RNA Worlds

From Life's Origins to Diversity
in Gene Regulation

ALSO FROM COLD SPRING HARBOR LABORATORY PRESS

SUBJECT COLLECTIONS FROM COLD SPRING HARBOR
PERSPECTIVES IN BIOLOGY

Cell–Cell Junctions
Generation and Interpretation of Morphogen Gradients
NF-κB: A Network Hub Controlling Immunity, Inflammation, and Cancer
The Origins of Life
Symmetry Breaking in Biology
The p53 Family

RELATED LABORATORY MANUALS

RNA: A Laboratory Manual

OTHER RELATED TITLES

A Short Guide to the Human Genome
Epigenetics
Translational Control in Biology and Medicine

RNA Worlds

From Life's Origins to Diversity in Gene Regulation

EDITED BY

John F. Atkins
University of Utah
University College Cork
Trinity College Dublin

Raymond F. Gesteland
University of Utah

Thomas R. Cech
Howard Hughes Medical Institute
University of Colorado

www.cshperspectives.org

COLD SPRING HARBOR LABORATORY PRESS
Cold Spring Harbor, New York • www.cshlpress.com

RNA Worlds: From Life's Origins to Diversity in Gene Regulation
A Subject Collection from *Cold Spring Harbor Perspectives in Biology*
Articles online at www.cshperspectives.org

All rights reserved
© 2011 by Cold Spring Harbor Laboratory Press, Cold Spring Harbor, New York
Printed in the United States of America

Publisher	John Inglis
Acquisition Editor	Richard Sever
Director of Development, Marketing & Sales	Jan Argentine
Project Manager	Inez Sialiano
Permissions Coordinator	Carol Brown
Production Editor	Kaaren Kockenmeister
Production Manager	Denise Weiss
Book Marketing Manager	Ingrid Benirschke
Sales Account Managers	Jane Carter and Elizabeth Powers
Cover Designers	Susan Roberts and John Atkins

Front cover artwork: The ancient RNA World is represented by RNA self-replication at the center of the spiral with the diversity of modern RNA represented at the near end of the spiral. Some RNA structures are shown in layers at the sides.

Library of Congress Cataloging-in-Publication Data

RNA worlds : from life's origins to diversity in gene regulation / edited by John F. Atkins, Raymond F. Gesteland, Thomas R. Cech.
    p. cm.
  Includes bibliographical references and index.
  ISBN 978-0-87969-946-8 (hardcover : alk. paper)
 1. Genetic regulation. I. Atkins, John F. II. Gesteland, Raymond F. III. Cech, Thomas. IV. Title.

QH450.R635 2011
572.8'8--dc22

2010027850

10 9 8 7 6 5 4 3 2 1

All World Wide Web addresses are accurate to the best of our knowledge at the time of printing.

Authorization to photocopy items for internal or personal use, or the internal or personal use of specific clients, is granted by Cold Spring Harbor Laboratory Press, provided that the appropriate fee is paid directly to the Copyright Clearance Center (CCC). Write or call CCC at 222 Rosewood Drive, Danvers, MA 01923 (978-750-8400) for information about fees and regulations. Prior to photocopying items for educational classroom use, contact CCC at the above address. Additional information on CCC can be obtained at CCC Online at http://www.copyright.com/.

All Cold Spring Harbor Laboratory Press publications may be ordered directly from Cold Spring Harbor Laboratory Press, 500 Sunnyside Blvd., Woodbury, New York 11797-2924. Phone: 1-800-843-4388 in Continental U.S. and Canada. All other locations: (516) 422-4100. FAX: (516) 422-4097. E-mail: cshpress@cshl.edu. For a complete catalog of all Cold Spring Harbor Laboratory Press publications, visit our website at http://www.cshlpress.com/.

Contents

Preface, vii

Foreword to the First Edition of *The RNA World*, ix
Francis Crick

Prologue to the First Edition of *The RNA World*, xi
James D. Watson

The RNA Worlds in Context, 1
Thomas R. Cech

THE EARLY RNA WORLD

Setting the Stage: The History, Chemistry, and Geobiology behind RNA, 7
Steven A. Benner, Hyo-Joong Kim, and Zunyi Yang

The Origins of the RNA World, 21
Michael P. Robertson and Gerald F. Joyce

Getting Past the RNA World: The Initial Darwinian Ancestor, 43
Michael Yarus

The Origins of Cellular Life, 51
Jason P. Schrum, Ting F. Zhu, and Jack W. Szostak

WHAT RNA CAN DO BY ITSELF

Riboswitches and the RNA World, 63
Ronald R. Breaker

Riboswitches: Structures and Mechanisms, 79
Andrew D. Garst, Andrea L. Edwards, and Robert T. Batey

Small Self-cleaving Ribozymes, 93
Adrian R. Ferré-D'Amaré and William G. Scott

Group II Introns: Mobile Ribozymes that Invade DNA, 103
Alan M. Lambowitz and Steven Zimmerly

EXIT THE RNA WORLD: PROTEIN SYNTHESIS ON RIBOSOMES

The Roles of RNA in the Synthesis of Protein, 123
Peter B. Moore and Thomas A. Steitz

Evolution of Protein Synthesis from an RNA World, 141
Harry F. Noller

The Ribosome: Some Hard Facts about Its Structure and Hot Air about Its Evolution, 155
V. Ramakrishnan

RNP-ZYMES: WHAT RNA CAN DO IN COLLABORATION WITH PROTEIN

Noncoding RNPs of Viral Origin, 165
Joan Steitz, Sumit Borah, Demian Cazalla, Victor Fok, Robin Lytle, Rachel Mitton-Fry, Kasandra Riley, and Tasleem Samji

Spliceosome Structure and Function, 181
Cindy L. Will and Reinhard Lührmann

Telomerase: An RNP Enzyme Synthesizes DNA, 205
Elizabeth H. Blackburn and Kathleen Collins

RNA REGULATED GENE EXPRESSION

Bacterial Small RNA Regulators: Versatile Roles and Rapidly Evolving Variations, 215
Susan Gottesman and Gisela Storz

RNA in Defense: CRISPRs Protect Prokaryotes against Mobile Genetic Elements, 231
Matthijs M. Jore, Stan J.J. Brouns, and John van der Oost

Ancestral Roles of Small RNAs: An Ago-Centric Perspective, 243
Leemor Joshua-Tor and Gregory J. Hannon

RNA Interference and Heterochromatin Assembly, 255
Tom Volpe and Robert A. Martienssen

The X as Model for RNA's Niche in Epigenomic Regulation, 267
Jeannie T. Lee

The Long Arm of Long Noncoding RNAs: Roles as Sensors Regulating Gene Transcriptional Programs, 279
Xiangting Wang, Xiaoyuan Song, Christopher K. Glass, and Michael G. Rosenfeld

Contents

TOOLS FOR RNA SCIENCE

Folding and Finding RNA Secondary Structure, 293
David H. Mathews, Walter N. Moss, and Douglas H. Turner

Predicting and Modeling RNA Architecture, 309
Eric Westhof, Benoît Masquida, and Fabrice Jossinet

RNA Reactions One Molecule at a Time, 321
Ignacio Tinoco, Gang Chen, and Xiaohui Qu

Aptamers and the RNA World, Past and Present, 333
Larry Gold, Nebojsa Janjic, Thale Jarvis, Dan Schneider, Jeffrey J. Walker, Sheri K. Wilcox, and Dom Zichi

In Vivo RNAi: Today and Tomorrow, 343
Norbert Perrimon, Jian-Quan Ni, and Lizabeth Perkins

Index, 355

Preface

THE TITLE OF THIS BOOK COULD HAVE BEEN "The RNA World, 4th edition," because it is very much the descendent of three previous volumes published by Cold Spring Harbor Laboratory (CSHL) Press. Instead, we've chosen "RNA Worlds" to reflect the book's dual purpose. On the one hand, the volume covers the exciting diversity of form and function of RNA in the present day world, including RNA functions that have been discovered only recently and are still emerging. On the other hand, the volume maintains a major focus on the ancient RNA world that is thought to have predated genetically encoded proteins, DNA, and organisms we know about. As documented here, evidence for the reality of such a primordial role for RNA is increasing, although challenges remain in demonstrating model RNA-based replicases and in uncovering feasible origins of precursors for RNA synthesis. Nevertheless the plasticity of RNA stemming from its 2′ hydroxyl group and single-stranded nature, which permits diverse folding in contrast to its storage molecule cousin, DNA, continues to fascinate.

Even as we see health and economic benefits of biological research, including RNA research, pursuit of intellectual curiosity remains fundamentally important for the human psyche as it was for explorers in previous eras. Curiosity about our origins seems particularly deep-seated, as reflected by the complex tapestry of explanations in the religious heritages of diverse societies. But the scientific explanations of life's origins seem more ennobling to us than, for instance, the idea of the original human female being derived from a rib of a man.

The evidence is overwhelming that not only all humans but all known life on earth had a common origin. This would not be a surprise to Darwin who wrote in his "Origin of Species" book one and a half centuries ago that "probably all organic beings which have ever lived on this earth have descended from some one primordial form". To what extent these conclusions have been providing a bulwark against the intolerance of fundamentalist beliefs is, of course, part of a "well-worn" debate.

These considerations contribute to the broad interest in theories and experiments concerning a primordial RNA world. How best to focus interest on the key roles that RNA played in the origins of life and continues to play in present day species diversity and function? It is crucial to continue the present research directions in biology to reveal the bonanza yet in store and to encourage its distillation and dissemination to a wider audience.

Even the pioneers of the RNA world concept did not foresee that ribozymes had survived to the present day. Is it conceivable that some RNA-based organisms currently exist on our planet? After all, without ribosomes, such an "organism" may be even smaller than organisms that rely on protein synthesis. Such an organism might be uniquely present in some minute deep-rock niches, and maybe in environments that could not support predatory DNA-based organisms. Such a ribo-organism might even have evolved strategies to resist predation. While microorganisms are now known to live deeper in the earth and in more places than previously appreciated, the possibility of RNA-based organisms whose ancestor arose after the general advent of protein-based life does not seem to have been widely considered. Search of stable shales with appropriate pore sizes would likely yield new organisms with an unquantifiable possibility of the discovery of RNA-based organisms that would even require modification of the statement above about common ancestry. As referenced in the chapter by Benner and colleagues, this is just one of the possibilities for radically different life on earth whose potential existence merits investigation. Such a project would be much cheaper than the quest to find life on Mars and, if successful, just as wonderful scientifically.

Since the third edition of "The RNA World" book, a key pioneer of RNA world studies and author in three previous editions of the book, Leslie Orgel, has passed away. We will not forget his rigorous approach to origin-of-life experiments and his warm humor.

We thank Richard Sever and John Inglis of CSHL Press for wise advice and continuing support. It is a pleasure also to acknowledge the care and understanding of the project coordinator at CSHL, Inez Sialiano, in making this book a reality, and Susan Roberts for help with the cover graphics.

J.F. ATKINS
R.F. GESTELAND
T.R. CECH

Foreword to the First Edition of *The RNA World*

The term "RNA world" originally referred to a hypothetical time in the evolution of earthly life when there was no elaborate mechanism for protein synthesis such as we have today. As described by Joyce and Orgel (Chapter 1), there were speculations in the 1960s that RNA catalysts existed at that stage in evolution, that RNA was the sole genetic material, and that the standard Watson-Crick pairing was the basis of genetic replication. None of these early authors was smart enough to suggest that relics of these hypothetical catalytic RNAs might still be around today. Indeed, it was speculated that the original ribosomes might have been made solely of RNA but not that their main catalytic activity (the formation of the peptide bond) might still be performed today by RNA alone, as recent evidence seems to suggest.

This hypothesis of an RNA world without protein was largely forgotten but has now become fashionable again because of the remarkable discoveries by Altman and by Cech (described by Cech, Chapter 11) of RNA molecules that do indeed have catalytic activity of one sort or another. These discoveries have removed the chief objection to the RNA world—that RNA by itself cannot act catalytically.

This volume on the RNA world covers a somewhat wider field. It was realized in the 1960s that not all present-day RNA was messenger RNA or viral RNA. The ribosome was known to contain several distinct types of "structural" RNA molecules; tRNAs were familiar, and there were isolated examples of other small RNA molecules. What was not realized was the richness of the present-day RNA world, including snRNPs (as described by Baserga and Steitz, Chapter 14) and now many others. We have come to realize that RNA molecules can occur with many different shapes and can thus have many different functions. Several of the chapters in this book deal with what these structures are, what their functions may be, what makes them stable, and how to predict their structure. Just how many of these RNAs are, by themselves, truly catalytic remains to be seen. Some of them may use divalent cations to produce both their folding and their catalytic activities, as discussed by Pan et al. (Chapter 12).

Some of the properties of present-day RNA molecules were quite unexpected. One might perhaps have guessed that some limited form of messenger-RNA editing could occur very rarely as a freak, but who would have predicted that such editing would take so many distinct forms and occur as widely as Bass describes in Chapter 15?

That a retrovirus contains a sequence rather like part of a tRNA molecule was a surprise when it was first discovered. In 1987, Maizels and Weiner proposed that such structures tagged RNA genomes for replication in the early RNA world. In Chapter 23, they explore the many possible ramifications of this genome tag hypothesis. In particular, they propose that both tRNA and tRNA-aminoacylation activity were subject to selection before protein synthesis arrived on the scene. Whether one can safely extrapolate from a present-day property of rapidly evolving RNA viruses to as far back in time as the early RNA world remains to be seen.

A good case can be made (Gold et al., Chapter 19) that small lengths of RNA can often form relatively rigid structures more easily than can polypeptides of similar length. New techniques are now being used (Chapter 19 and Chapter 20 [Szostak and Ellington]) to explore this vast "space" of small RNA sequences for interesting properties such as the binding of one specific substrate or another and, in some cases, a particular type of catalytic activity.

The recombinant-DNA revolution of the 1970s was possible because molecular biologists could use the sophisticated products of billions of years of natural evolution—the replicases, restriction enzymes, and so forth—as precise and delicate chemical tools. These molecules were all proteins that acted on nucleic acids in one way or another. Now experimentalists (Chapters 19 and 20) are using a combination of biochemical processes, related to nucleic acid replication, which together embody the mechanism of natural selection. These powerful tools can provide them with new and desirable RNA molecules in the laboratory while they wait. Because these methods do not need whole cells, but only cellular components, they can handle many more individuals at one time and so explore the RNA "space" more quickly. The message is clear: Where possible, let Nature do the work, and if she proves a little slow, take your whip to her.

The exact nature of the early RNA world is now a matter of active debate. What can we learn from the present "relics" and, in particular, can we deduce the composition and nature of each particular type of early RNA from the detailed study of its many descendants today? Can we learn anything from small RNA-related molecules (such as coenzymes) about the general nature of the early RNA world? Early speculations had assumed that the RNA world was rather simple and rather inaccurate. Benner and his colleagues (Chapter 2) now believe it was much more complex and more accurate. As more sequence data accumulate, one would expect these suggestions to become less speculative.

It may be possible to deduce something about some of the activities in the RNA world by a detailed study of the present mechanisms of protein synthesis. For example, Weiss and Cherry (Chapter 3) have proposed an ingenious model for the origins of the smaller ribosomal subunit. They suggest that it was originally an RNA replicase that used oligonucleotides as a substrate, cannibalizing triplets from them for its own replication.

We may, in time, arrive at a rather plausible picture of this early stage in evolution, even if true molecular fossils (that is, actual specimens of the molecules that then existed) are forever unavailable to us because of the ceaseless battering of thermal motion over billions of years.

The details of the transition from the RNA world to the present protein-dominated world are also debatable. For example, was there some sort of protein synthesis before the existence of a primitive ribosome? Did these first ribosomes have some protein, or did the ribosomal proteins arrive a little later?

It may turn out that we will eventually be able to see how this RNA world got started. At present, the gap from the primal "soup" to the first RNA system capable of natural selection looks forbiddingly wide. Was there perhaps a pre-RNA world, with replication based on some even more primitive system?

Such a pre-RNA world might have been of several types. It would presumably have been based on some form of genetic polymer or sheet that was formed more easily than RNA under prebiotic conditions. It might eventually have transcribed its detailed genetic information directly onto RNA to form the RNA world. Alternatively, it might have acted solely as a midwife, setting up the basis of RNA synthesis for some use of its own, with RNA replication then taking over and replacing it. Cairns-Smith has already suggested that this pre-RNA system might have been based on clays, or on some organic polymer (Cairns-Smith, A.G., Genetic Takeover and the Mineral Origins of Life. Cambridge University Press [1982]). This time we should be smart enough to ask if there are any relics of this pre-RNA world still with us today.

Finally, a word of heresy: As Moore has pointed out (Chapter 5), the real fossil record suggests that our present form of protein-based life was already in existence 3.6 billion years ago and evolved rather slowly for a billion or so years after that. This leaves an astonishingly short time to get life started. Moreover, the three main lines of descent (see Fig. 3 in Woese and Pace, Chapter 4) seem a very long distance from their hypothetical common ancestor.

The rather far-fetched hypothesis of Directed Panspermia would predict that life was sent here in the form of "bacteria" suitable for growth in anaerobic conditions, and that several somewhat different forms would probably have been sent at the same time, in the hope that at least one would survive. All are likely to have evolved originally from a common ancestor (on another planet) that existed some billions of years before the formation of our solar system. Therefore, the final question about the RNA world and the pre-RNA world (if it existed) is: Where did it occur? Are we totally confident that our form of life started here, or did it perhaps originate elsewhere in the universe? It might have been easier to start elsewhere because, for example, the atmosphere there was more reducing than the Earth's early atmosphere appears to have been.

The lively and authoritative chapters of this book deal comprehensively both with the hypothetical RNA world and with the complexities of RNA structure and function we find around us today. I recommend it to all molecular biologists and especially to anyone fascinated by the baroque complexity of the nucleic acids and of RNA in particular.

FRANCIS CRICK

Prologue to the First Edition of *The RNA World*
Early Speculations and Facts about RNA Templates

RNA FIRST CAME ALIVE TO ME DURING THE FALL OF 1947 at Indiana University when I took Salvador Luria's course on viruses. There I first learned that whereas the then-known phage, pox and papilloma viruses, contained DNA, this molecule was totally absent in several purified plant viruses as well as in the viruses that caused encephalitis and polio, which instead contained RNA. Apparently a given virus had either RNA or DNA, in contrast to cells which contained both. But whether it was the nucleic acid component that carried their genetic specificity was still unclear. At that time, most scientists wanted Avery, MacLeod, and McCarty's experiment on pneumococcal transformation by purified DNA to be extended to other life forms before jumping on the nucleic acid bandwagon. Then in the spring of 1952 came the report from Al Hershey that the DNA component of phage T2 carried genetic specificity. This immediately thrilled me, but I remember well the audience's indifference when in mid-April I read Hershey's letter at an Oxford meeting of The Society for General Microbiology.

When I had arrived at Cambridge in the fall of 1951, I started taking seriously the work of Brachet and his collaborators in Brussels, who emphasized the correlation between the RNA content and the protein-synthesizing capacity of cells. Those cells making large amounts of protein possessed large numbers of virus-sized ribonucleoprotein particles, known initially as microsomal particles but since 1958 as ribosomes. Most importantly, these particles had been pinpointed as the actual sites of protein synthesis by means of the then just-developed cell-free systems for protein synthesis. Here the key lab was that of Paul Zamecnik at Massachusetts General Hospital. Equally important was Brachet and Chantrenne's demonstration that the nucleus, and hence DNA, had no direct participation in protein synthesis. To show this, they cut the giant alga Acetabularia in half and observed that the half without a nucleus could maintain almost normal protein synthesis for more than a month. Yet from the one geneone enzyme (protein) results of Beadle and Tatum, the ultimate source of the genetic information that specifies the amino acid sequences of proteins had to be the genes found in the nucleus. I thus postulated a two-stage scheme for protein synthesis in which DNA first serves as a template for nucleus-located synthesis of RNA, and this RNA then in turn moves to the cytoplasm where it functions as the template for protein synthesis.

No one then had any compelling reason to take my hypothesis seriously, but by November 1952 I liked it well enough to print (DNA→RNA→protein on a small piece of paper which I taped on the wall above my writing table in my rooms at Clare College. From the day of our first meeting, Francis Crick and I thought it highly likely that the genetic information of DNA is conveyed by the sequence of its four bases, but we knew it was premature to promote this idea before the structure of DNA was known. However, from the moment we first saw how to build a double helix out of the four base pairs, it was clear that the essential uniqueness of a gene must reside in its respective sequence of base pairs. Moreover, not only could base pairing provide the way for genes to be copied exactly during gene duplication, but it was also very likely to underlie the process by which the genetic information of a DNA molecule is transferred to its RNA product.

Still totally unclear, however, was how RNA might serve as the template for ordering the amino acids in their respective polypeptide products. Emboldened by our fantastic good luck in so simply finding the structural essence of the gene duplication process, I saw no reason not to take on the challenge of finding out what RNA molecules looked like in three dimensions. Such knowledge, I felt, would be indispensable to understanding how they functioned in protein synthesis. I took on this task when I moved to Pasadena in the fall of 1953. There I joined forces with Alex Rich, who had started working on DNA just before Francis and I found the double helix. There was no difficulty in getting him to move on to a better pasture, and soon I was collecting RNA samples and drawing them into fibers that Alex exposed to X-ray beams. But despite much travail, even those fibers displaying high birefringence never gave rise to ordered diffraction patterns like those of DNA. Although we thought we saw reflections that might have come from short sections of double helices, we could never be sure and saw no way to decide whether RNA was a one- or two-chain molecule. Here the base composition was a bad tease. Viral RNAs clearly did not show equivalence of A with U or G with C, but the RNA from cells had A/U and G/C ratios sometimes closely approaching 1/1. But we could see no difference in the general features of the X-ray diagrams from viral or cellular RNAs. To our annoyance, RNA, no matter from what source, showed identical X-ray diagrams characterized by strong reflections at 3.36 Å and 4.00 Å. Clearly, there was some ordered structure in RNA, but we saw no way to get to it. After some six months of such frustration, we gave up.

A very different approach to understanding protein synthesis came from the very clever Russian-born theoretical physicist George Gamow, who was struck by the fact that the 3.5-Å distance between adjacent amino acids in extended polypeptide chains was very similar to the 3.4-Å separation between adjacent base pairs in the B form of the double helix. Not then cognizant of RNA's primary role in protein synthesis, Gamow proposed that amino acids are directly ordered by contacts with the DNA base pairs, with the polypeptide products containing the same number of amino acids as there are base pairs in their DNA templates. To deal with the fact that DNA has only 4 letters in its alphabet, whereas 20 different letters (amino acids) are used to specify proteins, Gamow assumed that each amino acid must be coded by several adjacent base pairs. Because there are only 16 (4×4) combinations of the four bases, taken two at a time, Gamow made the assumption that each amino acid must be specified by groups of three adjacent base pairs (a codon) along DNA chains. To deal with the fact that there are 64 ($4 \times 4 \times 4$) such triplets, he assumed that many amino acids must be specified by more than one triplet (redundant triplets). In such an "overlapping" code, adjacent amino acid codons have two out of the three base pairs in common, thereby restricting which amino acids can lie adjacent to each other. Gamow's overlapping code was only the first of several that would later be devised, each leading to different combinations of forbidden amino acid neighbors.

When George first told me his scheme, I quickly dismissed it since DNA was not the template that ordered the amino acids. But he was having combinatorial fun, and besides I could not rule out the possibility that some RNA molecules were double helices. With time, moreover, I realized there was a real virtue to Gamow-like codes. In disproving them, the possibility of overlapping codes could be ruled out, pointing the way to codes in which adjacent groups of most likely three bases specified successive amino acids along a polypeptide chain. In fact, the first known amino acid sequence, that determined by Sanger for insulin, disproved the first code, although Gamow did not at first realize this because his initial list of the 20 amino acids had several embarrassing mistakes (e.g., he included both cystine and cysteine). Other such overlapping codes devised over the next year were also ruled out as more proteins became sequenced. It was during the first coding rush that Leslie Orgel and I, on a trip to Berkeley where Gamow was spending the spring of 1954, suggested that we form a club of 20 members whose purpose was to crack the RNA structure and in so doing reveal how the genetic code operated. Soon to be known as the "RNA Tie Club," with its members reflecting Gamow's eclectic taste, it never had a formal meeting, nor did all its members ever cough up the money to purchase their RNA ties and tiepins bearing their respective amino acid code letters. Gamow's tiepin sported ALA (alanine), and mine, PRO (proline). Much more important in the long run was the opportunity it provided to exchange ideas about the code through "Notes to the RNA Tie Club."

Several of these communications, the most important of which came from Francis Crick, Leslie Orgel, and Sydney Brenner, later became incorporated into published manuscripts. Among these was the definitive disproof by Brenner of any form of overlapping code, which he wrote in Johannesburg in the fall of 1956 just before he returned to England to join Francis Crick. Equally important was the communication by Crick, John Griffiths, and Orgel suggesting a comma-less code as a device to let non-overlapping triplets be read in the appropriate reading frame. But the most influential paper intellectually was the RNA Tie Club's first Note, written by Francis Crick and sent out early in 1955 under the title On Degenerate Templates and the Adaptor Hypothesis. After spending the previous August in Woods Hole batting about potential codes with Gamow and Brenner, Crick began questioning the basic assumption that a nucleic acid template provided specific cavities complementary in shape and charge to the amino acid side groups. Here he argued that the nucleic acid bases want to hydrogen-bond and are not at all suitable for forming cavities that could attract the hydrophobic side groups of amino acids such as valine, leucine, or isoleucine. Equally tricky to imagine was any structural basis for degenerative codes in which many amino acid side groups are specified by more than one set of triplets. Faced with what he considered insuperable obstacles, Crick made the radical proposal that prior to peptide bond formation, amino acids are enzymatically attached to small "adaptor" molecules that have surface specifically tailored to bond to nucleic acid triplets. Here Crick suggested that the "adaptor" molecules might be tiny polynucleotides that base pair to RNA templates.

My reaction to the adaptor hypothesis was initially very negative, even though I had spent the fall of 1954 in futile efforts trying to fold RNA chains into shapes bearing cavities appropriate for the amino acid side groups.

The adaptor idea seemed much too complicated to me to have ever gotten started at the origin of life several billion years ago. Francis, too, had his moments of doubt; he even concluded his now famous Tie Club Note with the phrase, "In the comparative isolation of Cambridge I must confess there are times when I have no stomach for decoding." In fact, by the time I returned to Cambridge for the year beginning in June 1955, Francis and Alex Rich were immersed in building new three-dimensional models for collagen to compete with one earlier proposed by Linus Pauling, this despite the fact that Francis and I had often referred to collagen as the most boring of macromolecules.

Equally frustrated, I reverted to thinking about plant viruses, in particular tobacco mosaic virus (TMV), whose helical construction I had worked out in the spring of 1952 after Francis and I had been told by Sir Lawrence Bragg to stop trying to work out the structure of DNA through model building. Always troublesome to me was the apparent necessity to postulate both genetic and protein synthesis roles for RNA. Knowing that only a tiny fraction of TMV particles are actually infectious, I speculated whether in fact these rare particles contained DNA, not RNA, chains. But this was ruled out when it became possible to reconstitute infectious TMV particles from their purified RNA and protein components. This experiment, first successfully accomplished in the spring of 1955 in Berkeley by Heinz Fraenkel-Conrat and Robley Williams, generated much newspaper publicity which gave uninformed readers the idea that life itself had been created. Francis, however, put the matter in its proper light, being quoted in the English press as saying that this was a finding he had anticipated.

Reconstitution by itself, however, did not answer the question of whether the protein component played any more than a protective-coat role for the genetic-information-bearing RNA component. Less than a year later, however, Alfred Gierer, working in Gerhard Schramm's lab in Tübingen, clearly showed that the RNA alone was infectious. This primacy of nucleic acids as bearers of genetic information then lay at the heart of the way Francis and I thought about cells and viruses. But this was far from an acceptable paradigm for many of the attendees at the key, late March 1956, CIBA Foundation meeting on the Structure of Viruses. They were not at home with the concept that information flows unidirectionally from nucleic acids to proteins and never backward. This was awkwardly shown when André Lwoff and I passed on to Robley Williams a telegram message supposedly from Wendell Stanley reading "TMV protein infectious—be cautious!" To our amazement, Robley didn't question the result until we revealed the hoax.

At this gathering of some 30 scientists, Francis presented our ideas on why the protective coats of viruses are made up of protein subunits. In our view, it was a consequence of the fact that no viral nucleic acid had sufficient coding capacity to specify a single polypeptide chain large enough to surround a more centrally located core of nucleic acid. The 2-million-molecular-weight RNA of TMV, for example, contains only 6000 bases and, assuming a coding ratio of three bases per amino acid, is only capable of specifying a 2000-amino-acid polypeptide or about 230,000 molecular weight. To make up the 38-million-molecular-weight protein coat, at least 165 subunits would be needed. In fact, the TMV subunit contains only some 150 amino acids, suggesting that the TMV RNA codes for several proteins or that the coding ratio is very much larger than three. Initially, we thought that tomato bushy stunt virus (TBSV), with a much higher RNA content, might be more useful in giving a realistic value for the coding ratio. It contains only four bases for every amino acid in the crystallographic subunit. Later, direct analysis of its protein subunit size suggested that each crystallographic repeat contains some five protein subunits, very likely implying that TBSV RNA, like TMV RNA, also codes for several proteins.

After that 1956 meeting, I again had a go at the RNA structure, taking advantage of the newly discovered enzyme polynucleotide phosphorylase that Marianne Grunberg-Manago and Severo Ochoa found could be used to make synthetic RNA molecules. By then back at the National Institutes of Health, Alex Rich had shown the month before that random AU and AGCU co-polymers gave X-ray diffraction patterns similar to those we had obtained using purified cellular and viral RNAs. I focused instead on poly(A) (adenine) fibers drawn from material prepared in the Molteno Institute by Roy Markham and David Lipkin. To my delight, they generated clean helical X-ray diagrams that were best interpreted as base-paired double helices built up from two parallel poly(A) chains. Initially, I was disturbed by the fact that many of the key reflections overlapped with those generated by purified RNA, which by then we had every reason to believe was single-chained. Later, this apparent paradox was possibly resolved by the finding that sections of hydrogen-bonded hairpins form along most RNA chains. Conceivably, it is these short sections of imperfect double helices that generate the DNA-like feature of the RNA X-ray patterns.

By then discouraged that the study of purified RNA would lead toward understanding how protein synthesis occurs, I decided to concentrate on the structure of ribosomal particles, at that time believing that they must carry the genetic information for ordering amino acids in proteins. In the spring of 1956 I convinced Alfred Tissières, then a Fellow at Kings College working in the Molteno Institute on oxidative phosphorylation, to join me at Harvard, where I would be moving in the fall of 1956. Alfred had in fact already done some preliminary experiments on the then-called microsomal particles and was keen to follow them up.

I arrived at Harvard some six months before Alfred but was preoccupied most of this time trying to be an effective teacher for the seniors and beginning graduate students who were taking my course on viruses. As soon as my lectures were under control, however, I went across the Charles River to see Paul Zamecnik whom I had first met the year before while briefly stopping at Harvard on my way back to England. There, in the building where Fritz Lipmann also had his lab, I first appreciated the importance of the discovery made there a year earlier by Mahlon Hoagland of the activated high-energy acyl-amino acid intermediates for protein synthesis. Even more important, I first learned of the more recent observation by Mahlon, Mary Stephenson, and Paul of a soluble RNA (sRNA) fraction to which the activated amino acids are transferred prior to protein synthesis. Quickly, I realized that these sRNA molecules might be the polynucleotide adaptors postulated two years before by Francis in his first Note to the RNA Tie Club. Prior to my visit, Crick's ideas were unknown to the Massachusetts General group and they eagerly sought out Francis when they came together at the 1957 Gordon Conference on Nucleic Acids and Proteins.

Tissières and I commenced our molecular characterization of *Escherichia coli* ribosomes in the spring of 1957, suspecting that they might have structural plans like those of the small RNA viruses, whose isocahedral-shaped protein shells are formed by the regular aggregation of a single protein building block. We could not have been more wrong about how they are organized. To start with, we found that the *E. coli* ribosomes, like those from all other organisms, are formed by the aggregation of two RNA-containing subunits, the larger 50S subunit approximately twice the size of the smaller 30S subunit. Each subunit contains a single major RNA chain, with the 50S ribosomal subunits possessing 23S RNA chains and the 30S subunits possessing 16S RNA chains. Moreover, both subunits contain a large number of different small proteins that for the most part are subunit-specific. At low Mg^{++} levels, the 30S and 50S subunits do not associate with each other, but when the Mg^{++} concentration is raised, they come together to form the 70S complex that subsequent work has shown to be the ribosomal form that carries out protein synthesis.

Naively, at first, we assumed that either the 16S RNA, or the 23S RNA, or both, were the actual templates for protein synthesis. Puzzling to us, however, was why the templates existed in only two size classes while there was great variation in the sizes of their putative polypeptide products.

Equally disturbing was why the base composition of ribosomal RNAs barely varied between bacterial species with highly different AT/GC ratios. A priori we had expected to find that the base compositions of the RNA templates would reflect those of their DNA templates. Luckily, there was one powerful exception. In 1956 the phage T2-specific RNA made after T2 infection of *E. coli* was shown by Volkan and Astrachan to have a T2 DNA-like base composition. Moreover, in contrast to the metabolically very stable ribosomal RNA chains, T2 RNA had been found to have a half-life of only several minutes.

In retrospect, Tissières and I should have gravitated early to T2 RNA, but in fact not until the fall of 1959 were its molecular properties investigated anywhere. Then, Masayasu Nomura and Ben Hall, working with Sol Spiegelman at the University of Illinois, provided tentative evidence for the incorporation of T2 RNA into abnormally small ribosomes. In trying to follow up this observation early in 1960, my graduate student Bob Risebrough came to a radically different conclusion. After T2 RNA is synthesized it does not become part of a ribosomal subunit by aggregating with newly made ribosomal proteins. What in fact happens is that in the presence of Mg^{++}, the T2 RNA becomes attached to the smaller 30S ribosomal subunit, which in turn binds the larger 50S subunit to form the >70S complex that actually carries out protein synthesis. This result instantly changed the way we visualized protein synthesis. Instead of serving template roles, ribosomes function as stable assemblage sites for protein synthesis. The true template had to be a new RNA class unknown until that moment both because it comprises such a small percentage of the total RNA and because it is heterogeneous in length. Later, these metabolically unstable templates, whose amounts respond to cellular needs, would be named messenger RNA (mRNA) by Jacques Monod and François Jacob.

The first lab outsider to whom I revealed this conceptual breakthrough was Leo Szilard, whom I had gone down to see in New York where he was successfully plotting the radiation therapy that would cure his bladder cancer. Leo's reaction was entirely negative, not being convinced that we had the right interpretation for Risebrough's experimental results. Predictably, he wanted us to show that mRNA existed in uninfected *E. coli* cells before he would change his mind-set. These were experiments that we had in fact already planned to start several weeks later, as soon as François Gros arrived from Paris to spend several months working with Alfred and me. Also soon to join this effort was Wally Gilbert, then still teaching theoretical physics to Harvard students, but who was increasingly tempted to move into molecular biology by our excitement about mRNA. By the time of the June Gordon Conference on Nucleic Acids we were virtually convinced that *E. coli* mRNA also existed, and by the summer's end we had data for a convincing publication. Already at the Gordon Conference we had heard rumors that Sydney Brenner and François Jacob, using very different arguments, had also postulated the existence of mRNA and that their idea was being tested by Sydney that week in Matt Meselson's lab at Caltech. Eventually, we were to publish our independent proofs for mRNA's existence early in 1961 in back-to-back articles in *Nature*.

With the basic scheme for how RNA participates in protein synthesis known, the path became open for definitive experiments on the exact nature of the genetic code. Using enzymatically synthesized RNA as messengers for in vitro protein synthesis, the correct assignments for all of the triplet codons were determined by early 1966. With this major goal achieved, the time had clearly come to ask how the DNA→RNA→protein flow of information had ever gotten started. Here, Francis was again far ahead of his time. In 1968 he argued that RNA must have been the first genetic molecule, further suggesting that RNA, besides acting as a template, might also act as an enzyme and, in so doing, catalyze its own self-replication. As the chapters in this book will show, how right he was!

J.D. WATSON

RNA Worlds

From Life's Origins to Diversity
in Gene Regulation

The RNA Worlds in Context

Thomas R. Cech

Department of Chemistry and Biochemistry, Howard Hughes Medical Institute, University of Colorado, Boulder, Colorado 80309-0215

Correspondence: thomas.cech@colorado.edu

SUMMARY

The chapters in this collection discuss not one RNA world, but two. The first is the primordial RNA world, a hypothetical era when RNA served as both information and function, both genotype and phenotype. The second RNA world is that of today's biological systems, where RNA plays active roles in catalyzing biochemical reactions, in translating mRNA into proteins, in regulating gene expression, and in the constant battle between infectious agents trying to subvert host defense systems and host cells protecting themselves from infection. This second RNA world is not at all hypothetical, and although we do not have all the answers about how it works, we have the tools to continue our interrogation of this world and refine our understanding. The fun comes when we try to use our secure knowledge of the modern RNA world to infer what the primordial RNA world might have looked like.

Outline

1 The primordial RNA world

2 The contemporary RNA world

3 The world of RNA technology and medical applications

References

1 THE PRIMORDIAL RNA WORLD

The term "RNA world" was first coined by Gilbert (1986), who was mainly interested in how catalytic RNA might have given rise to the exon–intron structure of genes. But the concept of RNA as a primordial molecule is older, hypothesized by Crick (1968), Orgel (1968), and Woese (1967). Noller subsequently provided evidence that ribosomal RNA is more important than ribosomal proteins for the function of the ribosome, giving experimental support to these earlier speculations (Noller and Chaires 1972; Noller 1993). The discovery of RNA catalysis (Kruger et al. 1982; Guerrier-Takada et al. 1983) provided a much firmer basis for the plausibility of an RNA world, and speculation was rekindled. The ability to find a broad range of RNA catalysts by selection of RNAs from large random-sequence libraries (SELEX) (Ellington and Szostak 1990; Tuerk and Gold 1990; Wright and Joyce 1997) fueled the enthusiasm, and made it possible to conceive of a ribo-organism that carried out complex metabolism (Benner et al. 1989). The widely accepted order of events for the evolution of an RNA world and from the RNA world to contemporary biology is summarized in Figure 1.

Did an RNA world exist? Some of the most persuasive arguments in favor of an RNA world are as follows. First, RNA is both an informational molecule and a biocatalyst—both genotype and phenotype—whereas protein has extremely limited ability to transmit information (as with prions). Thus, RNA should be capable of replicating itself, and indeed RNA can perform the sort of chemistry required for RNA replication (Cech 1986). Second, it is more parsimonious to conceive of a single type of molecule replicating itself than to posit that two different molecules (such as a nucleic acid and a protein capable of replicating that nucleic acid) were synthesized by random chemical reactions in the same place at the same time. Third, the ribosome uses RNA catalysis to perform the key activity of protein synthesis in all extant organisms, so it must have done so in the Last Universal Common Ancestor (LUCA). Fourth, other catalytic activities of RNA—activities that RNA would need in an RNA world but that have not been found in contemporary RNAs—are generally already present in large combinatorial libraries of RNA sequences and can be discovered by SELEX. Fifth, RNA clearly preceded DNA, because multiple enzymes are dedicated to the biosynthesis of the ribonucleotide precursors of RNA, whereas deoxyribonucleotide biosynthesis is derivative of ribonucleotide synthesis, requiring only two additional enzymatic activities (thymidylate synthase and ribonucleotide reductase.) Finally, a primordial RNA world has the attractive feature of continuity; it could evolve into contemporary biology by the sort of events that are well precedented, whereas it is unclear how a self-replicating system based on completely unrelated chemistry could have been supplanted by RNA.

Opinions vary, however, as to whether RNA comprised the first autonomous self-replicating system or was a derivative of an earlier system. Benner et al. (this collection) and Robertson and Joyce (this collection) are circumspect, noting that the complexity and the chiral purity of modern RNA create challenges for thinking about it arising de novo. On the other hand, the recent finding that activated pyrimidine ribonucleotides can be synthesized under plausible prebiotic conditions (Powner et al. 2009) means that it is premature to dismiss the RNA-first scenarios. Yarus (this collection), an unabashed enthusiast for an RNA world, argues for a closely related replicative precursor. In vitro evolution studies directed towards an RNA replicase ribozyme continue apace and are of great importance in establishing the biochemical plausibility of RNA-catalyzed RNA replication (Johnston et al. 2001; Zaher and Unrau 2007; Lincoln and Joyce 2009; Shechner et al. 2009).

What might the first ribo-organism have looked like? Schrum et al. (this collection) describe progress in achieving replication of simple nucleic acid-like polymers within lipid envelopes, thereby constituting "protocells." These liposomes can grow and upon agitation can divide to give daughter protocells, carrying newly replicated nucleic acids. Whether by lipids or other means, some form of encapsulation must have been a key early step in life. Encapsulation can protect the genome from degradation and

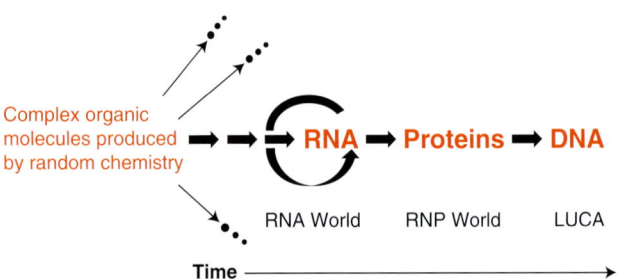

Figure 1. An RNA world model for the successive appearance of RNA, proteins, and DNA during the evolution of life on Earth. Many isolated mixtures of complex organic molecules failed to achieve self-replication, and therefore died out (indicated by the arrows leading to extinction.) The pathway that led to self-replicating RNA has been preserved in its modern descendants. Multiple arrows to the left of self-replicating RNA cover the likely self-replicating systems that preceded RNA. Proteins large enough to self-fold and have useful activities came about only after RNA was available to catalyze peptide ligation or amino acid polymerization, although amino acids and short peptides were present in the mixtures at far left. DNA took over the role of genome more recently, although still >1 billion years ago. LUCA (Last Universal Common Ancestor) already had a DNA genome and carried out biocatalysis using protein enzymes as well as RNP enzymes (such as the ribosome) and ribozymes.

predation, allows useful small molecules to be concentrated for the cell's use, and enables natural selection by ensuring that the benefit of newly derived functions accrues to the organism that stumbled across them.

2 THE CONTEMPORARY RNA WORLD

Today, RNA is the central molecule in gene expression in all extant life, serving as the messenger. It is also central to biocatalysis, seen dramatically in the ribosome but also in ribozymes and RNPzymes such as telomerase and the signal recognition particle. More recently, its diverse roles in regulation of (DNA) gene expression have been discovered. It is useful to organize the discussion of contemporary RNA activities as a spectrum, going from those activities that are so RNA-centered that one could conceive of them having operated in a primordial RNA world very much as they do today, to those that rely more and more on collaboration with proteins, to those RNAs that work on DNA (Fig. 1).

What can RNA do by itself? It can bind small metabolites (such as guanine, S-adenosylmethionine, and lysine) with exquisite specificity, and then use this binding energy to switch from one RNA structure to another. These riboswitches are common regulators of gene expression in Gram-positive bacteria, and are also found in other organisms including plants (Breaker [this collection]; Garst et al. [this collection]). Furthermore, even very small RNAs can act as ribozymes, accomplishing sequence-specific self-cleavage (Ferré-D'Amaré and Scott [this collection]). These self-cleavers can be easily re-engineered into multiple-turnover RNA-cleaving enzymes, so it is straightforward to imagine that they could have served such a function in a primordial RNA world. Larger ribozymes can accomplish sophisticated RNA splicing reactions, as described for group II introns by Lambowitz and Zimmerly (this collection). There are a number of similarities, both mechanistic and structural, between group II intron self-splicing and spliceosomal splicing of mRNA introns, providing a plausible continuum from the RNA world to post-protein contemporary biology.

Although RNA can perform many activities by itself, in modern cells RNA more often works in concert with proteins. The ribosome uses both RNA and protein to catalyze message-encoded protein synthesis. Yet the heart of the peptidyl transferase center is a ribozyme, and other fundamental activities such as mRNA start-site selection, codon–anticodon interaction, and decoding involve direct RNA–RNA interactions, so the RNA world ancestry of the ribosome is apparent (Moore and Steitz, Noller, Ramakrishnan [all in this collection]). The same can be said of the spliceosome (Will and Lührmann [this collection]).

Although a detectable level of catalysis of an isolated step of RNA splicing can be achieved with pure snRNAs (Valadkhan et al. 2009), the efficient and regulated splicing of an entire genome's collection of primary transcripts requires the collaboration of almost 200 proteins and five snRNAs in the modern spliceosome. Telomerase represents another paradigm, as it includes a canonical protein enzyme (TERT) that operates in intimate collaboration with RNA (Blackburn and Collins [this collection])—so it appears to derive from more recent evolution, after protein enzymes and DNA chromosomes were well established.

It seems likely that the most recently evolved functions of RNA involve regulation of DNA—because there would have been no DNA to regulate in a primordial RNA world! Nevertheless, similar principles could have been active in an RNA world. Gottesman and Storz (this collection) describe RNA regulation in bacteria, which occurs through a range of mechanisms ranging from the simple "antisense RNA" principle of inhibition by complementary base-pairing to RNA–protein interactions. In eukaryotes, several classes of noncoding RNAs perform diverse functions in the regulation of gene expression. Small double-stranded RNAs (for example, the 21-bp small-interfering RNAs and the microRNAs) regulate the stability or the translatability of mRNAs (Joshua-Tor and Hannon [this collection]). Here the RNA provides such a simple function—recognition of complementary sequences on the mRNA target—that the authors choose to organize their discussion according to subfamilies of the Argonaute proteins that bind the small RNAs. The RNA interference (RNAi) pathway is involved not only in mRNA-level events, but also in the regulation of chromatin structure as described by Volpe and Martienssen (this collection). Maintenance of the highly condensed heterochromatin found at chromosome centromeres depends on this RNAi activity. Finally, long noncoding RNAs usually acting *in cis* (on the chromosome or the local region where they were synthesized) can turn off gene expression by attracting proteins that modify chromatin structure. The effect can spread to an entire chromosome, in the case of the Xist RNA that condenses one of the two X chromosomes in female mammals and thereby gives gene dosage compensation (Lee [this collection]). In other cases, the effect is more local, affecting transcription of a single gene or a group of genes (Wang et al. [this collection]). These recently discovered activities of RNA show that the RNA world never stopped (and has not stopped) evolving.

Diverse viral encoded ncRNAs are used as weapons either to circumvent host defense or otherwise manipulate host cellular machinery for their own purposes (Steitz et al. [this collection]). Although several of the classes of viral ncRNAs are counterparts of cellular equivalents, some are distinctive. Bacteria have evolved the CRISPR (Clustered Regularly Interspaced Short Palindromic Repeat) defense system to protect themselves from alien DNA such as that injected by bacteriophages (Wang et al. [this collection]). Here, the information identifying the invading genome is stored in the form of DNA, but it is subsequently converted to small guide RNAs that recognize and interfere with subsequent invaders. Although there is a clear analogy between CRISPR and eukaryotic RNAi, the two systems appear to have evolved completely independently.

3 THE WORLD OF RNA TECHNOLOGY AND MEDICAL APPLICATIONS

I oversimplified when I said that there were two RNA worlds. There is in fact a third—the world of RNA research and development. This third RNA world should be of special interest to students, because this RNA world offers opportunities for gainful employment!

RNA function depends on its structure—it is the seemingly limitless variety of structures that allows so many diverse functions. We can now predict RNA secondary structure quite well (Mathews et al. [this collection]) and see much progress on predicting 3D structure (Westhof et al. [this collection]). Remarkably, we can now watch molecules of RNA fold and unfold and switch from one state to another in "single-molecule experiments" (Tinoco et al. [this collection]). We can use double-stranded RNAs and the intrinsic RNAi machinery present in organisms to do genome-wide knock-downs of gene function (Perrimon et al. [this collection]). Finally, RNA science is poised to make an impact on medicine. For example, aptamers can monitor the concentrations of many of the proteins in human serum, which has diagnostic applications because the presence of many proteins is correlated with health and disease (Gold et al. [this collection]). In addition, both microRNAs and antisense nucleic acids that inhibit miRNAs have pharmaceutical potential, which is under development in numerous biotechnology and pharmaceutical companies.

Thus, the authors of this collection take us on a fascinating journey through three RNA worlds. The primordial RNA world (ca. four billion years ago) relied on the dual ability of RNA to serve as both informational molecule and biocatalyst, providing a self-replicating system. Coupled with other ribozymes that carried out complex metabolism and encapsulated in some sort of envelope, self-replicating RNA constituted an early life form that was the ancestor of contemporary biology. The second RNA world is that of contemporary biology, where RNA occasionally acts by itself (ribozymes and riboswitches) but more often acts in concert with proteins. The ribosome and the spliceosome still "remember" their ribozyme heritage, whereas telomerase and the signal recognition particle have moved on to incorporate canonical protein enzymes. The RNA interference system and CRISPR have gone further, reducing the role of the RNA to that of a simple guide sequence. Finally, the third RNA world—that of RNA technology and medical applications—is a baby compared to even the second RNA world, because it arose only in the past half-century. Although this last RNA world is only perhaps one millionth of one per cent as old as the primordial RNA world, it is a vibrant community, and my co-editors John Atkins and Ray Gesteland and I feel privileged to be part of it.

REFERENCES

Benner SA, Ellington AD, Tauer A. 1989. Modern metabolism as a palimpsest of the RNA world. *Proc Natl Acad Sci* **86**: 7054–7058.

Benner SA, Kim H-J Yang Z. 2010. Setting the Stage: The history, chemistry, and geobiology behind RNA. *Cold Spring Harb Perspect Biol* doi: 10.1101/cshperspect.a003541.

Blackburn EH, Collins K. 2010. Telomerase: An RNP enzyme synthesizes DNA. *Cold Spring Harb Perspect Biol* doi: 10.1101/cshperspect.a003558.

Breaker RR. 2010. Riboswitches and the RNA world. *Cold Spring Harb Perspect Biol* doi: 10.1101/cshperspect.a003566.

Cech TR. 1986. A model for the RNA-catalyzed replication of RNA. *Proc Natl Acad Sci* **83**: 4360–4363.

Crick FH. 1968. The origin of the genetic code. *J Mol Biol* **38**: 367–379.

Ellington AD, Szostak JW. 1990. In vitro selection of RNA molecules that bind specific ligands. *Nature* **346**: 818–822.

Ferré-D'Amaré AR, Scott WG. 2010. Small self-cleaving ribozymes. *Cold Spring Harb Perspect Biol* doi: 10.1101/cshperspect.a003574.

Garst AD, Edwards AL, Batey RT. 2010. Ribswitches: Structures and mechanisms. *Cold Spring Harb Perspect Biol* doi: 10.1101/cshperspect.a003533.

Gilbert W. 1986. Origin of life: The RNA world. *Nature* **319**: 618.

Gold L, Janjic N, Jarvis T, Schneider D, Walker JJ, Wilcox SK, Zichi D. 2010. Aptamers and the RNA world, past and present. *Cold Spring Harb Perspect Biol* doi: 10.1101/cshperspect.a003582.

Gottesman S, Storz G. 2010. Bacterial small RNA regulators: Versatile roles and rapidly evolving variations. *Cold Spring Harb Perspect Biol* doi: 10.1101/cshperspect.a003798.

Guerrier-Takada C, Gardiner K, Marsh T, Pace N, Altman S. 1983. The RNA moiety of ribonuclease P is the catalytic subunit of the enzyme. *Cell* **35**: 849–857.

Johnston WK, Unrau PJ, Lawrence MS, Glasner ME, Bartel DP. 2001. RNA-catalyzed RNA polymerization: Accurate and general RNA-templated primer extension. *Science* **292**: 1319–1325.

Joshua-Tor L, Hannon GJ. 2010. Ancestral roles of small RNAs: An agocentric perspective. *Cold Spring Harb Perspect Biol* doi: 10.1101/cshperspect.a003772.

Kruger K, Grabowski PJ, Zaug AJ, Sands J, Gottschling DE, Cech TR. 1982. Self-splicing RNA: Autoexcision and autocyclization of the ribosomal RNA intervening sequence of *Tetrahymena*. *Cell* **31**: 147–157.

Lambowitz AM, Zimmerly S. 2010. Group II introns: Mobile ribozymes that invade DNA. *Cold Spring Harb Perspect Biol* doi: 10.1101/cshperspect.a003616.

Lee JT. 2010. The X as model for RNA's niche in epigenomic regulation. *Cold Spring Harb Perspect Biol* doi: 10.1101/cshperspect.a003749.

Lincoln TA, Joyce GF. 2009. Self-sustained replication of an RNA enzyme. *Science* **323**: 1229–1232.

Mathews DH, Moss WN, Turner DH. 2010. Folding and finding RNA secondary structure. *Cold Spring Harb Perspect Biol* doi: 10.1101/cshperspect.a003665.

Moore PB, Steitz TA. 2010. The roles of RNA in the synthesis of protein. *Cold Spring Harb Perspect Biol* doi: 10.1101/cshperspect.a003780.

Noller HF. 2010. Evolution of protein synthesis from an RNA world. *Cold Spring Harb Perspect Biol* doi: 10.1101/cshperspect.a003681.

Noller HF. 1993. Peptidyl transferase: Protein, ribonucleoprotein, or RNA? *J Bacteriol* **175**: 5297–5300.

Noller HF, Chaires JB. 1972. Functional modification of 16S ribosomal RNA by kethoxal. *Proc Natl Acad Sci* **69**: 3115–3118.

Orgel LE. 1968. Evolution of the genetic apparatus. *J Mol Biol* **38**: 381–393.

Perrimon N, Ni J-Q, Perkins L. 2010. In vivo RNAi: Today and tomorrow. *Cold Spring Harb Perspect Biol* doi: 10.1101/cshperspect.a003640.

Powner MW, Gerland B, Sutherland JD. 2009. Synthesis of activated pyrimidine ribonucleotides in prebiotically plausible conditions. *Nature* **459**: 239–242.

Ramakrishnan V. 2010. The ribosome: Some hard facts about its structure and hot air about its evolution. In *RNA Worlds* (eds. JF Atkins, RF Gesteland, TR Cech). Cold Spring Harbor Laboratory Press, Cold Spring Harbor, NY.

Robertson MP, Joyce GF. 2010. The origins of the RNA world. *Cold Spring Harb Perspect Biol* doi: 10.1101/cshperspect.a003608.

Schrum JP, Zhu TF, Szostak JW. 2010. The origins of cellular life. In *RNA Worlds* (eds. JF Atkins, RF Gesteland, TR Cech). Cold Spring Harbor Laboratory Press, Cold Spring Harbor, NY.

Shechner DM, Grant RA, Bagby SC, Koldobskaya Y, Piccirilli JA, Bartel DP. 2009. Crystal structure of the catalytic core of an RNA-polymerase ribozyme. *Science* **326**: 1271–1275.

Steitz J, Borah S, Cazalla D, Fok V, Lytle R, Mitton-Fry R, Riley K, Samji T. 2010. Noncoding RNPs of viral origin. *Cold Spring Harb Perspect Biol* doi: 10.1101/cshperspect.a005165.

Tinoco I, Chen G, Qu X. 2010. RNA reactions one molecule at a time. *Cold Spring Harb Perspect Biol* doi: 10.1101/cshperspect.a003624.

Tuerk C, Gold L. 1990. Systematic evolution of ligands by exponential enrichment: RNA ligands to bacteriophage T4 DNA polymerase. *Science* **249**: 505–510.

Valadkhan S, Mohammadi A, Jaladat Y, Geisler S. 2009. Protein-free small nuclear RNAs catalyze a two-step splicing reaction. *Proc Natl Acad Sci* **106**: 11901–11906.

Volpe T, Martienssen RA. 2010. RNA interference and heterochromatin assembly. *Cold Spring Harb Perspect Biol* doi: 10.1101/cshperspect.a003731.

Wang X, Song X, Glass CK, Rosenfeld MG. 2010. The long arm of long noncoding RNAs: Roles as sensors regulating gene transcriptional programs. *Cold Spring Harb Perspect Biol* doi: 10.1101/cshperspect.a003756.

Westhof E, Masquida B, Jossinet F. 2010. Predicting and modeling RNA architecture. *Cold Spring Harb Perspect Biol* doi: 10.1101/cshperspect.a003632.

Will CL, Lührmann R. 2010. Spliceosome structure and function. *Cold Spring Harb Perspect Biol* doi: 10.1101/cshperspect.a003707.

Woese CR. 1967. The genetic code: The molecular basis for genetic expression. p. 186. Harper & Row.

Wright MC, Joyce GF. 1997. Continuous in vitro evolution of catalytic function. *Science* **276**: 614–617.

Yarus M. 2010. Getting past the RNA world: The initial Darwinian ancestor. *Cold Spring Harb Perspect Biol* doi: 10.1101/cshperspect.a003590.

Zaher HS, Unrau PJ. 2007. Selection of an improved RNA polymerase ribozyme with superior extension and fidelity. *RNA* **13**: 1017–1026.

Setting the Stage: The History, Chemistry, and Geobiology behind RNA

Steven A. Benner, Hyo-Joong Kim, and Zunyi Yang

Foundation for Applied Molecular Evolution, Gainesville, Florida 32601

Correspondence: sbenner@ffame.org

SUMMARY

No community-accepted scientific methods are available today to guide studies on what role RNA played in the origin and early evolution of life on Earth. Further, a definition-theory for life is needed to develop hypotheses relating to the "RNA First" model for the origin of life. Four approaches are currently at various stages of development of such a definition-theory to guide these studies. These are (a) paleogenetics, in which inferences about the structure of past life are drawn from the structure of present life; (b) prebiotic chemistry, in which hypotheses with experimental support are sought that get RNA from organic and inorganic species possibly present on early Earth; (c) exploration, hoping to encounter life independent of terran life, which might contain RNA; and (d) synthetic biology, in which laboratories attempt to reproduce biological behavior with unnatural chemical systems.

Outline

1 Introduction
2 Moving backward in time
3 Moving forward in time on a rocky earth
4 Exploration
5 Synthetic biology
6 The state of the RNA world: 2009
References

1 INTRODUCTION

Most scientists have been taught that science develops models for reality by systematically observing, analyzing observations, and designing experiments to test hypotheses that emerge from that analysis. Those making a career in science soon realize that things are not so simple, especially as some of the most interesting questions in science do not lend themselves to such an approach. These include questions about the current reality ("Does alien life exist in the cosmos?") and future reality ("Will human carbon dioxide emissions trigger catastrophic global warming?"). We cannot observe directly most of the cosmos. We cannot observe directly the future. Other approaches are needed to address such questions.

Other approaches are also needed to address equally interesting questions about the historical past. Going backward in time, these include questions like: "How did *Homo sapiens* originate?" "How did multicellular life emerge?" "How was the Earth formed?"

Fortunately, past reality is more accessible than future or distant realities. Ever since the Enlightenment, natural historians have developed strategies to generate models for the past using observations made today, combined with models that can be experimentally tested that (one hopes) apply to all realities at all times. Using these strategies, we today have good models for how the Earth and its Sun were formed, plausible models for how humankind came into being, and rudimentary models for how multicellularity arose.

Unfortunately, one of the most interesting questions about the past has resisted these approaches: "How did life originate?" The best evidence suggests that the antiquity of the key events is greater than the antiquity of any surviving physical record. Hard work in the Olduvai George or elsewhere on Earth does not seem likely to find fossils relevant to this question. Nor can we (yet) constrain our models by observing the formation of life on planets orbiting distant stars that are today undergoing biogenesis.

Still worse, the problem is associated with a resilient semantic question: "What is 'life'?" The question has been long discussed (Koshland 2002), often avoided (Baross and Benner 2007), and infrequently addressed, as when a NASA panel concluded that "life is a self-sustaining chemical system capable of Darwinian evolution" (Joyce et al. 1994).

It is clear that any definition of life must incorporate a "theory of life" (Cleland and Chyba 2000; Benner 2009). The "NASA definition" did so, excluding, for example, nonchemical and Lamarckian systems (Benner et al., 2004). But synthetic biologists attempting to make artificial systems "capable of Darwinian evolution" also find that things are not so simple. Synthetic biologists are today rearranging natural genes in unnatural arrangements to form a cell whose genes come entirely from elsewhere (Lartigue et al. 2009). Others have already generated artificial chemical systems capable of Darwinian evolution by rearranging atoms in terran genetic molecular systems (Yang et al. 2009). Still others have found cellular systems that can grow, divide, and collect material (Yarus 2010). We now know that such advances simply move the bar, because few in the community are prepared to call these "artificial life."

The need to codevelop a definition-theory of life concomitant with our development of a model for its earliest forms on Earth makes understanding life's origins one of the most intellectually interesting challenges in contemporary science. Many approaches might meet this challenge (Fig. 1).

One of the most direct comes from the tradition of natural history, which includes fields like geology, paleontology, and molecular evolution. In this tradition, we start with the life that we know on Earth and whatever physical record that we have about past life (including fossils) and work *backward* in time to create models for simpler forms of life, coupling this with what we know about the chemistry of life (Benner et al. 2002). Conjectural models for past biology can be confirmed in part using the emerging field of paleogenetics, which resurrects genes and proteins from extinct organisms for study in the laboratory (Liberles 2007). Paleogenetics allows experiments to be performed directly on biomolecules that vanished long ago, bringing the power of the experimental method directly to bear on historical questions (Benner 2007).

The concept of an "RNA world" (Gilbert 1986) is itself an outcome of this backwards-in-time process (Rich 1962; Crick 1968; White 1976; Visser and Kellogg 1978). The most direct evidence for an RNA World, crystallographic evidence that the RNA in the ribosome catalyzes peptide bond formation (Harms et al. 2001; Moore and Steitz 2003), is applied within a historical argument, one that infers that all ribosomes in terran biology use RNA to synthesize peptide bonds, and therefore the last common ancestor of all ribosomes used RNA to synthesize peptide bonds. Extrapolating back in time using genomic data has offered us a glimpse of the metabolic complexity of the RNA world (Benner et al. 1989; Benner et al. 1993; Koonin 2003).

Alternatively, we might work forwards in time. We can today make an inventory of molecules observed in nearby regions of our galaxy that are right now forming stars, planets, and (perhaps) life, adding molecules found in comets and meteorites that may have delivered to a nascent Earth (Pizzarello 2004; Chyba and Sagan 1992). We might complete our inventory by guessing what other molecules might have been generated on primitive Earth, exploiting

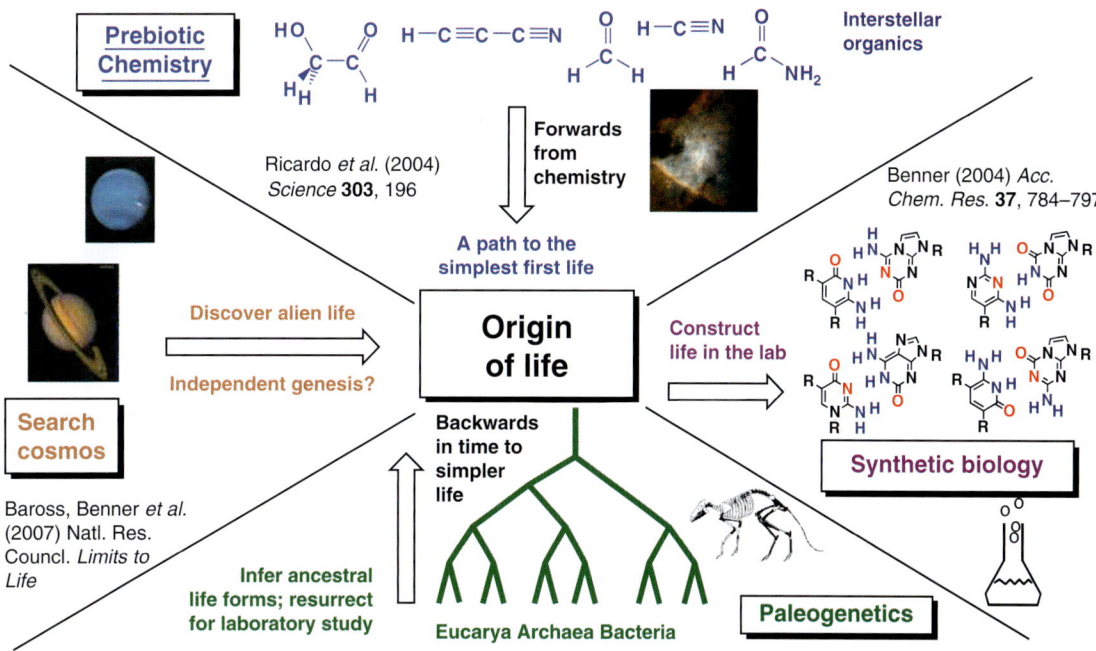

Figure 1. The origin of life as a historical question cannot be studied directly. There are, however, many indirect ways to approach the question. Four of these are illustrated here. The bottom wedge represents approaches that work backwards in time from contemporary biology to more ancient forms of life, The top wedge represents approaches that work forward in time from organic species presumably available on early Earth to the first Darwinian chemical systems. The left wedge represents approaches that hope to discover an alien or weird form of life by exploration, a form of life whose structure might constrain models for how terran life emerged. The right wedge represents efforts in the laboratory to create artificial Darwinian systems, systems that might further constrain models of how terran life emerged, even if they are not constrained by current models for the environment on early Earth. Benner et al. (2007) *Adv Enzymol Mol Biol Protein Evol* **75**: 1–132.

our best models for its minerals, atmosphere, and ocean at that time. Given this inventory, experiments can be imagined whereby we create a chemical system capable of Darwinian evolution (perhaps RNA) from this inventory (Benner et al. 2006; Rich 1962).

A third approach, exploration, has driven discovery and paradigm change throughout human existence. We might go to other planets to see what we can find. If, for example, Mars or Titan hold life forms having histories independent of life on Earth, observation there might jolt our concept of what life is, how it might emerge, and whether RNA is universally a key component.

Alternatively, an alternative form of life might exist here in Earth in a "shadow biosphere," evading our detection so far because we have not looked in the right place or in the right way (Benner 1999; Cleland and Copley 2005; Davies et al. 2009). A shadow biosphere might even hold RNA-only organisms (Benner 1999), which would not necessarily be detectable using probes that look for ribosomal RNA. Unfortunately, exploration on other planets remains expensive, slow, and dangerous, as terran exploration has always been. Nor can exploration easily find what we do not know how to seek or how to recognize should we encounter it.

Therefore, a fourth approach might return to synthetic biology (Benner and Sismour 2005). Failing to find a second example of life, we might create one in a laboratory on Earth. Again, we might start with nucleic acids as catalysts, focusing on a cartoon of how RNA life might have emerged on Earth. In this cartoon, a nucleic acid emerged on early Earth able to catalyze the template-directed synthesis of its copy (Rich 1962). We might attempt in the laboratory to create such a nucleic acid.

Unfortunately, confident as we are of RNA catalysis and the *possibility* that RNA could catalyze its own template-directed replication (which, if accompanied by strand displacement, might meet a definition for life), no one has actually *showed* that RNA can do so. The closest appears to be work by Bartel, Zaher, Unrau and others (Zaher and Unrau 2007), much discussed elsewhere in this collection, which has yielded an RNA molecule able to add about two dozen monomers to a primer before it falls apart in the solutions containing the high concentrations of Mg^{2+} for its activity.

This sets the stage for two questions whose answers would forward the "RNA first" model for the origin of life: (a) How is functional behavior distributed within RNA sequence space? and (b) How might RNA molecules within that space emerge spontaneously?

The second question requires that we address the challenge of creating oligomeric nucleic acids prebiotically, solving various associated problems (chirality, destruction by water, entropy of polymer assembly from dilute building blocks; Shapiro 2007). The first question tells us how grim the challenge is. If only one in 10^{30} RNA molecules 100 nucleotides in length can spark Darwinian evolution, the challenge is bigger than if 10% of all RNA molecules 20 nucleotides in length can do so.

Both of these questions are experimentally accessible today, and are being pursued (Rajamani et al. 2008), perhaps without the full intensity that one might desire. The two ingredients for success in science (funding and enthusiasm) are presently missing in much of the scientific community to pursue such questions with intensity. Even the Howard Hughes Medical Institute and the Templeton Foundation, long sources of support for fundamental questions of these types, have turned their attention elsewhere in recent years.

These four approaches set the stage for a collection that places RNA at the center of our model for life as a universal. As the Table of Contents shows, Earth today is still very much an "RNA world." More than half of the pages of this collection describe what RNA does today in the terran biosphere, especially with the proteins that it encodes and synthesizes, the DNA that instructs it and the metabolism that supports it. These can be approached by scientific methods that we learned in school. Accordingly, this article will emphasize the historical, alien, and synthetic parts of the RNA world, approachable only by more exotic methods.

2 MOVING BACKWARD IN TIME

The RNA World model may be examined using just about every strategy available to draw inferences about the historical past through observations made on modern biosphere. Perhaps the most remarkable consequence of those observations is that all known life on Earth is built from the same encoded biopolymers (RNA, DNA, and proteins) built from the same building blocks. This historical reality allows us to align the sequences of these biopolymers, first to ask whether they are "homologous" (related by common ancestry) and, if they are, to construct evolutionary trees that show their family relationships.

These alignments led to the discovery that ribosomal RNA molecules from all known life on Earth were homologous, and that their alignment could guide the construction of a "universal" tree of life (Woese 1998). Conversely, the ribosomal RNA tree can be used to defines species, even in microorganisms in which classical definitions of species fail (Woese 2004). By extension, all ribosomes on Earth descended from a common ancestor. The antiquity of the ribosome is strong evidence for the use of RNA in early terran life (see Lambowitz and Zimmerly 2010; Moore and Steitz 2010).

Evidence for an earlier biosphere on Earth that relied on RNA before the emergence of encoded proteins comes from the structure of the ribosome. An RNA component of the ribosome appears to be in direct contact with reacting atoms as the peptide bond is formed (Moore and Steitz 2003). Given the homology among all ribosomal RNA, one can infer from the ribosomal structures from even a few organisms that the synthesis of *all* peptide bonds on Earth is catalyzed by RNA (Harms et al. 2001), and that their common ancestral ribosome also used RNA to catalyze peptide bond synthesis. This completes the use of structural biology to define the concept of the RNA world, a use that began with the structure of transfer RNA (Kim et al. 1974), which prompted Francis Crick to comment that tRNA looked like a molecule trying to be a catalyst (Crick 1968).

Paleogenetic resurrections have enriched this concept. For example, the structures of many proteins that interact with the ribosome and ribosomal RNA are also homologous in many forms of life. Elongation factors, which present charged tRNA to the ribosome, are an example. Evolutionary trees based on their alignment are models for the familial history of elongation factors. Taking the next step, once the those trees are available, it is possible to infer the sequence of ancestral elongation factors and, through the magic of recombinant DNA biotechnology, bring them back for study in the laboratory.

Gaucher et al. did exactly this for bacterial elongation factors dating back perhaps three billion years (Gaucher et al. 2003; Gaucher et al. 2008). Inferring the sequences of ancestral elongation factors, these authors inferred the sequences of various candidate ancestral elongation factors, synthesized genes encoding these ancient proteins, and resurrected the now-extinct elongation factors in the laboratory. These resurrected proteins are, of course, physical manifestations of a hypothesis that these sequences actually existed three billion years ago and worked with a functioning ribosome that also existed at that time.

Paleogenetics experiments allowed these historical hypotheses to be tested in the laboratory, first by seeing if the hypothetical ancestral elongation factors actually work. The result was positive: The putative ancestral elongation factors work.

This, in turn, allowed a discovery: The temperature optima for the ancestral elongation factors deep in the eubacterial tree was rather high, about 65°C (Gaucher et al. 2003). The temperature optimum of *modern* elongation factors, measured in vitro, is the same as the ambient temperature at which their hosts live. Exploiting a favorite axiom from natural history ("The present is the key to the past"), these observations with resurrected elongation factors implies that ancestral bacteria deep in the bacterial tree living billions of years ago lived at 65°C.

Further resurrections of ancestral elongation factors throughout the eubacterial tree inferred a temperature history of life, all assuming a functioning, homologous ribosome (Gaucher et al. 2008). Perhaps not so far in the future, entire ancestral ribosomes will be resurrected, taking paleogenetics back in time about as far as the contemporary record might allow. If we are more fortunate, we might be able to infer the sequence of membrane proteins that diverged before the last common ancestor (Linkkila and Gogarten 1991) and use paleogenetics strategies to go still farther back in time.

Experiments and observations with the ribosome were not the only ones to drive biological chemists to conclude that an RNA world was a historical reality. Already in 1976, White noted that RNA fragments attached to various cofactors (Fig. 2, magenta moieties) were widely distributed in modern terran life, and were most likely present in the last common ancestor of life on Earth (White 1976). White suggested that these RNA fragments were remnants of an RNA world. Just two years later, Visser and Kellogg (1978) used these observations to account for the reactivity and distribution of biotin in modern metabolism.

By the end of the 1980s, sufficient sequence data were available to take the next step. Here, comparative genomics was used to infer the entire genetic complement of the organism that is represented in the universal tree as the organism connecting the archeal, bacterial, and eucaryal kingdoms (Benner et al. 1989; Benner et al. 1993; Anantharaman et al. 2002; Koonin 2003). Correlations between these with biosignatures in the geological record, including the emergence of atmospheric dioxygen, were used to infer a metabolism encoded by the "protogenome" found in that organism (Benner et al. 1989). From there, it was argued that the noninvolvement of the RNA portions of the RNA cofactors in the core of their functional chemical reactivity implied that they arose at a time when RNA was the only biopolymer available. This, in turn, was taken to imply that the RNA world was metabolically complex, containing many RNA enzymes able to catalyze, for example, the transfer phosphate groups, reduction and oxidation reactions, and the formation of carbon–carbon bonds.

3 MOVING FORWARD IN TIME ON A ROCKY EARTH

Such narratives nicely illustrate why many are confident that an RNA World existed as a historical reality, and why

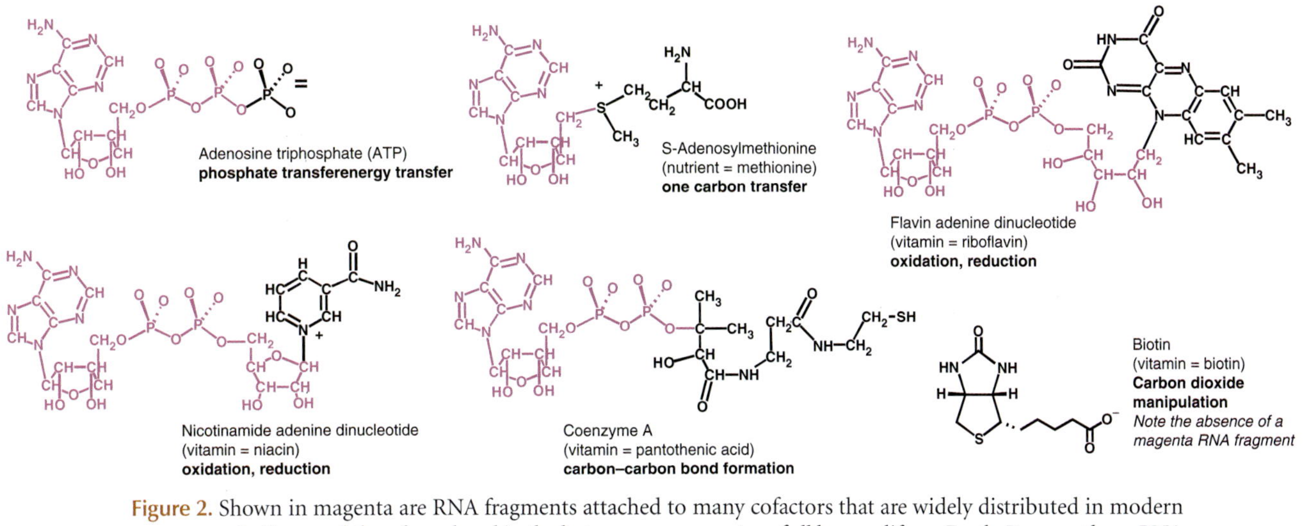

Figure 2. Shown in magenta are RNA fragments attached to many cofactors that are widely distributed in modern terran metabolism, and therefore placed in the last common ancestor of all known life on Earth. Because those RNA fragments do not participate in the chemistry of the metabolic reaction, they are not likely to have arisen convergently, but rather reflect an episode of life on Earth when RNA was the only encoded component of biocatalysis, and used these fragments as "handles." Under this hypothesis, the *absence* of a magenta RNA cofactor on biotin implies that it arose *after* the RNA World.

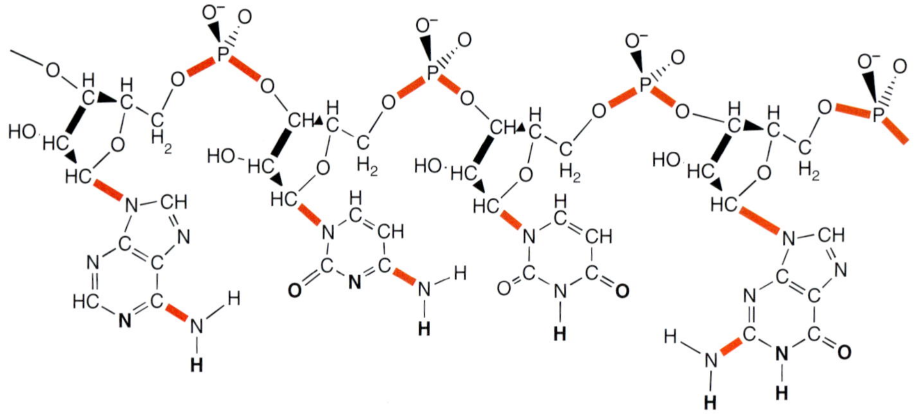

Figure 3. The red bonds in this general structure for RNA are all thermodynamically unstable with respect to hydrolysis in water, and suffer from a standard organic reaction mechanism by which they can hydrolyze.

some agree that metabolism was complex in the RNA biosphere. However, no matter how convincing these narratives, they do not force the conclusion that RNA was the *first* encoded biopolymer on Earth. To support this conclusion, one must find a model that accounts for the emergence of polymeric RNA without the intervention of prior biology.

The community is today twice divided. The first divide separates some who attempt to find such a model from others who consider that such a model cannot be found (Shapiro 2007). At the center of those who doubt that RNA emerged prebiotically are many experiments defining the intrinsic instability of RNA in water. For example, the red bonds in Figure 3 are all thermodynamically unstable with respect to hydrolysis in water, a substance often believed essential for life (Baross and Benner 2007). Even if prebiotic processes managed to assemble (up a free energy gradient) several dozen ribonucleotides of the same chirality, the product RNA might have promptly fallen apart. Still worse, reactivity intrinsic in the 2′-OH group of RNA will lead to cleavage even in the absence of water. Indeed, the most successful ribozymal RNA polymerases to date (Lawrence and Bartel 2003; Zaher and Unrau 2007) do not indefinitely replicate RNA because they themselves fall apart in water in the presence of divalent magnesium cations, which are required at substantial concentration for their reactivity. It is not surprising that Joyce and Orgel called RNA a "prebiotic chemist's nightmare" (Joyce and Orgel 1999; Joyce and Orgel 2006).

These observations alone are sufficient for some thoughtful authors to abandon RNA as a candidate for the first genetic molecule (Larralde et al. 1995; Shapiro 2007). Some have abandoned *all* genetic biopolymers, suggesting instead that something like Darwinian evolution must have been supported by a set of small organic molecules dissipating free energy in a cycle that, through its operation, can adapt to changing conditions and evolve in a Darwinian sense (Kauffman 1986; Smith and Morowitz 2004). Others, notably the late Leslie Orgel, disputed this view (Orgel 2008). This topic is being debated at a special session in the April 2010 Astrobiology Science Conference in Houston, a debate that has an unpredictable outcome.

Unfortunately, RNA encounters prebiotic synthesis problems long before it becomes an oligomer. As discussed in Robertson and Joyce 2010, approaches exist to generate various nucleobases from precursor molecules that the community has come to view as "plausibly prebiotic." Unfortunately, three of these nucleobases (not uracil) suffer hydrolytic deamination in water (Levy and Miller 1998).

Still worse, ribose is unstable. Just a decade ago, Stanley Miller and his group quantitated the instability of ribose by measuring the rate of decomposition of ribose under a variety of conditions. At pH 7 and 100°C, ribose decomposes with a half-life of only 73 minutes; its half-life is 44 year at 0°C (Larralde et al. 1995). These observations led Miller to "preclude the use of ribose and other sugars as prebiotic reagents except under very special conditions. It follows that ribose and other sugars were not components of the first genetic material, and that other possibilities, such as the peptide nucleic acids, and other non-sugar based backbones, should be examined."

Unfortunately, uncharged nonsugar backbones do not appear to be able to support Darwinian evolution (Richert et al. 1996; Benner and Hutter 2002). Other carbohydrates, such as threose, can scaffold a biopolymer with pairing properties (Horhota et al. 2005), but suffer from the same instability problems as ribose. The hunt is on for other biopolymers that might both support RNA-like replication and are stable under the conditions in which they are formed.

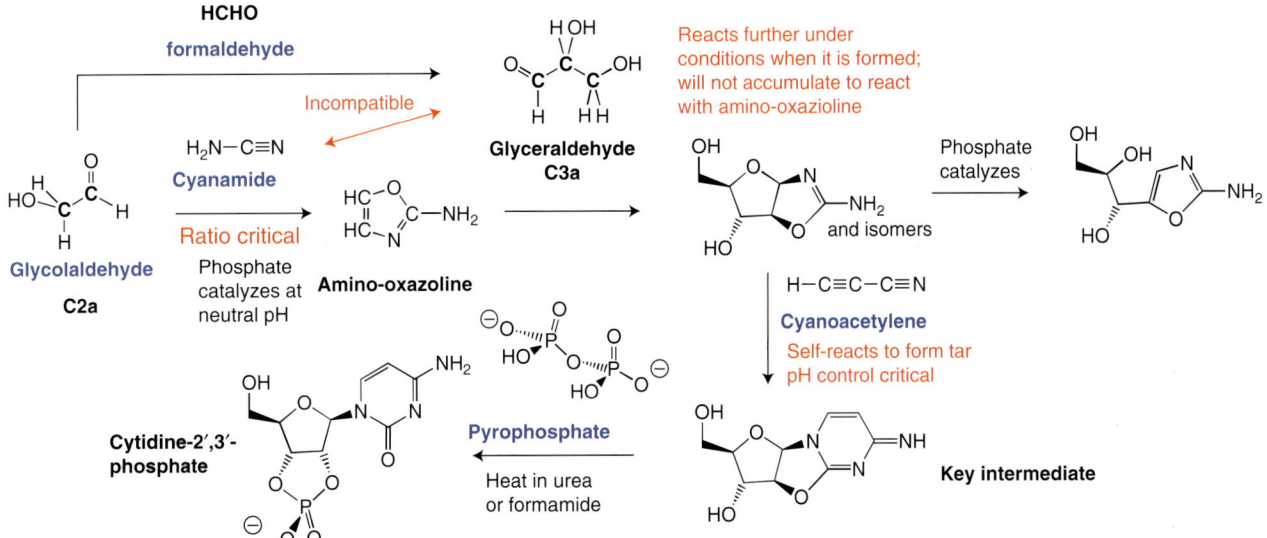

Figure 4. A scheme to obtain one of the red bonds in Figure 3 (Powner et al. 2009). For those accepting Shapiro's critique of an "RNA first" model for life's origin (Shapiro 2007), this scheme offers much to criticize. Compounds such as glyceraldehyde are not formed under conditions in which they accumulate. Glyceraldehyde and cyanamide are incompatible. Hands-on intervention by intelligent chemists must control their ratios and availabilities. To get the 2′,3′-cyclic phosphate of cytidine, pyrophosphate and the key intermediate must be heated in urea or formamide (certainly available on early Earth), but as a different reaction medium. Some regard this as an example of "synthetic organic chemistry," not "prebiotic chemistry," others disagree (Benner 2009).

Even the community that has *not* abandoned RNA as the first Darwinian molecule is divided, however. The division is illustrated by a line of recent work of Sutherland and his coworkers, illustrated in Figure 4 (Powner et al. 2009). Their approach focused on generating in water the bond that joins ribose to a cytosine heterocycle, one of the red bonds in Figure 3. They developed a laboratory route that forms this bond early, with later completion of the synthesis of both the heterocyclic and ribose rings. In this route, 2-aminooxazole and glyceraldehyde react in water, freshly prepared cyanoacetylene is added, and the product mixture is recovered and suspended in a urea solution with pyrophosphate, which is then placed on a filter, evaporated, and heated. Phosphate serves as a catalyst and buffer critical for several of these steps.

Even among those searching for a prebiotic origin of RNA, Sutherland's approach is not universally regarded as satisfactory. It requires unstable molecules such as glyceraldehyde as precursors; glyceraldehyde might be made from the condensation of glycolaldehyde and formaldehyde, which is indisputably prebiotic, but reacts further under conditions in which it is formed. The solution of cyanoacetylene must be freshly prepared, as cyanoacetylene polymerizes to form tar. Even if the precursors are stipulated, several of them are incompatible; for example, cyanamide destroys glyceraldehyde. Further, the ratio of glycolaldehyde and cyanamide is critical to the success of the first step.

Attempting to manage these issues, Sutherland suggested that 2-aminooxazole might have been formed in a location different from glyceraldehyde, sublimed from that location into a planetary atmosphere, to be later rained into a separate pond holding the glyceraldehyde in the correct ratio (Anastasi et al. 2007). Critics of this type of prebiotic chemistry find this an excessive *Deus ex machina*, but also note that if amino-oxazole is to be formed in a different locale than glyceraldehyde, the effort to find pH neutral conditions to form amino-oxazole (recognizing the instability of glyceraldehyde at high pH) was unnecessary. If this were not sufficient, the final step requires yet another change of environments, to urea or formamide as a reaction mixture, not water.

Shapiro criticized such syntheses, noting that they are analogous to a golfer "who having played a golf ball through an 18-hole course, then assumed that the ball could also play itself around the course in his absence" (Shapiro 2007). Sutherland himself recognized the potential power of such criticisms, writing: "One can imagine a number of scenarios that would result in heating, and progressive dehydration followed by cooling, rehydration, and irradiation" to get the desired cytidine. "Comparative assessment of these scenarios is beyond the scope of this work" (Powner et al. 2009). To which Shapiro might reply: "Indeed."

Thus, the remaining part of the community still considering an "RNA first" model for life's origins is looking

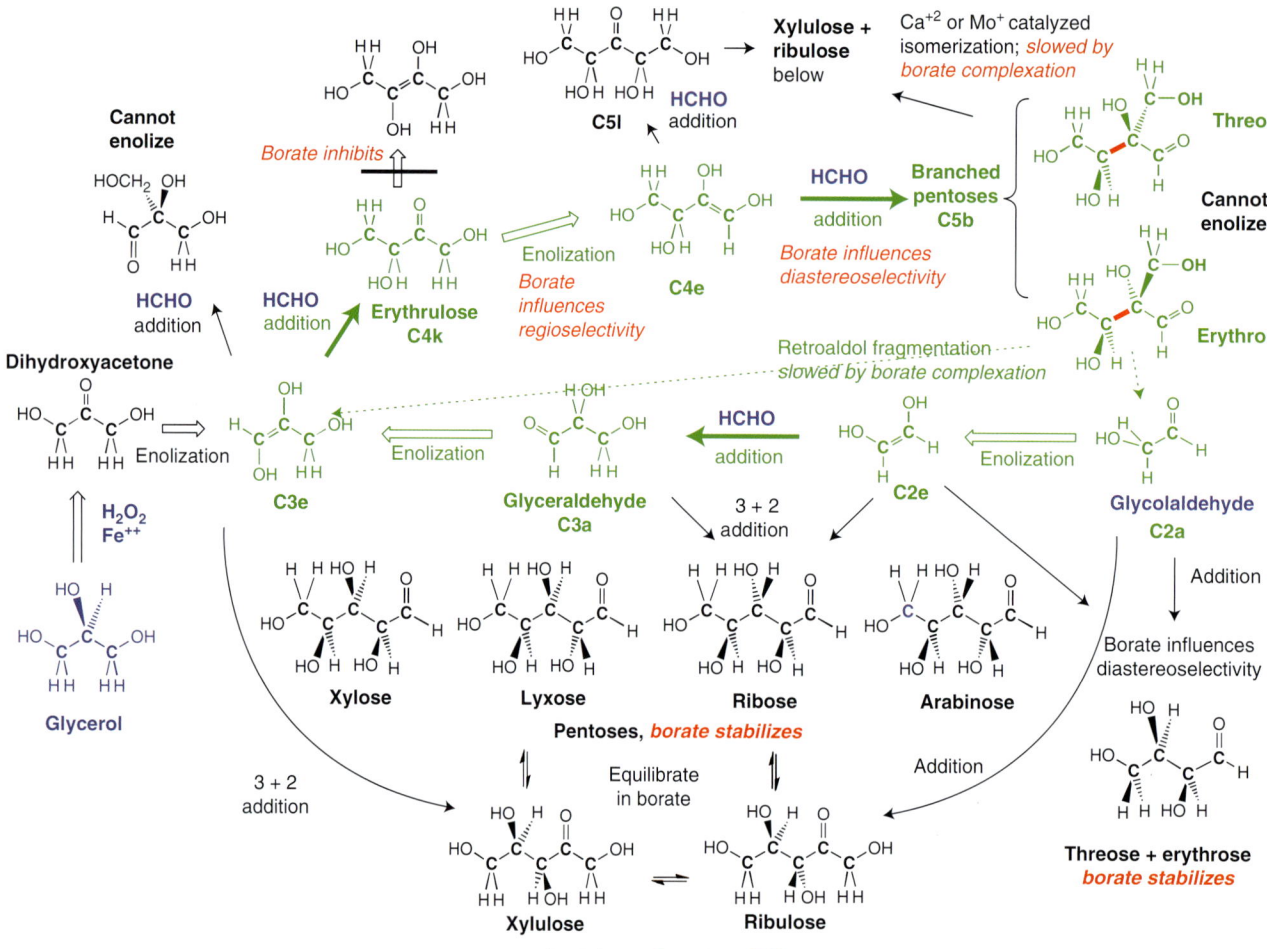

Figure 5. In an effort to find a mineral-stabilized route to pentoses and pentuloses, HJ Kim et al. (in prep.) propose a "premetabolic cycle" under the control of mineral borate. Compounds and reactions involved in the cycle are shown in green; the cycle fixes formaldehyde (HCHO) operating clockwise. Prebiotic compounds in blue feed the cycle. Leakage from the cycle indicated by black arrows emerging from green compounds; in each case but one, these lead directly or indirectly to pentoses and pentuloses, which are stabilized by borate. Whether this represents long sought experimental support for a "metabolism first" model for the origin of life, or the way in which pentoses and pentuloses emerged prebiotically, is disputable.

for ways to stabilize unstable intermediates, especially if such stabilization also guides productively their further reaction and avoids reactions that lead to tar. To these ends, some have added moieties, such as phosphate, to ribose (Xiang et al. 1994; Muller et al. 1990). Others propose organic scaffoldings (Persil and Hud 2007). Still others propose that minerals stabilize reactive intermediates, including minerals that contain borate (Ricardo et al. 2004).

Although these efforts have focused on making RNA under a "gene first" model for the origin of life, they curiously have come as close as any to providing a working example of a cycle reminiscent of models that put "metabolism first." For example, Figure 5 shows a cycle that starts with glycolaldehyde, abundant in the cosmos (Hollis et al. 2001), and fixes formaldehyde (formed by electrical discharge through an atmosphere of moist carbon dioxide) in the presence of borate minerals to give stereoisomeric branched pentoses. These branched pentoses are themselves unable to enolize, and form stable complexes with borate. They can, however, undergo calcium-catalyzed rearrangement to give pentuloses directly, or suffer retroaldolcleavage to give glycolaldehyde and glyceraldehyde, which can either assemble in the presence of borate to form pentoses, which are themselves stabilized by borate, or add more formaldehyde to form more branched pentose.

Experimental work with this cycle has uncovered several of the challenges that must be met for such cycles to support prebiotic synthesis, let alone to serve as a "metabolism first" evolving system. Key to these is "leakage," the loss of material from the cycle to give undesired products. If carbon leaks from the cycle faster than it is fixed, the cycle disappears.

In the cycle in Figure 5, a principal source of leakage arises from the addition of formaldehyde to the enediol of erythrulose at the less hindered center. In this particular case, experiments show that the leakage is productive; the resulting 3-pentulose isomerizes to give pentuloses (ribulose, xylulose) that are stabilized by borate as well. Leakage through the reaction of two glycolaldehyde molecules gives threose, an alternative genetic carbohydrate, which is also stabilized. These are both fortunate outcomes, as all of these carbohydrates have potential as building blocks for genetic materials. Other cycles need not be so fortunate.

4 EXPLORATION

Since the third edition of *The RNA World* appeared (Gesteland et al. 2006), exploration of our solar system has discovered substantial amounts of perchlorate on Mars and oceans of methane on Titan, the largest moon of Saturn. The first is interesting as it offers one explanation for the failure of previous landers on Mars, including the 1976 Viking lander, to detect organic species (Benner et al. 2000). As a first step in its analysis of Martian soil, Viking heated a sample to 500°C. Organics heated with perchlorate at this temperature ignite, a property useful in fireworks. At Martian temperatures, however, perchlorate is entirely compatible with organic species. Thus, optimists still note that organics may exist in the surface of Mars, and RNA is not excluded from these.

Another dramatic climax in our exploration of the Solar System was the Cassini-Huygens landing on Titan. There, at 94 K (-179°C), oceans of methane were observed. Unfortunately, RNA is unlikely to dissolve in liquid methane (although the experiment seems not to have been performed). The alkalinity of the alternative fluid proposed for Titan, subsurface water-ammonia eutectics, would make an RNA world unlikely there as well.

The search for remnants of the RNA world on Earth (Benner 1999; Cleland and Copley 2005) has also not advanced much. Several authors have now given serious consideration to pursuing it further (Davies et al. 2009), and the first ideas suggesting how to explore for a "shadow biosphere" have been presented. For example, as 70% of the volume of a typical eubacterial cell is filled with the machinery to make proteins, a surviving rioorganism might be considerably smaller than the smallest modern bacteria. It may, therefore, be found in environments where size is a constraint.

5 SYNTHETIC BIOLOGY

Since the third edition of *The RNA World*, synthetic biological approaches, represented by the right wedge in Figure 1, have made progress (Benner 2009). As discussed in other articles in this collection, experimentalists are changing the structure of RNA to help manage parts of its unfortunate reactivities. Amino groups replacing the ribose hydroxyl groups are being used to facilitate template-directed polymerization (Stutz et al. 2007) (Fig. 6). Threose is being contemplated as an alternative to ribose (Schoning et al. 2000). Many in the community have recognized that the RNA World model might be supported if nucleic acids of *any* kind can support Darwinian evolution in the laboratory.

One limitation of the RNA World has been hypothesized to be the paucity of building blocks in the RNA biopolymer (Benner et al. 1999). Certainly, proteins, with 20 amino acids, have a richer diversity of functionality than RNA, and this functionality is useful for binding and catalysis. With these thoughts in mind, several groups have sought to increase the number of nucleotides in nucleic acids (Hirao 2006; Benner 2004; Henry and Romesberg 2003).

The challenge of synthesizing components of an artificially expanded genetic information systems and showing that they bind to each other with expanded Watson-Crick specificity has been met for several systems (Henry and Romesberg 2003; Benner 2004). Accordingly, the bar has moved. Today's challenge has been to develop enzymatic processes that allow artificial genetic systems to be copied, and their copies to be copied indefinitely in a PCR format, without unidirectional loss of the unnatural components of the molecule (Johnson et al. 2004). This process of repeated copying, with imperfections, with the imperfections themselves being copy-able, is the essence of Darwinism at the molecular level.

Today, three examples are now known in which six nucleotide letters can be incorporated generally into PCR without substantial loss of the components with essentially independent selection of sequence (Sismour et al. 2004; Sismour et al. 2005; Yang et al. 2007). In principle, any of these systems can support in vitro evolution.

A particularly successful example of these is the six letter PCR incorporating 6-amino-5-nitropyridin-2-one and its complement imidazo(1,2-c)pyrimidin-5(1H)-one, trivially designated Z and P (Fig. 6). Protein polymerases support "six letter PCR" with this system. Further, mechanisms have been found by which Z:P pairs are mutated to give C:G pairs and C:G pairs are mutated back to Z:P pairs. As with natural mutation, this process involves

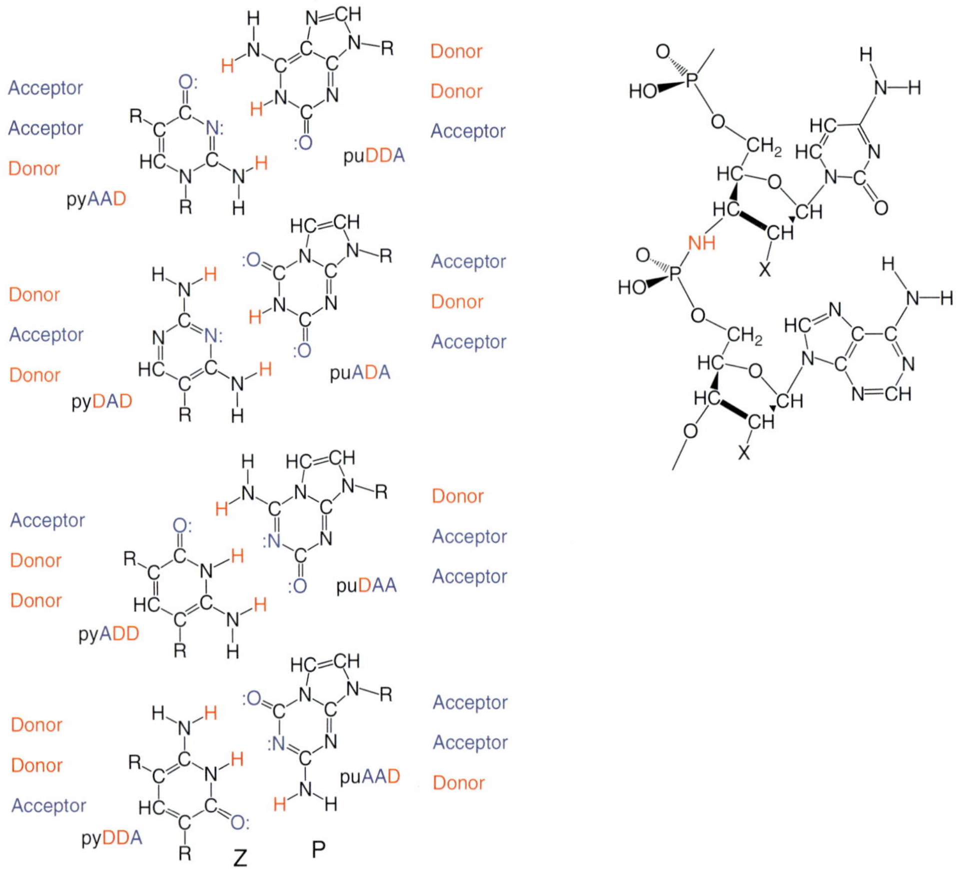

Figure 6. Some unnatural nucleic acid structures that have been developed by synthetic biologists as RNA-like (but not RNA) molecules possibly capable of supporting Darwinian evolution, including the expanded genetic alphabet (left) that supports six-letter PCR and the phosphoramidate linkage that supports enzyme-free primer extension (Stutz et al. 2007).

the protonation and deprotonation of the nucleobase heterocycles. Thus, this artificial genetic system is capable of supporting Darwinian evolution, and has been shown to do so in the laboratory (Fig. 7) (Yang and Benner 2009).

Although this system certainly can support the test of the hypothesis that nucleic acids having more letters can deliver more functional diversity than nucleic acids having fewer (Reader and Joyce 2002), it does not constitute a "nucleic acids only" system; it depends on polymerases that have been delivered by four billion years of natural biological evolution. Joseph Piccirilli suggested some time ago that perhaps the RNA World had more nucleotides of this type, and therefore had access to greater catalytic diversity than it presently seems to have access to with just adenine, guanine, cytosine, and uracil. In this model, the genetic biopolymer evolved to lose building blocks, becoming more streamlined, perhaps reflecting the fact that four nucleotides are optimal in a genetic system (Szathmary 1999).

6 THE STATE OF THE RNA WORLD: 2009

As subsequent articles in this collection show, we today live in an RNA World. Further, it seems very likely that catalytic RNA played a pervasive role in earlier life. These inferences make even more intriguing the failure of prebiotic experiments to generate RNA easily, as well as our failure to have tripped across RNA-only organisms in a shadow biosphere, and the difficulty in getting synthetic systems based on RNA or RNA-like to catalyze their own reproduction via monomer addition. We are missing something in our models for reality.

What might be wrong? For example, we may have misread the historical import of the modern record. The synthesis of oligonucleotides by monomer addition, for example, is difficult in the laboratory, even though it is today "universal" in terran biology. But perhaps the first RNA replication did not use monomer addition. Ligation, recombination (Hayden et al. 2005), and other processes are certainly available to obtain longer RNA molecules

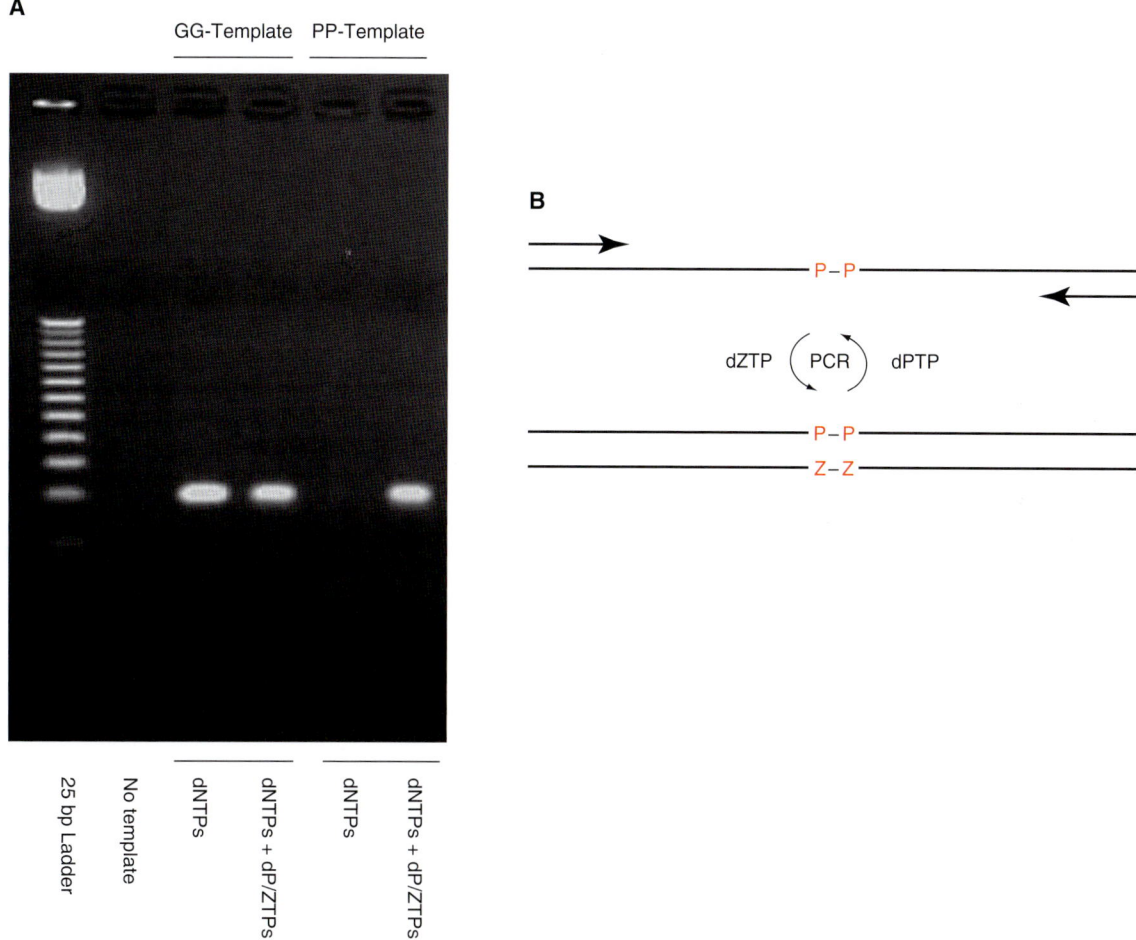

Figure 7. "Six letter PCR" showing products of amplification of an amplicon containing two adjacent nonstandard nucleotides.

from shorter ones, which seem at present to be much easier to obtain prebiotically. These alternative processes may have been lost in all lineages of terran life that we know about.

It is also, of course, possible that we have not modeled correctly the early environment on Earth. Hot versus cold, water or not, the nature of available minerals, and places on the planet (surface or deep oceans) that could possibly have cradled life's formation are all uncertain. For example, solvents like formamide, which lack certain of the RNA-destroying properties of water, may have dominated a largely water-free environment where the first RNA molecules were assembled.

For these reasons, synthetic biology might offer the most constructive paradigm for future effort. Synthetic biology sets a grand challenge, here the construction of an RNA system that catalyzes its own replication in the context of chosen metabolism, isolation systems, and environmental conditions. Pursuit of that challenge drags scientists across uncharted terrain where they must ask and answer unscripted questions. If theory driving that pursuit is inadequate, the synthesis fails, and fails in a way that cannot be ignored. In contrast, if observations contradict an accepted theory, they are (as often as not) ignored (Benner 2009). For this reason, synthesis drives discovery and paradigm change in ways that analysis cannot. And the incremental advances in the historical, alien, origins parts of the RNA world since the third edition of this collection, suggest that discovery and paradigm change both will be needed before the RNA World becomes established as a statement about the historical past.

REFERENCES

Anantharaman V, Koonin EV, Aravind L. 2002. Comparative genomics and evolution of proteins involved in RNA metabolism. *Nucl Acids Res* **30:** 1427–1464.

Anastasi C, Crowe MA, Sutherland JD. 2007. Two-step potentially prebiotic synthesis of r-D-cytidine-5'-phosphate from D-glyceraldehyde-3-phosphate. *J Am Chem Soc* **129:** 24–25.

Baross J, Benner SA, Cody GD, Copley SD, Pace NR, Scott JH, Shapiro R, Sogin ML, Stein JL, Summons R, et al. 2007. *The limits of organic life in planetary systems.* The National Academies Press, Washington DC.

Benner SA. 1999. How small can a microorganism be? *Size limits of very small microorganisms: proceedings of a workshop, steering group on astrobiology of the space studies board.* National Research Council, pp. 126–135.

Benner SA. 2004. Understanding nucleic acids using synthetic chemistry. *Acc Chem Res* **37**: 784–797.

Benner SA. 2007. The early days of paleogenetics. Connecting molecules to the planet. *Experimental Paleogenetics*, (ed. D.A., Liberles), pp. 3–19, Academic Press.

Benner SA. 2009. *Life, the universe and the scientific method*, Gainesville FL, FfAME Press.

Benner SA, Hutter D. 2002. Phosphates, DNA, and the search for nonterrean life: A second generation model for genetic molecules. *Bioorg Chem* **30**: 62–80.

Benner SA, Sismour AM. 2005. Synthetic biology. *Nature Rev Genetics* **6**: 533–543.

Benner SA, Ellington AD, Tauer A. 1989. Modern metabolism as a palimpsest of the RNA world. *Proc Nat Acad Sci* **86**: 7054–7058.

Benner SA, Ricardo A, Carrigan MA. 2004. Is there a common chemical model for life in the universe? *Curr Opinion Chem Biol* **8**: 672–689.

Benner SA, Sassi SO, Gaucher EA. 2007. Molecular paleosciences. Systems biology from the past. *Adv Enzymol Related Areas Mol Biol Protein Evol* **75**: 1–132.

Benner SA, Burgstaller P, Battersby TR, Jurczyk S. 1999. Did the RNA world exploit an expanded genetic alphabet? In *The RNA world*, 2nd ed. (ed. RF Gesteland et al.), pp. 163–181. Cold Spring Harbor Laboratory Press, Cold Springs Harbor, NY.

Benner SA, Caraco MD, Thomson JM, Gaucher EA. 2002. Planetary biology. Paleontological, geological, and molecular histories of life. *Science* **293**: 864–868.

Benner SA, Carrigan MA, Ricardo A, Frye F. 2006. Setting the stage: The history, chemistry and geobiology behind RNA. In The *RNA world* 3rd ed. (ed. R.F. Gesteland et al.), pp. 1–21. Cold Spring Harbor Laboratory Press, Cold Springs Harbor, NY.

Benner SA, Cohen MA, Gonnet GH, Berkowitz DB, Johnsson K. 1993. Reading the palimpsest. Contemporary biochemical data and the RNA world. In *The RNA world* (ed. R. Gesteland, J. Atkins), pp. 27–70. Cold Spring Harbor Laboratory Press, Cold Springs Harbor, NY.

Benner SA, Devine KG, Matveeva LN, Powell DH. 2000. The missing organic molecules on Mars. *Proc Natl Acad Sci* **97**: 2425–2430.

Chyba C, Sagan C. 1992. Endogenous production, exogenous delivery and impact-shock synthesis of organic-molecules: An inventory for the origins of life. *Nature* **355**: 125–132.

Cleland CE, Chyba CF. 2000. Defining 'life'. *Orig Life Evol Biosphere* **32**: 387–393.

Cleland CE, Copley SD. 2005. The possibility of alternative microbial life on Earth. *Int J Astrobiol* **4**: 165–173.

Crick FHC. 1968. The origin of the genetic code. *J Mol Biol* **38**: 367–379.

Davies PCW, Benner SA, Cleland CE, Lineweaver CH, McKay CP, Wolfe-Simon F. 2009. Signatures of a shadow biosphere. *Astrobiol* **9**: 241–249.

Gaucher EA, Govindarajan S, Ganesh OK. 2008. Palaeotemperature trend for Precambrian life inferred from resurrected proteins. *Nature* **451**: 704–U2.

Gaucher EA, Thomson JM, Burgan MF, Benner SA. 2003. Inferring the paleoenvironment during the origins of bacteria based on resurrected ancestral proteins. *Nature* **425**: 285–288.

Gesteland R.F, Cech T.R, Atkins J.F. editors. 2006. *The RNA World* 3rd Edition. Cold Spring Harbor Laboratory Press, Cold Springs Harbor, NY.

Gilbert W. 1986. The RNA World. *Nature* **319**: 818.

Harms J, Schluenzen F, Zarivach R, Bashan A, Gat S, Agmon I, Bartels H, Franceschi F, Yonath A. 2001. High resolution structure of the large ribosomal subunit from a Mesophilic Eubacterium. *Cell* **107**: 679–688.

Hayden EJ, Riley CA, Burton AS, et al. 2005. RNA-directed construction of structurally complex and active ligase ribozymes through recombination. *RNA* **11**: 1678–1687.

Henry AA, Romesberg FE. 2003. Beyond A, C, G and T: Augmenting Nature's alphabet. *Curr Opin Chem Biol* **7**: 727–733.

Hirao I. 2006. Unnatural base pair systems for DNA/RNA-based biotechnology. *Curr Opin Chem Biol* **10**: 622–627.

Hollis JM, Vogel SN, Snyder LE, Jewell PR, Lovas FJ. 2001. The spatial scale of glycolaldehyde in the galactic center. *Astrophys J* **554**: L81–L85 Part 2.

Horhota A, Zou K, Ichida JK, Yu B, McLaughlin LW, Szostak JW, Chaput JC. 2005. Kinetic analysis of an efficient DNA-dependent TNA polymerase. *J Am Chem Soc* **127**: 7427.

Jermann TM, Opitz JG, Stackhouse J, Benner SA. 1995. Reconstructing the evolutionary history of the artiodactyl ribonuclease superfamily. *Nature* **374**: 57–59.

Johnson SC, Sherrill CB, Marshall DJ, Moser MJ, Prudent JR. 2004. A third base pair for the polymerase chain reaction. Inserting isoC and isoG. *Nucl Acids Res* **32**: 1937–1941.

Joyce GF. 1994. Foreward. In *Origins of life: The central concepts* (ed. D.W. Deamer et al.), pp. xi–xii. Jones & Bartlett, Boston.

Joyce GF, Orgel LE. 2006. in *The RNA world* (ed. R.F. Gesteland et al.), pp. 23–56, Cold Spring Harbor Laboratory Press, Cold Spring Harbor, New York.

Joyce GF, Orgel LE. 1999. Prospects for understanding the origin of the RNA world. in *The RNA World*, 2nd edition (ed. R.F. Gesteland et al.), pp. 49–77, Cold Spring Harbor Laboratory Press, Cold Spring Harbor, New York.

Kauffman SA. 1986. Autocatalytic sets of proteins. *J Theor Biol* **119**: 1–24.

Kim SH, Suddath FL, Quigley GJ, Mcpherson A, Sussman JL, Wang AHJ, Seeman NC, Rich A. 1974. 3-Dimensional tertiary structure of yeast phenylalanine transfer-RNA. *Science* **185**: 435–440.

Koonin EV. 2003. Comparative genomics, minimal gene-sets and the last universal common ancestor. *Nature Reviews Microbiol* **1**: 127–136.

Koshland DE. 2002. The seven pillars of life. *Science* **295**: 2215.

Lambowitz AM, Zimmerly S. 2010. Group II introns: mobile ribozymes that invade DNA. *Cold Spring Harb Perspect Biol* **2**: a003616.

Larralde R, Robertson MP, Miller SL. 1995. Rates of decomposition of ribose and other sugars. Implications for chemical evolution. *Proc Natl Acad Sci* **92**: 8158–8160.

Lartigue C, Vashee S, Algire MA, Chuang RY, Benders GA, Ma L, Noskov VN, Denisova EA, Gibson DG, Assad-Garcia N, et al. 2009. Creating bacterial strains from genomes that have been cloned and engineered in yeast. *Science*, published online.

Lawrence MS, Bartel DP. 2003. Processivity of ribozyme-catalyzed RNA polymerization. *Biochemistry* **42**: 8748–8755.

Levy M, Miller S. 1998. The stability of the RNA bases: Implications for the origin of life. *Proc Natl Acad Sci* **95**: 7933–7938.

Liberles D.A. editor 2007. *Experimental PALEOGENETICS*, Academic Press.

Linkkila TP, Gogarten JP. 1991. Tracing origins with molecular sequences – rooting the universal tree of life. *Trends in Biochemical Sciences* **16**: 287–288.

Moore PB, Steitz TA. 2003. The structural basis of large ribosomal subunit function. *Annual Review Biochem* **72**: 813–850.

Moore PB, Steitz TA. 2010. The roles of RNA in the synthesis of protein. *Cold Spring Harb Perspect Biol* **2**: a003780.

Muller D, Pitsch S, Kittaka A, Wagner E, Wintner CE, Eschenmoser A. 1990. Chemistry of α-aminonitriles—aldomerisation of glycolaldehyde phosphate to rac-hexose 2,4,6-triphosphates and (in presence of formaldehyde) rac-pentose 2,4-diphosphates—rac-allose 2,4,6-triphosphate and rac-ribose 2,4-diphosphate are the main reaction-products. *Helv Chim Acta* **73:** 1410–1468.

Orgel LE. 2008. On the implausibility of metabolic cycles on prebiotic Earth. *PLOS Biol* **6:** e18.

Persil O, Hud NV. 2007. Harnessing DNA intercalation. *Trends in Biotech* **25:** 433–436.

Pizzarello S. 2004. Chemical evolution and meteorites: An update. *Origins Life Evol Biosphere* **34:** 25–34.

Powner MW, Gerland B, Sutherland JD. 2009. Synthesis of activated pyrimidine ribonucleotides in prebiotically plausible conditions. *Nature* **459:** 239–242.

Rajamani S, Vlassov A, Benner S, Coombs A, Olasagasti F, Deamer D. 2008. Lipid-assisted synthesis of RNA-like polymers from mononucleotides. *Orig Life Evol Biosphere* **38:** 57–74.

Reader JS, Joyce GF. 2002. A ribozyme composed of only two different nucleotides. *Nature* **470:** 841–844.

Ricardo A, Carrigan MA, Olcott AN, Benner SA. 2004. Borate minerals stabilize ribose. *Science* **303:** 196.

Rich A. 1962. On the problems of evolution and biochemical information transfer. in *Horizons In Biochemistry* (eds. M. Kasha, B. Pullmann), pp. 103–126. Academic Press, New York.

Richert C, Roughton AL, Benner SA. 1996. Nonionic analogs of RNA with dimethylene sulfone bridges. *J Am Chem Soc* **118:** 4518–4531.

Robertson MP, Joyce GF. 2010. The origins of the RNA world. *Cold Spring Harb Perspect Biol* **2:** a003608.

Schoning KU, Scholz P, Guntha S, Wu X, Krishnamurthy R, Eschenmoser A. 2000. Chemical etiology of nucleic acid structure: The α-threofuranosyl-(3′→2′) oligonucleotide system. *Science* **290:** 1347–1351.

Shapiro R. 1988. Prebiotic ribose synthesis: A critical analysis, *Origins of Life* **18:** 71–85.

Shapiro R. 2000. A replicator was not involved in the origin of life. *IUBMB Life* **49:** 173–176.

Shapiro R. 2007. A simpler origin for life. *Scientific American* **296:** 46–53.

Sismour AM, Benner SA. 2005. The use of thymidine analogs to improve the replication of an extra DNA base pair: A synthetic biological system. *Nucl Acids Res* **33:** 5640–5646.

Sismour AM, Lutz S, Park J-H, Lutz MJ, Boyer PL, Hughes SH, Benner SA. 2004. PCR amplification of DNA containing non-standard base pairs by variants of reverse transcriptase from human immunodeficiency virus-1. *Nucl Acids Res* **32:** 728–735.

Smith E, Morowitz HJ. 2004. Universality in intermediary metabolism. *Proc Natl Acad Sci* **101:** 13168–13173.

Stutz JAR, Kervio E, Deck C, Richert C. 2007. Chemical primer extension: Individual steps of spontaneous replication. *Chemisty & Biodiversity* **4:** 784–802.

Szathmary E. 1999. The origin of the genetic code: Amino acids as cofactors in an RNA world. *Trends Genet* **15:** 223–229.

Visser CM, Kellog RM. 1978. Biotin. Its place in evolution. *J Mol Evol* **11:** 171–178.

White HBIII. 1976. Coenzymes as fossils of an earlier metabolic state. *J Mol Evol* **7:** 101.

Woese C. 1967. The evolution of the genetic code. In *The genetic code*, pp. 179–195. Harper & Row, New York.

Woese CR. 1998. The universal ancestor. *Proc Natl Acad Sci* **95:** 6854–6859.

Woese CR. 2004. A new biology for a new century. *Microb Mol Biol Rev* **68:** 173–186.

Xiang Y-B, Drenkard S, Baumann K, Hickey D, Eschenmoser A. 1994. Chemistry of α-amino nitriles. 12 Exploratory experiments on thermal-reactions of α-amino nitriles. *Helv Chim Acta,* **77:** 2209.

Yang Z, Benner SA. 2009. Darwinian systems based on an expanded genetic alphabet. *Nature Chem Biol* submitted.

Yang Z, Sismour AM, Sheng P, Puskar NL, Benner SA. 2007. Enzymatic incorporation of a third nucleobase pair. *Nucl Acids Res* **35:** 4238–4249.

Yarus M. 2010. Getting past the RNA world: the initial Darwinian ancestor. *Cold Spring Harb Perspect Biol* **2:** a003590.

Zaher HS, Unrau PJ. 2007. Selection of an improved RNA polymerase ribozyme with superior extension and fidelity. *RNA* **13:** 1017–1026.

The Origins of the RNA World

Michael P. Robertson and Gerald F. Joyce

Departments of Chemistry and Molecular Biology and The Skaggs Institute for Chemical Biology, The Scripps Research Institute, La Jolla, California 92037

Correspondence: gjoyce@scripps.edu

SUMMARY

The general notion of an "RNA World" is that, in the early development of life on the Earth, genetic continuity was assured by the replication of RNA and genetically encoded proteins were not involved as catalysts. There is now strong evidence indicating that an RNA World did indeed exist before DNA- and protein-based life. However, arguments regarding whether life on Earth began with RNA are more tenuous. It might be imagined that all of the components of RNA were available in some prebiotic pool, and that these components assembled into replicating, evolving polynucleotides without the prior existence of any evolved macromolecules. A thorough consideration of this "RNA-first" view of the origin of life must reconcile concerns regarding the intractable mixtures that are obtained in experiments designed to simulate the chemistry of the primitive Earth. Perhaps these concerns will eventually be resolved, and recent experimental findings provide some reason for optimism. However, the problem of the origin of the RNA World is far from being solved, and it is fruitful to consider the alternative possibility that RNA was preceded by some other replicating, evolving molecule, just as DNA and proteins were preceded by RNA.

Outline

1 Introduction

2 An "RNA-first" view of the origin of life

3 An "RNA-later" view of the origin of life

4 Concluding remarks

References

1 INTRODUCTION

The general idea that, in the development of life on the Earth, evolution based on RNA replication preceded the appearance of protein synthesis was first proposed over 40 yr ago (Woese 1967; Crick 1968; Orgel 1968). It was suggested that catalysts made entirely of RNA are likely to have been important at this early stage in the evolution of life, but the possibility that RNA catalysts might still be present in contemporary organisms was overlooked. The unanticipated discovery of ribozymes (Kruger et al. 1982; Guerrier-Takada et al. 1983) initiated extensive discussion of the role of RNA in the origins of life (Sharp 1985; Pace and Marsh 1985; Lewin 1986) and led to the coining of the phrase "the RNA World" (Gilbert 1986).

"The RNA World" means different things to different investigators, so it would be futile to attempt a restrictive definition. All RNA World hypotheses include three basic assumptions: (1) At some time in the evolution of life, genetic continuity was assured by the replication of RNA; (2) Watson-Crick base-pairing was the key to replication; (3) genetically encoded proteins were not involved as catalysts. RNA World hypotheses differ in what they assume about life that may have preceded the RNA World, about the metabolic complexity of the RNA World, and about the role of small-molecule cofactors, possibly including peptides, in the chemistry of the RNA World.

There is now strong evidence indicating that an RNA World did indeed exist on the early Earth. The smoking gun is seen in the structure of the contemporary ribosome (Ban et al. 2000; Wimberly et al. 2000; Yusupov et al. 2001). The active site for peptide-bond formation lies deep within a central core of RNA, whereas proteins decorate the outside of this RNA core and insert narrow fingers into it. No amino acid side chain comes within 18 Å of the active site (Nissen et al. 2000). Clearly, the ribosome is a ribozyme (Steitz and Moore 2003), and it is hard to avoid the conclusion that, as suggested by Crick, "the primitive ribosome could have been made entirely of RNA" (1968).

A more tenuous argument can be made regarding whether life on Earth began with RNA. In what has been referred to as "The Molecular Biologist's Dream" (Joyce and Orgel 1993), one might imagine that all of the components of RNA were available in some prebiotic pool, and that these components could have assembled into replicating, evolving polynucleotides without the prior existence of any evolved macromolecules. However, a thorough consideration of this "RNA-first" view of the origin of life inevitably triggers "The Prebiotic Chemist's Nightmare", with visions of the intractable mixtures that are obtained in experiments designed to simulate the chemistry of the primitive Earth. Perhaps this continuing nightmare will eventually have a happy ending, and recent experimental findings provide some reason for optimism. However, the problem of the origin of the RNA World is far from being solved, and it is fruitful to consider the alternative possibility that RNA was preceded by some other replicating, evolving molecule, just as DNA and proteins were preceded by RNA.

2 AN "RNA-FIRST" VIEW OF THE ORIGIN OF LIFE

2.1 Abiotic Synthesis of Polynucleotides

This section considers the synthesis of oligonucleotides from ß-D-nucleoside 5′-phosphates, leaving aside for now the question of how the nucleotides became available on the primitive Earth. Two fundamentally different chemical reactions are involved. First, the nucleotide must be converted to an activated derivative, for example, a nucleoside 5′-polyphosphate. Next the 3′-hydroxyl group of a nucleotide or oligonucleotide molecule must be made to react with the activated phosphate group of a monomer. Synthesis of oligonucleotides from nucleoside 3′-phosphates will not be discussed because activated nucleoside 2′- or 3′-phosphates in general react readily to form 2′,3′-cyclic phosphates. These cyclic phosphates are unlikely to oligomerize efficiently because the equilibrium constant for dimer formation is only of the order of 1.0 L/mol (Erman and Hammes 1966; Mohr and Thach 1969). In the presence of a complementary template somewhat larger oligomers might be formed because the free energy of hybridization would help to drive forward the chain extension reaction.

In enzymatic RNA and DNA synthesis, the nucleoside 5′-triphosphates (NTPs) are the substrates of polymerization. Polynucleotide phosphorylase, although it is a degradative enzyme in nature, can be used to synthesize oligonucleotides from nucleoside 5′-diphosphates. Nucleoside 5′-polyphosphates are, therefore, obvious candidates for the activated forms of nucleotides. Although nucleoside 5′-triphosphates are not formed readily, the synthesis of nucleoside 5′-tetraphosphates from nucleotides and inorganic trimetaphosphate provides a reasonably plausible prebiotic route to activated nucleotides (Lohrmann 1975). Other more or less plausible prebiotic syntheses of nucleoside 5′-polyphosphates from nucleotides have also been reported (Handschuh et al. 1973; Osterberg et al. 1973; Reimann and Zubay 1999). Less clear, however, is how the first phosphate would have been mobilized to convert the nucleosides to 5′-nucleotides. Nucleoside 5′-polyphosphates are high-energy phosphate esters, but are relatively unreactive in aqueous solution. This may be advantageous for enzyme-catalyzed polymerization, but is a severe obstacle for the nonenzymatic polymerization of nucleoside 5′-polyphosphates, which would occur far more slowly than the hydrolysis of the resulting polynucleotide.

In a different approach to the activation of nucleotides, the isolation of an activated intermediate is avoided by using a condensing agent such as a carbodiimide (Khorana 1961). This is a popular method in organic synthesis, but its application to prebiotic chemistry is problematic. Potentially prebiotic molecules such as cyanamide and cyanoacetylene activate nucleotides in aqueous solution, but the subsequent condensation reactions are inefficient (Lohrmann and Orgel 1973).

Most attempts to study nonenzymatic polymerization of nucleotides in the context of prebiotic chemistry have used nucleoside 5'-phosphoramidates, particularly nucleoside 5'-phosphorimidazolides. Although phosphorimidazolides can be formed from imidazoles and nucleoside 5'-polyphosphates (Lohrmann 1977), they are only marginally plausible as prebiotic molecules. They were chosen because they are prepared easily and react at a convenient rate in aqueous solution.

Nucleotides contain three principal nucleophilic groups: the 5'-phosphate, the 2'-hydroxyl, and the 3'-hydroxyl group, in order of decreasing reactivity. The reaction of a nucleotide or oligonucleotide with an activated nucleotide, therefore, normally yields 5',5'-pyrophosphate-, 2',5'-phosphodiester-, and 3',5'-phosphodiester-linked adducts (Fig. 1A), in order of decreasing abundance (Sulston et al. 1968). Thus the condensation of several monomers would likely yield an oligomer containing one pyrophosphate and a preponderance of 2',5'-phosphodiester linkages (Fig. 1B). There is little chance of producing entirely 3',5'-linked oligomers from activated nucleotides unless a catalyst can be found that increases the proportion of 3',5'-phosphodiester linkages. Several metal ions, particularly Pb^{2+} and UO_2^{2+}, catalyze the formation of oligomers from nucleoside 5'-phosphorimidazolides (Sleeper and Orgel 1979; Sawai et al. 1988). The Pb^{2+}-catalyzed reaction is especially efficient when performed in eutectic solutions of the activated monomers (in concentrated solutions obtained by partial freezing of more dilute solutions). Substantial amounts of long oligomers are formed under eutectic conditions, but the product oligomers always contain a large proportion of 2',5'-linkages (Kanavarioti et al. 2001; Monnard et al. 2003).

What kinds of prebiotically plausible catalysts might lead to the production of 3',5'-linked oligonucleotides directly from nucleoside 5'-phosphorimidazolides or other activated nucleotides? It is unlikely, but not impossible, that a metal ion or simple acid-base catalyst would provide

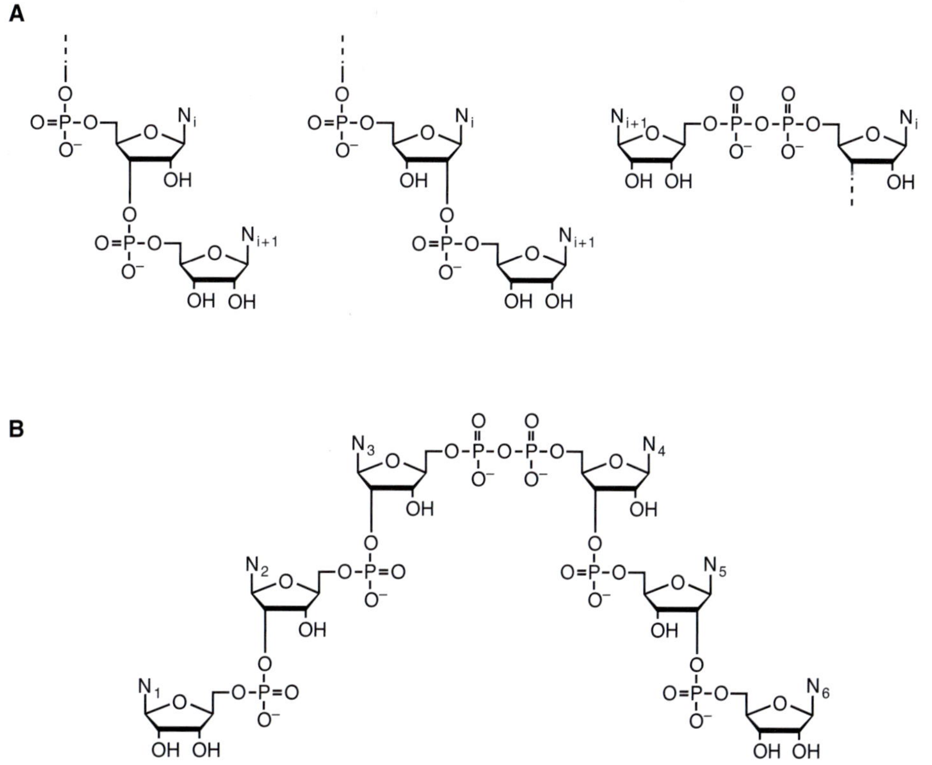

Figure 1. Phosphodiester linkages resulting from chemical condensation of nucleotides. (A) Reaction of an activated mononucleotide (N_{i+1}) with an oligonucleotide ($N_1–N_i$) to form a 3',5'-phosphodiester (*left*), 2',5'-phosphodiester (*middle*), or 5',5'-pyrophosphate linkage (*right*). (B) Typical oligomeric product resulting from chemical condensation of activated mononucleotides.

sufficient regiospecificity. The most attractive of the other hypotheses is that adsorption to a specific surface of a mineral might orient activated nucleotides rigidly and thus catalyze a highly regiospecific reaction.

The work of Ferris and coworkers provides support for this hypothesis (Ferris et al. 2004; Ferris 2006). They have studied the oligomerization of nucleoside 5′-phosphorimidazolides and related activated nucleotides on the clay mineral montmorillonite (Ferris and Ertem 1993; Kawamura and Ferris 1994; Miyakawa and Ferris 2003). Some samples of the mineral are effective catalysts, promoting the formation of oligomers even from dilute solutions of activated nucleotide substrates. Furthermore, the mineral profoundly affects the regiospecificity of the reaction. The oligomerization of adenosine 5′-phosphorimidazolide, for example, gives predominantly 3′,5′-linked products in the presence of montmorillonite, but predominantly 2′,5′-linked products in aqueous solution (Ding et al. 1996; Kawamura and Ferris 1999). Once short oligomers have been synthesized, they can be further extended by adsorbing them on either montmorillonite or hydroxylapatite and repeatedly adding activated monomers, resulting in the accumulation of mainly 3′,5′-linked oligoadenylates up to 40–50 subunits in length (Ferris et al. 1996; Ferris 2002). However, even when adsorbed on montmorillonite, the phosphorimidazolides of the pyrimidine nucleosides yield oligomers that are predominantly 2′,5′-linked.

Long oligomers have also been obtained from monomers in a single step using a different activated nucleotide in which imidazole is replaced by 1-methyl-adenine (Prabahar and Ferris 1997; Huang and Ferris 2003). Using the 1-methyl-adenine derivative of adenylate or uridylate, oligomers containing up to 40 subunits were produced, consisting of ~75% 3′,5′-linkages for oligoadenylate and ~60% 3′,5′-linkages for oligouridylate (Huang and Ferris 2006). Oligomerization of the 1-methyl-adenine derivative of guanylate or cytidylate was less efficient, but all four activated monomers could be co-incorporated, to at least a modest extent, within abiotically synthesized oligonucleotides.

Detailed analysis of this work on catalysis by montmorillonite suggests that oligomerization occurs at a limited number of structurally specific active sites within the interlayers of the clay (Wang and Ferris 2001). These sites must not be saturated with sodium ions, which appear to block access of the activated nucleotides (Joshi et al. 2009). Several different samples of montmorillonite have proven to be good catalysts, in part depending on their proton versus sodium ion content. It will be important to determine if there are other types of minerals that are comparably efficient catalysts of oligonucleotide synthesis, and if so, to study the regiospecificity and sequence generality of the reactions they catalyze.

2.2 Nonenzymatic Replication of RNA

If a mechanism existed on the primitive Earth for the polymerization of activated nucleotides, it would have generated a complex mixture of product oligonucleotides that differed in both length and sequence. The next stage in the emergence of an RNA World would have been the replication of some of these molecules, so that a process equivalent to natural selection could begin. The reaction central to replication of nucleic acids is template-directed synthesis, that is, the synthesis of a complementary oligonucleotide under the direction of a preexisting oligonucleotide. A good deal of work has already been performed on this aspect of nonenzymatic replication. This work has been reviewed elsewhere (Joyce 1987; Orgel 2004a), so only a summary of the results will be given here.

The first major conclusion is that most activated nucleotides do not undergo efficient, regiospecific, template-directed reactions in the presence of an RNA or DNA template. In general, only a small proportion of template molecules succeed in directing the synthesis of a complete complement, and the complement usually contains a mixture of 2′,5′- and 3′,5′-phosphodiester linkages. After a considerable search, a set of activated nucleotides was found that undergo efficient and highly regiospecific template-directed reactions. Working with guanosine 5′-phospho-2-methylimidazolide (2-MeImpG), it was shown that poly(C) can direct the synthesis of long oligo(G)s in a reaction that is highly efficient and highly regiospecific (Inoue and Orgel 1981). If poly(C) is incubated with an equimolar mixture of the four 2-MeImp$N$s (N = G, A, C, or U), less than 1% of the product consists of noncomplementary nucleotides (Inoue and Orgel 1982). Subsequent experiments suggested that this and the related reactions discussed later occur preferentially within the context of double helices that have a structure resembling the A form of RNA (Kurz et al. 1997, 1998; Kozlov et al. 1999, 2000).

Random copolymers containing an excess of C residues can be used to direct the synthesis of products containing G and the complements of the other bases present in the template (Inoue and Orgel 1983). The reaction with a poly(C,G) template is especially interesting because the products, like the template, are composed entirely of C and G residues. If these products in turn could be used as templates, it might allow the emergence of a self-replicating sequence. Self-replication, however, is unlikely, mainly because poly(C,G) molecules that do not contain an excess of C residues tend to form stable self-structures that prevent them from acting as templates (Joyce and Orgel 1986). The self-structures are of two types: (1) the standard Watson-Crick variety based on C•G pairs, and (2) a quadrahelix structure that results from the association of four

G-rich sequences. As a consequence, any C-rich oligonucleotide that can serve as a good template will give rise to G-rich complementary products that tend to be locked in self-structure and so cannot act as templates. Overcoming the self-structure problem using the standard C and G nucleotides is very difficult because it requires the discovery of conditions that favor the binding of mononucleotides to allow template-directed synthesis to occur, but suppress the formation of long duplex regions that would exclude activated monomers from the template.

Some progress has been made in discovering defined-sequence templates that are copied faithfully to yield complementary products (Inoue et al. 1984; Acevedo and Orgel 1987; Wu and Orgel 1992a). Successful templates typically contain an excess of C residues, with A and U residues isolated from each other by at least three C residues. Runs of G residues are copied into runs of C residues, so long as the formation of self-structures by G residues can be avoided (Wu and Orgel 1992b). In light of the available evidence, it seems unlikely that a pair of complementary sequences can be found, each of which facilitates the synthesis of the other using nucleoside 5′-phospho-2-methylimidazolides as substrates. Some of the obstacles to self-replication may be attributable to the choice of reagents and reaction conditions, but others seem to be intrinsic to the template-directed condensation of activated mononucleotides.

A related nonenzymatic replication scheme involves synthesis by the ligation of short 3′,5′-linked oligomers (James and Ellington 1999). This is certainly an attractive possibility, made more plausible by the discovery of analogous ribozyme-catalyzed reactions (Bartel and Szostak 1993), but it faces two major obstacles. The first is the difficulty of obtaining the substrates in the first place. The second is concerned with fidelity. Pairs of oligonucleotides containing a single base mismatch, particularly if the mismatch forms a G•U wobble pair, still hybridize as efficiently as fully complementary oligomers, except in a temperature range very close to the melting point of the perfectly paired structure. Maintaining fidelity would therefore be difficult under any plausible temperature regime.

Despite these problems, template-directed ligation of short oligonucleotides may be a viable alternative to the oligomerization of activated monomers. Ferris' work discussed above suggests that predominantly 3′,5′-linked oligonucleotides might form spontaneously from activated nucleotides on some variety of montmorillonite (Ferris et al. 1996) or on some other mineral. Oligonucleotide 5′-triphosphates undergo slow but remarkably 3′,5′-regiospecific ligation in the presence of a complementary template (Rohatgi et al. 1996a,b). The combination of some such pair of reactions might provide a replication scheme for polynucleotides starting with an input of activated monomers.

There also are efforts in what is sometimes termed "synthetic biology" to achieve nonenzymatic replication with molecules that resemble biological nucleic acids, but are not constrained by considerations of plausible prebiotic chemistry. For example, the 2′- and 3′-hydroxyl groups of activated mononucleotides can be replaced by an amino group at either position, providing enhanced nucleophilicity and resulting in more rapid template-dependent (and template-independent) oligomerization (Lohrmann and Orgel 1976; Zielinski and Orgel 1985). Dinucleotide building blocks, consisting of 3′-amino, 3′-deoxynucleotide analogs can also be oligomerized in the presence of a suitable condensing agent (Zielinski and Orgel 1987). With additional modification of the nucleotide bases, it has been possible to carry out the template-directed copying of nucleic acid sequences that contain of all four bases (Schrum et al. 2009). These efforts, although not explaining the origin of the RNA World, contribute to understanding the chemical challenges that must be overcome in achieving the nonenzymatic replication of RNA.

2.3 The First RNA Replicase

The notion of the RNA World places emphasis on an RNA molecule that catalyzes its own replication. Such a molecule must function as an RNA-dependent RNA polymerase, acting on itself (or copies of itself) to produce complementary RNAs, and acting on the complementary RNAs to produce additional copies of itself. The efficiency and fidelity of this process must be sufficient to produce viable "progeny" RNA molecules at a rate that exceeds the rate of decomposition of the "parents." Beyond these requirements, the details of the replication process are not highly constrained.

The RNA-first view of the origin of life assumes that a supply of activated ß-D-nucleotides was available by some as yet unrecognized abiotic process. Furthermore, it assumes that a means existed to convert the activated nucleotides to an ensemble of random-sequence polynucleotides, a subset of which had the ability to replicate. It seems to be implicit in the model that such polynucleotides replicate themselves but, for whatever reason, do not replicate unrelated neighbors. It is not clear whether replication involves one molecule copying itself (and its complement) or a family of molecules that together copy each other. These questions are set aside for the moment in order to first consider the question of whether an RNA molecule of reasonably short length can catalyze its own replication with sufficiently high fidelity.

Accuracy and Survival. The concept of an error threshold, that is, an upper limit to the frequency of copying errors that can be tolerated by a replicating macromolecule, was first introduced by Eigen (1971). This important idea has been extended in a series of mathematically sophisticated papers by McCaskill, Schuster, and others (McCaskill 1984a; Eigen et al. 1988; Schuster and Swetina 1988). Here only a brief summary of the subject is provided.

Eigen's model (1971) envisages a population of replicating polynucleotides that draw on a limited supply of activated mononucleotides to produce additional copies of themselves. In this model, the rate of synthesis of new copies of a particular replicating RNA is proportional to its concentration, resulting in autocatalytic growth. The net rate of production is the difference between the rate of formation of error-free copies and the rate of decomposition of existing copies of the RNA. For an advantageous RNA to outgrow its competitors, its net rate of production must exceed the mean rate of production of all other RNAs in the population. Only the error-free copies of the advantageous RNA contribute to its net rate of production, but all the copies of the other RNAs contribute to their collective production. Thus the relative advantage enjoyed by the advantageous individual compared with the rest of the population (often referred to as the "superiority" of the advantageous individual) must exceed the probability of producing an error copy of that advantageous individual.

The proportion of copies of an RNA that are error free is determined by the fidelity of the component condensation reactions that are required to produce a complete copy. For simplicity, consider a self-replicating RNA that is formed by n condensation reactions, each having mean fidelity q. The probability of obtaining a completely error-free copy is given by q^n, which is the product of the fidelity of the component condensation reactions. If an advantageous individual is to outgrow its competitors, q^n must exceed the superiority, s, of that individual. Expressed in terms of the number of reactions required to produce the advantageous individual,

$$n < |\ln s|/|\ln q|.$$

For $s > 1$ and $q > 0.9$, this equation simplifies to

$$n < \ln s/(1-q).$$

This is the "error threshold," which describes the inverse relationship between the fidelity of replication, q, and the maximum allowable number of component condensation reactions, n. The maximum number of component reactions is highly sensitive to the fidelity of replication, but depends only weakly on the superiority of the advantageous individual. For a self-replicating RNA that is formed by the template-directed condensation of activated mononucleotides, a total of $2n - 2$ condensation reactions are required to produce a complete copy. This takes into account the synthesis of both a complementary strand and a complement of the complement.

It should be recognized that a marked superiority of one sequence over all other sequences could not be maintained over evolutionary time because novel variants would soon arise to challenge the dominant species. However, a marked initial superiority may be important in allowing an efficient self-replicating RNA to emerge from a pool of less efficient replicators. In the absence of other efficient replicators, a primitive self-replicating RNA that operates with low fidelity may gain a foothold by taking advantage of a somewhat less stringent error threshold. Whether or not this can occur depends on its superiority. For example, an RNA that replicates 10-fold more efficiently than its competitors and does so with 90% fidelity could be no longer than 12 nucleotides, and a similarly advantageous RNA that replicates with 70% fidelity could be no longer than four nucleotides. It seems highly unlikely than *any* of the 17 million possible RNA dodecamers is able to catalyze its own replication with 90% fidelity, and even less likely that a tetranucleotide could catalyze its own replication with 70% fidelity. However, an RNA that replicates 10^6-fold more efficiently than its competitors and does so with 90% fidelity could be as long as 67 nucleotides, and one that replicates with 70% fidelity could be as long as 20 nucleotides.

When self-replication is first established, fidelity is likely to be poor and there is strong selection pressure favoring improvement of the fidelity. As fidelity improves, a larger genome can be maintained. This allows exploration of a larger number of possible sequences, some of which may lead to further improvement in fidelity, which in turn allows a still larger genome size, and so on. Once the evolving population has achieved a fidelity of about 99%, a genome length of about 100 nucleotides can be maintained, even for modest superiority values. This would allow RNA-based life to become firmly established. Until that time, it is a race between evolutionary improvement in the context of a sloppy self-replicating system and the risk of delocalization of the genetic information because of overstepping the error threshold. If the time required to bootstrap to high fidelity and large genomes is too long, there is a risk that the population will succumb to an environmental catastrophe before it has had the chance to develop appropriate countermeasures.

It is difficult to state with certainty the minimum possible size of an RNA replicase ribozyme. An RNA consisting of a single secondary structural element, that is, a small stem-loop containing 12–17 nucleotides, would not be expected to have replicase activity, whereas a double stem-loop, perhaps forming a "dumbbell" structure or a pseudoknot, might just be capable of a low level of activity. A triple stem-loop structure, containing 40–60 nucleotides, offers a reasonable hope of functioning as a replicase ribozyme. One could, for example, imagine a molecule consisting of a pseudoknot and a pendant stem-loop that forms a cleft for template-dependent replication.

Suppose there is some 40-mer that enjoys a superiority of 10^3-fold and replicates with 90% fidelity. This should be regarded as a highly optimistic but not outrageous view of what is possible for a minimum replicase ribozyme. Would such a molecule be expected to occur within a population of random-sequence RNAs? A complete library consisting of one copy each of all 10^{24} possible 40-mers would weigh about 1 kg. There may be many such 40-mers, encompassing both distinct structural motifs and, more importantly, a large number of equivalent representations of each motif. As a result, even a small fraction of the total library, consisting of perhaps 10^{20} sequences and weighing about 1 g, might be expected to contain at least one self-replicating RNA with the requisite properties. It is not sufficient, however, that there be just one copy of a self-replicating RNA. The above calculations assume that a self-replicating RNA can copy itself (or that a fully complementary sequence is automatically available, as will be discussed later). If two or more copies of the same 40mer RNA are needed, then a much larger library, consisting of 10^{48} RNAs and weighing 10^{28} g would be required. This amount is comparable to the mass of the Earth.

At first sight, it might seem that one way to ease the error threshold would be for the replicase ribozyme to accept dinucleotide or trinucleotide substrates, so that copies of the RNA could be formed by fewer condensation reactions. Calculations show that, over a broad range of superiority values, RNAs that are required to replicate with 90% fidelity when using mononucleotide substrates would be required to replicate with roughly 80% fidelity when using dinucleotide substrates or roughly 70% fidelity when using trinucleotide substrates. Thus the use of short oligomers offers only a modest advantage because of lessening of the error threshold, which likely would be outweighed by the greater difficulty of achieving high fidelity when discriminating among the 16 possible dinucleotide or 64 possible trinucleotide substrates, rather than among the four mononucleotides.

If one accepts the RNA-first view that there was a prebiotic pool of random-sequence RNAs, and if one assumes that the pool included a replicase ribozyme containing, say, 40 nucleotides and replicating itself with about 90% fidelity, then it is not difficult to imagine how RNA-based evolution might have started. During the initial period a successful clone would have expanded in the absence of competition. As competition for substrates intensified there would have followed a succession of increasingly more advantageous individuals, each replicating within its error threshold. After a period of intensifying competition, the single most advantageous species would have been replaced by a "quasispecies," that is, a mixture of the most advantageous individual and substantial amounts of closely related individuals that replicate almost as fast and almost as faithfully as the most advantageous one (Eigen and Schuster 1977; Eigen et al. 1988). Under these conditions the persistence of a particular advantageous individual is no longer the problem, but one must understand the evolution of the composition of the quasispecies and the conditions for its persistence. This difficult problem has been partially solved by McCaskill (1984b). The general form of the solution is very similar to the error threshold described by Eigen (1971), but with different values for the constant in the inequality. Thus concerns about the error threshold apply to the quasispecies as well as to the succession of individuals. Practically speaking, however, once a quasispecies distribution of sophisticated replicators had emerged, the RNA World would have been on solid footing and, barring an environmental catastrophe, unlikely to lose the ability to maintain genetic information over time.

Another Chicken-and-Egg Paradox. The previous discussion has tried mightily to present the most optimistic view possible for the emergence of an RNA replicase ribozyme from a soup of random-sequence polynucleotides. It must be admitted, however, that this model does not appear to be very plausible. The discussion has focused on a straw man: The myth of a small RNA molecule that arises de novo and can replicate efficiently and with high fidelity under plausible prebiotic conditions. Not only is such a notion unrealistic in light of current understanding of prebiotic chemistry (Joyce 2002), but it should strain the credulity of even an optimist's view of RNA's catalytic potential. If you doubt this, ask yourself whether you believe that a replicase ribozyme would arise in a solution containing nucleoside 5′-diphosphates and polynucleotide phosphorylase!

If one accepts the notion of an RNA World, one is faced with the dilemma of how such a genetic system came into existence. To say that the RNA World hypothesis "solves the paradox of the chicken-and-the-egg" is correct if one means that RNA can function both as a genetic molecule and as a catalyst that promotes its own replication. RNA-catalyzed RNA replication provides a chemical basis for

Darwinian evolution based on natural selection. Darwinian evolution is a powerful way to search among vast numbers of potential solutions for those that best address a particular problem. Selection based on inefficient RNA replication, for example, could be used to search among a population of RNA molecules for those individuals that promote improved RNA replication. But here one encounters another chicken-and-egg paradox: Without evolution it appears unlikely that a self-replicating ribozyme could arise, but without some form of self-replication there is no way to conduct an evolutionary search for the first, primitive self-replicating ribozyme.

One way that RNA evolution may have gotten started without the aid of an evolved catalyst might be by using nonenzymatic template-directed synthesis to permit some copying of RNA before the appearance of the first replicase ribozyme. Suppose that the initial ensemble of monomers was not produced by random copolymerization, but rather by a sequence of untemplated and templated reactions (Fig. 2), and further suppose that members of the initial ensemble of multiple stem-loop structures could be replicated, albeit inefficiently, by the template-directed process. This would have two important consequences. First, any molecule with replicase function that appeared in the mixture would likely find in its neighborhood similar molecules and their complements, related by descent, thus eliminating the requirement for two unrelated replicases to meet. Second, a majority of molecules in the mixture would contain stem-loop structures. If it is true that ribozyme function is favored by stable self-structure, and if the base-sequences of the stems in stem-loop structures are relatively unimportant for function, this model might provide an economical way of generating a relatively small ensemble of sequences that is enriched with catalytic sequences.

How plausible is the assumption that replicases could act on sequences similar to themselves, while ignoring unrelated sequences? This selectivity could be ensured by segregating individual molecules (or clonal lines) on the surface of mineral grains, on the surface of micelles, or within membranes. Closely related molecules might be segregated as a group through specific hydrogen-bonding interactions (the family that sticks together, replicates together). For any segregation mechanism, weak selection would result if the replicating molecules are sufficiently dispersed that diffusion over their intermolecular distance is slow compared with replication. Computer simulations have shown that under such conditions of segregation, evolutionary bootstrapping can occur, resulting in progressively larger genomes that are copied with progressively greater fidelity (Szabó et al. 2002). Alternatively, the

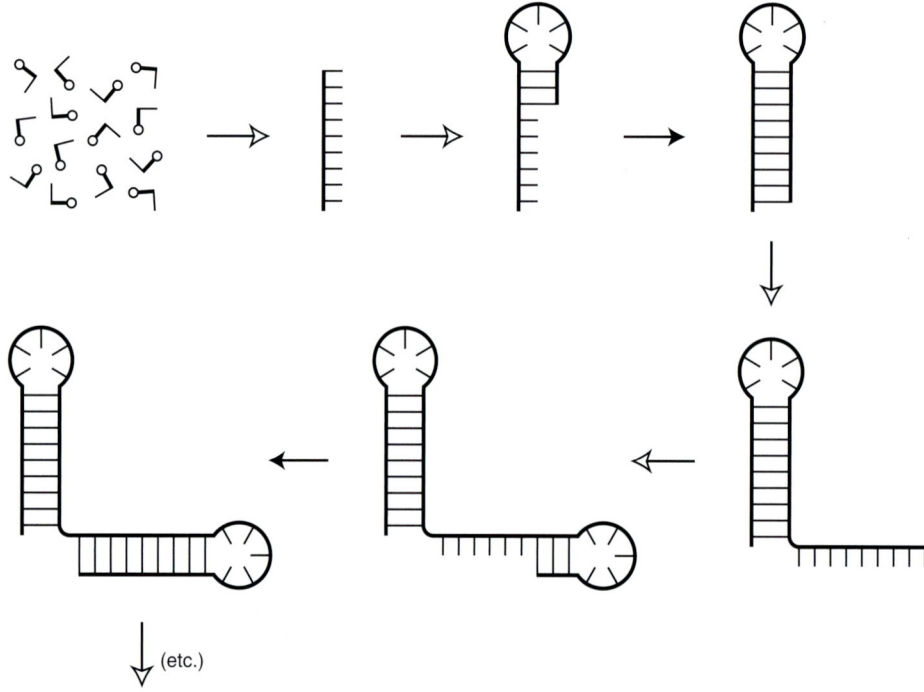

Figure 2. Nonenzymatic synthesis of multi-stem-loop structures as a result of untemplated (open arrowhead) and templated (filled arrowhead) reactions. Template-directed synthesis is assumed to occur rapidly whenever a template, activated monomers, and a suitable primer are available. Once the complementary strand is completed, additional residues are added slowly in a random-sequence manner.

requirement for replication of related, but not unrelated, sequences might be met through the use of "genomic tags" (Weiner and Maizels 1987). Among self-replicating sequences, it is plausible that some are restricted to copying molecules with a particular 3′-terminal subsequence. A replicator that happened by chance to carry a terminal sequence that matched the preference of its active site would replicate itself while ignoring its neighbors.

Another resolution of the paradox of how RNA evolution was initiated without the aid of an evolved ribozyme is to abandon the RNA-first view of the origin of life and suppose that RNA was *not* the first genetic molecule (Cairns-Smith 1982; Shapiro 1984; Joyce et al. 1987; Joyce 1989, 2002; Orgel 1989, 2004a). Perhaps RNA replication arose in the context of an evolving system based on something other than RNA (see the section "Alternative Genetic Systems"). Even if this is true, all of the arguments concerning the relationship between the fidelity of replication and the maximum allowable genome length would still apply to this earlier genetic system. Of course, the challenge to those who advocate the RNA-later approach is to show that there is an informational entity that is prebiotically plausible and is capable of initiating its own replication without the aid of a sophisticated catalyst.

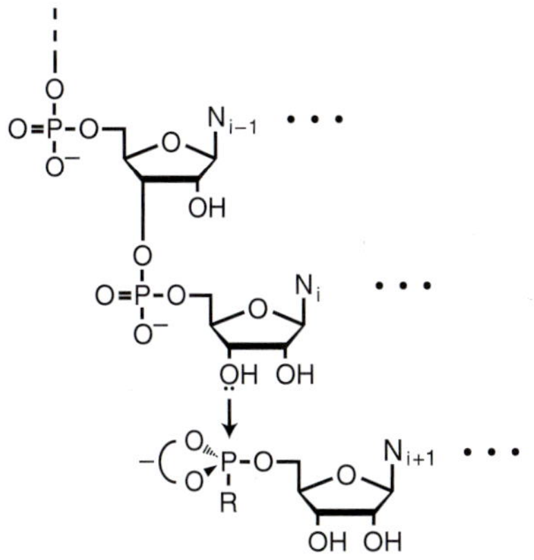

2.4 Replicase Function in the Evolved RNA World

Although it is difficult to say how the first RNA replicase ribozyme arose, it is not difficult to imagine how such a molecule, once developed, would function. The chemistry of RNA replication would involve the template-directed polymerization of mono- or short oligonucleotides, using chemistry in many ways similar to that used by contemporary group I ribozymes (Cech 1986; Been and Cech 1988; Doudna and Szostak 1989). One important difference is that, unlike group I ribozymes, which rely on a nucleoside or oligonucleotide leaving group, an RNA replicase would more likely make use of a different leaving group that provides a substantial driving force for polymerization and that, after its release, does not become involved in some competing phosphoester transfer reaction.

From Ligases to Polymerases. The polymerization of activated nucleotides proceeds via nucleophilic attack by the 3′-hydroxyl of a template-bound oligonucleotide at the α-phosphorus of an adjacent template-bound nucleotide derivative (Fig. 3). The nucleotide is "activated" for attack by the presence of a phosphoryl substituent, for example a phosphate, polyphosphate, alkoxide, or imidazole group. As discussed previously, polyphosphates, such as inorganic pyrophosphate, are the most obvious candidates for the leaving group. The condensation reaction could be assisted

Figure 3. Nucleophilic attack by the 3′-hydroxyl of a template-bound oligonucleotide (N_1-N_i) on the α-phosphorus of an adjacent template-bound mononucleotide (N_{i+1}). Dotted lines indicate base pairing to a complementary template. R is the leaving group.

by favorable orientation of the reactive groups, deprotonation of the nucleophilic 3′-hydroxyl, stabilization of the trigonal-bipyramidal transition state, and charge neutralization of the leaving group. All of these tasks might be performed by RNA (Narlikar and Herschlag 1997; Emilsson et al. 2003), acting either alone (Orteleva-Donnelly and Strobel 1999) or with the help of a suitably positioned metal cation or other cofactor (Shan et al. 1999; Shan et al. 2001).

The possibility that an RNA replicase ribozyme could have existed has been made abundantly clear by work involving ribozymes that have been developed in the laboratory through in vitro evolution (Bartel and Szostak 1993; Ekland et al. 1995; Ekland and Bartel 1996; Robertson and Ellington 1999; Jaeger et al. 1999; Rogers and Joyce 2001; Johnston et al. 2001; McGinness and Joyce 2002; Ikawa et al. 2004; Fujita et al. 2009). Bartel and Szostak (1993), for example, began with a large population of random-sequence RNAs and evolved the "class I" RNA ligase ribozyme, an optimized version of which is about 100 nucleotides in length and catalyzes the joining of two template-bound oligonucleotides. Condensation occurs between the 3′-hydroxyl of one oligonucleotide and the 5′-triphosphate of another, forming a 3′,5′-phosphodiester linkage and releasing inorganic pyrophosphate. This reaction is classified as ligation because of the nature of the oligonucleotide substrates, but involves the same chemical transformation as is catalyzed by modern RNA polymerase enzymes.

X-ray crystal structures of two RNA ligase ribozymes, the L1 and above-mentioned class I ligases, have been determined, providing a glimpse into the mechanistic strategies that these two structurally and evolutionarily distinct ribozymes use to catalyze the same reaction (Robertson and

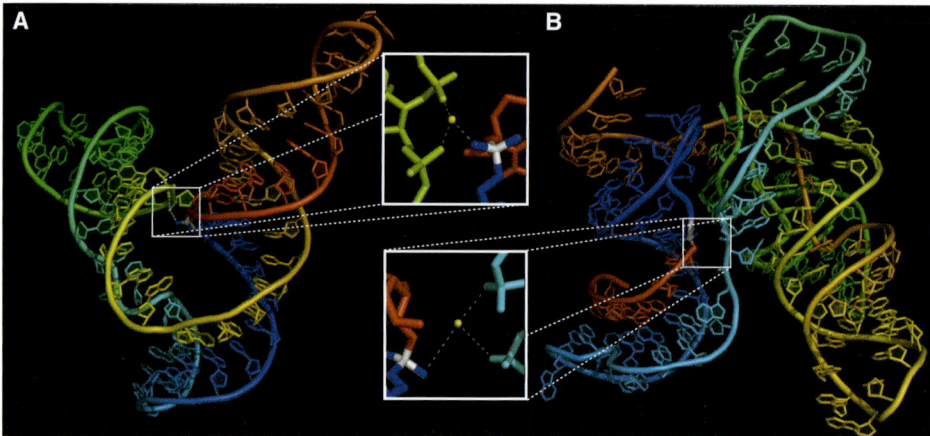

Figure 4. X-ray crystal structure of the (A) L1 ligase and (B) class I ligase ribozymes. Insets show the putative magnesium ion binding sites at the respective ligation junctions. The structures are rendered in rainbow continuum, with the 5′-triphosphate-bearing end of the ribozyme colored violet and the 3′-hydroxyl-bearing end of the substrate colored red. The phosphate at the ligation junction is shown in white, and the proximate magnesium ion (modeled for the class I ligase) is shown as a yellow sphere, with dashed lines indicating coordination contacts.

Scott 2007; Shechner et al. 2009) (Fig. 4). Both crystal structures capture the product of the ligation reaction, and consequently offer an incomplete view of the reaction pathway. For example, the pyrophosphate leaving group is absent from the structures, so no conclusions can be drawn regarding potential ribozyme-assisted orientation of the reactive triphosphate or charge neutralization of the pyrophosphate leaving group. There is, however, information regarding other aspects of the reaction mechanism that can be inferred from the product structures.

Both the L1 and class I ligases are dependent on the presence of magnesium ions for their activity. A prominent feature of the L1 structure (Fig. 4A) involves a bound metal ion in the active site, coordinated by three nonbridging phosphate oxygens, one of which belongs to the newly formed phosphodiester linking what originally were the two substrates. This magnesium ion is favorably positioned to help neutralize the increased negative charge of the transition state and, potentially, to activate the 3′-hydroxyl nucleophile and to help orient the α-phosphate for a more optimal in-line alignment. In the case of the class I structure (Fig. 4B), no catalytic metal ions appear to have been retained in the vicinity of the active site, although two magnesium ions are observed to participate in crucial structural interactions that help shape the active site architecture. Despite the lack of direct observation of a catalytic metal at the active site, there is what appears to be an empty metal binding site formed by two nonbridging phosphate oxygens, positioned directly opposite the ligation junction in a manner similar to that observed for the magnesium binding site of the L1 ligase and reminiscent of the arrangement seen in protein polymerases. The lack of a metal in the crystal structure may simply be an artifact of the crystallization process or may imply a local conformational change in the product that disfavors retention of the bound metal. These structures show that, despite some remaining gaps in the detailed understanding of how these ribozymes function, the available information points to a universal catalytic strategy, very similar to that used by modern protein-based RNA polymerases.

Subsequent to its isolation as a ligase, the class I ribozyme was shown to catalyze a polymerization reaction in which the 5′-triphosphate-bearing oligonucleotide is replaced by one or more NTPs (Ekland and Bartel 1996). This reaction proceeds with high fidelity ($q = 0.92$), but the reaction rate drops sharply with successive nucleotide additions.

Bartel and colleagues performed further in vitro evolution experiments to convert the class I ligase to a bona fide RNA polymerase that operates on a separate RNA template (Johnston et al. 2001). To the 3′ end of the class I ligase they added 76 random-sequence nucleotides that were evolved to form an accessory domain that assists in the polymerization of template-bound NTPs. The polymerization reaction is applicable to a variety of template sequences, and for well-behaved sequences proceeds with an average fidelity of 0.967. This would be sufficient to support a genome length of about 30 nucleotides, although the ribozyme itself contains about 190 nucleotides. The ribozyme has a catalytic rate for NTP addition of at least 1.5 min^{-1}, but its K_m is so high that, even in the presence of micromolar concentrations of oligonucleotides and millimolar concentrations of NTPs, it requires about 2 h to complete each NTP addition (Lawrence and Bartel 2003). The ribozyme operates best under conditions of high Mg^{2+} concentration, but becomes degraded under those conditions over 24 h, by which time it has added no more than 14 NTPs (Johnston et al. 2001).

Further optimization of the polymerase ribozyme using highly sophisticated in vitro evolution techniques has led to additional improvements in its biochemical properties. By directly selecting for extension of an external primer on a separate template, Zaher and Unrau (2007) were able to improve the maximum length of template-dependent polymerization to >20 nucleotides, with a rate that is ~threefold faster than that of the parent for the first nine monomer additions and up to 75-fold faster for additions beyond 10 nucleotides. In addition, although not rigorously quantitated, the new ribozyme displays significantly improved fidelity, particularly with respect to discrimination against G•U wobble pairs. It is this improved fidelity that appears to be the underlying source for the observed improvements in the maximum length of extension and the rate of polymerization.

A different RNA ligase ribozyme can operate on a separate RNA template in a largely sequence-general manner, and does so with a K_m that is at least 100-fold lower than that of the class-I-derived polymerase (McGinness and Joyce 2002). However, its catalytic rate is much lower as well, and it is unable to add more than a single NTP. Yet another RNA ligase ribozyme can operate on a separate template with the help of designed tertiary interactions that "clamp" the template–substrate complex to the ribozyme (Ikawa et al. 2004). But it too is a relatively slow catalyst that cannot add more than a single NTP.

A highly pessimistic view is that because there is no known polymerase ribozyme that combines all of the properties necessary to sustain its own replication, no such ribozyme is possible. A more balanced view is that RNA clearly is capable of greatly accelerating the template-dependent polymerization of nucleoside 5′-polyphosphates. Such catalytic RNAs can operate in a sequence-general manner and with reasonable fidelity. It seems only a matter of time (and likely considerable effort) before more robust polymerase ribozymes will be obtained. Nature did not have the opportunity to conduct carefully arranged evolution experiments using highly-purified reagents, but did have the luxury of much greater reaction volumes and much more time.

RNA Replication. Despite falling short of the ultimate goal of a general-purpose RNA polymerase ribozyme, a robust reaction system for RNA-catalyzed RNA replication has recently been shown. The system uses a pair of cross-replicating ligase ribozymes that each catalyze the formation of the other, using a mixture of four different substrate oligonucleotides (Lincoln and Joyce 2009). In reaction mixtures containing only these RNA substrates, $MgCl_2$, and buffer, a small starting amount of ribozymes gives rise to many additional ribozymes through a process of RNA-catalyzed exponential amplification. Whenever the substrates become depleted, the replication process can be restarted and sustained indefinitely by replenishing the supply of substrates.

Because the substrates are recognized by the ribozymes through specific Watson-Crick pairing interactions, evolution experiments can be performed by providing a variety of substrates that have different sequences in these recognition regions and different corresponding sequences in the catalytic domain of the ribozyme. RNA replication was performed with a library of 144 possible substrate combinations, resulting in the emergence of a set of highly advantageous replicators that included recombinants which were not present at the start of the experiment. Until the advent of a general-purpose RNA polymerase ribozyme, the system of cross-replicating ligases offers the best platform to study the biochemical properties and evolutionary behavior of an all-RNA replicative system.

Nucleotide Biosynthesis. RNA replicase activity is probably not the only catalytic behavior that was essential for the existence of the RNA World. Maintaining an adequate supply of the four activated nucleotides would have been a top priority. Even if the prebiotic environment contained a large reservoir of these compounds, the reservoir would eventually become depleted, and some capacity for nucleotide biosynthesis would have been required.

Ribozymes have been obtained, through in vitro evolution, that catalyze some of the steps of nucleotide biosynthesis. Unrau and Bartel (1998), for example, developed a ribozyme that catalyzes a reaction between 4-thiouracil and 5-phosphoribosyl-1-pyrophosphate (PRPP) to form 4-thiouridylate (Fig. 5A). The 4-thiouracil is provided free in solution and the PRPP is tethered to the 3′ end of the ribozyme. An optimized form of this ribozyme, containing 124 nucleotides, has an observed rate of 0.2 min^{-1} in the presence of 4 mM 4-thiouracil (Chapple et al. 2003). This is at least 10^7-fold faster than the uncatalyzed rate of reaction, which is too slow to measure. Unrau and colleagues used a similar approach to develop two different ribozymes that catalyze the formation of 6-thioguanylate from 6-thioguanine and tethered PRPP (Lau et al. 2004), as well as a third guanylate synthase ribozyme that arose as an unanticipated consequence of a related in vitro evolution experiment (Lau and Unrau 2009). The first two guanylate synthase ribozymes are slightly larger and have about twofold higher catalytic efficiency compared

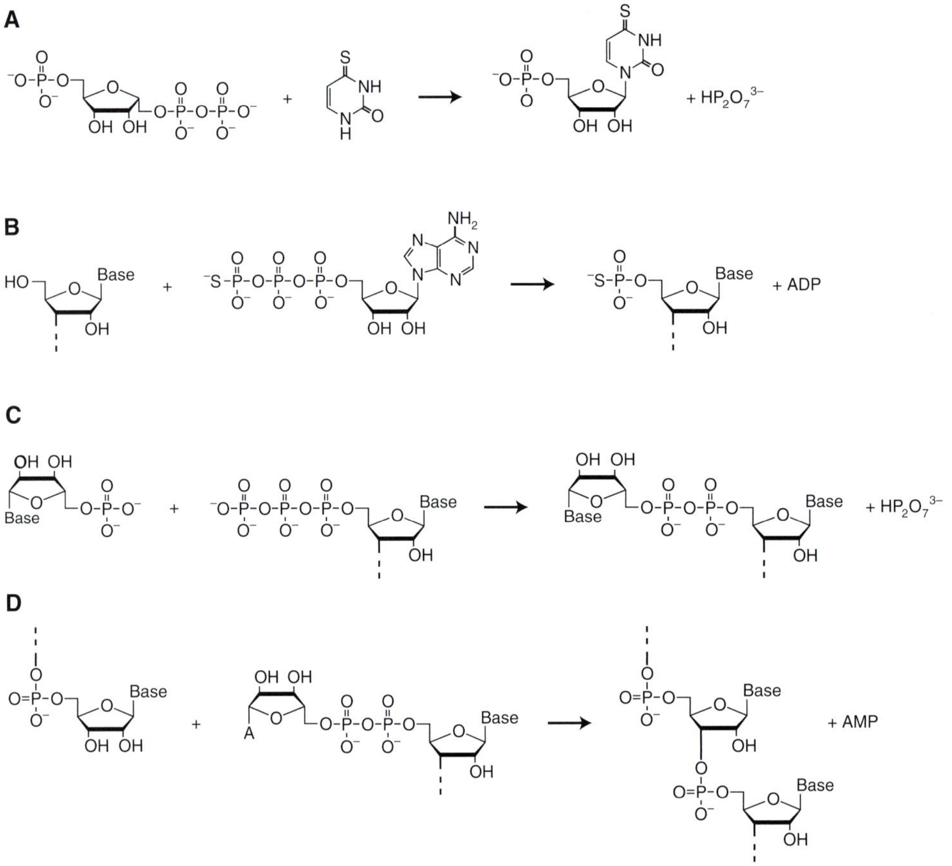

Figure 5. Known RNA-catalyzed reactions that are relevant to nucleotide biosynthesis. (*A*) Formation of 4-thiouridylate from free 4-thiouracil and ribozyme-tethered 5-phosphoribosyl-1-pyrophosphate. (*B*) 5′-phosphorylation of an oligonucleotide using γ-thio-ATP as the phosphate donor. (*C*) Activation of a nucleoside 5′-phosphate by formation of a 5′,5′-pyrophosphate linkage. (*D*) Template-directed ligation of RNA driven by release of a 5′,5′-pyrophosphate-linked adenylate.

with the uridylate synthase ribozyme, although guanylate synthesis is expected to have a much higher uncatalyzed rate of reaction.

RNA-catalyzed synthesis of PRPP has not been shown, but a ribozyme has been obtained that catalyzes the 5′-phosphorylation of oligonucleotides using γ-thio-ATP as the phosphate donor (Lorsch and Szostak 1994) (Fig. 5B). The ribozyme shows a rate enhancement of about 10^9-fold compared with the uncatalyzed rate of reaction. Once a nucleoside 5′-phosphate has been formed, it can phosphoryl be activated by another ribozyme that catalyzes the condensation of a nucleoside 5′-phosphate and a ribozyme-tethered nucleoside 5′-triphosphate (Huang and Yarus 1997) (Fig. 5C). This results in the formation of a 5′,5′-pyrophosphate linkage, which provides an activated nucleotide leaving group that can drive subsequent RNA-catalyzed, template-directed ligation of RNA (Hager and Szostak 1997) (Fig. 5D).

None of these four RNA-catalyzed reactions has precisely the right format for the corresponding reaction in a hypothetical nucleotide biosynthesis pathway in the RNA World. However, they show that RNA is capable of performing the relevant chemistry with substantial catalytic rate enhancement. It remains to be seen whether ribozymes can be developed that catalyze the formation of the fundamental building blocks of RNA, D-ribose and the four nucleotide bases, using starting materials that would have been abundant on the primitive Earth.

3 AN "RNA-LATER" VIEW OF THE ORIGIN OF LIFE

3.1 Abiotic Synthesis of Nucleotides

The RNA-first view of the origin of life proceeds from the assumption that pure ß-D-nucleotides were available in some prebiotic pool. How close to such a pool could one hope to get without magic (or evolved enzymes) on the primitive Earth? Could one hope to achieve replication in a pool containing a more realistic mixture of organic molecules, including, of course, ß-D-ribonucleotides? The synthesis of a nucleotide could occur in a number of ways. The simplest, conceptually, would be to synthesize a nucleoside base, couple it to ribose, and finally to phosphorylate the resulting nucleoside. However, a number of other routes are feasible, for example the assembly of the base on a preformed ribose or ribose phosphate, or the coassembly of the base and sugar-phosphate.

The classical prebiotic synthesis of sugars is by the polymerization of formaldehyde (the "formose" reaction). It yields a very complex mixture of products including only a small proportion of ribose (Mizuno and Weiss 1974). This reaction does not provide a reasonable route to the ribonucleotides. However, a number of more recent experimental findings, to some extent, address this deficiency.

The base-catalyzed aldomerization of glycoaldehyde phosphate in the presence of a half-equivalent of formaldehyde under strongly alkaline conditions gives a relatively simple mixture of tetrose- and pentose-diphosphates and hexose-triphosphates, of which ribose 2,4-diphosphate is the major component (Müller et al. 1990). Reactions of this kind proceed efficiently when 2 mM solutions of substrates are incubated at room temperature and pH 9.5 in the presence of layered hydroxides such as hydrotalcite (magnesium aluminum hydroxide) (Pitsch 1992; Pitsch et al. 1995a). The phosphates are absorbed between the positively charged layers of the mineral. The reaction proceeds under these milder conditions presumably because of the high concentration of substrates in the interlayer and because the positive charge on the metal hydroxide layers favors enolization of glycoaldehyde phosphate. A reaction between glycoaldehyde or glyceraldehyde and the amidotriphosphate ion provides an ingenious and prebiotically plausible route to glycoaldehyde phosphate and glyceraldehyde-2-phosphate, respectively, the two substrates in the above reactions (Krishnamurthy et al. 2000).

A number of other studies have addressed the problems presented by the lack of specificity of the formose reaction and by the instability of ribose. The Pb^{2+} ion is an excellent catalyst for the formose reaction and enables yields of the pentose sugars as high as 30% to be achieved (Zubay 1998). Furthermore, it seems likely that ribose is almost exclusively the first pentose product of the reaction and that the other pentoses are formed from it by isomerization. Other recent studies have addressed the problem presented by the instability of ribose. The four pentose sugars, including ribose, are all strongly stabilized in the presence of borate ions or calcium borate minerals (Ricardo et al. 2004). However, the effect of borate on the progress of the formose reaction has not been reported.

Many sugars, including the four pentoses, react readily with cyanamide to form stable bicyclic amino-oxazolines (Sanchez and Orgel 1970) (Fig. 6A). Strikingly, the ribose derivative crystallizes readily from aqueous solution even when complex mixtures of related molecules, including a mixture of the amino-oxazoline derivatives of the other three pentose sugars, are present (Springsteen and Joyce 2004). The crystals are multiply twinned, each crystal containing many small domains of each of the two enantiomorphs. Thus the reaction of a mixture of racemic sugars with cyanamide followed by crystallization might stabilize ribose, segregate it from other sugars, and present it in enantiospecific microdomains. Much remains to be shown, but the reactions described above suggest that ribose synthesis, although still problematic, may not be the intractable problem it once seemed.

The synthesis of the nucleoside bases is one of the success stories of prebiotic chemistry. Adenine is formed with remarkable ease from ammonia and hydrogen cyanide (Orò 1961). This synthesis has been described as "the rock of the faith" by Stanley Miller. Reasonably plausible syntheses of the other purine bases and of the pyrimidines have also been described (Sanchez et al. 1967; Ferris et al. 1968; Robertson and Miller 1995; Orgel 2004b; Saladino et al. 2004). The coupling of the purine bases with ribose or ribose-phosphate has been achieved under mild conditions, but in relatively low yield (Fuller et al. 1972). The corresponding reaction with pyrimidines does not occur.

There is a different potential route for the prebiotic synthesis of pyrimidine nucleotides via arabinose amino-oxazoline that first was explored nearly 40 yr ago (Tapiero and Nagyvary 1971) and in recent years has begun to look very persuasive (Ingar et al. 2003; Anastasi et al. 2007; Powner et al. 2009). The earlier studies began with arabinose 3-phosphate, which, like arabinose and other sugars, reacts with cyanamide to give the corresponding amino-oxazoline (Fig. 6B). This in turn reacts with cyanoacetylene to form a tricyclic intermediate that hydrolyzes to produce a mixture of cytosine arabinoside-3′-phosphate and cytosine 2′,3′-cyclic phosphate.

Sutherland and colleagues (Powner et al. 2009) have taken this approach further by starting simply with glycoaldehyde and cyanamide, which in the presence of 1 M phosphate at neutral pH gives 2-amino-oxazole in excellent yield (Fig. 6C). The phosphate both buffers and catalyzes the reaction, directing glycoaldehyde toward 2-amino-oxazole, rather than a complex mixture of aldomerization products. Glyceraldehyde is then added, resulting in formation of the various pentose amino-oxazolines, including

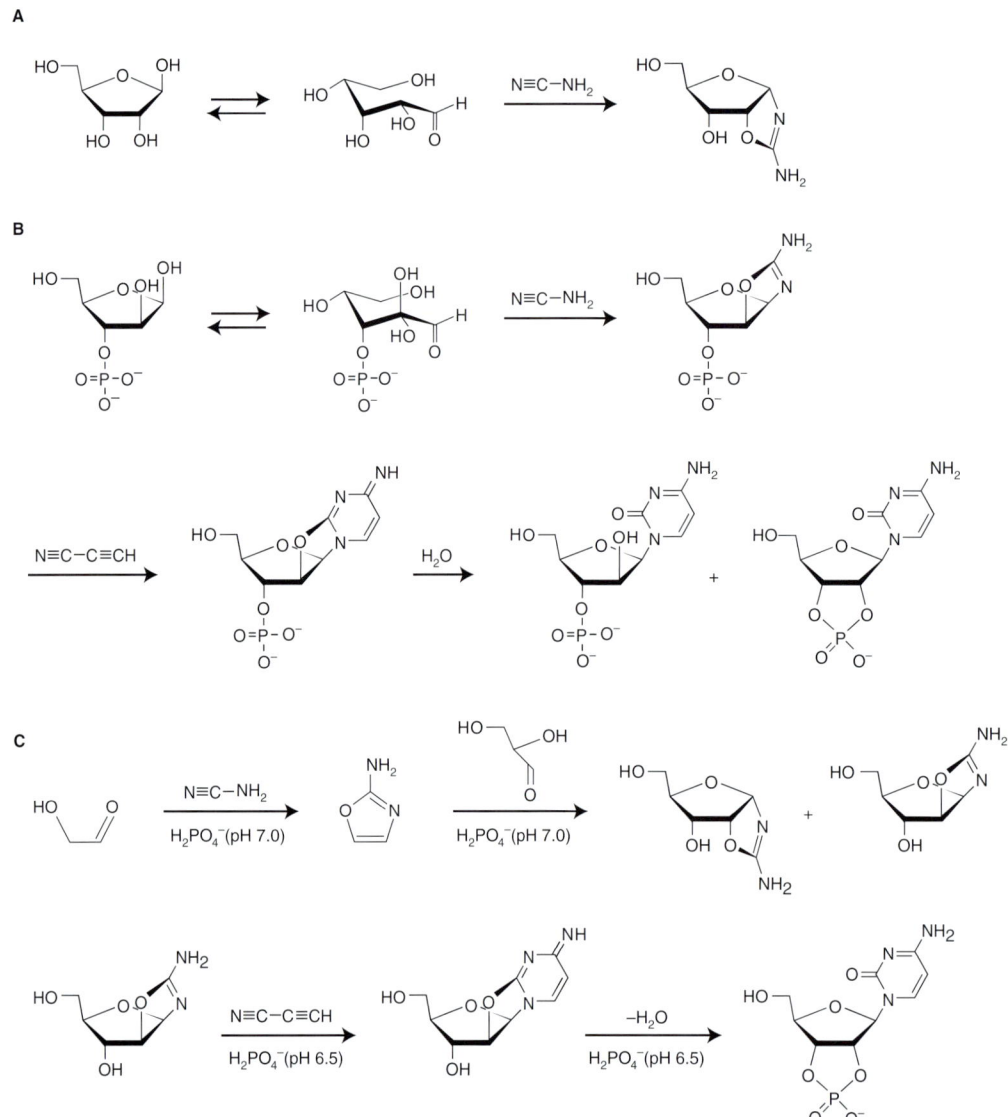

Figure 6. Potential prebiotic synthesis of pyrimidine nucleosides. (A) Reaction of ribose with cyanamide to form a bicyclic product, with cyanamide joined at both the anomeric carbon and 2-hydroxyl. (B) Analogous reaction of arabinose-3-phosphate to form a bicyclic product, which then reacts with cyanoacetylene to form a tricyclic intermediate that hydrolyzes to give a mixture of cytosine arabinoside-3′-phosphate and cytosine 2′,3′-cyclic phosphate. (C) Reaction of glycoaldehyde with cyanamide in neutral phosphate buffer, followed by addition of glyceraldehyde, to form ribose and arabinose amino-oxazoline (and lesser amounts of the xylose and lyxose compounds). Arabinose amino-oxazoline then reacts with cyanoacetylene to give cytosine 2′,3′-cyclic phosphate as the major product.

the arabinose compound. Arabinose amino-oxazoline, in turn, can react with cyanoacetylene, also in phosphate buffer, to form cytosine 2′,3′-cyclic phosphate as the major product. Perhaps equally intriguing, although given less emphasis in these studies, is that reaction of arabinose amino-oxazoline with cyanoacetylene also gives a substantial yield of cytosine 2′,3′-cyclic-5′-bisphosphate, which is more amenable to being converted to an activated monomer that would be suitable for polymerization.

There has been significant progress, especially recently, concerning the synthesis of the nucleosides and nucleotides from prebiotic precursors in reasonable yield. However, the story remains incomplete because these syntheses still require temporally separated reactions using high concentrations of just the right reactants, and would be disrupted by the presence of other closely related compounds. The reactions channel material toward the desired products, but other fractionation processes must be discovered that provide the correct starting materials at the requisite time and place. This "preprebiotic" chemistry likely would involve a series of reactions catalyzed by minerals or metal ions, coupled with a series of subtle fractionations of nucleotide-like materials based on adsorption on minerals, selective complex formation, crystallization, etc.

Even minerals could not achieve on a macroscopic scale one desirable separation, the resolution of D-ribonucleotides from their L-enantiomers. This is a serious problem because experiments on template-directed synthesis using poly(C) and the imidazolides of G suggest that the polymerization of the D-enantiomer is strongly inhibited by the L-enantiomer (Joyce et al. 1984). This difficulty may not be insuperable; perhaps with a different mode of phosphate activation, the inhibition would be less severe. However, enantiomeric cross-inhibition is certainly a serious problem if life arose in a racemic environment.

It is possible that the locale for life's origins was not racemic, even though the global chemical environment contained nearly equal amounts of each pair of stereoisomers. There likely were biases in the inventory of compounds delivered to the Earth by comets and meteorites. For example, some carbonaceous chondrite meteorites contain a significant enantiomeric excess of L-amino acids that are known to be indigenous to the meteorite (Engel and Macko 1997; Cronin and Pizzarello 1997; Pizzarello et al. 2003; Glavin and Dworkin 2009). These in turn could bias terrestrial syntheses, although the level of enantiomeric enrichment generally declines with successive chemical reactions. A special exception are a remarkable set of reactions and fractionation processes that amplify a slight chiral imbalance, even to the level of local homochirality (Kondepudi et al. 1990; Soai et al. 1995; Viedma 2005; Klussmann et al. 2006; Noorduin et al. 2008; Viedma et al. 2008). These systems have in common both a catalytic process for amplification of same-handed molecules and an inhibition process for suppression of opposite-handed molecules.

Some of the most appealing examples of chiral symmetry-breaking reactions involve saturating solutions of various amino acids that form an equilibrium between the liquid phase and solid phase. The solid phase consists of either racemic or enantiopure crystals, and the liquid phase reflects whatever enantiomeric excess exists at the eutectic point for the mixture. For some amino acids, such as serine and histidine, the enantiomeric excess at the eutectic is >90% (Klussmann et al. 2006). This means that, starting from a small concentration imbalance of D- and L-isomers, the imbalance is amplified as both isomers enter the solid phase and the solution phase approaches the eutectic equilibrium. This and related near-equilibrium mechanisms (Noorduin et al. 2008; Viedma et al. 2008) could provide a means to achieve high enantiomeric enrichment in a local environment. This in turn could bias the production of ribose and the derived nucleotides.

Scientists interested in the origins of life seem to divide neatly into two classes. The first, usually but not always molecular biologists, believe that RNA must have been the first replicating molecule and that chemists are exaggerating the difficulties of nucleotide synthesis. They believe that a few more striking chemical "surprises" will establish that a pool of racemic mononucleotides could have formed on the primitive Earth, and that further experiments with different activating groups, minerals, and chiral amplification processes will solve the enantiomeric cross-inhibition problem. The second group of scientists are much more pessimistic. They believe that the de novo appearance of oligonucleotides on the abiotic Earth would have been a near miracle. Time will tell which is correct.

3.2 Alternative Genetic Systems

The problems that arise when one tries to understand how an RNA World could have arisen de novo on the primitive Earth are sufficiently severe that one must explore other possibilities. What kind of alternative genetic systems might have preceded the RNA World? How could they have "invented" the RNA World? These topics have generated a good deal of speculative interest and some relevant experimental data.

Eschenmoser and colleagues have undertaken a systematic study of the properties of analogs of nucleic acids in which ribose is replaced by some other sugar, or in which the furanose form of ribose is replaced by the pyranose form (Eschenmoser 1999) (Fig. 7B). Strikingly, polynucleotides based on the pyranosyl analog of ribose (p-RNA) form Watson-Crick paired double helices that are more stable than RNA, and p-RNAs are less likely than the corresponding RNAs to form multiple-strand competing structures (Pitsch et al. 1993, 1995b, 2003). Furthermore, the helices twist much more gradually than those of standard nucleic acids, which should make it easier to separate strands of p-RNA during replication. Pyranosyl RNA appears to be an excellent choice as a genetic system; in some ways it seems an improvement compared with the standard nucleic acids. However, p-RNA does not interact with normal RNA to form base-paired double helices.

Most double-helical structures reported in the literature are characterized by a backbone with a six-atom repeat. Eschenmoser and colleagues made the surprising discovery that an RNA-like structure based on threose nucleotide analogs (TNA) (Fig. 7C), although it involves a five-atom repeat, can still form a stable double-helical structure with standard RNA (Schöning et al. 2000). This provides

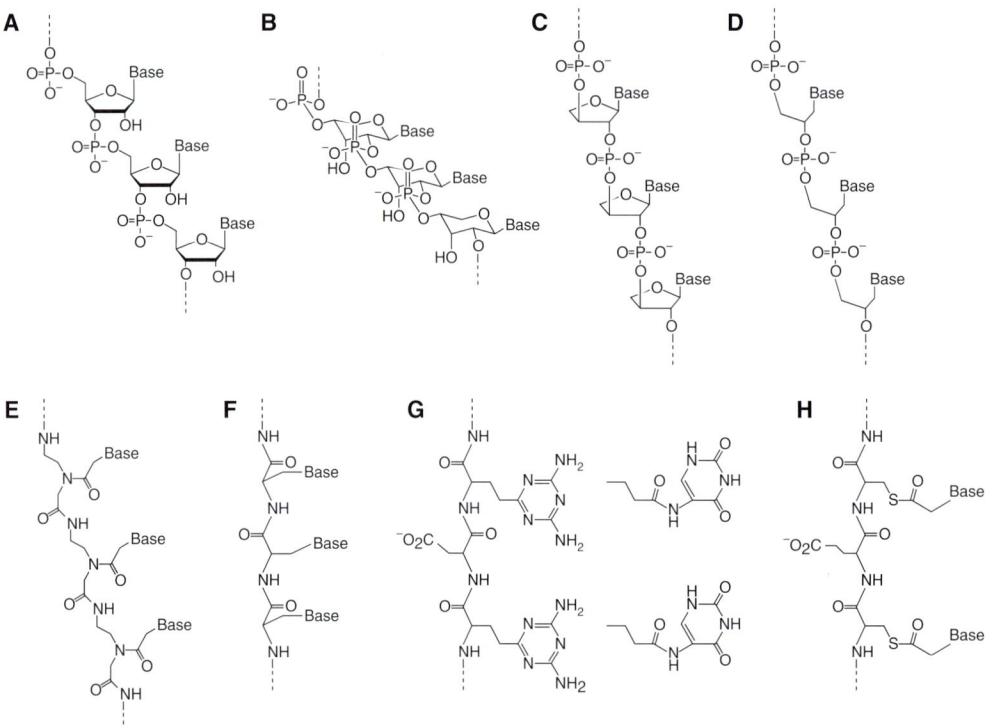

Figure 7. The structures of (*A*) RNA; (*B*) p-RNA; (*C*) TNA; (*D*) GNA; (*E*) PNA; (*F*) ANA; (*G*) diaminotriazine-tagged (*left*) and dioxo-5-aminopyrimidine-tagged (*right*) oligodipeptides; and (*H*) tPNA. ANA contains a backbone of alternating D- and L-alanine subunits. The diaminotriazine tags are shown linked to a backbone of alternating L-aspartate and L-glutamate subunits; the dioxo-5-aminopyrimidine tags (shown unattached) can be linked similarly. tPNA is shown with a backbone of alternating L-cysteine and L-glutamate subunits.

an example of a pairing system based on a sugar that could be formed more readily than ribose: Tetroses are the unique products of the dimerization of glycoaldehyde, whereas pentoses are formed along with tetroses and hexoses from glycoaldehyde and glyceraldehyde. A structural simplification of Eschenmoser's threose nucleic acid has been achieved by Meggers and colleagues (Zhang et al. 2005). They replaced threose by its open chain analogue, glycol, in the backbone of TNA, resulting in glycol nucleic acid (GNA) (Fig. 7D). Complementary oligomers of GNA form antiparallel, double-helices with surprisingly high duplex stabilities.

Peptide nucleic acid (PNA) is another nucleic acid analog that has been studied extensively (Fig. 7E). It was discovered by Nielsen and colleagues in the context of research on antisense oligonucleotides (Egholm et al. 1992, 1993; Wittung et al. 1994). PNA is an uncharged, achiral analog of RNA or DNA in which the ribose-phosphate backbone of the nucleic acid is replaced by a backbone held together by amide bonds. PNA forms very stable double helices with complementary RNA or DNA. Work in the Orgel laboratory has shown that information can be transferred from PNA to RNA, or from RNA to PNA, in template-directed reactions, and that PNA/DNA chimeras are readily formed on either DNA or PNA templates (Schmidt et al. 1997a,b; Koppitz et al. 1998). Thus it seems that a transition from a PNA World to an RNA World is possible.

The alanyl nucleic acids (ANA) are interesting for a different reason. They are polypeptides formed from nucleo amino acids (Fig. 7F), but pairing structures can be formed only if the two enantiomers of their constituent α-amino acids occur in a regular alternating sequence (Diederichsen 1996, 1997). Because abiotic syntheses of potentially chiral molecules would under almost all circumstances yield racemic products, pairing structures that can be formed from racemic mixtures are particularly attractive. The ANA-type backbone of alternating D- and L-amino acids could, in principle, support paired, double-stranded structures based on a variety of side-chain interactions.

Eschenmoser and colleagues have examined repeating homochiral dipeptide backbones that have either triazines or aminopyrimidines attached at alternating positions (Mittapalli et al. 2007a,b) (Fig. 7G). In this case, not only is the backbone a potential precursor to that of RNA, but also the bases have been replaced by potential precursors, such as 2,4-diaminotriazine (^{T}NN), 2,4-dioxotriazine (^{T}OO), 2,4-diamino-5-aminopyrimidine (^{AP}NN), and 2,4-dioxo-5-aminopyrimidine (^{AP}OO). Oligomers containing either ^{T}NN or ^{AP}OO subunits were found to pair strongly with complementary RNA, whereas oligomers containing either ^{T}OO or ^{AP}NN subunits did not. Not surprisingly, therefore, pairing between complementary substituted oligodipeptides of the same type (either oligo[^{T}NN]•oligo[^{T}OO] or oligo[^{AP}NN]•oligo[^{AP}OO]) also was weak. However, cross-pairing between oligo(^{T}NN) and oligo(^{AP}OO) was robust (Mittapalli et al. 2007a). This raises the intriguing possibility that an informational polymer could have a mixed composition of ^{T}NN and ^{AP}OO subunits, which would direct the synthesis of an opposing strand that has a complementary sequence as well as a "complementary" backbone composition.

Even more radical are thioester peptide nucleic acids (tPNA) (Fig. 7H), containing a repeating dipeptide backbone with cysteine residues at alternating positions, which are transiently linked via a thioester to a nucleic acid base (Ura et al. 2009). The bases are in dynamic equilibrium between the solution and cysteine positions along the backbone. Occupancy of a base at a particular position is enhanced by the presence of the complementary base on the opposing strand. In this way the informational polymer can self-assemble in a template-directed manner, with mismatched bases exchanging rapidly and matched bases remaining thioesterified for an extended period of time. Perhaps genetic information could be propagated in such a system, although the fidelity of replication, and therefore the maximum number of informational subunits, is likely to be modest.

The studies described previously suggest that there are many ways of linking together the nucleotide bases into chains that are capable of forming base-paired double helices. It is not clear that it is much easier to synthesize the monomers of p-RNA, TNA, GNA, PNA, ANA, or tPNA than to synthesize the standard nucleotides. However, it is possible that a base-paired structure of this kind will be discovered that can be synthesized readily under prebiotic conditions. The properties of the ^{T}NN- and ^{AP}OO-tagged oligodipeptides suggest that it may be fruitful to explore a broader range of potential precursors to RNA, changing the recognition elements as well as the backbone. A strong candidate for the first genetic material would be any informational macromolecule that is replicable in a sequence-general manner and derives from compounds that would have been abundant on the primitive Earth, and preferably has the ability to cross-pair with RNA.

The transition from an RNA-like World to the RNA World could take place in two ways. The transition might be continuous if the pre-RNA template could direct the synthesis of an RNA product with a complementary sequence. Such a transition, for example, from PNA to RNA, would preserve information. RNA could then act as a genetic material in a formerly PNA World. However, even if chimeras were involved in the transition, it is unlikely that the original function of a PNA catalyst could be retained throughout the transition because PNA and RNA have such different backbone structures. A direct and continuous transition from p-RNA to RNA would not be possible because p-RNA does not form complementary double helices with RNA, but this limitation does not apply to TNA and GNA.

The second type of transition can be described as a genetic takeover. A pre-existing self-replicating system evolves, for its own selective advantage, a mechanism for synthesizing and polymerizing the components of a completely different genetic system, and is taken over by it. Cairns-Smith (1982) has proposed that the first genetic system was inorganic, perhaps a clay, and that it "invented" a self-replicating system based on organic monomers. However, he clearly recognized the possibility of one organic genetic material replacing another (Cairns-Smith and Davies 1977). Genetic takeover does not require any structural relationship between the polymers of the two genetic systems. It suggests the possibility that the original genetic system may have been unrelated to nucleic acids.

The hypothesis of a genetic material completely different from nucleic acids has one enormous advantage—it opens up the possibility of using very simple, easily synthesized prebiotic monomers in place of nucleotides. However, it also raises two new and difficult questions. Which prebiotic monomers are plausible candidates as the components of a replicating system? Why would an initial genetic system invent nucleic acids once it had evolved sufficient synthetic know-how to generate molecules as complex as nucleotides?

A number of prebiotic monomers that might have made up a simple genetic material have already been suggested. They include hydroxy acids (Weber 1987), amino acids (Orgel 1968; Zhang et al. 1994), phosphomonoesters of polyhydric alcohols (Weber 1989), aminoaldehydes (Nelsestuen 1980), and molecules containing two sulfhydryl groups (Schwartz and Orgel 1985). The list could be expanded almost indefinitely. The discussion here concerns a small class of these monomers that appear to be particularly attractive in the light of recent work on enzyme mechanisms.

There is accumulating evidence that several enzymes that make or break phosphodiester bonds have two or three metal ions at their active sites (Cooperman et al. 1992; Sträter et al. 1996). In the case of the editing site for phosphodiester hydrolysis in the Klenow fragment of *Escherichia coli* DNA polymerase I, no other functional groups of the enzyme come close to the phosphodiester bond that is cleaved. This has led to the suggestion that the major role of the enzyme is to act as scaffolding on which to hang metal ions in precisely determined positions (Beese and Steitz 1991; Steitz 1998). A similar suggestion has been made for ribozymes on the basis of both indirect and direct evidence (Freemont et al. 1988; Yarus 1993; Steitz and Steitz 1993; Shan et al. 1999, 2001; Stahley and Strobel 2005).

Perhaps these observations can be extended to suggest that, if informational polymers preceded RNA, they may also have been dependent on metal ions for their catalytic activity. If so, the range of prebiotic monomers that needs to be considered is greatly reduced. In addition to the functional groups that react to form the backbone, the monomers must have carried metal-binding functional groups. If the metal ions involved were divalent ions such as Mg^{2+} and Ca^{2+}, the side groups are likely to have been carboxylate or phosphate groups. If transition metal ions were involved, sulfhydryl groups and possibly imidazole derivatives are likely to have been important.

Prebiotic monomers suitable for building polymers that bind Mg^{2+} or Ca^{2+} include aspartic acid, glutamic acid, and serine phosphate among biologically important amino acids. ß-amino acids, such as isoglutamic acid, hydroxydicarboxylic acids, such as α-hydroxysuccinic acid, and hydroxytricarboxylic acids, such as citric acid, are other possible candidates. A polymer containing D-aspartic acid, L-aspartic acid, and glycine as its subunits is typical of potentially informational co-polymers that might, in the presence of divalent metal ions, both replicate and function as a catalyst. Transition-metal ions might play a corresponding role for polymers containing cysteine or homocysteine. The present challenge is to show replication, or at least information transfer in template-directed synthesis, in some such system.

What selective advantage could a simpler, metabolically competent system derive from the synthesis of oligonucleotides? This is a baffling question. Most arguments that come to mind do not stand up to detailed analysis. If, for example, one postulates that nucleotides were first synthesized as parts of cofactors such as DPN, one must explain why the particular heterocyclic bases and sugars were chosen. Even if one supposes that among the many "experiments" in secondary metabolism performed by early organisms one happened by accident on a pair of complementary nucleotides that could form a replicating polymer, one must still explain how polymerization subsequently contributed to the success of the "inventor." Could oligonucleotides, by hybridization, have functioned at first as selective "glues" for tying pairs of macromolecules together? Could RNA have been invented by one organism as "antisense" against the genome of another?

The discussion so far, even though highly speculative, is still conservative in overall outlook. It supposes that the original information-accumulating system that led to the evolution of life on Earth was either RNA or some linear copolymer that replicated in an aqueous environment in much the same way as RNA. There remains a lingering doubt that the discussion is not on the right track at all; maybe the original system was not an organic copolymer (Cairns-Smith 1982), or maybe it replicated in a nonaqueous environment and RNA is an adaptation that permitted invasion of the oceans. Perhaps systems of high complexity can develop without any need for a genome in the usual sense (Dyson 1982; Kauffman 1986; Wächtershäuser 1988; De Duve 1991; Eschenmoser 2007). Perhaps...

Laboratory simulations of prebiotic chemistry are dependent on organic chemistry and can only explore the kinds of reactions understood by organic chemists. A good deal is known about reactions in aqueous solution, but less about reactions at the interface between water and inorganic solids. Very little is known about reactions in systems in which inorganic solids are depositing from aqueous solutions containing organic material. It is hard to see how speculative schemes involving heterogeneous aqueous systems can be tested until much more is known about the underlying branches of chemistry.

4 CONCLUDING REMARKS

After contemplating the possibility of self-replicating ribozymes emerging from pools of random polynucleotides and recognizing the difficulties that must have been overcome for RNA replication to occur in a realistic prebiotic soup, the challenge must now be faced of constructing a realistic picture of the origin of the RNA World. The constraints that must have been met in order to originate a self-sustained evolving system are reasonably well understood. One can sketch out a logical order of events, beginning with prebiotic chemistry and ending with DNA/protein-based life. However, it must be said that the details of this process remain obscure and are not likely to be known in the near future.

The presumed RNA World should be viewed as a milestone, a plateau in the early history of life on Earth. So too, the concept of an RNA World has been a milestone in the scientific study of life's origins. While this concept does not explain how life originated, it has helped to guide scientific thinking and has served to focus experimental efforts. Further progress will depend primarily on new experimental results, as chemists, biochemists, and molecular biologists work together to address problems concerning molecular replication, ribozyme enzymology, and RNA-based cellular processes.

ACKNOWLEDGMENTS

This work was supported by research grant NNX07AJ23G from the National Aeronautics and Space Administration. Previous versions of this article, which were published in the First (1993), Second (1999), and Third (2006) Editions of *The RNA World*, were coauthored by Leslie Orgel, who died on October 27, 2007. Many portions of the text have not been changed in the current edition because they remain an accurate reflection of current scientific understanding. The contributions of Leslie Orgel to this work and to the scientific literature of the origins of life are gratefully acknowledged.

REFERENCES

Acevedo OL, Orgel LE. 1987. Non-enzymatic transcription of an oligodeoxynucleotide 14 residues long. *J Mol Biol* **197:** 187–193.

Anastasi C, Crowe MA, Sutherland JD. 2007. Two-step potentially prebiotic synthesis of α-D-cytidine-5′-phosphate from D-glyceraldehyde-3-phosphate. *J Am Chem Soc* **129:** 24–25.

Ban N, Nissen P, Hansen J, Moore PB, Steitz TA. 2000. The complete atomic structure of the large ribosomal subunit at 2.4 Å resolution. *Science* **289:** 905–920.

Bartel DP, Szostak JW. 1993. Isolation of new ribozymes from a large pool of random sequences. *Science* **261:** 1411–1418.

Been MD, Cech TR. 1988. RNA as an RNA polymerase: Net elongation of an RNA primer catalyzed by the *Tetrahymena* ribozyme. *Science* **239:** 1412–1416.

Beese LS, Steitz TA. 1991. Structural basis for the 3′-5′ exonuclease activity of *Escherichia coli* DNA polymerase I: A two metal ion mechanism. *EMBO J* **10:** 25–33.

Cairns-Smith AG. 1982. *Genetic takeover and the mineral origins of life.* Cambridge University Press, Cambridge.

Cairns-Smith AG, Davies CJ. 1977. The design of novel replicating polymers. In *Encyclopaedia of ignorance* (eds. Duncan R., Weston-Smith M.), pp. 391–403. Pergamon Press, Oxford.

Cech TR. 1986. A model for the RNA-catalyzed replication of RNA. *Proc Natl Acad Sci* **83:** 4360–4363.

Chapple KE, Bartel DP, Unrau PJ. 2003. Combinatorial minimization and secondary structure determination of a nucleotide synthase ribozyme. *RNA* **9:** 1208–1220.

Cooperman BS, Baykov AA, Lahti R. 1992. Evolutionary conservation of the active site of soluble inorganic pyrophosphatase. *Trends Biochem Sci* **17:** 262–266.

Crick FHC. 1968. The origin of the genetic code. *J Mol Biol* **38:** 367–379.

Cronin JR, Pizzarello S. 1997. Enantiomeric excesses in meteoritic amino acids. *Science* **275:** 951–955.

De Duve C. 1991. *Blueprint for a cell: The nature and origin of life.* Neil Patterson Publishers, Burlington, North Carolina.

Diederichsen U. 1996. Pairing properties of alanyl peptide nucleic acids containing an amino acid backbone with alternating configuration. *Angew Chemie* **35:** 445–448.

Diederichsen U. 1997. Alanyl PNA: Evidence for linear band structures based on guanine–cytosine base pairs. *Angew Chemie* **36:** 1886–1889.

Ding ZP, Kawamura K, Ferris JP. 1996. Oligomerization of uridine phosphorimidazolides on montmorillonite: A model for the prebiotic synthesis of RNA on minerals. *Orig Life Evol Biosph* **26:** 151–171.

Doudna JA, Szostak JW. 1989. RNA-catalysed synthesis of complementary-strand RNA. *Nature* **339:** 519–522.

Dyson FJ. 1982. A model for the origin of life. *J Mol Evol* **18:** 344–350.

Egholm M, Buchardt O, Christensen L, Behrens C, Freier SM, Driver DA, Berg RH, Kim SK, Norden B, Nielson PE. 1993. PNA hybridizes to complementary oligonucleotides obeying the Watson-Crick hydrogen-bonding rules. *Nature* **365:** 566–568.

Egholm M, Buchardt O, Nielson PE, Berg RH. 1992. Peptide nucleic acids (PNA). Oligonucleotide analogues with an achiral peptide backbone. *J Am Chem Soc* **114:** 1895–1897.

Eigen M. 1971. Selforganization of matter and the evolution of biological macromolecules. *Naturwiss* **58:** 465–523.

Eigen M, Schuster P. 1977. The hypercycle: A principle of natural self-organization. Part A: Emergence of the hypercycle. *Naturwiss* **64:** 541–565.

Eigen M, McCaskill J, Schuster P. 1988. Molecular quasi-species. *J Phys Chem* **92:** 6881–6891.

Ekland EH, Bartel DP. 1996. RNA-catalysed RNA polymerization using nucleoside triphosphates. *Nature* **382:** 373–376.

Ekland EH, Szostak JW, Bartel DP. 1995. Structurally complex and highly active RNA ligases derived from random RNA sequences. *Science* **269:** 364–370.

Emilsson GM, Nakamura S, Roth A, Breaker RR. 2003. Ribozyme speed limits. *RNA* **9:** 907–918.

Engel M, Macko S. 1997. Isotopic evidence for extraterrestrial non-racemic amino acids in the Murchison meteorite. *Nature* **389:** 265–268.

Erman JE, Hammes GG. 1966. Relaxation spectra of ribonuclease. V. The interaction of ribonuclease with cytidine 2′:3′-cyclic phosphate. *J Am Chem Soc* **88:** 5614–5617.

Eschenmoser A. 1999. Chemical etiology of nucleic acid structure. *Science* **284:** 2118–2124.

Eschenmoser A. 2007. On a hypothetical generational relationship between HCN and constituents of the reductive citric acid cycle. *Chem Biodivers* **4:** 554–573.

Ferris JP. 2002. Montmorillonite catalysis of 30–50 mer oligonucleotides: Laboratory demonstration of potential steps in the origin of the RNA world. *Orig Life Evol Biosph* **32:** 311–332.

Ferris JP. 2006. Montmorillonite-catalyzed formation of RNA oligomers: The possible role of catalysis in the origins of life. *Phil Trans R Soc B* **361:** 1777–1786.

Ferris JP, Ertem G. 1993. Montmorillonite catalysis of RNA oligomer formation in aqueous solution. A model for the prebiotic formation of RNA. *J Am Chem Soc* **115:** 12270–12275.

Ferris JP, Sanchez RA, Orgel LE. 1968. Studies in prebiotic synthesis. III. Synthesis of pyrimidines from cyanoacetylene and cyanate. *J Mol Biol* **33:** 693–704.

Ferris JP, Hill AR, Liu R, Orgel LE. 1996. Synthesis of long prebiotic oligomers on mineral surfaces. *Nature* **381:** 59–61.

Ferris JP, Joshi PC, Wang KJ, Miyakawa S, Huang W. 2004. Catalysis in prebiotic chemistry: Application to the synthesis of RNA oligomers. *Adv Space Res* **33:** 100–105.

Freemont PS, Friedman JM, Beese LS, Sanderson MR, Steitz TA. 1988. Cocrystal structure of an editing complex of Klenow fragment with DNA. *Proc Natl Acad Sci* **85:** 8924–8928.

Fujita Y, Furuta H, Ikawa Y. 2009. Tailoring RNA modular units on a common scaffold: A modular ribozyme with a catalytic unit for β-nicotinamide mononucleotide-activated RNA ligation. *RNA* **15:** 877–888.

Fuller WD, Sanchez RA, Orgel LE. 1972. Studies in prebiotic synthesis. VI. Synthesis of purine nucleosides. *J Mol Biol* **67:** 25–33.

Gilbert W. 1986. The RNA world. *Nature* **319:** 618.

Glavin DP, Dworkin JP. 2009. Enrichment of the amino acid L-isovaline by aqueous alteration on CI and CM meteorite parent bodies. *Proc Natl Acad Sci* **106:** 5487–5492.

Guerrier-Takada C, Gardiner K, Marsh T, Pace N, Altman S. 1983. The RNA moiety of ribonuclease P is the catalytic subunit of the enzyme. *Cell* **35:** 849–857.

Hager AJ, Szostak JW. 1997. Isolation of novel ribozymes that ligate AMP-activated RNA substrates. *Chem Biol* **4:** 607–617.

Handschuh GJ, Lohrmann R, Orgel LE. 1973. The effect of Mg^{2+} and Ca^{2+} on urea-catalyzed phosphorylation reactions. *J Mol Evol* **2**: 251–262.

Huang F, Yarus M. 1997. Versatile 5' phosphoryl coupling of small and large molecules to an RNA. *Proc Natl Acad Sci* **94**: 8965–8969.

Huang W, Ferris JP. 2003. Synthesis of 35-40 mers of RNA oligomers from unblocked monomers. A simple approach to the RNA world. *Chem Commun* **12**: 1458–1459.

Huang W, Ferris JP. 2006. One-step, regioselective synthesis of up to 50-mers of RNA oligomers by montmorillonite catalysis. *J Am Chem Soc* **128**: 8914–8919.

Ikawa Y, Tsuda K, Matsumura S, Inoue T. 2004. De novo synthesis and development of an RNA enzyme. *Proc Natl Acad Sci* **101**: 13750–13755.

Ingar AA, Luke RWA, Hayter BR, Sutherland JD. 2003. Synthesis of a cytidine ribonucleotide by stepwise assembly of the heterocycle on a sugar phosphate. *Chembiochem* **4**: 504–507.

Inoue T, Orgel LE. 1981. Substituent control of the poly(C)-directed oligomerization of guanosine 5'-phosphoroimidazolide. *J Am Chem Soc* **103**: 7666–7667.

Inoue T, Orgel LE. 1982. Oligomerization of (guanosine 5'-phosphor)-2-methylimidazolide on poly(C). *J Mol Biol* **162**: 204–217.

Inoue T, Orgel LE. 1983. A nonenzymatic RNA polymerase model. *Science* **219**: 859–862.

Inoue T, Joyce GF, Grzeskowiak K, Orgel LE, Brown JM, Reese CB. 1984. Template-directed synthesis on the pentanucleotide CpCpGpCpC. *J Mol Biol* **178**: 669–676.

Jaeger L, Wright MC, Joyce GF. 1999. A complex ligase ribozyme evolved in vitro from a group I ribozyme domain. *Proc Natl Acad Sci* **96**: 14712–14717.

James KD, Ellington AD. 1999. The fidelity of template-directed oligonucleotide ligation and the inevitability of polymerase function. *Orig Life Evol Biosph* **29**: 375–390.

Johnston WK, Unrau PJ, Lawrence MS, Glasner ME, Bartel DP. 2001. RNA-catalyzed RNA polymerization: Accurate and general RNA-templated primer extension. *Science* **292**: 1319–1325.

Joshi PC, Aldersley MF, Delano JW, Ferris JP. 2009. Mechanism of montmorillonite catalysis in the formation of RNA oligomers. *J Am Chem Soc* **131**: 13369–13374.

Joyce GF. 1987. Non-enzymatic template-directed synthesis of informational macromolecules. *Cold Spring Harbor Symp Quant Biol* **52**: 41–51.

Joyce GF. 1989. RNA evolution and the origins of life. *Nature* **338**: 217–224.

Joyce GF. 2002. The antiquity of RNA-based evolution. *Nature* **418**: 214–221.

Joyce GF, Orgel LE. 1986. Non-enzymic template-directed synthesis on RNA random copolymers: Poly(C,G) templates. *J Mol Biol* **188**: 433–441.

Joyce GF, Orgel LE. 1993. Prospects for understanding the origin of the RNA world. In *The RNA world* (eds. Gesteland R.F., Atkins J.F.), pp. 1–25. Cold Spring Harbor Laboratory Press, Cold Spring Harbor, NY.

Joyce GF, Schwartz AW, Miller SL, Orgel LE. 1987. The case for an ancestral genetic system involving simple analogues of the nucleotides. *Proc Natl Acad Sci* **84**: 4398–4402.

Joyce GF, Visser GM, van Boeckel CAA, van Boom JH, Orgel LE, van Westrenen J. 1984. Chiral selection in poly(C)-directed synthesis of oligo(G). *Nature* **310**: 602–604.

Kanavarioti A, Monnard PA, Deamer DW. 2001. Eutectic phases in ice facilitate nonenzymatic nucleic acid synthesis. *Astrobiology* **1**: 271–281.

Kauffman SA. 1986. Autocatalytic sets of proteins. *J Theor Biol* **119**: 1–24.

Kawamura K, Ferris JP. 1994. Kinetic and mechanistic analysis of dinucleotide and oligonucleotide formation from the 5'-phosphorimidazolide of adenosine on Na^+-montmorillonite. *J Am Chem Soc* **116**: 7564–7572.

Kawamura K, Ferris JP. 1999. Clay catalysis of oligonucleotide formation: Kinetics of the reaction of the 5'-phosphorimidazolides of nucleotides with the non-basic heterocycles uracil and hypoxanthine. *Orig Life Evol Biosph* **29**: 563–591.

Khorana HG. 1961. *Some recent developments in the chemistry of phosphate esters of biological interest*, pp. 126–141. Wiley & Sons, New York.

Klussmann M, Iwamura H, Mathew SP, Wells Jr DH, Pandya U, Armstrong A, Blackmond DG. 2006. Thermodynamic control of asymmetric amplification in amino acid catalysis. *Nature* **441**: 621–623.

Kondepudi DK, Kaufman RJ, Singh N. 1990. Chiral symmetry breaking in sodium chlorate crystallization. *Science* **250**: 975–976.

Koppitz M, Nielsen PE, Orgel LE. 1998. Formation of oligonucleotide-PNA-chimeras by template-directed ligation. *J Am Chem Soc* **120**: 4563–4569.

Kozlov IA, Politis PK, Van Aerschot A, Busson R, Herdewijn P, Orgel LE. 1999. Nonenzymatic synthesis of RNA and DNA oligomers on hexitol nucleic acid templates: the importance of the A structure. *J Am Chem Soc* **121**: 2653–2656.

Kozlov IA, Zielinski M, Allart B, Kerremans L, Van Aerschot A, Busson R, Herdewijn P, Orgel LE. 2000. Nonenzymatic template-directed reactions on altritol oligomers, preorganized analogues of oligonucleotides. *Chem Eur J* **6**: 151–155.

Krishnamurthy R, Guntha S, Eschenmoser A. 2000. Regioselective α-phosphorylation of aldoses in aqueous solution. *Angew Chemie* **39**: 2281–2285.

Kruger K, Grabowski PJ, Zaug AJ, Sands J, Gottschling DE, Cech TR. 1982. Self-splicing RNA: Autoexcision and autocyclization of the ribosomal RNA intervening sequence of Tetrahymena. *Cell* **31**: 147–157.

Kurz M, Göbel K, Hartel C, Göbel MW. 1997. Nonenzymatic oligomerization of ribonucleotides on guanosine-rich templates: Suppression of the self-pairing of guanosine. *Angew Chemie* **36**: 842–845.

Kurz M, Göbel K, Hartel C, Göbel MW. 1998. Acridine-labeled primers as tools for the study of nonenzymatic RNA oligomerization. *Helv Chim Acta* **81**: 1156–1180.

Lau MWL, Cadieux KEC, Unrau PJ. 2004. Isolation of fast purine nucleotide synthase ribozymes. *J Am Chem Soc* **126**: 15686–15693.

Lau MWL, Unrau PJ. 2009. A promiscuous ribozyme promotes nucleotide synthesis in addition to ribose chemistry. *Chem Biol* **16**: 815–825.

Lawrence MS, Bartel DP. 2003. Processivity of ribozyme-catalyzed RNA polymerization. *Biochemistry* **42**: 8748–8755.

Lewin R. 1986. RNA catalysis gives fresh perspective on the origin of life. *Science* **231**: 545–546.

Lincoln TA, Joyce GF. 2009. Self-sustained replication of an RNA enzyme. *Science* **323**: 1229–1232.

Lohrmann R. 1975. Formation of nucleoside 5'-polyphosphates from nucleotides and trimetaphosphate. *J Mol Evol* **6**: 237–252.

Lohrmann R. 1977. Formation of nucleoside 5'-phosphoramidates under potentially prebiological conditions. *J Mol Evol* **10**: 137–154.

Lohrmann R, Orgel LE. 1973. Prebiotic activation processes. *Nature* **244**: 418–420.

Lohrmann R, Orgel LE. 1976. Template-directed synthesis of high molecular weight polynucleotide analogues. *Nature* **261**: 342–344.

Lorsch J, Szostak JW. 1994. In vitro evolution of new ribozymes with polynucleotide kinase activity. *Nature* **371**: 31–36.

McCaskill JS. 1984a. A stochastic theory of molecular evolution. *Biol Cybernetics* **50**: 63–73.

McCaskill JS. 1984b. A localization threshold for macromolecular quasispecies from continuously distributed replication rates. *J Chem Phys* **80**: 5194–5202.

McGinness KE, Joyce GF. 2002. RNA-catalyzed RNA ligation on an external RNA template. *Chem Biol* **9**: 297–307.

Mittapalli GK, Osornio YM, Guerrero MA, Reddy KR, Krishnamurthy R, Eschenmoser A. 2007a. Mapping the landscape of potentially primordial informational oligomers: Oligodipeptides tagged with 2,4-disubstituted 5-aminopyrimidines as recognition elements. *Angew Chemie* **46**: 2478–2484.

Mittapalli GK, Reddy KR, Xiong H, Munoz O, Han B, De Riccardis F, Krishnamurthy R, Eschenmoser A. 2007b. Mapping the landscape of potentially primordial informational oligomers: Oligodipeptides and oligodipeptoids tagged with triazines as recognition elements. *Angew Chemie* **46:** 2470–2477.

Miyakawa S, Ferris JP. 2003. Sequence- and regioselectivity in the montmorillonite-catalyzed synthesis of RNA. *J Am Chem Soc* **125:** 8202–8208.

Mizuno T, Weiss AH. 1974. Synthesis and utilization of formose sugars. *Adv Carbohyd Chem Biochem* **29:** 173–227.

Mohr SC, Thach RE. 1969. Application of ribonuclease T_1 to the synthesis of oligoribonucleotides of defined base sequence. *J Biol Chem* **244:** 6566–6576.

Monnard PA, Kanavarioti A, Deamer DW. 2003. Eutectic phase polymerization of activated ribonucleotide mixtures yields quasi-equimolar incorporation of purine and pyrimidine nucleobases. *J Am Chem Soc* **125:** 13734–13740.

Müller D, Pitsch S, Kittaka A, Wagner E, Wintner CE, Eschenmoser A. 1990. Chemie von α-aminonitrilen. Aldomerisierung von Glykolaldehydphosphat zu *racemischen* hexose-2,4,6-triphosphaten und (in gegenwart von formaldehyd) *racemischen* pentose-2,4-diphosphaten: *rac.*-allose-2,4,6-triphosphat und *rac.*-ribose-2,4-diphosphat sind die reaktionshauptprodukte. *Helv Chim Acta* **73:** 1410–1468.

Narlikar GJ, Herschlag D. 1997. Mechanistic aspects of enzymatic catalysis: Lessons from comparison of RNA and protein enzymes. *Annu Rev Biochem* **66:** 19–59.

Nelsestuen GL. 1980. Origin of life: Consideration of alternatives to proteins and nucleic acids. *J Mol Evol* **15:** 59–72.

Nissen P, Hansen J, Ban N, Moore PB, Steitz TA. 2000. The structural basis of ribosome activity in peptide bond synthesis. *Science* **289:** 920–930.

Noorduin WL, Izumi T, Millemaggi A, Leeman M, Meekes H, Van Enckevort WJP, Kellogg RM, Kaptein B, Vlieg E, Blackmond DG. 2008. Emergence of a single solid chiral state from a nearly racemic amino acid derivative. *J Am Chem Soc* **130:** 1158–1159.

Orgel LE. 1968. Evolution of the genetic apparatus. *J Mol Biol* **38:** 381–393.

Orgel LE. 1989. Was RNA the first genetic polymer? In *Evolutionary tinkering in gene expression* (eds. Grunberg-Manago M., Clark B.F.C., Zachau H.G.), pp. 215–224. Plenum, London.

Orgel LE. 2004a. Prebiotic chemistry and the origin of the RNA world. *Crit Rev Biochem Mol Biol* **39:** 99–123.

Orgel LE. 2004b. Prebiotic adenine revisited: Eutectics and photochemistry. *Orig Life Evol Biosph* **34:** 361–369.

Orò J. 1961. Mechanism of synthesis of adenine from hydrogen cyanide under plausible primitive earth conditions. *Nature* **191:** 1193–1194.

Osterberg R, Orgel LE, Lohrmann R. 1973. Further studies of urea-catalyzed phosphorylation reactions. *J Mol Evol* **2:** 231–234.

Pace NR, Marsh TL. 1985. RNA catalysis and the origin of life. *Orig Life Evol Biosph* **16:** 97–116.

Pitsch S. 1992. "Zur chemie von glykolaldehyd-phosphat: Seine bildung aus oxirancarbonitril und seine aldomerisierumg zu den (racemischen) pentose-2,4-diphosphaten und hexose-2,4,6-triphosphaten." Ph.D. Thesis, ETH, Zürich.

Pitsch S, Eschenmoser A, Gedulin B, Hui S, Arrhenius G. 1995a. Mineral induced formation of sugar phosphates. *Orig Life Evol Biosph* **25:** 297–334.

Pitsch S, Krishnamurthy R, Bolli M, Wendeborn S, Holzner A, Minton M, Lesueur C, Schlönvogt I, Jaun B, Eschenmoser A. 1995b. Pyranosyl-RNA ('p-RNA'): Base-pairing selectivity and potential to replicate. *Helv Chim Acta* **78:** 1621–1635.

Pitsch S, Wendeborn S, Jaun B, Eschenmoser A. 1993. Why pentose- and not hexose-nucleic acids? Pyranosyl-RNA ('p-RNA'). *Helv Chim Acta* **76:** 2161–2183.

Pitsch S, Wendeborn S, Krishnamurthy R, Holzner A, Minton M, Bolli M, Miculka C, Windhab N, Micura R, Stanek M, et al. 2003. Pentopyranosyl oligonucleotide systems. 9th communication. The ß-D-ribopyranosyl-(4′→2′)-oligonucleotide system ('pyranosyl-RNA'): Synthesis and resumé of base-pairing properties. *Helv Chim Acta* **86:** 4270–4363.

Pizzarello S, Zolensky M, Turk KA. 2003. Nonracemic isovaline in the Murchison meteorite: Chiral distribution and mineral association. *Geochim Cosmochim Acta* **67:** 1589–1595.

Prabahar KJ, Ferris JP. 1997. Adenine derivatives as phosphate-activating groups for the regioselective formation of 3′,5′-linked oligoadenylates on montmorillonite: Possible phosphate-activating groups for the prebiotic synthesis of RNA. *J Am Chem Soc* **119:** 4330–4337.

Powner MW, Gerland B, Sutherland JD. 2009. Synthesis of activated pyrimidine ribonucleotides in prebiotically plausible conditions. *Nature* **459:** 239–242.

Reimann R, Zubay G. 1999. Nucleoside phosphorylation: A feasible step in the prebiotic pathway to RNA. *Orig Life Evol Biosph* **29:** 229–247.

Ricardo A, Carrigan MA, Olcott AN, Benner SA. 2004. Borate minerals stabilize ribose. *Science* **303:** 196.

Robertson MP, Ellington AD. 1999. *In vitro* selection of an allosteric ribozyme that transduces analytes into amplicons. *Nature Biotechnol* **17:** 62–66.

Robertson MP, Miller SL. 1995. An efficient prebiotic synthesis of cytosine and uracil. *Nature* **375:** 772–774.

Robertson MP, Scott WG. 2007. The structural basis of ribozyme-catalyzed RNA assembly. *Science* **315:** 1549–1553.

Rogers J, Joyce GF. 2001. The effect of cytidine on the structure and function of an RNA ligase ribozyme. *RNA* **7:** 395–404.

Rohatgi R, Bartel DP, Szostak JW. 1996a. Kinetic and mechanistic analysis of nonenzymatic, template-directed oligoribonucleotide ligation. *J Am Chem Soc* **118:** 3332–3339.

Rohatgi R, Bartel DP, Szostak JW. 1996b. Nonenzymatic, template-directed ligation of oligoribonucleotides is highly regioselective for the formation of 3′–5′ phosphodiester bonds. *J Am Chem Soc* **118:** 3340–3344.

Saladino R, Crestini C, Costanzo G, DiMauro E. 2004. Advances in the prebiotic synthesis of nucleic acids bases: Implications for the origin of life. *Curr Org Chem* **8:** 1425–1443.

Sanchez RA, Orgel LE. 1970. Studies in prebiotic synthesis. V. Synthesis and photoanomerization of pyrimidine nucleosides. *J Mol Biol* **47:** 531–543.

Sanchez RA, Ferris JP, Orgel LE. 1967. Studies in prebiotic synthesis. II. Synthesis of purine precursors and amino acids from cyanoacetylene and cyanate. *J Mol Biol* **30:** 223–253.

Sawai H, Kuroda K, Hojo T. 1988. Efficient oligoadenylate synthesis catalyzed by uranyl ion complex in aqueous solution. In *Nucleic acids research symposium series* (ed. Hayatsu H.), vol. 19, pp. 5–7. IRL Press Limited, Oxford.

Shechner DM, Grant RA, Bagby SC, Koldobskaya Y, Piccirilli JA, Bartel DP. 2009. Crystal structure of the catalytic core of an RNA polymerase ribozyme. *Science* **326:** 1271–1275.

Schmidt JG, Christensen L, Nielsen PE, Orgel LE. 1997a. Information transfer from DNA to peptide nucleic acids by template-directed syntheses. *Nucleic Acids Res* **25:** 4792–4796.

Schmidt JG, Nielsen PE, Orgel LE. 1997b. Information transfer from peptide nucleic acids to RNA by template-directed syntheses. *Nucleic Acids Res* **25:** 4797–4802.

Schöning K, Scholz P, Guntha S, Wu X, Krishnamurthy R, Eschenmoser A. 2000. Chemical etiology of nucleic acid structure: The α-threofuranosyl-(3′→2′) oligonucleotide system. *Science* **290:** 1347–1351.

Schrum JP, Ricardo A, Krishnamurthy M, Blain JC, Szostak JW. 2009. Efficient and rapid template-directed nucleic acid copying using 2′-amino-2′,3′-dideoxyribonucleoside-5′-phosphorimidazolide monomers. *J Am Chem Soc* **131:** 14560–14570.

Schuster P, Swetina J. 1988. Stationary mutant distributions and evolutionary optimization. *Bull Math Biol* **50:** 635–660.

Schwartz AW, Orgel LE. 1985. Template-directed synthesis of novel, nucleic acid-like structures. *Science* **228:** 585–587.

Shan S, Kravchuk AV, Piccirilli JA, Herschlag D. 2001. Defining the catalytic metal ion interactions in the *Tetrahymena* ribozyme reaction. *Biochemistry* **40:** 5161–5171.

Shan S, Yoshida A, Sun S, Piccirilli JA, Herschlag D. 1999. Three metal ions at the active site of the *Tetrahymena* group I ribozyme. *Proc Natl Acad Sci* **96:** 12299–12304.

Shapiro R. 1984. The improbability of prebiotic nucleic acid synthesis. *Orig Life Evol Biosph* **14:** 565–570.

Sharp PA. 1985. On the origin of RNA splicing and introns. *Cell* **42:** 397–400.

Sleeper HL, Orgel LE. 1979. The catalysis of nucleotide polymerization by compounds of divalent lead. *J Mol Evol* **12:** 357–364.

Soai K, Shibata T, Morioka H, Choji K. 1995. Asymmetric autocatalysis and amplification of enantiomeric excess of a chiral molecule. *Nature* **378:** 767–768.

Springsteen G, Joyce GF. 2004. Selective derivatization and sequestration of ribose from a prebiotic mix. *J Am Chem Soc* **126:** 9578–9583.

Stahley MR, Strobel SA. 2005. Structural evidence for a two-metal-ion mechanism of group I intron splicing. *Science* **309:** 1587–1590.

Steitz TA. 1998. A mechanism for all polymerases. *Nature* **391:** 231–232.

Steitz TA, Moore PB. 2003. RNA, the first macromolecular catalyst: The ribosome is a ribozyme. *Trends Biochem Sci* **28:** 411–418.

Steitz TA, Steitz JA. 1993. A general two-metal-ion mechanism for catalytic RNA. *Proc Natl Acad Sci* **90:** 6498–6502.

Sträter N, Lipscomb WN, Klabunde T, Krebs B. 1996. Two-metal ion catalysis in enzymatic acyl- and phosphoryl-transfer reactions. *Angew Chemie* **35:** 2024–2055.

Strobel SA, Ortoleva-Donnelly L. 1999. A hydrogen-bonding triad stabilizes the chemical transition state of a group I ribozyme. *Chem Biol* **6:** 153–165.

Szabó P, Scheuring I, Czárán T, Szathmáry E. 2002. *In silico* simulations reveal that replicators with limited dispersal evolve towards higher efficiency and fidelity. *Nature* **420:** 340–343.

Sulston J, Lohrmann R, Orgel LE, Todd MH. 1968. Nonenzymatic synthesis of oligoadenylates on a polyuridylic acid template. *Proc Natl Acad Sci* **59:** 726–733.

Tapiero CM, Nagyvary J. 1971. Prebiotic formation of cytidine nucleotides. *Nature* **231:** 42–43.

Unrau PJ, Bartel DP. 1998. RNA-catalysed nucleotide synthesis. *Nature* **395:** 260–263.

Ura Y, Beierle JM, Leman LJ, Orgel LE, Ghadiri MR. 2009. Self-assembling sequence-adaptive peptide nucleic acids. *Science* **325:** 73–77.

Viedma C. 2005. Chiral symmetry breaking during crystallization: Complete chiral purity induced by nonlinear autocatalysis and recycling. *Phys Rev Lett* **94:** 065504.

Viedma C, Ortiz JE, de Torres T, Izumi T, Blackmond DG. 2008. Evolution of solid phase homochirality for a proteinogenic amino acid. *J Am Chem Soc* **130:** 15274–15275.

Wächtershäuser G. 1988. Before enzymes and templates: Theory of surface metabolism. *Microbiol Rev* **52:** 452–484.

Wang KJ, Ferris JP. 2001. Effect of inhibitors on the montmorillonite clay-catalyzed formation of RNA: Studies on the reaction pathway. *Orig Life Evol Biosph* **31:** 381–402.

Weber AL. 1987. The triose model: Glyceraldehyde as a source of energy and monomers for prebiotic condensation reactions. *Orig Life Evol Biosph* **17:** 107–119.

Weber AL. 1989. Model of early self-replication based on covalent complementarity for a copolymer of glycerate-3-phosphate and glycerol-3-phosphate. *Orig Life Evol Biosph* **19:** 179–186.

Weiner AM, Maizels N. 1987. tRNA-like structures tag the 3′ ends of genomic RNA molecules for replication: Implications for the origin of protein synthesis. *Proc Natl Acad Sci* **84:** 7383–7387.

Wimberly BT, Brodersen DE, Clemons WMJr, Morgan-Warren RJ, Carter AP, Vonrhein C, Hartsch T, Ramakrishnan V. 2000. Structure of the 30S ribosomal subunit. *Nature* **407:** 327–338.

Wittung P, Nielsen PE, Buchardt O, Egholm M, Norden B. 1994. DNA-like double helix formed by peptide nucleic acid. *Nature* **368:** 561–563.

Woese C. 1967. *The genetic code*, pp. 179–195. Harper and Row, New York.

Wu T, Orgel LE. 1992a. Nonenzymatic template-directed synthesis on oligodeoxycytidylate sequences in hairpin oligonucleotides. *J Am Chem Soc* **114:** 317–322.

Wu T, Orgel LE. 1992b. Nonenzymatic template-directed synthesis on hairpin oligonucleotides. II. Templates containing cytidine and guanosine residues. *J Am Chem Soc* **114:** 5496–5501.

Yarus M. 1993. How many catalytic RNAs? Ions and the Cheshire Cat conjecture. *FASEB J* **7:** 31–39.

Yusupov M, Yusupova G, Baucom A, Lieberman K, Earnest TN, Cate JH, Noller HF. 2001. Crystal structure of the ribosome at 5.5 Å resolution. *Science* **292:** 883–896.

Zhang S, Lockshin C, Cook R, Rich A. 1994. Unusually stable beta-sheet formation in an ionic self-complementary oligopeptide. *Biopolymers* **34:** 663–672.

Zhang L, Peritz A, Meggers E. 2005. A simple glycol nucleic acid. *J Am Chem Soc* **127:** 4174–4175.

Zielinski WS, Orgel LE. 1985. Oligomerization of activated derivatives of 3′-amino-3′-deoxyguanosine on poly(C) and poly(dC) templates. *Nucleic Acids Res* **13:** 2469–2484.

Zielinski WS, Orgel LE. 1987. Oligomerization of dimers of 3′-amino-3′-deoxy-nucleotides (GC and CG) in aqueous solution. *Nucleic Acids Res* **13:** 1699–1715.

Zubay G. 1998. Studies on the lead-catalyzed synthesis of aldopentoses. *Orig Life Evol Biosph* **28:** 13–26.

Getting Past the RNA World: The Initial Darwinian Ancestor

Michael Yarus

Department of Molecular, Cellular, and Developmental Biology, University of Colorado, Boulder, Colorado 80309-0347

Correspondence: yarus@colorado.edu

SUMMARY

A little-noted result of the confirmation of multiple premises of the RNA-world hypothesis is that we now know something about the dawn organisms that followed the origin of life, perhaps over 4 billion years ago. We are therefore in an improved position to reason about the biota just before RNA times, during the era of the first replicators, the first Darwinian creatures on Earth. An RNA congener still prominent in modern biology is a plausible descendent of these first replicators.

Outline

1. Introduction
2. A new vantage
3. Evidence for ancient RNA
4. Conservation and persistence
5. Persistence of CHNOPS
6. Persistence of RNA
7. The hypothesis
8. Earlier ideas about nucleotide cofactors
9. Pre-RNA world credentials
10. An old variant backbone
11. An augmented alphabet, an augmented manifold of reactions
12. Chickens and eggs
13. Continuity with an RNA world
14. Novel replication chemistry
15. Catch-22

References

1 INTRODUCTION

In the following discussion, we try to imagine the biological past over 4 Gya (Gigayears ago). Such speculation is worth new effort because of the success of the RNA-world hypothesis itself. The ultimate goal is a glimpse of the widely accepted, though hypothetical, initial replicator, whose biomolecular activity initiated Darwinian evolution on Earth. We are searching before the last universal common ancestor (LUCA), looking back toward the substantially older initial darwinian ancestor (IDA). That such a duplicating molecule existed (for example, Szathmary 2006) is therefore an indispensable assumption.

2 A NEW VANTAGE

Viewed from our time, it requires a huge leap to reach the IDA. The 4 Gy gap necessarily spanned is almost unthinkable, so great that it has carried most rock of the time away, consequently swallowing most hope of finding scientific evidence. But that is precisely the point; if 4 Gya was a time when oligoribonucleotides were prominent, we know something new, whose implications are usable as the crux of an inquiry into earlier times.

Furthermore, it requires only arithmetic to see that the RNA world can provide a decisive new vantage point. From the RNA world, the time to be spanned to the IDA is very much shortened. Because the solar system (and thus the Earth) congealed 4.5 Gya (Dalrymple 1991), in all likelihood an RNA world at 4 Gya is about 20-fold closer to the initiation of biological evolution than is the present. We want to know the Earth's biota a few hundred million years before the age of RNA. So, extrapolation from 4 Gya back to the IDA is 20-fold shorter than from today, probable molecular change is roughly 20-fold smaller, and the accuracy of any speculation roughly 20-fold greater than before acquisition of an RNA world standpoint. Instead of nearly completely spanning planetary history to reach the IDA, from the RNA world we span a time, for example, comparable to the history of Earth's mammals (Luo et al. 2001).

Moreover, the supporting data we need from the RNA world is minimal. Thus, the argument that follows does not use most of what we presently know with high probability. We need not assume that an organism existed that used RNA in every capacity for which it has been contemplated, or in even every capacity for which RNA's competence has been shown in selection experiments. Though RNA has surely proven its unanticipated versatility (Chen et al. 2007), we require only that a molecule made of something like modern ribonucleotides performed essential functions 4 Gya.

3 EVIDENCE FOR ANCIENT RNA

This indispensable contention rests on very robust experimental evidence, linking many independent observations. Such evidence is very frequent, for example, in studies of the apparent descent of translation—coded protein biosynthesis. If our immediate evolutionary predecessors were RNA-based organisms, then we plausibly require that they use RNA to invent translation and their successor catalysts, the peptides.

This idea in turn is strongly supported by selection of RNAs with the required translational capabilities. For example, the synthesis of aminoacyl-RNA (which activates amino acids and potentially links them to a matching coding triplet) is easily found within the RNA reaction repertoire, using amino acid adenylate (Illangasekare et al. 1995) (the universal biological substrate) or other carboxyl-activated amino acids (Lee et al. 2000) as precursor. Further, these RNA transacylators can be both faster and more accurate than modern proteins performing the same reaction (Illangasekare and Yarus 1999). In addition, aminoacyl group transfer is so simple a reaction for RNA that it can be performed by small RNAs, even in a ribozymic reaction center containing only three sparingly constrained nucleotides, or a ribozyme totaling five nucleotides in length (Turk et al. 2010). Such a catalyst would appear in an untemplated pool of ribonucleotide sequences consisting of only attograms (10^{-18} g) of RNA, unexpectedly easily reached by geochemical means. I argue, as previously (Yarus 2001; Yarus et al. 2005), that the unanticipated confirmation of translational activity, in fact, confirmation of the thorough competence of RNA in translational reactions, requires that the probability of an RNA invention of protein biosynthesis be strongly elevated. To do less than this is to ignore the theory of probabilities.

Alternatively, one could argue from the fact that the present peptidyl transferase is RNA, which exploits ribonucleotide properties of both peptidyl-tRNA (Weinger et al. 2004) and the rRNA (Nissen et al. 2000) to accelerate peptide transfer. The highly conserved rRNA peptidyl transferase sequence includes a ribonucleotide cradle whose sequence is particularly suited to poise the reacting amino acids (Welch et al. 1997), further supporting and illustrating its origin in the RNA world. Thus there is exceedingly strong support, all unexpected only a few years ago, for the origination of translation as a constellation of RNA activities. In fact, all chemical group transfers, and even the information transfers required for coded protein biosynthesis, have precedents within the chemical repertoire of pure small RNAs (Yarus 1991; Yarus et al. 2009). Any of the observations above, taken alone, is a strong argument for the existence of RNA-like molecules before the present nucleoprotein era. Taken together, these multiple experiments are very persuasive indeed.

4 CONSERVATION AND PERSISTENCE

So, having staked out a position for RNA 4 Gya, what can we consequently see? To see anything we need to define some evolutionary tendencies that span the RNA world.

One of these is continuity, the property that links predecessors with descendants by small genetic changes. Evolutionary descent follows a path that takes finite steps, but each descendant is continuously linked to its forbear by individual practical changes in a genetic text. One frequently noted consequence of continuity on large time scales is that successful adaptations tend to persist, and are rarely discarded. For example, it seems likely to many that iron, nickel, copper, manganese, and molybdenum clusters within modern protein enzymes are a retained adaptation from a mineral chemistry (Rees 2002) that was once even more useful to life. In fact, such mineral cofactors probably predate the RNA world, surviving even beyond the gigayears that stretch between us and the era of RNA dominance. They therefore illustrate the permanence of useful biochemical adaptations. Persistence and exaptation of successful evolutionary devices therefore is a rule and not an exception. We will depend on parallel ideas in the following discussion.

A second useful idea is Bayes' Theorem, which is a rigorous statement about the probability of ideas in the light of new evidence, like new evidence of an RNA world (Yarus 2001). Bayes' application here is that your successors are the most likely outcome of your existence, or alternatively, that your predecessors were those most likely to give rise to the present situation. Bayes' Theorem again strongly reinforces the preservation of prior adaptations, not simply because they passively preexist in almost their current form (as for continuity, earlier), but because persistence in descent naturally results from the prior success of an adaptation. Continuity implies that evolution has a path, and Bayes' implies that this path has a probable functional logic. Continuity and Bayes' together constitute a mutually reinforcing argument for the persistence of prior adaptations.

Now we aggressively combine these ideas with the newly plausible existence of RNA-like molecules 4 Gya, to consider what might have existed relatively shortly before, during the time of the IDA, the primordial replicator.

5 PERSISTENCE OF CHNOPS

Continuity should extend even down to atomic composition. Therefore we assume here that the primordial replicator was composed of the most prevalent biological atoms; C, H, N, O, P, and S. This notion at first looks bland and uncontroversial—but is worth emphasis because it wholly contradicts a frequently discussed idea, that of the "genetic takeover." The idea, owing to Cairns-Smith (Cairns-Smith 1982), is that another kind of replicator entirely, perhaps a layer of clay, not only preceded the rise of RNA but also templated the first RNA molecules. In this case, a clay "genetic material" with prevalent Al and Si centers would give way to a CHNOPS replicator, thereby violating this implementation of continuity.

However, although clays do have useful interactions with ribonucleotides (discussed later), it is unclear whether clays can replicate any of their properties. It has been claimed that the redox and binding affinities of montmorillonite clays can be propagated from preexisting to newly grown layers of clay (Weiss 1981). But efforts to define and reproduce these observations have not been successful (Arrhenius et al. 1986). Thus replication of clay "information" has remained hypothetical, and transfer of replicated clay properties to nucleic acids even more so. In materials of unlimited complexity, the only replication that we know takes place entirely within molecules that are CHNOPS, so this continuity assumption seems apt for the present purpose.

6 PERSISTENCE OF RNA

Focusing more closely, the primordial replicator must give way to RNA at 4 Gya. Therefore the immediate ancestor must be chemically related to RNA by a plausible transformation. At least the last in this chain of evolutionary transitions must be small, implying a predecessor anchored to RNA by change of a few atoms. Therefore it is economical of assumptions and consistent with continuity and Bayes' to extend the resemblance to RNA further back. Evolutionary succession would somehow conserve a predecessor's chemical proficiencies. It is difficult to see how to do this in a primitive context except by propagating the forerunner's RNA-like atoms and bonding. Is there a potential replicator that fits these criteria?

7 THE HYPOTHESIS

I suggest that AMP-containing enzymatic cofactors are the modern descendants of the IDA. An example (NADP, the virtually ubiquitous redox cofactor) is shown in Figure 1A. These structures meet the logic of continuity and Bayes' Theorem proposed above. NAD and NADP in particular are small 5'-5' linked ribonucleic acids with a potential stacked structure (Fig. 1B), which aligns their nucleotides for coordinated base-pairing interaction during a possible complementary replication (Fig. 1C). Their original complements/templates (Fig. 1C) may not presently be used by modern protein enzymes—it seems to

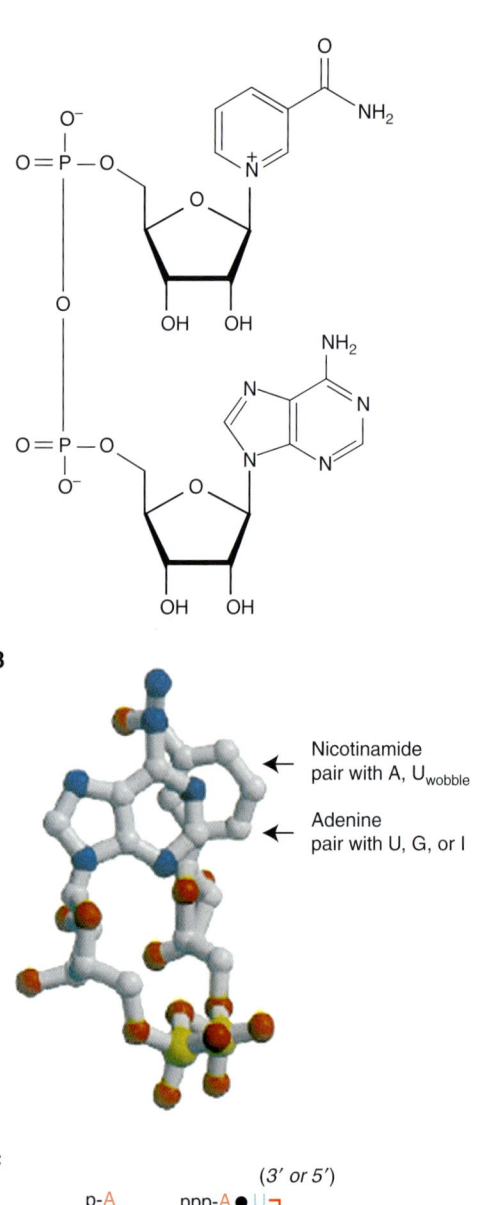

Figure 1. (A) Structure of NADP, the 3′ phosphorylated derivative of nicotinamide adenine diphosphate (NAD). The image is a public domain illustration from Wikipedia. (B) A solution structure for NAD, with the nucleobases at the *top*, ribose in the *middle*, and the pyrophosphate at the *bottom*. Possible base pairings for replication are indicated. The figure is derived from the model of (Smith and Tanner 2000). (C) Replication scheme for an AMP-containing cofactor, like NAD or FAD. For simplicity, the backbone linking chemistry is shown as that for RNA catalysis (Huang and Yarus 1997), but this might vary. The complementary "template" strand and its replication on an NAD-like template are unspecified, because it is unclear what polarity it would have, or if it would be 5′-5′ linked. However, 5′-5′ polarity appears simplest, and is adopted in the text for clarity. R is nicotinamide nucleotide or a congener and N is its hypothetical complement.

be hoping for too much to suppose that both the primordial cofactor and its template became protein cofactors that survive today. Therefore, we are somewhat uncertain what the complements to the cofactors were or even what template backbones were because these may now be lost (Fig. 1 legend). However, it is notable that frequently, as for the chemically active nucleotides of NAD and FAD, both chemical centers and of course AMP could be complementary to existing nucleotides, as suggested by the notations within Figure 1B. For conciseness, discussion below takes both strands to be 5′-5′ linked dinucleotides.

Although we came to this specific notion by a slightly exotic Continuity/Bayes' logic, the idea agrees broadly with a substantial literature about ancient molecules.

8 EARLIER IDEAS ABOUT NUCLEOTIDE COFACTORS

Casting NAD, FAD, and their related cofactors as the IDA is an extension of the influential notion of Harold White III (White 1976) that AMP-containing cofactors bear witness to the existence of a prior generation of RNA enzymes, whose AMP-containing reaction centers were appropriated by protein enzymes. In this article, we hypothetically extend the history of the cofactors further into the past and speculate that they may also have had a different and even more ancient role. More recently, Copley, Smith, and Morowitz (Copley et al. 2007) have emphasized that biocatalysis may have begun with small molecules, some of which lay on the evolutionary path to the RNA world. Though they were not thinking precisely of these same molecules, the present scheme is one implementation of such ideas.

9 PRE-RNA WORLD CREDENTIALS

The nicotinamide of NAD is easily reached by several synthetic routes beginning with plausibly prebiotic chemicals, like ethylene and ammonia (Friedmann et al. 1971) or aspartate and dihydroxyacetone phosphate (Cleaves and Miller 2001). NAD or a congener may have plausibly been one of the earliest cofactors to enter the biochemical inventory. It is therefore credible to consider nicotinamide-containing cofactors an early participant in biochemistry, before the rise of complex RNA catalysts.

10 AN OLD VARIANT BACKBONE

Desire for properties thought to be useful to primordial genetic materials, for example, novel chemical resistance or achirality, has led to varied ingenious proposals for alternate backbone composition and continuities (see Robertson and Joyce 2010; Benner et al. 2010). It is therefore doubly striking that there exists a natural alternative ribose/phosphodiester backbone universal in modern biochemistry, the 5′-5′ linked cofactors, that possesses frequently desired properties. For example, the prominent sensitivity of natural RNA to hydrolysis by any factor that stabilizes the ribose 2′ oxyanion (for example, bases and metals) does not exist for NAD and its congeners. It is also conceivable (see Fig. 1B) that the cofactors, because they do not repeatedly employ their chiral atoms directly in the backbone, may be more tolerant of variant sugars, as would be prevalent in a primitive milieu. For example, because the chiral sugars are at the ends rather than in the middle of the backbone, oligomerization of the 5′-5′ nucleotides may be less sensitive to sugar chirality than 5′-3′ RNA (Joyce et al. 1984). Finally, the cofactors are particularly ancient on the evidence of their own ubiquitous metabolic uses, but similar old structures also mark the 5′ terminus of all eukaryotic messages (reviewed in Schoenberg and Maquat 2009).

11 AN AUGMENTED ALPHABET, AN AUGMENTED MANIFOLD OF REACTIONS

Benner (Benner et al. 2010) has argued the benefits of a more diverse set of nucleobases. By synthesis of a third base pair, his, and other, laboratories have shown that a three complementary pair/six base system is capable of replication (Kimoto et al. 2009; Yang et al. 2007). It is striking that the 5′-linked dinucleotide cofactors deploy an expanded repertoire of nucleotide residues, and automatically use their more versatile reactivity. Thus it is manifest that such structural diversity enhances chemical versatility. In fact, it is quite frequent that free modern cofactors have a diminished version of the chemical activity that they display in conjunction with a protein enzyme. For example, NADH, the quintessential redox cofactor (here in its reduced form), will reduce nitro blue tetrazolium in the presence of phenazine methosulfate. This reaction underlies a classical NAD assay and cytochemical stain (Ponti et al. 1978). NADH-dye reduction goes via superoxide aerobically, but is almost as fast by direct means in the absence of oxygen (Compare Fig. 2C).

12 CHICKENS AND EGGS

The NAD reaction just mentioned would be particularly notable in a pre-RNA and RNA world, because of the complete absence of redox centers in four-nucleotide RNA (compare (Tsukiji et al. 2003)). Such new reactions are potentially crucial because they supply one key to the Darwinian evolution of the molecule. An appeal of the RNA world hypothesis is that it solves the "chicken and egg" problem; it shows that in an earlier, simplified biota the genotype/replicator and phenotype/catalyst could have been one and the same molecule. An exciting aspect of the current notion is that the genotype and phenotype could have been joined more simply, and earlier. The cofactors suggest that genotype/phenotype junction could exist in a tiny system consisting of two complementary two-nucleotide entities. Such IDAs would be selected for both replication and activity (would have undergone Darwinian evolution) (Fig. 2) because their biochemistry became useful via their own replication to effective levels.

13 CONTINUITY WITH AN RNA WORLD

Continuity for nucleoside diphosphate cofactors and RNA seems assured because RNA-cofactor binding and RNA-cofactor chemistry are already known in varied forms (reviewed in Jadhav and Yarus 2002b). Catalytic RNAs can synthesize A-containing cofactors—RNAs have been selected which form nucleoside 5′-5′ structures via attack

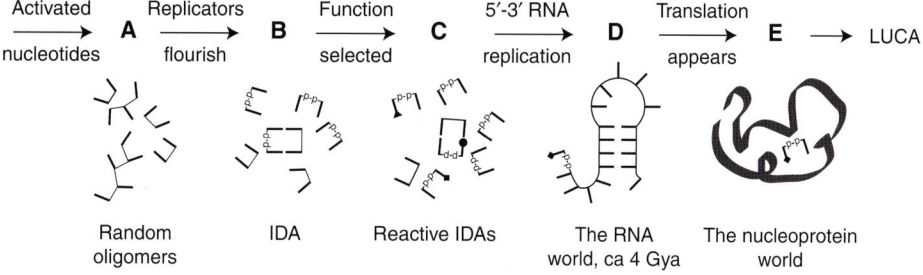

Figure 2. The IDA in context; the origin of life. (A) Activated nucleotides and compatible molecules oligomerize arbitrarily. (B) Replicators necessarily become abundant, by templating with minimal catalysis. (C) 5′-5′ replicators with reactive nucleotides are selected to participate in metabolism. (D) 5′-3′ RNA replicase creates an RNA world, ± 5′-5′ cofactor initiation and reactivity. (E) RNAs devise translation; 5′-5′ cofactors are adopted by peptide catalysts (ribbon). The pathway is initiated by its most complex event, geochemical creation of several activated nucleotide-like materials. Thus, while "simple" is a debatable evolutionary characterization, progress might be relatively simple once begun. After panel (B), all crucial transitions depend on somewhat similar selections for enhanced chemical proficiency.

by varied phosphorylated substrates (Huang and Yarus 1997); at the α-P of their own 5′ terminal triphosphate. When such 5′-5′ nucleoside-forming reactions are performed using RNAs initiated with 5′ ATP, attacks by phosphopantetheine, NMN and FMN create bona fide covalently linked CoA, NAD, and FAD as the 5′ terminus of the catalyst (Huang et al. 2000). The ambit of RNA activities also includes the synthesis of free 5′ diphosphate-linked ribonucleotides themselves, using free nucleotide substrates to form free small molecules with cofactor-like linkages (Huang et al. 1998).

5-prime cofactor RNAs may have been frequent agents in a diverse ribozyme metabolism (White 1976). Indeed, a subset of cofactors, like NADP (Fig. 1A) and CoA, have 3′ A phosphates, suggesting continuation by an RNA backbone downstream, now perhaps lost. Metabolically significant activities have been measured for such 5′ cofactor-RNAs. An RNA catalyst that uses a short recognition sequence near the 5′ terminus can attach coenzyme A *in trans* to a recognized RNA (Jadhav and Yarus 2002a). If the CoA-tagged RNA is itself a suitable catalyst (which can be isolated from preformed CoA-randomized RNA), it will synthesize acyl-CoA-RNAs, including acetyl-CoA-RNA and butyryl-CoA-RNAs (Jadhav and Yarus 2002a). An impressively complete case is that of a ribozyme that binds either free NAD or NADH and uses the cofactor to oxidize a linked benzyl alcohol (Tsukiji et al. 2003) or to reduce a linked benzaldehyde (Tsukiji et al. 2004).

The ready chemical synthesis of 5′-5′ nucleotides from activated ribonucleotides is often followed by their incorporation at the terminus of a longer 3′-5′ or 2′-5′ RNA backbone (Ferris and Ertem 1992). The same ease of 5′ incorporation applies, as just reviewed, to RNA-catalyzed incorporation of 5′-5′ structures at the RNA termini. In addition, terminal or noncovalently bound cofactor-RNAs are accessible for further biochemical transformation, in which a 5′-5′ linked cofactor performs RNA enzyme chemistry. Accordingly, any activities evolved for cofactors before the rise of RNA catalysts could be quickly adopted into an emergent RNA metabolism.

14 NOVEL REPLICATION CHEMISTRY

But might 5′-5′ cofactors preexist to join the RNA world as replicators? Cofactor chemistry seems favorable to the hypothesis, though decisive barriers remain.

Because of the superior nucleophilicity of the 5′ hydroxyl, incubation of 5′ activated nucleotides leads to abundant 5′-5′ products. AMP and rA with carbodimide activation give A5′pA (Sulston et al. 1968). AMP alone under the same conditions gives A5′p-pA as the majority product (Sulston et al. 1968; Robertson and Joyce 2010) which can then be incorporated at the 5′ terminus of a 5′-3′ polymer, for example, in the presence of montmorillonite clay (Ferris and Ertem 1992). Thus the nucleotide backbone in 5′-5′ linked cofactors like NAD is the one most readily formed by 5′ activated nucleotides.

NAD and its congeners are COSMIC LOPER, an acronym coined (Benner and Switzer 1999) to emphasize that a genetic material must have the unusual capability of tolerating changes in structure (mutations) without compromising the shared, essential ability of such varied molecules to replicate. This chemical property was envisioned as dependent on the dominance of the hydrophilic charged backbone of the nucleic acids. The 5′-diphosphate cofactors have, if anything, a more polar backbone than normal 5′-3′ RNA. This is illustrated by the existence of varied cofactors (NAD, FAD, CoA, and SAM), all retaining their solubility, somewhat similar form, and biochemical roles despite differences in chemical makeup.

Accelerated phosphodiester bond formation and even complete replication in an all-nucleic acid context have been broadly observed, and are not an unprecedented conjecture. Symmetrical RNA–RNA ligases with complementary overlaps will replicate exponentially without apparent limits, each ligating the complementary ligase together from its supplied substrate fragments (Lincoln and Joyce 2009). Complimentary deoxytrinucleotides will replicate a hexanucleotide template when activated by carbodiimide, and similar reactions extend to full replication of both strands (Sievers and Von Kiedrowski 1994). Although full replication of an oligomeric template by individual nucleotide addition has not been achieved, large ribozymes do accurately fill in small, primed templates (Zaher and Unrau 2007). More simply, activated ribonucleotides (e.g., GMP imidazolide) polymerize on a template like poly C (Inoue and Orgel 1983; Orgel 1992). However, extension to other systems than oligo C template-directing GMP imidazolide has proven difficult, and good enantiopurity is essential for chain growth (because one sugar enantiomer poisons chain growth by the other) (see Robertson and Joyce 2010) (Joyce et al. 1984).

15 CATCH-22

I have found no published evaluation of whether single activated nucleotides polymerize on a dinucleotide template (Fig. 1C), especially in 5′-5′ linkage. This is critically important because such chemistry potentially solves the evolutionary problem of needing a replicator to evolve an effective replicator (Robertson and Joyce 2010). That is, the miniaturization of the replicator to one internucleotide bond in each complement, as in the cofactors, may make it possible to envision replication using activated nucleotides and a template alone, removing or diminishing the need for a specific replication catalyst. Thus Darwinian evolution of RNA would no longer be caught in the replication Catch-22.

Despite the absence of dinucleotide replication data, in an Orgel system, with a polymeric template and complementary activated nucleotide substrates, the initial two nucleotides in a new 5′-nucleotide-initiated chain are successful in a reaction of similar molecularity. Further, whatever the result with imidazolides, there are effective, new and more "biological" activated nucleotides, like 3′-5′ cyclics, to try (Costanzo et al. 2009).

However, given that a 5′-5′ dimer template and potential substrate nucleotides are small molecules with enhanced conformational freedom, it may well be that such systems will need the stimulation observed for activated nucleotide polymerization in ice-bound eutectic solutions (Monnard et al. 2003) or facilitation by montmorillonite clay matrices (Ferris and Ertem 1992). They may also benefit from the support provided under eutectic conditions for the activities of fragmented RNA structures (Vlassov et al. 2004). Small size therefore decisively simplifies Darwinian evolution (see *Chickens and Eggs* and *Catch-22* earlier), even though it may also slow replication chemistry. However, there seems no reason why the IDA may not have enjoyed both the cold and the clay.

Replication thus remains an essential issue to be decided. And even beyond replication, there are many footholds for a critic in the ideas above. Moreover, even if these ideas are entirely correct, we will take a step back toward the IDA, but will still not know that it is the last step. Nonetheless, in my own defense, the argument's gaps seem experimentally amenable. There may well be a new world to find just beyond the RNA world we know.

ACKNOWLEDGMENTS

Many thanks to Gunter Von Kiedrowski and Hyman Hartman for a discussion of clay replication, and to Leslie Leinwand and Bill McClain for comments on a draft manuscript. During the preparation of this monograph, I am grateful for support from the National Institutes of Health (research grant GM 48080), NASA (Colorado University Astrobiology Center NCC2-1052) and the University of Colorado Council on Research and Creative Work.

REFERENCES

Arrhenius G, Cairns-Smith AG, Hartman H, Miller SL, Orgel LE. 1986. Remarks on the Review Article "Replication and Evolution in Inorganic Systems" by Armin Weiss. *Angew Chem Int Ed Engl* **25:** 658.

Benner SA, Switzer CY. 1999. Chance and necessity in biomolecular chemistry: Is life as we know it universal? In: H Frauenfelder, J Diesenhofer, PG Wolynes (eds) *Simplicity and complexity in proteins and nucleic acids*. Dahlem University Press, pp 339–363.

Benner S, Kim H-J, Yang Z. 2010. Setting the stage. The history, chemistry, and geobiology behind RNA. *Cold Spring Harb Perspect Biol* doi: 10.1101/cshperspect.a003541.

Cairns-Smith AG. 1982. *Genetic takeover and the mineral origins of life*. Cambridge University Press, Cambridge.

Chen X, Li N, Ellington AD. 2007. Ribozyme catalysis of metabolism in the RNA world. *Chem Biodivers* **4:** 633–655.

Cleaves HJ, Miller SL. 2001. The nicotinamide biosynthetic pathway is a by-product of the RNA world. *J Mol Evol* **52:** 73–77.

Copley SD, Smith E, Morowitz HJ. 2007. The origin of the RNA world: Co-evolution of genes and metabolism. *Bioorg Chem* **35:** 430–443.

Costanzo G, Pino S, Ciciriello F, Di Mauro E. 2009. Generation of long RNA chains in water. *J Biol Chem* **284:** 33206–33216.

Dalrymple GB. 1991. *The age of the Earth*. Stanford University Press, Palo Alto, CA.

Ferris JP, Ertem G. 1992. Oligomerization of ribonucleotides on montmorillonite: reaction of the 5′-phosphorimidazolide of adenosine. *Science* **257:** 1387–1389.

Friedmann N, Miller SL, Sanchez RA. 1971. Primitive earth synthesis of nicotinic acid derivatives. *Science* **171:** 1026–1027.

Huang F, Yarus M. 1997. 5′-RNA self-capping from guanosine diphosphate. *Biochemistry* **36:** 6557–6563.

Huang F, Bugg CW, Yarus M. 2000. RNA-Catalyzed CoA, NAD, and FAD synthesis from phosphopantetheine, NMN, and FMN. *Biochemistry* **39:** 15548–15555.

Huang F, Yang Z, Yarus M. 1998. RNA enzymes with two small-molecule substrates. *Chem Biol* **5:** 669–678.

Illangasekare M, Yarus M. 1999. Specific, rapid synthesis of Phe-RNA by RNA. *Proc Natl Acad Sci* **96:** 5470–5475.

Illangasekare M, Sanchez G, Nickles T, Yarus M. 1995. Aminoacyl-RNA synthesis catalyzed by an RNA. *Science* **267:** 643–647.

Inoue T, Orgel LE. 1983. A nonenzymatic RNA polymerase model. *Science* **219:** 859–862.

Jadhav VR, Yarus M. (2002a). Acyl-CoAs from coenzyme ribozymes. *Biochemistry* **41:** 723–729.

Jadhav VR, Yarus M. (2002b). Coenzymes as coribozymes. *Biochimie* **84:** 877–888.

Joyce GF, Visser GM, van Boeckel CA, van Boom JH, Orgel LE, van Westrenen J. 1984. Chiral selection in poly(C)-directed synthesis of oligo(G). *Nature* **310:** 602–604.

Kimoto M, Sato A, Kawai R, Yokoyama S, Hirao I. 2009. Site-specific incorporation of functional components into RNA by transcription using unnatural base pair systems. *Nucleic Acids Symp Ser (Oxf)*: 73–74.

Lee N, Bessho Y, Wei K, Szostak JW, Suga H. 2000. Ribozyme-catalyzed tRNA aminoacylation. *Nat Struct Biol* **7:** 28–33.

Lincoln TA, Joyce GF. 2009. Self-sustained replication of an RNA enzyme. *Science* **323:** 1229–1232.

Luo ZX, Crompton AW, Sun AL. 2001. A new mammaliaform from the early Jurassic and evolution of mammalian characteristics. *Science* **292:** 1535–1540.

Monnard P, Kanavarioti A, Deamer D. 2003. Eutectic phase polymerization of activated ribonucleotide mixtures yields quasi-equimolar incorporation of purine and pyrimidine nucleobases. *J Am Chem Soc* **125:** 13734–13740.

Nissen P, Hansen J, Ban N, Moore PB, Steitz TA. 2000. The structural basis of ribosome activity in peptide bond synthesis. *Science* **289:** 920–930.

Orgel LE. 1992. Molecular replication. *Nature* **358:** 203–209.

Ponti V, Dianzani MU, Cheeseman K, Slater TF. 1978. Studies on the reduction of nitroblue tetrazolium chloride mediated through the action of NADH and phenazine methosulphate. *Chem Biol Interact* **23:** 281–291.

Rees DC. 2002. Great metalloclusters in enzymology. *Annu Rev Biochem* **71:** 221–246.

Robertson MP, Joyce GF. 2010. The origins of the RNA world. *Cold Spring Harb Perspect Biol* doi: 10.1101/cshperspect.a003608.

Schoenberg DR, Maquat LE. 2009. Re-capping the message. *Trends Biochem Sci* **34**: 435–442.

Sievers D, Von Kiedrowski G. 1994. Self-replication of complementary nucleotide-based oligomers. *Nature* **369**: 221–224.

Smith PE, Tanner JJ. 2000. Conformations of nicotinamide adenine dinucleotide (NAD(+)) in various environments. *J Mol Recognit* **13**: 27–34.

Sulston J, Lohrmann R, Orgel LE, Miles HT. 1968. Nonenzymatic synthesis of oligoadenylates on a polyuridylic acid template. *Proc Natl Acad Sci* **59**: 726–733.

Szathmary E. 2006. The origin of replicators and reproducers. *Philos Trans R Soc Lond B Biol Sci* **361**: 1761–1776.

Tsukiji S, Pattnaik SB, Suga H. 2003. An alcohol dehydrogenase ribozyme. *Nat Struct Biol* **10**: 713–717.

Tsukiji S, Pattnaik SB, Suga H. 2004. Reduction of an aldehyde by a NADH/Zn2+-dependent redox active ribozyme. *J Am Chem Soc* **126**: 5044–5045.

Turk R, Chumachenko NV, Yarus M. 2010. Multiple translational products from a five-nucleotide ribozyme. *PNAS* **107**: 4585–4589.

Vlassov A, Johnston B, Landweber L, Kazakov S. 2004. Ligation activity of fragmented ribozymes in frozen solution: implications for the RNA world. *Nucleic Acids Res* **32**: 2966–2974.

Weinger JS, Parnell KM, Dorner S, Green R, Strobel SA. 2004. Substrate-assisted catalysis of peptide bond formation by the ribosome. *Nat Struct Mol Biol* **11**: 1101–1106.

Weiss A. 1981. Replication and evolution in inorganic systems. *Angewandte Chemie-International Edition in English* **20**: 850–860.

Welch M, Majerfeld I, Yarus M. 1997. 23S rRNA similarity from selection for peptidyl transferase mimicry. *Biochemistry* **36**: 6614–6623.

White HB, III. 1976. Coenzymes as fossils of an earlier metabolic state. *J Mol Evol* **7**: 101–104.

Yang Z, Sismour AM, Sheng P, Puskar NL, Benner SA. 2007. Enzymatic incorporation of a third nucleobase pair. *Nucleic Acids Res* **35**: 4238–4249.

Yarus M. 1991. An RNA-amino acid complex and the origin of the genetic code. *New Biol* **3**: 183–189.

Yarus M. 2001. On translation by RNAs alone. *Cold Spring Harb Symp Quant Biol* **66**: 207–215.

Yarus M, Caporaso JG, Knight R. 2005. Origins of the genetic code: The escaped triplet theory. *Annu Rev Biochem* **74**: 179–198.

Yarus M, Widmann JJ, Knight R. 2009. RNA-amino acid binding: A stereochemical era for the genetic code. *J Mol Evolution* **69**: 406–429.

Zaher HS, Unrau PJ. 2007. Selection of an improved RNA polymerase ribozyme with superior extension and fidelity. *RNA* **13**: 1017–1026.

The Origins of Cellular Life

Jason P. Schrum, Ting F. Zhu, and Jack W. Szostak

Howard Hughes Medical Institute, Department of Molecular Biology and the Center for Computational and Integrative Biology, Massachusetts General Hospital, Boston, Massachusetts 02114

Correspondence: szostak@molbio.mgh.harvard.edu

SUMMARY

Understanding the origin of cellular life on Earth requires the discovery of plausible pathways for the transition from complex prebiotic chemistry to simple biology, defined as the emergence of chemical assemblies capable of Darwinian evolution. We have proposed that a simple primitive cell, or protocell, would consist of two key components: a protocell membrane that defines a spatially localized compartment, and an informational polymer that allows for the replication and inheritance of functional information. Recent studies of vesicles composed of fatty-acid membranes have shed considerable light on pathways for protocell growth and division, as well as means by which protocells could take up nutrients from their environment. Additional work with genetic polymers has provided insight into the potential for chemical genome replication and compatibility with membrane encapsulation. The integration of a dynamic fatty-acid compartment with robust, generalized genetic polymer replication would yield a laboratory model of a protocell with the potential for classical Darwinian biological evolution, and may help to evaluate potential pathways for the emergence of life on the early Earth. Here we discuss efforts to devise such an integrated protocell model.

Outline

1. Introduction
2. Background
3. Recent results
4. Challenges and future research directions

References

1 INTRODUCTION

The emergence of the first cells on the early Earth was the culmination of a long history of prior chemical and geophysical processes. Although recognizing the many gaps in our knowledge of prebiotic chemistry and the early planetary setting in which life emerged, we will assume for the purpose of this review that the requisite chemical building blocks were available, in appropriate environmental settings. This assumption allows us to focus on the various spontaneous and catalyzed assembly processes that could have led to the formation of primitive membranes and early genetic polymers, their coassembly into membrane-encapsulated nucleic acids, and the chemical and physical processes that allowed for their replication. We will discuss recent progress toward the construction of laboratory models of a protocell (Fig. 1), evaluate the remaining steps that must be achieved before a complete protocell model can be constructed, and consider the prospects for the observation of spontaneous Darwinian evolution in laboratory protocells. Although such laboratory studies may not reflect the specific pathways that led to the origin of life on Earth, they are proving to be invaluable in uncovering surprising and unanticipated physical processes that help us to reconstruct plausible pathways and scenarios for the origin of life.

The term protocell has been used loosely to refer to primitive cells or to the first cells. Here we will use the term protocell to refer specifically to cell-like structures that are spatially delimited by a growing membrane boundary, and that contain replicating genetic information. A protocell differs from a true cell in that the evolution of genomically encoded advantageous functions has not yet occurred. With a genetic material such as RNA (or perhaps one of many other heteropolymers that could provide both heredity and function) and an appropriate environment, the continued replication of a population of protocells will lead inevitably to the spontaneous emergence of new coded functions by the classical mechanism of evolution through variation and natural selection. Once such genomically encoded and therefore heritable functions have evolved, we would consider the system to be a complete, living biological cell, albeit one much simpler than any modern cell (Szostak et al. 2001).

2 BACKGROUND

2.1 Membranes as Compartment Boundaries

All biological cells are membrane-bound compartments. The cell membrane fulfills the essential function of creating an internal environment within which genetic materials can reside and metabolic activities can take place without being lost to the environment. Modern cell membranes are composed of complex mixtures of amphiphilic molecules such as phospholipids, sterols, and many other lipids as well as diverse proteins that perform transport and enzymatic functions. Phospholipid membranes are stable under a wide range of temperature, pH, and salt concentration conditions. Such membranes are extremely good permeability barriers, so that modern cells have complete control over the uptake of nutrients and the export of wastes through the specialized channel, pump and pore proteins embedded in their membranes. A great deal of complex biochemical machinery is also required to mediate the growth and division of the cell membrane during the cell cycle. The question of how a structurally simple protocell could accomplish these essential membrane functions is a critical aspect of understanding the origin of cellular life.

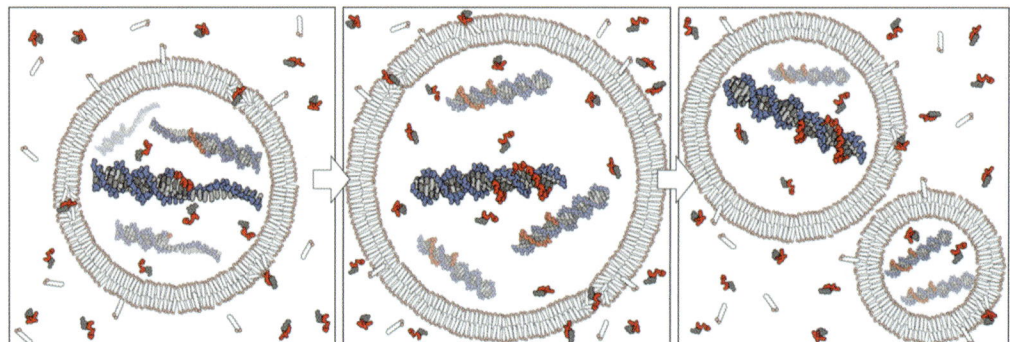

Figure 1. A simple protocell model based on a replicating vesicle for compartmentalization, and a replicating genome to encode heritable information. A complex environment provides lipids, nucleotides capable of equilibrating across the membrane bilayer, and sources of energy (*left*), which leads to subsequent replication of the genetic material and growth of the protocell (*middle*), and finally protocellular division through physical and chemical processes (*right*). (Reproduced from Mansy et al. 2008 and reprinted with permission from Nature Publishing ©2008.)

Vesicles formed by fatty acids have long been studied as models of protocell membranes (Gebicki and Hicks 1973; Hargreaves and Deamer 1978; Walde et al. 1994a). Fatty acids are attractive as the fundamental building block of prebiotic membranes in that they are chemically simpler than phospholipids. Fatty acids with a saturated acyl chain are extremely stable compounds and therefore might have accumulated to significant levels, even given a relatively slow or episodic synthesis. Moreover, the condensation of fatty acids with glycerol to yield the corresponding glycerol esters provides a highly stabilizing membrane component (Monnard et al., 2002). Finally, phosphorylation and the addition of a second acyl chain yields phosphatidic acid, the simplest phospholipid, thus providing a conceptually simple pathway for the transition from primitive to more modern membranes. The prebiotic chemistry leading to the synthesis of fatty acids and other amphiphilic compounds is treated in more detail in Mansy (2010).

The best reason for considering fatty acids as fundamental to the nature of primitive cell membranes is not, however, their chemical simplicity. Rather, fatty-acid molecules in membranes have dynamic properties that are essential for both membrane growth and permeability. Because fatty acids are single chain amphiphiles with less hydrophobic surface area than phospholipids, they assemble into membranes only at much higher concentrations. This equilibrium property is mirrored in their kinetics: Fatty acids are not as firmly anchored within the membrane as phospholipids; they enter and leave the membrane on a time scale of seconds to minutes (Chen and Szostak 2004). Fatty acids can also exchange between the two leaflets of a bilayer membrane on a subsecond time scale. Rapid flip-flop is essential for membrane growth when new amphiphilic molecules are supplied from the environment. New molecules enter the membrane primarily from the outside leaflet, and flip-flop allows the inner and outer leaflet areas to equilibrate, leading to uniform growth.

Considering that protocells on the early Earth did not, by definition, contain any complex biological machinery, they must have relied on the intrinsic permeability properties of their membranes. Membranes composed of fatty acids are in fact reasonably permeable to small polar molecules and even to charged species such as ions and nucleotides (Mansy et al. 2008). This appears to be largely a result of the ability of fatty acids to form transient defect structures and/or transient complexes with charged solutes, which facilitate transport across the membrane. The subject of the permeability of fatty-acid based membranes is dealt with in greater detail by Mansy (2010).

Prebiotic vesicles were almost certainly composed of complex mixtures of amphiphiles. Amphiphilic molecules isolated from meteorites (Deamer 1985; Deamer and Pashley 1989) as well as those synthesized under simulated prebiotic conditions (McCollum et al. 1999; Dworkin et al. 2001; Rushdi and Simoneit 2001) are highly heterogeneous, both in terms of acyl chain length and head group chemistry. Membranes composed of mixtures of amphiphiles often have superior properties to those composed of single pure species. For example, mixtures of fatty acids together with the corresponding alcohols and/or glycerol esters generate vesicles that are stable over a wider range of pH and ionic conditions (Monnard et al. 2002), and are more permeable to nutrient molecules including ions, sugars and nucleotides (Chen et al. 2004; Sacerdote and Szostak 2005; Mansy et al. 2008). This is in striking contrast to the apparent requirement for homogeneity in the nucleic acids, where even low levels of modified nucleotides can be destabilizing or can block replication.

3 RECENT RESULTS

3.1 Pathways for Vesicle Growth

Fatty-acid vesicle growth has been shown to occur through at least two distinct pathways: growth through the incorporation of fatty acids from added micelles, and growth through fatty-acid exchange between vesicles. The growth of membrane vesicles from micelles has been observed following the addition of micelles or fatty-acid precursors to pre-formed vesicles (Walde et al. 1994a, 1994b; Berclaz et al. 2001). When initially alkaline fatty-acid micelles are mixed with a buffered solution at a lower pH, the micelles become thermodynamically unstable. As a consequence, the fatty-acid molecules can either be incorporated into pre-existing membranes, leading to growth (Berclaz et al. 2001), or can self-assemble into new vesicles (Blochliger et al. 1998; Luisi et al. 2004). These pioneering studies were done by cryo-TEM, which does not allow growth to be followed in real time, and by light scattering, which is difficult to interpret in the case of samples with heterogeneous size distributions. We therefore adapted a fluorescence assay based on FRET (Förster resonance energy transfer) to measure changes in membrane area in real time. This assay is based on the distance dependence of energy transfer between donor and acceptor fluorescent dyes; thus when a membrane grows in area by incorporating additional fatty-acid molecules, the dyes are diluted and the efficiency of FRET decreases. Studies on small (typically 100 nm in diameter) unilamellar fatty-acid vesicles using this assay showed that the slow addition of fatty-acid micelles led to vesicle growth with an efficiency of $\sim$90% (Hanczyc et al. 2003).

The real-time FRET assay allowed for a kinetic dissection of the growth process, revealing a surprisingly complex series of events after the rapid addition of micelles (Chen and Szostak 2004). Two major processes were observed. The first fast phase resulted in membrane area growth that was limited to ~40% increase in area, independent of the amount of added micelles. A second much slower phase led to a further increase in membrane area that varied with the amount of added micelles. We interpreted the fast phase as reflecting the rapid assembly of a layer of adhering micelles around the pre-formed vesicles, with rapid monomer exchange resulting in the efficient incorporation of this material into the pre-formed membrane. We interpreted the slow phase as the consequence of micelle–micelle interactions leading to the assembly of intermediate structures that could partition between two pathways—with some monomers dissociating and contributing to membrane growth and the remainder ultimately assembling into new membrane vesicles. Although these interpretations are consistent with our data, the experiments are rather indirect, and further exploration of the mechanism of membrane growth is certainly desirable.

A second, distinct pathway for vesicle growth involves fatty-acid exchange between vesicles. Under certain conditions this exchange can lead to growth of a subpopulation of vesicles at the expense of their surrounding neighbors. Within populations of osmotically relaxed vesicles, such exchange processes do not result in significant changes in size distribution with time. Similarly, a population of uniformly osmotically swollen vesicles does not change in size distribution, but such vesicles are in equilibrium with a lower solution concentration of fatty acids because the tension in the membrane of the swollen vesicles makes it more energetically favorable for fatty-acid molecules to reside in membrane. When osmotically swollen vesicles are mixed with osmotically relaxed (isotonic) vesicles, rapid fatty-acid exchange processes result in growth of the swollen vesicles and corresponding shrinkage of the relaxed vesicles (Chen et al. 2004). Because vesicles can be osmotically swollen as a result of the encapsulation of high concentrations of nucleic acids such as RNA, this process allows for the growth of vesicles containing genetic polymers at the expense of empty vesicles (or vesicles that contain less internal nucleic acid). Because faster replication would increase the internal nucleic acid concentration, this pathway of competitive vesicle growth provides the potential for a direct physical link between the rate of replication of an encapsulated genetic polymer and the rate of growth of the protocell as a whole.

Assuming that the division of osmotically swollen vesicles could occur either stochastically or at some threshold size, protocells that developed some heritable means of faster replication and growth would have a shorter cell cycle, on average, and would therefore gradually take over the population. This simple physical mechanism might therefore lead to the emergence of Darwinian evolution by competition at the cellular level. However, if replication is limited by the rapid reannealing of complementary strands (see later discussion), it may be difficult to reach osmotically significant concentrations. Furthermore, osmotically swollen vesicles are intrinsically difficult to divide owing to the energetic cost of reducing the volume of a spherical vesicle to that of two daughter vesicles of the same total surface area. One possibility is that osmotically driven competitive growth might alternate with the faster membrane growth that follows micelle addition. If new fatty-acid material was only available sporadically, rapid membrane growth might follow an influx of fresh fatty acids, facilitating division (see following discussion).

All of the experiments discussed earlier were done with small unilamellar vesicles prepared by extrusion through 100 nm pores in filters. In contrast, fatty-acid vesicles that form spontaneously by rehydrating dry fatty-acid films tend to be several microns in diameter and multilamellar (Hargreaves and Deamer 1978; Hanczyc et al. 2003). Such large multilamellar fatty-acid vesicles are so heterogeneous that quantitative studies of growth and division are difficult. We have recently developed a simple procedure for the preparation of micron-sized, monodisperse (homogeneous in size) multilamellar vesicles by large-pore dialysis (Zhu and Szostak 2009b). The preparation of large monodisperse multilamellar vesicles has allowed us to directly observe an unusual mode of vesicle growth (Zhu and Szostak 2009a). We showed that feeding a micron-sized multilamellar fatty-acid vesicle with fatty-acid micelles results in the formation of a thin membranous protrusion which extends from the side of the initially spherical parental vesicle. Over time, this thin membrane tubule elongates and thickens, gradually incorporating more and more of the parental vesicle, until eventually the entire vesicle is transformed into a long, hollow threadlike vesicle (Fig. 2). This pathway occurs with vesicles ranging in size from 1 to at least 10 μm in diameter, composed of a variety of different fatty acids and related amphiphiles. Only multilamellar vesicles grow in this manner and only when vesicle volume increases slowly (relative to surface area growth) because of a relatively impermeable buffer solute. Confocal microscopy has provided insight into the mechanism of this mode of growth: The outermost membrane layer grows first, and because there is little volume between it and the next membrane layer, and that volume cannot increase on the same time scale, the extra membrane area is forced into the form of a thin tubule. Over time, this

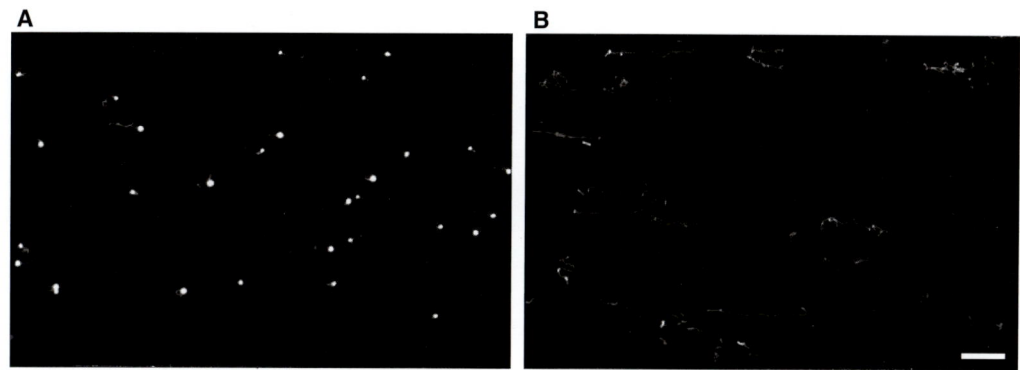

Figure 2. Vesicle shape transformations during growth. All vesicles are labeled with 2 mM encapsulated HPTS, a water-soluble fluorescent dye, in their internal aqueous space. (A) 10 min and (B) 30 min after the addition of five equivalents of oleate micelles to oleate vesicles (in 0.2 M Na-bicine, pH 8.5). Scale bar: 50 μm. (Reproduced from Zhu and Szostak 2009a and reprinted with permission from ACS Publications ©2009.)

tubule grows, and as a result of poorly understood exchange processes, the entire original vesicle is ultimately transformed into a long thread-like hollow vesicle (Fig. 3).

3.2 Pathways for Vesicle Division

Vesicle division by the extrusion of large vesicles through small pores is a way in which mechanical energy can be used to drive division (Hanczyc et al. 2003). Vesicle growth by micelle feeding followed by division by extrusion can be performed repetitively, resulting in cycles of growth and division in which both membrane material and vesicle contents are distributed to daughter vesicles in each cycle. However, division by extrusion results in the loss of 30%–40% of the encapsulated vesicle contents to the environment during each cycle (Hanczyc et al. 2003; Hanczyc and Szostak 2004). Most of this loss is a result of the unavoidable geometric constraint of dividing a spherical vesicle into two spherical (or subspherical) daughter vesicles with conservation of surface area; some additional loss may occur as a result of pressure-induced membrane rupture. Although extrusion is a useful laboratory model for vesicle division, an analogous extrusion process appears unlikely to occur in a prebiotic scenario on the early Earth because vesicle extrusion from the flow of suspended vesicles through a porous rock would require both the absence of any large pores or channels and a very high pressure gradient (Zhu and Szostak 2009a).

The above problems stimulated a search for a more realistic pathway for vesicle division. The possible spontaneous division of small unilamellar vesicles after micelle addition has been reported (Luisi et al. 2004; Luisi 2006), and electron microscopy has revealed structures that are possible intermediates in growth and division, notably pairs of vesicles joined by a shared wall (Stano et al. 2006). However, the mechanism of the proposed division as well as the nature of the energetic driving force remain unclear. Additional studies are required to clearly distinguish between the vesicle-stimulated assembly of new vesicles, and the more biologically relevant processes of growth and division.

We have recently found that the growth of large multilamellar vesicles into long threadlike vesicles, described above, provides a pathway for coupled vesicle growth and division (Zhu and Szostak 2009a). The long threadlike vesicles are extremely fragile, and divide spontaneously

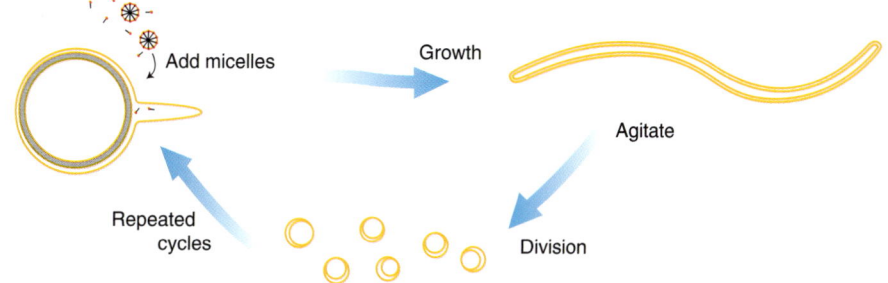

Figure 3. Schematic diagram of coupled vesicle growth and division. (Reproduced from Zhu and Szostak 2009a and reprinted with permission from the Journal of the American Chemical Society ©2009.)

into multiple daughter vesicles in response to modest shear forces. In an environment of gentle shear, growth and division become coupled processes because only the filamentous vesicles can divide (Fig. 3). If the initial parental vesicle contains encapsulated genetic polymers such as RNA, these molecules are distributed randomly to the daughter vesicles and are thus inherited. The robustness and simplicity of this pathway suggests that similar processes might have occurred under prebiotic conditions. The mechanistic details of this mode of division remain unclear. One possibility, supported by some microscopic observations, is that the long thin membrane tubules are subject to the "pearling instability" (Bar-Ziv and Moses 1994), and minimize their surface energy by spontaneously transforming from a cylindrical shape to a string of beads morphology. The very thin tether joining adjacent spherical beads may be a weak point that can be easily disrupted by shear forces.

3.3 RNA-Catalyzed RNA Replication on the Early Earth and the Modern Laboratory

A core assumption of the RNA world hypothesis is that the RNA genomes of primitive cells were replicated by a ribozyme RNA polymerase (Gilbert 1986). The idea of RNA-catalyzed RNA replication provides a solution to the apparent paradox of DNA replication catalyzed by proteins that are encoded by DNA. This simplification of early biochemistry gained instant plausibility from the discovery of catalytic RNAs almost 30 years ago (Kruger et al. 1982; Guerrier-Takada et al. 1983). In the time after the discovery of the first ribozymes, the RNA World hypothesis has continued to gain support, most dramatically from the discovery that the ribosome is a ribozyme (Nissen et al. 2000), and that all proteins are assembled through the catalytic activity of the ribosomal RNA. Support has also come from the in vitro evolution of a wide range of new ribozymes, including bona fide RNA polymerases made of RNA (Johnston et al. 2001). On the other hand, no ribozyme polymerase yet comes close to being a self-replicating RNA, like the replicase envisaged as the core of the RNA World biochemistry.

Why has the in vitro evolution of an RNA replicase been so much more difficult than originally expected? It is clear that the problem is not with catalysis of the chemical step, even with catalytically demanding triphosphate substrates. Evolutionarily optimized versions of the Class I ligase carry out multiple-turnover ligation reactions at >1 s^{-1}, with over 50,000 turnovers overnight (Ekland et al. 1995), and optimized versions of the smaller, simpler DSL ligase carry out sustained multiple turnover ligation reactions at rates $>1/\text{min}$ (Voytek and Joyce 2007). These catalytic rates are more than sufficient to carry out the replication of a 100–200 nt ribozyme in minutes to hours, if these rates could be maintained in the context of a polymerase reaction using monomer substrates. However, even the best available polymerase ribozyme requires 1–2 days to copy 10–20 nucleotides of a template strand, apparently as a consequence of poor binding to both the ribonucleoside triphosphate (NTP) monomers and the primer-template substrate. The need to overcome the electrostatic repulsion between negatively charged NTP and RNA substrates and ribozyme is thought to contribute to the very high Mg^{++} requirement for the polymerase ribozyme (Glasner et al. 2000). Such high levels of Mg^{++} lead to hydrolytic degradation of the ribozyme, and are also not compatible with known fatty-acid based membranes because of crystallization of the fatty acid-magnesium salt. In addition, fatty-acid membranes are almost impermeable to NTPs (Mansy et al. 2008).

The incompatibility between currently available ribozyme polymerases and fatty-acid based vesicles suggests either that early replicases were quite different, or that RNA replication in early cells proceeded in a very different manner. For example, less charged and more activated nucleotides might be easier for a ribozyme polymerase to bind, with little or no Mg^{++}. Many potential leaving groups, such as imidazole, adenine or 1-Me-adenine have been examined in template-directed and nontemplated polymerization reactions, but have not yet been tested as substrates for ribozyme polymerases (Prabahar and Ferris 1997; Huang and Ferris 2006). The tethering of ribozyme and primer-template to hydrophobic aggregates has been examined as a way of increasing local substrate concentration (Müller and Bartel 2008). However, this approach did not lead to a dramatic improvement in the extent of template copying, apparently as a result of ribozyme inhibition at high effective RNA concentrations. It may be possible to evolve polymerases that operate well at high RNA concentrations, but membrane localization by chemical derivatization adds further complexity to a replication pathway because of the need for a specific catalyst for the derivatization step. Alternative means of facilitating the interaction of a ribozyme with a primer-template substrate, such as the presence of basic peptides or other cofactors, might overcome this problem. Finally it is noteworthy that very small self-aminoacylating ribozymes have been obtained using RNA libraries that have an unconstrained sequence at their 3′-end (Chumachenko et al. 2009). This strategy may also be fruitful in selections for ribozyme polymerases.

Given the above constraints and uncertainties, what can we say about the emergence of RNA-catalyzed RNA replication in the origin of life? It is still possible that under the proper conditions, and using the right substrates, a small simple ribozyme could effectively catalyze RNA replication. However, a replicase must do more than catalyze a simple phospho-transfer reaction. Binding in a nonsequence-specific manner to a primer-template complex, facilitating binding of the proper incoming monomer, catalyzing primer extension, and repeating this process until the end of the template (or set of templates for a multi-component replicase) is reached might require a complex replicase structure. Such a replicase would presumably be rare in collections of random RNA sequences. If life required a very special sequence to get started, then the origin of life on earth could have been a low probability event and life on other earth-like planets might be very rare. If, on the other hand, RNA-catalyzed RNA replication could have emerged gradually in a series of simpler steps, it might have been easier and thus more likely for life to begin, and life elsewhere might be common. For this reason, we now turn to a consideration of nonenzymatic template-directed replication chemistry.

3.4 Chemical Template Replication Revisited

The nonenzymatic template-directed polymerization of activated ribonucleotides was studied in depth by Leslie Orgel, together with his students and colleagues, over a period of several decades (Orgel 2004). Here, the template itself acts as a catalyst by helping to align and orient the monomers so that they are pre-organized for polymerization. The main lesson from this work is that spontaneous chemical copying of RNA sequences is indeed possible, but is subject to several important constraints and limitations. The constraints make template-directed RNA copying incompatible with currently available membrane vesicle systems, and the limitations have, so far at least, made it impossible to obtain repeated cycles of RNA replication through chemical copying.

To obtain reasonable reaction rates, Orgel made use of nucleoside monophosphates activated with a good leaving group such as imidazole. The ribonucleotide 5′-phosphorimidazolides spontaneously assemble on a template oligonucleotide and polymerize over several days, generating a complementary strand. However, monomer binding to RNA templates is weak and concentrations on the order of 0.1 M were required for optimal copying. In addition, a Mg^{++} concentration of ∼0.1 M, which as noted above is incompatible with the presence of fatty-acid based membranes, is required for optimal polymerization. Even under these rather extreme conditions, polymerization proceeds at only 1–2 nucleotides per day, and is therefore limited by monomer hydrolysis. Beyond these constraints, three aspects of the copying reaction present major hurdles to multiple rounds of replication. First, the chemical structure of RNA results in a problem of regiospecificity, because new linkages can be either 3–5′ or 2′–5′ phosphodiester bonds. Surprisingly, under most conditions, it is the 2′–5′ phosphodiester bonds that are most common in polymerization products. This problem can be ameliorated by the choice of ions and leaving groups; for unknown reasons, Zn^{2+} ions and the 2-methylimidazole leaving group favor the synthesis of 3′–5′ linkages (Lohrmann et al. 1980; Inoue and Orgel 1982). Studies of oligonucleotide ligation showed that the helical context of an extended RNA duplex also favors the formation of 3′–5′ linkages (Rohatgi et al. 1996). However, it remains difficult to obtain a homogeneous RNA backbone without enzymatic catalysis. Second, adenine (A) residues in the template are difficult to copy, and two or more As in succession block chain growth (Inoue and Orgel 1983), presumably as a result of the poor base-stacking propensity of the incoming U monomers. It has recently been found that template copying at subzero temperatures can proceed past multiple A residues in the template (Vogel and Richert 2007), but this required sequential additions of oxyazabenzotriazole-activated monomer together with a series of helper oligos. Third, there is the issue of fidelity. The copying of G and C residues is remarkably accurate, with error rates estimated at 0.5% or less. However, the addition of A and U residues causes problems, most significantly the formation of G:U wobble base-pairs (Wu and Orgel 1992), which would lead to significant error rates in a four base system. Might an alternative genetic polymer, perhaps even a close relative of RNA, overcome these problems and enable chemical replication? The identification of such a system might ultimately lead to the discovery of plausible progenitors of RNA, or, alternatively, to the discovery of new replication strategies that allow for the chemical replication of RNA itself.

3.5 Phosphoramidate Nucleic Acids

Early studies of template copying using more reactive nucleotide derivatives were performed by Orgel et al., who examined both 2′- and 3′-amino ribonucleotide 5′-phosphorimidazolides (Lohrmann and Orgel 1976; Zielinksi and Orgel 1985; Tohidi et al. 1987). Polymerization of these nucleotides yields phosphoramidate nucleic acids, which are generally similar to standard phosphodiester linked nucleic acids except that the phosphoramidate linkage is more acid labile. As expected, replacing a sugar hydroxyl in the monomer with a more nucleophilic amino group resulted in a large increase in monomer reactivity. The activated 3′-amino ribonucleotides participated in rapid copying of short oligonucleotide templates 5-13 nucleotides in length, yielding N3′→P5′ linked complementary oligonucleotides (Tohidi et al. 1987). The increased reactivity also led to faster intramolecular monomer cyclization, which depleted the template copying reactions of activated substrate molecules (Hill et al. 1988).

More recently, the Richert group has begun to explore the potential of 3′-amino- nucleotide analogs in template-directed condensation reactions. Deoxyribonucleotide monomers, activated with an oxyazabenzotriazole leaving group, completed a template-directed reaction with a 3′-amino-terminated primer in seconds (Rothlingshofer et al. 2008). The fidelity of this reaction is sufficient to allow for sequencing by nonenzymatic primer extension. However, the monomer is rapidly consumed by internal cyclization.

The high intrinsic reactivity of the amino-sugar modified nucleotides suggested to us that an alternative phosphoramidate nucleic acid might act as a good platform for chemical self-replication. Our group has therefore started to study a series of phosphoramidate nucleic acids with sugar phosphate backbones that vary in their degree of conformational flexibility or constraint. We are currently focusing on the 2′-5′ linked phosphoramidate analog of DNA, and the corresponding monomers, the 2′-amino dideoxyribonucleotide 5′-phosphorimidazolides (Fig. 4). The 2′-amino ImpddNs are advantageous as monomers because they cannot undergo intramolecular cyclization because of the steric constraint of the ribose ring; they are only depleted during polymerization reactions by competing hydrolysis.

Our first experiments with 2′-amino ImpddG led to rapid and efficient primer-extension across a dC_{15} template, generating full-length product in ∼6 hour (Mansy et al. 2008). Encouraged by the rapid copying of oligo-dC templates by 2′-amino ImpddG, we performed a more extensive study of the copying of templates with differing sugar-phosphate backbones, lengths and sequences (Schrum et al. 2009). The most important property contributing to good template activity appears to be preorganization in the form of an A-type helix. Thus, RNA templates were uniformly superior to DNA templates of the same sequence, and LNA (locked nucleic acid) templates, which are chemically locked in a C3′-endo sugar conformation, were superior to RNA templates. This result is consistent with previous observations that under most conditions RNA-template directed polymerization of ribonucleotides leads to a majority of 2′-5′ linkages; it appears that an A-type helical geometry generally favors polymerization through attack by a 2′ nucleophile. A second key factor is that enhanced monomer affinity for the template increases the reaction efficiency. Thus G:C base-pairs lead to efficient copying, whereas A:U base-pairs were very poorly copied. Replacing A with D (diaminopurine) results in a D:U base-pair with three hydrogen bonds, and slightly improved primer-extension. However, the poor base stacking of U residues must also be improved to obtain efficient template copying. When we replaced U with C5-propynyl-U, the resulting A:U^P base-pairs led to improved copying, but the D:U^P combination had a clearly synergistic effect (Fig. 5). In the context of a DNA duplex, the D:U^P base-pair has previously been shown to be energetically almost equivalent to a C:G base-pair (Chaput et al. 2002). When we used G, C, D, and U^P as the four monomer and template bases, we were able to copy mixed-sequence RNA templates over 15 nts in length in about a day. This represents a

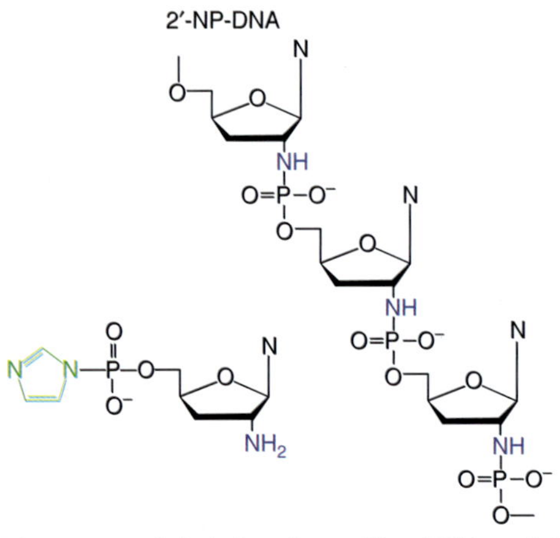

Figure 4. Structures of 2′-5′ phosphoramidate DNA, and the corresponding activated monomers.

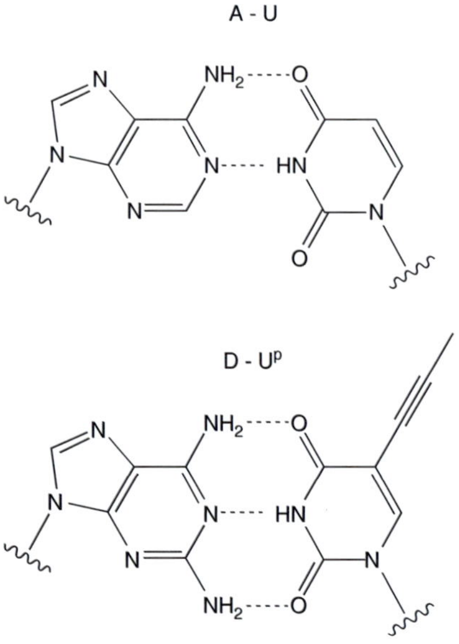

Figure 5. Watson Crick base-pairs. *Top*: Standard A:U base-pair. *Bottom*: Alternative diaminopurine (D): C5-propynyl-uracil (U^P) base-pair.

significant step toward developing a robust, generalized chemical replication system.

This system is, however, far from ideal, and there are strong indications that the fidelity of template copying becomes an issue when all four nucleotides are present, largely because of the formation of G:U wobble base-pairs. Given that primer-extension on 2′–5′ linked DNA templates is approximately similar to that on the corresponding RNA templates, we are currently synthesizing 2′–5′ linked phosphoramidate DNA templates to assess self-replication in this system. In light of the efficient templating of LNA oligonucleotides, we are also interested in exploring prebiotically plausible nucleic acids that are more conformationally constrained than RNA. A particularly interesting candidate is TNA (threose nucleic acid) (Schöning et al. 2000) and its 2′-amino substituted phosphoramidate version, NP-TNA (Wu et al. 2002). These are both base-pairing systems that form standard Watson-Crick duplexes, despite having only five atoms per backbone repeat unit (vs. six for RNA and DNA). It will be of great interest to see if the resulting decrease in flexibility leads to increased fidelity in template copying reactions.

3.6 Protocell Assembly

In principle, protocell-like objects could form spontaneously as new membranes self-assemble and encapsulate genetic molecules in solution. Recently, simple physical processes that would enhance the efficiency of the coassembly of nucleic acids and membrane vesicles have been proposed. One such alternative scenario is based on the fact that the clay mineral montmorillonite is not only a catalyst of RNA polymerization (Ferris et al. 1996; Huang and Ferris 2006) but also catalyzes membrane assembly (Hanczyc et al. 2003). Experiments with clay particles containing surface-adsorbed RNA showed that such particles stimulated vesicle assembly, and frequently became trapped inside the vesicles whose assembly they had catalyzed. Thus, a common mineral can catalyze both the assembly of a genetic polymer and the assembly of a membrane vesicle, and bring these two components together to generate a protocell-like structure consisting of a genetic polymer trapped within a membrane compartment. Although the effectiveness of this process is attractive, some means of releasing at least some of the bound nucleic acid from the mineral surface, and/or replicating it on the surface, would be necessary for subsequent replication to occur.

More recent experiments suggest a very different geochemical scenario leading to the assembly of similar protocell-like structures. The hollow channels within the rocks of the alkaline off-axis hydrothermal vents provide a protected compartmentalized environment where it has been suggested that primitive metabolic activities might have originated (Martin and Russell 2003). Recent theoretical studies suggested that the strong thermal gradients present in hydrothermal vents, together with the thin channels produced by mineral precipitation, could greatly concentrate small organic molecules such as nucleotides as well as larger nucleic acids from a very dilute external reservoir (Baaske et al. 2007). Work from our laboratory (Budin et al. 2009) has confirmed the predicted concentration effect, and has also shown that subcritical concentrations of fatty acids can be concentrated to the extent that they self-assemble into vesicles at the bottom of the capillary channels. Moreover, DNA oligonucleotides can also be greatly concentrated and can become encapsulated within the vesicles, resulting in the spontaneous assembly of protocell-like structures.

3.7 Encapsulated Template Replication: Emergence of a Protocell

The experiments discussed earlier suggest that the assembly of protocell-like structures is not that difficult, because it appears to be possible through multiple distinct mechanisms. The more challenging question is, how could such a structure replicate? We have already considered the replication of the protocell membrane and the genetic material as separate entities. To address the question of their replication as a combined structure we must consider in more detail the molecular constituents of the protocell membrane and the molecular nature of the encapsulated genetic material.

Genome replication within a protocell can only occur if the building blocks used to copy template strands are able to enter the fatty-acid vesicle compartment. Early work using phospholipid-based vesicles and protein enzymes showed the feasibility of constructing primitive cell-like compartments (Chakrabarti et al. 1994; Luisi et al 1994). More recent permeability studies (Mansy et al. 2008; see also Mansy 2010) showed that nucleotides could spontaneously diffuse across simple fatty-acid membranes, but that net negative charge is a critical determinant of permeability. Thus, nucleotides that are chemically activated, e.g., by conversion of the 5′-phosphate to a 5′-phosphorimidazolide, equilibrate across vesicle membranes much more rapidly on account of the reduction of the net negative charge. In addition, mixtures of fatty acids with their glycerol esters generate membranes that are more permeable to polar and charged molecules. By combining these observations, we were able to show that activated 2′-amino-2′,3′-dideoxyguanosine-5′-phosphorimidazolide, the same nucleotide previously shown to rapidly copy oligo-dC templates, could be added to the outside of fatty-acid vesicles containing an encapsulated primer-template complex, and copy the internal dC_{15} template. Copying of the encapsulated dC_{15} template by primer-extension reached >95% completion in 12–24 hour (Mansy et al. 2008), compared with 6–12 hour in free solution. The longer time required for copying encapsulated templates reflects the time required for entry of external nucleotides to the interior of the vesicles. Importantly, the presence of high concentrations (5–10 mM) of highly reactive activated nucleotides did not have any disruptive effects on the integrity of the vesicle membrane, as no leakage of encapsulated primer-template complexes was observed. It is also important to note that control experiments with phospholipid membranes showed no copying of internal template, because the activated nucleotides could not enter the vesicle; similarly, "modern" activated nucleotides such as nucleoside triphosphates cannot cross fatty-acid based membranes. Successful copying of encapsulated templates therefore requires both "primitive" nucleotides with reduced charge, and "primitive" membranes composed of single chain amphiphiles.

The copying of a genetic polymer inside a membrane compartment is an important step toward the realization of a self-replicating system capable of Darwinian evolution. What, then, are the remaining barriers to the assembly of such a system? The copying of a single-stranded template produces a double-stranded product; these strands would have to separate before a second cycle of genome replication could begin. Separate follow-up experiments by our group showed that some fatty-acid based vesicles are able to retain encapsulated DNA and RNA oligonucleotides over a temperature range of 0 °C to 100 °C (Mansy and Szostak 2008). As with permeability, mixtures of amphiphiles lead to improved thermostability, with glycerol esters being particularly stabilizing, possibly because of the additional hydrogen bond donors and acceptors provided by the glycerol head group. Furthermore, we found that encapsulated double-stranded DNA could be denatured at elevated temperatures, with the strands reannealing once the temperature was lowered (Mansy and Szostak 2008). This implies that thermal fluctuations could provide a mechanism for strand separation that is compatible with the integrity of fatty-acid vesicles, potentially allowing for complete cycles of replication of encapsulated genetic polymers. The mutual compatibility of nucleic acid replication and fatty-acid compartment growth is very encouraging because it alleviates concerns related to the permeability and stability of membrane vesicles. Vesicles therefore do seem to be a physically plausible way to segregate and spatially localize genomes, keep emergent catalytic polynucleotides physically close to their encoding genome, and protect the nascent evolving system from parasitic polymers.

4 CHALLENGES AND FUTURE RESEARCH DIRECTIONS

4.1 Prospects for a Complete Protocell Model

Although considerable progress has been made toward the assembly of model protocells, several remaining issues must be solved before multiple cycles of protocell replication can be achieved in the laboratory. These factors are also relevant to protocell replication on the early Earth. The most important factor at this time appears to be the competition between strand reannealing and strand copying, after thermal strand separation. PCR reactions generally plateau at about 1-μM DNA strand concentration, which is the concentration at which strand reannealing and strand copying occur on a similar time scale of about 1 minute However, nonenzymatic template copying requires on the order of a day for completion, which implies that either template copying must be much faster, or reannealing must be much slower. One way to make reannealing sufficiently slow is to keep strand concentrations subnanomolar. Low strand concentrations are possible in large vesicles, but it is hard to see how a few molecules of a genetic polymer could have any significant phenotypic effect on a large vesicle composed of millions of amphiphilic molecules. The emergence of metabolic ribozymes would be more plausible if nucleic acid strand concentrations were much higher, so that a catalyst of modest efficiency could generate enough product to influence cell properties.

What other factors might affect the rate of strand reannealing? Perhaps the most obvious possibility is that secondary structure, which can form extremely rapidly because the interactions are intramolecular, could greatly slow down strand annealing. This phenomenon is essential to the viability of single-stranded RNA phage such as Qβ• (Axelrod et al. 1991). Significant intrastrand secondary structure would also be an expected consequence of selection for sequences that fold into functional shapes with catalytic activity. On the other hand, chemical replication through dense secondary structure would probably be much slower than replication of an unfolded, open template. The outcome of simultaneous selection for an open, accessible template sequence, and a folded functional structure remains unclear. An alternative but even more speculative solution might result from the rapid binding by base-pairing of short oligonucleotides or even monomers to freshly separated strands. If a template strand was largely occupied by monomers or short oligomers, even if these were in rapid exchange, the strand might be prevented from annealing to a complementary strand. This possibility has the advantage that it need not block or slow the copying reaction, however its effectiveness remains to be tested experimentally.

Another challenge faced by replicating protocells, whether on the early earth or in the modern laboratory, is the continuous dilution of protocells through the competing formation of new empty vesicles. When new fatty acids are supplied as micelles, the efficiency of incorporation into pre-formed vesicles (or protocells) can be quite high, but some new vesicles are always formed. Thus over time, the descendants of a given protocell will gradually be diluted out by the continuous formation of these new vesicles. To avoid extinction by dilution, the protocells must out-compete other vesicles either by having a more rapid cell cycle, thereby generating more progeny during division, or by surviving destructive processes more efficiently. We have previously proposed (Chen et al. 2004) that faster growth, driven by the osmotic pressure of encapsulated nucleic acid, could lead to an effectively shorter cell cycle for protocells that contain high copy numbers of their replicating genome. However, in light of the problems associated with this approach, it is of considerable interest to explore new ways in which a protocell genome could lead to faster growth or growth that occurs at the expense of empty vesicles. An alternative strategy for surviving dilution would be for a protocell genome to colonize empty vesicles. This could occur through a low level of stochastic vesicle–vesicle fusion events, possibly catalyzed by low levels of divalent cations such as Ca^{2+}. Systematic efforts to measure vesicle fusion frequencies under different environmental conditions could therefore be quite useful. Finally, it is possible that this problem could be circumvented entirely if early life was discontinuous (Budin et al. 2009). For example, protocells could be occasionally disrupted by drying, or simply dissolve as a result of dilution with water to a level below the critical aggregate concentration; subsequent re-hydration or concentration would result in reformation of vesicles encapsulating genomic nucleic acids, thus generating a new "randomized" set of protocells. As long as such events were fairly uncommon, and assuming that genomic replication had kept ahead of vesicle replication so that each vesicle contained multiple genome copies prior to disruption, this process of disruption and reformation would lead to the spread of evolving genomic sequences through the "new" vesicles.

Laboratory models of protocell systems should be helpful in modeling many of the above scenarios. Assuming that protocell reproduction can be achieved, and made efficient enough to continue through many generations, it should then be possible to observe the spontaneous evolution of adaptive innovations in this relatively simple chemical system. The nature of such adaptations may provide clues as to how modern cells evolved from their earliest ancestors. Ultimately this line of research may also tell us whether the conserved biochemistry of life is driven by chemical necessity, or whether biochemically very different forms of life are also possible.

ACKNOWLEDGMENTS

We thank Itay Budin, Matt Powner, and other members of the Szostak lab for helpful discussions.

REFERENCES

Axelrod VD, Brown E, Priano C, Mills DR. 1991. Coliphage Q β RNA replication: RNA catalytic for single-strand release. *Virology* **184:** 595–608.

Baaske P, Weinert FM, Duhr S, Lemke KH, Russell MJ, Braun D. 2007. Extreme accumulation of nucleotides in simulated hydrothermal pore systems. *Proc Natl Acad Sci* **104:** 9346–9351.

Bar-Ziv R, Moses E. 1994. Instability and "pearling" states produced in tubular membranes by competition of curvature and tension. *Phys Rev Lett* **73:** 1392–1395.

Berclaz N, Muller M, Walde P, Luisi PL. 2001. Growth and transformation of vesicles studied by ferritin labeling and cryotransmission electron microscopy. *J Phys Chem B* **105:** 1056–1064.

Blochliger E, Blocher M, Walde P, Luisi PL. 1998. Matrix effect in the size distribution of fatty acid vesicles. *J Phys Chem B* **102:** 10383–10390.

Budin I, Bruckner R, Szostak JW. 2009. Formation of protocell-like vesicles in a thermal diffusion column. *J Am Chem Soc* **131:** 9628–9629.

Chakrabarti AC, Breaker RR, Joyce GF, Deamer DW. 1994. Production of RNA by a polymerase protein encapsulated within phospholipid vesicles. *J Mol Evol* **39:** 555–559.

Chaput JC, Sinha S, Switzer C. 2002. 5-propynyluracil.diaminopurine: An efficient base-pair for non-enzymatic transcription of DNA. *Chem Commun* **15:** 1568–1569.

Chen IA, Roberts RW, Szostak JW. 2004. The emergence of competition between model protocells. *Science* **305:** 1474–1476.

Chen IA, Szostak JW. 2004. A kinetic study of the growth of fatty acid vesicles. *Biophys J* **87:** 988–998.

Chumachenko NV, Novikov Y, Yarus M. 2009. Rapid and simple ribozymic aminoacylation using three conserved nucleotides. *J Am Chem Soc* **131:** 5257–5263.

Deamer DW. 1985. Boundary structures are formed by organic components of the Murchison carbonaceous chondrite. *Nature* **317:** 792–794.

Deamer DW, Pashley RM. 1989. Amphiphilic components of the Murchison carbonaceous chondrite: Surface properties and membrane formation. *Orig Life Evol Biosph* **19:** 21–38.

Dworkin J, Deamer D, Sandford S, Allamandola L. 2001. Self-assembling amphiphilic molecules: Synthesis in simulated interstellar/precometary ices. *Proc Natl Acad Sci* **98:** 815–819.

Ekland EH, Szostak JW, Bartel DP. 1995. Structurally complex and highly active RNA ligases derived from random RNA sequences. *Science* **269:** 1319–1325.

Ferris JP, Hill AR Jr, Liu R, Orgel LE. 1996. Synthesis of long prebiotic oligomers on mineral surfaces. *Nature* **381:** 59–61.

Gebicki JM, Hicks M. 1973. Ufasomes are stable particles surrounded by unsaturated fatty acid membranes. *Nature* **243:** 232–234.

Gilbert W. 1986. The RNA World. *Nature* **319:** 618.

Glasner ME, Yen CC, Ekland EH, Bartel DP. 2000. Recognition of nucleoside triphosphates during RNA-catalyzed primer extension. *Biochemistry* **39:** 15556–15562.

Guerrier-Takada C, Gardiner K, Marsh T, Pace N, Altman S. 1983. The RNA moiety of ribonuclease P is the catalytic subunit of the enzyme. *Cell* **35:** 849–857.

Hanczyc MM, Szostak JW. 2004. Replicating vesicles as models of primitive cell growth and division. *Curr Opin Chem Biol* **8:** 660–664.

Hanczyc MM, Fujikawa SM, Szostak JW. 2003. Experimental models of primitive cellular compartments: Encapsulation, growth, and division. *Science* **302:** 618–622.

Hargreaves WR, Deamer DW. 1978. Liposomes from ionic, single-chain amphiphiles. *Biochemistry* **17:** 3759–3768.

Hill AR Jr, Nord LD, Orgel LE, Robins RK. 1988. Cyclization of nucleotide analogues as an obstacle to polymerization. *J Mol Evol* **28:** 170–171.

Huang W, Ferris JP. 2006. One-step, regioselective synthesis of up to 50-mers of RNA oligomers by montmorillonite catalysis. *J Am Chem Soc* **128:** 8914–8919.

Inoue T, Orgel LE. 1982. Oligomerization of (guanosine 5′-phosphor)-2-methylimidazolide on poly(C). An RNA polymerase model. *J Mol Biol* **162:** 201–217.

Inoue T, Orgel LE. 1983. A nonenzymatic RNA polymerase model. *Science* **219:** 859–862.

Johnston WK, Unrau PJ, Lawrence MS, Glasner ME, Bartel DP. 2001. RNA-catalyzed RNA polymerization: Accurate and general RNA-templated primer extension. *Science* **292:** 1319–1325.

Kruger K, Grabowski PJ, Zaug AJ, Sands J, Gottschling DE, Cech TR. 1982. Self-splicing RNA: Autoexcision and autocyclization of the ribosomal RNA intervening sequence of Tetrahymena. *Cell* **31:** 147–157.

Lohrmann R, Orgel LE. 1976. Template-directed synthesis of high molecular weight polynucleotide analogues. *Nature* **261:** 342–344.

Lohrmann R, Bridson PK, Orgel LE. 1980. Efficient metal-ion catalyzed template-directed oligonucleotide synthesis. *Science* **208:** 1464–1465.

Luisi PL. 2006. *The emergence of life: From chemical origins to synthetic biology*, Cambridge University Press, Cambridge.

Luisi PL, Walde P, Oberholzer T. 1994. Enzymatic RNA synthesis in self-reproducing vesicles: An approach to the construction of a minimal synthetic cell. *Ber Bunsenges Phys Chem* **98:** 1160–1165.

Luisi PL, Stano P, Rasi S, Mavelli F. 2004. A possible route to prebiotic vesicle reproduction. *Artif Life* **10:** 297–308.

Mansy SS. 2010. Membrane transport in primitive cells. *Cold Spring Harb Perspect Biol* **2:** a002188.

Mansy SS, Szostak JW. 2008. Thermostability of model protocell membranes. *Proc Natl Acad Sci* **105:** 13351–13355.

Mansy SS, Schrum JP, Krishnamurthy M, Tobé S, Treco DA, Szostak JW. 2008. Template-directed synthesis of a genetic polymer in a model protocell. *Nature* **454:** 122–125.

Martin W, Russell MJ. 2003. On the origins of cells: A hypothesis for the evolutionary transitions from abiotic geochemistry to chemoautotrophic prokaryotes, and from prokaryotes to nucleated cells. *Philos Trans R Soc Lond B Biol Sci* **358:** 59–83.

McCollom TM, Ritter G, Simoneit BR. 1999. Lipid synthesis under hydrothermal conditions by Fischer-Tropsch-type reactions. *Orig Life Evol Biosph* **29:** 153–166.

Monnard PA, Apel CL, Kanavarioti A, Deamer DW. 2002. Influence of ionic inorganic solutes on self-assembly and polymerization processes related to early forms of life: Implications for a prebiotic aqueous medium. *Astrobiology* **2:** 139–152.

Müller UF, Bartel DP. 2008. Improved polymerase ribozyme efficiency on hydrophobic assemblies. *RNA* **14:** 552–562.

Nissen P, Hansen J, Ban N, Moore PB, Steitz TA. 2000. The structural basis of ribosome activity in peptide bond synthesis. *Science* **289:** 920–930.

Orgel LE. 2004. Prebiotic chemistry and the origin of the RNA world. *Crit Rev Biochem Mol Biol* **39:** 99–123.

Prabahar KJ, Ferris JP. 1997. Adenine derivatives as phosphate-activating groups for the regioselective formation of 3′,5′-linked oligoadenylates on montmorillonite: Possible phosphate-activating groups for the prebiotic synthesis of RNA. *J Am Chem Soc* **119:** 4330–4337.

Rohatgi R, Bartel DP, Szostak JW. 1996. Nonenzymatic, template-directed ligation of oligoribonucleotides is highly regioselective for the formation of 3′-5′ phosphodiester bonds. *J Am Chem Soc* **118:** 3340–3344.

Röthlingshöfer M, Kervio E, Lommel T, Plutowski U, Hochgesand A, Richert C. 2008. Chemical primer extension in seconds. *Angew Chem Int Ed Engl* **47:** 6065–6068.

Rushdi AI, Simoneit BR. 2001. Lipid formation by aqueous Fischer-Tropsch-type synthesis over a temperature range of 100 to 400 degrees C. *Orig Life Evol Biosph* **31:** 103–118.

Sacerdote MG, Szostak JW. 2005. Semi-permeable lipid bilayers exhibit diastereoselectivity favoring ribose; implications for the origins of life. *Proc Natl Acad Sci* **102:** 6004–6008.

Schöning K, Scholz P, Guntha S, Wu X, Krishnamurthy R, Eschenmoser A. 2000. Chemical etiology of nucleic acid structure: The α-threofuranosyl-(3′→2′) oligonucleotide system. *Science* **290:** 1347–1351.

Schrum JP, Ricardo A, Krishnamurthy K, Blain JC, Szostak JW. 2009. Efficient and rapid template-directed nucleic acid copying using 2′-amino-2′, 3′-dideoxyribonucleoside-5′-phosphorimidazolide monomers. *J Am Chem Soc* **31:** 14560–14570.

Stano P, Wehrli E, Luisi PL. 2006. Insights into the self-reproduction of oleate vesicles. *J Phys: Condens Matter* **18:** S2231–S2238.

Szostak JW, Bartel DP, Luisi PL. 2001. Synthesizing life. *Nature* **409:** 387–390.

Tohidi M, Zielinski WS, Chen CH, Orgel LE. 1987. Oligomerization of 3′-amino-3′deoxyguanosine-5′phosphorimidazolidate on a d(CpCpCpCpC) template. *J Mol Evol* **25:** 97–99.

Vogel SR, Richert C. 2007. Adenosine residues in the template do not block spontaneous replication steps of RNA. *Chem Commun* **19:** 1896–1898.

Voytek SB, Joyce GF. 2007. Emergence of a fast-reacting ribozyme that is capable of undergoing continuous evolution. *Proc Natl Acad Sci* **104:** 15288–15293.

Walde P, Wick R, Fresta M, Mangone A, Luisi PL. 1994a. Autopoietic self-reproduction of fatty acid vesicles. *J Am Chem Soc* **116:** 11649–11654.

Walde P, Goto A, Monnard P-A, Wessicken M, Luisi PL. 1994b. Oparin's reactions revisited: Enzymatic synthesis of poly(adenylic acid) in micelles and self-reproducing vesicles. *J Am Chem Soc* **116:** 7541–7547.

Wu T, Orgel LE. 1992. Nonenzymatic template-directed synthesis on hairpin oligonucleotides. 3. Incorporation of adenosine and uridine residues. *J Am Chem Soc* **114:** 7963–7969.

Wu X, Guntha S, Ferencic M, Krishnamurthy R, Eschenmoser A. 2002. Base-pairing systems related to TNA: α-Threofuranosyl oligonucleotides containing phosphoramidate linkages. *Org Lett* **4:** 1279–1282.

Zhu TF, Szostak JW. 2009a. Coupled growth and division of model protocell membranes. *J Am Chem Soc* **131:** 5705–5713.

Zhu TF, Szostak JW. 2009b. Preparation of large monodisperse vesicles. *PLoS ONE* **4:** e5009.

Zielinski WS, Orgel LE. 1985. Oligomerization of activated derivatives of 3′-amino-3′-deoxyguanosine on poly(C) and poly(dC) templates. *Nucleic Acids Res* **13:** 2469–2484.

Riboswitches and the RNA World

Ronald R. Breaker

Department of Molecular, Cellular and Developmental Biology; Department of Molecular Biophysics and Biochemistry; Howard Hughes Medical Institute, Yale University, New Haven, Connecticut 06520-8103

Correspondence: ronald.breaker@yale.edu

SUMMARY

Riboswitches are structured noncoding RNA domains that selectively bind metabolites and control gene expression (Mandal and Breaker 2004a; Coppins et al. 2007; Roth and Breaker 2009). Nearly all examples of the known riboswitches reside in noncoding regions of messenger RNAs where they control transcription or translation. Newfound classes of riboswitches are being reported at a rate of about three per year (Ames and Breaker 2009), and these have been shown to selectively respond to fundamental metabolites including coenzymes, nucleobases or their derivatives, amino acids, and other small molecule ligands.

The characteristics of some riboswitches suggest they could be modern descendents of an ancient sensory and regulatory system that likely functioned before the emergence of enzymes and genetic factors made of protein (Nahvi et al. 2002; Vitreschak et al. 2004; Breaker 2006). If true, then some of the riboswitch structures and functions that serve modern cells so well may accurately reflect the capabilities of RNA sensors and switches that existed in the RNA World. This article will address some of the characteristics of modern riboswitches that may be relevant to ancient versions of these metabolite-sensing RNAs.

Outline

1. Riboswitches and Their Moving Parts
2. Modern Mechanisms for Riboswitch-Mediated Gene Control
3. Possible Mechanisms for Riboswitch Control of Ribozymes
4. Ligand Binding Affinity and Kinetics of Riboswitch Function
5. How Many Riboswitch Classes Currently Exist?
6. Something Special About SAM?
7. Increasing Capabilities by Stacking Riboswitch Components
8. The Origin of Riboswitches
9. Conclusions

References

1 RIBOSWITCHES AND THEIR MOVING PARTS

The term riboswitch was established to define RNAs that control gene expression by binding metabolites without the need for protein factors. More recently, the name has begun to be used for riboswitch-like RNAs that respond to changes in temperature (Johansson 2009; Klinkert and Narberhaus 2009), tRNA binding (Gutiérrez-Preciado et al. 2009), or metal ion binding (Cromie et al. 2006; Dann et al. 2007). Although the functions of the RNAs encompassed by this expanded definition certainly would have been useful in an RNA World, the discussion hereafter will be focused on riboswitches that have evolved to respond to small organic compounds.

Riboswitches need to form molecular architectures with sufficient complexity to carry out two main functions: molecular recognition and conformational switching. Simple riboswitches each carry one aptamer that senses a single ligand and one expression platform that usually controls gene expression via a single mechanism. Because only four types of monomers are used by RNA to form selective binding pockets for target metabolites, aptamer sequences and structures tend to be strikingly well conserved over great evolutionary distances (e.g., see Grundy and Henkin 1998; Gelfand et al. 1999; Sudarsan et al. 2003; Nahvi et al. 2004). This sequence and structure conservation serves as the basis for assigning riboswitch representatives to specific classes (Fig. 1). Aspects of the tertiary structure folds used

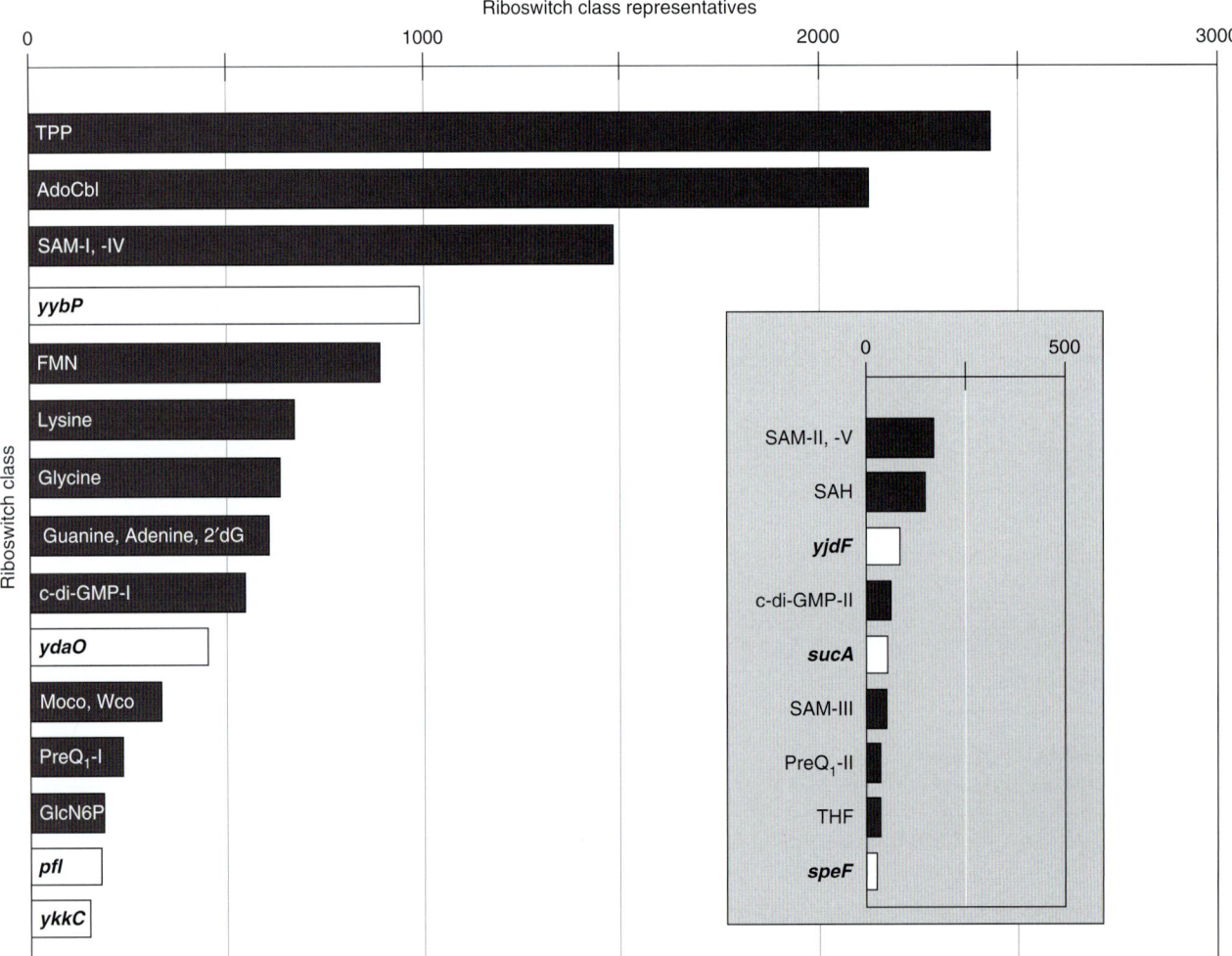

Figure 1. Classes of validated and candidate riboswitches plotted in descending order relative to their frequency in the genomes of ~700 bacterial species. Classes with at least some biochemical or genetic validation (filled bars) are named for the ligand that is most tightly bound. Multiple compound names indicate that binding site variants exist with altered ligand specificity. Candidate riboswitches whose ligand-binding functions have not been validated (open bars) are named for genes that are commonly associated with the motif.

by riboswitch aptamers are presented in greater detail in Garst et al. (2010).

The expression platform of each bacterial riboswitch usually is located downstream of the aptamer, where it assesses the ligand binding status of the RNA and regulates gene expression accordingly (Mandal and Breaker 2004a; Barrick and Breaker 2007). Alternately folding structures are common for RNA, and folding differences can be harnessed to influence several different processes that contribute to gene expression efficiency (Fig. 2). Therefore, expression platforms tend to be far less conserved through evolution compared to aptamer domains.

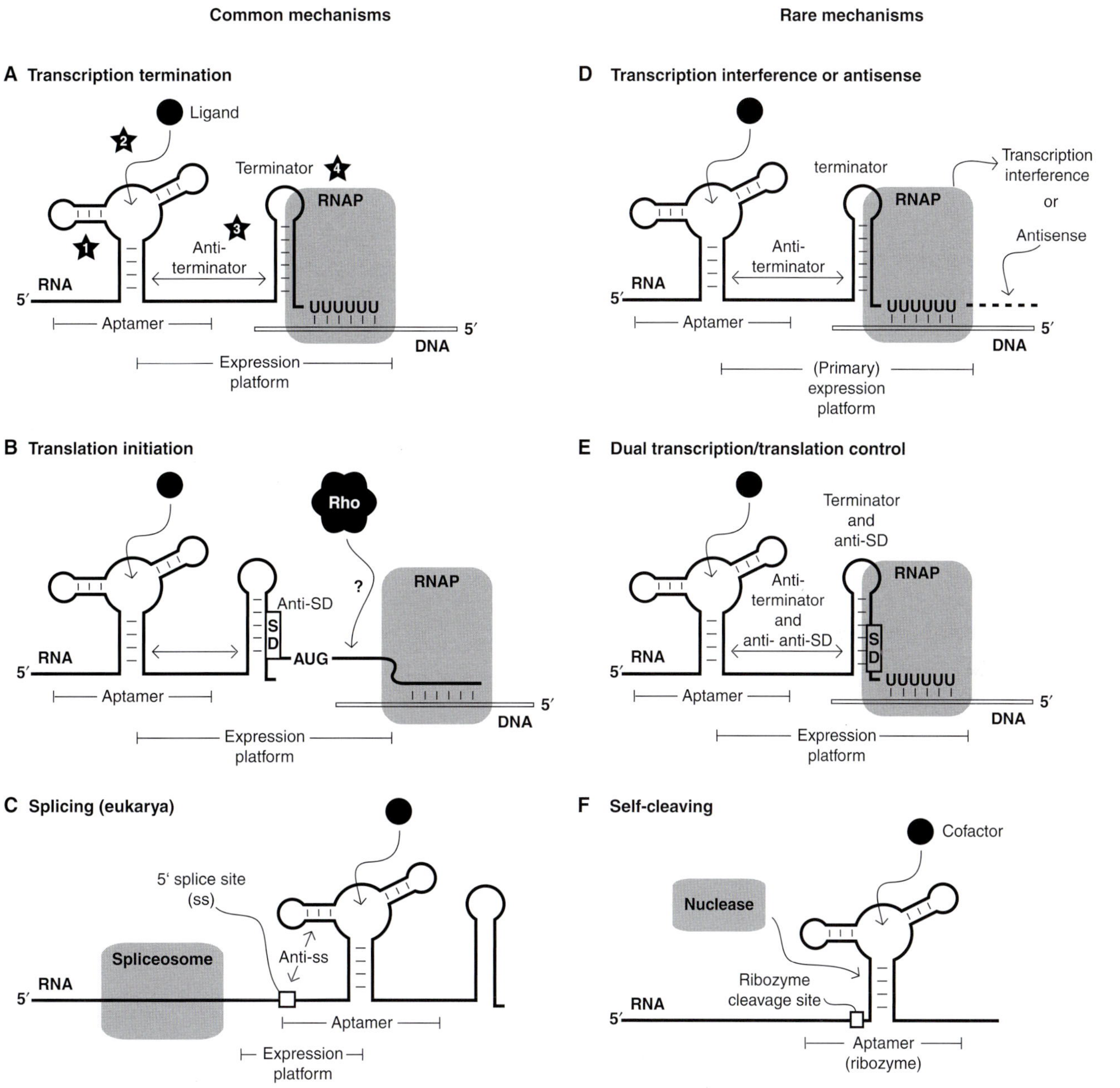

Figure 2. Established or predicted mechanisms of riboswitch-mediated gene regulation. The most common mechanisms are (A) transcription termination, (B) translation initiation, and (C) splicing control (in eukaryotes). More rare mechanisms observed or predicted in some bacterial species include (D) transcription interference or possibly antisense action, (E) dual transcription and translation control, and (F) ligand-dependent self-cleaving ribozyme action. The numbers in a identify steps that are important for kinetically driven riboswitches (Wickiser et al. 2005a,b; Gilbert et al. 2006). Numbers represent (1) folding of the aptamer, (2) docking of the ligand, (3) folding of the expression platform, and (4) speed of RNA polymerase (RNAP). In B, Rho represents the transcriptional terminator protein (Skordalakes and Berger 2003).

Members of all experimentally validated riboswitch classes can bind their target ligand without the need for protein factors. Also, at least some representatives of each class have predicted or experimentally proven RNA structures residing in expression platforms that could independently control gene expression. In other words, the RNA alone is sufficient to carry out molecular recognition and gene control actions without the obligate assistance of protein factors. As with ancient ribozymes, any RNA World riboswitches would have probably had to function in an environment devoid of genetically encoded proteins. However, proteins could be directly involved in the folding and action of some modern riboswitches, much like protein factors assist large ribozymes such as RNase P (Sharin et al. 2005) and self-splicing ribozymes (Halls et al. 2007).

Aptamers can be structurally preorganized to various extents in the absence of ligand, but ligand binding induces at least some structural reorganization or stabilization of aptamer substructures that consequently influences the folding and function of the adjoining expression platform. It is this metabolite-dependent interplay between aptamer and expression platform that is harnessed by the riboswitch to modulate gene expression through various mechanisms.

2 MODERN MECHANISMS FOR RIBOSWITCH-MEDIATED GENE CONTROL

One of the most common mechanisms used by bacterial riboswitches involves the modulation of transcription termination (Fig. 2A) (Barrick and Breaker 2007). The formation of a strong stem followed by a run of uridine residues constitutes an intrinsic transcription terminator (Gusarov and Nudler 1999; Yarnell and Roberts 1999), which causes RNA polymerase to stall transcription and eventually to release the DNA template and nascent RNA product. Ligand binding to the aptamer controls the formation of the terminator stem usually by regulating the formation of a competing secondary structure or anti-terminator. Similarly, mutually-exclusive base-paired structures are exploited by riboswitches to control ribosome access to the ribosome binding site or Shine-Dalgarno (SD) sequence, thereby regulating translation initiation (Fig. 2B). This mechanism could be used to regulate translation from full-length mRNAs, but perhaps the transcription terminator protein Rho (Skordalakes and Berger 2003) will recognize nascent mRNAs that are not being actively translated because of SD access control by a riboswitch, thus causing transcription termination as well.

The most widespread riboswitch class discovered to date responds to the coenzyme thiamin pyrophosphate (TPP) (Mironov et al. 2002; Winkler et al. 2002b; Barrick and Breaker 2007). This riboswitch is quite common in plants and fungi (Sudarsan et al. 2003) where representatives have been shown to control splicing (Fig. 2C) (Kubodera et al. 2003; Cheah et al. 2007; Croft et al. 2007; Bocobza et al. 2007; Wachter et al. 2007). Additional data is needed to validate the proposed existence of a eukaryotic riboswitch class responsive to arginine (Borsuk et al. 2007), but it seems likely that more metabolite-binding riboswitches will be discovered in eukaryotes. Because eukaryotic TPP riboswitches appear to most commonly control splicing, introns may be excellent vehicles for riboswitch expression and regulatory function, and therefore may yield future riboswitch discoveries.

In a few instances in bacteria, a riboswitch aptamer resides upstream of an intrinsic transcription terminator stem that is formed using SD nucleotides (Fig. 2D). This arrangement should allow metabolite-controlled transcription termination as well as allow control of translation initiation for any mRNAs whose transcription progress extends into the coding region.

Also, riboswitches do not necessarily function to exclusively control the expression of an adjoining ORF, but could act in a bimolecular fashion on other RNAs, even after they may have modulated expression of a local gene. One possible example of this is a SAM-I riboswitch identified by bioinformatics (Rodionov et al. 2004) that is transcribed in the opposite direction of the genes it controls. Experimental analysis of this riboswitch (André et al. 2008) confirms the production of antisense RNA in response to changing sulfur levels. However the true mechanism of gene control may not be because of base pairing between sense and antisense transcripts. Rather, competition between RNA polymerases initiating at convergent promoters for the transcription of the mRNA versus the riboswitch-linked RNA may be the dominant control mechanism (Fig. 2E).

One of the most interesting mechanisms for riboswitch-mediated gene control integrates ligand binding and ribozyme activities (Fig. 2F). Representatives of the *glmS* riboswitch class function as metabolite-responsive self-cleaving ribozymes (Winkler et al. 2004). The ligand for this ribozyme-riboswitch is glucosamine-6-phosphate (GlcN6P), which is the metabolic product of the protein whose expression is controlled by the RNA. A *glmS* ribozyme construct from *Bacillus subtilis* cleaves with an GlcN6P-induced rate constant of nearly 100/min (Brooks and Hampel 2009), which is an enhancement of more than 1 billion-fold above the rate constant for spontaneous RNA cleavage by phosphoester transfer. Ribozyme self-cleavage initiates rapid destruction of its associated mRNA coding region via selective nuclease degradation (Collins et al. 2007).

3 POSSIBLE MECHANISMS FOR RIBOSWITCH CONTROL OF RIBOZYMES

It seems reasonable to speculate that metabolite-regulated ribozymes may have been prevalent in RNA World organisms. However, some features of the riboswitch mechanisms described above would not be applicable in a hypothetical RNA World cell. Perhaps a mechanism involving intrinsic transcription termination hairpins existed, but these hairpins would have needed to control transcription by an RNA polymerase ribozyme instead of protein-based polymerase enzymes. Of course, expression platforms that control ribosome access to the SD sequence could not have been relevant until at least a primitive translation apparatus was established.

RNA World organisms could have easily made use of numerous allosteric ribozymes with architectural features that were similar to those created in the laboratory by fusing aptamers to ribozymes (Breaker 2002; Silverman 2003). Such allosteric ribozymes could have emerged through RNA genome rearrangements that placed aptamers near ribozymes, so that ligand-mediated structure modulation affects a neighboring structure critical for ribozyme activity. Alternatively, the ligand could have functioned as a cofactor to directly participate in accelerating the chemical step of the ribozyme, rather than functioning as an allosteric effector. A modern example of cofactor participation is seen with *glmS* ribozymes, wherein GlcN6P actively participates in *glmS* ribozyme self-cleavage (Cochrane et al. 2009).

Self-splicing ribozymes also may have had functional homologs in the RNA World that were regulated by small molecule ligands. Grafting of an aptamer to a structurally sensitive region of a group I ribozyme (Hougland et al. 2006) has been shown to yield ligand-dependent splicing and gene control in cells (Thompson et al. 2002). Given that group I ribozymes use guanosine or one of its phosphorylated derivatives to initiate the first step of splicing, perhaps at least some representatives of these RNAs serve as metabolite-responsive riboswitches in modern cells. This function would be consistent with the fact that many group I ribozymes are associated with similar genes, which implies a functional linkage like that seen between riboswitches and the genes they control. If true, group I ribozymes may be one of the oldest and most widespread of all the known riboswitch classes.

4 LIGAND BINDING AFFINITY AND KINETICS OF RIBOSWITCH FUNCTION

Riboswitch aptamers can tightly bind their target ligands with values for dissociation constants (K_D) ranging from the mid micromolar as measured for GlcN6P binding by *glmS* ribozymes (Winkler et al. 2004) to the mid picomolar range as observed for some riboswitch representatives that bind TPP (Welz and Breaker 2007), FMN (Lee et al. 2009) and c-di-GMP (Sudarsan et al. 2008). For *Escherichia coli*, the presence of a single molecule per cell corresponds to a concentration in the low nanomolar range. Therefore, picomolar K_D values for aptamers are at least two orders of magnitude better than required for a riboswitch to detect compounds as rare as one per cell!

There are several possible explanations for this K_D and metabolite concentration paradox (Ames and Breaker 2009). Perhaps, the conditions used to estimate ligand affinities are not a good approximation of cellular conditions, and the actual K_D values may be much poorer. It is also known that nucleotides flanking the aptamer domains can diminish measured K_D values, presumably by forming alternative folds that compete with the ligand-receptive structure. Thus the affinities of riboswitch aptamers may be poorer for nascent transcripts as they emerge from RNA polymerase. Another hypothesis that would resolve this contradiction is that some riboswitches may not reach equilibrium with their target metabolite, but they rely on the kinetics of RNA folding and ligand binding to properly modulate gene expression. In other words, the speed of ligand association, rather than the equilibrium constant reflecting ligand affinity, may be the critical determinant for the concentration of ligand needed to trigger riboswitch action. Unless the ligand binds before RNA polymerase passes beyond the terminator stem, transcription will progress to completion even if ligand binding eventually occurs.

Evidence for the importance of ligand-binding kinetics has emerged from an analysis of an FMN riboswitch from the *ribD* operon of *B. subtilis* (Wickiser et al. 2005b). This riboswitch operates via transcription termination whereby ligand binding permits formation of a terminator stem (Fig. 3A). The study revealed that the concentration required to trigger efficient transcription termination is more than 10-fold higher than the K_D value measured for the minimal aptamer domain. Furthermore, conducting transcription reactions under conditions that accelerate the speed of RNA polymerization (e.g., increasing NTP concentrations or using mutations to remove RNA polymerase pause sites) creates a demand for even higher FMN concentration to trigger termination.

This and related data (Wickiser et al. 2005a; Gilbert et al. 2006) indicate that at least some riboswitches are not thermodynamically driven, but rely on the kinetics of transcription, RNA folding and metabolite binding to tune the concentration of ligand needed to trigger genetic control. Interestingly, this characteristic may give riboswitches an advantage over protein genetic factors in some instances

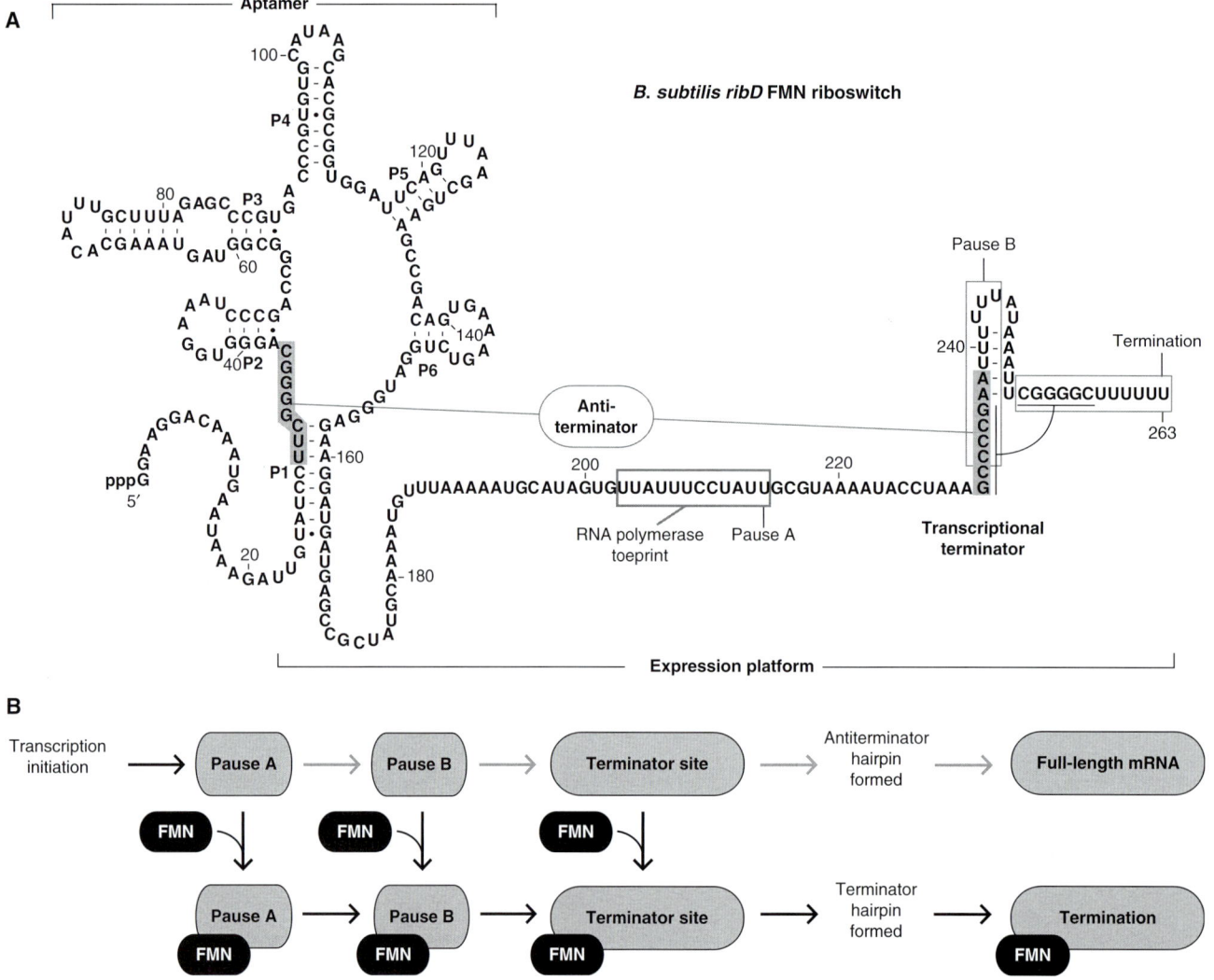

Figure 3. Kinetic function of an FMN riboswitch. (A) Sequence and secondary structure models for the *ribD* FMN riboswitch from *B. subtilis* (Winkler et al. 2002a). The RNA functions as a genetic "OFF" switch wherein FMN binding stabilizes P1 formation, precludes the formation of an antiterminator stem, and permits the formation of a terminator stem that represses gene expression. (B) Simplified kinetic scheme for the function of the *ribD* FMN riboswitch depicted in A. Steps represented by black and gray arrows lead to termination and full-length mRNA production, respectively. See elsewhere for details (Wickiser et al. 2005b).

(Ames and Breaker 2009). A kinetically driven riboswitch can be tuned to respond to a different concentration of metabolite by accruing mutations in the aptamer that change the rate constant for ligand association. Perhaps more likely to occur are mutations in the span of nucleotides linking aptamer to expression platform that change the time needed for RNA polymerase to reach the terminator stem. Kinetically driven riboswitches could lower or raise the concentration needed to trigger function by simply inserting or deleting nucleotides to this linker, respectively. RNA World riboswitches may have exploited similar characteristics to experience a more smooth evolutionary landscape where functional tuning could occur by mutations outside the binding and catalytic sites of receptors and ribozymes.

5 HOW MANY RIBOSWITCH CLASSES CURRENTLY EXIST?

Precisely defining what constitutes a distinctive riboswitch class can be difficult because of variation in aptamer sequence, structure, and function. For example, one or a few mutations to the core of guanine riboswitch aptamers can change ligand specificity to adenine (Mandal and Breaker 2004b) or 2′-deoxyguanosine (Kim et al. 2007), respectively. These mutations result in changes to the binding pocket, but the global architecture of the RNA remains largely unchanged (Serganov et al. 2004; Edwards and Batey 2009). Alternatively, some riboswitch aptamers carry substantial differences in sequence and structural elements that are distal from their near-identical ligand binding cores (see SAM riboswitch discussion later). Comparative sequence analysis algorithms cluster these RNAs into different groups despite the fact that they use similar binding sites to bind the same ligand (Weinberg et al. 2008).

A total of 17 riboswitch classes with at least some experimental validation have been identified based on a conservative approach to classification, wherein representatives are grouped either by similar global architecture or by similar binding site architecture (Fig. 1). At least seven candidate riboswitch classes also have been identified, although evidence for metabolite binding has not been reported for these candidates. Analysis of the numbers of representatives in each class can be conducted to determine if nearly all the riboswitch classes have been reported, or whether there are many more left to be discovered (Ames and Breaker 2009).

There are several reasons why it is important to estimate how many riboswitch classes still may await discovery in modern cells. From an RNA World vantage point, each new riboswitch discovery provides additional insights into the structural potential that primitive organisms may have harnessed long ago. Of course it may be difficult, or even impossible, to prove that a given riboswitch class is a direct descendant of one that existed in a ribo-organism. Regardless, these discoveries also provide a more complete understanding of the role that metabolite-sensing RNAs serve in modern cells. Locating riboswitch sequences in the genomes of organisms reveals new genetic regulons for compounds that are sensed by RNA (e.g., see Sudarsan et al. 2008). Associating genes of unknown function to specific metabolites via riboswitch regulation networks provides clues that are useful to establish protein function. Moreover, if some of these riboswitches control key metabolic processes in pathogenic organisms, then they may serve as new targets for the development of antibacterial compounds (Blount and Breaker 2006; Breaker 2009).

A plot of the riboswitch classes in descending order of frequency of incidence (Fig. 1) reveals that the numbers of representatives for these classes do not sort randomly. Assuming bioinformatics search algorithms used to identify new classes are revealing most common examples in sequence databases, then there naturally appears to be an increasing number of classes that have ever fewer representatives. This distribution is similar to that expected for phenomena that follow a power law relationship (Ames and Breaker 2009). If this relationship holds, then we speculate that more than 100 novel riboswitch classes may exist even in the ~700 bacterial species whose genomes have been fully sequenced. The actual number of bacterial riboswitch classes that ultimately will be discovered may be far greater because of the vast amount of novel genomic DNA information that remains to be sequenced.

6 SOMETHING SPECIAL ABOUT SAM?

S-adenosylmethionine (SAM; Fig. 4A) is a coenzyme made by fusing the amino acid L-methionine with ATP such that the triphosphate moiety of ATP is removed. Compounds like SAM and other nucleotide-like coenzymes have been proposed to be relics from an RNA World (White III 1976). RNA World organisms would therefore have needed ribozymes to synthesize this compound, probably used aptamers to sense this compound, and used ribozymes that used the compound as a methylating agent. When SAM is used as a coenzyme for methylation reactions, the resulting byproduct after methyl group transfer is S-adenosylhomocysteine (SAH; Fig. 4A). This compound is similar in structure to SAM, which creates a considerable molecular recognition challenge for enzymes and receptors that need to selectively bind only SAM or only SAH. However, this precise molecular discrimination is important, because SAH is toxic to cells and needs to be selectively recycled to regenerate SAM.

Interestingly, at least five distinct classes of riboswitches exist that bind SAM or its derivative SAH (Wang and Breaker 2008). SAM-I riboswitches (Fig. 4B) strongly discriminate between SAM and SAH by using carbonyl oxygen atoms (O2) of two U residues to create a partial negative-charged surface that attracts the distinguishing positive charge present on the sulfur of SAM (Montange and Batey 2006). The conserved nucleotides that form the binding pocket of SAM-I aptamers are also present in two additional structural families SAM-IV (Weinberg et al. 2008) and SAM-I/SAM-IV (Weinberg et al. 2010). Although these three structural families were discovered at different times because they are clustered separately by bioinformatics algorithms, they appear to use different accessory structures to form ligand-binding cores that are identical.

A similar situation is observed for SAM-II and SAM-V riboswitches (Fig. 4C). SAM-II riboswitch aptamers (Corbino et al. 2005) form a binding pocket that selectively recognizes SAM in part by exploiting the carbonyl oxygen atoms (O4) to bind the positive charge on sulfur (Gilbert et al. 2008). SAM-V aptamers (Poiata et al. 2009) represent a different structural family member but representatives carry distinctive sequences that flank a putative ligand-binding core that appears to be identical to that formed by SAM-II. It is not clear whether members of the

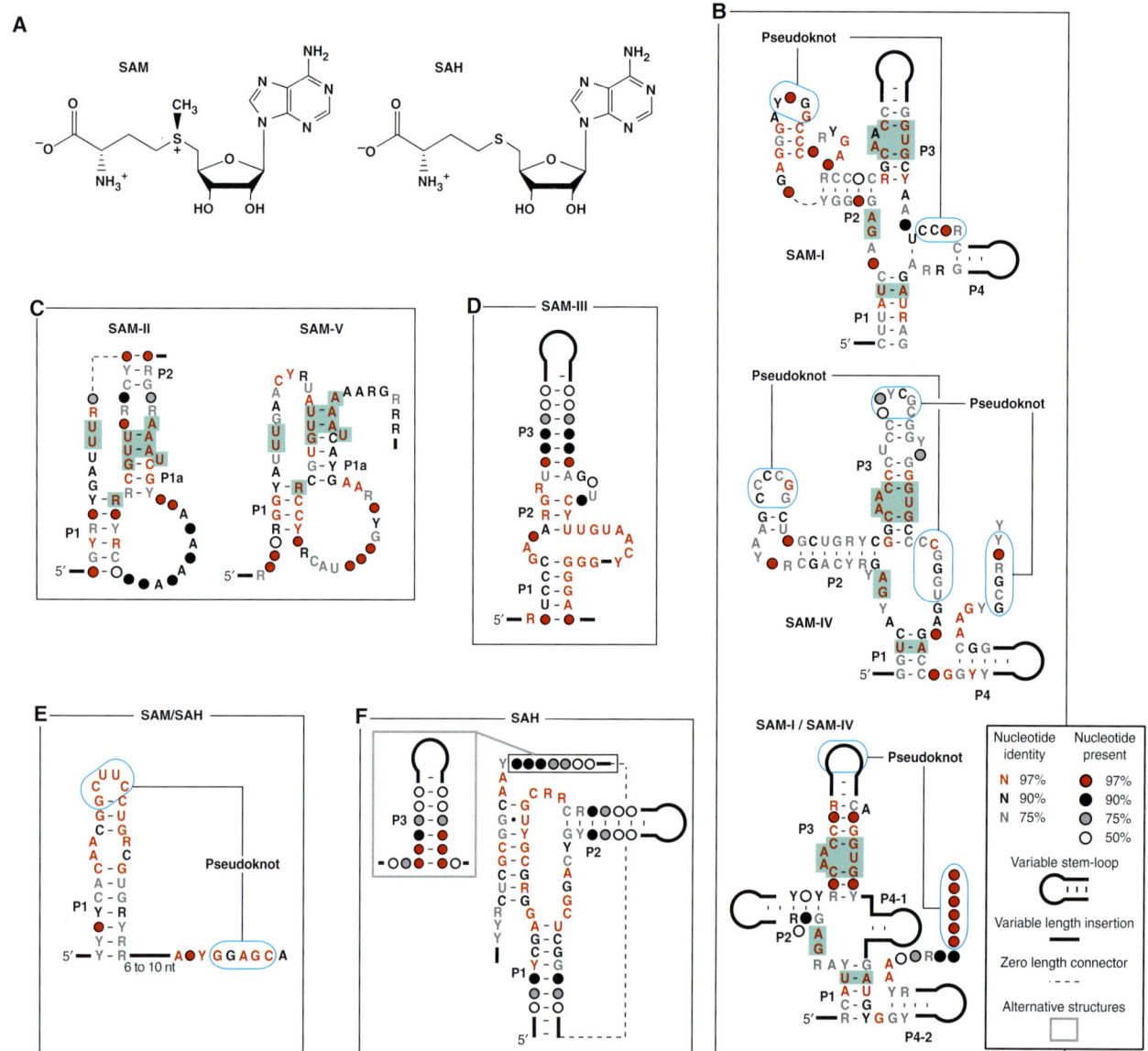

Figure 4. Riboswitch aptamers that respond to the coenzyme SAM or its metabolic derivative SAH. (A) The chemical structures of SAM and SAH. (B) Three variations of the SAM-I aptamer class. Despite substantial differences in sequence and structural subdomains, SAM-I (Grundy and Henkin 1998), SAM-IV (Weinberg et al. 2008), and SAM-I/SAM-IV (Weinberg et al. 2010) aptamer families carry a common ligand binding core (green-shaded nucleotides), including the identities of nucleotides that directly contact SAM (red bold letters). (C) A second superfamily of SAM-specific riboswitches use either the SAM-II (Corbino et al. 2005) or SAM-V (Poiata et al. 2009) aptamer families, which also carry a common ligand binding core. Annotations are as described for B. (D) Sequence and structural features of SAM-III aptamers (Fuchs et al. 2006), representing the third class of SAM-specific riboswitches. (E) Sequence and structural features of SAM/SAH aptamers (Weinberg et al. 2010), which genetically track like SAM riboswitches but bind SAH more tightly. (F) A distinct class of aptamers that genetically track as SAH riboswitches and selectively bind SAH while strongly discriminating against SAM (Wang et al. 2008).

structural superfamilies most commonly represented by SAM-I and SAM-II riboswitches have emerged via divergent or convergent evolution, but their existence helps showcase the diversity of structures that can be used by RNA to support the formation of identical metabolite binding pockets.

SAM-III or SMK riboswitches (Fuchs et al. 2006) represent a third distinct class of SAM-binding RNAs. Yet again, SAM is bound more tightly than SAH in large part by exploiting the positive charge on sulfur, which is recognized by a carbonyl oxygen (O4) of a U residue (Lu et al. 2008). The ligand binding core is otherwise distinct from the two classes discussed previously. It is interesting to note that riboswitch aptamers examined in detail so far rely only on carbonyl oxygen atoms from U residues to interact with the positive charge on sulfur. Phosphate oxygen atoms appear ideally suited to associate with positive-charged ligands, and yet no SAM riboswitches exploiting this possible interaction have been discovered to date.

A fourth distinct SAM riboswitch class exists, but representatives of this class do not strongly discriminate between SAM and SAH. The SAM/SAH aptamers (Weinberg et al. 2010) are exceptionally small for natural metabolite-binding RNAs (as short as 43 nucleotides), and actually bind SAH at least fivefold more tightly than SAM. However, the motif is always genetically associated with genes for SAM biosynthesis, suggesting that SAM is the biologically relevant ligand. Most likely, SAH does not attain a concentration needed to inappropriately trigger riboswitch function.

Much greater discrimination in favor of SAH and against SAM is achieved by a fifth distinct riboswitch class in this series (Wang et al. 2008). SAH riboswitches are genetically associated with genes for enzymes that degrade SAH to prevent its toxic build-up and that recycle components for the regeneration of SAM. Given that SAM concentrations are likely to be higher that SAH concentrations, SAH riboswitches must strongly discriminate against SAM. Indeed, a representative SAH riboswitch appears to discriminate against SAM by approximately 1000 fold. This discrimination conceivably could be achieved by forming an SAH binding pocket that creates a steric block of the extra methyl group on SAM.

Based on current data, there does not appear to be anything that is particularly special about SAM as a riboswitch ligand. The existence of a diversity of riboswitch aptamers for SAM and SAH appears to be because of the importance that cells place on sensing these ligands, and perhaps due more to the rich structural potential that RNA can rely on to form selective binding pockets for metabolites. As evidence of this structural potential, we are discovering that multiple riboswitch classes have evolved to sense other metabolites, including preQ1 (Roth et al. 2007; Meyer et al. 2008). This same structural potential of course would also have been available to organisms from the RNA World.

7 INCREASING CAPABILITIES BY STACKING RIBOSWITCH COMPONENTS

The architecture of simple riboswitches composed of a single aptamer and a single expression platform has very limited functional capabilities. Most obvious is the fact that the riboswitch will respond only to its target ligand, or perhaps also to a close chemical analog that may also be present in a cell. Furthermore, the dose-response curve for a simple riboswitch can do no better than conform to the functional optimum for a one-to-one interaction between a receptor and its ligand (Fig. 5). Specifically, a simple riboswitch that functions to perfection will require an 81-fold change in ligand concentration to progress from 10% to 90% gene modulation (Fig. 5B). If some of the riboswitch RNAs being made fail to function properly, because of folding problems for example, then the dynamic range for gene control will be reduced and the dynamic range for ligand sensing may be expanded.

These biochemical certainties for simple riboswitches place severe limitations on the performance characteristics of metabolite-sensing RNAs. However, modern organisms (and likely their RNA World ancestors) found ways to overcome the biochemical limitations of simple riboswitches. For example, riboswitches for the amino acid glycine commonly carry two aptamers and only a single expression platform (Fig. 6A). These aptamers function cooperatively, such that glycine binding by one aptamer increases the affinity for glycine binding to the adjacent aptamer. This cooperative binding function allows the riboswitch to show a far narrower ligand sensing dynamic range, which yields a genetic switch that has more digital character (Fig. 5B).

Ligand binding by a glycine riboswitch from *Vibrio cholerae* that includes two cooperative aptamers shows a Hill coefficient of 1.64 (Mandal et al. 2004). This value does not attain perfection, but must give cells that carry cooperative aptamers an evolutionary advantage. Many glycine riboswitches control expression of genes whose protein products form the glycine cleavage system, which allows cells to use excess glycine as an energy source. Apparently some cells take advantage of the cooperative riboswitch to more quickly produce proteins for glycine cleavage when the amino acid is in slight excess, and then turn off production of the glycine cleavage system before glycine concentrations drop to a point that may jeopardize synthesis of proteins.

Other tandem-arranged riboswitch configurations also exist, and these further expand the functional characteristics of metabolite-binding RNAs. For example, a type of tandem riboswitch architecture has been observed that can produce a more digital genetic response without cooperative ligand binding. Two independently functioning riboswitches from the same class (or that sense the same ligand) can occur in the same mRNA to produce a dose-response curve that is steeper than that for a single riboswitch (Fig. 6b). An example of this type of gene control system in *Bacillus anthracis* is composed of two TPP riboswitches (Sudarsan et al. 2006; Welz and Breaker 2007). If

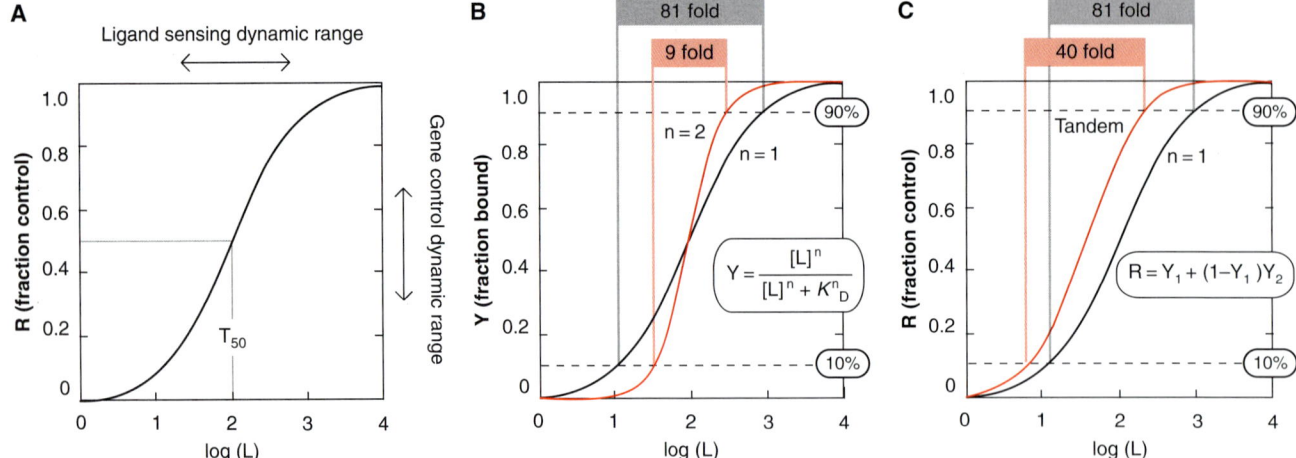

Figure 5. Binding kinetics for simple riboswitches, or complex riboswitches that use cooperative binding or a tandem architecture using riboswitches of from the same class. (A) Dose-response curve for a typical riboswitch carrying a single aptamer that functions perfectly. Note that the plot represents the performance of a population of individual riboswitch molecules where [R] represents the fraction of riboswitches causing gene expression change on ligand binding. Ligand concentration [L] is in arbitrary units, and T_{50} (Wickiser et al. 2005b) represents the concentration of ligand needed to half-maximally modulate gene expression. (B) Comparison of the dose-response curve for a simple riboswitch (one aptamer and one expression platform) versus a cooperative riboswitch (two aptamers and one expression platform). The curve for the cooperative riboswitch reflects perfect cooperativity between the aptamers and a Hill coefficient (n) of two. Note that [Y], the fraction of riboswitches bound by ligand, is equivalent to [R] if ligand binding always triggers a change in gene expression. (C) Comparison of the dose-response curve for a simple riboswitch versus a tandem arrangement of independently functioning riboswitches of the same class and near identical T_{50} values. Other annotations are as described in A and B. (Adapted, with permission, from Welz and Breaker 2007.)

the concentrations of ligand needed to half-maximally modulate gene expression (T_{50} values) are near identical, then only a 40-fold change in ligand concentration is needed to progress from 10% to 90% gene modulation.

A very similar type of tandem riboswitch architecture has recently been observed in *Candidatus pelagibacter ubique* (Poiata et al. 2009). This arrangement is composed of representatives of SAM-II and SAM-V riboswitch classes (Fig. 6C). The SAM-II riboswitch occurs first in the nascent mRNA, and has an expression platform that includes an intrinsic terminator stem. The SAM-V riboswitch occurs second, and likely controls gene expression by regulating access to the ribosome binding site. This arrangement could yield a more digital dose-response curve like the tandem TPP riboswitches described above. In addition, the two distinct expression platforms that control either transcription or translation provide a particularly economical genetic switch. If SAM concentrations are adequate, newly initiated transcripts are terminated by the SAM-II riboswitch before the coding region is produced. If SAM concentrations become adequate only after the full mRNA is produced, then the SAM-V riboswitch can repress translation without requiring that the mRNA be degraded. This capability could be particularly advantageous for organisms that live in environments that are constantly limited for nutrients.

Although simple riboswitches only respond to one ligand type, this restriction in signaling complexity can be overcome by stacking tandem riboswitches from different ligand-binding classes such that gene expression is responsive to more than one chemical signal (Fig. 6D). Indeed, a natural example of such a two-input Boolean logic gate has been observed in the *metE* mRNA from the bacterium *Bacillus clausii* (Sudarsan et al. 2006). A SAM-I riboswitch aptamer precedes an aptamer for AdoCbl, and each aptamer is associated with its own intrinsic terminator stem. The binding of either SAM or AdoCbl causes transcription termination before the coding region of the mRNA is made, and thus the tandem arrangement functions as Boolean NOR gate.

The examples described above are undoubtedly just a small sample of the ways that riboswitches can be conjoined or combined with other functional domains to create even more complex RNA devices. Many types of RNA switches are being created (or recreated) in laboratories by molecular engineers (Breaker 2004; Silverman 2003). For example, integration of multiple aptamers into a structurally sensitive site of a hammerhead ribozyme produced a cooperative

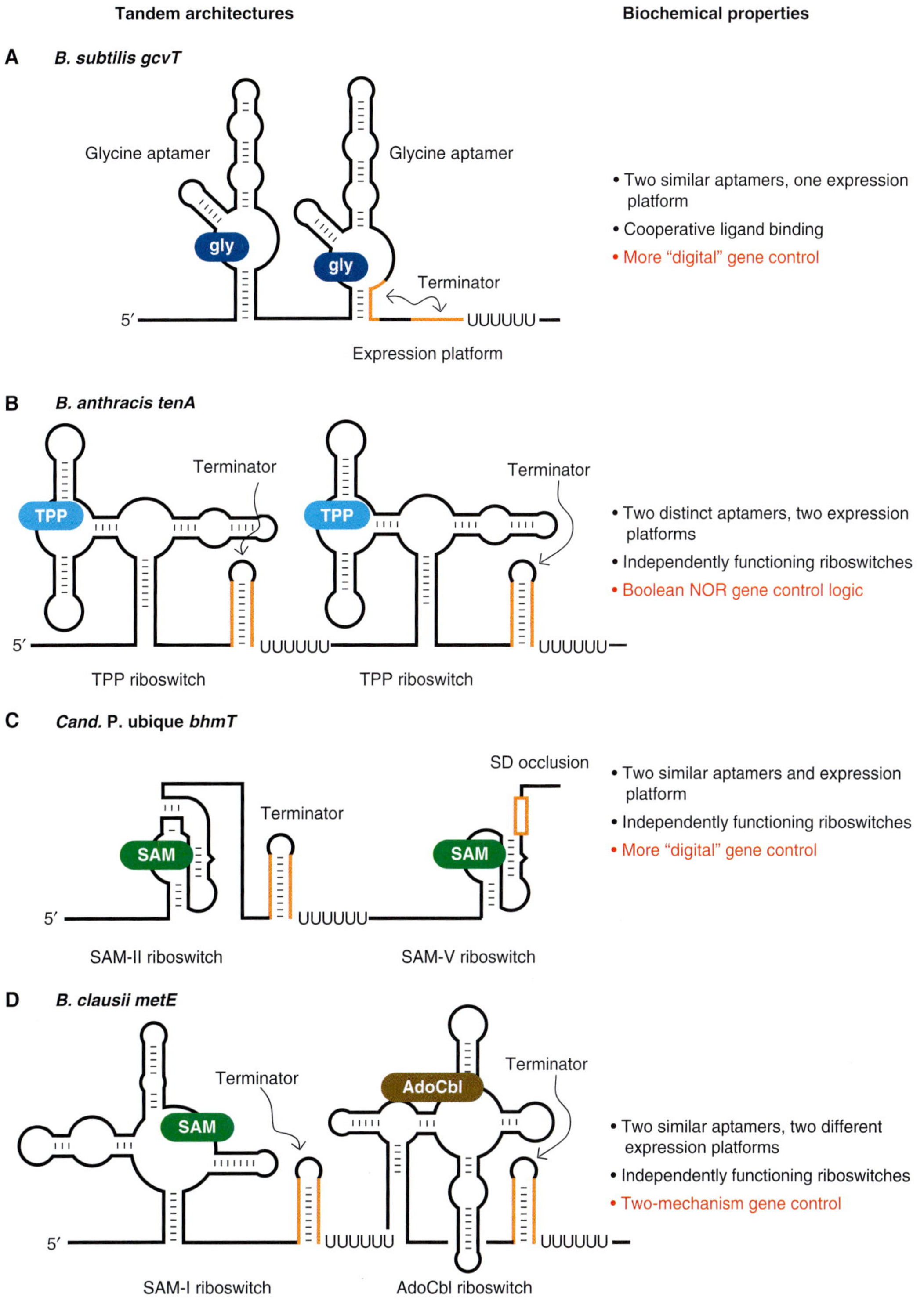

Tandem architectures

A *B. subtilis gcvT*

Glycine aptamer

Glycine aptamer

gly

gly

Terminator

5′

UUUUUU

Expression platform

B *B. anthracis tenA*

Terminator

Terminator

TPP

TPP

5′

UUUUUU

UUUUUU

TPP riboswitch

TPP riboswitch

C *Cand. P. ubique bhmT*

SD occlusion

Terminator

SAM

SAM

5′

UUUUUU

SAM-II riboswitch

SAM-V riboswitch

D *B. clausii metE*

Terminator

AdoCbl

Terminator

SAM

5′

UUUUUU

UUUUUU

SAM-I riboswitch

AdoCbl riboswitch

Biochemical properties

- Two similar aptamers, one expression platform
- Cooperative ligand binding
- More "digital" gene control

- Two distinct aptamers, two expression platforms
- Independently functioning riboswitches
- Boolean NOR gene control logic

- Two similar aptamers and expression platform
- Independently functioning riboswitches
- More "digital" gene control

- Two similar aptamers, two different expression platforms
- Independently functioning riboswitches
- Two-mechanism gene control

Figure 6. Architectures and functions of tandem riboswitch arrangements. (*A*) Cooperative glycine riboswitch system that yields a more digital genetic response in numerous Gram positive bacteria including in the 5′ UTR of the *B. subtilis gcvT* gene (Mandal et al. 2004). (*B*) Tandem TPP riboswitches from the 5′ UTR of the thiamin metabolism gene *tenA* from *Bacillus anthracis* (Welz and Breaker 2007). (*C*) Tandem SAM-II and SAM-V riboswitches identified in ocean bacteria such as "*Cand.* P. ubique" (Poiata et al. 2009). (*D*) A two-input Boolean NOR logic gate composed of tandem riboswitches for SAM and AdoCbl located in the *metE* gene from *Bacillus clausii* (Sudarsan et al. 2006).

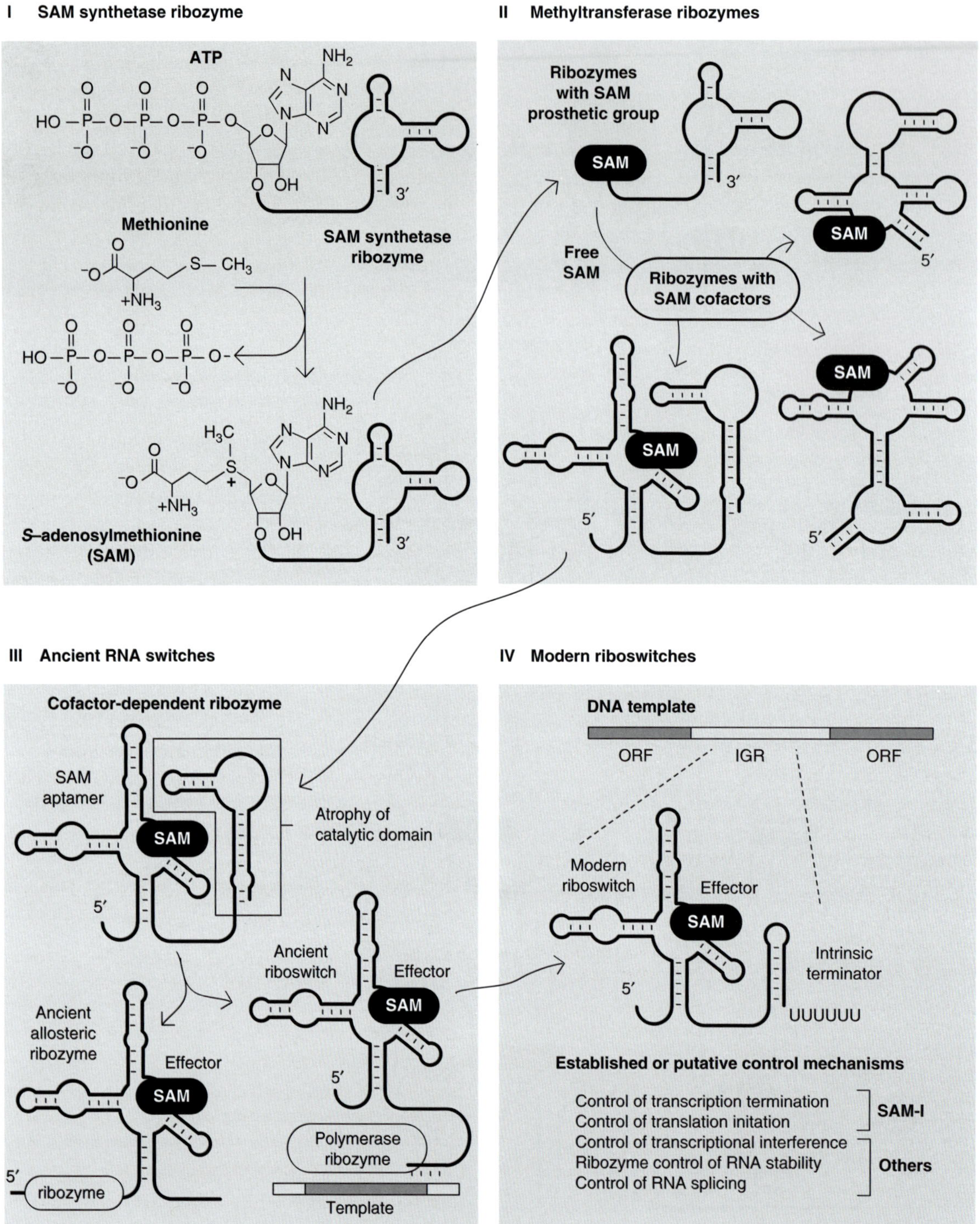

Figure 7. A possible path for the descent of modern riboswitches from RNA World ribozymes. SAM-dependent RNAs were arbitrarily chosen as an example. (*Stage I*) Emergence of ribozymes that synthesize SAM. Ribozymes might have coupled methionine to a 5′-terminal ATP moiety or coupled methionine to a free ATP substrate. (*Stage II*) Utilization of SAM by diverse methyltransferase ribozymes. In a complex RNA World, various methyltransferase ribozymes might have been present that used SAM as a prosthetic group or as a diffusible cofactor. (*Stage III*) Co-opting ribozyme subdomains to create RNA switches. Allosteric systems that controlled ribozyme function or that controlled the biosynthesis of new RNAs could have made use of variant SAM-binding domains of certain methyltransferase ribozymes. (*Stage IV*) Perseverance of ancient SAM-binding aptamers and their use in modern riboswitches. (Adapted, with permission, from Breaker 2006.)

allosteric ribozyme that functions as a Boolean AND gate (Jose et al. 2001). This RNA construct incorporates several features of natural riboswitches with ribozyme function to provide just one example of the incredible complexity of function that could be achieved by RNA. Although there may be little demand in modern cells for controlling ribozyme functions using aptamers as allosteric binding sites, RNA World organisms could have made extensive use of ribozymes whose reaction kinetics were controlled by metabolite binding aptamers integrated into their structures. Of course riboswitches from the RNA World would not necessarily have carried expression platforms like those from modern cells, but could have used other ligand-mediated structural changes to regulate early metabolic processes.

8 THE ORIGIN OF RIBOSWITCHES

The fact that riboswitches are entrusted by some modern cells to sense many compounds and regulate key metabolic genes strongly validates the hypothesis that RNA World organisms could have used similar RNA-based sensors and switches. However, we cannot presume that all modern riboswitches are direct descendants from homologous riboswitches from the RNA World. Biological innovation with RNA polymers would not have ceased with the emergence of proteins, and so new or reinvented riboswitch classes are probably continuing to appear.

Some riboswitch classes, particularly those that sense TPP, AdoCbl, and FMN are exceptionally widespread (Barrick and Breaker 2007). Moreover, these riboswitch ligands are proposed to be molecular relics from an RNA World (White 1976, Benner et al. 1989). Such characteristics are precisely what would be expected for riboswitches that have an ancient origin. Furthermore, the most widespread riboswitch classes have some of the most complex aptamer structures, and these information-rich structures are unlikely to emerge frequently during evolution. Rather, widespread complex riboswitches most likely have an early and perhaps RNA World origin.

In contrast, the more narrowly distributed riboswitches that have less-complex structures could represent more recent evolutionary inventions. Even relatively widespread riboswitches with simplistic aptamers, such as those that bind purines (Kim and Breaker 2008) or preQ$_1$ (Roth et al. 2007) could be reinvented rather than directly descend from primordial homologs. Perhaps modern classes of riboswitch aptamers have mixed beginnings; with some having emerged relatively recently to satisfy new regulatory needs whereas others representing an unbroken structural and functional lineage from RNAs that have changed little since the RNA World.

One intriguing hypothesis is that some riboswitch aptamers, particularly those that bind coenzymes, may be descendent from ancient ribozymes that synthesized these coenzymes or used them to catalyze metabolic reactions. A coenzyme-dependent ribozyme and a coenzyme-responsive riboswitch both would need to form a selective binding pocket for the compound, and therefore it is likely that some RNA sequences that function as ribozymes could be very close in sequence-space to those that function as riboswitches (Fig. 7).

9 CONCLUSIONS

The existence of so many riboswitch classes that are involved in key regulatory processes provides support for the hypothesis that RNA World organisms could have harnessed the structural and functional potential of aptamers to create sophisticated sensory and regulatory networks without the need for proteins. Also as seen with modern organisms, when more complex biochemical functions are demanded for the regulatory task, multiple riboswitches can be assembled in different architectures to provide more sophisticated RNA devices. Tandem aptamers that yield more digital genetic responses, multi-mechanism gene control, or Boolean logic gate function result from simply grafting the parts from two or more riboswitches together. Because examples of sophisticated riboswitches are successfully competing with proteins for demanding roles in modern cells, there is reason to believe that RNA World organisms could have used similar RNAs to regulate very complex metabolic states before the emergence of proteins.

There also appears to be considerable potential for exploiting riboswitches for practical applications. Existing riboswitches can be fused to coding regions to create ligand-controlled genetic constructs. Reverse engineering of natural riboswitches could help guide the creation of designer riboswitches to expand the tools available for genetics studies (Suess and Weigand 2008). Also, analogs of riboswitch ligands can be used to deregulate key metabolic pathways in bacteria, and thus may serve as new classes of antibiotics (Blount and Breaker 2006; Breaker 2009). Therefore the continued discovery and analysis of riboswitches should help efforts to create molecular tools and drug targets from RNA, as well as provide modern examples of RNAs whose functions may closely mimic those from the RNA World.

ACKNOWLEDGMENTS

RNA research in the Breaker laboratory is supported by the Howard Hughes Medical Institute and by grants from the National Institutes of Health.

REFERENCES

André G, et al. 2008. S-box and T-box riboswitches and antisense RNA control a sulfur metabolic operon of *Clostridium acetobutylicum*. *Nucleic Acids Res* **36:** 5955–5969.

Ames TD, Breaker RR. 2009. Bacterial riboswitch discovery and analysis. In: *The chemical biology of nucleic acids*, Meyer G., ed., John Wiley & Sons, Ltd (in press).

Barrick JE, Breaker RR. 2007. The distributions, mechanisms, and structures of metabolite-binding riboswitches. *Genome Biol* **8:** R239.

Benner SA, Ellington AD, Tauer A. 1989. Modern metabolism as a palimpsest of the RNA world. *Proc Natl Acad Sci USA* **86:** 7054–7058.

Blount KF, Breaker RR. 2006. Riboswitches as antibacterial drug targets. *Nat Biotechnol* **24:** 1558–1564.

Bocobza S, Adato A, Mandel T, Shapira M, Nudler E, Aharoni A. 2007. Riboswitch-dependent gene regulation and its evolution in the plant kingdom. *Genes Dev* **21:** 2874–2879.

Borsuk P, Przykorska A, Blachnio K, Koper M, Pawlowicz JM, Pekala M, Weglenski P. 2007. L-arginine influences the structure and function of arginase mRNA in *Aspergillus nidulans*. *Biol Chem* **388:** 135–144.

Breaker RR. 2002. Engineered allosteric ribozymes as biosensor components. *Curr Opin Biotechnol* **13:** 31–39.

Breaker RR. 2004. Natural and engineered nucleic acids as tools to explore biology. *Nature* **432:** 838–845.

Breaker RR. 2006. Riboswitches and the RNA world. In: *The RNA world*, 3rd ed., Gesteland R.F., Cech T.R., Atkins J.F., eds., Cold Spring Harbor Laboratory Press, pp. 89–107.

Breaker RR. 2009. Riboswitches: from ancient gene-control systems to modern drug targets. *Future Microbiol* **4:** 771–773.

Brooks KM, Hampel KJ. 2009. A rate-limiting conformational step in the catalytic pathway of the glmS ribozyme. *Biochemistry* **48:** 5669–5678.

Cheah MT, Wachter A, Sudarsan N, Breaker RR. 2007. Control of alternative RNA splicing and gene expression by eukaryotic riboswitches. *Nature* **447:** 497–500.

Cochrane JC, Lipchock SV, Smith KD, Strobel SA. 2009. Structural and chemical basis for glucosamine-6-phosphate binding and activation of the glmS ribozyme. *Biochemistry* **48:** 3239–3246.

Collins JA, Irnov I, Baker S, Winkler WC. 2007. Mechanism of mRNA destabilization by the glmS ribozyme. *Genes Dev* **21:** 3356–3368.

Coppins RL, Hall KB, Groisman EA. 2007. The intricate world of riboswitches. *Curr Opin Microbiol* **10:** 176–181.

Corbino KA, Barrick JE, Lim J, Welz R, Tucker BJ, Puskarz I, Mandal M, Rudnick ND, Breaker RR. 2005. Evidence for a second class of S-adenosylmethionine riboswitches and other regulatory RNA motifs in α-proteobacteria. *Genome Biol* **6:** R70.

Croft MT, Moulin M, Webb ME, Smith AG. 2007. Thiamine biosynthesis in algae is regulated by riboswitches. *Proc Natl Acad Sci* **104:** 20770–20775.

Cromie MJ, Shi Y, Latifi T, Groisman EA. 2006. An RNA sensor for intracellular Mg^{2+}. *Cell* **125:** 71–84.

Dann CE 3rd, Wakeman CA, Sieling CL, Baker SC, Ironv I, Winkler WC. 2007. Structure and mechanism of a metal-sensing regulatory RNA. *Cell* **130:** 878–892.

Edwards AL, Batey RT. 2009. A structural basis for the recognition of 2′-deoxyguanosine by the purine riboswitch. *J Mol Biol* **385:** 938–948.

Fuchs RT, Grundy FJ, Henkin TM. 2006. The S_{MK} box is a new SAM-binding RNA for translational regulation of SAM synthase. *Nat Struct Mol Biol* **13:** 226–233.

Garst AD, Edwards AL, Batey RT. 2010. Riboswitches: structures and mechanisms. *Cold Spring Harb Perspect Biol* doi: 10.1101.cshperspect.a003533.

Gelfand MS, Mironov AA, Jomantas J, Kozlov YI, Perumov DA. 1999. A conserved RNA structure element involved in the regulation of bacterial riboflavin synthesis genes. *Trends Genet* **15:** 439–442.

Gilbert SD, Rambo RP, Van Tyne D, Batey RT. 2008. Structure of the SAM-II riboswitch bound to S-adenosylmethionine. *Nat Struct Mol Biol* **15:** 177–182.

Gilbert SD, Stoddard CD, Wise SJ, Batey RT. 2006. Thermodynamic and kinetic characterization of ligand binding to the purine riboswitch aptamer domain. *J Mol Biol* **359:** 754–768.

Grundy FJ, Henkin TM. 1998. The S box regulon: A new global transcription termination control system for methionine and cysteine biosynthesis genes in gram-positive bacteria. *Mol Microbiol* **30:** 737–749.

Gusarov I, Nudler E. 1999. The mechanism of intrinsic transcription termination. *Mol Cell* **3:** 495–504.

Gutiérrez-Preciado A, Henkin TM, Grundy FJ, Yanofsky C, Merino E. 2009. Biochemical features and functional implications of the RNA-based T-box regulatory system. *Microbiol Mol Biol Rev* **73:** 36–61.

Halls C, Mohr S, Del Campo M, Yang Q, Jankowsky E, Lambowitz AM. 2007. Involvement of DEAD-box proteins in group I and group II intron splicing. Biochemical characterization of Mss116p, ATP hydrolysis-dependent and -independent mechanisms, and general RNA chaperone activity. *J Mol Biol* **365:** 835–855.

Hougland JL, Piccirilli JA, Forconi M, Lee J, Herschlag D. 2006. How the group I intron works: a case study of RNA structure and function. In: *The RNA world*, 3rd ed., Gesteland R.F., Cech T.R., Atkins J.F., eds., Cold Spring Harbor Laboratory Press, pp. 133–205.

Johansson J. 2009. RNA thermosensors in bacterial pathogens. *Contrib Microbiol* **16:** 150–160.

Jose A, Soukup GA, Breaker RR. 2001. Cooperative binding of effectors by an allosteric ribozyme. *Nucleic Acids Res* **29:** 1631–1637.

Kim JN, Breaker RR. 2008. Purine sensing by riboswitches. *Biol Cell* **100:** 1–11.

Kim JN, Roth A, Breaker RR. 2007. Guanine riboswitch variants from *Mesoplasma florum* selectively recognize 2′-deoxyguanosine. *Proc Natl Acad Sci* **104:** 16092–16097.

Klinkert B, Narberhaus F. 2009. Microbial thermosensors. *Cell Mol Life Sci* **66:** 2661–2676.

Kubodera T, et al. 2003. Thiamine-regulated gene expression of *Aspergillus oryzae thiA* requires splicing of the intron containing a riboswitch-like domain in the 5′ UTR. *FEBS Lett* **555:** 516–520.

Lee ER, Blount KF, Breaker RR. 2009. Roseoflavin is a natural antibacterial compound that binds to FMN riboswitches and regulates gene expression. *RNA Biol* **6:** 187–194.

Lu C, Smith AM, Fuchs RT, Ding F, Rajashankar K, Henkin TM, Ke A. 2008. Crystal structures of the SAM-III/S_{MK} riboswitch reveal the SAM-dependent translation inhibition mechanism. *Nat Struct Mol Biol* **15:** 1076–1083.

Mandal M, Breaker RR. 2004a. Gene regulation by riboswitches. *Nat Rev Mol Cell Biol* **5:** 451–463.

Mandal M, Breaker RR. 2004b. Adenine riboswitches and gene activation by disruption of a transcription terminator. *Nat Struct Mol Biol* **11:** 29–35.

Mandal M, Lee M, Barrick JE, Weinberg Z, Emilsson GM, Ruzzo WL, Breaker RR. 2004. A glycine-dependent riboswitch that uses cooperative binding to control gene expression. *Science* **306:** 275–279.

Meyer MM, Roth A, Chervin SM, Garcia GA, Breaker RR. 2008. Confirmation of a second natural preQ$_1$ aptamer class in Streptococcaceae bacteria. *RNA* **14:** 685–695.

Mironov AS, Gusarov I, Rafikov R, Lopez LE, Shatalin K, Kreneva RA, Perumov DA, Nudler E. 2002. Sensing small molecules by nascent RNA: a mechanism to control transcription in bacteria. *Cell* **111:** 747–756.

Montange RK, Batey RT. 2006. Structure of the S-adenosylmethionine riboswitch regulatory mRNA element. *Nature* **441:** 1172–1175.

Nahvi A, Barrick JE, Breaker RR. 2004. Coenzyme B$_{12}$ riboswitches are widespread genetic control elements in prokaryotes. *Nucleic Acids Res* **32:** 143–150.

Nahvi A, Sudarsan N, Ebert MS, Zou X, Brown KL, Breaker RR. 2002. Genetic control by a metabolite binding mRNA. *Chem Biol* **9:** 1043–1049.

Poiata E, Meyer MM, Ames TD, Breaker RR. 2009. A variant riboswitch aptamer class for S-adenosylmethionine common in marine bacteria. *RNA* **15:** 2046–2056.

Rodionov DA, Vitreschak AG, Mironov AA, Gelfand MS. 2004. Comparative genomics of the methionine metabolism in Gram-positive bacteria: a variety of regulatory systems. *Nucleic Acids Res* **32**: 3340–3353.

Roth A, Breaker RR. 2009. The structural and functional diversity of metabolite-binding riboswitches. *Annu Rev Biochem* **78**: 305–334.

Roth A, Winkler WC, Regulski EE, Lee BW, Lim J, Jona I, Barrick JE, Ritwick A, Kim JN, Welz R, et al. 2007. A riboswitch selective for the queuosine precursor preQ$_1$ contains an unusually small aptamer domain. *Nat Struct Mol Biol* **14**: 308–317.

Sharin E, Schein A, Mann H, Ben-Asouli Y, Jarrous N. 2005. RNase P: Role of distinct protein cofactors in tRNA substrate recognition and RNA-based catalysis. *Nucleic Acids Res* **33**: 5120–5132.

Silverman SK. 2003. Rube Goldberg goes (ribo)nuclear? Molecular switches and sensors made from RNA. *RNA* **4**: 377–383.

Skordalakes E, Berger JM. 2003. Structure of the Rho transcription terminator: Mechanism of mRNA recognition and helicase loading. *Cell* **114**: 135–146.

Sudarsan N, Barrick JE, Breaker RR. 2003. Metabolite-binding RNA domains are present in the genes of eukaryotes. *RNA* **9**: 644–647.

Sudarsan N, Hammond MC, Block KF, Welz R, Barrick JE, Roth A, Breaker RR. 2006. Tandem riboswitch architectures exhibit complex gene control functions. *Science* **314**: 300–304.

Sudarsan N, Lee ER, Weinberg Z, Moy RH, Kim JN, Link KH, Breaker RR. 2008. Riboswitches in eubacteria sense the second messenger cyclic di-GMP. *Science* **321**: 411–413.

Suess B, Weigand JE. 2008. Engineered riboswitches: overview, problems and trends. *RNA Biol* **5**: 24–29.

Serganov A, Yuan YR, Pikovskaya O, Polonskaia A, Malinina L, Phan AT, Hobartner C, Micura R, Breaker RR, Patel DJ. 2004. Structural basis for discriminative regulation of gene expression by adenine- and guanine-sensing mRNAs. *Chem Biol* **11**: 1729–1741.

Thompson KM, Syrett HA, Knudsen SM, Ellington AD. 2002. Group I aptazymes as genetic regulatory switches. *BMC Biotechnol* **2**: 21.

Wachter A, Tunc-Ozdemir M, Grove BC, Green PJ, Shintani DK, Breaker RR. 2007. Riboswitch control of gene expression in plants by splicing and alternative 3′ processing of mRNAs. *Plant Cell* **19**: 3437–3450.

Wang JX, Breaker RR. 2008. Riboswitches that sense S-adenosylmethionine and S-adenosylhomocysteine. *Biochem Cell Biol* **86**: 157–168.

Wang JX, Lee ER, Morales DR, Lim J, Breaker RR. 2008. Riboswitches that sense S-adenosylhomocysteine and activate genes involved in coenzyme recycling. *Mol Cell* **28**: 691–702.

Weinberg Z, Regulski EE, Hammond MC, Barrick JE, Yao Z, Ruzzo WL, Breaker RR. 2008. The aptamer core of SAM-IV riboswitches mimics the ligand-binding site of SAM-I riboswitches. *RNA* **14**: 822–828.

Weinberg Z, Wang JX, Bogue J, Yang J, Corbino K, Moy R, Breaker RR. 2010. Comparative genomics reveals 104 candidate structured RNAs from bacteria, archaeal and their metagenomes. *Genome Biol* (submitted).

Welz R, Breaker RR. 2007. Ligand binding and gene control characteristics of tandem riboswitches in *Bacillus anthracis*. *RNA* **13**: 573–582.

White HB, III. 1976. Coenzymes as fossils of an earlier metabolic state. *J Mol Evol* **7**: 101–104.

Wickiser JK, Cheah MT, Breaker RR, Crothers DM. 2005a. The kinetics of ligand binding by an adenine-sensing riboswitch. *Biochemistry* **44**: 13404–13414.

Wickiser JK, Winkler WC, Breaker RR, Crothers DM. 2005b. The speed of RNA transcription and metabolite binding kinetics operate an FMN riboswitch. *Mol Cell* **18**: 49–60.

Winkler WC, Cohen-Chalamish S, Breaker RR. 2002a. An mRNA structure that controls gene expression by binding FMN. *Proc Natl Acad Sci* **99**: 15908–15913.

Winkler W, Nahvi A, Breaker RR. 2002b. Thiamine derivatives bind messenger RNAs directly to regulate bacterial gene expression. *Nature* **419**: 952–956.

Winkler WC, Nahvi A, Roth A, Collins JA, Breaker RR. 2004. Control of gene expression by a natural metabolite-responsive ribozyme. *Nature* **428**: 281–286.

Vitreschak AG, Rodionov DA, Mironov AA, Gelfand MS. 2004. Riboswitches: The oldest mechanism for the regulation of gene expression? *Trends Genet* **20**: 44–50.

Yarnell WS, Roberts JW. 1999. Mechanism of intrinsic transcription termination and antitermination. *Science* **284**: 611–615.

Riboswitches: Structures and Mechanisms

Andrew D. Garst, Andrea L. Edwards, and Robert T. Batey

Department of Chemistry and Biochemistry, University of Colorado at Boulder, Boulder, Colorado 80309-0215

Correspondence: robert.batey@colorado.edu

SUMMARY

A critical feature of the hypothesized RNA world would have been the ability to control chemical processes in response to environmental cues. Riboswitches present themselves as viable candidates for a sophisticated mechanism of regulatory control in RNA-based life. These regulatory elements in the modern world are most commonly found in the 5′-untranslated regions of bacterial mRNAs, directly interacting with metabolites as a means of regulating expression of the coding region via a secondary structural switch. In this review, we focus on recent insights into how these RNAs fold into complex architectures capable of both recognizing a specific small molecule compound and exerting regulatory control over downstream sequences, with an emphasis on transcriptional regulation.

Outline

1 Introduction
2 Constructing an RNA receptor: The basic architectural building blocks
3 Folding pathways of the purine riboswitch aptamer
4 Recognition of effector molecules by riboswitch receptors
5 Ligand binding induces conformational changes in the receptor
6 Temporal effects of transcription
7 Models for structural switching
8 Riboswitches in the RNA world
References

1 INTRODUCTION

Life in an RNA world would have relied on RNA as both a medium for heritable genetic information and chemical catalysis. In addition to these functions, life would have had to react to changing environmental conditions—that is, be capable of regulating biological functions. Insights into how RNA can accomplish this crucial task have been revealed through recent discoveries that this molecule accomplishes a wide variety of regulatory functions in modern biology. One of the most striking recent examples of how RNA regulates gene expression was revealed by the discovery of *riboswitches*, a common means of genetic regulation at the mRNA level in the bacterial kingdom (Barrick and Breaker 2007).

Riboswitches are elements commonly found in the 5′-untranslated region (UTR) of mRNAs that exert their regulatory control over the transcript in a *cis*-fashion by directly binding a small molecule ligand (McDaniel et al. 2003; Mironov et al. 2002; Nahvi et al. 2002; Winkler et al. 2002a; Winkler et al. 2002b). The typical riboswitch contains two distinct functional domains (Fig. 1A). The effector molecule is recognized by an *aptamer domain*, which adopts a compact three-dimensional fold to scaffold the ligand binding pocket (Winkler and Breaker 2003). As with proteins, these RNA receptors must discriminate between chemically related metabolites with high selectivity to elicit the appropriate regulatory response. A second domain, the *expression platform*, contains a secondary structural switch that interfaces with the transcriptional or translational machinery. Regulation is achieved by virtue of a region of overlap between these two domains, known as the *switching sequence*, whose pairing directs folding of the RNA into one of two mutually exclusive structures in the expression platform that represent the on and off states of the mRNA (Fig. 1B).

The above model of the riboswitch immediately suggests two fundamental questions regarding their regulatory mechanism. First, how does RNA create a binding pocket that achieves both high affinity and specificity for the effector molecule? This requires knowledge of the atomic-level structure of the RNA-ligand complex. The second pertinent question is: How is effector binding communicated to the expression platform to yield a regulatory outcome? This requires knowledge both of potential ligand-induced conformational changes in the aptamer and the cotranscriptional folding pathway of the mRNA. Given that transcriptional attenuation is the predominant mode of regulation for riboswitches (Barrick and Breaker 2007), they must rapidly adopt the requisite conformations to act before escape of polymerase beyond its 3′-boundary. Thus, the above questions must be considered in the context of RNA folding. although most of this article will focus on the mechanism of transcriptional regulation, it has been suggested that tight coupling of transcription and translation in bacteria may impart the same temporal constraints on translation (Link and Breaker 2009).

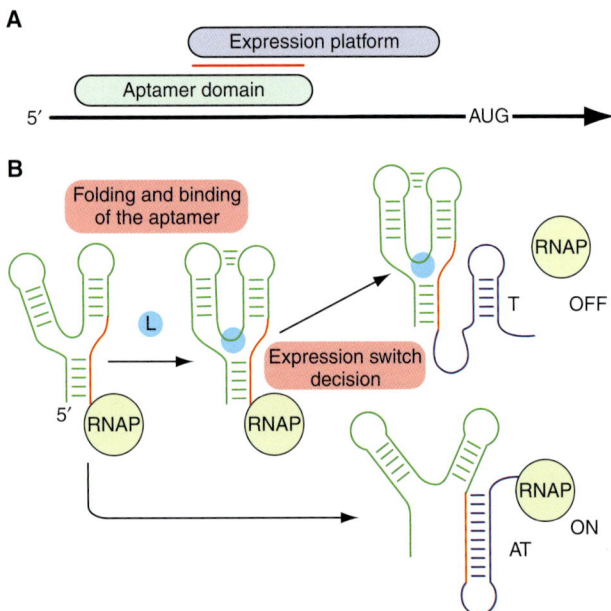

Figure 1. Organization and mechanism of the typical riboswitch. (*A*) Arrangement of riboswitch structural elements in primary sequence of the mRNA transcript. The aptamer domain (green) and the expression platform (purple) overlap through a sequence that can base pair with either domain (red bar). (*B*) During transcription of the riboswitch, several events occur to elicit the appropriate regulatory outcome. Early events during transcription include the folding of the aptamer domain and potential binding of the effector (L, cyan). Depending on whether effector is bound, the RNA adopts one of two potential folds in the expression platform (an antiterminator, AT, or rho-independent terminator, T) that determines the regulatory response.

2 CONSTRUCTING AN RNA RECEPTOR: THE BASIC ARCHITECTURAL BUILDING BLOCKS

Sequence and structural analyses of over ten riboswitch aptamer domains in complex with their effectors (Table 1) have revealed these RNAs are organized using many of the same architectural principles observed in other large RNAs (Hendrix et al. 2005; Leontis et al. 2006). The sequence of the aptamer domain of the purine riboswitch, one of the simplest aptamers, contains three conserved helical elements (P1, P2, and P3) (Fig. 2A) that are present in all ∼500 identified sequences (Mandal et al. 2003). These helices are organized into higher order structures by two principles (Batey et al. 1999; Holbrook, 2008). First, they form sets of coaxial stacks in which two or more individual

Table 1. Structures of riboswitches

Riboswitch	PDB	References
Purine		
Guanine	1U8D, 1Y27	(Batey et al. 2004; Serganov et al. 2004)
Adenine	1Y26	(Serganov et al. 2004)
2′-deoxyguanosine	3DS7	(Edwards and Batey 2009)
S-adenosylmethionine (SAM)		
SAM-I	2GIS	(Montange and Batey 2006)
SAM-II	2QWY	(Gilbert et al. 2008)
SAM-III	3E5C	(Lu et al. 2008)
Thiamine pyrophosphate (TPP)		
Bacterial (*Escherichia coli*)	2GDI, 2HOJ	(Edwards and Ferre-D'Amare 2006; Serganov et al. 2006)
Eukaryotic (*A. thaliana*)	2CKY	(Thore et al. 2006)
Lysine	3DOU, 3DIL	(Garst et al. 2008; Serganov et al. 2008)
Flavin mononucleotide (FMN)	3F2Q	(Serganov et al. 2009)
Magnesium	2QBZ	(Dann et al. 2007)
glmS (ribozyme)	2NZ4, 2Z75	(Cochrane et al. 2007; Klein and Ferré-D'Amaré 2006)
Pre-Q_1	3FU2, 2KFC, 3GCA	(Kang et al. 2009; Klein et al. 2009; Spitale et al. 2009)
Cyclic-di-GMP	3IRW, 3IWN	(Kulshina et al. 2009; Smith et al. 2009)

helices are arranged in a single pseudocontinuous helix, exemplified by the P1 and P3 stack (Fig. 2B) (Batey et al. 2004; Serganov et al. 2004). Second, helices and helical stacks tend to be arranged in a parallel configuration (Fig. 2B). This architectural behavior is manifest in almost all RNAs containing three or more helices (Holbrook, 2008). A notable exception to this is tRNA, in which the two coaxial stacks are arranged perpendicularly via the interaction of the D- and T-loops.

RNA helices are further organized through sequence motifs that adopt a variety of secondary structural elements: terminal loops, internal bulges and loops, and multi-helix junctions (Batey et al. 1999; Hendrix et al. 2005). One of the most common RNA motifs is the GA_3 tetraloop, a terminal loop whose first nucleotide is a guanosine followed by three adenosine residues (Varani 1995). This tetraloop presents a stack of the adenosines for recognition by other RNA motifs, such as the tetraloop receptor (Cate et al. 1996; Jaeger et al. 1994; Murphy and Cech 1994) (Fig. 3A), which was recently identified in a subclass of the di-cyclic guanosine monophosphate riboswitches (Sudarsan et al. 2008). Interaction between the loop and the receptor serves to anchor the two helices together in a side-by-side arrangement, facilitating parallel packing. Variations of this terminal loop-internal loop theme are observed in the structures of the lysine (Fig. 3B) and thiamine pyrophosphate (TPP) riboswitches to achieve similar structural outcomes (Garst et al. 2008; Serganov et al. 2008; Serganov et al. 2006; Thore et al. 2006). Other common RNA building blocks that have been identified in riboswitches include kink-turns (K-turns) (Blouin and Lafontaine 2007; Winkler et al. 2001), kissing-loop interactions (Blouin and Lafontaine 2007), sarcin-ricin loops (Grundy et al. 2003; Sudarsan et al. 2003), T-loops (Barrick and Breaker 2007), and pseudoknots (McDaniel et al. 2005). Multi-helix junctions also represent an important component of many riboswitches but their sequences and structures are diverse, reflecting unique requirements for positioning the flanking helices or hosting sites of activity (Montange and Batey 2008).

The lysine riboswitch, one of the largest of the known riboswitches, illustrates how these architectural themes are used to establish a complex global fold. The aptamer domain of the *Thermotoga maritima* lysine riboswitch is organized as a bundle of three coaxially stacked sets of helices: P1/P2a, P2b/P3, and P4/P5 (Garst et al. 2008; Serganov et al. 2008) (Fig. 3C). The helical stacks are arranged in a parallel fashion by variants of the kink-turn, kissing-loop, tetraloop-tetraloop receptor and sarcin-ricin loop motifs (Fig. 3D). The P2 helix contains two internal loops: a canonical sarcin-ricin loop in the center of P2 and an unconventional kink-turn motif. The sarcin-ricin loop serves to help establish a receptor for the docking of the terminal loop L4, which forms a tetraloop-like structure (Fig. 3B), thereby arranging P4 parallel to P2. The noncanonical kink turn near the terminus of P2 bends the helix to direct the terminal loop L2 toward L3, which creates a kissing-loop interaction. This brings the P3 helix in parallel to P2 and P4, establishing the overall structure.

The importance of these architectural motifs for function is highlighted by studies in which altering these sequences renders a riboswitch nonfunctional under physiological conditions. Deletion of the L2–L3 interaction in the purine riboswitch results in a ~1000-fold decrease in affinity for guanine (Gilbert et al. 2007; Lemay et al. 2006; Stoddard et al. 2008). Siimilarly, mutation of the kink-turn or kissing-loop interaction in the *B. subtilis lysC* lysine riboswitch debilitates ligand-dependent transcription termination in vitro (Blouin and Lafontaine 2007). Studies of the SAM-I riboswitch show similar results (Heppell and Lafontaine 2008; McDaniel et al. 2005; Winkler et al. 2003).

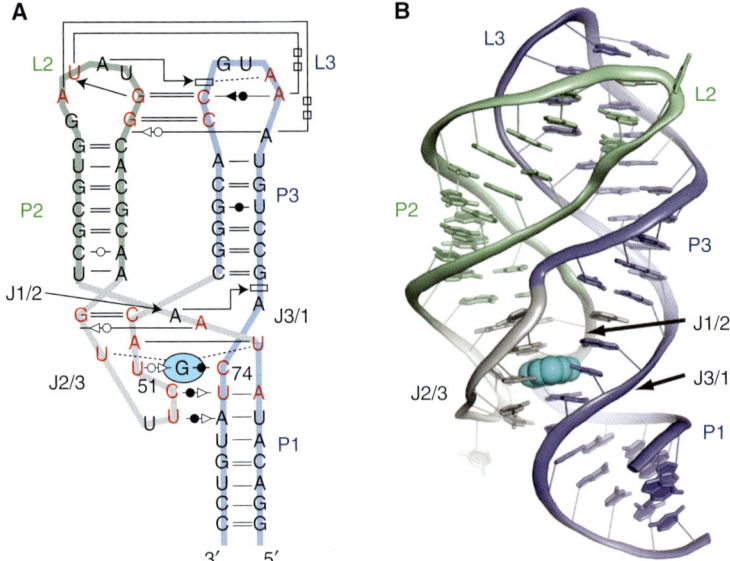

Figure 2. Global architectures of the purine riboswitch. (A) The secondary structure of the aptamer domain of the *Bacillus subtilis xpt-pbuX* guanine riboswitch, drawn to reflect its tertiary architecture. Base interactions are depicted using the nomenclature of Leontis and Westhof (Leontis and Westhof 2001). Nucleotides that are >95% conserved in phylogeny are highlighted in red. (B) Crystal structure of the guanine riboswitch showing the organization of paired (P), loop (L), and joining (J) regions. The ligand is shown in cyan.

3 FOLDING PATHWAYS OF THE PURINE RIBOSWITCH APTAMER

Crystal structures of riboswitch aptamer domains are static pictures that do not lend insight into the process by which these RNAs acquire their three-dimensional structures. For riboswitches to function during transcription, they must rapidly form the tertiary interactions described earlier and avoid kinetic folding traps commonly found in RNA (Pan and Sosnick 1997; Treiber and Williamson 1999). In other words, slow folding or misfolding will impede their ability to elicit a proper regulatory response during the short

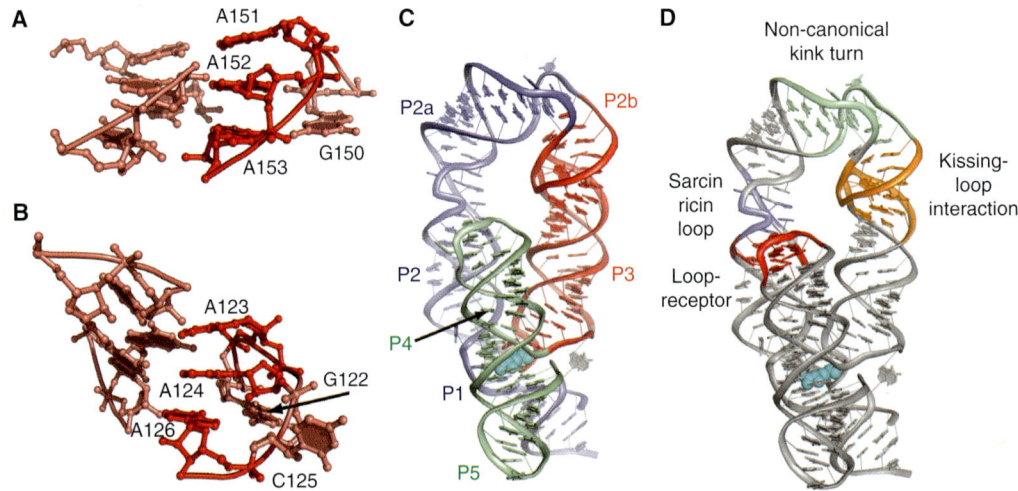

Figure 3. Organization of the lysine riboswitch. (A) The GAAA tetraloop-tetraloop receptor motif found across phylogeny in diverse RNAs, emphasizing the docking of the three adenosine residues into the minor groove of the receptor. (B) Mimicry of the tetraloop-tetraloop receptor interaction by the conserved pentaloop-helix interaction found in the lysine riboswitch. (C) The lysine riboswitch structure emphasizing the three-helix bundle formed by the P1/P2a, P2b/P3, and P4/P4 coaxial stacks (blue, red, and green, respectively). Lysine is shown in cyan. (D) Four recurrent structural motifs found in the lysine riboswitch that serve to organize the three-helix bundle.

temporal window of transcription of the 5′-UTR. This folding process can be monitored by a range of methods including classical approaches that use temperature and chemical denaturants to measure relative stability of structural elements, and more recently developed single molecule techniques that monitor folding transitions in real time. Because of its simple architecture, the purine riboswitch has emerged as the dominant model system for understanding aptamer domain folding.

A technique that uses the reactivity of 2′-hydroxyl groups to N-methylisatoic anhydride (NMIA) as a probe of RNA folding (Wilkinson et al. 2006) was recently used to study folding of the B. subtilis xpt-pbuX guanine riboswitch (Stoddard et al. 2008). An advantageous feature of this chemical probing method—commonly called "SHAPE"—is that it can be performed over a broad temperature range (Merino et al. 2005). This allows for determination of the melting temperature (T_m, defined as the temperature at which the RNA is half folded and half unfolded) of every nucleotide in the RNA of interest (Merino et al. 2005). Applied to the purine aptamer, at high temperatures (>70°C), only the helical elements of the aptamer are structured, as expected from the large free energy associated with RNA secondary structure formation (Fig. 4A) (Brion and Westhof 1997). As the temperature drops below 60°C, the L2–L3 interaction forms along with nucleotides becoming protected in J3/1, indicating that the peripheral loop–loop interaction promotes partial preorganization of the three-way junction via coaxial stacking of P1 and P3. Notably, J1/2 and J2/3 remain conformationally flexible in the absence of ligand, which would allow the effector access to the core of the RNA (see next section).

The previous analysis using SHAPE chemistry agrees well with single molecule fluorescence resonance energy transfer (smFRET) experiments with the B. subtilis pbuE adenine riboswitch (Lemay et al. 2006). In this approach, folding transitions are reported by changes in the FRET value of single molecules in which L2 and L3 were labeled with donor and acceptor fluorophores, respectively (Fig. 4A). Importantly, the smFRET data provide additional insight into the rates of folding and the influences of magnesium (Mg^{++}) that is not possible using ensemble techniques. Addition of Mg^{++} above 50 μM was shown to significantly increase the folding rate of the loop-loop

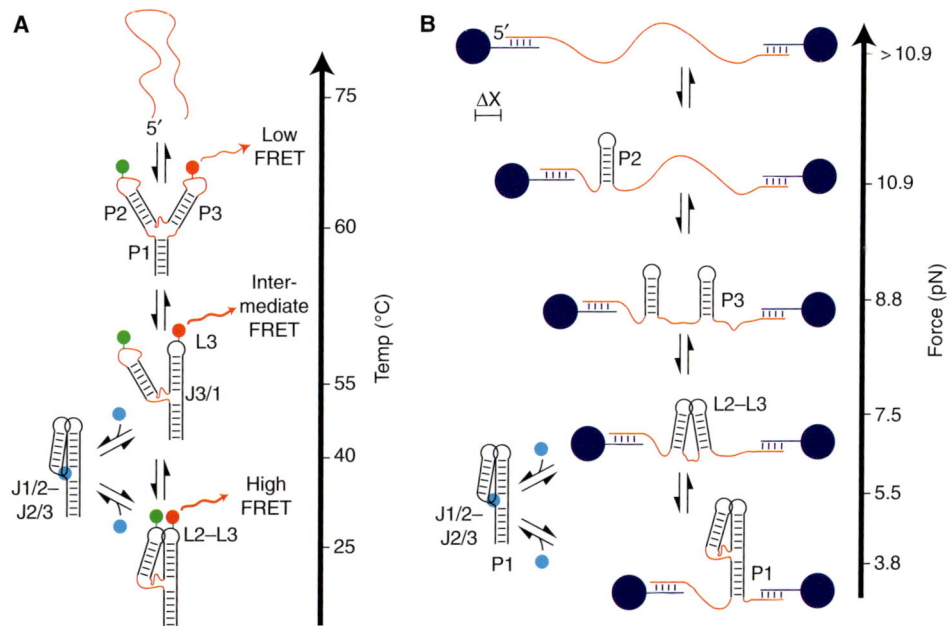

Figure 4. Folding of the purine riboswitch aptamer domain. (A) Thermal denaturation and smFRET of the purine riboswitch reveal similar heirarchical folding landscapes. The T_m of each structural element (denoted by red strands) of the xpt-pbuX purine riboswitch reveals folding intermediates similar to those observed in smFRET experiments of the pbuE riboswitch at ambient temperature. smFRET experiments consisted of a fluorescein donor (green) and Cy3 acceptor (red) placed at L2 and L3 respectively allow observation of the loop–loop interaction in real time. (B) Unfolding the pbuE riboswitch by force measured as changes in distance (ΔX). This experiment reveals a pathway in which individual helices form sequentially from 5′ to 3′. Although denaturation studies show that ligand (cyan sphere) binding stabilizes the three way junction by ~15°C, force spectroscopy supports a model in which P1 is also stabilized by ligand binding.

interaction; at physiologically relevant Mg^{++} concentrations, the global fold of the aptamer is adopted within ~1 s (Lemay et al. 2006). Considering that bacterial polymerases have elongation rates of around ~45 nucleotides per second at 25°C (Ryals et al. 1982), transcription of this 70-nucleotide aptamer would require ~1.5 s. Thus, folding and transcription occur on similar timescales, as would be expected for a cotranscriptional process.

Another method for monitoring folding processes uses force, which has gained popularity because of technological advances that allow higher spatial and temporal resolution (King et al. 2009; Li et al. 2008). Application of constant force to the ends of a macromolecule alters the energy landscape such that specific structural elements become bistable (equally occupying the folded and unfolded states) at a specific applied force, known as a $F_{1/2}$ (King et al. 2009; Li et al. 2008). Like smFRET, this technique simultaneously measures thermodynamic and kinetic parameters of folding and unfolding. Observations of *Vibrio vulnificus add* adenine riboswitch folding by force spectroscopy revealed that as the applied force was decreased, extension changes corresponding to formation of P2 occurred, followed by P3, the loop-loop interaction, and finally P1 (Fig. 4B) (Greenleaf et al. 2008). By fully extending the RNA and allowing it to refold for predetermined time intervals before reapplying force, the authors obtained estimates for the rate of aptamer domain folding similar to those derived from smFRET measurements.

The folding model obtained by force spectroscopy is distinct from that obtained using other techniques (Greenleaf et al. 2008). Although chemical probing and smFRET propose a classic hierarchical folding pathway in which all secondary structure is acquired before tertiary structure, the folding pathway proposed using force spectroscopy proceeds from the 5′ to 3′ direction. Such a model is more consistent with expectations for folding during transcription. Nevertheless, both models illustrate the importance of tertiary structure formation before ligand binding, as this has cooperative influences on ligand binding by preorganizing the binding site.

4 RECOGNITION OF EFFECTOR MOLECULES BY RIBOSWITCH RECEPTORS

Following proper folding, the aptamer domain must accomplish two interconnected tasks: specific recognition of the effector and coupling binding to a regulatory outcome. The first objective requires that the RNA effectively discriminates between closely related compounds in the cell. For example, the purine riboswitches achieve over a 10,000-fold level of discrimination between guanine and adenine (Mandal et al. 2003; Mandal and Breaker 2004), whereas the lysine riboswitch discriminates between lysine and ornithine, amino acids that differ by a single methylene group in their side chain, by at least 5000-fold (Garst and Batey, unpublished data).

The purine riboswitch family is divided into three classes that are defined by their effector molecules: guanine/hypoxanthine, adenine, or 2′-deoxyguanosine (Kim and Breaker 2008). This family of aptamers is defined by a common conserved secondary and tertiary structure. In each class, the effector is bound using a pocket formed by the RNA three-way junction (Kim et al. 2007; Mandal et al. 2003; Mandal and Breaker 2004). Within the junction are five base triples between nucleotides that define the local architecture of the binding pocket (Fig. 2A). At the center is a base triple involving the ligand and two pyrimidine residues that serve to specifically recognize the nucleobase (Fig. 5A) (Batey et al. 2004; Noeske et al. 2005; Serganov et al. 2004). Every functional group in the ligand is directly

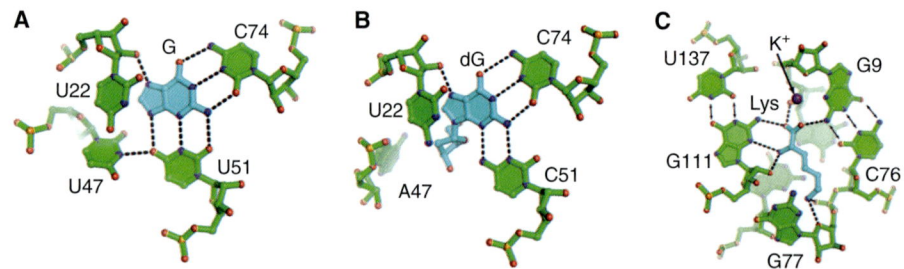

Figure 5. Details of ligand-riboswitch interactions. (A) Recognition of guanine (cyan) by a pocket of conserved pyrimidine residues (green). The ligand interacts with the RNA primarily through hydrogen bonding interactions (black dashed lines) between pyrimidine 51 and 74. The binding pocket for adenine-binding riboswitches is nearly identical except that the ligand interacts with a uridine residue at position 74. (B) 2′-deoxyguanosine (cyan) is recognized through interactions with two cytosine residues (green). Note the shift of C51 relative to U51 to accommodate the 2′-deoxyribose moiety (C) Lysine (cyan) recognition by nucleotides in the the lysine riboswitch binding pocket (green). A potassium ion (purple sphere) partially mediates the interaction between the carboxylate group and the RNA.

recognized through hydrogen bonding interactions, in part explaining the ability of the RNA to achieve a high specificity for this compound. Surprisingly, the nucleobase does not completely stack on the bases above or below it, a feature common to other riboswitches that recognize nucleobase-containing effectors (Gilbert et al. 2008; Montange and Batey, 2006; Serganov et al. 2009; Serganov et al. 2006; Thore et al. 2006).

Purine specificity is achieved through the identity of two pyrimidine residues that participate in a base triple with the ligand. Nucleotide 74 forms a Watson-Crick pair with the ligand, allowing the RNA to discriminate between adenine and guanine/2′-deoxyguanosine (Batey et al. 2004; Mandal et al. 2003; Mandal and Breaker 2004; Serganov et al. 2004). This position is always uridine in adenine riboswitches and always cytosine in guanine/2′-deoxyguanosine riboswitches (Fig. 5A). Similarly, the Watson-Crick face of nucleotide 51 is used to discriminate between nucleobases (adenine and guanine) and nucleosides (Edwards and Batey 2009). Residue 51 is always uridine in guanine and adenine riboswitches and cytosine in 2′-deoxyguanosine riboswitches (Fig. 5B). This cytosine shifts toward nucleotide 74 such that it opens up the binding pocket to allow for the presence of the 2′-deoxyribose sugar moiety (Edwards and Batey, 2009). Thus, small sequence changes to the aptamer can generate riboswitches that respond to chemically distinct effectors.

Like the purine riboswitch, the lysine riboswitch fully encapsulates its ligand within a binding pocket that directly interacts with all of the effector's polar functional groups (Garst et al. 2008; Serganov et al. 2008). The main chain atoms of lysine are recognized through minor groove interactions with invariant G-C and G•U pairs, and by a potassium ion bound near the carboxyl group (Fig. 5C). The positively charged amino group of the side chain is recognized through an electrostatic interaction with a nonbridging phosphate oxygen positioned at the opposite end of the pocket. Discrimination between lysine and ornithine is likely caused by the inability of a shorter ornithine side chain to position its amino group adjacent to the nonbridging phosphate group. The ability of the effector to form this electrostatic interaction is also the basis for discrimination against other abundant amino acids such as serine and aspartate that could sterically fit in this pocket.

A distinct challenge faced by RNA is recognition of negatively charged moieties. Instead of directly contacting negative charges presented by ligands, aptamers tend to use metal cations as a bridge. For example, the phosphate groups of TPP and flavin mononucleotide (FMN) are coordinated by Mg^{++} ions to mediate RNA contacts (Serganov et al. 2009; Serganov et al. 2006; Thore et al. 2006). Similarly, the carboxylic acid moiety of lysine interacts with the RNA via a potassium cation (Fig. 5C) (Serganov et al. 2008). The SAM-I and SAM-II riboswitches represent an exception in which the carboxylic acid moiety interacts directly with the Watson-Crick face of a guanine and adenine residue, respectively (Gilbert et al. 2008; Montange and Batey 2006).

These structures highlight what appears to be a general feature of the majority of riboswitch aptamer-ligand interactions: The ligand is highly solvent inaccessible, buried within the core of the RNA. On average, 90% of the metabolite surface area is solvent inaccessible, which is greater than that of ligands bound to artificial aptamers (71%) (Edwards et al. 2007). At first glance, this might be attributed to the need to discriminate between closely related metabolites in the cell (Gilbert and Batey 2005). However, in vitro selected RNAs can easily achieve comparable specificities for their ligands without the need for extensive burial. The theophylline aptamer is capable of binding its cognate purine with an affinity that rivals the adenine riboswitch, capable of discriminating against caffeine by >10,000-fold (Jenison et al. 1994). Like many riboswitches, the theophylline aptamer undergoes a significant ligand-dependent conformational change (Zimmermann et al. 1997; Zimmermann et al. 2000), indicating that folding and a high degree of ligand burial are not necessarily correlated phenomena.

5 LIGAND BINDING INDUCES CONFORMATIONAL CHANGES IN THE RECEPTOR

A more likely reason for extensive ligand burial is related to interdomain communication in the riboswitch. Solvent inaccessibility of the effector molecule implies that the aptamer must adopt a flexible "open" state that allows the ligand entry into the pocket, followed by a conformational change that encapsulates the ligand. This phenomenon, referred to as an induced-fit binding mechanism, is a common feature of RNA binding reactions, such as in RNP assembly (Williamson 2000). Conformational changes in the riboswitch aptamer domain are used to couple ligand binding to folding of downstream structural switch in the expression platform (see next section) that in turn instructs the transcriptional or translational machinery.

The inherent flexibility of riboswitches in the absence of a ligand typically precludes crystallographic examination of these states. Instead, they are studied in solution using techniques such as chemical probing and NMR spectroscopy, which provide information on conformational dynamics at the global and local levels. NMR studies of both the *pbuE* and *xpt-pbuX* purine aptamers show that in the absence of ligand, nucleotides in the three-way junction are completely disordered even at ambient temperatures as indicated by the absence of peaks for corresponding nucleotides (Noeske et al. 2007a; Noeske et al. 2007b). This is consistent with the high degree of reactivity in these regions identified by in-line probing (Mandal et al. 2003) and SHAPE chemistry (Stoddard et al. 2008). Both NMR and chemical probing data indicate the J1/2 and J2/3 regions of the aptamer domain are disordered in the absence of the effector (Fig. 4A) (Noeske et al. 2007b; Ottink et al. 2007; Stoddard et al. 2008), suggesting that these elements form a flexible lid that closes around the docked ligand.

Addition of the appropriate purine nucleobase induces structure in the three-way junction, as evidenced both by the appearance of new peaks in NMR spectra (Noeske et al. 2005) and a $>15°C$ increase in the T_m of these nucleotides by NMIA probing (Fig. 4A) (Stoddard et al. 2008). Along with formation of intermolecular interactions between the ligand and RNA, these data reveal formation of RNA–RNA interactions in the junction. Importantly, a subset of these ligand-dependent interactions involves formation of two base triples between J2/3 and the 3′-side of the P1 helix (Stoddard et al. 2008). Formation of these interactions was proposed to stabilize incorporation of the switching sequence into the receptor domain, serving as the basis for the communication of binding to the signal transduction domain (Batey et al. 2004). This was further supported by force spectroscopy of the *pbuE* aptamer, in which an additional ~2.2 pN of force was required to make the P1 helix bistable in the presence of adenine, providing direct evidence for adenine induced stabilization of this helix (Fig. 4B) (Greenleaf et al. 2008).

Ligand-induced folding of the lysine riboswitch is similar to that of the purine riboswitches in many respects. Small angle X-ray scattering (SAXS), which provides a measure of the global conformation in solution (Lipfert and Doniach 2007; Putnam et al. 2007), demonstrates that the lysine aptamer is largely preorganized by Mg^{++}, whereas NMIA chemical probing data reveal that a subset of nucleotides in the five-way junction remains disordered under these conditions (Garst et al. 2008). In the presence of saturating lysine concentrations, these nucleotides become nonreactive to NMIA, indicating that lysine induces local order to the junction. Interestingly, this RNA crystallizes well in the absence of lysine, allowing the free state of this aptamer to be characterized (Garst et al. 2008; Serganov et al. 2008). In contrast to the solution studies, the crystal structures of the free and bound state were nearly identical. This suggests that in solution the aptamer exists as an ensemble of interconverting structures and that lysine recognizes a subpopulation of binding-competent RNAs. Although the aptamer can adopt a "bound-like" state in the absence of ligand, it requires lysine to fully stabilize the crystallographically observed structure.

Although the purine and lysine aptamers appear to undergo localized conformational changes on binding ligand, others show more dramatic structural organizations. For example, studies of the TPP riboswitch using site-specific labeling with the fluorescent reporter 2-aminopurine (2AP) revealed both the ligand binding pocket and a terminal loop-helix interaction fold with comparable rates on addition of ligand (Lang et al. 2007). However, a sevenfold disparity between the folding rate of the loop-helix interaction and the top of the P1 helix was measured, indicating that the free state of the TPP aptamer is largely disordered and ligand-induced organization of the RNA propagates from the periphery to the central junction. Similarly, fluorescence and NMR spectroscopic analyses of the preQ$_1$ (a biosynthetic precursor of the queuosine nucleoside found in the anticodon of many tRNAs) riboswitch revealed that the free state of this RNA folds into a hairpin structure with a highly disordered single stranded 3′-tail (Kang et al. 2009; Rieder et al. 2009). Addition of preQ$_1$ is required for association of the hairpin loop with the 3′-tail to complete the pseudoknot structure.

6 TEMPORAL EFFECTS OF TRANSCRIPTION

To this point, we have focused on studies of isolated riboswitch aptamers, which typically bind their effector ligands with nanomolar equilibrium dissociation constants (K_D). However, a number of studies have shown higher ligand concentrations are required to effectively regulate gene expression. The FMN, SAM-I, and lysine riboswitches require ligand concentrations $\sim100–1000$-fold above the K_D of the aptamer to achieve half maximal transcription termination (T_{50}) (Blouin and Lafontaine 2007; Tomsic et al. 2008; Wickiser et al. 2005b). This is a hallmark of a kinetically controlled process in which the aptamer domain has insufficient time to equilibrate with the cellular environment before the expression platform commits the RNA to an alternative folding route that may be largely irreversible. Studies of a number of riboswitches have found that transcripts encompassing the entire riboswitch are deficient in effector binding, strongly arguing for irreversible folding (Lemay et al. 2006; Mironov et al. 2002; Winkler et al. 2002a).

The time sensitivity of riboswitch regulation is explained in part by the slow kinetics of ligand binding, an inherent feature of induced fit processes. Measurement of the association kinetics of 2-aminopurine to the *xpt* and *pbuE* purine riboswitches yielded rate constants on the order of 10^5 $M^{-1}s^{-1}$ for both aptamers (Gilbert et al. 2006; Wickiser et al. 2005a). Similar values were obtained for the FMN and TPP riboswitches (Wickiser et al. 2005b). These rates suggest that at 1 μM intracellular ligand concentration, at least 10 s would be required for effector to fully saturate the aptamer. If transcription occurred at a constant rate for this duration, a bacterial polymerase would transcribe nearly 500–1000 nucleotides, well beyond the boundaries of even the largest riboswitches, making regulation by the riboswitch impossible.

One means by which riboswitches deal with this time limitation is by manipulating the rate of transcription using programmed pause sites embedded within the expression platform. Two such sites were identified by synchronized transcription assays of the FMN aptamer that have pause lifetimes of 1 and 10 s respectively (Wickiser et al. 2005b). Mutations that ablate pausing in these sites cause significant elevation of the T_{50} values in the context of the same aptamer, as expected when transcription proceeds through the riboswitch more rapidly. Uridine-rich tracts in the expression platform of the *pbuE* riboswitch have also been suggested to serve as transcriptional pause sites, though they remain to be experimentally validated (Wickiser et al. 2005a). Although the general importance of pausing remains to be addressed for the majority of aptamers, this phenomenon may provide an important mechanism for tuning the response range of a riboswitch to ligand concentrations relevant for the cell.

Pausing may also provide time at important points during transcription to allow structural rearrangements that guide more efficient folding of the downstream sequence. For example, RNAs such as RNase P, tmRNA, and SRP RNAs all form labile, nonnative structural intermediates at the point of transcriptional pauses (Wong et al. 2007). These intermediates sequester upstream portions of long-range helices in which the 5′ and 3′ sides are separated in primary sequence by more than 50 nucleotides. It has been proposed that these intermediates provide the RNA with a mechanism for preventing misfolding and enhancing the rate of the overall folding reaction in vivo. By analogy, the long range P1 helix in riboswitch aptamers may be formed as part of a labile folding intermediate that is trapped by ligand binding.

7 MODELS FOR STRUCTURAL SWITCHING

Riboswitches, as well as a number of other types of RNA regulatory elements in bacteria and eukarya, use mutually exclusive secondary structures to direct expression machinery. As these processes are often cotranscriptional, it is important to consider the rate at which RNA transcribes and folds, as well as the relative thermodynamic stability of competing secondary structures. As opposed to the majority of studies that observe folding of a fully synthesized RNA (see previous discussion), the directionality of transcription imprints order that biases the final conformation because upstream sequences can fold before transcription of downstream sequences. For riboswitches, this means that folding of the aptamer can occur in the absence of the expression platform, and the antiterminator can form before synthesis of the terminator stem-loop.

The effects of transcription order have been studied for many RNAs, but perhaps most clearly so using a simple bistable switch that mimics many of the features of naturally occurring riboswitches (Fig. 6) (Xayaphoummine et al. 2007). This RNA was engineered to fold into mutually exclusive branched or rodlike structures that can be differentiated using native gel electrophoresis. Transcription of the "forward sequence" (e.g., *abc . . . xyz*) of the switch (Fig. 6A) yields exclusively the branched structure, whereas transcription of the reversed sequence (e.g., *zyx . . . cba*) leads to the alternative rodlike structure with ~90% frequency. In contrast, heat renaturation of the RNA leads to an equal distribution of the two states, which do not significantly interchange at room temperature. In other words, RNA can readily adopt nonequilibrium structures during cotranscriptional folding because of high activation energies associated with their interconversion (Crothers et al. 1974). This result emphasizes that traditional methods for studying RNA folding might fail to capture some of the most crucial features of cotranscriptional folding which are inherently tied to the mechanism of riboswitch regulation.

A second key conclusion gained from studies of the engineered switch is that the relative stability of helices ($P_a > P_c$, and $P_c > P_b$) is sufficient to encode a precise folding pathway (Xayaphoummine et al. 2007). Similar models have been devised to explain the variable regulatory activities of purine riboswitches within the *B. subtilis* genome (Mulhbacher and Lafontaine 2007). In cases where the relative stability of the antiterminator hairpin (ΔG_{AT}) is greater than that of the terminator hairpin (ΔG_T) (Fig. 6C), reporter gene assays demonstrate that there is a significant amount of transcription read through in vivo, even at saturating ligand concentrations in the growth media. In contrast, increasing the stability of the terminator hairpin is correlated with the enhanced ability of the riboswitch to promote transcription termination under high ligand concentrations (Mulhbacher and Lafontaine 2007). Similar factors may account for the large range of behaviors observed for the 12 SAM-I riboswitches in the *B. subtilis* genome (Tomsic et al. 2008).

The analogy between the engineered and natural riboswitches can be extended by considering self-induced folding transitions occurring during folding of the reverse sequence (Fig. 6B). In the absence of a competitor oligonucleotide, the reverse switch sequence destabilizes the P_b

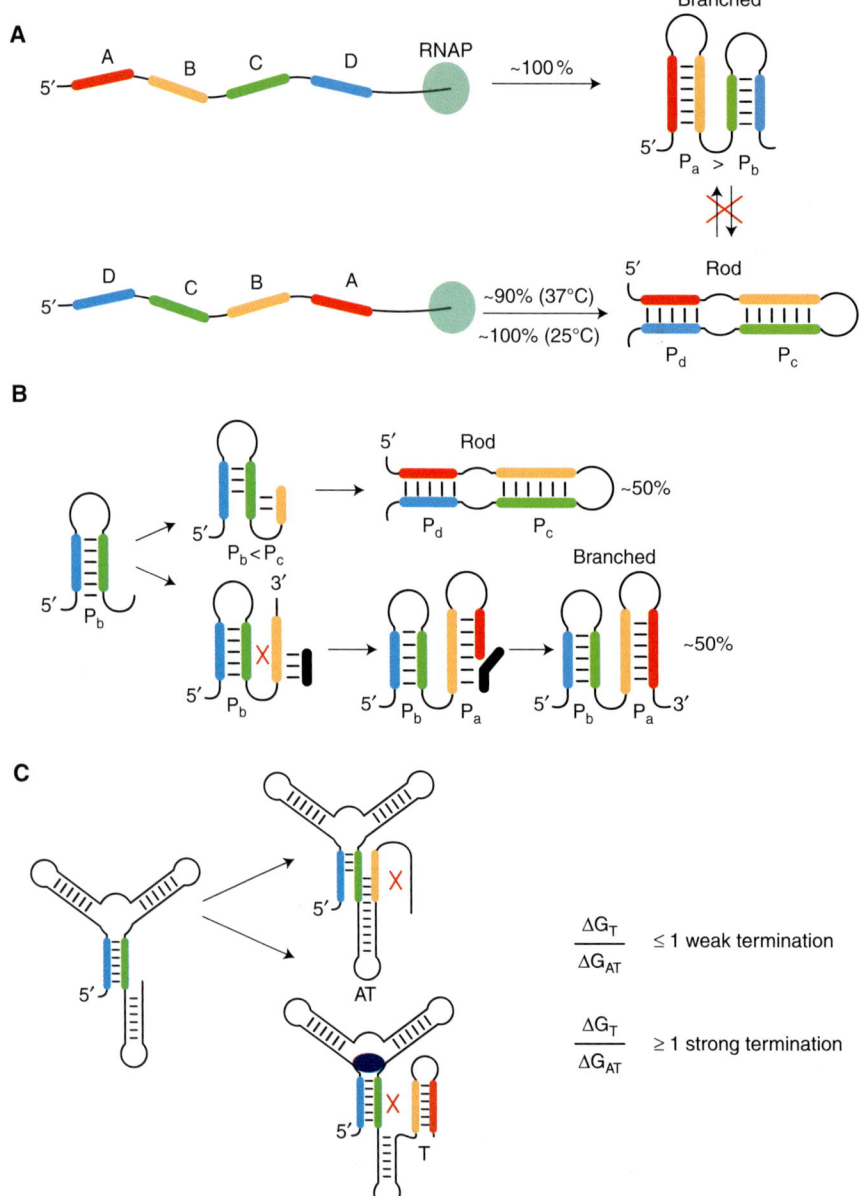

Figure 6. Cotranscriptional folding of synthetic RNA switches. (*A*) A synthetically designed RNA sequence transcribed in forward and reverse directions illustrating the influence of transcription order on the folding outcome. The free energy of each sequential helix dictates the outcome of folding. (*B*) A model of the reverse synthetic sequence reveals that structural transitions can be redirected by addition of a competitor oligonucleotide (black) to the transcription reaction. (*C*) Regulatory efficiency of purine riboswitches correlates with the relative free energy of the terminator (T) and anti terminator (AT) stems, pointing to similarities between riboswitches and their synthetic counterparts.

stem in favor of the P_c stem; lower temperatures bias the structure further toward P_c because of slower transcription rates. This refolding process can be disrupted by introduction of a competitor oligonucleotide (Fig. 6B, black line) that pairs with the P_c stem early in transcription. The oligonucleotide efficiently redirects folding into the branched structure, in the same fashion as ligand binding prevents self-induced transitions during riboswitch folding that would lead to aberrant antitermination (Fig. 6C). Elements of the unbound aptamer can be thought of as behaving like the antisense oligonucleotide by disrupting terminator stem formation while simultaneously nucleating the antiterminator stem. This comparison warrants further investigation of the simple secondary structures in both domains of the RNA that must interchange during transcription to elicit regulatory control.

8 RIBOSWITCHES IN THE RNA WORLD

Although the exact nature of the RNA world cannot be reconstructed, riboswitches present themselves as an elegant solution to the problem of biological regulation. These RNAs are capable of recognizing and responding to compounds that were likely to have been metabolites in the RNA world, such as the purine nucleobases. In addition, the secondary structural switch that interfaces with the expression machinery can function by directing RNA–RNA interactions. In modern riboswitches that act at the translational level, effector binding to the aptamer domain occludes the Shine-Dalgarno sequence that recruits the 30S subunit by pairing with a site near the 3′-end of the 16S ribosomal RNA. Thus, riboswitches contain the essential features to efficiently function in the hypothesized RNA world.

If riboswitches are truly biological fossils of an ancient world, it is even more impressive that they have survived billions of years of evolutionary pressure to serve in modern organisms. Some riboswitches that are widely distributed in phylogeny, such as the thiamine pyrophosphate and adenosylcobalamin classes, might have survived because they occupy a niche in metabolic regulation that perhaps is better suited to RNA than protein. They are an energetically cost-effective means for genetic regulation compared with protein synthesis, as only mRNA transcription needs to occur. Furthermore, riboswitches offer an immediate feedback response as riboswitch regulation involves fewer steps compared with that of regulatory proteins. Thus, modern organisms continue to capitalize on the unique abilities of RNA to fine-tune gene expression levels using a mechanism that could have easily found a home in the RNA world.

ACKNOWLEDGMENTS

A.D.G. and A.L.E. contributed equally to this work. We thank members of the Batey laboratory and Jennifer Pfingsten for careful reading of this manuscript. This work was supported by grants from the National Institutes of Health.

REFERENCES

Barrick JE, Breaker RR. 2007. The distributions, mechanisms, and structures of metabolite-binding riboswitches. *Genome Biol* **8:** R239.

Batey RT, Gilbert SD, Montange RK. 2004. Structure of a natural guanine-responsive riboswitch complexed with the metabolite hypoxanthine. *Nature* **432:** 411–415.

Batey RT, Rambo RP, Doudna JA. 1999. Tertiary Motifs in RNA Structure and Folding. *Angew Chem Int Ed Engl* **38:** 2326–2343.

Blouin S, Lafontaine DA. 2007. A loop loop interaction and a K-turn motif located in the lysine aptamer domain are important for the riboswitch gene regulation control. *RNA* **13:** 1256–1267.

Brion P, Westhof E. 1997. Hierarchy and dynamics of RNA folding. *Annu Rev Biophys Biomol Struct* **26:** 113–137.

Cate JH, Gooding AR, Podell E, Zhou K, Golden BL, Kundrot CE, Cech TR, Doudna JA. 1996. Crystal structure of a group I ribozyme domain: Principles of RNA packing. *Science* **273:** 1678–1685.

Cochrane JC, Lipchock SV, Strobel SA. 2007. Structural investigation of the GlmS ribozyme bound to its catalytic cofactor. *Chem Biol* **14:** 97–105.

Crothers DM, Cole PE, Hilbers CW, Shulman RG. 1974. The molecular mechanism of thermal unfolding of *Escherichia coli* formylmethionine transfer RNA. *J Mol Biol* **87:** 63–88.

Dann CE3rd, Wakeman CA, Sieling CL, Baker SC, Irnov I, Winkler WC. 2007. Structure and mechanism of a metal-sensing regulatory RNA. *Cell* **130:** 878–892.

Edwards AL, Batey RT. 2009. A structural basis for the recognition of 2′-deoxyguanosine by the purine riboswitch. *J Mol Biol* **385:** 938–948.

Edwards TE, Ferre-D'Amare AR. 2006. Crystal structures of the thi-box riboswitch bound to thiamine pyrophosphate analogs reveal adaptive RNA-small molecule recognition. *Structure* **14:** 1459–1468.

Edwards TE, Klein DJ, Ferre-D'Amare AR. 2007. Riboswitches: Small-molecule recognition by gene regulatory RNAs. *Curr Opin Struct Biol* **17:** 273–279.

Garst AD, Heroux A, Rambo RP, Batey RT. 2008. Crystal structure of the lysine riboswitch regulatory mRNA element. *J Biol Chem* **283:** 22347–22351.

Gilbert SD, Batey RT. 2005. Riboswitches: natural SELEXion. *Cell Mol Life Sci* **62:** 2401–2404.

Gilbert SD, Love CE, Edwards AL, Batey RT. 2007. Mutational analysis of the purine riboswitch aptamer domain. *Biochemistry* **46:** 13297–13309.

Gilbert SD, Rambo RP, Van Tyne D, Batey RT. 2008. Structure of the SAM-II riboswitch bound to S-adenosylmethionine. *Nat Struct Mol Biol* **15:** 177–182.

Gilbert SD, Stoddard CD, Wise SJ, Batey RT. 2006. Thermodynamic and kinetic characterization of ligand binding to the purine riboswitch aptamer domain. *J Mol Biol* **359:** 754–768.

Greenleaf WJ, Frieda KL, Foster DA, Woodside MT, Block SM. 2008. Direct observation of hierarchical folding in single riboswitch aptamers. *Science* **319:** 630–633.

Grundy FJ, Lehman SC, Henkin TM. 2003. The L box regulon: lysine sensing by leader RNAs of bacterial lysine biosynthesis genes. *Proc Natl Acad Sci U S A* **100:** 12057–12062.

Hendrix DK, Brenner SE, Holbrook SR. 2005. RNA structural motifs: building blocks of a modular biomolecule. *Q Rev Biophys* **38:** 221–243.

Heppell B, Lafontaine DA. 2008. Folding of the SAM aptamer is determined by the formation of a K-turn-dependent pseudoknot. *Biochemistry* **47:** 1490–1499.

Holbrook SR. 2008. Structural principles from large RNAs. *Annu Rev Biophys* **37:** 445–464.

Jaeger L, Michel F, Westhof E. 1994. Involvement of a GNRA tetraloop in long-range RNA tertiary interactions. *J Mol Biol* **236:** 1271–1276.

Jenison RD, Gill SC, Pardi A, Polisky B. 1994. High-resolution molecular discrimination by RNA. *Science* **263:** 1425–1429.

Kang M, Peterson R, Feigon J. 2009. Structural Insights into riboswitch control of the biosynthesis of queuosine, a modified nucleotide found in the anticodon of tRNA. *Mol Cell* **33:** 784–790.

Kim JN, Breaker RR. 2008. Purine sensing by riboswitches. *Biol Cell* **100:** 1–11.

Kim JN, Roth A, Breaker RR. 2007. Guanine riboswitch variants from *Mesoplasma florum* selectively recognize 2′-deoxyguanosine. *Proc Natl Acad Sci* **104:** 16092–16097.

King GM, Carter AR, Churnside AB, Eberle LS, Perkins TT. 2009. Ultrastable atomic force microscopy: Atomic-scale stability and registration in ambient conditions. *Nano Lett* **9:** 1451–1456.

Klein DJ, Ferré-D'Amaré AR. 2006. Structural basis of glmS ribozyme activation by glucosamine-6-phosphate. *Science* **313:** 1752–1756.

Klein DJ, Edwards TE, Ferre-D'Amare AR. 2009. Cocrystal structure of a class I preQ1 riboswitch reveals a pseudoknot recognizing an essential hypermodified nucleobase. *Nat Struct Mol Biol* **16:** 343–344.

Kulshina N, Baird NJ, Ferre-D'Amare AR. 2009. Recognition of the bacterial second messenger cyclic diguanylate by its cognate riboswitch. *Nat Struct Mol Biol* **16:** 1212–1217.

Lang K, Rieder R, Micura R. 2007. Ligand-induced folding of the thiM TPP riboswitch investigated by a structure-based fluorescence spectroscopic approach. *Nucleic Acids Res* **35:** 5370–5378.

Lemay JF, Penedo JC, Tremblay R, Lilley DM, Lafontaine DA. 2006. Folding of the adenine riboswitch. *Chem Biol* **13:** 857–868.

Leontis NB, Westhof E. 2001. Geometric nomenclature and classification of RNA base pairs. *RNA* **7:** 499–512.

Leontis NB, Lescoute A, Westhof E. 2006. The building blocks and motifs of RNA architecture. *Current Opinion in Structural Biology* **16:** 279–287.

Li PT, Vieregg J, Tinoco IJr, 2008. How RNA unfolds and refolds. *Annu Rev Biochem* **77:** 77–100.

Link KH, Breaker RR. 2009. Engineering ligand-responsive gene-control elements: Lessons learned from natural riboswitches. *Gene Ther*.

Lipfert J, Doniach S. 2007. Small-angle X-ray scattering from RNA, proteins, and protein complexes. *Annu Rev Biophys Biomol Struct* **36:** 307–327.

Lu C, Smith AM, Fuchs RT, Ding F, Rajashankar K, Henkin TM, Ke A. 2008. Crystal structures of the SAM-III/S(MK) riboswitch reveal the SAM-dependent translation inhibition mechanism. *Nat Struct Mol Biol* **15:** 1076–1083.

Mandal M, Breaker RR. 2004. Adenine riboswitches and gene activation by disruption of a transcription terminator. *Nat Struct Mol Biol* **11:** 29–35.

Mandal M, Boese B, Barrick JE, Winkler WC, Breaker RR. 2003. Riboswitches control fundamental biochemical pathways in *Bacillus subtilis* and other bacteria. *Cell* **113:** 577–586.

McDaniel BA, Grundy FJ, Henkin TM. 2005. A tertiary structural element in S box leader RNAs is required for S-adenosylmethionine-directed transcription termination. *Mol Microbiol* **57:** 1008–1021.

McDaniel BA, Grundy FJ, Artsimovitch I, Henkin TM. 2003. Transcription termination control of the S box system: Direct measurement of S-adenosylmethionine by the leader RNA. *Proc Natl Acad Sci* **100:** 3083–3088.

Merino EJ, Wilkinson KA, Coughlan JL, Weeks KM. 2005. RNA structure analysis at single nucleotide resolution by selective 2′-hydroxyl acylation and primer extension (SHAPE). *J Am Chem Soc* **127:** 4223–4231.

Mironov AS, Gusarov I, Rafikov R, Lopez LE, Shatalin K, Kreneva RA, Perumov DA, Nudler E. 2002. Sensing small molecules by nascent RNA: a mechanism to control transcription in bacteria. *Cell* **111:** 747–756.

Montange RK, Batey RT. 2006. Structure of the S-adenosylmethionine riboswitch regulatory mRNA element. *Nature* **441:** 1172–1175.

Montange RK, Batey RT. 2008. Riboswitches: emerging themes in RNA structure and function. *Annu Rev Biophys* **37:** 117–133.

Mulhbacher J, Lafontaine DA. 2007. Ligand recognition determinants of guanine riboswitches. *Nucleic Acids Res* **35:** 5568–5580.

Murphy FL, Cech TR. 1994. GAAA tetraloop and conserved bulge stabilize tertiary structure of a group I intron domain. *J Mol Biol* **236:** 49–63.

Nahvi A, Sudarsan N, Ebert MS, Zou X, Brown KL, Breaker RR. 2002. Genetic control by a metabolite binding mRNA. *Chem Biol* **9:** 1043.

Noeske J, Schwalbe H, Wohnert J. 2007b. Metal-ion binding and metal-ion induced folding of the adenine-sensing riboswitch aptamer domain. *Nucleic Acids Res* **35:** 5262–5273.

Noeske J, Buck J, Furtig B, Nasiri HR, Schwalbe H, Wohnert J. 2007a. Interplay of 'induced fit' and preorganization in the ligand induced folding of the aptamer domain of the guanine binding riboswitch. *Nucleic Acids Res* **35:** 572–583.

Noeske J, Richter C, Grundl MA, Nasiri HR, Schwalbe H, Wohnert J. 2005. An intermolecular base triple as the basis of ligand specificity and affinity in the guanine- and adenine-sensing riboswitch RNAs. *Proc Natl Acad Sci* **102:** 1372–1377.

Ottink OM, Rampersad SM, Tessari M, Zaman GJ, Heus HA, Wijmenga SS. 2007. Ligand-induced folding of the guanine-sensing riboswitch is controlled by a combined predetermined induced fit mechanism. *RNA* **13:** 2202–2212.

Pan T, Sosnick TR. 1997. Intermediates and kinetic traps in the folding of a large ribozyme revealed by circular dichroism and UV absorbance spectroscopies and catalytic activity. *Nat Struct Biol* **4:** 931–938.

Putnam CD, Hammel M, Hura GL, Tainer JA. 2007. X-ray solution scattering (SAXS) combined with crystallography and computation: Defining accurate macromolecular structures, conformations and assemblies in solution. *Q Rev Biophys* **40:** 191–285.

Rieder U, Lang K, Kreutz C, Polacek N, Micura R. 2009. Evidence for pseudoknot formation of class I preQ1 riboswitch aptamers. *Chembiochem* **10:** 1141–1144.

Ryals J, Little R, Bremer H. 1982. Temperature dependence of RNA synthesis parameters in *Escherichia coli*. *J Bacteriol* **151:** 879–887.

Serganov A, Huang L, Patel DJ. 2008. Structural insights into amino acid binding and gene control by a lysine riboswitch. *Nature* **455:** 1263–1267.

Serganov A, Huang L, Patel DJ. 2009. Coenzyme recognition and gene regulation by a flavin mononucleotide riboswitch. *Nature* **458:** 233–237.

Serganov A, Polonskaia A, Phan AT, Breaker RR, Patel DJ. 2006. Structural basis for gene regulation by a thiamine pyrophosphate-sensing riboswitch. *Nature* **441:** 1167–1171.

Serganov A, Yuan YR, Pikovskaya O, Polonskaia A, Malinina L, Phan AT, Hobartner C, Micura R, Breaker RR, Patel DJ. 2004. Structural basis for discriminative regulation of gene expression by adenine- and guanine-sensing mRNAs. *Chem Biol* **11:** 1729–1741.

Smith KD, Lipchock SV, Ames TD, Wang J, Breaker RR, Strobel SA. 2009. Structural basis of ligand binding by a c-di-GMP riboswitch. *Nat Struct Mol Biol* **16:** 1218–1223.

Spitale RC, Torelli AT, Krucinska J, Bandarian V, Wedekind JE. 2009. The structural basis for recognition of the PreQ0 metabolite by an unusually small riboswitch aptamer domain. *J Biol Chem* **284:** 11012–11016.

Stoddard CD, Gilbert SD, Batey RT. 2008. Ligand-dependent folding of the three-way junction in the purine riboswitch. *RNA* **14:** 675–684.

Sudarsan N, Lee ER, Weinberg Z, Moy RH, Kim JN, Link KH, Breaker RR. 2008. Riboswitches in eubacteria sense the second messenger cyclic di-GMP. *Science* **321:** 411–413.

Sudarsan N, Wickiser JK, Nakamura S, Ebert MS, Breaker RR. 2003. An mRNA structure in bacteria that controls gene expression by binding lysine. *Genes Dev* **17:** 2688–2697.

Thore S, Leibundgut M, Ban N. 2006. Structure of the eukaryotic thiamine pyrophosphate riboswitch with its regulatory ligand. *Science* **312:** 1208–1211.

Tomsic J, McDaniel BA, Grundy FJ, Henkin TM. 2008. Natural variability in S-adenosylmethionine (SAM)-dependent riboswitches: S-box elements in *Bacillus subtilis* exhibit differential sensitivity to SAM In vivo and in vitro. *J Bacteriol* **190:** 823–833.

Treiber DK, Williamson JR. 1999. Exposing the kinetic traps in RNA folding. *Curr Opin Struct Biol* **9:** 339–345.

Varani G. 1995. Exceptionally stable nucleic acid hairpins. *Annu Rev Biophys Biomol Struct* **24:** 379–404.

Wickiser JK, Cheah MT, Breaker RR, Crothers DM. 2005a. The kinetics of ligand binding by an adenine-sensing riboswitch. *Biochemistry* **44**: 13404–13414.

Wickiser JK, Winkler WC, Breaker RR, Crothers DM. 2005b. The speed of RNA transcription and metabolite binding kinetics operate an FMN riboswitch. *Mol Cell* **18**: 49–60.

Wilkinson KA, Merino EJ, Weeks KM. 2006. Selective 2′-hydroxyl acylation analyzed by primer extension (SHAPE): Quantitative RNA structure analysis at single nucleotide resolution. *Nat Protoc* **1**: 1610–1616.

Williamson JR. 2000. Induced fit in RNA-protein recognition. *Nat Struct Biol* **7**: 834–837.

Winkler WC, Breaker R. 2003. Genetic control by metabolite-binding riboswitches. *Chembiochem* **4**: 1024–1032.

Winkler W, Nahvi A, Breaker RR. 2002a. Thiamine derivatives bind messenger RNAs directly to regulate bacterial gene expression. *Nature* **419**: 952–956.

Winkler WC, Cohen-Chalamish S, Breaker RR. 2002b. An mRNA structure that controls gene expression by binding FMN. *Proc Natl Acad Sci U S A* **99**: 15908–15913.

Winkler WC, Grundy FJ, Murphy BA, Henkin TM. 2001. The GA motif: An RNA element common to bacterial antitermination systems, rRNA, and eukaryotic RNAs. *RNA* **7**: 1165–1172.

Winkler WC, Nahvi A, Sudarsan N, Barrick JE, Breaker RR. 2003. An mRNA structure that controls gene expression by binding S-adenosylmethionine. *Nat Struct Biol* **10**: 701–707.

Wong TN, Sosnick TR, Pan T. 2007. Folding of noncoding RNAs during transcription facilitated by pausing-induced nonnative structures. *Proc Natl Acad Sci* **104**: 17995–18000.

Xayaphoummine A, Viasnoff V, Harlepp S, Isambert H. 2007. Encoding folding paths of RNA switches. *Nucleic Acids Res* **35**: 614–622.

Zimmermann GR, Jenison RD, Wick CL, Simorre JP, Pardi A. 1997. Interlocking structural motifs mediate molecular discrimination by a theophylline-binding RNA. *Nat Struct Biol* **4**: 644–649.

Zimmermann GR, Wick CL, Shields TP, Jenison RD, Pardi A. 2000. Molecular interactions and metal binding in the theophylline-binding core of an RNA aptamer. *RNA* **6**: 659–667.

Small Self-cleaving Ribozymes

Adrian R. Ferré-D'Amaré[1] and William G. Scott[2]

[1]Howard Hughes Medical Institute and Division of Basic Sciences, Fred Hutchinson Cancer Research Center, Seattle, Washington 8109-1024

[2]The Center for the Molecular Biology of RNA and The Department of Chemistry and Biochemistry, University of California, Santa Cruz, Santa Cruz, California 95064

Correspondence: aferre@fhcrc.org and wgscott@ucsc.edu

SUMMARY

The hammerhead, hairpin, hepatitis delta virus (HDV), Varkud Satellite (VS), and *glmS* ribozymes catalyze sequence-specific intramolecular cleavage of RNA. They range between 50 and 150 nucleotides in length, and are known as the "small self-cleaving ribozymes." Except for the *glmS* ribozyme that functions as a riboswitch in Gram-positive bacteria, they were originally discovered as domains of satellite RNAs. However, recent studies show that several of them are broadly distributed in genomes of organisms from many phyla. Each of these ribozymes has a unique overall architecture and active site organization. Crystal structures have revealed how RNA active sites can bind preferentially to the transition state of a reaction, whereas mechanistic studies have shown that nucleobases can efficiently perform general acid–base and electrostatic catalysis. This versatility explains the abundance of ribozymes in contemporary organisms and also supports a role for catalytic RNAs early in evolution.

Outline

1. Introduction
2. The internal transesterification reaction and its catalysis
3. Overall structures of self-cleaving ribozymes
4. Active site structures and catalytic mechanisms
5. The hairpin ribozyme: Analogy to RNase A
6. The hammerhead ribozyme: Variation on a theme
7. The *glmS* ribozyme: Co-opting a cofactor
8. The HDV ribozyme: Nucleobase and metal-ion catalysis
9. How widespread are the self-cleaving ribozymes?
10. Concluding remarks

References

1 INTRODUCTION

Sequence-specific cleavage and ligation are fundamental nucleic acid reactions. Five ribozymes that catalyze them are known from present-day organisms: the hammerhead (Prody et al. 1986), hairpin (Buzayan et al. 1986a), hepatitis delta virus (HDV) (Sharmeen et al. 1988), Varkud satellite (VS) (Saville and Collins, 1990), and glmS (Winkler et al. 2004) ribozymes. In nature, these small, self-cleaving ribozymes catalyze intramolecular reactions. Although the evolutionary origin of these ribozymes is unknown, they have been central to our understanding of the structural, biochemical, and biological versatility of RNA, and thus to the epistemological construction of the RNA World hypothesis. They provided the first high-resolution glimpses of RNA catalysts (Pley et al. 1994; Scott et al. 1995), and have yielded the most detailed structural descriptions of ribozyme-catalyzed chemical reactions (Chi et al. 2008; Klein et al. 2007a; Rupert et al. 2002). The five ribozymes each possess unique folds and active sites, enabling comparison of structurally distinct RNAs that catalyze the same chemical reaction. These studies led to a paradigm shift from the view that all ribozymes are metalloenzymes (Pyle, 1993) to one in which RNA plays a direct role in chemical catalysis (Bevilacqua and Yajima, 2006; Murray et al. 1998). Discovery (Winkler et al. 2004) and characterization (Klein and Ferré-D'Amaré, 2006) of the glmS ribozyme revealed the first natural ribozyme with a coenzyme. Finally, genomic sequencing and analysis are revealing that ribozymes are far more widespread in nature than originally thought (Martick et al. 2008; Salehi-Ashtiani and Szostak, 2001; Webb et al. 2009; Winkler et al. 2004). This article reviews mechanistic insights derived from the study of the small self-cleaving RNAs, whose importance to ribozyme enzymology is analogous to that of serine proteases, lysozyme, and RNase A to protein enzymology.

2 THE INTERNAL TRANSESTERIFICATION REACTION AND ITS CATALYSIS

RNA undergoes nonspecific base-catalyzed degradation through an internal transesterification reaction wherein the 2′-O of a ribose attacks the adjacent 3′ phosphate; cleavage products have a 2′,3′-cyclic phosphate and a 5′-OH, respectively (Fig. 1). This reaction is catalyzed by deprotonation of the 2′-OH; hence, its acceleration with increasing pH. The five ribozymes discussed in this article employ the same chemical mechanism, but are highly sequence-specific. RNase A catalyzes RNA cleavage through the same reaction, thus providing a frame of reference. (Unlike the ribozymes, the protein enzyme hydrolyzes the cyclic phosphate in a subsequent step.) Although RNase A shows only modest sequence specificity, it achieves a rate enhancement of $\sim 10^{11}$ over the uncatalyzed reaction. The principal factors responsible are schematized in Figure 2 (Raines 1998). Like all enzymes, RNase A overcomes the entropic and steric penalties of bringing the reactants into the active site in a productive conformation through binding energy. The transesterification reaction proceeds through a concerted S_N2-like mechanism wherein the 2′ ribose oxygen, the phosphorus, and the 5′ ribose oxygen are aligned. Richards

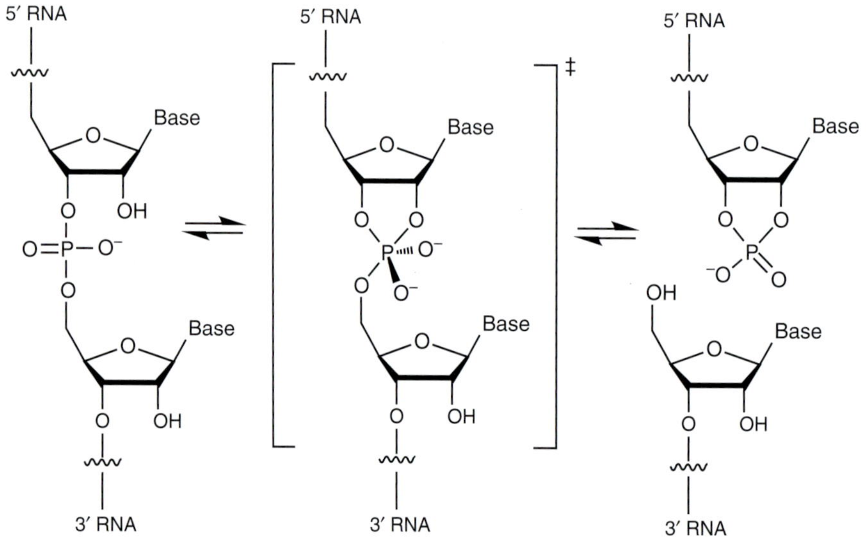

Figure 1. Internal transesterification reaction catalyzed by the hammerhead, hairpin, HDV, VS, and glmS ribozymes. The concerted cleavage reaction proceeds without intermediates. The hammerhead, hairpin, and VS ribozymes also catalyze the ligation reaction.

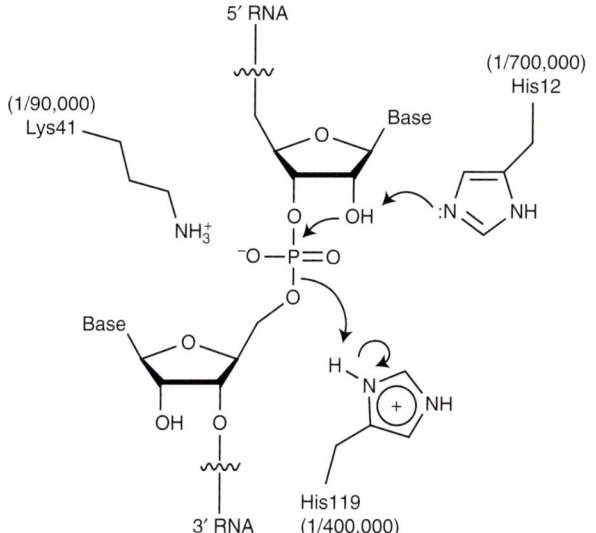

Figure 2. RNase A active site structure and catalytic mechanism. The degree of impairment resulting from site-directed mutations of the catalytic residues (Raines 1998) is indicated in parentheses.

et al. (1971) determined the structure of the enzyme bound to a dinucleotide mimic inhibitor and discovered that RNase A splays apart the nucleotides flanking the scissile phosphate to achieve alignment. The structure revealed the location of two catalytic histidines and one lysine. His12 functions as a general base catalyst, deprotonating the 2′-OH, and His114 functions as a general acid, protonating the 5′-oxo leaving group. The reaction proceeds through a trigonal bipyramidal oxyphosphorane transition-state whose excess negative charge is stabilized by the ammonium group of Lys41. Mutating these catalytic amino acids greatly impairs RNase A (Fig. 2).

3 OVERALL STRUCTURES OF SELF-CLEAVING RIBOZYMES

How is an enzyme active site constructed entirely of RNA? Crystal structures of the hairpin, hammerhead, HDV and *glmS* ribozymes reveal four unique answers (Fig. 3). Each positions the substrate inside an active-site cleft, surrounding it with nucleotides distant in primary sequence. Ribozymes are similar to protein enzymes in that the need to converge multiple functional groups from different parts of the nucleic acid at the active site places a lower bound on their length (for a first-principles discussion about proteins, see de Gennes 1990). RNA is primarily helical; active site formation requires a multi-helical fold, either head-to-tail (coaxial) (Quigley and Rich 1976), or side-by-side. The prevalence of the latter fold is counterintuitive, as it brings the negatively charged backbones of helices into close apposition (Murthy and Rose 2000). Like other structured

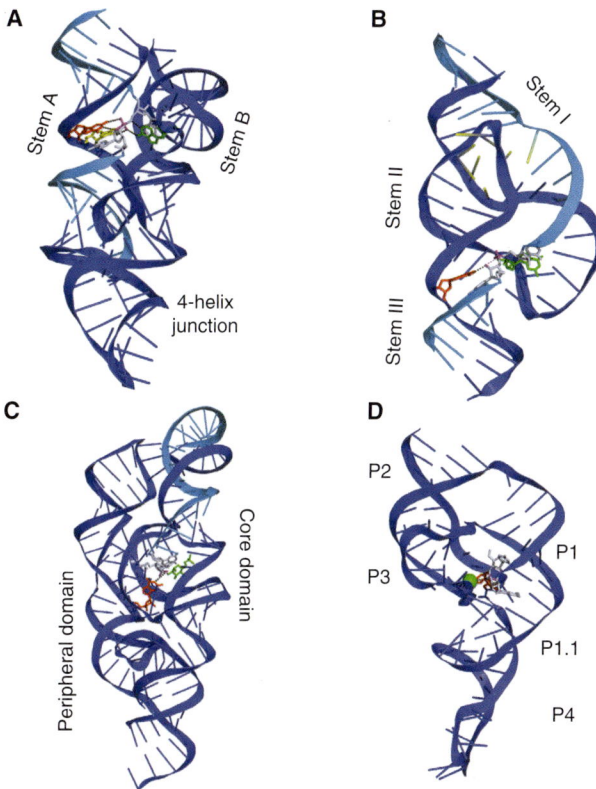

Figure 3. Cartoon representations of the overall structures of four self-cleaving ribozymes. (*A*) The hairpin ribozyme, (*B*) the hammerhead ribozyme, (*C*) the *glmS* ribozyme-riboswitch, and (*D*) the hepatitis delta virus (HDV) ribozyme. Residues implicated in general acid and base catalysis in the cleavage reaction are green and red, respectively. The scissile phosphate and nucleophilic 2′-O atoms are magenta, flanked by substrate residues shown in light blue.

RNAs, the architecture of self-cleaving ribozymes consists of multihelical junctions, interactions of nonhelical elements (helix-terminal loops and internal bulges), and pseudoknotting.

Multihelical junctions organize the hammerhead, hairpin, and VS ribozymes. The active site of the hairpin ribozyme is formed by apposition of the minor grooves of two irregular helices (stems A and B) (Rupert and Ferré-D'Amaré 2001). One of the strands of stem A carries the scissile phosphate. Although these two helices suffice for activity in vitro (Butcher et al. 1995; Shin et al. 1996), in the natural genomic context of the hairpin ribozyme, they are joined by a four-helix junction that stabilizes the active structure (Murchie et al. 1998) (Fig. 3A). The hammerhead ribozyme is comprised of three helices (stems I, II, and III) arranged into an approximate "γ" shape, with stems II and III stacking coaxially, and stem I packing against stem II (Martick and Scott 2006). In addition to the junction, interactions of nonhelical elements at the tip of stem II and the distal portion of stem I are critical for organizing the RNA (Fig. 3B). The substrate strand is part of stem I, and the active site lies at the three-helix junction (Haseloff and Gerlach 1989; Ruffner et al. 1990; Symons 1997; Uhlenbeck 1987). No crystal structure is available for the VS ribozyme. Modeling based on biochemistry and small-angle-X ray scattering indicates that it is comprised of two three-helix junctions and that it is also stabilized by a long-range interaction between two loops (Lilley 2004). The hammerhead, hairpin and VS ribozymes share the manner in which they bind their substrate. All three have sequence elements that form canonical base pairs with nucleotides on either side of the scissile phosphate on the substrate strand (although not with nucleotides immediately adjacent to the cleavage site). As a result, the substrate is bound as part of an irregular helix that docks with the rest of the ribozyme to form the active molecule. An important consequence of this mode of substrate binding is that after cleavage, both of the product RNA strands may remain bound by the ribozyme (as each forms a short A-form helix with parts of the ribozyme), so that depending on the conditions, these ribozymes can efficiently catalyze the ligation of substrates containing the correct functional groups (a 2′,3′-cyclic phosphate and a 5′-OH) on their termini (Blount and Uhlenbeck 2002; Canny et al. 2007; Fedor 1999; Jones et al. 2001; Saville and Collins 1991).

Pseudoknots organize the structures of the HDV and *glmS* ribozymes. A pseudoknot is a nucleic acid structure formed by nucleotides in the loop of a stem-loop base pairing with nucleotides outside of the stem-loop (Pleij 1990). The HDV ribozyme has five helices (P1–P4 and P1.1) connected as a nested double pseudoknot (Ferré-D'Amaré et al. 1998; Perrotta and Been 1991). The intricate connectivity of the RNA chain allows this compact (minimal forms are ∼60 nt) and remarkably stable (it remains active in 5M urea) ribozyme to pack two helices stably side-by-side to form the active site (Fig. 3D). The core of the *glmS* ribozyme is also a nested double pseudoknot (Klein and Ferré-D'Amaré 2006). In addition, this larger ribozyme (∼150 nt) has a peripheral domain that itself contains a pseudoknot. This peripheral domain packs against one of the sides of the core double pseudoknot (Fig. 3C). The HDV and *glmS* ribozymes position their substrates similarly. Unlike the hammerhead, hairpin, and VS ribozymes, these two ribozymes only base pair with the segment 3′ of the scissile phosphate of their substrates. There are a few non-Watson-Crick interactions between the nucleotide immediately 5′ of the scissile phosphate and the ribozyme core, but no helix is formed between the ribozyme and substrate nucleotides further 5′. As a result, product dissociation occurs readily. Hence, as with RNase A, neither HDV nor *glmS* ribozymes can relegate cleaved RNA.

4 ACTIVE SITE STRUCTURES AND CATALYTIC MECHANISMS

Unlike proteins, RNA has no functional groups that are positively charged at neutral pH. The groups with pKa's closest to neutrality are the N1 nitrogens of purines (3.5 and 9.2 for A and G, respectively) and the N3 nitrogens of pyrimidines (4.2 and 9.2 for C and U, respectively). As a polyanion, however, RNA binds cations. In principle, these can function not only as electrostatic catalysts that stabilize negative charges (analogous to Lys41 of RNase A) (Fig. 2), but also as either Lewis acids that perturb the pKa of RNA functional groups bound to them, or as Brønsted acids that exist as hydrates in solution, lowering the pKa of the cation-coordinated waters. Thus, in addition to providing a localized positive charge, tightly bound cations might assist in catalysis by providing hydroxide or hydronium ions to function as reactants or specific base/acid catalysts. Because the Group I intron has been shown to employ tightly bound Mg^{2+} ions for electrostatic transition-state stabilization and as Lewis acids, it was widely believed that all catalytic RNAs would employ bound metal ions as cofactors (reviewed in Pyle 1993). In all but one of the ribozymes considered here, that assumption proved unnecessary.

5 THE HAIRPIN RIBOZYME: ANALOGY TO RNase A

The hairpin ribozyme was the first self-cleaving ribozyme for which strong evidence was gathered against direct participation of divalent metal ions in catalysis. Like all highly structured RNAs, this ribozyme requires divalent cations for folding under physiological conditions. However, it was found that the hairpin ribozyme remains fully active if magnesium ion is replaced entirely with cobalt (III) hexammine (Hampel and Cowan 1997; Nesbitt et al. 1997; Young et al. 1997). $Co(NH_3)_6^{3+}$ is isosteric with $Mg(H_2O)_6^{2+}$, but unlike the water ligands of Mg^{2+}, the amino ligands cannot readily dissociate; therefore, the metal ion cannot chemically participate in catalysis, nor can its ligands be replaced by RNA functional groups to form direct cation–RNA coordinations. Thus, the full activity of the hairpin ribozyme in $Co(NH_3)_6^{3+}$ in the absence of Mg^{2+} implies that the RNA, rather than acting as a scaffold for binding divalent cations, must itself instead be directly participating in catalysis. The structure of the ribozyme unambiguously corroborated this prediction.

The structure of the hairpin ribozyme was determined in uncleaved, intermediate, and cleaved states. An uncleavable substrate analog was used to obtain the initial-state structure, and an RNA-vanadate complex to mimic the intermediate or transition-state (Rupert and Ferré-D'Amaré 2001; Rupert et al. 2002). The hairpin ribozyme active site possesses some striking similarities to that of RNase A. As in the protein, the substrate nucleotides flanking the scissile phosphate are splayed apart. The nucleotide preceding the scissile phosphate forms a noncanonical pair within stem A, whereas the nucleotide that follows it flips out of stem A and base pairs with an unpaired nucleotide in stem B. This has the effect of bringing the three reactive atoms into a near-in-line conformation (Fig. 4A). Two conserved and functionally important purines, G8 and A38, are positioned on either side of the scissile phosphate, in locations analogous to those of the catalytic histidines of RNase A. Another conserved nucleobase, A9, lies in a position analogous to that of Lys41.

The hairpin ribozyme transition-state analog structure implicates G8 and A39 as the general base and general acid, respectively, in the cleavage reaction (their roles would be reversed in the ligation reaction). It also indicates that the exocyclic amines of A9 and A38 play a role analogous to the amine of Lys41 in electrostatic transition-state stabilization. Modeling of the pH-dependence of the hairpin ribozyme reaction, and comparison of it to that of the RNase A reaction, not only corroborates this assessment but permits assignment of slightly perturbed pKa values to the general base and general acid of 9.0 and 5.0, respectively (Bevilacqua 2003), consistent with their identification in the crystal structures as a G and an A (but also see Liu et al. 2009). Hence the hairpin ribozyme structure, in a manner consistent with the observed biochemistry, revealed that the RNA

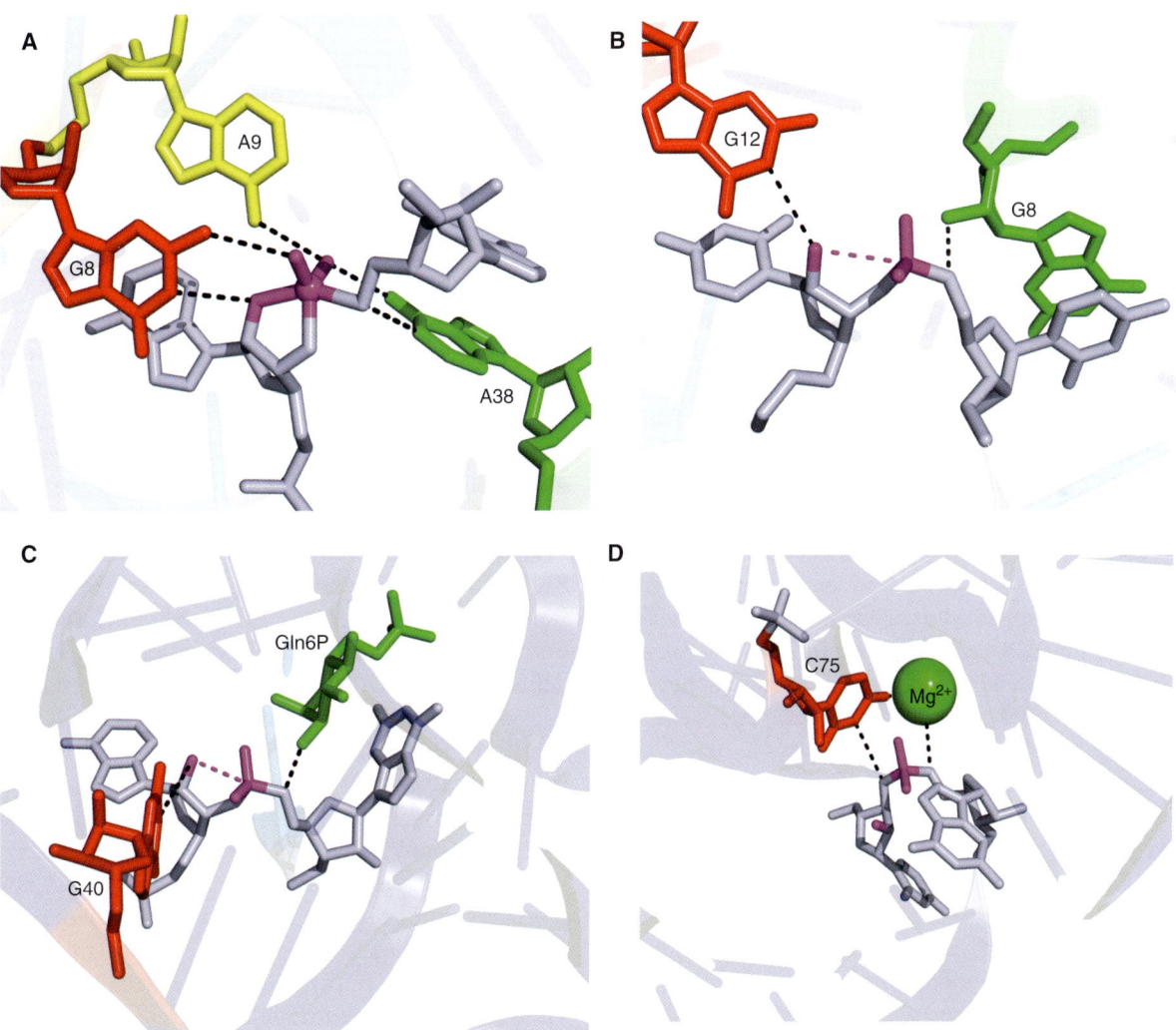

Figure 4. Active sites of the (*A*) hairpin, (*B*) hammerhead, (*C*) *glmS*, and (*D*) HDV ribozymes color-coding as in Figure 3. The moieties thought to contribute to catalysis are labeled.

itself possesses the necessary structural and chemical capacity to catalyze acid-base phosphodiester chemistry.

6 THE HAMMERHEAD RIBOZYME: VARIATION ON A THEME

Although the hammerhead has long been the prototype ribozyme and was the first catalytic RNA whose structure was determined, its history has been fraught with experimental discord (Blount and Uhlenbeck 2005). Between its discovery (Prody et al. 1986) and 2003, most attention was focused on fifteen near-invariant nucleotides at the three-helix junction core of a "minimal" hammerhead (Blount and Uhlenbeck 2005; McKay 1996; Scott 1999; Wedekind and McKay 1998). The crystal structures revealed an active site that offered no clear explanation for the invariance of many of the core residues (McKay 1996). The hammerhead ribozyme was originally thought to be a metalloenzyme. Although no obvious mechanism was suggested by metal ions present in the crystal structures of the minimal hammerhead ribozyme, divalent cations were still considered to be the most likely acid/base catalytic components. In 1998, it was shown that the hammerhead, hairpin, and VS ribozymes, unlike the HDV ribozyme, did not require divalent metal ions for catalysis, but instead were catalytically proficient in high concentrations of monovalent cations, including the nonmetallic ammonium ion (Murray et al. 1998). In other words, folding and charge stabilization were sufficient to support catalytic activity of these ribozymes.

In 2003, two groups revealed that "full-length" hammerhead ribozymes derived from natural sequences, which included a previously neglected tertiary contact between helices I and II in a region with little sequence conservation, were up to 1000-fold more active than minimal hammerheads (De la Peña et al. 2003; Khvorova et al. 2003). Subsequent crystal structures of precleavage and postcleavage full-length hammerhead ribozymes that include these tertiary contacts revealed a much more informative active site that is both reminiscent of the hairpin ribozyme active site and intriguingly different (Chi et al. 2008; Martick and Scott 2006).

Crystal structures of full-length hammerhead ribozymes revealed an arrangement of residues in the active site that makes their participation in catalysis clear (Fig. 4B) The invariant G12 hydrogen-bonds to the nucleophilic 2′-O; this conserved guanine is likely the general base in the self-cleavage reaction. A hydrogen bond between a the 2′-OH of a second invariant nucleotide, G8, and the 5′-oxo leaving group was also identified. This implicates the 2′-OH of the ribose of G8 as a general acid catalyst. (The requirement for G at position 8 is due to a Watson-Crick tertiary base pair it forms with the equally invariant C3.) In addition, the crystal structure of a very slowly cleaving G12A hammerhead (whose core structure is essentially identical but whose purine general base has a pKa lowered by ~5 units) revealed, in the product structure, a potential additional interaction between A9 and the scissile phosphate, reminiscent of the coincidently named A9 of the hairpin ribozyme. However, whether A9 or another entity engages in transition-state stabilization in the hammerhead ribozyme remains speculative.

7 THE glmS RIBOZYME: CO-OPTING A COFACTOR

This ribozyme was discovered as an RNA domain that catalyzes site-specific RNA cleavage in the presence of the small-molecule metabolite glucosamine-6-phosphate (GlcN6P) (Winkler et al. 2004). In the absence of GlcN6P, the rate of cleavage of the ribozyme is indistinguishable from that of background hydrolysis. Addition of GlcN6P results in as much as 10^7-fold increase of reactivity at the specific cleavage site. The glmS ribozyme is not a metalloenzyme, because it is fully active with $Co(NH_3)_6^{3+}$ (Roth et al. 2006). In principle, GlcN6P could function either as an allosteric activator or as a coenzyme. The structure reveals that the N1 of the conserved nucleotide G40 lies close to the 2′-OH nucleophile, well positioned to serve as a general base. GlcN6P binds on the opposite side of the scissile phosphate, with its amine group in van der Waals contact with the 5′-O leaving group, an arrangement compatible with its function as a general acid (Fig. 4C). If GlcN6P binds to the ribozyme in its protonated (ammonium) form, a positive charge would lie adjacent to the scissile phosphate, where it could stabilize the transition state. Therefore, the structure of the active site of the glmS ribozyme is consistent with a coenzyme function for GlcN6P. Moreover, structures of the glmS ribozyme in different states (precleavage, precleavage bound to GlcN6P, transition-state mimic, postcleavage) do not show any evidence of conformational change of the RNA resulting from GlcN6P binding (Klein and Ferré-D'Amaré 2006; Klein et al. 2007b).

A remarkable feature of the glmS ribozyme is the absolute requirement for both G40 and GlcN6P for catalytic activity. The ribozyme is inactive in the absence of GlcN6P, but is also inactive in the presence of GlcN6P if a G40A mutation is introduced into the ribozyme. Structure determination shows that the mutant adopts the wild-type conformation, and that GlcN6P binds in precisely the same location as to the wild-type (Klein et al. 2007a). Thus, lack of activity of the G40A mutant is not caused by misfolding. The strict requirement for both catalytic functional groups contrasts with the behavior of other enzymes. For instance, site-directed mutation of either catalytic histidine of RNase A to alanine results in an enzyme that is still active, although impaired by 400,000 to 700,000-fold (Raines 1998). Analogously, abasic substitutions of the three catalytic purines in the active site of the hairpin ribozyme results in 9350- and 14,000-fold loss in activity (for A9, G8, and A38, respectively) but each mutant retains activity (Kuzmin et al. 2005; Lebruska et al. 2002). The simultaneous requirement for G40 and GlcN6P by the glmS ribozyme implies that the active-site guanine and the small molecule coenzyme mutually tune each other's chemical properties such that neither is catalytically proficient in the absence of the other.

8 THE HDV RIBOZYME: NUCLEOBASE AND METAL-ION CATALYSIS

Of the five known self-cleaving ribozymes, the HDV ribozyme is the only one that employs an active-site divalent cation for catalysis. Ironically, this was also the first ribozyme for which compelling evidence for nucleobase participation in catalysis was gathered. Structure determination of the postcleavage form of the HDV ribozyme revealed that the N3 imine of the functionally essential residue C75 (C76 in the antigenomic HDV ribozyme) hydrogen bonds to the 5′-OH leaving group of the reaction. In addition, C75 was found in a region of strongly negative electrostatic potential. These observations suggested that C75 may have an altered pKa, and function as a general acid/base catalyst (Ferré-D'Amaré et al. 1998). This was reinforced by several observations, including the rescue of a C75U mutant by exogenous imidazole, and the shift of the reaction pKa in C76A mutant that matches the difference in pKa between C and A (Perrotta et al. 1999). Kinetic experiments cannot distinguish between general acid and general base functions for C75/76. Strong evidence for the former comes from Das and Piccirilli (2005), who synthesized chemically modified HDV ribozymes in which the leaving group oxygen of the reaction was replaced with a sulfur. These RNAs were expected lose their dependence on C75 if the role of this nucleobase was to protonate the leaving group (stabilizing it), but not if C75 functions as a general base, deprotonating the 2′-OH. Kinetic characterization revealed that the sulfur substitution limits the need for C75. The most definitive evidence for divalent cation participation in HDV ribozyme catalysis comes from kinetic analysis, and from the inactivity of the ribozyme in the presence of $Co(NH_3)_6^{3+}$ (Nakano et al. 2000). Thus, in the active site of the HDV ribozyme, a magnesium ion-activated water molecule functions as a general base, and the N3 imine of C75 serves as a general acid (Fig. 4D).

9 HOW WIDESPREAD ARE THE SELF-CLEAVING RIBOZYMES?

The hammerhead and hairpin ribozymes were discovered embedded within the sense and antisense strands, respectively, of the satellite RNA of tobacco ringspot virus (Buzayan et al. 1986b; Prody et al. 1986), and subsequently in other plant satellite virus RNAs and viroids. These viruslike RNAs are single-stranded, covalently closed circular molecules that are replicated via the rolling-circle mechanism (Fig. 5). The ribozyme domains occur at the interface of two monomeric segments of a concatamer, and catalyze the self-cleavage and self-ligation of the genomic and antigenomic RNA segments. Two variants of the HDV ribozyme were subsequently discovered in the sense and antisense strands of HDV, a single-stranded circular RNA virus that is a satellite of the hepatitis B (DNA) virus and a human pathogen. HDV also replicates by the rolling-circle mechanism, and the ribozymes are responsible for generating unit-length genomes. The VS ribozyme was discovered as part of an abundant mitochondrial RNA of natural isolates of Neurospora. The RNA is encoded by a retroplasmid, and the function of the ribozyme is to process into unit length molecules the RNA concatamers that

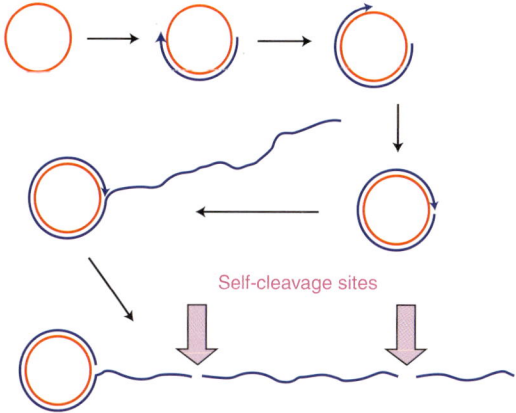

Figure 5. Rolling-circle replication of satellite RNAs. Single-stranded covalently closed circles of sense (or genomic) RNA (red circle) are replicated by an RNA polymerase of the host, which transcribes processively, generating a linear multimeric concatamer (blue) that is complementary to the monomeric satellite RNA sense-strand. The multimeric complementary copy is then cleaved in to monomeric fragments, and these antigenomic RNAs circularize to produce templates for the second half of the replicative cycle, in which copies of the genomic sense strand are ultimately produced. The cleavage and ligation sites are often comprised of conserved ribozyme sequences that catalyze both phosphodiester bond cleavage and, in the case of circularization of the monomeric fragments, phosphodiester bond ligation. In the case of the satellite RNA of tobacco ringspot virus, the cleavage site in the sense strand is a hammerhead ribozyme, and that in the antisense strand is the hairpin ribozyme. HDV ribozyme sequences are present in both the sense and antisense strands of HDV.

result from transcription of the plasmid (Collins 2002). Because the biological contexts of the hammerhead, hairpin, HDV, and VS ribozymes are very similar, it was thought that self-cleaving ribozymes would be restricted to this specialized evolutionary niche.

Hints that these ribozymes might be more widespread in nature began to emerge in the 1990s. Circularizing RNA transcripts of repetitive satellite "junk" DNAs from newt (Epstein and Pabon-Pena 1991) and *Schistosoma* (Ferbeyre et al. 2000) were found to contain hammerhead ribozymes; the latter were found to be exceptionally active. Although genomic scans for known hammerhead sequences aided in discovery of the *Schistosoma* hammerhead (Bourdeau et al. 1999; Ferbeyre et al. 1998), it turned out that the search criteria were too restrictive. If the implicit requirement that the hammerhead ribozyme sequence be continuous is relaxed, a whole set of discontinuous ribozymes that closely resemble the *Schistosoma* sequence can be found in the 3′-UTRs of mature mRNAs that code for the clec2 family of proteins in mammals ranging from platypus to mice (Martick et al. 2008). These hammerhead ribozymes are located immediately downstream from the stop codon, are quite active, and capable of deactivating translation in vivo. These findings suggested self-cleaving RNAs that regulate gene expression may be widespread in the genomes of free-living organisms. Moreover, in vitro selection experiments revealed that the minimal hammerhead ribozyme commonly arises, suggesting it might have evolved multiple times independently (Salehi-Ashtiani and Szostak 2001).

The *glmS* ribozyme was discovered through a bioinformatic search for riboswitches in bacterial genomes (Barrick et al. 2004). Riboswitches, which are not typically catalytic, are gene-regulatory RNAs that bind small molecules and modulate the expression of mRNAs of which they are part (see Breaker 2010 and Garst et al. 2010). The *glmS* ribozyme-riboswitch is present in the 5′-UTRs of mRNAs that encode the essential enzyme GlcN6P synthase in Gram-positive bacteria. The 5′ termini of mRNAs are capped by a triphosphate in Gram-positive bacteria. GlcN6P-activated self-cleavage of the ribozyme domain exposes a 5′-OH, and this recruits an RNase protein that degrades the ribozyme-cleaved mRNA. Because GlcN6P synthase is an unstable protein, degradation of the mRNA completes a negative feedback loop that modulates intracellular GlcN6P concentration (Collins et al. 2007). The *glmS* ribozyme is widespread in Gram-positive bacteria, including important human pathogens. It is at present unknown in Gram-negative bacteria, archaea or Eukarya.

The most recent expansion of the known range of self-cleaving ribozymes comes from searches for HDV-like ribozymes. Salehi-Ashtiani et al. (2006) discovered several self-cleaving RNAs in the human genome through in vitro selection. One of these is conserved throughout mammals in an intron of the *CPEB3* gene. This ribozyme is closely related to the HDV ribozyme. When Lupták and coworkers performed bioinformatic searches for additional HDV/*CPEB3* ribozymes using a structure-based template, they discovered variants of the ribozymes scattered throughout phylogeny, in viruses, bacteria, protists, plants, fungi, and animal phyla including platyhelmynths, nematodes, molluscs, insects, echinoderms, cephalochordates, and fish (Webb et al. 2009). All these RNAs preserve the core structure and catalytic elements of the HDV ribozyme, and many are active in vitro and in vivo. An HDV-like ribozyme from was found to be differentially active in different developmental stages of the African mosquito *Anopheles gambiae*. HDV-like ribozymes occur in dozens of loci in several organisms. However, the genomic location and numbers of ribozymes vary greatly in sister species. Perhaps related to this is that the cleavage sites of several of them occur at the 5′ termini of transposable elements. Although it is too early to draw conclusions about the origins of these RNAs, it is clear now that self-cleaving ribozymes are common elements of genomes throughout the biosphere.

10 CONCLUDING REMARKS

The study of the small self-cleaving ribozymes has revealed that remarkably small RNA units can achieve high sequence specificity and catalytic efficiency. The information content required for activity is even lower than indicated by the length of the ribozymes, because the only requirement for some tracts of their sequences is that A-form double helices be able to form. That is, the composition of many of the helical segments of small ribozymes has little effect on catalytic activity. Thus, RNAs capable of biochemical catalysis are likely to be quite common in sequence space, and even a modest repertory of RNAs of random sequence present at the outset of the RNA world may have contained ribozymes. This conclusion is supported by in vitro selection experiments (e.g., Salehi-Ashtiani and Szostak 2001). In addition, the limited sequence requirements of ribozymes would have been important for primordial RNA organisms whose replicative polymerases may have been error-prone (see Robertson and Joyce 2010). In vitro selection experiments also show that small ribozymes can catalyze a broad range of biochemical transformations (reviewed in Wilson and Szostak 1999), as would have been required in the RNA world if ribozymes were responsible for all biochemical catalysis. Thus far, however, all compact ribozymes known from nature catalyze the same overall transformation, RNA cleavage by internal transesterification. At this point in time, it is unclear if this indicates that small ribozymes that catalyze other reactions remain to be discovered in present-day organisms, or that ribozymes with other biochemical functions have become extinct.

If the advent of biochemical catalysis predates the transition from an RNA world to one in which protein enzymes play the predominant role, it is fair to ask what forms of catalysis might be of such a fundamental nature as to be shared by both RNA and protein enzymes. Metal-ion assisted catalysis and general acid–base catalysis both emerge as contenders for this distinction. Group I and Group II self-splicing introns (see Lambowitz and Zimmerly 2010) share with protein enzymes that polymerize nucleic acids the use of catalytic Mg^{2+} ions (Steitz and Steitz 1993). On the other hand, the similar general acid–base catalytic strategies, employed by the small self-cleaving RNAs and analogous protein enzymes, such as RNase A, are a recurring theme. What is remarkable, in the context of small self-cleaving ribozymes, is the variation within shared catalytic strategies. The hairpin ribozyme is perhaps the most RNase A-like, in that purines play roles analogous to both catalytic histidines of the protein, and both the RNA and the protein employ amine groups to aid in transition-state stabilization. Nonetheless, this is clearly not the only way, and the variability of the acid catalyst in particular indicates a remarkable ability for evolution to fine-tune a ribozyme's catalytic strategy to meet what are presumably different selective pressures. The accelerating pace of discovery of self-cleaving ribozymes in the genomes of contemporary free-living organisms indicates that far from being molecular living fossils, these RNAs are active players in nucleic acid metabolism. Perhaps this is less surprising now, in light of the catalytic versatility of RNA revealed by the study of self-cleaving ribozymes.

ACKNOWLEDGMENTS

The authors dedicate this review to the memory of Frederic M. Richards (1925-2009), and gratefully acknowledge the efforts of past and present members of their research groups. Work in the authors' laboratories summarized here was funded in part by the National Institutes of Health (AI043393 to WGS and GM63576 to ARF) the W.M. Keck Foundation (ARF and WGS) and the Howard Hughes Medical Institute (ARF).

REFERENCES

Barrick JE, Corbino KA, Winkler WC, Nahvi A, Mandal M, Collins J, Lee M, Roth A, Sudarsan N, Jona I, et al. 2004. New RNA motifs suggest an expanded scope for riboswitches in bacterial genetic control. *Proc Natl Acad Sci* **101:** 6421–6426.

Bevilacqua PC. 2003. Mechanistic considerations for general acid-base catalysis by RNA: revisiting the mechanism of the hairpin ribozyme. *Biochemistry* **42:** 2259–2265.

Bevilacqua PC, Yajima R. 2006. Nucleobase catalysis in ribozyme mechanism. *Current Op Chem Biol* **10:** 455–464.

Blount KF, Uhlenbeck OC. 2002. Internal equilibrium of the hammerhead ribozyme is altered by the length of certain covalent cross-links. *Biochemistry* **41:** 6834–6841.

Blount KF, Uhlenbeck OC. 2005. The structure-function dilemma of the hammerhead ribozyme. *Annu Rev Biophys Biomol Struct* **34:** 410–440.

Bourdeau V, Ferbeyre G, Pageau M, Paquin B, Cedergren R. 1999. The distribution of RNA motifs in natural sequences. *Nucleic Acids Res* **27:** 4457–4467.

Butcher SE, Heckman JE, Burke JM. 1995. Reconstitution of hairpin ribozyme activity following separation of functional domains. *J Biol Chem* **270:** 29648–29651.

Buzayan JM, Gerlach WL, Bruening G. 1986a. Non-enzymatic cleavage and ligation of RNAs complementary to a plant virus satellite RNA. *Nature* **323:** 349–353.

Buzayan JM, Hampel A, Bruening G. 1986b. Nucleotide sequence and newly formed phosphodiester bond of spontaneously ligated satellite tobacco ringspot virus RNA. *Nucleic Acids Res* **14:** 9729–9743.

Breaker RR. 2010. Riboswitches and the RNA world. *Cold Spring Harb Perspect Biol* doi: 10.1101.cshperspect.a003566.

Canny MD, Jucker FM, Pardi A. 2007. Efficient ligation of the *Schistosoma* hammerhead ribozyme. *Biochemistry* **46:** 3826–3834.

Chi Y-I, Martick M, Lares M, Kim R, Scott WG, Kim S-H, Joyce GF. 2008. Capturing hammerhead ribozyme structures in action by modulating general base catalysis. *Plos Biol* **6:** e234.

Collins R. 2002. The *Neurospora* varkud satellite ribozyme. *Biochem Soc Trans* **30:** 1122–1126.

Collins JA, Irnov I, Baker S, Winkler WC. 2007. Mechanism of mRNA destabilization by the *glmS* ribozyme. *Genes Dev* **21:** 3356–3368.

Das SR, Piccirilli JA. 2005. General acid catalysis by the hepatitis delta virus ribozyme. *Nature Chem Biol* **1:** 45–52.

de Gennes P-G. 1990. Minimum numer of aminoacids required to build up a specific receptor with a folded polypeptide chain. In *Introduction to polymer dynamics* Cambridge, Cambridge University Press, pp. 17–26.

De la Peña M, Gago S, Flores R. 2003. Peripheral regions of natural hammerhead ribozymes greatly increase their self-cleavage activity. *EMBO J* **22:** 5561–5570.

Epstein LM, Pabon-Pena LM. 1991. Alternative modes of self-cleavage by newt satellite 2 transcripts. *Nucleic Acids Res* **19:** 1699–1705.

Fedor MJ. 1999. Tertiary structure stabilization promotes hairpin ribozyme ligation. *Biochemistry* **38:** 11040–11050.

Ferbeyre G, Smith JM, Cedergren R. 1998. Schitosome satellite DNA encodes active hammerhead ribozymes. *Mol Cell Biol* **18:** 3880–3888.

Ferré-D'Amaré AR, Zhou K, Doudna JA. 1998. Crystal structure of a hepatitis delta virus ribozyme. *Nature* **395:** 567–574.

Ferbeyre G, Bourdeau V, Pageau M, Miramontes P, Cedergren R. 2000. Distribution of hammerhead and hammerhead-like RNA motifs through the GenBank. *Genome Res* **10:** 1011–1019.

Garst AD, Edwards AL, Batey RT. 2010. Riboswitches: structures and mechanisms. *Cold Spring Harb Perspect Biol* doi: 10.1101.cshperspect.a003533.

Hampel A, Cowan JA. 1997. A unique mechanism for RNA catalysis: The role of metal cofactors in hairpin ribozyme cleavage. *Chem Biol* **4:** 513–517.

Haseloff J, Gerlach WL. 1989. Sequences required for self-catalyzed cleavage of the satellite RNA of tobacco ringspot virus. *Gene* **82:** 43–52.

Jones FD, Ryder SP, Strobel SA. 2001. An efficient ligation reaction promoted by a Varkud Satellite ribozyme with extended 5′- and 3′-termini. *Nucleic Acids Res* **29:** 5115–5120.

Khvorova A, Lescoute A, Westhof E, Jayasena S. 2003. Sequence elements outside the hammerhead ribozyme catalytic core enable intracellular activity. *Nat Struct Biol* **10:** 708–712.

Klein DJ, Ferré-D'Amaré AR. 2006. Structural basis of *glmS* ribozyme activation by glucosamine-6-phosphate. *Science* **313:** 1752–1756.

Klein DJ, Been MD, Ferré-D'Amaré AR. 2007a. Essential role of an active-site guanine in glmS ribozyme catalysis. *J Am Chem Soc* **129**: 14858–14859.

Klein DJ, Wilkinson SR, Been MD, Ferré-D'Amaré AR. 2007b. Requirement of helix P2.2 and nucleotide G1 for positioning of the cleavage site and cofactor of the glmS ribozyme. *J Mol Biol* **373**: 178–189.

Kuzmin YI, Da Costa CP, Cottrell JW, Fedor MJ. 2005. Role of an active site adenine in hairpin ribozyme catalysis. *J Mol Biol* **349**: 989–1010.

Lambowitz AM, Zimmerly S. 2010. Group II introns: mobile ribozymes that invade DNA. *Cold Spring Harb Perspect Biol* doi: 10.1101/cshperspect.a003616.

Lebruska LL, Kuzmine II, Fedor MJ. 2002. Rescue of an abasic hairpin ribozyme by cationic nucleobases. Evidence for a novel mechanism of RNA catalysis. *Chem Biol* **9**: 465–473.

Lilley DM. 2004. The Varkud satellite ribozyme. *RNA* **10**: 151–158.

Liu L, Cottrell JW, Scott LG, Fedor MJ. 2009. Direct measurement of the ionization state of an essential guanine in the hairpin ribozyme. *Nature Chem Biol* **5**: 351–357.

Martick M, Scott WG. 2006. Tertiary contacts distant from the active site prime a ribozyme for catalysis. *Cell* **126**: 309–320.

Martick M, Horan LH, Noller HF, Scott WG. 2008. A discontinuous hammerhead ribozyme embedded in a mammalian messenger RNA. *Nature* **454**: 899–902.

McKay DB. 1996. Three-dimensional structure of the hammerhead ribozyme. *Nucleic Acids Mol Biol* **10**: 161–172.

Murchie AIH, Thomson JB, Walter F, Lilley DMJ. 1998. Folding of the hairpin ribozyme in its natural conformation achieves close physical proximity of the loops. *Mol Cell* **1**: 873–881.

Murray JB, Seyhan AA, Walter NG, Burke JM, Scott WG. 1998. The hammerhead, hairpin and VS ribozymes are catalytically proficient in monovalent cations alone. *Chem Biol* **5**: 587–595.

Murthy VL, Rose GD. 2000. Is counterion delocalization responsible for collapse in RNA folding. *Biochemistry* **39**: 14365–14370.

Nakano S-I, Chadalavada DM, Bevilacqua PC. 2000. General acid-base catalysis in the mechanism of a hepatitis delta virus ribozyme. *Science* **287**: 1493–1497.

Nesbitt S, Hegg LA, Fedor MJ. 1997. An unusual pH-independent and metal-ion independent mechanism for hairpin ribozyme catalysis. *Chem Biol* **4**: 619–630.

Perrotta AT, Been MD. 1991. A pseudoknot-like structure required for efficient self-cleavage of hepatitis delta virus RNA. *Nature* **350**: 434–436.

Perrotta AT, Shih I, Been MD. 1999. Imidazole rescue of a cytosine mutation in a self-cleaving ribozyme. *Science* **286**: 123–126.

Pleij CW. 1990. Pseudoknots: A new motif in the RNA game. *Trends Biochem Sci* **15**: 143–147.

Pley HW, Flaherty KM, McKay DB. 1994. Three-dimensional structure of a hammerhead ribozyme. *Nature* **372**: 68–74.

Prody GA, Bakos JT, Buzayan JM, Schneider IR, Bruening G. 1986. Autolytic processing of dimeric plant virus satellite RNA. *Science* **231**: 1577–1580.

Pyle AM. 1993. Ribozymes: A distinct class of metalloenzymes. *Science* **261**: 709–714.

Quigley GJ, Rich A. 1976. Structural domains of transfer RNA molecules. *Science* **194**: 796–806.

Raines RT. 1998. Ribonuclease A. *Chem Rev* **98**: 1045–1065.

Richards FM, Wyckoff HW, Carlson WD, Allewell NM, Lee B, Mitsui Y. 1971. Protein structure, ribonuclease-S, and nucleotide interactions. *Cold Spring Harbor Symp Quant Biol* **36**: 35–43.

Robertson MP, Joyce GF. 2010. The origins of the RNA world. *Cold Spring Harb Perspect Biol* doi: 10.1101.cshperspect.a003608.

Roth A, Nahvi A, Lee M, Jona I, Breaker RR. 2006. Characteristics of the glmS ribozyme suggest only structural roles for divalent metal ions. *RNA* **12**: 607–619.

Ruffner DE, Stormo GD, Uhlenbeck OC. 1990. Sequence requirements of the hammerhead RNA self-cleavage reaction. *Biochemistry* **29**: 10695–10702.

Rupert PB, Ferré-D'Amaré AR. 2001. Crystal structure of a hairpin ribozyme-inhibitor complex with implications for catalysis. *Nature* **410**: 780–786.

Rupert PB, Massey AP, Sigurdsson ST, Ferré-D'Amaré AR. 2002. Transition state stabilization by a catalytic RNA. *Science* **298**: 1421–1424.

Salehi-Ashtiani K, Szostak JW. 2001. In vitro evolution suggests multiple origins for the hammerhead ribozyme. *Nature* **414**: 82–84.

Salehi-Ashtiani K, Lupták A, Litovchick A, Szostak JW. 2006. A genome-wide search for ribozymes reveals an HDV-like sequence in the human CPEB3 gene. *Science* **313**: 1788–1792.

Saville BJ, Collins RA. 1990. A site-specific self-cleavage reaction performed by a novel RNA in *Neurospora* mitochondria. *Cell* **61**: 685–696.

Saville BJ, Collins RA. 1991. RNA-mediated ligation of self-cleavage products from a *Neurospora* mitochondrial plasmid transcript. *Proc Natl Acad Sci USA* **88**: 8826–8830.

Scott WG. 1999. Biophysical and biochemical investigations of RNA catalysis in the hammerhead ribozyme. *Q Rev Biophys* **32**: 241–284.

Scott WG, Finch JT, Klug A. 1995. The crystal structure of an all-RNA hammerhead ribozyme: A proposed mechanism for RNA catalytic cleavage. *Cell* **81**: 991–1002.

Sharmeen L, Kuo MY-P, Dinter-Gottlieb G, Taylor J. 1988. Antigenomic RNA of human hepatitis delta virus can undergo self-cleavage. *J Virol* **62**: 2674–2679.

Shin C, Choi JN, Song SI, Song JT, Ahn JH, Lee JS, Choi YD. 1996. The loop B domain is physically separable from the loop A domain in the hairpin ribozyme. *Nucleic Acids Res* **24**: 2685–2689.

Steitz TA, Steitz JA. 1993. A general two-metal-ion mechanism for catalytic RNA. *Proc Natl Acad Sci* **90**: 6498–6502.

Symons RH. 1997. Plant pathogenic RNAs and RNA catalysis. *Nucl Acids Res* **25**: 2683–2689.

Uhlenbeck OC. 1987. A small catalytic oligoribonucleotide. *Nature* **328**: 596–600.

Webb C-HT, Riccitelli NJ, Ruminski DJ, Lupták A. 2009. Widespread occurrence of self-cleaving ribozymes. *Science* **326**: 953.

Wedekind JE, McKay DB. 1998. Crystallographic structures of the hammerhead ribozyme: relationship to ribozyme folding and catalysis. *Annu Rev Biophys Biomol Struct* **27**: 475–502.

Wilson DS, Szostak JW. 1999. In vitro selection of functional nucleic acids. *Annu Rev Biochem* **68**: 611–647.

Winkler WC, Nahvi A, Roth A, Collins JA, Breaker RR. 2004. Control of gene expression by a natural metabolite-responsive ribozyme. *Nature* **428**: 281–286.

Young KJ, Gill F, Grasby JA. 1997. Metal ions play a passive role in the hairpin ribozyme catalyzed reaction. *Nucleic Acids Res* **25**: 3760–3766.

Group II Introns: Mobile Ribozymes that Invade DNA

Alan M. Lambowitz[1] and Steven Zimmerly[2]

[1]Institute for Cellular and Molecular Biology, Department of Chemistry and Biochemistry, and Section of Molecular Genetics and Microbiology, School of Biological Sciences, University of Texas at Austin, Austin, Texas 78712

[2]Department of Biological Sciences, University of Calgary, Calgary, Alberta T2N 1N4, Canada

Correspondence: lambowitz@mail.utexas.edu

SUMMARY

Group II introns are mobile ribozymes that self-splice from precursor RNAs to yield excised intron lariat RNAs, which then invade new genomic DNA sites by reverse splicing. The introns encode a reverse transcriptase that stabilizes the catalytically active RNA structure for forward and reverse splicing, and afterwards converts the integrated intron RNA back into DNA. The characteristics of group II introns suggest that they or their close relatives were evolutionary ancestors of spliceosomal introns, the spliceosome, and retrotransposons in eukaryotes. Further, their ribozyme-based DNA integration mechanism enabled the development of group II introns into gene targeting vectors ("targetrons"), which have the unique feature of readily programmable DNA target specificity.

Outline

1. Introduction
2. Characteristics of group II introns
3. Reactions catalyzed by group II intron RNAs
4. Three-dimensional structure of group II intron ribozymes
5. The involvement of proteins in group II intron splicing
6. Group II intron mobility
7. Group II intron evolution

References

1 INTRODUCTION

Group II introns are mobile genetic elements that are found in bacterial and organellar genomes and are thought to be ancestors of spliceosomal introns and retrotransposons in eukaryotes. They consist of a catalytically active intron RNA ("ribozyme") and an intron-encoded protein (IEP), whose combined activities enable intron proliferation within genomes. The group II intron RNA catalyzes its own splicing via transesterification reactions that are the same as those of spliceosomal introns, yielding spliced exons and an excised intron lariat RNA. The IEP is a multifunctional reverse transcriptase (RT), which is related to non-LTR-retrotransposon RTs and assists splicing by stabilizing the catalytically active RNA structure. It then remains bound to the excised intron RNA in an RNP that invades DNA sites. DNA invasion occurs by a remarkable mechanism in which the intron RNA uses its ribozyme activity to reverse splice directly into a DNA strand, after which it is reverse transcribed back into DNA by the IEP. Cycles of RNA splicing and reverse splicing enable the introns to proliferate to new DNA sites, while minimally impairing gene expression.

The characteristics of group II introns, including their splicing and mobility mechanisms, active-site structure, and naturally occurring variants that are split into two or more functionally reassociating segments, suggest an evolutionary scenario for the origin of introns, the spliceosome, and retrotransposons in eukaryotes. An evolutionary relationship between group II and spliceosomal introns appears increasingly plausible in light of newly obtained structural information, and indeed, a recent hypothesis asserts that group II intron invasion was the major driving force for the emergence of eukaryotes (Martin and Koonin 2006). Group II introns have also found practical applications as novel gene targeting vectors ("targetrons"), whose ribozyme-based DNA-integration mechanism enables their ready reprogramming to insert into desired DNA sites with high efficiency and specificity.

Here, we present an overview of group II intron structure and their splicing and mobility mechanisms, including recent insights from X-ray crystal structures of a group II intron RNA. We incorporate what has been learned into an evolutionary framework that considers the origin of group II introns, their structural variations, and how they may have evolved into spliceosomal introns and retrotransposons in eukaryotes.

2 CHARACTERISTICS OF GROUP II INTRONS

2.1 Phylogenetic Distribution

Group II introns have been found in bacteria and in the mitochondrial (mt) and chloroplast (cp) genomes of fungi, plants, protists, and an annelid worm (Belfort et al. 2002; Lambowitz and Zimmerly 2004; Toro et al. 2007; Vallès et al. 2008). Group II introns are rare in archaea, with the few found there likely acquired from eubacteria by relatively recent horizontal transfers (Rest and Mindell 2003). In eubacteria, group II introns are present in ~25% of sequenced genomes, generally in small numbers, and typically as active retroelements with functional ribozyme and RT components. By contrast, group II introns in organelles frequently have degenerate RNA structures and either lack ORFs or encode degenerate IEPs that no longer promote intron mobility (Michel and Ferat 1995; Bonen 2008; Barkan 2009). Such immobile group II introns are inherited vertically and rely on host-encoded splicing factors (see later). Group II introns have not been found in the nuclear genomes of eukaryotes, but their hypothesized descendants, spliceosomal introns and retrotransposons, are highly abundant in eukaryotes, together comprising more than half of the human genome.

2.2 Group II Intron RNAs

Group II intron RNAs are characterized by a conserved secondary structure, which spans 400-800 nts and is organized into six domains, DI-VI, radiating from a central "wheel" (Fig. 1) (Michel and Ferat 1995; Qin and Pyle 1998; Pyle and Lambowitz 2006). These domains interact to form a conserved tertiary structure that brings together distant sequences to form an active site. The active site binds the splice sites and branch-point nucleotide residue and uses specifically bound Mg^{++} ions to activate the appropriate bonds for catalysis. DV is the heart of the active site and contains the so-called catalytic triad AGC and an AY bulge, both of which bind catalytically important Mg^{++} ions. DI is the largest domain, with upper and lower halves separated by the κ and ζ motifs. The lower half contains the ε′ motif, which is associated with the active site, while the upper half contains sequence elements that bind the 5′ and 3′ exons at the active site. DVI contains the branch-point nucleotide, generally a bulged A. DII and III are smaller domains that contribute structurally, with DIII also acting as a "catalytic effector" to accelerate the catalytic step (Fedorova et al. 2003). The DIV loop encodes the IEP, with subdomain DIVa near its 5′ end containing a high-affinity binding site for the IEP. Group II intron RNAs also have conserved 5′- and 3′-end sequences, GUGYG and AY, respectively, which resemble those of spliceosomal introns (GU...AG) and which are bound at the active site by the ε–ε′ and γ–γ′ interactions.

Critical for the folding of group II intron RNAs are a series of conserved motifs involved in long-range tertiary interactions (Greek letters, EBS (exon-binding site), or

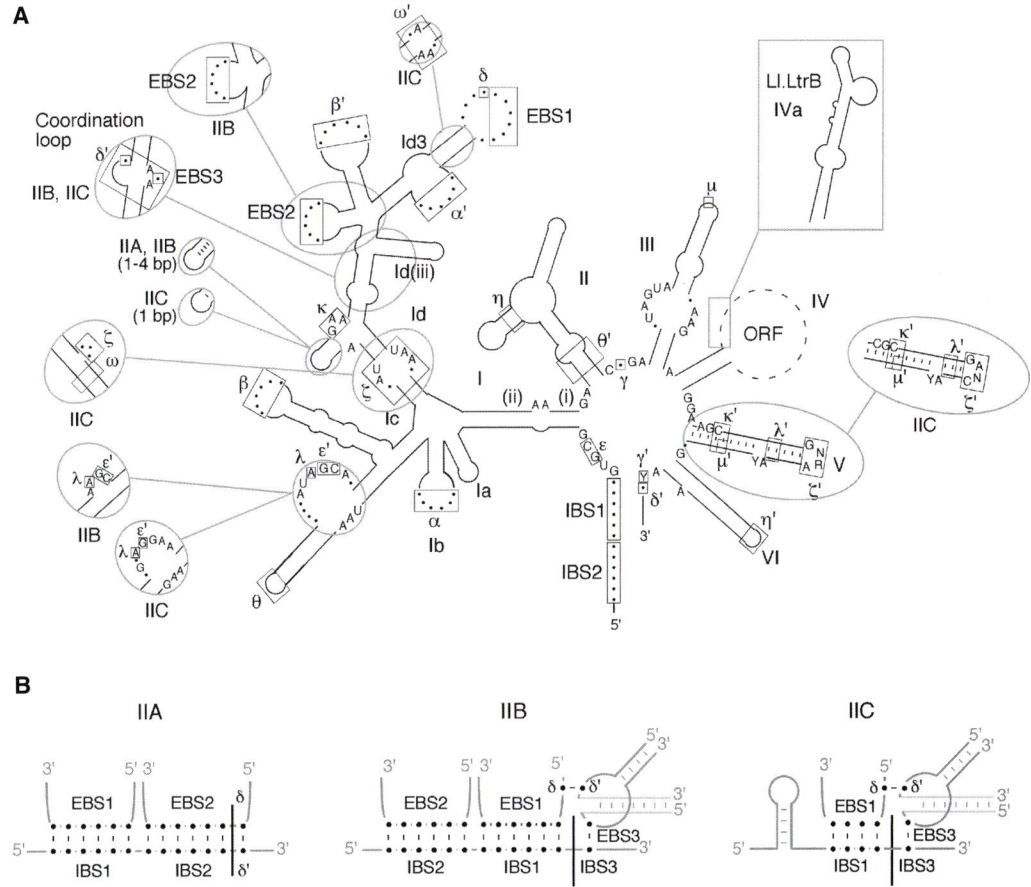

Figure 1. Group II intron RNA secondary structure. (A) Structure of a representative bacterial IIA1 intron (not to scale), with notable variations in IIB and IIC introns shown in circles. Boxes indicate sequences involved in tertiary interactions (Greek letters, EBS, IBS). The "loop" of DIV, which encodes the IEP, is depicted by dashed lines, with a box showing the location and structure of DIVa of the *Lactococcus lactis* Ll.LtrB intron, a high-affinity binding site for the IEP. Subdomains discussed in the text are labeled, with base pairs (dashes) shown only for DV and the κ-stem-loop. Compared to IIA introns, major differences in other subgroups include structural features of DV (IIC introns); different ε′ motifs (IIB, IIC); the number of base pairs in the κ-stem-loop (IIC); a coordination loop containing EBS3 and δ′ (IIB, IIC); the absence of the DId(iii) stem-loop (IIB, IIC); the absence of a stem in the EBS2 motif (IIB, IIC); a unique ζ–ζ′ motif (IIC); and the ω–ω′ interaction (IIC, some IIB). (B) Base-pairing interactions used by IIA, IIB, and IIC introns to bind the exons at the active site. EBS, exon-binding site; IBS, intron-binding site.

IBS (intron-binding site) (Fig. 1). Such interactions have been identified by systematic covariation, genetic, and biochemical analyses. Some involve Watson-Crick base pairing (IBS1-EBS1, IBS2-EBS2, IBS3-EBS3, α–α′, β–β′, δ–δ′, ε–ε′, and γ–γ′), whereas others are tetraloop-receptor interactions (ζ–ζ′, θ–θ′, and η–η′) or other types of non-Watson-Crick interactions (κ–κ′, λ–λ′, and μ–μ′) (Qin and Pyle 1998; Costa et al. 2000; Boudvillain et al. 2000; Fedorova and Pyle 2005). The tertiary interactions between domains make it possible for group II intron RNAs to be split readily into different *trans*-splicing segments (Belhocine et al. 2008; Glanz and Kück 2009) and for some domains (DIc, DIII, DV, DI/II/III/IV) to act in *trans* to promote the splicing of introns lacking them (Jarrell et al. 1988a; Goldschmidt-Clermont et al. 1991; Suchy and Schmelzer 1991). Such fragmented group II introns and *trans*-acting segments occur naturally and underlie evolutionary scenarios for the origin of snRNAs (see later).

Although all group II introns have similar overall secondary structures, three major subgroups, denoted IIA, IIB, and IIC, and further subdivisions (A1, A2, B1, B2) are distinguished by specific variations (Fig. 1A; Michel et al. 1989; Toor et al. 2001; Dai et al. 2008). Unlike group I introns, in which subgroups differ mainly in peripheral structures, the differences between group II intron subgroups extend to the active site. One defining difference involves the interactions that bind the exons at the active site, which are critical for both RNA splicing and DNA target site recognition during intron mobility (Fig. 1B). IIA and IIB introns recognize their flanking exons via three base-pairing interactions (IBS1/EBS1 and IBS2/EBS2 for the 5′ exon and δ–δ′ (IIA) or IBS3/EBS3 (IIB) for the 3′ exon), while IIC introns use only two of these interactions (IBS1/EBS1 and IBS3/EBS3) and may also recognize a stem-loop derived from a transcription terminator or *attC* site in the 5′ exon (Fig. 1B) (Toor et al. 2006). Group II intron subgroups also differ in DV and in tertiary interaction motifs (see Fig. 1 legend). The latter include the "coordination loop" containing EBS3 and δ′, which is present in IIB and IIC but not IIA introns. The coordination loop may also position the branch-point A at the active site (Hamill and Pyle 2006), but this proposal has been questioned (Michel et al. 2009). A notable subfamily of IIB introns contains two additional stem-loops between DVI and the 3′-splice site (Stabell et al. 2009).

2.3 Degenerate, Twintron, and *Trans*-Splicing Group II Introns

Many organellar group II introns have structural defects that impair ribozyme activity. Plant mt and cp DNAs, for example, each encode about twenty group II introns, none of which is self-splicing or mobile (Barkan 2004). Their structural deviations include mispairs, insertions and deletions in DV and DVI, the absence of the bulged A in DVI, and subdomains that are unrecognizable or absent (Michel et al. 1989; Bonen 2008). The most extreme examples are found in *Euglena* cp DNA, which contains ~150 introns lacking various domains. The smallest, referred to as group III introns, are ~100 nt and contain only DI-like and DVI-like structures, lacking even DV, which is catalytically essential (Copertino and Hallick 1993). The splicing of these highly degenerate introns presumably requires *trans*-acting RNAs or proteins that compensate for the missing RNA structures.

Group II introns sometimes form twintrons in which one intron has inserted into another, reflecting their activity as mobile elements (Copertino and Hallick 1993). In some cases, the invading intron disrupts a critical feature of the outer intron and must be spliced first (Drager and Hallick 1993), while in other cases, the invading intron need not be spliced before the outer intron (Nakamura et al. 2002). Some twintrons form large nested arrays by repeated insertion of one group II intron into another (Dai and Zimmerly 2003).

Finally, *trans*-splicing group II introns are variants caused by genomic rearrangements that split an intron into two or more separately transcribed segments (Glanz and Kück 2009). These segments reassociate via tertiary contacts between group II intron domains and *trans*-splice the exons to produce a functional mRNA. *Trans*-splicing group II introns split at different locations have arisen naturally many times, particularly in plant mitochondria and chloroplasts (Qiu and Palmer 2004; Bonen 2008). One example is mt *nad1*-I4, which is continuous in some flowering plants, but *trans*-splicing in others, the only significant difference in the intron being its division into two independently transcribed segments (Bonen 1993). More extreme examples are cp *psaA*-I2 in *Chlamydomonas* spp. and mt *nad5*-I3 in some angiosperms, which are transcribed in three segments, with the middle segment beginning in DI and ending in DIV (Goldschmidt-Clermont et al. 1991; Knoop et al. 1997).

2.4 Group II Intron-Encoded Proteins

Most group II introns in bacteria and about half in mitochondria and chloroplasts encode an IEP in the loop of DIV. The best characterized IEP is the LtrA protein encoded by the *Lactococcus lactis* Ll.LtrB intron (Fig. 2A). LtrA has four domains: RT, reverse transcriptase; X/thumb; D, DNA binding; and En, DNA endonuclease. The RT domain is characterized by seven conserved sequence blocks (RT1-7) that form the fingers and palm regions of retroviral RTs, with RT5 containing the highly conserved sequence YADD that is part of the RT active site (Xiong and Eickbush 1990). Although presumably having a similar fold, group II intron and non-LTR-retrotransposon RTs are larger than those of retroviral RTs because of an amino-terminal extension (RT0) and "insertions" between the RT sequence blocks (RT2a, 3a, 4a, 7a in LtrA; cf., HIV-1 RT in Fig. 2F). Some of these insertions have conserved structural features and may be functionally important (Malik et al. 1999; Blocker et al. 2005).

Domain X is sometimes referred to as the "maturase domain" because it was identified as a site of mutations affecting RNA splicing activity (Mohr et al. 1993; Moran et al. 1994). It is characterized by conserved sequences and three predicted α-helices that are structurally analogous to the thumb domain of retroviral RTs (Blocker et al. 2005). In addition to reverse transcription, the RT and X/thumb domains function together in specific binding of the intron RNA, which both promotes formation of the active ribozyme structure and positions the protein to initiate cDNA synthesis (Saldanha et al. 1999; Wank et al. 1999; Cui et al. 2004).

The carboxy-terminal D and En domains interact with the target DNA during intron mobility. Domain D contributes to DNA binding, whereas the En domain is a Mg^{++}-dependent DNA endonuclease of the H-N-H family that cleaves a target DNA strand to generate the primer for

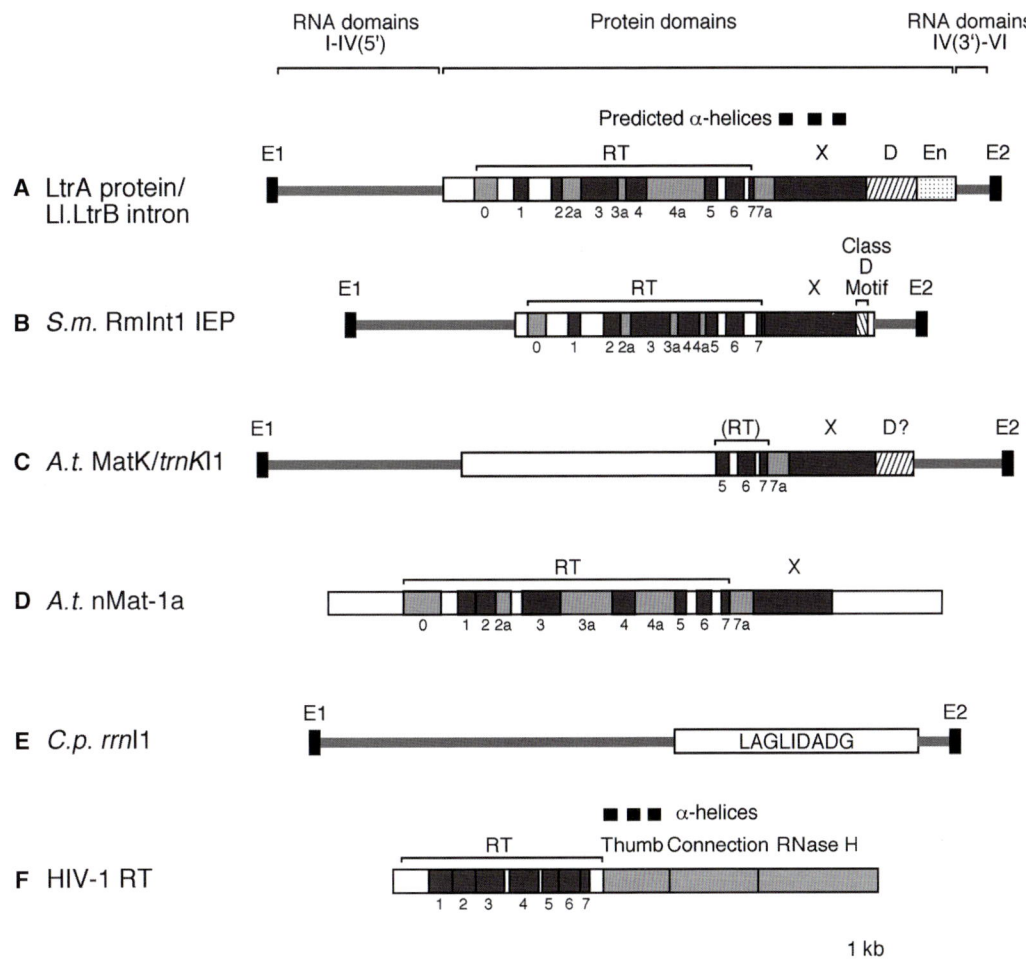

Figure 2. Group II intron IEPs and related proteins. (A) LtrA protein encoded by the *L. lactis* Ll.LtrB intron. (B) IEP lacking an En domain encoded by the *Sinorhizobium meliloti* RmInt1 intron, which belongs to bacterial lineage D (see Fig. 3). The "class D motif" at the carboxy-terminus is a conserved sequence that is required for splicing and mobility functions in lineage D IEPs (Molina-Sánchez et al. 2010). (C) MatK protein encoded by the *Arabidopsis thaliana* trnKI1 intron. MatK proteins retain conserved sequence blocks RT5-7 and domain X, but their amino-terminal halves have diverged from those of canonical group II IEPs, and they lack an En domain (Mohr et al. 1993). (D) nMat-1a protein encoded by a nuclear gene in *Arabidopsis thaliana*. nMat-1 proteins contain complete RT and X domains, but have mutations expected to inhibit RT activity; nMat-2 proteins (not shown) also contain an En domain, but with mutations expected to inhibit En activity (Mohr and Lambowitz 2003). (E) LAGLIDADG protein encoded by *Cryphonectria parasitica* rrnI1. (F) HIV-1 RT. Schematics of introns and ORFs are to scale. Insertions between RT sequence blocks are denoted 2a, 3a, 4a, and 7a. The locations of the three-predicted α-helices characteristic of thumb domains are shown above domain X in LtrA (cf. with HIV-1 RT in panel F).

reverse transcription (San Filippo and Lambowitz 2002). Some bacterial group II introns, the best studied of which is *Sinorhizobium meliloti* RmInt1, encode IEPs that lack an En domain and use a different mechanism to prime reverse transcription (Fig. 2B) (see later).

Many group II IEPs have lost conserved sequences required for RT activity, but continue to function in RNA splicing, which is essential for gene expression after intron insertion. Some of these degenerate IEPs have small changes (e.g., mutations in the conserved YADD motif at the RT active site), whereas others are highly degenerate.

The latter include two prominent examples of plant IEPs that have evolved to splice multiple organellar group II introns: cp MatK proteins (Fig. 2C) (Mohr et al. 1993) and nuclear-encoded nMat proteins (nMat-1a, -1b, -2a, -2b) (Fig. 2D) (Mohr and Lambowitz 2003) (see later).

Finally, a small group of fungal mt group II introns encode IEPs belonging to a family of DNA endonucleases characterized by the conserved motif LAGLIDADG (Fig. 2E) (Toor and Zimmerly 2002). These enzymes are often encoded in group I introns, where they function to promote intron homing by double-strand-break stimulated homologous recombination and sometimes act as maturases to promote RNA splicing (Belfort et al. 2002). Whether they function similarly in group II intron splicing and mobility is unknown.

2.5 Group II Intron Lineages

Group II intron ribozymes and IEPs function together as RNPs, with each IEP binding specifically to the intron RNA that encodes it. As a result, the intron RNAs and IEPs have coevolved over long times to form phylogenetic lineages of mobile introns (Fontaine et al. 1997; Toor et al. 2001). This situation again contrasts with that of group I introns, whose IEPs generally act independently as homing endonucleases and are frequently exchanged among introns (Belfort et al. 2002). Phylogenetic analyses identified eight lineages of group II intron IEPs, termed bacterial classes A-F, ML (mitochondrial-like) and CL (chloroplast-like), the latter because they are the major lineages in mitochondria and chloroplasts, respectively (Zimmerly et al. 2001; Simon et al. 2008) (Fig. 3). Each IEP lineage is associated with a specific RNA subgroup: ML with IIA, bacterial class C with IIC, and the remainder with IIB RNAs. CL IEPs are associated with IIB1 and IIB2 RNAs, while bacterial A, B, D, E, and F IEPs are associated with less typical IIB structures (Simon et al. 2009). Notably, bacteria contain all group II intron lineages, while mitochondria and chloroplasts contain both ML and CL introns but not other lineages. This distribution may reflect that ML and CL introns were present in bacterial endosymbionts that colonized eukaryotes and then exchanged between the two organelles.

3 REACTIONS CATALYZED BY GROUP II INTRON RNAs

Group II ribozymes catalyze their own splicing via two sequential transesterification reactions (Fig. 4A). In the first step, the 2′ OH of the bulged A in DVI acts as the nucleophile to attack the 5′-splice site, producing an intron lariat/3′-exon intermediate. In the second step, the 3′ OH of the cleaved 5′ exon is the nucleophile and attacks the 3′-splice site, resulting in exon ligation and excision of an intron lariat RNA. Some group II introns self-splice in vitro, but the reaction is generally slow (k_{obs}= 0.2 − 1.0 × 10^{-2}/min) and requires nonphysiological conditions—e.g., high concentrations of monovalent salt and/or Mg^{++} (Jarrell et al. 1988b; Daniels et al. 1996; Hiller et al. 2000), reflecting that proteins are needed to help fold group II intron RNAs into the catalytically active structure for efficient splicing. An important variation of the splicing reaction, termed "hydrolytic splicing," involves the use of

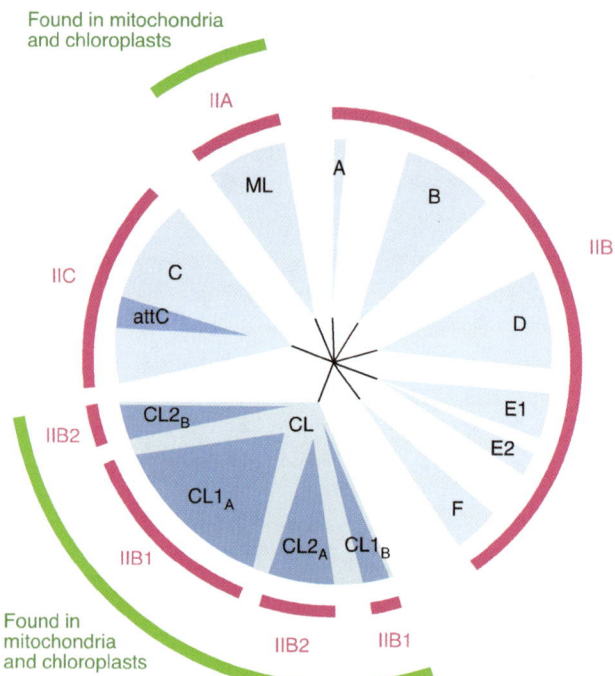

Figure 3. Group II intron lineages. The major lineages of group II intron IEPs, denoted CL (chloroplast-like), ML (mitochondrial-like), and bacterial classes A-F, are shown as blue sectors. Notable sublineages, including four subdivisions of CL and a subclass of IIC introns that inserts after *attC* sites, are shown as darker blue sectors within the major lineages. RNA structural subgroups that correspond to IEP lineages are shown in magenta. All group II intron lineages and RNA types are found in bacteria. Lineages and RNA types also found in organelles are delineated in green (outer circle). Note that there may be limited exceptions to the overall pattern of coevolution within the CL group, with different sublineages possibly having exchanged IIB RNA structures (Simon et al. 2009). An alternate nomenclature for group II lineages has been proposed, which does not distinguish between IEP and ribozyme lineages or take into account exceptions to their coevolution (Toro et al. 2002).

water as the nucleophile for the first transesterification reaction, rather than the 2′ OH of the bulged A of DVI (Fig. 4B) (van der Veen et al. 1987; Jarrell et al. 1988b). Some group II introns can splice exclusively via this pathway in vivo (Podar et al. 1998a; Bonen 2008).

The active site for the splicing reaction contains at least two specifically bound Mg^{++} ions, which appear to be associated with the AGC triad and AY bulge in DV based on thio substitution/rescue and metal ion cleavage experiments (Chanfreau and Jacquier 1994; Sigel et al. 2000; Gordon and Piccirilli 2001; Gordon et al. 2007). For group I introns, *R*p- and *S*p-phosphorothioate substitutions at the splice sites have opposite effects on the two transesterification steps, which are simple reversals of each other at the same active site (McSwiggen and Cech 1989). By contrast, group II and spliceosomal introns show strong sensitivity to *R*p but not *S*p substitutions for both steps (Moore

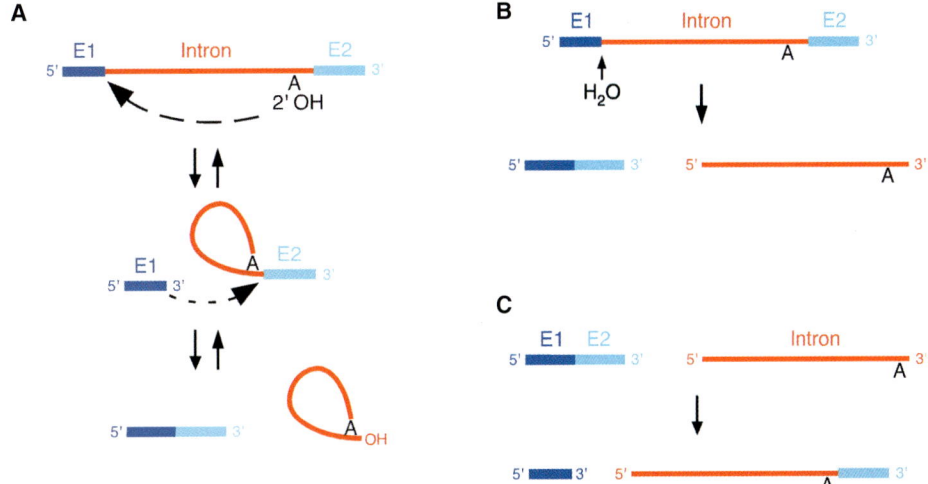

Figure 4. Reactions catalyzed by group II introns RNAs. (*A*) Forward and reverse splicing. (*B*) Hydrolytic splicing. The initial step is hydrolytic cleavage at the 5′ splice site. The second step leading to exon ligation (not shown) is the same as for splicing via lariat formation (panel A). (*C*) Partial reverse splicing by linear intron RNA, leading to ligation of the 3′ end of the intron RNA to the 5′ end of the 3′ exon. Intron RNA, red; 5′ and 3′ exons (E1 and E2), dark and light blue, respectively.

and Sharp 1993; Padgett et al. 1994; Podar et al. 1995; Podar et al. 1998b). These and other findings suggest that group II introns either use separate active sites with different Mg^{++} ions to catalyze the two splicing steps or a single active site, which is rearranged between steps (discussed in Pyle and Lambowitz (2006)). The active sites for the two steps at least partially overlap, as all of the reacting groups for both steps are in close proximity prior to splicing (de Lencastre et al. 2005) and mutations to key residues block both steps (Chanfreau and Jacquier 1994). A conformational change that occurs after the first step involves the formation of the η–η′ interaction between the distal loops of DII and DVI and is thought to remove the branch-point A from the active site (Chanfreau and Jacquier 1996). Other conformational changes may be linked to exon binding and release (Costa and Michel 1999).

Because transesterification reactions are reversible and energetically neutral, excised group II intron RNAs can reverse splice into ligated exons, guided by the same EBS/IBS and δ–δ′ base pairing interactions between the intron and flanking exon sequences used for RNA splicing. Lariat RNAs can carry out both reverse splicing steps, resulting in insertion of the intron RNA between the 5′ and 3′ exons (Fig. 4A, reverse arrows), while linear intron RNAs can carry out only the first step, ligation of the 3′ end of the intron to the 3′ exon (Fig. 4C) (Mörl and Schmelzer 1990; Mörl et al. 1992; Roitzsch and Pyle 2009). Reverse splicing of lariat or linear introns occurs into DNA as well as RNA exons, with a relatively small decrease of reactivity, reflecting a modest contribution of the 2′ OHs of the RNA exons (Mörl et al. 1992; Griffin et al. 1995). The ability to reverse splice efficiently into DNA underlies group II intron mobility.

Group II intron RNAs can carry out a number of other reactions in vitro, but their biological significance is unclear. These include: spliced-exon reopening (SER) in which the lariat intron hydrolytically cleaves the exon junction in RNA or DNA (Jarrell et al. 1988b; Mörl et al. 1992); RNA cleavage after 5′-exon sequences (Müller et al. 1988); reverse branching in which excised lariat ligates itself to 5′ exon RNA sequence through the reverse of the first splicing step (Chin and Pyle 1995); and circle formation in which an intron's 3′ end attacks the 5′ exon–intron junction, releasing the 5′ exon and an intron RNA cyclized by a 2′-5′ bond (Murray et al. 2001). Circular group II intron RNAs have also been reported in vivo (Li-Pook-Than and Bonen 2006; Molina-Sánchez et al. 2006).

4 THREE-DIMENSIONAL STRUCTURE OF GROUP II INTRON RIBOZYMES

A recent milestone in understanding group II ribozymes is the 3.1-Å X-ray crystal structure of the catalytic core of an *Oceanobacillus iheyensis* IIC intron (Toor et al. 2008a). The construct used for crystallization was a 412-nt RNA that was deleted for the distal stems of DII, III and VI, as well as the ORF of DIV (Fig. 5A). The intron was crystallized after hydrolytic splicing in vitro and hence is not branched. The 5′ and 3′ termini and DVI were not visible in the structure, and DVI was later found to be degraded (Toor et al.

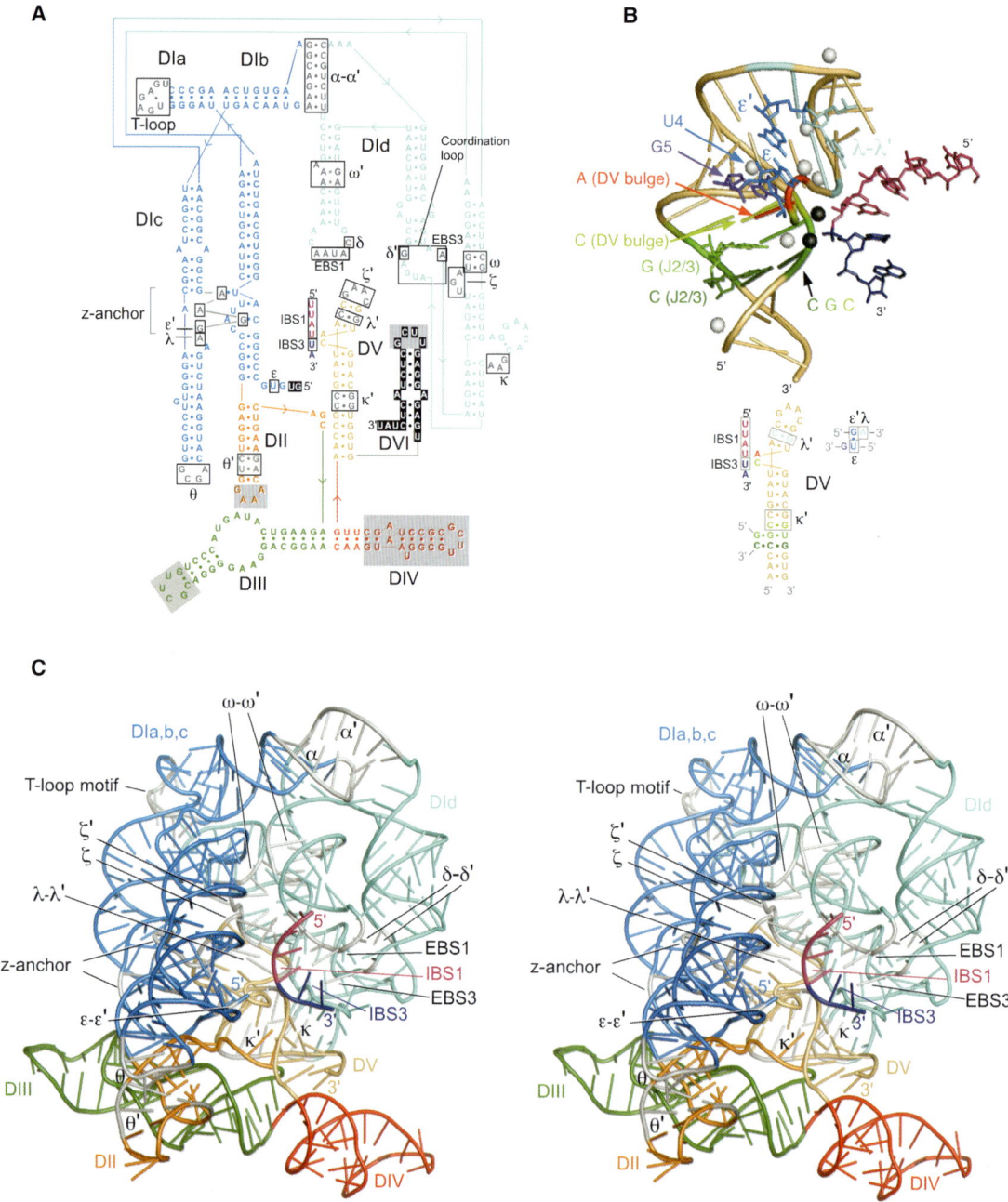

Figure 5. Crystal structure of the *Oceanobacillus iheyensis* group IIC intron. (*A*) Sequence and secondary structure of the crystallized RNA. Boxes indicate motifs involved in tertiary interactions. Solid gray boxes in DII, DIII, and DIV indicate regions deleted from the crystallization construct and replaced with sequences not present in the wild-type intron. Nucleotide residues in DI and DVI shown as white letters on a black background are not visible in the crystal structure. (*B*) Structure of the active-site region, with a corresponding color-coded secondary structure below. DV is a beige tube helix, with a bound RNA modeled as the 5′ and 3′ exons (pink and indigo, respectively; Toor et al. 2008b; Toor et al. 2010). The triple interactions in the triple helix stack between the CGC triad, the CG of J2/3, and the C of the AC bulge are shown in dark green, green and yellow-green. The three-base stack consisting of the A of the AC bulge, G5, and U4 of the ε–ε′ interaction are in red, purple and blue, respectively, while the λ–λ′ interaction is cyan. Metal ions bound to DV in the crystal are indicated by spheres, with black spheres representing the proposed active-site Mg^{++} ions, which were identified by binding of Yb^{3+} in the crystal derivatives. (*C*) X-ray crystal structure. A stereoview is shown, with domains colored as in (*A*) and regions involved in tertiary interactions colored gray (Toor et al. 2008a,b). Note added in proof: The conserved single base pair in the κ stem-loop, which we noted was missing in the original structure (Toor et al. 2008a), is present in the recently corrected and refined structure (Toor et al. 2010).

2010). In a follow-up structure, what was initially thought to be a 6-nt exon RNA bound at the active site (Toor et al. 2008b) was found to be an unidentified RNA fragment, possibly derived from the degraded DVI (Toor et al. 2010).

The structures provide critical insights into the nature of the active site (Fig. 5B). The catalytic center, DV, whose structure in isolation had been determined by crystallography and NMR (Zhang and Doudna 2002; Sigel et al. 2004), assumes a very different structure in the context of the ribozyme. The two helices of DV, rather than stacking coaxially, are bent severely to bring the AC bulge near the CGC triad, thus juxtaposing the phosphate backbones of the most conserved DV sequences. Nine potential Mg^{++}-binding sites were assigned in or near DV, of which two bound Yb^{3+} ions in heavy atom derivatives. These two metal ions are coordinated to the phosphate backbones of the CGC triad and AC bulge, which were identified as catalytic metal-binding sites in other group II introns (see above), and are separated by a distance consistent with a two-metal-ion model of catalysis (Steitz and Steitz 1993).

The folded structure of DV is stabilized by the close packing of other conserved elements (Fig. 5B,C). The two bases just upstream of the γ-nucleotide in J2/3 form major groove base triples with the CG of the catalytic triad. The C of the AC bulge stacks on top of these two triple pairs to form a third triple base pair with the final C of the CGC triad. The A of the AC bulge protrudes in a different direction and stacks below G5 and U4 of the ε–ε′ interaction, which is in turn positioned by the λ–λ′ interaction between DIc and the upper stem of DV.

The remainder of the structure is dominated by DI, which along with DV comprises the minimal catalytic core (Fig. 5C). The two halves of DI wrap around and cap DV, anchored by the ζ–ζ′ and κ–κ′ interactions. On the opposite side of DV, the α–α′ pairing connects the halves. One half of DI, which consists of DI(i,ii)/Ia,/Ib/Ic, is a structural unit that positions the ε–ε′ pairing at the core of the molecule. Key to this positioning is the z-anchor, a zig-zagging series of four bases in the ε′ motif, which alternately form pairings with ε and the bulge loop of DI(i). Another key organizing feature is the 5-way junction formed by the loop of DIa interacting with the helices of DI(ii), DIb, DIc and DId, similar to the T-loop in tRNAs (Michel et al. 2009).

The other half of DI consists of nearly all of DId and is the exon-binding region. An important organizing center here is the ω–ω′ interaction, a ribose zipper that anchors the EBS1 stem near the ζ receptor, thus helping to position the 5′ exon. EBS3 and the coordination loop are positioned under EBS1 near the active site. Completing the structure are DII, which stacks coaxially with DI(i) and helps to dock DIc through the θ–θ′ interaction, and DIII and DIV, which are stacked coaxially along the bottom of the RNA.

The structure contains most of the previously predicted tertiary interactions, including IBS1-EBS1, α–α′, ε–ε′, δ–δ′, θ–θ′, ζ–ζ′, κ–κ′, and λ–λ′. Interactions not observed are either missing from IIC introns (β–β′, IBS2-EBS2), deleted from the crystallized construct (η–η′, μ–μ′), or unresolved (γ–γ′). However, some features of the structure are not consistent with previous data, particularly the position of DIII, which is pointed away from the ribozyme in the crystal structure, but whose conserved internal loop is predicted to be in the core of the molecule in contact with the ε motif, based on both chemogenetic and cross-linking experiments (Dai et al. 2008; Fedorova and Pyle 2008). In addition, the κ stem-loop does not form the single base pair that is phylogenetically conserved in IIC introns (Figs. 1A, 5A), and IBS3 and EBS3, while near each other, are not paired, possibly reflecting binding of a noncomplementary RNA fragment (Toor et al. 2010) (see earlier). Thus, it is not certain that the entire RNA has crystallized in a native conformation. IIA and IIB introns are expected to have active-site structures similar to that determined for the IIC intron, but contain a number of motifs not found in IIC introns and vice versa and thus must stabilize this structure in different ways.

5 THE INVOLVEMENT OF PROTEINS IN GROUP II INTRON SPLICING

As for other large RNAs, proteins play a major role in folding group II introns into their active three-dimensional structure in vivo (Lambowitz et al. 1999). Some of these proteins are RNA splicing factors that stabilize the active structure, while others are RNA chaperones that resolve stable, inactive structures ("kinetic traps") that limit the rate of productive RNA folding.

5.1 Group II Intron-Encoded RTs are Intron-Specific RNA Splicing Factors

A key splicing factor for mobile group II introns is the IEP, which promotes splicing of the intron that encodes it. This splicing function (maturase activity) was first shown genetically for the yeast mt aI1 and aI2 IEPs (Carignani et al. 1983; Moran et al. 1994) and has been studied biochemically for the *L. lactis* Ll.LtrB IEP (LtrA protein) (Fig. 2A) (Matsuura et al. 1997; Saldanha et al. 1999). All three of these important model introns are members of subgroup IIA and encode ML lineage proteins. Purified recombinant LtrA can by itself promote splicing of the

Ll.LtrB intron at near-physiological Mg^{++} concentrations (5 mM), at which the intron cannot self-splice efficiently. To promote splicing, LtrA binds tightly and specifically to the intron RNA to stabilize the active RNA structure, which reverts rapidly to the inactive structure when the protein is removed with protease (Matsuura et al. 2001; Noah and Lambowitz 2003; Mohr et al. 2006). Although largely monomeric in solution, LtrA binds the intron RNA with a stoichiometry of 2:1, suggesting that it functions as a dimer, the same active form as HIV-1 RT (Saldanha et al. 1999; Rambo and Doudna 2004).

Biochemical studies showed that LtrA has a high-affinity binding site in DIVa, an idiosyncratic stem-loop structure at the beginning of the LtrA ORF, accounting in part for its intron-specificity (Fig. 1A, inset) (Wank et al. 1999). The binding of LtrA to DIVa involves recognition of specific bases in the terminal loop and helical bulges, as well as their secondary structure context via phosphate-backbone interactions, analogous to classical phage coat-protein stem-loop interactions (Watanabe and Lambowitz 2004). Because DIVa in the Ll.LtrB intron contains the Shine-Dalgarno sequence and start codon of the IEP, the binding of LtrA to DIVa down-regulates its own translation (Singh et al. 2002; Cui et al. 2004).

Anchored by its tight binding to DIVa, LtrA makes additional contacts with conserved core regions of the intron RNA that stabilize the active RNA structure (Wank et al. 1999; Matsuura et al. 2001; Dai et al. 2008). If DIVa is deleted, LtrA can still promote splicing at reduced efficiency by binding directly to these core regions. The yeast aI2 IEP, which is translated as a fusion protein with the upstream exon and then proteolytically processed, also binds both DIVa and catalytic core regions, suggesting a conserved mode of interaction even when DIVa does not contain the start codon for the intron ORF (Huang et al. 2003).

LtrA-binding sites on the intron RNA were identified by RNA-footprinting (Matsuura et al. 2001) and by using circularly permuted intron RNAs to position fluorescence quenching and cross-linking reagents at different sites within the intron (Dai et al. 2008). Mapping of these sites onto a three-dimensional model of the intron RNA revealed a large continuous binding surface, which extends from DIVa across contiguous regions of DI, II, and VI. This binding surface suggests that LtrA promotes splicing by stabilizing interactions between these RNA domains, as well as the folded structure of DI. After splicing, the IEP remains tightly bound to excised intron and presumably uses most or all of the same interactions to stabilize the ribozyme structure for reverse splicing into DNA during intron mobility. Despite the structural differences between IIA, IIB, and IIC introns, modeling suggests that the IEP could bind and act similarly in all three cases (Dai et al. 2008).

Regions of the LtrA protein required for splicing were identified by using a high throughput mutagenesis method and mapped onto a three-dimensional model of the protein (Cui et al. 2004; Blocker et al. 2005). The results suggest that the RNA-binding surface extends across the RT and X/thumb domains and includes the amino-terminal extension, template-primer binding track, and patches on the back of the protein. Biochemical and genetic experiments suggest that LtrA's amino-terminal extension binds DIVa, whereas other regions of the RT and X domains bind the intron core (Cui et al. 2004; Gu et al. 2010). Unlike retroviral RTs, non-LTR-retroelement RTs bind specifically to their template RNAs for initiation of cDNA synthesis. Thus, the splicing function of group II intron RTs may have evolved from interactions that were used initially for RNA template recognition (Kennell et al. 1993).

5.2 Evolution of IEPs to Function in Splicing Multiple Group II Introns

Although most IEPs splice only the intron that encodes them, some have evolved to splice multiple introns, providing a common splicing apparatus. In the simplest cases, several bacterial introns that have proliferated within a genome are spliced by a single IEP, enabling all but one intron to lose its own ORF (Dai and Zimmerly 2003; Meng et al. 2005). Other IEPs, however, have expanded their splicing function to include more distantly related introns. Thus, the cp MatK protein, which is encoded by a *trnK*I1 intron in land plants and by a free-standing ORF in some nonphotosynthetic plants, is a highly degenerate IEP that has lost mobility functions but acquired the ability to bind and splice multiple cp IIA introns that lack ORFs (Fig. 2C; Ems et al. 1995; Vogel et al. 1999; Zoschke et al. 2010). This more general splicing function presumably reflects the loss of intron-specific interactions, such as those involving DIVa, and increased reliance on interactions with structural features shared by IIA introns. A further development is illustrated by the nMat proteins (nMat-1a, -1b, 2a, and 2b) of flowering plants (Fig. 2D; Mohr and Lambowitz 2003). These proteins, which evolved from mt group II intron IEPs, are encoded by nuclear genes and are transported into organelles to promote the splicing of group II introns that lack ORFs (Nakagawa and Sakurai 2006; Keren et al. 2009). This evolutionary progression in which an IEP loses mobility functions, evolves to splice multiple introns, and is ultimately subsumed into the host genome, limits potentially deleterious intron mobility and facilitates host regulation of RNA splicing in concert with other cellular processes.

5.3 Recruitment of Host-Encoded Proteins to Function in Group II Intron Splicing

Most mt and cp group II introns lack ORFs and rely on host-encoded proteins to promote their splicing. This situation is similar to that for group I introns, where a variety of host proteins, such aminoacyl-tRNA synthetases, have been recruited to function in splicing different introns (Lambowitz et al. 1999). Host-encoded splicing factors have been studied in the greatest detail for plant cp group II introns. Genetic analysis supported by in vivo binding data indicate that these cp group II introns utilize a battery of splicing factors, including proteins belonging to three large plant-specific families of RNA-binding proteins, CRM (chloroplast RNA splicing and ribosome maturation), POROR (plant RNA recognition), and pentatricopeptide repeat (PPR) proteins (Stern et al. 2010). These proteins function combinatorially to promote the splicing of different introns, with most introns requiring multiple splicing factors that associate with the RNA to form large complexes (Till et al. 2001; Kroeger et al. 2009). In the green alga *Chlamydomonas reinhardti*, the *trans*-splicing of cp *psaA*-I1 and -I2 requires a different set of host-encoded proteins, which also appear to function in large complexes, some of which may be membrane bound (Merendino et al. 2006; Perron et al. 1999). The different host-encoded splicing factors for plant and algal cp group II introns presumably reflect that their splicing function evolved relatively recently, after the divergence of plants and green algae.

5.4 DEAD-Box Proteins Function as RNA Chaperones in Group II Intron Splicing

In addition to the splicing factors that stabilize the active RNA structure, the efficient splicing of group II introns requires RNA chaperones to resolve stable inactive or intermediate structures that limit the rate of productive RNA folding (Huang et al. 2005; Köhler et al. 2010). In *S. cerevisiae* mitochondria, a key protein that plays this role is Mss116p, a member of the DEAD-box family of ATP-dependent RNA helicases, which function in diverse RNA structural rearrangements in all organisms (Séraphin et al. 1989; Huang et al. 2005). *MSS116* null mutants are defective in splicing all four mt group II introns. Two of these, aI1 and aI2, are mobile IIA introns that encode RT/maturase proteins that stabilize their active structures, while the other two, aI5γ and bI1, are small IIB introns that do not encode proteins. Significantly, these group II intron splicing defects are leaky, and *MSS116* mutants are also defective in splicing all mt group I introns, translation of some mt mRNAs, and other RNA processing reactions. Further, Mss116p can be replaced in group I or group II intron splicing reactions in vitro or in vivo by CYT-19, a *N. crassa* DEAD-box protein that functions as an RNA chaperone in mt group I intron splicing (Huang et al. 2005) or by other DEAD-box proteins, which do not ordinarily function in group I and II intron splicing (e.g., *S. cerevisiae* Ded1p and *Escherichia coli* SrmB; Del Campo et al. 2009 and refs. therein). Thus, Mss116p and these other DEAD-box proteins appear to interact nonspecifically with structurally diverse RNA and RNP substrates, as expected for general RNA chaperones. Biochemical studies confirmed these nonspecific interactions, as well as a correlation with the DEAD-box protein's RNA-unwinding activity in all cases, but leave open the possibility that other DEAD-box protein activities, such as strand annealing or RNA structural stabilization, also contribute to the splicing of some group II introns (Mohr et al. 2006; Solem et al. 2006; Halls et al. 2007; Del Campo et al. 2007, 2009). The involvement of DEAD-box proteins in group II intron splicing is an important parallel to the spliceosome, where DEAD-box proteins and related RNA helicases function at multiple steps to accelerate specific structural transitions.

6 GROUP II INTRON MOBILITY

Group II intron mobility occurs by a novel mechanism in which the excised intron RNA uses its ribozyme activity to reverse splice directly into a DNA strand where it is reverse transcribed by the IEP. The introns use different variations of this mechanism both to "retrohome" to specific DNA target sites at frequencies up to 100% and to "retrotranspose" to ectopic sites that resemble the normal homing sites at low frequencies (10^{-4} to 10^{-6}). Retrohoming was first observed for yeast mt introns, which were found to invade intronless alleles during genetic crosses (Meunier et al. 1990). The high efficiency and specificity of retrohoming underlie the use of group II introns in gene targeting.

6.1 Group II Intron Retrohoming by Reverse Splicing into DNA

The mechanism of group II intron retrohoming was elucidated by studies of the yeast mt aI1 and aI2 and *L. lactis* Ll.LtrB introns (reviewed in Lambowitz and Zimmerly 2004). Retrohoming is mediated by the RNP that is formed during RNA splicing and consists of the IEP and excised intron lariat RNA (Fig. 6A). RNPs initiate retrohoming by using both the IEP and intron RNA to recognize DNA target sequences. The IEP recognizes specific bases or structural features of the DNA target site, which differ for each intron, and this interaction helps separate the DNA strands, enabling the intron RNA to base pair to the 5′ and 3′ DNA exons. Importantly, this base pairing involves

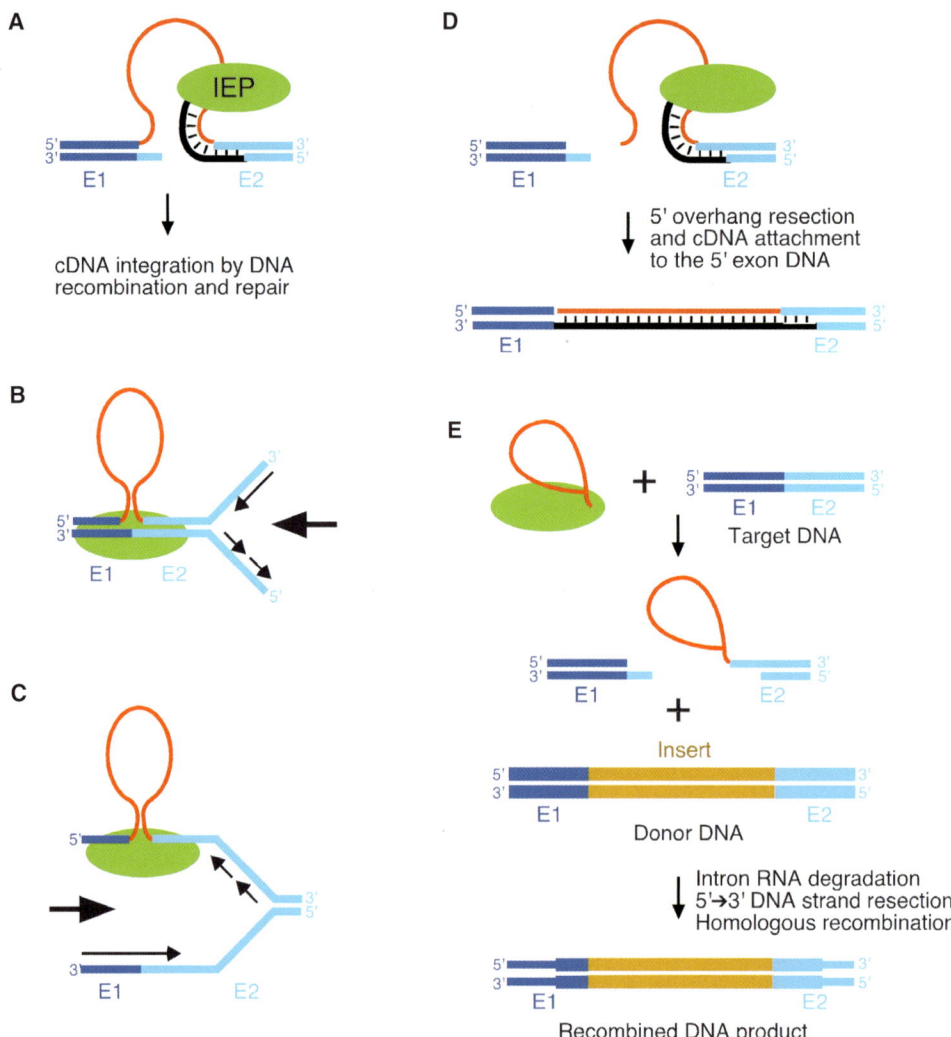

Figure 6. Group II intron mobility mechanisms. (*A*) Retrohoming via reverse splicing of the intron RNA into double-stranded DNA. After reverse splicing of the intron RNA into the top strand, the bottom strand is cleaved by the En domain of the IEP, and the 3′ end at the cleavage site is used as a primer for reverse transcription of the inserted intron RNA. The resulting intron cDNA is integrated by cellular DNA recombination and/or repair mechanisms. (*B*) Reverse splicing of the intron RNA into double-stranded DNA, with priming by the nascent leading strand of the DNA replication fork. (*C*) Reverse splicing of the intron RNA into single-stranded DNA, with priming by the nascent lagging strand of the DNA replication fork. (*D*) Retrohoming of linear intron RNA by the first step of reverse splicing, bottom-strand cleavage, reverse transcription, and attachment of the free cDNA end to the 5′ exon DNA likely by NHEJ (Zhuang et al. 2009b). (*E*) Use of group II introns to introduce a targeted double-strand break that stimulates gene targeting by homologous recombination. The top-strand break by the first step of reverse splicing can be made either by lariat RNA as shown in the figure or by linear intron RNA (not shown; Mastroianni et al. 2008). Recombination results in the precise insertion of a novel DNA sequence (gold) from the donor DNA into the target DNA. The target and donor DNAs are shown with different widths to illustrate the origin of different DNA segments in the recombined DNA product. Intron RNA, red; 5′ and 3′ exons (E1 and E2), dark and light blue, respectively; IEP, green. In (*B*) and (*C*), large arrows indicate the direction of the replication fork, and small arrows indicate the direction of DNA synthesis.

the same subgroup-specific EBS/IBS and δ–δ′ interactions used for RNA splicing (see earlier; Guo et al. 1997; Eskes et al. 1997; Mohr et al. 2000; Jiménez-Zurdo et al. 2003; Robart et al. 2007; Zhuang et al. 2009a). By using the same base-pairing interactions for both RNA splicing and DNA integration, the intron ensures that it inserts only at sites from which it can subsequently excise by RNA splicing.

After base pairing, the intron reverse splices into the DNA strand, resulting in the insertion of linear intron

RNA between the two DNA exons. The En domain of the IEP cleaves the opposite DNA strand a short distance downstream (position +10 for aI1 and aI2 and +9 for Ll.LtrB), and the 3′ OH of the cleaved DNA is used as a primer for reverse transcription of the inserted intron RNA, a process referred to as "target-primed reverse transcription" (TPRT). For yeast aI1 and aI2, cDNA integration generally occurs by a mechanism in which a nascent cDNA initiates recombination with an intron-containing allele, leading to transfer of the intron plus a variable length of the upstream exon (Eskes et al. 1997, 2000). By contrast, cDNA integration for the Ll.LtrB intron in both *L. lactis* and *E. coli* occurs by a DNA repair mechanism without coconversion of flanking exon sequences (Cousineau et al. 1998). In *E. coli*, this repair mechanism appears to involve synthesis of a full-length intron cDNA, removal of the RNA strand by the host RNase H, second-strand synthesis by the host replicative DNA polymerase III, and the use of host DNA ligase to seal nicks (Smith et al. 2005).

6.2 DNA Target Site Recognition By Group II Intron RNPs

The mechanism of DNA target site recognition by group II intron RNPs has been studied in most detail for the *L. lactis* Ll.LtrB group II intron, taking advantage of an efficient *E. coli* expression system to obtain large amounts of purified RNPs (Saldanha et al. 1999). Ll.LtrB RNPs bind to DNA nonspecifically and scan for target sites by facilitated diffusion, analogous to site-specific DNA binding proteins (Aizawa et al. 2003). The IEP is thought to first recognize a small number of specific bases in the distal 5′-exon region of the DNA target site via major groove interactions, most critically T-23, G-21, and A-20 (Singh and Lambowitz 2001). These initial base interactions, bolstered by phosphate backbone and possibly IEP interactions in the adjacent minor groove (positions-17 to -13), trigger local DNA melting, enabling the intron RNA's EBS2, EBS1, and δ sequences to base pair to the IBS2, IBS1, and δ′ sequences for reverse splicing. Intron RNA base pairing may occur concomitantly with and help drive DNA melting. Bottom-strand cleavage occurs at position +9 from the insertion site and requires additional interactions between the IEP and the 3′ exon, the most critical being recognition of T+5 (Singh and Lambowitz 2001). Binding of the RNP to the 5′ and 3′ DNA exons bends the target DNA, with the bend angle increasing when the cleaved 3′ end is repositioned from the En to RT active site for initiation of reverse transcription (Noah et al. 2006). Because reverse splicing into the DNA target site is reversible and the equilibrium favors intron excision, the reaction must be driven forward by the initiation of cDNA synthesis, which blocks the 3′-splice site and prevents excision (Aizawa et al. 2003). Other group II introns use the same mode of DNA target site recognition, but the 5′- and 3′-exon sequences recognized by the IEP differ even for closely related introns, suggesting rapid evolution of IEP specificity (Zhuang et al. 2009a and references therein).

6.3 Variations of the Retrohoming Mechanism

Group II introns whose IEPs lack an En domain or whose En activity is inactivated by mutation use a variation of the retrohoming pathway in which a nascent strand at a DNA replication fork primes reverse transcription (Fig. 6B,C; Zhong and Lambowitz 2003; Martínez-Abarca et al. 2004). Group II introns retrohome via this mechanism either by reverse splicing into double-stranded DNA with preferential use of leading strand DNA primers, or by reverse splicing into transiently single-stranded DNA at a DNA replication fork or transcription bubble with preferential use of lagging strand DNA primers. The latter pathway appears to be favored by naturally occurring group II introns whose IEPs lack an En domain (Ichiyanagi et al. 2002), including IIC introns, where single-stranded DNA facilitates formation of the DNA stem-loop of a transcription terminator or *attC* site that is recognized by the intron RNP (Robart et al. 2007; Léon and Roy 2009).

Linear group II intron RNAs, which may be generated by hydrolytic splicing or debranching of lariats, use another retrohoming variation, which was uncovered by injecting Ll.LtrB RNPs containing linear intron RNA into *Xenopus laevis* oocyte nuclei or *Drosophila melanogaster* embryos (Fig. 6D; Zhuang et al. 2009b). Here, the linear intron RNA carries out the first step of reverse splicing into the DNA target, thereby ligating the RNA to the 3′ but not the 5′ exon. The intron RNA is then reverse transcribed by the IEP, and the free end of the cDNA is linked to the 5′ exon DNA, likely via nonhomologous end-joining (NHEJ), a widely studied DNA repair process. Although inefficient, this mechanism could be used for retrohoming of linear RNAs, not only in eukaryotes but also in many prokaryotes, which have analogous NHEJ machinery (Bowater and Doherty 2006).

6.4 Retrotransposition of Group II Introns to New Sites

Retrotransposition to ectopic sites that resemble the normal homing site occurs at low frequency providing a means of group II intron dispersal to new genomic locations. In all cases examined, retrotransposition occurs by reverse splicing of the intron RNA into a DNA site, but with different variations of the pathway favored depending on the organism and growth conditions (Yang et al. 1998; Martínez-Abarca and Toro 2000; Dickson et al. 2001; Ichiyanagi et al. 2002). In *L. lactis*, retrotransposition of the Ll.LtrB intron occurs primarily by reverse splicing into transiently single-stranded DNA with priming by the nascent lagging strand (Fig. 6C) (Ichiyanagi et al. 2002). In *E. coli*, however, retrotransposition of this intron occurs both by the above mechanism and by inaccurate reverse splicing into double-stranded DNA, with or without En cleavage of the opposite strand (Fig. 6B,C) (Coros et al. 2005). As expected, pathways using nascent DNA strands as primers are favored under rapid growth conditions, which lead to an increased frequency of replication forks (Coros et al. 2005). Retromobility of the Ll.LtrB intron is also influenced in interesting ways by cellular interactions, host factors, and stress responses (Beauregard et al. 2008; Zhao et al. 2008; Coros et al. 2009).

6.5 Targetrons

Because group II introns recognize DNA target sites largely by base pairing of the intron RNA to the DNA target sequence, it is possible to retarget them to insert into desired DNA sites simply by modifying the base pairing sequences in the intron RNA (Guo et al. 2000; Karberg et al. 2001). This feature, combined with the high efficiency and specificity of the retrohoming reaction, enabled the development of group II introns into gene targeting vectors ("targetrons"; Perutka et al. 2004). A targetron based on the Ll.LtrB intron is sold commercially and widely used for gene targeting in bacteria. Recently, other group II introns have also been adapted for gene targeting, expanding the range of accessible target sites and providing other useful properties (Zhuang et al. 2009a).

Targetrons are also being developed for use in eukaryotes, where the obstacles include nuclear accessibility of RNPs and suboptimal Mg^{++} concentrations (Mastroianni et al. 2008). In *X. laevis* oocyte nuclei and *D. melanogaster* embryos, microinjected group II intron RNPs retrohome efficiently if additional Mg^{++} is provided (Mastroianni et al. 2008; Zhuang et al. 2009b). A more general solution is to identify or select introns that function at lower Mg^{++} concentrations. Importantly, targetrons can be used not only for site-specific DNA integration, but also to generate a targeted double-strand DNA break that stimulates targeted DNA integration by homologous recombination. The double-strand break results from the initial partial reverse splicing and second-strand cleavage reactions of the group II intron RNP (Fig. 6E; Karberg et al. 2001; Mastroianni et al. 2008). The introduction of a recombinogenic double-strand break by a protein endonuclease, such as a Zn-finger nuclease or a meganuclease, is a favored mode of gene targeting in higher organisms (Porteus and Carroll 2005), for which targetrons have an inherent advantage in the ease of retargeting breaks to desired sites.

7 GROUP II INTRON EVOLUTION

Group II introns are thought to have played a major role in eukaryotic genome evolution as ancestors of both spliceosomal introns and non-LTR-retrotransposons (Sharp 1985; Cech 1986; Zimmerly et al. 1995). An evolutionary relationship between group II and spliceosomal introns is suggested by their identical splicing pathways, similar boundary sequences, and structural similarities between key regions of group II intron domains and spliceosomal RNAs (Madhani and Guthrie 1992; Shukla and Padgett 2002; Keating et al. 2010). The latter include: divalent metal-ion binding sites in DV and U6 snRNA, which may contribute to catalysis; the similar branch site motifs of DVI and the U2-intron pairing, in which an equivalent adenosine is the branch site for the first step of splicing; the similarity between the ϵ–ϵ' interaction of group II introns and the pairing between the ACAGAGA sequence of U6 and the 5′ region of the intron; and recognition of 5′ and 3′ exons, in which the IBS1-EBS1 and δ–δ' motifs of IIA introns are analogous to the U5 snRNA stem-loop (Fig. 7). Confidence in these similarities has been bolstered by the recent IIC intron crystal structure (Toor et al. 2008a) and by increasing evidence that snRNAs can catalyze reactions related to RNA splicing (Valadkhan et al. 2009). An evolutionary relationship between group II introns and non-LTR-retrotransposons is suggested by the similarity of their RT sequences (Xiong and Eickbush 1990; Blocker et al. 2005) and TPRT mechanisms, in which a cleaved DNA target site is used as the primer for reverse transcription of a specifically bound RNA template (Luan et al. 1993; Zimmerly et al. 1995).

The phylogenetic distribution of group II introns, which are common in eubacteria and eukaryotic organelles but rare in archaebacteria, suggests a scenario in which mobile group II introns originated in eubacteria and were transmitted to eukaryotes, possibly via endosymbiotic bacteria that gave rise to mitochondria and chloroplasts (Cavalier-Smith 1991; Palmer and Logsdon 1991). The idea that group II introns originated as retroelements in bacteria (the "retroelement ancestor hypothesis") is supported by the observation that bacterial group II introns include all known lineages and generally behave as retroelements, whereas organellar introns belong to only the ML or CL lineages, and frequently lack ORFs and/or have degenerate IEP or RNA structures (Toor et al. 2001). The ancestral eubacterial retroelement might have arisen by invasion of a self-splicing ribozyme by an RT (Wank et al.

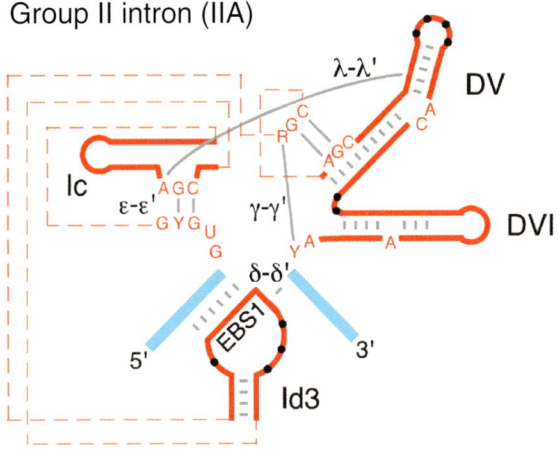

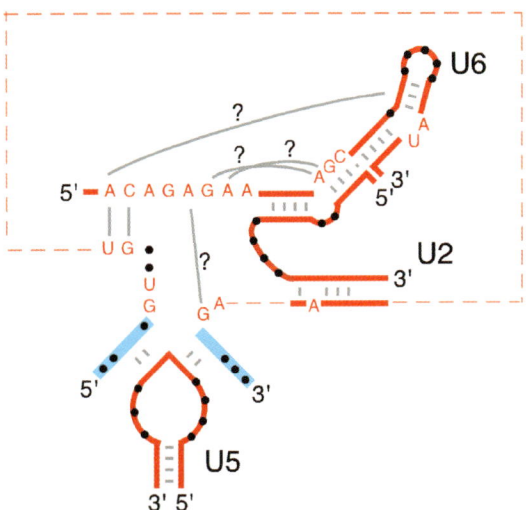

Figure 7. Similarities between the active site of group II introns and the putative active site of the spliceosome. Group II intron RNA and spliceosomal snRNA segments are shown in red, and exons are shown in blue. Base-pairing interactions that are similar for group II and spliceosomal introns are shown by gray bars, and unpaired bases at similar positions are shown by black dots. Dashed lines indicate connecting sequence of unspecified length. Question marks indicate hypothetical interactions that may occur in the spliceosome, based on interactions found in group II intron RNAs (Boudvillain et al. 2000; Toor et al. 2008a). The similarity between DId3 and the U5 snRNA is closest for IIA introns, while the ε-ε' and DV/U6 similarities are closest for IIA and IIB introns (see Fig. 1).

1999) or from a retroelement that evolved a self-splicing RNA at its termini to ameliorate damage to the host (Curcio and Belfort 1996). While IIC introns may be the earliest branching class (i.e., most similar to the last common ancestor), statistical support for this position is weak (Rest and Mindell 2003; Simon et al. 2009).

In eukaryotes, group II introns are thought to have invaded the nucleus and proliferated to many genomic sites, after which the ribozyme structure degenerated and fragmented into snRNAs that function in *trans* in a common splicing apparatus (Sharp 1991). A recent hypothesis suggests that the introduction of group II introns by a bacterial endosymbiont was the driving force for a fundamental step in the evolution of eukaryotes, the formation of the nuclear membrane, which separates transcription from translation and thus helps prevent translation of incompletely spliced RNAs (Martin and Koonin 2006). Regardless of its origin, the separation of transcription and translation by the nuclear membrane prevents immediate access of the IEP to the intron RNA, and necessitates the evolution of splicing factors that function in *trans*. In flowering plants, host-encoded splicing factors for cp group II introns evolved by the expansion and diversification of different families of RNA-binding proteins, with different proteins required to splice different introns. Such a solution would be impractical for the large numbers of introns in eukaryotic nuclear genomes, which instead evolved snRNAs derived from group II intron domains into a common RNA-based catalytic machinery replacing that in individual introns. These snRNAs continue to recognize the introns via conserved 5' and 3' sequences and a branch-point nucleotide similar to those of group II introns and catalyze splicing by the same transesterification reactions. This byzantine RNA-based spliceosomal machinery is arguably the strongest evidence that the eukaryotic splicing apparatus evolved from group II introns or their close relatives rather than de novo.

ACKNOWLEDGMENTS

We thank Alice Barkan, Marlene Belfort, Mark Del Campo, Georg Mohr, Sabine Mohr, and Rick Russell for comments on the manuscript. Work in the authors' laboratories was supported by National Institutes of Health grants GM37949 and GM37951 and Welch Foundation grant F-1607 to A.M.L. and CIHR grant MOP-93662 and NSERC grant RGP 2003717-02 to S.Z.

REFERENCES

Aizawa Y, Xiang Q, Lambowitz AM, Pyle AM. 2003. The pathway for DNA recognition and RNA integration by a group II intron retrotransposon. *Mol Cell* **11:** 795–805.

Barkan A. 2004. Intron splicing in plant organelles. In *Molecular biology and biotechnology of plant organelles* (eds H. Daniell, C. Chase), pp. 281–308. Kluwer Academic Publishers, Dordrecht.

Barkan A. 2009. Genome-wide analysis of RNA-protein interactions in plants. *Methods Mol Biol* **553:** 13–37.

Beauregard A, Curcio MJ, Belfort M. 2008. The take and give between retrotransposable elements and their hosts. *Ann Rev Genet* **42:** 587–617.

Belfort M, Derbyshire V, Parker MM, Cousineau B, Lambowitz AM. 2002. Mobile introns: Pathways and proteins. In *Mobile DNA II* (eds N. L. Craig, R. Craigie, M. Gellert, A. M. Lambowitz), pp. 761–783. ASM Press, Washington D.C.

Belhocine K, Mak AB, Cousineau B. 2008. Trans-splicing versatility of the Ll.LtrB group II intron. *RNA* **14:** 1782–1790.

Blocker FJ, Mohr G, Conlan LH, Qi L, Belfort M, Lambowitz AM. 2005. Domain structure and three-dimensional model of a group II intron-encoded reverse transcriptase. *RNA* **11:** 14–28.

Bonen L. 1993. *Trans*-splicing of pre-mRNA in plants, animals, and protists. *FASEB J* **7:** 40–46.

Bonen L. 2008. *Cis*- and *trans*-splicing of group II introns in plant mitochondria. *Mitochondrion* **8:** 26–34.

Boudvillain M, de Lencastre A, Pyle AM. 2000. A tertiary interaction that links active-site domains to the 5′ splice site of a group II intron. *Nature* **406:** 315–318.

Bowater R, Doherty AJ. 2006. Making ends meet: Repairing breaks in bacterial DNA by non-homologous end-joining. *PLoS Genet* **2:** e8.

Carignani G, Groudinsky O, Frezza D, Schiavon E, Bergantino E, Slonimski PP. 1983. An mRNA maturase is encoded by the first intron of the mitochondrial gene for the subunit I of cytochrome oxidase in *S. cerevisiae*. *Cell* **35:** 733–742.

Cavalier-Smith T. 1991. Intron phylogeny: A new hypothesis. *Trends Genet* **7:** 145–148.

Cech TR. 1986. The generality of self-splicing RNA: Relationship to nuclear mRNA splicing. *Cell* **44:** 207–210.

Chanfreau G, Jacquier A. 1994. Catalytic site components common to both splicing steps of a group II intron. *Science* **266:** 1383–1387.

Chanfreau G, Jacquier A. 1996. An RNA conformational change between the two chemical steps of group II self-splicing. *EMBO J* **15:** 3466–3476.

Chin K, Pyle AM. 1995. Branch-point attack in group II introns is a highly reversible transesterification, providing a potential proofreading mechanism for 5′-splice site selection. *RNA* **1:** 391–406.

Copertino D, Hallick R. 1993. Group II and group III introns of twintrons: potential relationships with nuclear pre-mRNA introns. *Trends Biochem Sci* **18:** 467–471.

Coros CJ, Landthaler M, Piazza CL, Beauregard A, Esposito D, Perutka J, Lambowitz AM, Belfort M. 2005. Retrotransposition strategies of the *Lactococcus lactis* Ll.LtrB group II intron are dictated by host identity and cellular environment. *Mol Microbiol* **56:** 509–524.

Coros CJ, Piazza CL, Chalamcharla VR, Smith D, Belfort M. 2009. Global regulators orchestrate group II intron retromobility. *Mol Cell* **34:** 250–256.

Costa M, Michel F. 1999. Tight binding of the 5′ exon to domain I of a group II self-splicing intron requires completion of the intron active site. *EMBO J* **18:** 1025–1037.

Costa M, Michel F, Westhof E. 2000. A three-dimensional perspective on exon binding by a group II self-splicing intron. *EMBO J* **19:** 5007–5018.

Cousineau B, Smith D, Lawrence-Cavanagh S, Mueller JE, Yang J, Mills D, Manias D, Dunny G, Lambowitz AM, Belfort M. 1998. Retrohoming of a bacterial group II intron: mobility via complete reverse splicing, independent of homologous DNA recombination. *Cell* **94:** 451–462.

Cui X, Matsuura M, Wang Q, Ma H, Lambowitz AM. 2004. A group II intron-encoded maturase functions preferentially in *cis* and requires both the reverse transcriptase and X domains to promote RNA splicing. *J Mol Biol* **340:** 211–231.

Curcio MJ, Belfort M. 1996. Retrohoming: cDNA-mediated mobility of group II introns requires a catalytic RNA. *Cell* **84:** 9–12.

Dai L, Zimmerly S. 2003. ORF-less and reverse-transcriptase-encoding group II introns in archaebacteria, with a pattern of homing into related group II intron ORFs. *RNA* **9:** 14–19.

Dai L, Chai D, Gu SQ, Gabel J, Noskov SY, Blocker FJ, Lambowitz AM, Zimmerly S. 2008. A three-dimensional model of a group II intron RNA and its interaction with the intron-encoded reverse transcriptase. *Mol Cell* **30:** 472–485.

Daniels DL, Michels WJ Jr, Pyle AM. 1996. Two competing pathways for self-splicing by group II introns: a quantitative analysis of *in vitro* reaction rates and products. *J Mol Biol* **256:** 31–49.

de Lencastre A, Hamill S, Pyle AM. 2005. A single active-site region for a group II intron. *Nat Struct Mol Biol* **12:** 626–627.

Del Campo M, Mohr S, Jiang Y, Jia H, Jankowsky E, Lambowitz AM. 2009. Unwinding by local strand separation is critical for the function of DEAD-box proteins as RNA chaperones. *J Mol Biol* **389:** 674–693.

Del Campo M, Tijerina P, Bhaskaran H, Mohr S, Yang Q, Jankowsky E, Russell R, Lambowitz AM. 2007. Do DEAD-box proteins promote group II intron splicing without unwinding RNA? *Mol Cell* **28:** 159–166.

Dickson L, Huang HR, Liu L, Matsuura M, Lambowitz AM, Perlman PS. 2001. Retrotransposition of a yeast group II intron occurs by reverse splicing directly into ectopic DNA sites. *Proc Natl Acad Sci* **98:** 13207–13212.

Drager RG, Hallick RB. 1993. A complex twintron is excised as four individual introns. *Nucleic Acids Res* **21:** 2389–2394.

Ems SC, Morden CW, Dixon CK, Wolfe KH, dePamphilis CW, Palmer JD. 1995. Transcription, splicing and editing of plastid RNAs in the non-photosynthetic plant *Epifagus virginiana*. *Plant Mol Biol* **29:** 721–733.

Eskes R, Liu L, Ma H, Chao MY, Dickson L, Lambowitz AM, Perlman PS. 2000. Multiple homing pathways used by yeast mitochondrial group II introns. *Mol Cell Biol* **20:** 8432–8446.

Eskes R, Yang J, Lambowitz AM, Perlman PS. 1997. Mobility of yeast mitochondrial group II introns: Engineering a new site specificity and retrohoming via full reverse splicing. *Cell* **88:** 865–874.

Fedorova O, Pyle AM. 2005. Linking the group II intron catalytic domains: Tertiary contacts and structural features of domain 3. *EMBO J* **24:** 3906–3916.

Fedorova O, Pyle AM. 2008. A conserved element that stabilizes the group II intron active site. *RNA* **14:** 1048–1056.

Fedorova O, Mitros T, Pyle AM. 2003. Domains 2 and 3 interact to form critical elements of the group II intron active site. *J Mol Biol* **330:** 197–209.

Fontaine JM, Goux D, Kloareg B, Loiseaux-de Goër S. 1997. The reverse-transcriptase-like proteins encoded by group II introns in the mitochondrial genome of the brown alga *Pylaiella littoralis* belong to two different lineages which apparently coevolved with the group II ribozyme lineages. *J Mol Evol* **44:** 33–42.

Glanz S, Kück U. 2009. *Trans*-splicing of organelle introns–a detour to continuous RNAs. *Bioessays* **31:** 921–934.

Goldschmidt-Clermont M, Choquet Y, Girard-Bascou J, Michel F, Schirmer-Rahire M, Rochaix JD. 1991. A small chloroplast RNA may be required for trans-splicing in Chlamydomonas reinhardtii. *Cell* **65:** 135–143.

Gordon PM, Piccirilli JA. 2001. Metal ion coordination by the AGC triad in domain 5 contributes to group II intron catalysis. *Nat Struct Biol* **8:** 893–898.

Gordon PM, Fong R, Piccirilli JA. 2007. A second divalent metal ion in the group II intron reaction center. *Chem Biol* **14:** 607–612.

Griffin EA Jr, Qin Z, Michels WJ Jr, Pyle AM. 1995. Group II intron ribozymes that cleave DNA and RNA linkages with similar efficiency, and lack contacts with substrate 2′-hydroxyl groups. *Chem Biol* **2:** 761–770.

Gu S-Q, Cui X, Mou S, Mohr S, Yao J, Lambowitz AM. 2010. Genetic identification of potential RNA-binding regions in a group II intron-encoded reverse transcriptase. *RNA* **16:** 732–747.

Guo H, Karberg M, Long M, Jones JP 3rd, Sullenger B, Lambowitz AM. 2000. Group II introns designed to insert into therapeutically relevant DNA target sites in human cells. *Science* **289:** 452–457.

Guo H, Zimmerly S, Perlman PS, Lambowitz AM. 1997. Group II intron endonucleases use both RNA and protein subunits for recognition of specific sequences in double-stranded DNA. *EMBO J* **16:** 6835–6848.

Halls C, Mohr S, Del Campo M, Yang Q, Jankowsky E, Lambowitz AM. 2007. Involvement of DEAD-box proteins in group I and group II intron splicing: Biochemical characterization of Mss116p, ATP hydrolysis-dependent and -independent mechanisms, and general RNA chaperone activity. *J Mol Biol* **365:** 835–855.

Hamill S, Pyle AM. 2006. The receptor for branch-site docking within a group II intron active site. *Mol Cell* **23:** 831–840.

Hiller R, Hetzer M, Schweyen RJ, Mueller MW. 2000. Transposition and exon shuffling by group II intron RNA molecules in pieces. *J Mol Biol* **297:** 301–308.

Huang HR, Chao MY, Armstrong B, Wang Y, Lambowitz AM, Perlman PS. 2003. The DIVa maturase binding site in the yeast group II intron aI2 is essential for intron homing but not for in vivo splicing. *Mol Cell Biol* **23:** 8809–8819.

Huang HR, Rowe CE, Mohr S, Jiang Y, Lambowitz AM, Perlman PS. 2005. The splicing of yeast mitochondrial group I and group II introns requires a DEAD-box protein with RNA chaperone function. *Proc Natl Acad Sci* **102:** 163–168.

Ichiyanagi K, Beauregard A, Lawrence S, Smith D, Cousineau B, Belfort M. 2002. Retrotransposition of the Ll.LtrB group II intron proceeds predominantly via reverse splicing into DNA targets. *Mol Micoibiol* **46:** 1259–1272.

Jarrell KA, Dietrich RC, Perlman PS. 1988a. Group II intron domain 5 facilitates a *trans*-splicing reaction. *Mol Cell Biol* **8:** 2361–2366.

Jarrell KA, Peebles CL, Dietrich RC, Romiti SL, Perlman PS. 1988b. Group II intron self-splicing: Alternative reaction conditions yield novel products. *J Biol Chem* **263:** 3432–3439.

Jiménez-Zurdo JI, García-Rodríguez FM, Barrientos-Durán A, Toro N. 2003. DNA target site requirements for homing *in vivo* of a bacterial group II intron encoding a protein lacking the DNA endonuclease domain. *J Mol Biol* **326:** 413–423.

Karberg M, Guo H, Zhong J, Coon R, Perutka J, Lambowitz AM. 2001. Group II introns as controllable gene targeting vectors for genetic manipulation of bacteria. *Nat Biotech* **19:** 1162–1167.

Keating KS, Toor N, Perlman PS, Pyle AM. 2010. A structural analysis of the group II intron active site and implications for the spliceosome. *RNA* **16:** 1–9.

Kennell JC, Moran JV, Perlman PS, Butow RA, Lambowitz AM. 1993. Reverse transcriptase activity associated with maturase-encoding group II introns in yeast mitochondria. *Cell* **73:** 133–146.

Keren I, Bezawork-Geleta A, Kolton M, Maayan I, Belausov E, Levy M, Mett A, Gidoni D, Shaya F, Ostersetzer-Biran O. 2009. AtnMat2, a nuclear-encoded maturase required for splicing of group-II introns in *Arabidopsis* mitochondria. *RNA* **15:** 2299–2311.

Knoop V, Altwasser M, Brennicke A. 1997. A tripartite group II intron in mitochondria of an angiosperm plant. *Mol Gen Genetics* **255:** 269–276.

Köhler D, Schmidt-Gattung S, Binder S. 2010. The DEAD-box protein PMH2 is required for efficient group II intron splicing in mitochondria of *Arabidopsis thaliana*. *Plant Mol Biol* **72:** 459–467.

Kroeger TS, Watkins KP, Friso G, van Wijk KJ, Barkan A. 2009. A plant-specific RNA-binding domain revealed through analysis of chloroplast group II intron splicing. *Proc Natl Acad Sci* **106:** 4537–4542.

Lambowitz AM, Zimmerly S. 2004. Mobile group II introns. *Ann Rev Genet* **38:** 1–35.

Lambowitz AM, Caprara MG, Zimmerly S, Perlman PS. 1999. Group I and group II ribozymes as RNPs: Clues to the past and guides to the future. In *The RNA world* (eds R. F. Gesteland, T. R. Cech, J. F. Atkins), pp. 451–485. Cold Spring Harbor Laboratory Press, Cold Spring Harbor, NY.

Li-Pook-Than J, Bonen L. 2006. Multiple physical forms of excised group II intron RNAs in wheat mitochondria. *Nucleic Acids Res* **34:** 2782–2790.

Léon G, Roy PH. 2009. Group IIC intron mobility into attC sites involves a bulged DNA stem-loop motif. *RNA* **15:** 1543–1553.

Luan DD, Korman MH, Jakubczak JL, Eickbush TH. 1993. Reverse transcription of R2Bm RNA is primed by a nick at the chromosomal target site: A mechanism for non-LTR retrotransposition. *Cell* **72:** 595–605.

Madhani HD, Guthrie C. 1992. A novel base-pairing interaction between U2 and U6 snRNAs suggests a mechanism for the catalytic activation of the spliceosome. *Cell* **71:** 803–817.

Malik HS, Burke WD, Eickbush TH. 1999. The age and evolution of non-LTR retrotransposable elements. *Mol Biol Evol* **16:** 793–805.

Martin W, Koonin EV. 2006. Introns and the origin of nucleus-cytosol compartmentalization. *Nature* **440:** 41–45.

Martínez-Abarca F, Toro N. 2000. RecA-independent ectopic transposition *in vivo* of a bacterial group II intron. *Nucleic Acids Res* **28:** 4397–4402.

Martínez-Abarca F, Barrientos-Durán A, Fernández-López M, Toro N. 2004. The RmInt1 group II intron has two different retrohoming pathways for mobility using predominantly the nascent lagging strand at DNA replication forks for priming. *Nucleic Acids Res* **32:** 2880–2888.

Mastroianni M, Watanabe K, White TB, Zhuang F, Vernon J, Matsuura M, Wallingford J, Lambowitz AM. 2008. Group II intron-based gene targeting reactions in eukaryotes. *PLoS ONE* **3:** e3121. doi:10.1371/journal.pone.0003121.

Matsuura M, Noah JW, Lambowitz AM. 2001. Mechanism of maturase-promoted group II intron splicing. *EMBO J* **20:** 7259–7270.

Matsuura M, Saldanha R, Ma H, Wank H, Yang J, Mohr G, Cavanagh S, Dunny GM, Belfort M, Lambowitz AM. 1997. A bacterial group II intron encoding reverse transcriptase, maturase, and DNA endonuclease activities: Biochemical demonstration of maturase activity and insertion of new genetic information within the intron. *Genes Dev* **11:** 2910–2924.

McSwiggen JA, Cech TR. 1989. Stereochemistry of RNA cleavage by the *Tetrahymena* ribozyme and evidence that the chemical step is not rate-limiting. *Science* **244:** 679–683.

Meng Q, Wang Y, Liu XQ. 2005. An intron-encoded protein assists RNA splicing of multiple similar introns of different bacterial genes. *J Biol Chem* **280:** 35085–35088.

Merendino L, Perron K, Rahire M, Howald I, Rochaix JD, Goldschmidt-Clermont M. 2006. A novel multifunctional factor involved in *trans*-splicing of chloroplast introns in *Chlamydomonas*. *Nucleic Acids Res* **34:** 262–274.

Meunier B, Tian G-L, Macadre C, Slonimski PP, Lazowska J. 1990. Group II introns transpose in yeast mitochondria. In *Structure function and biogenesis of energy transfer systems* (eds E. Quagliariello, S. Papa, F. Palmieri, C. C Saccone), pp 169–174. Elsevier Scientific Publishers, Amsterdam.

Michel F, Ferat JL. 1995. Structure and activities of group II introns. *Ann Rev Biochem* **64:** 435–461.

Michel F, Costa M, Westhof E. 2009. The ribozyme core of group II introns: A structure in want of partners. *Trends Biochem Sci* **34:** 189–199.

Michel F, Umesono K, Ozeki H. 1989. Comparative and functional anatomy of group II catalytic introns—a review. *Gene* **82:** 5–30.

Mohr G, Lambowitz AM. 2003. Putative proteins related to group II intron reverse transcriptase/maturases are encoded by nuclear genes in higher plants. *Nucleic Acids Res* **31:** 647–652.

Mohr G, Perlman PS, Lambowitz AM. 1993. Evolutionary relationships among group II intron-encoded proteins and identification of a conserved domain that may be related to maturase function. *Nucleic Acids Res* **21:** 4991–4997.

Mohr G, Smith D, Belfort M, Lambowitz AM. 2000. Rules for DNA target-site recognition by a lactococcal group II intron enable retargeting of the intron to specific DNA sequences. *Genes Dev* **14:** 559–573.

Mohr S, Matsuura M, Perlman PS, Lambowitz AM. 2006. A DEAD-box protein alone promotes group II intron splicing and reverse splicing by acting as an RNA chaperone. *Proc Natl Acad Sci* **103:** 3569–3574.

Molina-Sánchez MD, Martínez-Abarca F, Toro N. 2006. Excision of the *Sinorhizobium meliloti* group II intron RmInt1 as circles *in vivo*. *J Biol Chem* **281:** 28737–28744.

Molina-Sánchez MD, Martínez-Abarca F, Toro N. 2010. Structural features in the C-terminal region of the *Sinorhizobium meliloti* RmInt1 group II intron-encoded protein contribute to its maturase and intron DNA-insertion function. *FEBS J* **277**: 244–255.

Moore MJ, Sharp PA. 1993. Evidence for two active sites in the spliceosome provided by stereochemistry of pre-mRNA splicing. *Nature* **365**: 364–368.

Moran JV, Mecklenburg KL, Sass P, Belcher SM, Mahnke D, Lewin A, Perlman P. 1994. Splicing defective mutants of the *COXI* gene of yeast mitochondrial DNA: Initial definition of the maturase domain of the group II intron AI2. *Nucleic Acids Res* **22**: 2057–2064.

Mörl M, Schmelzer C. 1990. Integration of group II intron bI1 into a foreign RNA by reversal of the self-splicing reaction in vitro. *Cell* **60**: 629–636.

Mörl M, Niemer I, Schmelzer C. 1992. New reactions catalyzed by a group II intron ribozyme with RNA and DNA substrates. *Cell* **70**: 803–810.

Müller MW, Schweyen RJ, Schmelzer C. 1988. Selection of cryptic 5′ splice sites by group II intron RNAs in vitro. *Nucleic Acids Res* **16**: 7383–7395.

Murray HL, Mikheeva S, Coljee VW, Turczyk BM, Donahue WF, Bar-Shalom A, Jarrell KA. 2001. Excision of group II introns as circles. *Mol Cell* **8**: 201–211.

Nakagawa N, Sakurai N. 2006. A mutation in At-nMat1a, which encodes a nuclear gene having high similarity to group II intron maturase, causes impaired splicing of mitochondrial NAD4 transcript and altered carbon metabolism in *Arabidopsis thaliana*. *Plant Cell Physiol* **47**: 772–783.

Nakamura Y, Kaneko T, Sato S, Ikeuchi M, Katoh H, Sasamoto S, Watanabe A, Iriguchi M, Kawashima K, Kimura T, et al. 2002. Complete genome structure of the thermophilic cyanobacterium *Thermosynechococcus elongatus* BP-1. *DNA Res* **9**: 123–130.

Noah JW, Lambowitz AM. 2003. Effects of maturase binding and Mg^{2+} concentration on group II intron RNA folding investigated by UV cross-linking. *Biochemistry* **42**: 12466–12480.

Noah JW, Park S, Whitt JT, Perutka J, Frey W, Lambowitz AM. 2006. Atomic force microscopy reveals DNA bending during group II intron ribonucleoprotein particle integration into double-stranded DNA. *Biochemistry* **45**: 12424–12435.

Padgett RA, Podar M, Boulanger SC, Perlman PS. 1994. The stereochemical course of group II intron self-splicing. *Science* **266**: 1685–1688.

Palmer JD, Logsdon JMJr, 1991. The recent origins of introns. *Curr Opinion Genet Dev* **1**: 470–477.

Perron K, Goldschmidt-Clermont M, Rochaix JD. 1999. A factor related to pseudouridine synthases is required for chloroplast group II intron trans-splicing in *Chlamydomonas reinhardtii*. *EMBO J* **18**: 6481–6490.

Perutka J, Wang W, Goerlitz D, Lambowitz AM. 2004. Use of computer-designed group II introns to disrupt *Escherichia coli* DExH/D-box protein and DNA helicase genes. *J Mol Biol* **336**: 421–439.

Podar M, Perlman PS, Padgett RA. 1995. Stereochemical selectivity of group II intron splicing, reverse splicing, and hydrolysis reactions. *Mol Cell Biol* **15**: 4466–4478.

Podar M, Perlman PS, Padgett RA. 1998b. The two steps of group II intron self-splicing are mechanistically distinguishable. *RNA* **4**: 890–900.

Podar M, Chu VT, Pyle AM, Perlman PS. 1998a. Group II intron splicing in vivo by first-step hydrolysis. *Nature* **391**: 915–918.

Porteus MH, Carroll D. 2005. Gene targeting using zinc finger nucleases. *Nat Biotechnol* **23**: 967–973.

Pyle AM, Lambowitz AM. 2006. Group II introns: Ribozymes that splice RNA and invade DNA. In *The RNA world*, 3rd ed. (eds R. F. Gesteland, T. R. Cech, J. F. Atkins), pp. 469–506. Cold Spring Harbor Laboratory Press, Cold Spring Harbor, NY.

Qiu YL, Palmer JD. 2004. Many independent origins of *trans* splicing of a plant mitochondrial group II intron. *J Mol Evol* **59**: 80–89.

Qin PZ, Pyle AM. 1998. The architectural organization and mechanistic function of group II intron structural elements. *Curr Opinion Struct Biol* **8**: 301–308.

Rambo RP, Doudna JA. 2004. Assembly of an active group II intron-maturase complex by protein dimerization. *Biochemistry* **43**: 6486–6497.

Rest JS, Mindell DP. 2003. Retroids in archaea: Phylogeny and lateral origins. *Mol Biol Evol* **20**: 1134–1142.

Robart AR, Seo W, Zimmerly S. 2007. Insertion of group II intron retroelements after intrinsic transcriptional terminators. *Proc Natl Acad Sci* **104**: 6620–6625.

Roitzsch M, Pyle AM. 2009. The linear form of a group II intron catalyzes efficient autocatalytic reverse splicing, establishing a potential for mobility. *RNA* **15**: 473–482.

Saldanha R, Chen B, Wank H, Matsuura M, Edwards J, Lambowitz AM. 1999. RNA and protein catalysis in group II intron splicing and mobility reactions using purified components. *Biochemistry* **38**: 9069–9083.

San Filippo J, Lambowitz AM. 2002. Characterization of the C-terminal DNA-binding/DNA endonuclease region of a group II intron-encoded protein. *J Mol Biol* **324**: 933–951.

Séraphin B, Simon M, Boulet A, Faye G. 1989. Mitochondrial splicing requires a protein from a novel helicase family. *Nature* **337**: 84–87.

Sharp PA. 1985. On the origin of RNA splicing and introns. *Cell* **42**: 397–400.

Sharp PA. 1991. "Five easy pieces". *Science* **254**: 663.

Shukla GC, Padgett RA. 2002. A catalytically active group II intron domain 5 can function in the U12-dependent spliceosome. *Mol Cell* **9**: 1145–1150.

Sigel RK, Vaidya A, Pyle AM. 2000. Metal ion binding sites in a group II intron core. *Nature Struct Biol* **7**: 1111–1116.

Sigel RKO, Sashital DG, Abramovitz DL, Palmer AGIII, Butcher SE, Pyle AM. 2004. Solution structure of domain 5 of a group II intron ribozyme reveals a new RNA motif. *Nature Struct Mol Biol* **11**: 187–192.

Simon DM, Kelchner SA, Zimmerly S. 2009. A broad-scale phylogenetic analysis of group II intron RNAs and intron-encoded reverse transcriptases. *Mol Biol Evol* **26**: 2795–2808.

Simon DM, Clarke NA, McNeil BA, Johnson I, Pantuso D, Dai L, Chai D, Zimmerly S. 2008. Group II introns in Eubacteria and Archaea: ORF-less introns and new varieties. *RNA* **14**: 1704–1713.

Singh NN, Lambowitz AM. 2001. Interaction of a group II intron ribonucleoprotein endonuclease with its DNA target site investigated by DNA footprinting and modification interference. *J Mol Biol* **309**: 361–386.

Singh RN, Saldanha RJ, D'Souza LM, Lambowitz AM. 2002. Binding of a group II intron-encoded reverse transcriptase/maturase to its high affinity intron RNA binding site involves sequence-specific recognition and autoregulates translation. *J Mol Biol* **318**: 287–303.

Smith D, Zhong J, Matsuura M, Lambowitz AM, Belfort M. 2005. Recruitment of host functions suggests a repair pathway for late steps in group II intron retrohoming. *Genes Dev* **19**: 2477–2487.

Solem A, Zingler N, Pyle AM. 2006. A DEAD protein that activates intron self-splicing without unwinding RNA. *Mol Cell* **24**: 611–617.

Stabell FB, Tourasse NJ, Kolstø AB. 2009. A conserved 3′ extension in unusual group II introns is important for efficient second-step splicing. *Nucleic Acids Res* **37**: 3202–3214.

Steitz TA, Steitz JA. 1993. A general two-metal-ion mechanism for catalytic RNA. *Proc Natl Acad Sci* **90**: 6498–6502.

Stern DB, Goldschmidt-Clermont M, Hanson MR. 2010. Chloroplast RNA metabolism. *Ann Rev Plant Biol* (in press).

Suchy M, Schmelzer C. 1991. Restoration of the self-splicing activity of a defective group II intron by a small *trans*-acting RNA. *J Mol Biol* **222**: 179–187.

Till B, Schmitz-Linneweber C, Williams-Carrier R, Barkan A. 2001. CRS1 is a novel group II intron splicing factor that was derived from a domain of ancient origin. *RNA* **7**: 1227–1238.

Toor N, Zimmerly S. 2002. Identification of a family of group II introns encoding LAGLIDADG ORFs typical of group I introns. *RNA* **8**: 1373–1377.

Toor N, Hausner G, Zimmerly S. 2001. Coevolution of group II intron RNA structures with their intron-encoded reverse transcriptases. *RNA* **7**: 1142–1152.

Toor N, Keating KS, Fedorova O, Rajashankar K, Wang J, Pyle AM. 2010. Tertiary architecture of the *Oceanobacillus iheyensis* group II intron. *RNA* **16**: 57–69.

Toor N, Keating KS, Taylor SD, Pyle AM. 2008a. Crystal structure of a self-spliced group II intron. *Science* **320**: 77–82.

Toor N, Rajashankar K, Keating KS, Pyle AM. 2008b. Structural basis for exon recognition by a group II intron. *Nat Struct Mol Biol* **15**: 1221–1222.

Toor N, Robart AR, Christianson J, Zimmerly S. 2006. Self-splicing of a group IIC intron: 5′ exon recognition and alternative 5′ splicing events implicate the stem-loop motif of a transcriptional terminator. *Nucleic Acids Res* **34**: 6461–6471.

Toro N, Jiménez-Zurdo JI, García-Rodríguez FM. 2007. Bacterial group II introns: Not just splicing. *FEMS Microbiol Rev* **31**: 342–358.

Toro N, Molina-Sánchez M, Fernández-López M. 2002. Identification and characterization of bacterial class E group II introns. *Gene* **299**: 245–250.

Valadkhan S, Mohammadi A, Jaladat Y, Geisler S. 2009. Protein-free small nuclear RNAs catalyze a two-step splicing reaction. *Proc Natl Acad Sci* **106**: 11901–11906.

Vallès Y, Halanych KM, Boore JL. 2008. Group II introns break new boundaries: Presence in a bilaterian's genome. *PLoS ONE* **3**: e1488. doi:10.1371/journal.pone.0001488.

van der Veen R, Kwakman JH, Grivell LA. 1987. Mutations at the lariat acceptor site allow self-splicing of a group II intron without lariat formation. *EMBO J* **6**: 3827–3831.

Vogel J, Börner T, Hess WR. 1999. Comparative analysis of splicing of the complete set of chloroplast group II introns in three higher plant mutants. *Nucleic Acids Res* **27**: 3866–3874.

Wank H, San Filippo J, Singh RN, Matsuura M, Lambowitz AM. 1999. A reverse transcriptase/maturase promotes splicing by binding at its own coding segment in a group II intron RNA. *Mol Cell* **4**: 239–250.

Watanabe K, Lambowitz AM. 2004. High-affinity binding site for a group II intron-encoded reverse transcriptase/maturase within a stem-loop structure in the intron RNA. *RNA* **10**: 1433–1443.

Xiong Y, Eickbush T. 1990. Origin and evolution of retroelements based upon their reverse transcriptase sequences. *EMBO J* **9**: 3353–3362.

Yang J, Mohr G, Perlman PS, Lambowitz AM. 1998. Group II intron mobility in yeast mitochondria: Target DNA-primed reverse transcription activity of aI1 and reverse splicing into DNA transposition sites *in vitro*. *J Mol Biol* **282**: 505–523.

Zhang L, Doudna JA. 2002. Structural insights into group II intron catalysis and branch-site selection. *Science* **295**: 2084–2088.

Zhao J, Niu W, Yao J, Mohr S, Marcotte EM, Lambowitz AM. 2008. Group II intron protein localization and insertion sites are affected by polyphosphate. *PLoS Biol* **6**: e150.

Zhong J, Lambowitz AM. 2003. Group II intron mobility using nascent strands at DNA replication forks to prime reverse transcription. *EMBO J* **22**: 4555–4565.

Zhuang F, Karberg M, Perutka J, Lambowitz AM. 2009a. EcI5, a group IIB intron with high retrohoming frequency: DNA target site recognition and use in gene targeting. *RNA* **15**: 432–449.

Zhuang F, Mastroianni M, White TB, Lambowitz AM. 2009b. Linear group II intron RNAs can retrohome in eukaryotes and may use non-homologous end-joining for cDNA ligation. *Proc Natl Acad Sci* **106**: 18189–18194.

Zimmerly S, Hausner G, Wu X. 2001. Phylogenetic relationships among group II intron ORFs. *Nucleic Acids Res* **29**: 1238–1250.

Zimmerly S, Guo H, Perlman PS, Lambowitz AM. 1995. Group II intron mobility occurs by target DNA-primed reverse transcription. *Cell* **82**: 545–554.

Zoschke R, Nakamura M, Liere K, Sugiura M, Börner T, Linneweber-Schmitz C. 2010. An organellar maturase associated with multiple group II introns. *Proc Natl Acad Sci* **107**: 3245–3250

The Roles of RNA in the Synthesis of Protein

Peter B. Moore[1,2] and Thomas A. Steitz[1,2,3]

[1]Departments of Molecular Biophysics and Biochemistry, Yale University, New Haven, Connecticut 208114
[2]Departments of Chemistry, Yale University, New Haven, Connecticut 208107
[3]Howard Hughes Medical Institute, New Haven, Connecticut 209812

Correspondence: peter.moore@yale.edu

SUMMARY

The crystal structures of ribosomes that have been obtained since 2000 have transformed our understanding of protein synthesis. In addition to proving that RNA is responsible for catalyzing peptide bond formation, these structures have provided important insights into the mechanistic details of how the ribosome functions. This review emphasizes what has been learned about the mechanism of peptide bond formation, the antibiotics that inhibit ribosome function, and the fidelity of decoding.

Outline

1. Introduction
2. The role of the ribosome in protein synthesis
3. Ribosome crystal structures
4. The PTC is made of RNA
5. What does the RNA in the PTC do to promote peptide bond formation?
6. Complexes with both A- and P-site substrates bound
7. The transition state intermediate and its possible stabilization
8. Metal ions in the PTC?
9. Protection of peptidyl-tRNA from hydrolysis and induced-fit activation of peptide synthesis
10. Release factors are tRNA mimics
11. Binding of the CCA end of tRNA to E site of the large subunit
12. Inhibition of the peptidyl transferase reaction by antibiotics
13. Translational fidelity
14. The tRNA binding sites in the small ribosomal subunit
15. The role of rRNA in mRNA decoding
16. Structural basis of elongation factor functions
17. Concluding comments

References

1 INTRODUCTION

In December, 1962, James Watson delivered a Nobel Prize lecture in Stockholm entitled "Involvement of RNA in the Synthesis of Protein" (Watson 1963) in which he described what was then known about protein synthesis. When one reads that text today, one is struck by how much had already been learned; the picture Watson painted for his audience was correct in its essence. However, there were still many missing pieces. For example, in 1962 no one knew how many tRNA binding sites there are on the ribosome, nor even whether all the ribosomes in a cell are the same.

In 1962, only three kinds of RNA were known: transfer RNAs, which were then called soluble RNAs, messenger RNAs, which had just been discovered, and ribosomal RNAs. It was understood that mRNAs convey sequence information from the genome to the protein synthetic apparatus, and that tRNAs are carrier molecules for amino acids that perform two functions. First, the esterification of amino acids to tRNAs, which is driven by ATP hydrolysis, activates them; the formation of polypeptides from aminoacyl tRNAs is spontaneous under intracellular conditions, but the formation of polypeptides from free amino acids is not. Second, tRNAs are adaptor molecules. There is at least one tRNA for every amino acid the cell uses for protein synthesis, and an enzyme that specifically aminoacylates each tRNA with its cognate amino acid. Furthermore, base pairing interactions between tRNA bases, i.e., between tRNA anticodons and mRNA codons, determine protein sequences. The RNAs whose role(s) in protein synthesis Watson could not explain were the ribosomal RNAs. No one knew why the ribosome, the enzyme that catalyzes peptide bonds formation, unlike any other enzyme then known, is made primarily of RNA.

The reason for writing an essay on the role of RNA in protein synthesis today is we can now answer the one major question that Watson could not. Ribosomes contain rRNA because the ribosome is a ribozyme. Even before the first ribozymes were discovered in the early 1980s, it had been suspected that rRNA might be the active principle in the ribosome (Crick 1968; see Noller 1991). The validity of that surmise was proven by the atomic resolution crystal structures of ribosomes that began appearing in the summer of 2000 (Fig. 1).

Here we review the crystallographic information on which the conclusion that the ribosome is a ribozyme is based, stressing the role of rRNA as a mediator of both peptide bond formation, and mRNA/tRNA interactions, the two aspects of ribosome function now best understood. In addition we will comment on some of the crystal structures that have been obtained of ribosomes with factors bound, and summarize what has been learned about the antibiotic inhibitors of large subunit function. Rather than discussing all of the ribosome crystal structures now available, we will concentrate on those obtained using the large ribosomal subunit of *Haloarcula marismortui* (Hma), and both the 70S ribosome and the small ribosomal subunit of *Thermus thermophilus* (Tth). Those interested in what has been learned from other ribosomal crystal structures should consult the appropriate reviews (e.g., Yonath and Bashan 2004; Berk and Cate 2007).

2 THE ROLE OF THE RIBOSOME IN PROTEIN SYNTHESIS

In all organisms, the ribosome consists of two ribonucleoprotein subunits, one about twice the size of the other. Prokaryotic large ribosomal subunits sediment at ~50S, and have masses of ~1.5 megadaltons. In prokaryotes, the small ribosomal subunit sediments at ~30S, and has a mass of ~0.8 megadaltons. The object that catalyzes protein synthesis is a 1:1 complex of the two subunits that sediments at ~70S, and it is about two-thirds RNA by weight. The large subunit contains 34 proteins and two RNAs, 23S

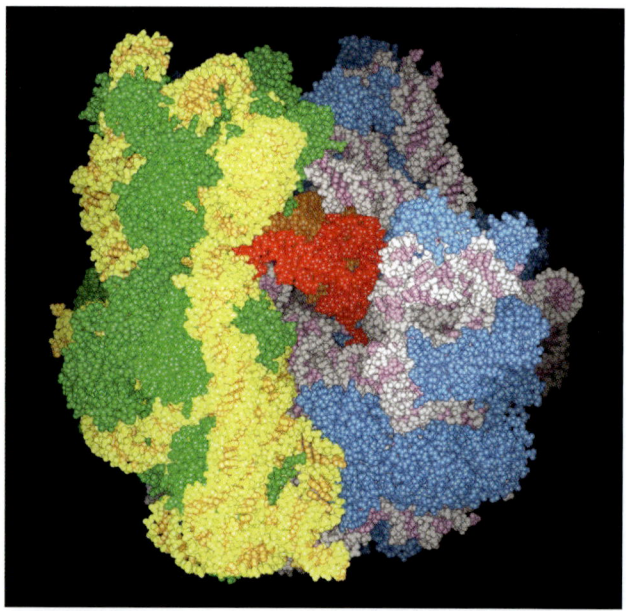

Figure 1. Spacefilling model of the 70S ribosome. The structure of the Hma large subunit (Ban et al. 2000) and that of the *T. thermophilus* small subunit (Wimberly et al. 2000) were docked using the rRNA from the Noller model (Yusupov et al. 2001) of the 70S as well as the A, P, and E site tRNAs from that model. The 23S rRNA and 5S rRNA are in pink and white whereas the 16S rRNA is in light and dark yellow. rProteins of the large subunit are blue and those of the small subunit are green. The A-site tRNA with its 3' end extending into the peptidyl transferase cavity is red and the P-site tRNA is yellow.

rRNA and 5S rRNA, whereas the small subunit consists of 21 proteins and a single RNA, 16S rRNA. With the exception of a single large subunit protein, L7/L12, which is present in four copies per ribosome, the ribosome is a 1:1 complex of all its components (Wittmann-Liebold 1986). The cytoplasmic ribosomes in eukaryotes are homologous to prokaryotic ribosomes, but are bigger and have more components (Wool 1986; Wool et al. 1995).

The roles of the two subunits of the ribosome in protein synthesis are different (Green and Noller 1997). The small ribosomal subunit mediates the interactions between mRNA codons and tRNA anticodons that determine the order in which amino acids are assembled into protein. The large ribosomal subunit contains the peptidyl transferase center, the site where peptide bond formation occurs. In isolation, both subunits can perform functions closely related to those they play in the intact 70S ribosome. By itself, the small subunit binds mRNA, and when mRNA is bound, it will bind tRNAs in a codon-specific manner. By itself, the large subunit will catalyze the formation of peptide bonds between aminoacyl tRNA-like substrates.

Both subunits have three tRNA binding sites: an A site that binds aminoacyl-tRNAs, a P site that interacts with peptidyl-tRNAs, and an E site through which deacylated tRNAs pass as they are discharged from the ribosome. The tRNA sites in the large subunit interact primarily with the CCA terminus of tRNAs, but also with their acceptor stems. The anticodon stems of tRNAs interact with the A, P, and E sites of the small ribosomal subunit. In a functioning ribosome, mRNA also binds to the A and P sites of the small subunit, and the interaction between mRNA codons and tRNA anticodons in the A site ensure that the correct aminoacyl tRNA is selected from the cytoplasmic pool at each step in chain elongation.

The two aspects of the mechanism of ribosome function of primary interest here, peptide bond formation and mRNA decoding, are critical for the elongation phase of protein synthesis, i.e., the stage in protein synthesis during which nascent polypeptides are elongated amino-terminal to carboxy-terminal, one amino acid at a time. Elongation is promoted by two protein factors. One facilitates the delivery of aminoacyl-tRNAs to the ribosome, and the other promotes translocation, the complex process that ultimately enables the ribosome to read the next codon of the mRNA bound to it. In prokaryotes, the protein factor responsible for tRNA delivery is called EF-Tu, and the protein that promotes translocation is called EF-G. It is important to note that ribosomes can catalyze mRNA-dependent polypeptide synthesis without the assistance of factors, but factor-free translation is slow, and inefficient (Pestka 1974; Gavrilova et al. 1976; Southworth et al. 2002).

The elongation cycle starts with a ribosome that has a mRNA bound to its small subunit, a peptidyl tRNA bound to the P site of both the large and small ribosomal subunits that makes the expected codon–anticodon interaction with the mRNA triplet exposed in the P site, an A site that has the mRNA codon 3' to the one in the P site bound to it, but no tRNA, and an E site that contains a deacylated tRNA. Ribosomes in this state interact preferentially with aminoacyl tRNA ternary complexes, which is to say aminoacyl tRNA·EF-Tu·GTP complexes. Base pairing interactions between the A-site mRNA codon and ternary complex anticodons determine which ternary complex will be selected from the mixture present in the cell. When a cognate ternary complex has been selected, the anticodon stem of its tRNA occupies the A-site of the small subunit, but not the A site of the large subunit. Hydrolysis of the GTP in the ternary complex ensues, EF-Tu·GDP is released from the ribosome, and the bound aminoacyl tRNA reorients so that its aminoacyl-CCA end can enter the large subunit's A site. This reorientation is called "accommodation." (At about this stage, the deacylated tRNA in the E site leaves the ribosome.) The peptidyl transferase reaction ensues quickly thereafter; the nascent peptide that was esterified to the 3' terminal ribose of the tRNA in the P site is transferred to the amino group of the aminoacyl tRNA bound to the A site, which elongates it by one amino acid. The tRNA in the P site is left deacylated. In the final step of the cycle, EF-G·GTP binds to the same site on the ribosome where EF-Tu binds, and facilitates translocation. The deacylated tRNA in the P site moves to the E site, the peptidyl tRNA in the A site moves to the P site, and the ribosome advances along its mRNA by 1 codon. Hydrolysis of the GTP bound to EF-G follows, and EF-G·GDP leaves the ribosome, returning it to its initial state.

3 RIBOSOME CRYSTAL STRUCTURES

No single experimental technique can ever provide all the information needed to understand the mechanism of a complex biological process, but history shows that mechanisms cannot be understood unless the structures of the participating macromolecules are known at atomic resolution. Fortunately, in 2000, the ribosome community began to obtain the high-resolution structural information about the ribosome it had needed for so long, and not surprisingly, the field has been utterly transformed by it.

A monograph could be written on the history of ribosome crystallography, and it would be a good read because the scientific hurdles that had to be overcome were substantial, and the personalities involved were/are colorful. In the end, once a strategy had been devised for phasing ribosome diffraction patterns (Ban et al. 1998), a burst of atomic resolution structures appeared. A 2.4 Å resolution structure of the large ribosomal subunit from Hma (Ban et al. 2000) appeared in August, 2000. A month later, a 3.05 Å resolution structure appeared of the 30S subunit from Tth (Wimberly et al. 2000), and it followed by only a week or two the publication of an independently determined, but less accurate, and lower resolution (3.4 Å) version of the same structure (Schluenzen et al. 2000), the faults of which were later corrected (Pioletti et al. 2001). The big surprise for the cognoscenti was the 3.0 Å resolution crystal structure of the *Deinococcus radiodurans* large ribosomal subunit, which appeared at the end of 2001 (Harms et al. 2001). The paper describing that structure was also the paper in which crystallization of those ribosomes was first announced. Implicit in this "coincidence" was the message that the technical problems that for so long plagued ribosome crystallography had been overcome.

Although the structures of the large and small subunits mentioned earlier were extraordinarily illuminating, they left unanswered a host of questions related to the way subunits interact in the 70S ribosome, but the crystal structures needed to address those issues began appearing not long thereafter. The first atomic model of the structure of 70S ribosome published was derived from a 5.5 Å resolution electron density map of the ribosome from *T. thermophilus*, which was interpreted using the structures of the two subunits that had already been determined at higher resolution (Yusupov et al. 2001). The principle drawback of such models is that they are least informative where information is most needed, namely in those regions where structures used to guide interpretation do not explain electron density. Fortunately, crystal structures of 70S ribosomes are now available that have resolutions high enough to eliminate such ambiguities. In 2005 Schuwirth et al. produced a 3.5 Å resolution structure for the 70S ribosome from *Escherichia coli*. A year later a 2.8 Å resolution structure was obtained for a new crystal form of the *T. thermophilus* 70S ribosome (Selmer et al. 2006), and shortly thereafter a higher resolution (3.7 Å) version of the original Yusupov structure appeared (Korostelev et al. 2006). These crystal structures have been the "parent structures" for scores of additional structures that show ribosomes and ribosomal subunits bound to substrates, substrate analogs, proteins factors, and antibiotic inhibitors.

4 THE PTC IS MADE OF RNA

The first atomic resolution structures of the large ribosomal subunit and its substrate complexes proved that rRNA is responsible for peptidyl transferase activity of the ribosome (Nissen et al. 2000). The evidence is straightforward. It has long been known that large subunits catalyze the formation of peptide bonds in the absence of mRNA, protein factors, or small ribosomal subunits when provided with low molecular weight analogs of the CCA-ends of peptidyl tRNA and aminoacyl tRNA (Monro et al. 1969). For that reason, structures were obtained of the complexes the Hma large subunit forms both with an A-site analog, and with a putative transition state analog (Nissen et al. 2000). They revealed that peptidyl tRNA mimics bind to the large subunit with their C74 and C75 analogs base-paired with two G residues belonging to the P-loop of 23S rRNA, and that aminoacyl tRNA mimics bind with their C75 analogs base-paired with a G residue belonging to the A-loop of 23S rRNA, as earlier biochemical and genetic experiments had indicated they should (Samaha et al. 1995; Kim and Green 1999). Thus there could be no doubt that these analog structures identified the site where peptide bond formation occurs in the ribosome, i.e., the peptidyl transferase center (PTC). Inspection of the surrounding region showed that there is no protein in the vicinity, and hence that protein cannot be directly involved in peptide bond formation. Although it was subsequently discovered that there is some poorly ordered protein in the vicinity of the PTC that had not been taken into account in these first large subunit crystal structures (see Klein et al. 2004), the conclusions still stands.

5 WHAT DOES THE RNA IN THE PTC DO TO PROMOTE PEPTIDE BOND FORMATION?

Enzymes can facilitate chemical reactions in three general ways: (1) substrate orientation, (2) specific chemical catalysis, and (3) transition state stabilization. Substrate orientation is a major contributor to the catalytic power of most enzymes. By binding substrates with their orbitals oriented properly for reaction, enzymes reduce the entropic barrier to reaction. Page and Jencks (1971) concluded that the rate enhancement caused by orientation can exceed 10^7-fold. As expected, the PTC avails itself of this important source of catalytic power by positioning the α-amino group of the aminoacyl moiety of the aminoacyl tRNA bound to the A site so that it is close to and pointed toward the carbonyl carbon of the ester that links the peptidyl moiety to the CCA portion of the pepetidyl tRNA bound in the P site. Thus, substrate orientation makes a major contribution to the ribosome's catalytic power, as pointed out by Nissen et al. (2000). Indeed, it was proposed some two decades ago that substrate orientation alone might accelerate the rate of peptide bond formation sufficiently to account for the catalytic activity of the ribosome (Nierhaus et al. 1980), and similar arguments have been advanced more recently (Sievers et al. 2004). Nevertheless, whether and to what extent the ribosome uses additional mechanisms to enhance the rate of peptide bond formation was still an open question. Does it use RNA, with or without the assistance of metal ions, to assist catalysis chemically, and does the ribosome stabilize the oxyanion intermediate in the synthesis reaction as proteases do when catalyzing the peptide hydrolysis reaction? These questions have now been largely answered.

Initial insights into the substrate complexes that form in the PTC were obtained from the structures of separate complexes with A-site analogs, e.g., C puromycin (Cpmn), and the P-site substrates analog, CCA-phe-caproic acid-biotin (CCApcb) plus sparsomycin (Hansen et al. 2002). An approximation to the structure of the large subunit containing both substrates was achieved initially by superimposing the structures of these separately determined A-site and P-site substrate complexes. In this hypothetical, two-substrate complex the α-amino group of the A-site amino acid is adjacent to the ester linked carbonyl carbon of the peptidyl-tRNA it is to attack. The orientations of the two single-stranded CCA sequences bound in these two sites are related by a two-fold rotation axis despite the fact that the tRNA molecules to which they are attached are related to each other by a translation. The proposal that this difference in the orientations of the 3′ termini of the two tRNA molecules may facilitate their translocation after peptide bond formation is as yet untested (Nissen et al. 2000). In any case the relative orientation of the aminoacyl and peptidyl CCAs correctly positions the attacking αNH$_2$ and the ester linked carbonyl carbon, more or less.

6 COMPLEXES WITH BOTH A- AND P-SITE SUBSTRATES BOUND

The structures obtained of Hma large subunit complexes with analogs of A- and P-site substrates bound simultaneously (Fig. 2) both limited the number of ways RNA might promote peptide bond formation, and showed that premature peptidyl-tRNA hydrolysis is suppressed by an induced-fit mechanism. Two A-site substrates were prepared that differed in whether or not a C74 mimic was included: CC-hydroxypuromycin (CChPmn) or CPmn. Both were studied in combination with a P-site substrate, CCApcb. The structures of these complexes confirmed that only the N3 of A2486 (2451 *E. coli*) and the 2′OH of A76 of the P-site substrate contact the attacking α-amino group of the aminoacyl-tRNA, as had earlier been concluded (Hansen et al. 2002), and thus only they could possibly play a direct, chemical role in catalysis (Fig. 2).

The possibility that the N3 of A2486 (2451) might serve as a general base to activate the attacking αNH$_2$, despite its normally low pKa (Nissen et al. 2000), was ruled out by the genetic and biochemical experiments reported by Green

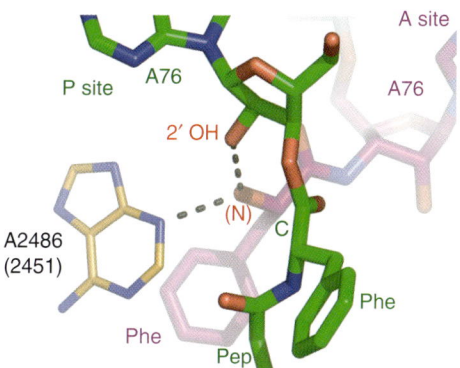

Figure 2. Preattack conformation of the substrates of the peptidyl transferase reaction. The hydroxyl group representing the α-amino group of the A-site substrate, CChPmn (purple) is in position to attack the ester group of the P-site substrate CCApcb (green). It is within hydrogen bonding distance of the N3 of A2486 (2451), and the 2′ hydroxyl group of the P-site substrate. In the ground state, the two reactive groups are 3.7 Å apart.

and coworkers, who showed that mutation of A2486 to any other base has no effect on the rate of peptide bond formation when full tRNA substrates are used with the 70s ribosome; these mutations do, however, inhibit the termination step, accounting for their dominant lethal phenotype (Youngman et al. 2004).

In contrast, the 2′OH of the P-site A76 remains an interesting candidate for catalytic involvement as both a general acid and general base; numerous biochemical experiments employing a deoxA76 in the P-site substrate have suggested that the 2′OH of A76 plays an important role in the PTC. Barta and colleagues proposed a mechanism in which the 2′OH serves simultaneously as a general base to activate the nucleophilic attack of the α-amino group and as a general acid to facilitate the leaving of the 3′ linked peptide ester (Dorner et al. 2003). Their proposal was based on experiments which showed that when acetylated aminoacyl deoxA is used as a P-site substrate and Pmn employed as an A-site substrate, the rate of peptide bond formation is reduced at least 100-fold, and on the structure-based model of Hansen et al. (2002), which suggested an interaction between this 2′OH and the α-NH$_2$ group. A more quantitative assessment of the magnitude of the effect of a deoxy-A76 on the rate of peptide bond formation was obtained using full-length tRNA substrates by Strobel and coworkers who found that the rate of peptide bond formation for peptidyl-tRNAs containing a 2′ deoxy A76 is at least 10^6-fold slower than normal (Weinger et al. 2004). This rate reduction also occurs if a 2′ fluoro analog rather than a 2′ deoxy analog is used. However, it is not obvious that this entire rate enhancement should be credited to the 2′OH group. A 100-fold rate enhancement would result even if the reaction were uncatalyzed simply because of the vicinal effect of the 2′OH. Also, in the substrate complex, water is sterically excluded from donating protons to the 3′ ester-linked hydroxyl (Schmeing et al. 2005a).

7 THE TRANSITION STATE INTERMEDIATE AND ITS POSSIBLE STABILIZATION

Structures have been obtained of several of the complexes the Hma large subunit forms with analogs of the expected intermediate in the peptide synthesis reaction in which either a phosphodiester or triester is used to mimic the tetrahedral carbon transition state (Fig. 3). The first such transition state analog (TSA) used to examine the structure of a complex was CCA-phosphate puromycin (Nissen et al. 2000). A 3.2 Å resolution map was interpreted as indicating that one of its nonbridging phosphoryl oxygens was interacting with the N3 of A2486 (2451) and represented the oxyanion mimic of the tetrahedral carbon. However, subsequent modeling of the tetrahedral intermediate from the structures of the A- and P-site substrate complexes implied that the oxyanion points away from A2486, and the pH dependence of the TSA binding was not consistent with the proposed interaction with A2486 (Parnell et al. 2002). The structures of numerous additional analogs of the transition state intermediate, which were determined at resolutions between 2.4 Å and 2.8 Å, clearly showed that the oxyanion of the intermediate points away from A2486 (2451), and interacts with a water molecule that is positioned by its interactions with several adjacent nucleotides. The polarity of this water molecule may increase the rate of peptide bond formation by stabilizing the oxyanion, but the magnitude of its contribution is difficult to estimate.

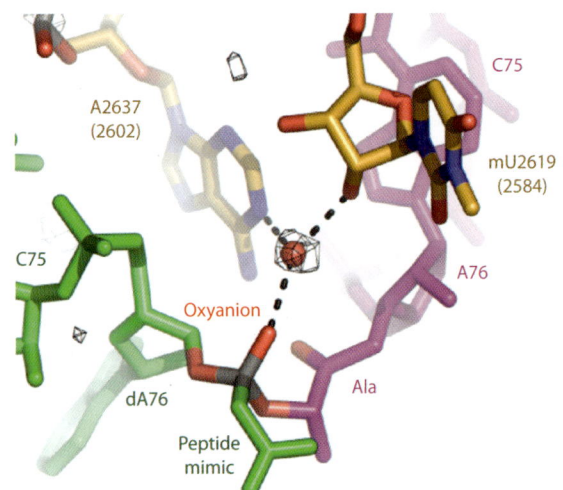

Figure 3. Structure of an analog of a transition site intermediate in the peptidyl transferase reaction showing the oxyanion bound to a water molecule and pointing away from A2486.

8 METAL IONS IN THE PTC?

The observation that the 2′OH of the A76 of the P-site bound tRNA enhances the rate of peptidyl-transferase activity by 10^6-fold or more raises the question of why it is so effective. Difference electron density maps of TSA complexes showed density peaks near this 2′OH, opening the possibility that there might be a Mg^{++} or Na^+ ion present, even though the peak was 2.8 Å from the A76 2′OH. The possibility that a Mg^{++} ion might be binding to this 2′OH, or anywhere else at the site of reaction, was eliminated by examining crystals of TSA complexes with the large subunit in which the Mg^{++} in the crystals was exchanged for Mn^{++}, because no peaks appeared in this region in anomalous difference Fourier maps. Likewise, the possibility that there might be a monovalent metal ion binding in the vicinity was eliminated by examining crystals containing 1.5-M potassium. There is no structural or biochemical evidence that metal ions play a direct role in peptide bond formation.

The analysis of the mechanism of peptide bond formation presented here, which derives from the structures of complexes of the large ribosomal subunit from Hma with substrate analogs, as well as numerous biochemical experiments, has not gone unchallenged. For example, Korostelev et al. (2006) concluded from their 3.7 Å resolution structure of the Tth 70S ribosome with tRNAs bound in both the P- and E-sites that the structure of the PTC in isolated 50S subunits is not the same as it is in 70S ribosomes. However, subsequent recalculations of the Korostelev electron density map (Simonovic and Steitz 2008), and some recently reported crystal structures of 70S ribosomes with tRNAs bound in both the A- and P-sites (Voorhees et al. 2009; Blaha et al. 2009) indicate that this is not the case. These new structures support the conclusions about the mechanism of peptide bond formation that were reached using Hma structures. Nevertheless, anyone concerned that peace is about to break out should consult (Gindulyte et al. 2006) and (Korostelev et al. 2009).

9 PROTECTION OF PEPTIDYL-tRNA FROM HYDROLYSIS AND INDUCED-FIT ACTIVATION OF PEPTIDE SYNTHESIS

Another long-standing question that can now be answered is how the ribosome prevents the premature hydrolysis of peptidyl-tRNAs. If the PTC can activate α-NH_2 groups for reaction, why does it not activate water molecules when the A-site is empty? The reason is that the ester-linked carbonyl carbon to be attacked is sterically protected by the PTC until the CCA of an aminoacyl-tRNA binds to the A-site and induces a (modest) conformational change in the PTC rRNA, which repositions the carbonyl group so that it can be attacked by an appropriately positioned α-amino group (Schmeing et al. 2005b). In short, this property of the PTC depends on induced fit, a mechanism responsible for enhancing the substrate specificity of many protein enzymes, that was first advanced by Koshland (1959) to explain why hexokinase does not hydrolyze ATP in the absence of glucose, and then proven to be correct by Bennett and Steitz (1976).

The structural evidence for an induced fit conformational change in the large ribosomal subunit derives from structures of the Hma 50S subunit complexed with CCApcb in the P-site, and simultaneously with either CChPmn or ChPmn in the A-site (Schmeing et al. 2005b). When ChPmn is in the A-site the ester group of CCApcb is both positioned and sequestered by the bases of U2620 (2585) on one side and A2486 (2451) and C2104 (2063) on the other so that the carbonyl oxygen of its ester group points toward the α-amino group of the A-site substrate, and in this orientation hydrolysis is prevented because water molecules cannot access its carbonyl carbon. However, when CChPmn binds to the A-site, it induces a series of conformational changes in the bases forming the PTC; the base of U2620 (2582) moves away from the ester group of the peptidyl-tRNA mimic, allowing its carbonyl group to reorient in a manner that is favorable for nucleophilic attack by the nearby α-hydroxl or α-amino group (Fig. 4). These conformational changes also explain why the rate of hydrolysis of peptidyl tRNAs is 100 times faster in the presence of CCA than it is in the presence of CA (Caskey et al. 1971).

10 RELEASE FACTORS ARE tRNA MIMICS

The synthesis of nascent polypeptide chains ends when translocation brings the stop codon of a mRNA into the A site of the ribosome, and the reason synthesis ceases at that junction is that cells do not (normally) contain tRNAs that are cognate to stop codons. Based on what has been said so far one might think that ribosomal complexes of this sort would be very stable because in the absence of A-site bound tRNAs, the conformation adopted by the PTC protects the ester bonds that link nascent polypeptides to P-site bound tRNAs from hydrolysis. However, in vivo, polypeptides are rapidly released from these complexes because of the action of proteins called release factors, of which there are two in bacteria, RF1 and RF2. RF1 is specific for 70S ribosomes carrying nascent polypeptides that have stopped synthesis at UAG or UAA codons, and RF2 is specific for UGA and UAA codons. Both recognize stop codons when they are exposed in the decoding centers of ribosomes, and make the PTC catalyze a reaction

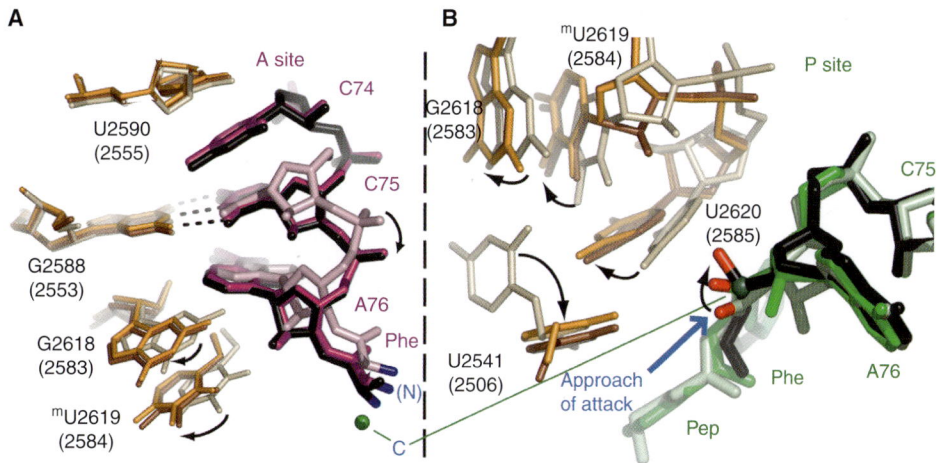

Figure 4. Movements of rRNA and peptidyl-tRNA are induced by proper binding of the A-site substrate. (*A*) A view of the A-site of the three complexes. Without C74, ChPmn (pink, with rRNA colored wheat) is positioned higher in the A-site. In substrates with C74, both CChPmn (purple, with orange rRNA) and the TSA (black, with brown rRNA), C74 stacks with rRNA base U2590 (2555), shifting the substrates down, and the α-amino group closer to the ester carbon of the P-site substrate (green). rRNA base G2618 (2583) shifts to maintain the A-minor interaction, causing methylU2619 (2584) to also move. (*B*) A view of the P-site of the same three complexes. The movement induced in G2618 (2583) by A-site substrate binding breaks its G-U wobble pair with U2451 (2506), which swings 90°. MethylU2619 (2584) and U2620 (2585) also shift to allow the ester group to move from the position it occupies when CCApcb (light green) is bound together with ChPmn, to that when CCApcb (medium green) is bound with CChPmn, and finally to that when it has been attacked by the A-site substrate, as shown with the TSA (black).

it otherwise promotes very inefficiently, namely the hydrolysis of the ester bonds that link peptides to tRNAs bound in the P site.

Atomic resolution structures have recently been obtained of 70S ribosome/tRNA complexes with RF1 bound (Lauerberg et al. 2008), or RF2 bound (Korostelev et al. 2008; Weixlbaumer et al. 2008). These structures show that RFs bind to the ribosome in much the same way that aminoacyl tRNAs bind before peptide bond formation. Specific interactions between the ends of RFs that bind to the A site of the small subunit decoding center and stop codons stabilize RFs in a conformation which ensures that their far ends will insert into PTC, where they occupy the same space as the CCA end of a A-site bound tRNA. The conformation of the PTC changes in response to RF binding in much the same way it changes when aminoacyl tRNAs bind to the A site, rendering the ester bond in the P-site of the PTC accessible to water. The GGQ sequence found in all RFs is positioned so that it can assist with the attack of a water molecule on that ester bond.

11 BINDING OF THE CCA END OF tRNA TO E SITE OF THE LARGE SUBUNIT

The exit site, or E-site, binds deacylated tRNAs as they transit from the P-site, after peptide bond formation, back to solution. Although proteins are found in the present day E-site of archaea (L44e and L15e) and eubacteria (L33), it appears that prior to the divergence of archaea and eubacteria, the binding site for CCA in the E-site on the large subunit was entirely rRNA. Not only are the sequence-specific interactions that ensure that only deacylated tRNAs bind to the E-site mediated by RNA, but the proteins that contact the CCA in eubacteria and archebacteria are not homologous, implying that they were added to the ribosome subsequent to the divergence of the two kingdoms (Schmeing et al. 2003) (Fig. 5). Furthermore, the conformation of the CCA end of a tRNA bound to the E site of an archaeal ribosome is not the same as it is when it is bound to the E site of a eubacterial ribosome (Schmeing et al. 2003; Selmer et al. 2006). Only A76 binds the same way to the E site of both types of ribosome, where it interacts with universally conserved bases.

12 INHIBITION OF THE PEPTIDYL TRANSFERASE REACTION BY ANTIBIOTICS

Microorganisms synthesize a wide variety of secondary metabolites that seem to function at low concentrations as cell-signaling molecules (Yim et al. 2007). At higher concentrations, some of these compounds are antibiotics because they inhibit ribosome function. The species specificities

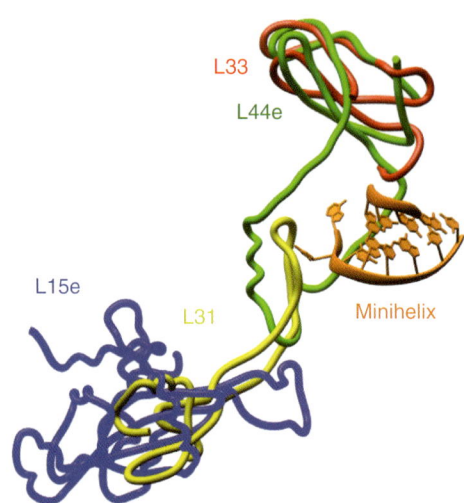

Figure 5. Proteins in the vicinity of the E site of the large ribosomal subunit. Oligonucleotide analogs of the acceptor stem of a tRNA (orange) will bind to the E site of the large ribosomal subunit from *H. marismortui*. The proteins in the immediate neighborhood, L44e (green) and L15e (blue), have no homologs in the eubacterial ribosome. When the structure of the large ribosomal subunit from *D. radiodurans* (Harms et al. 2001) is superimposed on that of *H. marismortui*, it is found that L33 (red) and L31 (yellow) occupy suggestively similar positions. (Reproduced with permission from Schmeing et al. (2003).)

of these ribosome inhibitors vary a lot, but some are sufficiently specific for eubacterial ribosomes to be useful for treating human diseases.

Crystal structures have been obtained of dozens of the complexes antibiotics and other inhibitors of protein synthesis form with the large ribosomal subunit (e.g., Fig. 6). So far, almost all of the anti-ribosomal antibiotics studied crystallographically bind to sites that are composed entirely of RNA, and most of those that target the large ribosomal subunit inhibit its activity in two different ways (Schleunzen et al. 2001; Hansen et al. 2002; Hansen et al. 2003). One class, the macrolides (e.g., erythromycin, tylosin, and azythromycin), bind to a site in the proximal part of the polypeptide exit tunnel adjacent to the peptidyl-transferase center, and although some macrolides also block PTC activity sterically, they all inhibit protein synthesis by interfering with the passage of nascent polypeptide down the exit tunnel. In the presence of macrolides ribosomes tend to synthesize short peptides rather than complete proteins (Tenson et al. 2003). The hypothesis that macrolides occlude the tunnel like the stopper in a bottle, which explains a lot of the phenomenology of macrolides, may be an oversimplification (Tu et al. 2005; Voss et al. 2006), and there is abundant evidence that the efficiency with which nascent polypeptides pass through the exit tunnel, which is the property of the protein synthesis system that macrolides

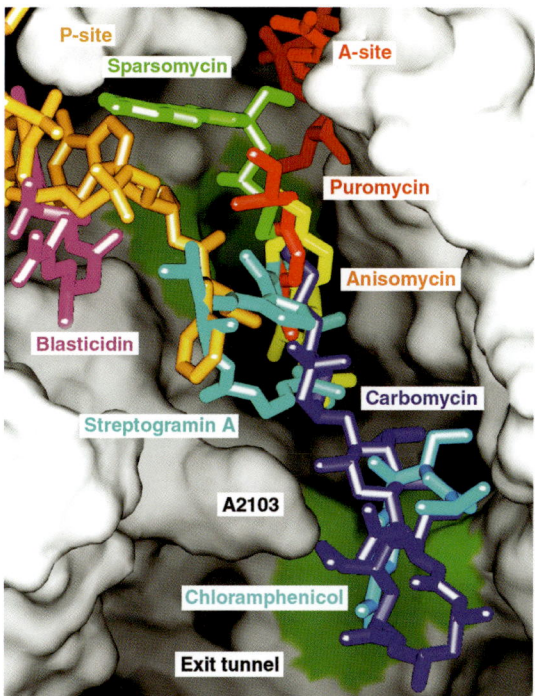

Figure 6. The positions of seven antibiotics and A-site (red) plus P-site (yellow) substrates bound to the peptidyl transferase center. The ribosome has been split open to reveal the lumen of the exit tunnel and adjacent regions of the peptidyl transferase site. Ribosomal components are depicted as a continuous surface that is colored green at two positions where splayed out bases provide hydrophobic binding sites for small molecules. Seven independently determined cocrystal structures have been aligned by superimposing the 23S rRNA in each complex.

alter, is affected by amino acid sequence (Nakatogawa and Ito 2004). In any case, all the macrolides studied so far bind to the *H. marismortui* 50S subunit (Hansen et al. 2003) in the same general location that macrolides bind to the large ribosomal subunit from the eubacterium *D. radiodurans* (Schleunzen et al. 2001).

The second set of antibiotics we will discuss is much more diverse chemically. Its members bind to either the A-site or the P-site of the PTC, and appear to obstruct the binding of either the A-site or P-site tRNA to the PTC, or both, consistent with their being competitive inhibitors of peptide bond formation (Fig. 6). For example, anisomycin bound to Hma large subunit (Hansen et al. 2003) and chloramphenicol bound to *D. radiodurans* 50S subunit (Schleunzen et al. 2001) both occupy a hydrophobic crevice formed by two splayed-out bases that also provides the binding site for the tyrosine side chain of an A-site bound substrate analog. Linezolid, a completely synthetic protein synthesis inhibitor, binds to ribosomes the same way (Ippolito et al. 2008). (At mM concentrations chloramphenicol binds to a second hydrophobic crevice in Hma subunits where macrolide antibiotics also bind.) Virginiamycin M occupies portions of both the A-site and P-site whereas blastocydin S exploits another strategy by mimicking C74 and C75 of the P-site bound tRNA and base-pairing with the P-loop.

Recent structural studies of mutant Hma large subunit have provided insights into the structural basis of the resistance that mutations confer to the macrolides, and to anisomycin (Blaha et al. 2008; Gurel et al. 2009). It has been known for years that the mutation A2058G (*E. coli*) in 23S rRNA makes eubacterial ribosomes highly resistant to macrolides. The nucleotide at that location in the Hma ribosome is a G, and as expected, erythromycin does not bind to crystals of the Hma large subunit, even at a concentration of 3 mM, and other macrolides, e.g., azithromycin, bind only weakly. However, when that G is mutated to an A, the affinity of the Hma ribosome for erythromycin increases by $>10^4$-fold, and it binds at <1 µM concentrations (Tu et al. 2005). Azythromycin also binds more tightly to the mutant, but in much the same orientation that it binds to wild-type ribosomes (Fig. 7A). The reduction in binding constant caused by the A2058G mutation in eubacterial ribosomes reflects the energetic cost of desolvating and burying the N2 of that G, and the less snug packing of the macrolide ring on the tunnel wall its presence requires (Fig. 7B). Dimethylation of the N6 of A2058 would also sterically interfere with macrolide binding. Erythromycin, telithromycin and azythromycin bind identically to the G2099A mutant subunit, and very similarly to the way 16-membered macrolides bind to wild-type 50S subunit. (Initial reports indicated that the orientation of the macrolide ring of erythromycin bound to the Hma 50S subunit is orthogonal to that observed when it binds to the large subunit from Dra (Schleunzen et al. 2001), but subsequent studies performed with Dra have indicated that this not the case (Wilson et al. 2005).)

Anisomycin is one of three low molecular weight antibiotics that inhibit the large ribosomal subunit by binding to its A-site cleft. These compounds are interesting because even though their binding site, i.e., the A-site cleft and its immediate surround, is highly conserved and their chemical structures quite similar, they are species-specific. Two members of this group, chloramphenicol and linezolid, are specific for eubacteria, whereas the third, anisomycin, is much more toxic to eukaryotes and archaea than it is to eubacteria. What is the structural basis for this difference?

No single mutation makes Hma highly resistant to anisomycin the way A2058G (*E. coli*) makes eubacterial highly resistant to macrolides. Nevertheless, when the structures of the Hma large ribosomal subunits containing mutations that confer comparatively weak resistance to anisomycin are considered as a group, an interesting picture of the

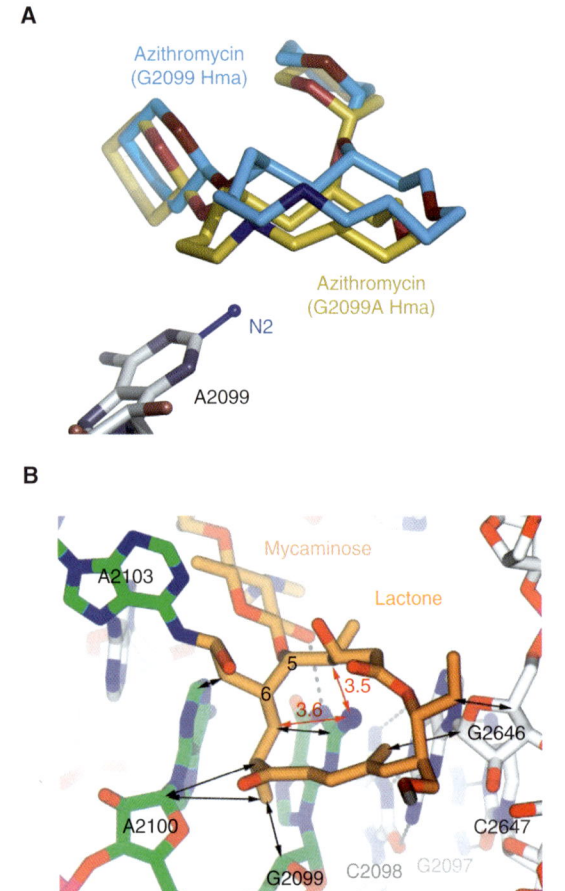

Figure 7. Macrolide interactions with the large ribosomal subunit of *H. marismortui*. (*A*) Comparison of the location and conformation of azithromycin bound to wild-type *H. marismortui* large ribosomal subunits, and G2099A mutant large ribosomal subunits from the same organism. Structures of both forms of the subunit with the drug bound have been superimposed. The drug bound to G2099A subunits is shown in brown, and the drug bound to wild-type is in cyan. The location that would be occupied by the N2 of a G residue at position 2099 is modeled in. (Reproduced with permission from Tu et al. 2005 [© Elsevier].) (*B*) Intrusion of the N2 of G2099 into the lactone ring of macrolide antibiotics. The structure of the 16-membered macrolide tylosin is shown bound to wild-type *H. marismortui* large ribosomal subunits. The N2 of G2099 lies under the hydrophobic surface of the lactone ring of the drug requiring it to move away from the base where it would otherwise make a number of close contacts that would be unfavorable for binding. In addition, when the drug is bound, the N2 of G2099 is desolvated and unable to make any hydrogen bonds. (Reproduced with permission from Hansen et al. 2002 [© Elsevier].)

These conformational states are in equilibrium with each other, and may vary in free energy by as much as 3 or 4 kcal/mol at 37°C. Any given substrates, substrate analog, or inhibitor of the PTC binds only when it is in a compatible conformational state, and thus that conformation will dominate when the ligand in question is present. For example, when anisomycin is bound to the Hma ribosome the conformation of its PTC is not the same as it is in the absence of anisomycin (Blaha et al. 2008). The conformational change that explains why the PTC catalyzes nucleophilic attacks on the peptidyl tRNA only when proper A-site substrates are present is undoubtedly a manifestation of this same phenomenon (Schmeing et al. 2005b).

The structural data support the hypothesis that mutations in Hma that cause resistance to anisomycin either stabilize the apo- conformation of the PTC or destabilize its anisomycin-bound conformation (Blaha et al. 2008); thermodynamically it does not matter which happens. Because the conformation of the center of the PTC, which includes the A-site cleft, is stabilized by its interactions with more peripheral nucleotides, mutations in nucleotides that do not contact anisomycin when it is bound to the ribosome can alter its binding constant, as observed. They also determine which conformations of the PTC will be favored in the ribosomes from any given species in the absence of PTC ligands. It follows that the conformation of the PTC in apo-structures of the ribosome should vary somewhat between species, and they do.

Comparison of the structures of large ribosomal subunits from several species, with and without A-site cleft antibiotics of several kinds bound strongly suggests that the position of a single, highly conserved nucleotide that neighbors the A-site cleft (U2504 in *E. coli*) controls the response of the ribosome to A-site cleft antibiotics (Gurel et al. 2009). In eubacteria its placement restricts the volume antibiotics can occupy on one side of the A-site cleft. In Hma, and one would surmise eukaryotes as well, its placement is more permissive allowing larger molecules can bind to the cleft. Because the position of this nucleotide is determined by its interactions with nucleotides that are even further removed from the center of the PTC and the identities of which vary from one group of organisms to the next, its conformation provides a satisfying explanation for species specificities of drugs of this sort.

Mutations in ribosomal proteins can also cause antibiotic resistance. A deletion of three residues in a long, antiparallel protein loop of L22 causes resistance to erythromycin, but does not prevent the drug from binding. The structure of Hma large subunit containing such a deletion mutation in protein L22 shows that the protein loop that lies on the wall of the wild-type ribosomal tunnel has moved from that position, making the tunnel wider by

conformational properties of the PTC emerges that may do more than explain the phenotypes of those mutations.

It appears likely that there are several different conformations possible for the PTC region of the ribosome that differ modestly in RNA backbone trajectory, but more conspicuously in the orientations of their nucleotides bases.

several Å just below the macrolide binding site (Tu et al. 2005). The resistance mechanism is less clear in this case, but may be related to the tunnel widening. Weak resistance to anisomycin also results from mutations in ribosomal protein L3 (Gurel, G., unpublished data).

13 TRANSLATIONAL FIDELITY

No aspect of protein synthesis has been more intensively studied than the fidelity with which mRNA sequences are translated into protein sequences. Fidelity is interesting because it cannot be explained by the energetics of the base pairing interactions on which it depends. The free energy of forming a three-base pair helix between an mRNA codon and the anticodon of an aminoacyl tRNA that are perfectly complementary, i.e., that are cognate, is in many instances only slightly more favorable than the free energy of the pairing of the same codon with tRNA codons that are not complementary at a single position, i.e., that are near cognate (Fig. 8). Thus if tRNA-mRNA interactions were all that counted, the error rate of protein synthesis might exceed one wrong amino acid for every 100 amino acids incorporated into protein. In fact, the error rate is about one in 1000–100,000 (see Gallant and Foley 1980). How is this level of accuracy achieved?

Two kinds of models have been advanced to explain the high fidelity of protein synthesis. Some have suggested that the ribosome might interact with the mRNA-tRNA complexes that form on its surface in such a way as to favor cognate complexes over near-cognate complexes energetically by enough to increase the fidelity of translation by 100–1000-fold. Others have proposed that the accuracy of translation is explained by proofreading. In order for the translation system to proofread, its duty cycle would have to be organized so that the free energy difference between cognate mRNA-tRNA pairings and near-cognate mRNA-tRNA pairings could be exploited (at least) twice every time the ribosome selects an aminoacyl tRNA from the cytoplasmic pool.

An historical comment is in order here. At the time proofreading mechanisms were first proposed for the translation system, it was believed that the free energy difference between cognate and near cognate pairings could explain a fidelity of 1 in 100, and thus if one iteration yielded a fidelity of 1 in 100, two iterations that were sequential and independent would yield 1 in 10,000, which is about what is needed. As already noted, more recent thermodynamic studies suggest that the free energy "penalty" associated with noncognate pairings is much smaller than first thought, perhaps so small that the error rates predicted from energy differences could be as high as 1 in 10 (!) (Xia et al. 1998). Thus by itself, proofreading would probably not be enough.

Our understanding of translational fidelity has been revolutionized by the small subunit crystal structures that have appeared since 2000. We now know that interactions between the small subunit and codon–anticodon complexes do indeed contribute substantially to the free energy advantage of cognate interactions, but that kinetic proof reading also contributes to fidelity.

14 THE tRNA BINDING SITES IN THE SMALL RIBOSOMAL SUBUNIT

Remarkably, the location of the A-, P-, and E sites on the small subunit, and much of what we now know about how they interact with tRNAs was revealed by a structure of the Tth small subunit obtained from crystals that contained no tRNA (Wimberly et al. 2000). The reason this structure is informative is that the Tth small subunit has a protruding "spur" (helix 6 of 16S rRNA), the conformation of which is essentially the same as that of the anticodon stem-loop of a tRNA. Fortuitously, in the crystals studied, the spur of each subunit is inserted into the P site of a neighbor, where its "anticodon" sequence makes a noncognate interaction with a three-base sequence at the 5′ end of the host subunit's 16S rRNA, which happens to fold back into the P-site so that it can act like a mRNA (Carter et al. 2000). The accuracy with which this crystal packing interaction mimics P-site tRNA binding could be shown by superimposing the small subunit structure in question on a lower resolution structure of the Tth 70S ribosome that did have tRNAs and mRNA bound (Cate et al. 1999). In addition to validating the P-site interactions seen in the tRNA-free small subunit crystals in question, this superposition identified the small subunit's A and E

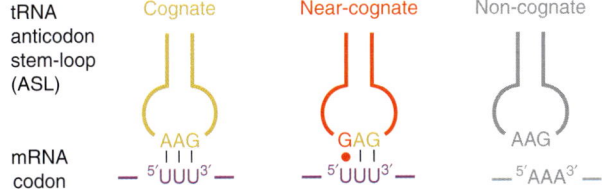

Figure 8. Codon–anticodon interactions. *Left* panel: a cognate codon–anticodon interaction. These is strict Watson-Crick base pairing between the first two bases in the mRNA codon and the corresponding anticodon bases, but wobble pairing is acceptable in the third position. *Middle* panel: a near-cognate codon–anticodon interaction. The distinction between this interaction and a cognate interaction is that a wobble pairing occurs in the first or second position. *Right* panel: a noncognate interaction. In a noncognate interaction no Watson-Crick or wobble pairs form between the codon and the anticodon.

sites in enough detail so that inferences could be drawn about the way tRNAs interact with those sites also.

Compositional arguments like the one that proves that peptide bond formation is RNA-catalyzed cannot be used to show that small subunit function is RNA-driven. Although the 3 tRNA binding sites in the 30S ribosomal subunit all contain 16S rRNA sequences, none of them is protein-free. However, there is a suggestive compositional gradient. Protein is least abundant in the A site, which is arguably the most important site on the small subunit because it is where decoding occurs; only a short segment of protein S12 interacts with tRNAs bound there. By contrast, tRNA-protein interactions dominate in the small subunit's E site, the site that appears the least important of the three in the sense that one could imagine a ribosome that lacks an E site, but not one that lacks an A- or P site (Carter et al. 2000). This compositional gradient is even more pronounced in the large ribosomal subunit; its E site includes protein (Schmeing et al. 2003), but its A- and P-sites are protein-free (Nissen et al. 2000).

Only in the case of the P site have experiments been performed that directly test the functional importance of its protein components (Hoang et al. 2004). The P site of the small subunit includes sequences from the C-terminal tails of proteins S9 and S13, and cells that have S9 and S13 genes that lack their normal C-terminal tail sequences are viable, but grow slowly. Thus the proteins in the P site optimize it, but do not give it functional properties it would otherwise lack. Until similar studies are performed on the A site, we will have to rely on purely structural evidence to tell us what RNA and protein contribute to its function.

15 THE ROLE OF rRNA IN mRNA DECODING

Much of what is now known about the way the A site operates comes from structures of the Tth small subunit with mRNAs and anticodon stem/loops bound in the A site (see Ogle and Ramakrishan, 2005). The first such structures showed that when cognate tRNA-mRNA complexes bind to the A site, three 16S rRNA bases, which genetic and biochemical experiments had earlier showed are vital for decoding (A1492, A1493 and G530), change conformation (Ogle et al. 2001). In the absence of A-site ligands, A1492 and A1493 are buried in the interior of helix 44, the secondary structure element of which they are a part, and G530 is in the *syn* conformation. When cognate tRNA-mRNA complexes bind to the A site, the three adopt positions that permit them to interact with the minor groove of the three-base-pair helix formed by mRNA codons and tRNA anticodons. A1493 swings out of the interior of helix 44, and makes a type I A-minor interaction (Nissen et al. 2001) with the base pair that includes the first base of the codon. The three hydrogen bonds so created stabilize the interaction between the small subunit and the A-site complex, and the reason this interaction is so significant for fidelity is that *any* Watson-Crick pair between the first base of a mRNA codon and a tRNA anticodon can interact with A1493 in exactly the same way, but non Watson-Crick base pairs cannot. Thus the A1493 interaction strongly favors cognate pairing over noncognate, or near-cognate pairing in the A site (Fig. 9A).

The minor groove face of the base pair formed by the second base of a codon and the second base of a tRNA anticodon is also "monitored." In this instance, G530 changes its conformation from *syn* to *anti*, and in that conformation, it interacts both with A1492, which like A1493 has swung out of helix 44, and the middle anticodon base in the helix. In addition to interacting with G530, A1492 makes a type II A-minor interaction with the middle codon base in the same base pair. Again the resulting 4 base complex will accommodate *any* codon-anticodon base pair that is a Watson-Crick base pair, but no non-Watson-Crick base pairs. Thus this interaction too favors cognate pairing over non-, or near-cognate pairings energetically.

The small subunit interacts with the third base pair in a codon-anticodon helix in a much less specific manner. It does not "measure" the width of the minor groove side of that pair the way it measures the width of the minor groove side of the other two base pairs. Thus not surprisingly sequence differences between codons in the third, or wobble, position are much less critical for coding than sequence differences in the first two, as has long been known. Non-Watson-Crick pairings are often acceptable in that position; GU pairings, for example, are almost as good as GC pairings.

The inferences drawn from the cognate A-site codon-anticodon structures have been tested, and extended using a series of small subunit structures in which the A site contains codon-anticodon complexes that have a non-cognate GU instead of a cognate GC at either the first or at the second codon position (Ogle et al. 2002; Ogle et al. 2003). Interestingly, the codon-anticodon interaction is so poor when there is a GU mismatch in the first position that the anticodon stem loop component of the complexes formed could not be visualized crystallographically unless the antibiotic paromomycin was included in the crystal to stabilize it. Paromomycin binds to helix 44 of 16S rRNA, filling space normally occupied by bases 1492 and 1493, and forcing them to adopt a conformation close to that seen when there is a cognate codon-anticodon pair in the A site. Given this extra "help," codon-anticodon complexes with GUs in the first position will bind in the A site, but the A minor-like interaction A1493 makes with them is imperfect. Fewer hydrogen bonds form, and

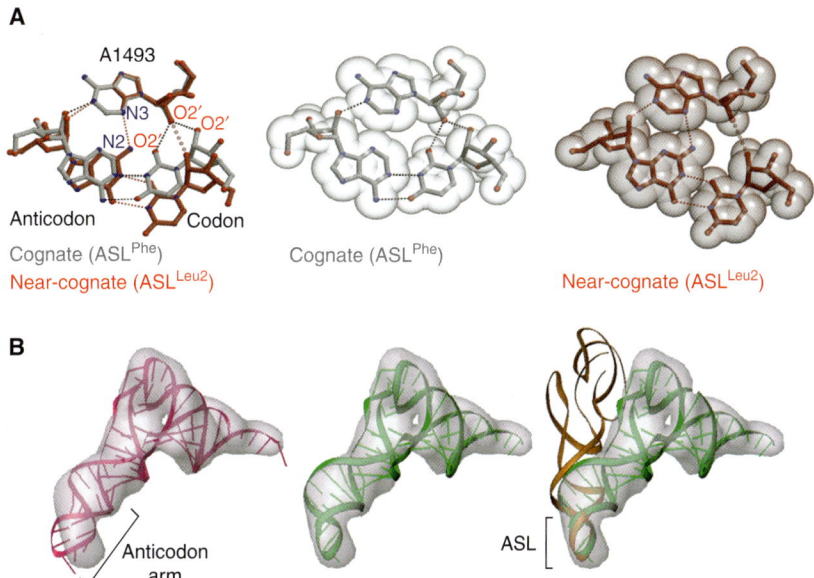

Figure 9. A-site tRNA interactions in the small ribosomal subunit. (*A*) interactions in the first position. *Left Panel*: The structure formed by a Watson-Crick pair in the first position with the ribosome compared with that formed when the base pair in the first position is a wobble base pair. In both cases A1493 of 16S rRNA inserts into the minor groove side of the base pair. More hydrogen bonds form and the fit is tighter if the pair is Watson-Crick than it does if it is a wobble pair. *Middle* Panel: A space-filling model of a cognate pair/A1493 interaction. *Right* Panel: A space-filling model of a wobble pair/A1493 interaction. (Reproduced with permission from Ogle *et al.* 2002 [© Elsevier].) (*B*). The conformation of aminoacyl tRNAs bound to the 70S ribosome in the pre-accommodation state. *Left* panel: Failure of the crystal structure of free tRNA to fit the cryoelectron microscopic electron density for tRNA in images of the 70S ribosome with aminoacyl tFNA together with EF-Tu bound in the pre-accommodation state. The tRNA crystal structure is shown in purple and the electron density in gray. *Middle* panel: Model obtained for tRNA in the pre-accommodated state by marrying the anticodon stem structure of a tRNA in the accommodated state to the crystal structure of a tRNA fit into the rest of tRNA electron density. The new model is green, and its anticodon stem is bent. *Right* Panel: Comparison of the orientation of a accommodated tRNA with a pre-accommodated tRNA The accommodated tRNA is brown and the pre-accommodated tRNA is green. (Reproduced with permission from Valle et al. (2003) [© Elsevier].)

the fit of the minor groove side of A1492 into the minor groove side of the GU pair is suboptimal. Worse, some hydrogen bond donors are left entirely without acceptors in these complexes, which imposes a large energetic cost. However, it is clear that the energetic penalty paid when there are first positions mismatches is reduced in the presence of the drug, and it should thus it makes sense that translational fidelity falls when cells are exposed to it.

GU mismatches in the second position are also problematic structurally. The electron density in the vicinity of the middle base pair of the codon–anticodon helix suggests that the small subunit is not able to find an energetically satisfactory way to accommodate GU mismatches. There appear to be two structures present in the crystal, which may be in equilibrium with each other. Sometimes the mRNA base occupies the position it would occupy if it were part of cognate base pair, and A1492 makes the expected type II A-minor interaction with it, but when this is so, the anticodon with which it is supposed to pair is positioned so that it cannot interact with G530 properly, which remains in the *syn* conformation. At other times, positions are reversed. The anticodon base occupies its cognate position, and interacts normally with an *anti* G530, but its codon mate is out of location, and does not interact properly with A1492. In either case, the full energetic value of the ribosome interactions that accompany cognate pairing is not realized.

The local conformational changes that occur when the A site of the small subunit interacts with cognate codon–anticodon complexes are accompanied by global changes in the over-all conformation of the small subunit that appear to be critical for fidelity (see the following) (Ogle et al. 2002). The three tRNA binding sites of the small ribosomal subunit are located at the interface between the subunit's head domain and its "shoulder" and "body" domains. When cognate interactions occur in the A site, the head rotates toward the large ribosomal subunit, clamping down on its tRNA binding sites, and the shoulder rotates in toward those sites also. These conformational changes do not occur when paromomycin binds to the subunit, nor when near-cognate codon–anticodon complexes occupy the A site. However, they do occur when near-cognate codon–anticodon complexes form in the presence of the drug. These observations indicate that these conformational changes are functionally significant. It is likely they contribute to the still poorly understood process that causes the rate of GTP hydrolysis by EF-Tu to accelerate when cognate interactions between mRNA and tRNA occur in the A site.

The decoding puzzle has one last piece that we have not yet considered. In all the codon–anticodon/small subunit complex structures examined at high resolution, the anticodon stem-loop in the A site is in its accommodated position, which is to say that when a 70S ribosome has an intact tRNA bound to its 30S A site in that orientation, its CCA end will occupy the A site in the large subunit's peptidyl transferase center. This is not the orientation of aminoacyl tRNAs on the ribosome when they first bind to it. As mentioned earlier, the tRNA-containing entity delivered to the ribosome is a ternary complex containing EF-Tu·GTP as well as aminoacyl tRNA, and while the anticodon loop of a ribosome-bound ternary complex occupies the A site, the rest of the molecule is in an entirely different orientation; their CCA ends are far from the peptidyl transferase center (Valle et al. 2002; Stark et al. 2002; Valle et al. 2003). tRNA reorientation, i.e., accommodation, occurs only after the GTP bound to the ternary complex has hydrolyzed, and EF-Tu·GDP has left the ribosome (Stark et al. 2002). Does the picture just provided for A-site interactions with codon and anticodons apply to the pre-accommodation state, or is the interaction completely different?

Amazingly, recent electron microscopic findings as well as a recent crystal structure suggest that as far as the anticodon stem-loop bound to the small subunit A site is concerned, the pre-accommodation state is the same as the accommodated state. It appears that the anticodon stems of tRNAs bound to ribosomes in the pre-accommodation state are bent so that their anticodon loops will fit into the A site in the orientation seen after accommodation (Valle et al. 2003; Li et al. 2008; Schmeing et al. 2009) (Fig. 9B). Accommodation appears to be driven by a springlike straightening out of the anticodon stem that becomes possible once a pre-accommodated tRNA has been released from its interactions with EF-Tu. In any case, it appears that tRNAs are bound to the ribosome only through the interactions they make with A-site components during accommodation. If that interaction is weak, as it will be if the codon–anticodon interaction is near-cognate, instead of cognate, the likelihood that accommodation will result in the dissociation of the tRNA from the ribosome will be high.

Recent kinetic studies have provided important insights into how the system just described actually works (Gromadski and Rodnina 2004; Blanchard et al. 2004). The binding of ternary complexes to the ribosome is a two-step process, the first of which is nonspecific. After a ternary complex has become engaged with the ribosome, a second, codon recognition step occurs that does depend on interactions. Kinetic data show that if the system were to reach equilibrium at that stage, which it normally does not, cognate interactions would be favored over near-cognate interactions by a factor of 300–400, which is almost enough by itself to account for the fidelity of translation. The high potential discrimination of this step unquestionably reflects the interactions 16S rRNA makes with complexes in the A site that are discussed earlier. However when the system operates normally, it does not come to equilibrium at this stage. The rates of subsequent steps are too fast, and hence it does not discriminate effectively between cognate and near-cognate interactions at this point. In fact, the fidelity of the tRNA selection phase of protein synthesis is determined mainly by the next step in the process, the activation of the GTPase of EF-Tu. The rate of activation is ~600 times faster if a cognate interaction has occurred in the A site of the 30S subunit than it is if the interaction in the A site is near-cognate. (All the steps that follow activation, including accommodation, are fast compared with cognate activation.) Furthermore, the rate of near-cognate activation is so slow compared with the rate of dissociation of near-cognate complexes from the ribosome that the probability of rejection of near-cognate complexes is very high at this step. Although the system discriminates quite effectively between cognate and near-cognate interactions during tRNA selection, the overall fidelity of the protein synthesizing system is higher than that process allows. It is the accommodation step that provides the additional fidelity required, and it does so because the probability of dissociation of near-cognate complexes during accommodation is higher than that of cognate complexes. Thus the system does proofread. It appears that the cognate/near-cognate discrimination provided by the tRNA selection phase we just discussed is about 60-fold, whereas that provided by the accommodation step is about 15-fold.

16 STRUCTURAL BASIS OF ELONGATION FACTOR FUNCTIONS

Although a detailed discussion of the crystal structures of 70S ribosomes with elongation factors bound is beyond the scope of this review, the factors being, after all "merely" proteins, a brief summary of the major conclusions derived from them seems in order. In the last year crystal structures have appeared of the Tth 70S ribosomes with three elongation factors bound: EF-P, EF-Tu and EF-G (Blaha et al. 2009; Schmeing et al. 2009; Gao et al. 2009). The least well known factor is EF-P, which stimulates the formation of the first peptide bond (Aoki et al. 1997; Glick and Ganoza 1975). The structure of this complex shows that it binds adjacent to a tRNA^{fmet} in the P site, on its E-site side (Blaha et al. 2009). It appears to stabilize that placement of fmet-tRNA^{fmet}, preventing both its back translocation and forward translocation. The structure of aminoacyl-tRNA-EF-Tu bound to the 70S ribosome with tRNAs in the P and E sites extends the resolution of previous cryo-EM studies to atomic resolution. The bend in the RNA of the anticodon stem and the interactions of the ribosome GTPase center with EF-Tu can be seen in this structure (Schmeing et al. 2009). The structure reported for the complex with EF-G is similarly revealing (Gao et al. 2009).

17 CONCLUDING COMMENTS

Many important aspects of ribosome function remain to be explained in molecular terms, but recent progress gives one hope that the crystal structures needed to address them may be obtained before long. Although this review has emphasized crystallographic findings, it is important to remember that information provided by genetics, biochemistry, and enzymology has been just as important to recent progress.

Finally, the long-standing hypothesis that the first ribosome-like object, which probably appeared on this planet around 3.5 billion years ago, was made entirely of RNA seems more likely than ever. Given what we now know about ribosome structure and function, it is easy to imagine an all-RNA particle having similar capabilities. That said, it would be foolish to assume that the biological world in which that first ribosome appeared was one in which RNA was the only polymer of consequence. Why would an RNA structure evolve that makes polypeptides if polypeptides did not already exist that would confer a selective advantage on the proto-organisms capable of synthesizing them?

ACKNOWLEDGMENTS

We thank Joachim Frank, Venki Ramakrishnan, Jeffrey Hansen, and Martin Schmeing for providing figures. The work was supported by a grant from the National Institutes of Health (GM022778) to P.B.M. and T.A.S.

REFERENCES

Aoki H, Dekany K, Adams SL, Ganoza MC. 1997. The gene encoding elongation factor protein P is essential for viability and it required for protein synthesis. *J Biol Chem* **272:** 32254–32259.

Ban N, Freeborn B, Nissen P, Penczek P, Grassucci RA, Sweet R, Frank J, Moore PB, Steitz TA. 1998. A 9 Å resolution X-ray crystallographic map of the large ribosomal subunit. *Cell* **93:** 1105–1115.

Ban N, Nissen P, Hansen J, Moore PB, Steitz TA. 2000. The complete atomic structure of the large ribosomal subunit at 2.4 A resolution. *Science* **289:** 905–920.

Bennett WS, Steitz TA. 1976. Glucose-induced conformational change in yeast hexokinase. *Proc Natl Acad Sci USA* **75:** 4848–4852.

Berk V, Cate JHD. 2007. Insights into protein biosynthesis from structures of bacterial ribosomes. *Curr Opin Struct Biol* **17:** 302–309.

Blaha G, Stanley RE, Steitz TA. 2009. Formation of the first peptide bond. *Science* **325:** 966–970.

Blaha G, Gurel G, Schroeder SJ, Moore PB, Steitz TA. 2008. Mutations outside the anisomycin-binding site can make ribosomes drug-resistant. *J Mol Biol* **379:** 505–519.

Blanchard SC, Gonzalez RL Jr, Kim HD, Chu S, Puglisi JD. 2004. tRNA selection and kinetic proofreading in translatioin. *Nature Struct Mol Biol* **11:** 1008–1014.

Carter AP, Clemons WM, Brodersen DE, Morgan-Warren RJ, Wimberly BT, Ramakrishnan V. 2000. Functional insights from the structure of the 30S ribosomal subunit and its interactions with antibiotics. *Nature* **407:** 340–348.

Caskey CT, Beaudet AL, Scolnick EM, Rosman M. 1971. Hydrolysis of fMet-tRNA by peptidyl transferase. *Proc Natl Acad Sci USA* **68:** 3163–3167.

Cate JH, Yusupov MM, Yusupova GZ, Earnest TN, Noller HF. 1999. X-ray crystal structures of 70S ribosome functional complexes. *Science* **285:** 2095–2104.

Crick FHC. 1968. The origin of the genetic code. *J Mol Biol* **38:** 367–379.

Dorner S, Panuschka F, Schmid W, Barta A. 2003 Mononucleoditde derivatives as ribosomal P-site substrates reveal an important conribution of the 2'-OH activity. *Nucl Acids Res* **31:** 6536–6542.

Gallant J, Foley D. 1980. On the causes and prevention of mistranslation. In *Ribosomes*, (ed. Chambliss G., Craven G.R., Davies J., Davis K., Kahan L., Nomura M.), pp. 615–638. Baltimore: University Park Press.

Gao Y.-G, Selmer M, Dunham CM, Weixlbaumer A, Kelley AC, Ramakrishnan V. 2009. The structure of the ribosome with elongation factor G trapped in the posttranslocational state. *Science* **326:** 694–699.

Gao H, Sengupta J, Valle M, Korostelev A, Eswar N, Stagg SM, Van Roey P, Agrawal RK, Harvey SC, Sali A et al. 2003. Study of the structural dynamics of the *E. coli* 70S ribosome using real-space refinement. *Cell* **113:** 789–801.

Gavrilova LP, Kostiashkina OE, Kotelianshy VE, Ruthkevitch HM, Spirin AS. 1976. Factor-free ("unenzymatic") and factor-dependent systems of translocation of polyuridylicacial by *Escherichia coli* ribosomes. *J Mol Biol* **101:** 537–552.

Gindulyte A, Bashan A, Agmon I, Massa L, Yonath A, Karle J. 2006. The transition state for the formation of the peptide bond. *Proc Natl Acad Sci USA* **103:** 13327–13332.

Glick BR, Ganoza MC. 1975. Identification of a soluble protein that stimulates peptide bond formation. *Proc Natl Acad Sci USA* **72:** 4257–4260.

Green R, Noller HF. 1997. Ribosomes and translation. *Annu Rev Biochem* **66:** 679–716.

Gromadski KB, Rodnina M. 2004. Kinetic determinants of high-fidelity tRNA discrimination on the ribosome. *Molec Cell* **13:** 191–200.

Gurel G, Blaha G, More PB, Steitz TA. 2009. U2504 determined the species specificity of the A-site cleft antibiotics. The structures of tiamulin, homoherringtonine and bruceantin bound to the ribosome. *J Mol Biol* **389:** 146–156.

Hansen JL, Moore PB, Steitz TA. 2003. Structure of five antibiotics bound at the peptidyl transferase center of the large ribosomal subunit. *J Mol Biol* **330:** 1061–1075.

Hansen JL, Schmeing TM, Moore PB, Steitz TA. 2002. Structural insights into peptide bond formation. *Proc Natl Acad Sci USA* **99:** 11670–11675.

Harms J, Schluenzen F, Zarivach R, Bashan A, Gat S, Agmon I, Bartels H, Francheschi F, Yonath A. 2001. High resolution structure of the large ribosomal subunit from a mesophilic bacterium. *Cell* **107:** 679–688.

Hoang L, Frederick K, Noller HF. 2004. Creating ribosomes with an all-RNA 30S subunit P site. *Proc Nat Acad Sci USA* **101:** 12439–12433.

Ippolito JA, Kanyo ZF, Wang D, Franeschi FJ, Moore PB, Steitz TA, Duffy EM. 2008. Crystal structure of the oxazolidinone antibiotic linezolid bound to the 50S ribosomal subunit. *J Med Chem* **51:** 3353–3356.

Kim DF, Green R. 1999. Base-pairing between 23S rRNA and tRNA in the ribosomal A site. *Molec Cell* **4:** 859–864.

Klein DJ, Moore PB, Steitz TA. 2004. The roles of ribosomal proteins in the structure, assembly and evolution of the large ribosomal subunit. *J Mol Biol* **340:** 141–177.

Korostelev A, Laurberg M, Noller HF. 2009. Multistart simulated annealing refinement of the crystal structure of the 70S ribosome. *Proc Natl Acad Sci USA* **106:** 18195–18200.

Korostelev A, Asahara H, Lancaster L, Laurberg M, Hirschi A, Zhu J, Trakhanov S, Scott WG, Noller HF. 2008. Crystal structure of a translation termination complex formed with RF2. *Proc Natl Acad Sci USA* **105:** 19684–19689.

Korostelev A, Trakhanov S, Laurberg M, Noller HF. 2006. Crystal structure of a 70S ribosome-tRNA complex reveals functional rearrangements. *Cell* **126:** 1065–1077.

Koshland DE, in *The Enzymes* (eds. Boyer P.D., Lardy H., Myrback K.) 305–346. (Academic Press, New York, 1959.

Lauerberg M, Asahara H, Korostelev A, Zhu J, Trakhanov S, Noller HF. 2008. Structural basis for translation termination of the 70S ribosome. *Nature* **454:** 852–857.

Li W, Agirrezabala X, Lei JL, Bouakaz L, Brunelle JL, Ortiz-Meoz RF, Green R, Sanyal S, Ehrenberg M, Frank J. 2008. Recognition of aminoacyl-tRNA: a common molecular mechanism revealed by cryo-EM. *EMBO J* **27:** 3322–3331.

Monro RE, Staehlin T, Celma ML, Vazquez D. 1969. The peptidyl transferase activity of ribosomes. *Cold Spring Harbor Symp Quant Biol* **34:** 357–366.

Nakatogawa H, Ito K. 2004. Intraribosomal regulation of expression and fate of proteins. *ChemBioChem* **5:** 48–51.

Nierhaus KH, Schulze H, Cooperman BS. 1980. Molecular mechanisms of the peptidyl transferase center. *Biochem Internat* **1:** 185–192.

Nissen P, Ban N, Hansen J, Moore PB, Steitz TA. 2000. The structural basis of ribosome activity in peptide bond synthesis. *Science* **289:** 920–930.

Nissen P, Ippolito JA, Ban N, Moore PB, Steitz TA. 2001. RNA tertiary interactions in the large ribosomal subunit: the A-minor motif. *Proc Natl Acad Sci USA* **98:** 4899–4903.

Noller HF. 1991. Ribosomal RNA and translation. *Annu Rev Biochem* **60:** 191–227.

Ogle JM, Ramakrishnan V. 2005. Structural insights into translational fidelity. *Annu Rev Biochem* **74:** 129–177.

Ogle JM, Carter AP, Ramakrishnan V. 2003. Insights into the decoding mechanism from recent ribosome structures. *Trends in Biochem Sci* **28:** 259–266.

Ogle JM, Brodersen DE, Clemons WM, Tarry MJ, Carter AP, Ramakrishnan V. 2001. Recognition of cognate transfer RNA by the 30S ribosomal subunit. *Science* **292:** 897–902.

Ogle JM, Murphy FVI, Tarry MJ, Ramakrishnan V. 2002. Selection of tRNA by the ribosome requires a transition from an open to a closed form. *Cell* **111:** 721–732.

Page MI, Jencks WP. 1971. Entropic contributions to rate acceleration in enzymatic and intramolecular reactions and the chelate effect. *Proc Natl Acad Sci USA* **68:** 1678–1683.

Parnell KM, Seila AC, Strobel SA. 2002. Evidence against stabilization of the transitionstate oxyanion by a pKa-perturbed RNA base in the peptidyl transferase center. *Proc Natl Acad Sci USA* **99:** 11658–11663.

Pestka S. 1974. Assay for nonenzymatic and enzymatic translocation with *Escherichia coli* ribosomes. *Meth Enzymol* **30:** 462–470.

Pioletti M, Schlunzen F, Harms J, Zarivach R, Glumann M, Avila H, Bashan A, Bartels H, Auerbach T, Jacobi C et al. 2001. Crystal structures of complexes of the small ribosomal subunit with tetracycline, edeine and IF3. *EMBO J* **20:** 1829–1839.

Samaha RR, Green R, Noller HF. 1995. A base pair between tRNA and 23S rRNA in the peptidyl transferase centre of the ribosome. *Nature* **377:** 309–314.

Schluenzen F, Tocilj A, Zarivach R, Harms J, Gluehmann M, Janell D, Bashan A, Bartles H, Agmon I, Franceschi F et al. 2000. Structure of functionally activated small ribosomal subunit at 3.3 A resolution. *Cell* **102:** 615–623.

Schleunzen F, Zarivach R, Harms J, Bashan A, Tocilj A, Albrecht R, Yonath A, Franceschi F. 2001. Structural basis for the interaction of antibiotics with the peptidyl transferase centre in eubacteria. *Nature* **413:** 814–821.

Schmeing TM, Moore PB, Steitz TA. 2003. Structure of deacylated tRNA mimics bound to the E site of the large ribosomal subunit. *RNA* **9:** 1345–1352.

Schmeing TM, Huang KS, Strobel SA, Steitz TA. 2005b. An induced-fit mechanism to promote peptide bond formation and exclude hydrolysis of peptidyl tRNA. *Nature* **438:** 520–524.

Schmeing TM, Kitchen D, Scaringe SA, Strobel SA, Steitz TA. 2005a. Structural insights into the roles of water and the 2′ hydroxyl of the P-site tRNA in the peptidyl transferase reaction. *Molec Cell* **20:** 437–448.

Schmeing TM, Voorhees RM, Kelley AC, Goa Y-G, Murphy FV IV, Weir JR, Ramakrishnan V. 2009. The crystal structure of the ribosome bound to EF-Tu and aminoacyl tRNA. *Science* **326:** 688–694.

Schuwirth BS, Borovinskaya MA, Hau CW, Zhang W, Vila-Sanjuro A, Holton JH, Doudna-Cate JH. 2005. Structure of the bacterial ribosome at 3.5 Å resolution. *Science* **310:** 827–834.

Selmer M, Dunham CM, Murphy FV IV, Weixlbaumer A, Petry S, Kelley AC, Weir JR, Ramakrishnan V. 2006. Structure of the 70S ribosome complexed with mRNA and tRNA. *Science* **3313:** 1935–1942.

Sievers A, Beringer M, Rodnina MV, Wolfenden R. 2004. The ribosome as an entropy trap. *Proc Natl Acad Sci USA* **101:** 7897–7901.

Simonovic M, Steitz TA. 2008. Cross-crystal averaging reveals that the structure of the peptidyl transferasemcneter is the same in the 70S ribosome and the 50S subunit. *Proc Natl Acad Sci USA* **105:** 500–505.

Southworth DR, Brunelle JL, Green R. 2002. EFT-independent translation of the mRNA-tRNA complexes promoted by modification of the ribosomes with thiol-specific reagents. *J Mol Biol* **324:** 611–623.

Stark H, Rodnina MV, Wieden H.-J., Zemlin F, Wintermeyer W, van Heel M. 2002. Ribosome interactions of aminoacyl-tRNA and elongation factor Tu in the codon-recognition complex. *Nature Struc Biol* **9:** 849–854.

Tenson T, Lovmar M, Ehrenberg M. 2003. The mechanism of action macrolides, lincosamides and streptogramin B reveals the nascent peptide exit path in the ribosome. *J Mol Biol* **330:** 1005–1014.

Tu D, Blaha G, Moore PB, Steitz TA. 2005 Structures of MLSbK antibiotics bound to mutated large ribosomal subunits provide a structural explanation for resistance. *Cell* **121:** 257–270.

Valle M, Sengupta J, Swami NK, Grassucci R, Burkhardt N, Nierhaus KH, Agrawal RK, Frank J. 2002. Cryo-EM reveals an active role for aminoacyl tRNA in the accommodation process. *EMBO J* **21:** 3557–3567.

Valle M, Zavialov AV, Li W, Stagg SM, Seugupta J, Nielsen RC, Nissen P, Harvey SC, Ehrenberg M, Frank J. 2003. Incorporation of aminoacyl-tRNA into the ribosome as seen by cryo-electron microscopy. *Nature Struc Biol* **10:** 899–906.

Voss NR, Gerstein M, Stetiz TA, Moore PB. 2006. The geometry of the ribosomal polypeptide exit tunnel. *J Mol Biol* **360:** 893–906.

Voorhees RM, Weixlbaumer A, Loakes D, Kelley AC, Ramakrishnan V. 2009 Insights into substrate stabilization from snapshots of the peptidyl transferase center of the intact 70S ribosome. *Nature Struc Biol* **16**: 528–533.

Watson JD. 1963. Involvement of RNA in the synthesis of proteins. *Science* **140**: 17–26.

Weinger JS, Parnell KM, Dorner S, Green R, Strobel SA. 2004. Substrate-assisted catalysis of peptide bond formation by the ribosome. *Nature Struc Biol* **11**: 1101–1106.

Weixlbaumer A, Jin H, Neubauer C, Voorhees RM, Petry S, Kelley AC, Ramakrishnan V. 2008. *Science* **322**: 953–956.

Wilson DN, Harms JM, Nierhaus KH, Schlunzen F, Fucini P. 2005. Species-specific antibiotic-ribosome interactions: implications for drug development. *Biol Chem* **386**: 1239–1252.

Wimberly BT, Brodersen DE, Clemons WM, Morgan-Warren RJ, Carter AP, Vonrhein C, Hartsch T, Ramakrishnan V. 2000. Structure of the 30S ribosomal subunit. *Nature* **407**: 327–339.

Wittmann-Liebold B. 1986. Ribosomal proteins: their structure and evolution. In *Structure, Function, and Genetics of Ribosomes* (ed. Hardesty B., Kramer G.), pp. 326–361. New York: Springer-Verlag.

Wool IG. 1986. Studies of the structure of eukaryotic (mammalian) ribosomes. In *Structure, Function, and Genetics of Ribosomes* (ed. Hardesty B., Kramer G.), pp. 391–411. New York: Springer-Verlag.

Wool IG, Chan Y.-L., Gluck A. 1995. Structure and evolution of mammalian ribosomal proteins. *Biochem Cell Biol* **73**: 933–947.

Xia T, SantaLucia J Jr, Burkard ME, Kierzek R, Schroeder SJ, Jiao C, Cox C, Turner DH. 1998. Thermodynamic parameters for an expanded nearest-neighbor model for formation of RNA duplexes with Watson-Crick base pairs. *Biochemistry* **37**: 14719–14735.

Yim G, Wang HH, Davies J. 2007. Antibiotics as signaling molecules. *Phil Transact Roy Soc B – Biol Sci* **362**: 1195–1200.

Yonath A, Bashan A. 2004. Ribosome crystallography: mutation, peptide bond formation, and amino acid polymerization are hampered by antibiotics. *Annu Rev Microbiol* **58**: 233–251.

Youngman EM, Brunelle JL, Kochaniak AB, Green R. 2004. The active site of the ribosome is composed of two layers of conserved nucleotides with distinct roles in peptide bond formation and peptide release. *Cell* **117**: 589–599.

Yusupov MM, Yusupova GZ, Baucom A, Lieberman K, Earnest TN, Cate JHD, Noller HF. 2001. Crystal structure of the ribosome at 5.5 A resolution. *Science* **292**: 883–896.

Evolution of Protein Synthesis from an RNA World

Harry F. Noller

Center for Molecular Biology of RNA and Department of Molecular, Cell, and Developmental Biology, Sinsheimer Laboratories, University of California at Santa Cruz, Santa Cruz, California 95064

Correspondence: harry@nuvolari.ucsc.edu

SUMMARY

Because of the molecular complexity of the ribosome and protein synthesis, it is a challenge to imagine how translation could have evolved from a primitive RNA World. Two specific suggestions are made here to help to address this, involving separate evolution of the peptidyl transferase and decoding functions. First, it is proposed that translation originally arose not to synthesize functional proteins, but to provide simple (perhaps random) peptides that bound to RNA, increasing its available structure space, and therefore its functional capabilities. Second, it is proposed that the decoding site of the ribosome evolved from a mechanism for duplication of RNA. This process involved homodimeric "duplicator RNAs," resembling the anticodon arms of tRNAs, which directed ligation of trinucleotides in response to an RNA template.

Outline

1. Introduction
2. Translation out of an RNA World
3. Peptidyl transferase: the ribosome is a ribozyme
4. Aminoacyl-tRNA selection: the 30S subunit a site
5. The 30S subunit p site: another function of rRNA
6. RNA molecular mechanics and translocation
7. What are ribosomal proteins for?
8. "Stop tRNAs" and the evolution of type I release factors
9. dRNA and the origins of the ribosomal decoding site
10. The driving force for evolution of translation from an RNA world
11. Conclusions

References

1 INTRODUCTION

Translation links the nucleotide sequences of genes to the amino acid sequences of proteins, establishing at the molecular level the correspondence between genotype and phenotype. The basic underlying mechanisms of translation must have arisen early in the history of molecular evolution, in some primitive form, before the existence of any genetically encoded protein. To understand how the ribosome, one of the most complex molecular structures in all of biology, and its associated translational ligands, could have emerged from an RNA world presents one of the most challenging problems in molecular evolution. Thanks to numerous fresh insights into the structure and function of ribosomes (and RNA in general), many of which are described in this collection, this once impenetrable problem can now be viewed as merely extraordinarily difficult. Among the central problems in reconstructing the molecular evolution of translation are : (1) The chicken-or-the-egg problem: If the ribosome requires proteins to function, where did the proteins come from to make the first ribosome and its translation factors? (2) What was the driving force for evolution of the ribosome? and (3) How did coding arise? Thanks to numerous advances in this field, we now have a likely answer to the first question, and a plausible answer to the second question (Noller 2004) Although the origins of coding remain a puzzle in spite of many decades of thought and speculation, a possible RNA World origin for the codon recognition function of the modern ribosome is suggested here. Another question, implicit in the RNA World hypothesis, is: (4) Can we account for all of the basic functions of translation in terms of RNA? The answer to this last question seems to be mainly "yes," although some proteins, such as the type I release factors, may have taken over functional roles that were once played by RNA.

2 TRANSLATION OUT OF AN RNA WORLD

We begin with the question of how the first translational system could have arisen without proteins, a question that was raised in the years following the elucidation of the genetic code and the discovery of the general properties of the translational apparatus (Woese 1967; Crick 1968; Orgel 1968). The simplest ribosomes (those from bacteria and archaea) contain about 50 different proteins and three rRNAs (16S, 23S, and 5S rRNAs) comprising about 4500 nucleotides and two-thirds of the mass of the ribosome. In addition to the ribosomal proteins, many nonribosomal protein factors are required for the steps of initiation, elongation, termination, and ribosome recycling. But how could the first ribosome have depended on proteins for its function? The overall process of translation was from the outset recognized to be centered around RNA—mRNA, tRNA, and the ribosome. In view of the fact that ribosomes contain large amounts of ribosomal RNA (rRNA), Crick asked whether the first ribosomes might have been made exclusively of RNA. Crick's conjecture notwithstanding, the overwhelming preponderance of opinion in the translation field was that the functions of the ribosome were determined by its proteins, and by the translation factors.

The first proteins shown to be dispensable were the translation factors. Polypeptide synthesis could be initiated in the absence of initiation factors, by manipulating the ionic conditions (Nirenberg and Leder 1964). Aminoacyl-tRNA could be bound to the ribosome in the absence of elongation factor EF-Tu, albeit at greatly reduced rates (Lill et al. 1986). Peptide bond formation was shown to be catalyzed by the large ribosomal subunit itself (Monro 1967). And translocation of tRNA could occur without EF-G (Pestka 1968; Gavrilova et al. 1976). The isolation of deletion mutants showed that at least 17 ribosomal proteins were individually dispensable (Dabbs 1986). Moreover, early in vitro reconstitution studies showed that many small-subunit ribosomal proteins could be singly omitted without abolishing function (Nomura et al. 1969). Conversely, although mutations in certain proteins were known to confer antibiotic resistance or affect translational accuracy (Davies and Nomura 1972), no examples were found in which mutation or chemical modification of a ribosomal protein caused loss of ribosome function.

Around the same time, findings from several laboratories began to point to the possibility of a functional role for rRNA. Inactivation of ribosomes on cleavage of a single phosphodiester bond of 16S rRNA by colicin E3 (Bowman et al. 1971; Senior and Holland 1971), resistance to the antibiotic kasugamycin conferred by the absence of methylation of two bases in 16S rRNA (Helser et al. 1972), inactivation of ribosomes by kethoxal modification of a few bases in 16S rRNA (Noller and Chaires 1972), and the unusually high conservation of sequences within the rRNAs (Woese et al. 1975) were early warning signs. Crosslinking of the anticodon and acceptor ends of tRNA with surprisingly high efficiency to 16S and 23S rRNA, respectively, placed the two most important functional features of tRNA in close proximity to universally conserved features of the two large rRNAs (Prince et al. 1982; Barta et al. 1984). Inactivation of ribosomes by cleavage of a single phosphodiester bond in the large subunit rRNA by α-sarcin (Endo and Wool 1982) and the dominant lethal phenotype of point mutations of G530 of 16S rRNA (Powers and Noller 1990) were more in keeping with the notion of a functional rRNA than of a mere structural scaffold. The technique of chemical footprinting of RNA quickly showed that tRNA, elongation factors, initiation factors and all major classes of ribosome-directed antibiotics interacted with 16S and/or 23S rRNA, often at universally conserved nucleotides [summarized in (Noller et al. 1990)].

In spite of the nearly overwhelming body of evidence, the idea that rRNA was a functional molecule, let alone *the* functional molecule of the ribosome, was met with widespread skepticism. The sole functional role for rRNA that was generally accepted was the Shine-Dalgarno mechanism for mRNA start-site selection (Shine and Dalgarno 1974), because of convincing supporting evidence (Steitz and Jakes 1975), but perhaps also because its straightforward base-pairing interactions put the mechanism in a comfortable context, in keeping with well-known properties of nucleic acids.

3 PEPTIDYL TRANSFERASE: THE RIBOSOME IS A RIBOZYME

To many outside observers of the field, the main function of the ribosome was considered to be the peptidyl transferase reaction, the sole chemical reaction known to be cayalyzed by the ribosome itself. Although other ribosomal functions, including the crucial processes of aminoacyl-tRNA selection and translocation would seem to merit at least as much mechanistic interest, peptide bond formation is also a symbolic event—the point of entry of an amino acid into the protein world. Footprinting and crosslinking of tRNA and its CCA end (Barta et al. 1984; Moazed and Noller 1989; Moazed and Noller 1991), and localization of the sites of interaction of several peptidyl transferase inhibitors (Moazed and Noller 1987), had unambiguously placed 23S rRNA at the "scene of the crime," although crosslinking studies had also shown that proteins L2 and L16 were nearby (reviewed in Wower et al. 1993). In vitro reconstitution experiments had eliminated all but a handful of large-subunit ribosomal proteins (Moore et al. 1975; Schulze and Nierhaus 1982). In one attempt to show the role of rRNA in catalysis of peptide bond formation, ribosomes were subjected to stringent protein-extraction procedures. *Thermus thermophilus* 50S subunits treated with 0.5% SDS and extensive digestion with protease K, followed by continuous vortexing for an hour or more with phenol, retained their full peptidyl transferase activity (Noller et al. 1992). Most, but not all, of the protein was removed by this procedure, leaving open the main question, but forcefully calling attention to the probable catalytic functionality of rRNA.

When the crystal structures of the ribosome and its subunits were solved, any remaining doubts about the functional role of rRNA were dispelled. Structures of the 30S and 50S subunits at 3.0 Å and 2.4 Å resolution, respectively, provided detailed descriptions of the folding of the RNA and protein components of the ribosome for the first time (Ban et al. 2000; Schluenzen et al. 2000; Wimberly et al. 2000). The 5.5 Å resolution structure of the complete 70S ribosome, with mRNA and tRNAs bound, showed how the subunits fit together, and revealed the interactions between the ribosome and the P- and E-site tRNAs (Yusupov et al. 2001) (Fig. 1). A complex containing the complete A-site tRNA bound to the 70S ribosome at 7.0 Å, and another complex of a tRNA anticodon stem-loop bound to the 30S subunit at 3.1–3.3 Å provided the details of the interactions between the ribosome and the A-site tRNA (Ogle et al. 2001). Several different complexes containing the 50S subunit bound with a variety of tRNA acceptor-end mimics, including a transition-state analogue, provided insight into how the aminoacyl and peptidyl ends of the tRNAs interact with the peptidyl transferase catalytic site (Nissen et al. 2000).

The distribution of the rRNA and protein moieties on the surface of the ribosomal subunits (Fig. 2) makes a clear case for the functional importance of rRNA. Proteins are distributed more or less evenly over the external surface of the ribosome, filling nooks and crannies in the rRNA (Fig. 2B,D), but the subunit interface surface, which contains the tRNA binding sites as well as other functional features, is made up mainly of rRNA (Fig. 2A,C). The overall impression is that of an RNA structure that has gradually incorporated a number of proteins over evolutionary time, but has not allowed them to impinge on its crucial functional centers.

The high-resolution structure of the 50S subunit (Ban et al. 2000), together with the knowledge of the positions of the acceptor ends of the tRNAs (Nissen et al. 2000; Yusupov et al. 2001), provided the first look at the structure of the peptidyl transferase center. No protein moieties were found with 17 Å of the catalytic site, definitively demonstrating that peptide bond formation is indeed catalyzed by RNA. Although a more recent high-resolution structure of the *T. thermophilus* 70S ribosome bound with tRNAs shows interactions between the amino-terminal tail of protein L27 and the backbone of the 3′ CCA end of the P-site tRNA (Selmer et al. 2006), no part of the protein is close enough to the catalytic site to play a direct (chemical) role in the reaction. Furthermore, early studies showed that *Escherichia coli* 50S subunits reconstituted in vitro without L27 were active in catalyzing peptide bond formation (Moore et al. 1975). Also, no counterpart to L27 is found in archaeal 50S subunits (Nissen et al. 2000). Comparison of structures of 50S complexes containing various susbstrate and transition-state analogs support an induced-fit model for catalysis of peptide bond formation based

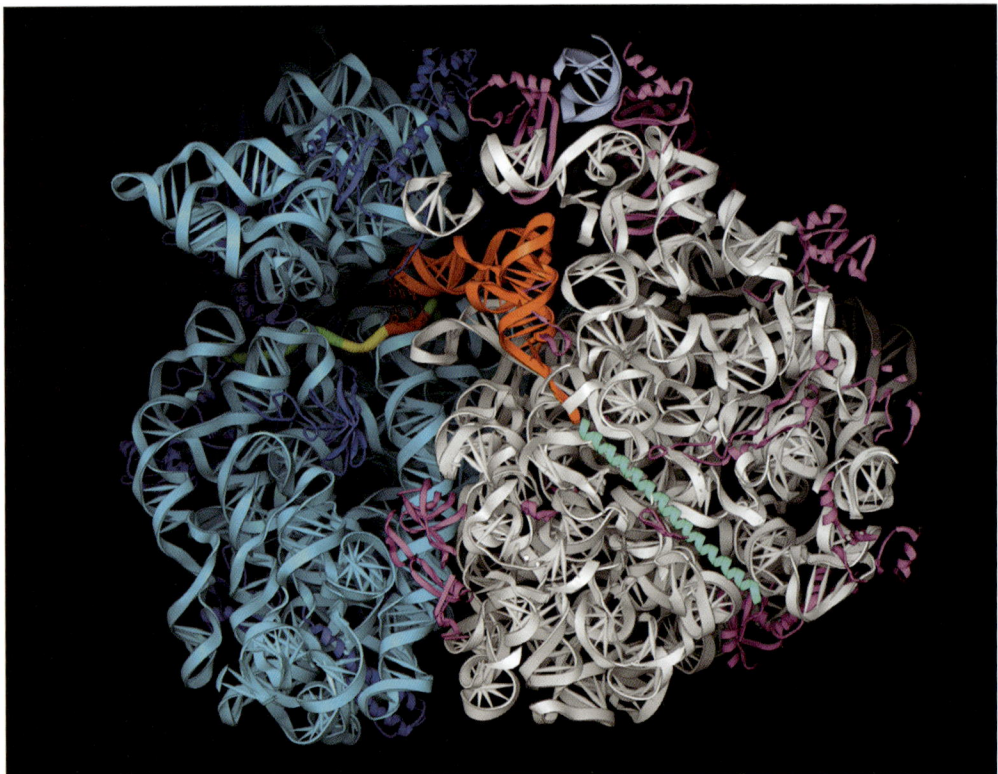

Figure 1. Cross-section of the crystal structure of the *T. thermophilus* 70S ribosome (Yusupov et al. 2001), with the 30S subunit on the *left* and the 50S subunit on the *right*. The locations of the peptidyl-tRNA (orange) at the subunit interface, the mRNA (green-yellow-red) wrapping around the neck of the 30S subunit, and a modeled α-helical nascent polypeptide chain (green) in the polypeptide exit tunnel are indicated. The 16S rRNA is shown in cyan, 23S rRNA in grey, 5S rRNA in grey-blue, 30S subunit proteins in dark blue, and 50S subunit proteins in magenta.

solely on RNA (Schmeing et al. 2005). The present state of our understanding of this fundamental biological function is reviewed in detail by Moore and Steitz (Moore and Steitz 2010).

4 AMINOACYL-tRNA SELECTION: THE 30S SUBUNIT A SITE

Among the other basic functions of the ribosome are the binding of tRNA to its A, P, and E sites. tRNA binding carries functional implications well beyond the simple positioning of substrates for the catalytic step. Binding to the A site (aminoacyl-tRNA selection) is an important determinant of translational accuracy (Kurland et al. 1990). The crystal structure of the 30S subunit in complex with a U_6 mRNA and a tRNA anticodon stem-loop analog of tRNA^{Phe} by Ramakrishnan and coworkers provided profound insight into the mechanism by which the ribosome mediates tRNA selection (Ogle et al. 2001). Binding of tRNA to the 30S subunit A site results in movement of three bases of 16S rRNA - G530, A1492, and A1493 into contact with the codon-anticodon duplex. These same three universally conserved bases had previously been identified in chemical footprinting experiments as the three main bases to interact with tRNA in the 30S A site (Moazed and Noller 1986; Moazed and Noller 1990); moreover, mutation of any one of them had been found to confer a dominant lethal phenotype (Powers and Noller 1990; Yoshizawa et al. 1999). The result of the tRNA-induced conformational rearrangement brings these bases into a remarkably close steric fit with the minor groove surfaces of the codon-anticodon base pairs, involving van der Waals contacts and hydrogen bonds to both the bases and backbone riboses (Fig. 3A). It is apparent that the close fit between 16S rRNA and the first two base pairs can be made only when perfect Watson-Crick pairing occurs. In contrast, only limited interactions are made with the third base pair, providing a structural explanation for the tolerance of noncanonical, or wobble pairing in the third position.

The full mechanistic significance of these interactions is not yet clear. Does the induced-fit interaction provide selectively enhanced thermodynamic stability to the cognate tRNA, thereby enhancing translational accuracy? Does the structural change in 16S rRNA initiate a signal that is

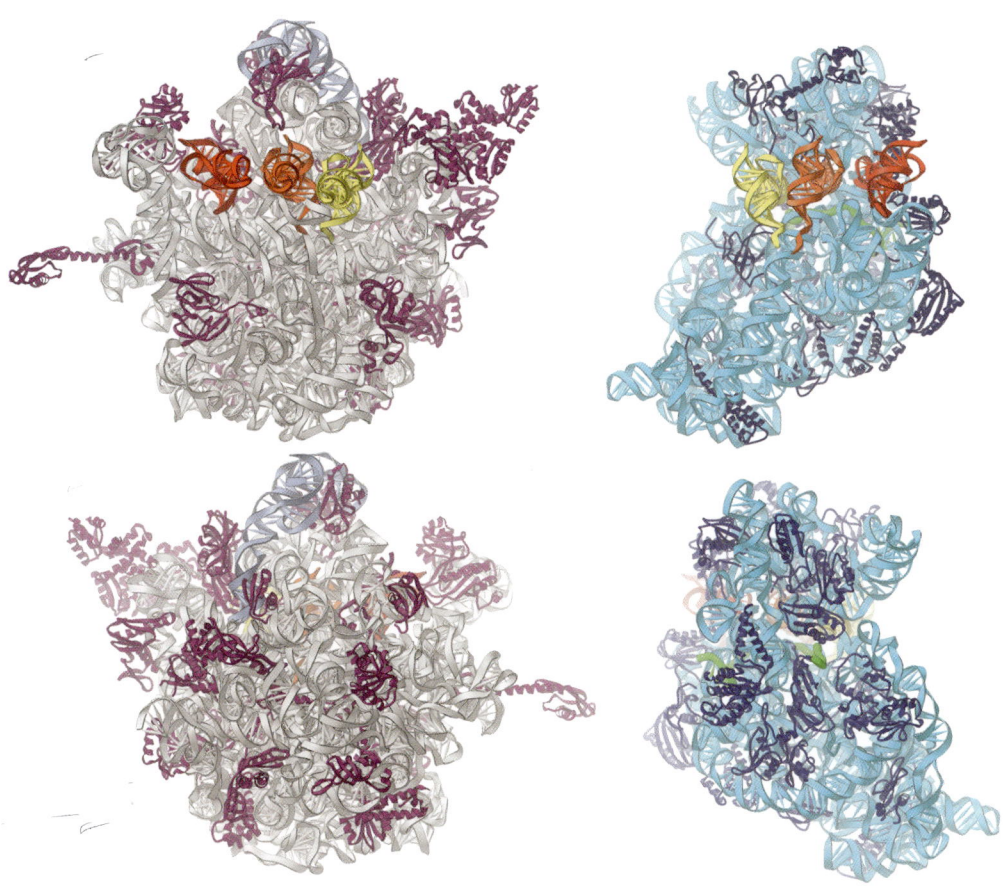

Figure 2. Interface and solvent views of the 30S and 50S subunits, as observed in the 70S ribosome crystal structure (Yusupov et al. 2001), showing the positions of the A-, P- and E-site tRNAs (yellow, orange, and red, respectively). *Top*, interface views of the 50S (*left*) and 30S (*right*) subunits, showing the relative absence of proteins surrounding the functional sites. *Bottom*, corresponding solvent surfaces of the subunits.

transmitted to the catalytic site of EF-Tu, accelerating hydrolysis of GTP? Or are both kinds of mechanisms involved? An additional observation explains the miscoding activity of aminoglycoside antibiotics. The crystal structure of the 30S subunit bound with paromomycin shows that this aminoglycoside causes A1492 and A1493 to rearrange from their normal locations stacked on the end of helix 44 to flip into almost exactly the positions induced by binding of the anticodon stem-loop (Carter et al. 2000; Ogle et al. 2001). Presumably, binding of noncognate tRNAs is stabilized by this largely prearranged conformational shift. Because the bases cannot form optimal minor groove interactions with the noncognate anticodon-codon complex, this result also suggests that there is more to this 16S rRNA interaction than simple thermodynamic stabilization of tRNA binding.

Most intriguing is that the ribosomal structure responsible for sensing whether true Watson-Crick pairs are made is formed from only three nucleotides of 16S rRNA. It is not difficult to imagine assembling such a mechanism from small, rudimentary RNAs of the kind that have been suggested to have populated the RNA World. Moreover, this simple steric minor-groove calibration could provide a general mechanism to monitor the accuracy of base pairing in a variety of functional contexts, such as RNA recombination (splicing) and RNA replication, as discussed later. In fact, the plausibility of such scenarios is made clear by the common occurrence of these kinds of interactions in the structures of the ribosomal and other RNAs. The Yale group has termed them "A-minor" interactions (Nissen et al. 2001), and has assigned them to three different structural classes, called types I, II, and III (Fig. 3B). More than 130 examples of type I and type II A-minor interactions are found in the *Haloarcula marismortui* 23S rRNA alone (Nissen et al. 2001). In the 30S A site, A1493 of 16S rRNA makes a Type I A-minor interaction with the first codon-anticodon base pair (Fig. 3A,B; *top*). A1492 makes a type II interaction with the mRNA nucleotide of the middle base pair, and pairs with G530, which itself interacts with the tRNA nucleotide of the middle base pair in a

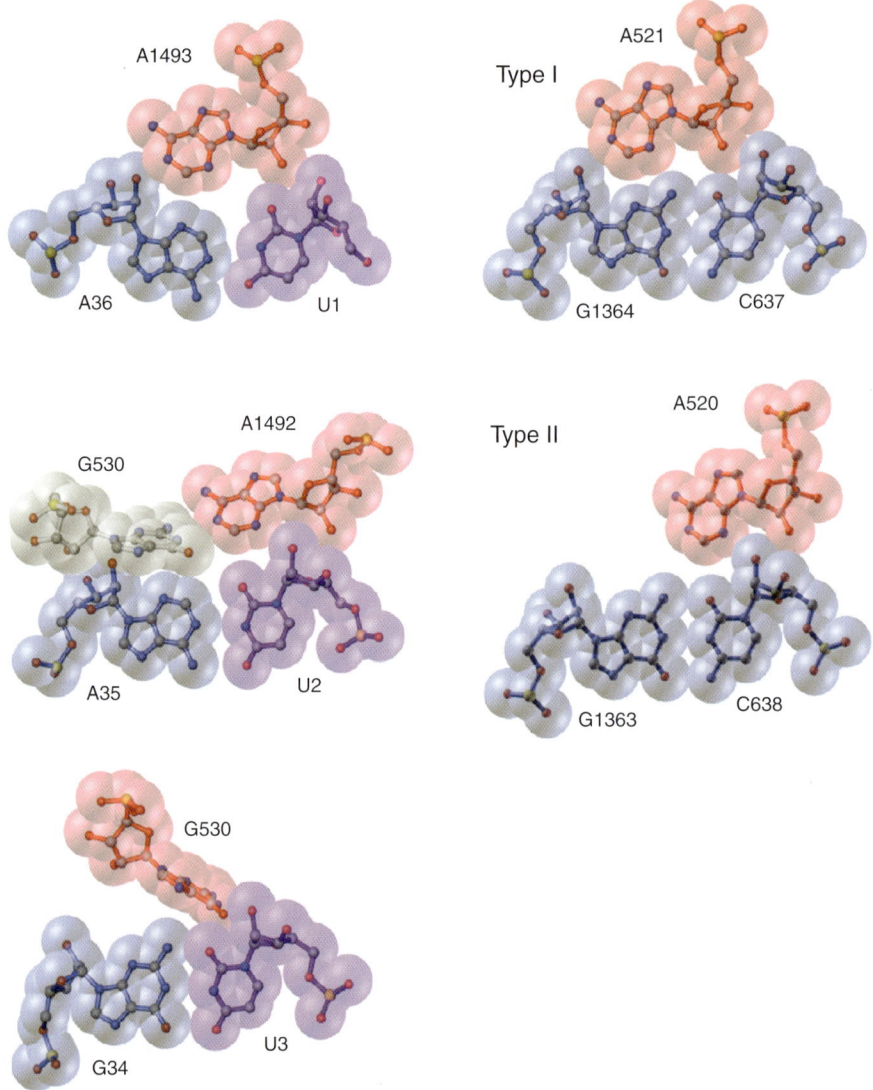

Figure 3. Steric minor-groove calibration of Watson-Crick codon-anticodon pairing by three conserved bases of 16S rRNA. (*left*) Contacts between G530, A1492, and A1493 of 16S rRNA and the codon-anticodon base pairs in the 30S A site (Ogle et al. 2001). (*right*) Type I and type II A-minor interactions (Nissen et al. 2001).

type II-like interaction (Fig. 3A,B; *middle*). The 30S subunit A site presents compelling evidence for the likelihood that the ribosome evolved from a purely RNA structure, and helps to explain why its function continues to be based on RNA.

5 THE 30S SUBUNIT P SITE: ANOTHER FUNCTION OF rRNA

One of the earliest indications of the functional role of rRNA was the implication of 16S rRNA in binding tRNA to the 30S P site. Inactivation of P-site binding by kethoxal modification of 16S rRNA (Noller and Chaires 1972), direct crosslinking of the wobble base of tRNA to C1400 of 16S rRNA (Prince et al. 1982), chemical footprinting of 16S rRNA by P-site tRNA (Moazed and Noller 1986; Moazed and Noller 1990) and modification-interference experiments (von Ahsen and Noller 1995) all pointed to the involvement of a constellation of 16S rRNA nucleotides in this function. The crystal structures directly showed the participation of 16S rRNA in tRNA binding to the 30S P site, providing further evidence for the RNA character of the ribosome (Wimberly et al. 2000; Yusupov et al. 2001). However, any hopes for a simple RNA-World picture were clouded by the intrusion of proteins S9 and S13 into the tRNA binding site (Wimberly et al. 2000). Both proteins contain extended carboxy-terminal tails, which contact phosphate 35 of the anticodon loop and the backbone of the anticodon stem, respectively. Both tails contain basic amino-acid side-chains that appear to make electrostatic interactions with the tRNA backbone.

An RNA-World impression of S9 and S13 is one of two proteins that landed on and took root in 16S rRNA as a later evolutionary refinement of the basic RNA structure of the subunit. The functional requirement for their C-terminal tails was tested directly by replacing the genomic copies of the E. coli genes encoding S9 and S13 with versions in which their carboxy-terminal tails were deleted (Hoang et al. 2004). The result was that mutant strains bearing the deleted versions of the two proteins were viable, including strains in which the tails of both proteins were deleted. The phenotypes were relatively mild, amounting to a 40% reduction in growth rate for the double deletion. In this strain, all of the cellular proteins are synthesized by ribosomes whose 30S P sites are composed purely of RNA. Thus, 16S rRNA is able to support all of the essential functions of the 30S P site, including translational initiation, P-site tRNA binding, and maintenance of the translational reading frame.

6 RNA MOLECULAR MECHANICS AND TRANSLOCATION

Perhaps the most demanding step of translation is the coupled movement of mRNA and tRNA, called translocation, which follows formation of each new peptide bond. This step depends on elongation factor EF-G and is coupled to hydrolysis of GTP. It is coupled to large-scale molecular movements, including relative rotation of the two ribosomal subunits, emphasizing the structural dynamics of the ribosome. Because the pioneer ribosomes must have been capable of translocation, we can ask: (a) How could such a fundamental process have operated in the absence of EF-G, which is commonly referred to as the "translocase" of protein synthesis? And, what was the source of energy to drive movement of mRNA and tRNA, and intersubunit rotation? Studies by Pestka (Pestka 1968) and Spirin (Gavrilova et al. 1976) showed many years ago that poly(U)-dependent synthesis of polyphenylalanine could proceed in the absence of EF-G, under certain in vitro conditions, or by modification of ribosomes with thiol-directed reagents. Criticisms that the observed synthesis might have involved some sort of "slippage" of the poly(U) mRNA were addressed by Green and coworkers using a defined mRNA (Southworth et al. 2002). The requirement for GTP was shown not to be absolute by the demonstration that the peptidyl transferase inhibitor sparsomycin can trigger a single round of translocation in vitro, with high efficiency and accuracy, in the complete absence of EF-G or GTP (Fredrick and Noller 2003). These studies suggest that translocation could have originated as a purely ribosomal, factor-independent process.

But in the absence of GTP hydrolysis, what is the source of free energy to drive the translocation reaction, and what keeps it from going backwards? The most obvious source of energy comes from peptide bond formation; peptidyl transfer results in formation of a peptide amide bond from an activated ribose ester linkage, accompanied by a large change in free energy. How can this free energy change be coupled to translocation? The answer most likely lies in the changing chemical nature of the acceptor end of the tRNA as it moves through the ribosome (Spirin 1985). It enters as an aminoacyl-tRNA, is transformed into a peptidyl-tRNA and then becomes completely deacylated. The 50S subunit contains three tRNA binding sites, the A, P, and E sites, which have specific affinities for the three forms of tRNA, providing at the same time a downhill energetic pathway and a unidirectional movement for the tRNA during translocation.

Translocation also appears to depend on rotation of the ~850 kDa 30S subunit relative to the 50S subunit for each step (Frank and Agrawal 2000; Frank et al. 2007; Horan and Noller 2007). In the absence of GTP hydrolysis, what is the source of energy to drive this massive intermolecular movement? Recent single-molecule FRET studies show that spontaneous intersubunit rotation can occur in a wide variety of mRNA-tRNA-ribosome complexes in the absence of EF-G or GTP, or even peptide bond formation (Cornish et al. 2008). This finding shows that thermal energy alone is sufficient to drive the intersubunit rotation underlying translocation. Translocation is also coupled to movement of a feature of the 50S subunit called the L1 stalk, which maintains contact with the elbow of the deacylated tRNA as it moves from the P/E to the E/E state. Single-molecule FRET experiments show that the L1 stalk can traverse through three different positions during translocation, corresponding to the P/E, E/E and vacant states of the E site (Cornish et al. 2009). Again, the L1 stalk was found to be able to move spontaneously in the absence of EF-G or GTP. These findings show that even the complex, large-scale molecular movements associated with translocation can be driven by thermal energy, obviating the need for special energy-generating steps in protein synthesis by the first ribosomes.

7 WHAT ARE RIBOSOMAL PROTEINS FOR?

If, as we suspect, the fundamental functions of the ribosome are based on its rRNA, why are there so many ribosomal proteins, some of which are highly conserved? So far, no one has reported observation of a ribosomal function being carried out by a rigorously protein-free preparation of rRNA. One explanation for this is that rRNA does not fold into its functional state in the absence of r-proteins (Nomura et al. 1969; Stern et al. 1989). Another reason for the presence of proteins in ribosomes is that they improve the efficiency and accuracy of translation. For example, mutations in proteins S4 and S5 have long been known to cause increased translational error frequencies, implying that they help to improve the accuracy of translation (Davies and Nomura 1972; Kurland et al. 1990). The presence of the carboxy-terminal tails of S9 and S13 is not essential for 30S P-site function, as discussed above, but improves tRNA binding and increases the growth rate of *E. coli* (Hoang et al. 2004). A further point is that even small improvements in the speed and accuracy of translation bring strong selective advantages.

In contrast to histones, for example, the structures of the ribosomal proteins are extremely heterogeneous, representing a large number of different domain types, including helical bundles, α/β RRM folds, all-β OB folds, and so on. Many contain long unstructured tails that penetrate the structure of the rRNA (Ban et al. 2000; Wimberly et al. 2000). Some are essential for correct overall folding and assembly, whereas others are not (Nomura et al. 1969). A few are positioned at or near the subunit interface, where they can influence ribosomal function, whereas the majority are located on the solvent surface, remote from any functional site. Thus, the ribosomal proteins clearly did not arise by duplication of one another, to play a single role, or related roles, but give the impression that they were added one at a time over the course of evolution, as incremental refinements of an essentially RNA-based ribosome.

8 "STOP tRNAs" AND THE EVOLUTION OF TYPE I RELEASE FACTORS

All but three of the 64 possible triplet codons are recognized by tRNAs. The remaining three, the stop codons UAG, UAA and UGA are recognized by the type I release factors (RF1 and RF2 in bacteria). Recent crystal structures of 70S ribosome termination complexes show that the stop codons are directly recognized by the release factors and suggest that catalysis of peptidyl-tRNA hydrolysis is catalyzed by the –NH group of the polypeptide backbone of a conserved Gln in the conserved GGQ motif (Korostelev et al. 2008; Laurberg et al. 2008; Weixlbaumer et al. 2008). The position of this –NH group superimposes with the position of the 3'-OH group of a deacylated tRNA bound to the A site. This raises the possibility that stop codons were originally recognized by deacylated tRNAs, which were replaced during evolution by the type I release factors (Laurberg et al., 2008). In fact, it has been shown that binding of deacylated tRNA to the A site catalyzes polypeptide release (Caskey et al. 1971). A potential shortcoming of a "stop tRNA" is that deacylated tRNAs are unable to bind elongation factor EF-Tu, which is critical for the accuracy of tRNA binding; the resulting high error frequency of translation termination may thus have driven the evolution of type I release factors.

9 dRNA AND THE ORIGINS OF THE RIBOSOMAL DECODING SITE

The molecular interactions involved in codon recognition shown in Figure 3 could, in principle, serve to monitor the accuracy of Watson-Crick base pairing in other RNA contexts. An obvious possible application of this type of quality control in the RNA World is the critical process of RNA replication. A-minor interactions are so simple and so widespread that it would be surprising if they were not put to use for this purpose. Is it possible that the 30S decoding site is a relic of an RNA replication mechanism from the RNA World? The following is a suggestion for how such a replication system may have worked, extrapolating from what we have learned from the ribosome.

Our starting assumption is that codon-anticodon interaction occurs at the site of the template–product interaction in an RNA World replicase, in which similar A-minor interactions were used to ensure the accuracy of RNA replication. The fact that the decoding site mediates triplet–triplet base pairing suggests that its RNA World role would have been to stabilize ligation of oligonucleotide triplets base-paired to a template; i.e., the replicase would have taken the form of an RNA ligase. The first critical question is whether the mRNA codon corresponds to the template (as it does in protein synthesis) or the product (in which case, the anticodon would be the template). Examination of the structure of the decoding site provides an unambiguous answer. The distance between phosphate and ribose groups of adjacent anticodons bound to the A and P sites is more than 30 Å, ruling out models in which the mRNA serves as template. Accordingly, we conclude that anticodons serve as short templates for positioning RNA triplets in the replicase active site.

This raises a second critical question: How can RNA be replicated using such short templates? Here, we introduce the idea of "duplicator RNA" (dRNA). dRNAs are small, tRNA-like structures that mediate *duplication* (as opposed to replication) of an RNA template (Fig. 4). dRNAs have a loop resembling an anticodon loop (here, it is shown as a seven-nucleotide loop, but loops of other sizes would be possible), and an unpaired self-complementary four-nucleotide tail, that allows dRNAs to form homodimers. One "anticodon" end of the dimer base pairs with a triplet

Evolution of Protein Synthesis from an RNA World

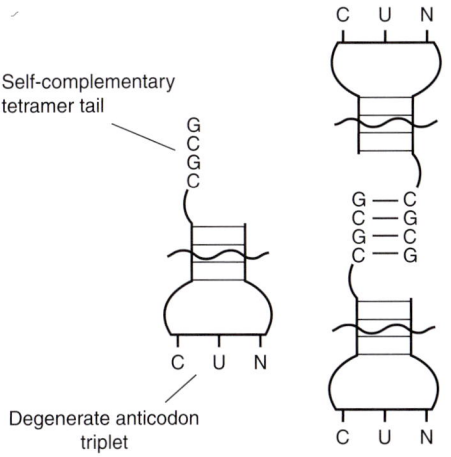

Figure 4. Duplicator RNA. (*left*) A schematic cartoon representing the structure of a duplicator RNA (dRNA) monomer, showing its two identity elements: A self-complementary tetramer tail and a degenerate anticodon triplet. There are 16 possible dRNAs. (*right*) A dRNA homodimer, formed by base pairing of its self-complementary tail. The wavy line indicates that other details of the structure between the anticodon and tail are not intended to be explicit.

sequence in the template RNA, and the other anticodon end pairs with either the nascent product RNA or the incoming triplet substrate (Fig. 5). The product interaction resembles the strong binding of the ribosomal P site and the substrate interaction corresponds to that of the decoding site. Ligation of the substrate oligomer to the growing product chain occurs at the junction between these two binding sites.

Because there are 16 possible self-complementary tetramers, there are 16 possible different dRNAs. Interestingly, the same number of dRNAs is predicted from a completely independent argument. The geometry of the A-minor interactions used in the decoding site creates the basis for the degeneracy of the genetic code (Fig. 3); adjacent base pairs can interact with adjacent adenosines to form type I and type II interactions, but this does not extend to a third base pair (which, in the decoding site is partially taken over by a separate guanine, G530). Degeneracy in the third position of the replicase triplet-triplet interaction would thus limit the number of dRNA anticodons (which could themselves be degenerate) to 16. Thus, each dRNA has two distinct identity elements—a self-complementary tetramer tail, and a (degenerate) triplet anticodon—which establish the link between the template and product RNAs. (Implicit in this discussion is the idea that dRNAs are precursors to tRNAs. Besides their anticodon-like features, their self-complementary tails are reminiscent of tRNA identity elements that are often present in the acceptor stems of tRNAs.)

There are several properties of this form of indirect templating that distinguish it from normal RNA replication, some of which seem especially well suited to the

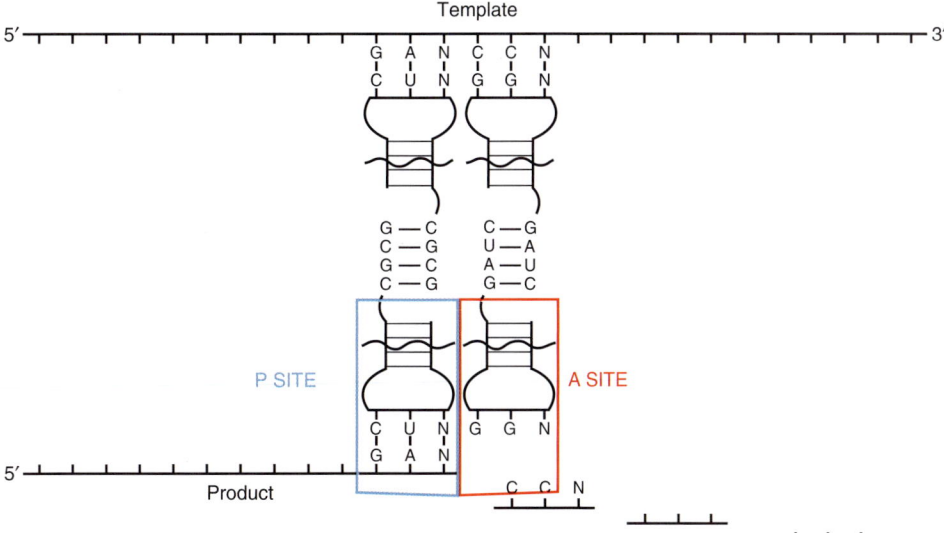

Figure 5. Indirect templating of RNA duplication mediated by dRNA homodimers. One dRNA dimer (*left*) is bound to the last triplet (GAN) in the product RNA, stabilized by a structure resembling the 30S ribosomal P site (blue box). A second dRNA dimer (*right*), bound to the next (CCN) triplet in the template by pairing with one of its GGN anticodons, binds the incoming substrate CCN trimer via base pairing with its other GGN anticodon. Discrimination of correct pairing with the incoming substrate trimer is promoted by A-minor interactions by a structure resembling the 30S ribosomal A site (red box). Binding of the upper anticodon triplets to the template RNA also uses structures resembling the 30S A and P sites (not shown).

challenges of the RNA World. First, the template is duplicated in a parallel fashion, eliminating the formation of an intermediate RNA duplex and the associated problem of unwinding a long duplex to allow release and folding of the product RNA. Second, the substrates are triplet oligonucleotides, which bind more stably to their templates than single nucleotides, yet are readily disrupted at ambient temperatures. These can be sourced from random-sequence triplet pools; because of third-position degeneracy, there are effectively only 16 different substrate oligomers.

The general scheme described here raises many detailed mechanistic questions. How are the substrate oligomers synthesized and activated? What is the mechanism of catalysis? How is the nascent chain translocated following addition of each oligomer? How is the secondary structure of the RNA template disrupted to allow dRNA binding? Would accidental binding of longer oligomers disrupt RNA duplication? Finally, the details of the structure of dRNA in the region linking the anticodon to the self-complementary tail (unspecified here) will be critical, so that the resulting geometry will allow binding of adjacent dRNAs to adjacent triplets and their parallel translocation with respect to the template and product RNAs at opposite ends of the dRNA dimers. These and other aspects of dRNA-mediated duplication are discussed elsewhere (Noller 2010 in preparation).

10 THE DRIVING FORCE FOR EVOLUTION OF TRANSLATION FROM AN RNA WORLD

It is challenging to ask how the structurally and functionally complex process of translation could have evolved from an RNA World. Most encouraging is the demonstration that small RNAs evolved in vitro are able to catalyze all of the principal chemical reactions of protein synthesis, including amino acid activation, aminoacylation of RNA, and peptidyl transfer (Zhang and Cech 1997; Illangasekare and Yarus 1999; Lee et al. 2000; Kumar and Yarus 2001). But quite apart from the mind-boggling prospect of evolving a structure as complex as the ribosome (even without its proteins), the probability of an early translational system producing a functionally capable protein, such as an enzyme, is vanishingly small (Woese 1967). Given these prospects, what was the driving force that led to the evolutionary selection of protein synthesis in the context of an RNA World? If we assume that some sort of Darwinian selection was in place, it must have provided a selective advantage to an RNA world system.

The diversity and efficiency of RNA function depends on the possible types of structure into which RNA can fold. Structural studies on naturally occurring and in vitro-selected RNAs have revealed a rich diversity of RNA folds. Nevertheless, it can already be seen that the range of RNA structures is limited, probably because of the relatively modest chemical differences between the four nucleotide monomers, compounded by the strong inherent tendency for ribonucleotides to adopt conformations resembling those that are found in A-type double helices (Saenger 1984). We would therefore expect an expansion of the range of possible RNA structures to confer a strong selective advantage.

Studies on the structures of RNA-ligand complexes show that the binding of even quite small molecules to RNA can cause large-scale structural changes. RNAs that appear to be unstructured adopt well-defined three-dimensional folds, and structured RNAs undergo conformational rearrangements in the presence of bound ligands. An example is the AMP-dependent structuring of an in vitro-selected aptamer, in which the 11-nucleotide RNA wraps around the nucleotide into an intricate, well-defined fold, such that the AMP plays the role of the conserved A in a GNRA tetraloop (Dieckmann et al. 1996). Another example is the discovery of "riboswitches," ligand-induced RNA structures that are found in naturally occurring mRNAs, influencing the expression of the mRNAs in which they are embedded (Mandal and Breaker 2004; Serganov et al. 2004; Breaker 2010). Thus, binding of small-molecule ligands is likely to have played an important part in expanding the structural repertoire of RNA.

Binding of peptides has also been shown to cause rearrangement of RNA structure. A vivid example is based on the HIV Tat-TAR complex, a protein-RNA interaction that is essential for viral function. Binding of the Tat protein to the TAR RNA could be mimicked by a nine-amino acid arginine-rich peptide derived from the binding site of the protein (Puglisi et al. 1992). Binding of this peptide causes a dramatic rearrangement of the structure of the TAR RNA, from a conventional hairpin stem-loop structure into a structure containing a bulge loop, resulting in formation of an A-U-U triple base pair tertiary interaction. At the core of the structure is an arginine side-chain that appears to play a crucial role in stabilizing the new fold. Remarkably, it was found that this same structural rearrangement occurred in the presence of a single argininamide (Puglisi et al. 1992) (Fig. 6), although the monomer bound with an affinity that was lower by five or six orders of magnitude. These findings show two points: (1) Simple peptides or even amino acid monomers can dramatically influence the structure of RNA, and (2) The extended structures provided by incorporating amino acids into peptides confer higher binding affinity (and concomitantly, increased specificity). Further examples of peptide-induced structural rearrangements of RNA have been observed for the HIV Rev-RRE, BIV Tat-TAR, and

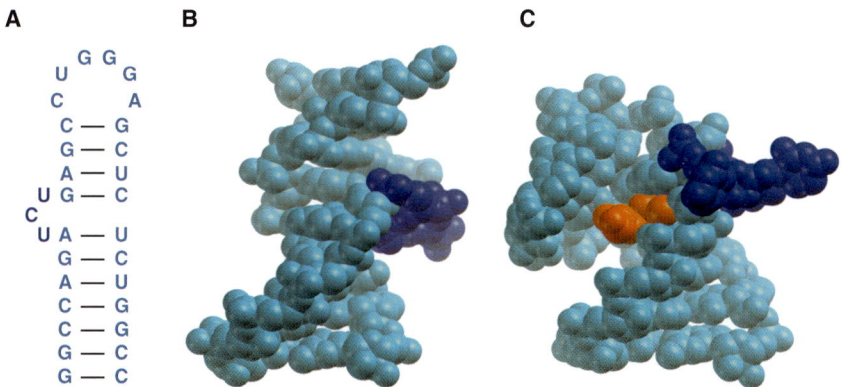

Figure 6. Influence of the HIV Tat peptide on the folding of TAR RNA (Puglisi et al. 1992). (*A*) Secondary structure of TAR RNA; (*B*) NMR structure of the free TAR RNA; (*C*) NMR structure of the TAR RNA bound to a nine-amino acid peptide from Tat protein or bound to a single argininamide (shown). The argininamide is shown in orange and the three-nucleotide bulge loop in dark blue.

HTLV-1 Rex-aptamer interactions (Puglisi et al. 1995; Battiste et al. 1996; Jiang et al. 1999). Accordingly, the ability to synthesize small peptides could have provided a strong selective advantage to an RNA World system possessing such a capability. The kernel of the ribosome may therefore have arisen as a relatively simple RNA that was able to catalyze formation of simple peptides, to help expand the structure space of RNA. Poole et al. (Poole et al. 1998) have proposed a related idea, that primitive peptides could have acted as chaperones, to assist the folding of RNAs.

The general idea that polypeptides promote formation of the active conformations of functional RNAs is supported by the fact that most, if not all, present-day functional RNAs are found associated with proteins in vivo. Ribonuclease P, spliceosomes and ribosomes, whose functions have been ascribed to their respective RNA moieties (Guerrier-Takada et al. 1983; Sharp 1991), nevertheless require proteins to function in their physiological states. In the case of 16S rRNA, assembly of ribosomal proteins has been shown not only to be important for formation of local RNA tertiary structure (Stern et al. 1989), but has also been found to influence the relative orientation of adjacent RNA helical elements, thereby helping to establish even the large-scale geometry of the RNA (Orr et al. 1998). RNase P is thought to use its protein component to help overcome electrostatic repulsion between its catalytic RNA subunit and its RNA substrate, another potential selective advantage for synthesis of (cationic) peptides (Reich et al. 1988).

11 CONCLUSIONS

Evolution of coding remains the most difficult step to explain. It is easiest to think of the evolution of translation as having begun with the synthesis of small peptides, possibly of random sequence. With short peptides containing a limited number of types of amino acids, useful amounts of peptides of defined sequence could be formed. In the absence of a coding mechanism, the substrates for the primitive peptidyl transferase would have been smaller proto-tRNAs, containing acceptor ends, but lacking anticodons (Maizels and Weiner 1993; Noller 1993; Schimmel et al. 1993). Coding would have evolved from a separate RNA-World mechanism, in which the Ramakrishnan A-minor calibration mechanism (Ogle et al. 2001) was used to check the accuracy of Watson-Crick base pairing in a completely different functional context, between short RNA sequences that were the counterparts of the codon and anticodon. At a later stage, the anticodon and proto-tRNA moieties would then be joined to form something resembling present-day tRNAs. It is interesting that the idea that the two halves of tRNA evolved separately has emerged independently, from three different laboratories, from three quite different lines of reasoning (Maizels and Weiner 1993; Noller 1993; Schimmel et al. 1993). How the coding of amino acids by specific nucleotide sequences emerged is harder to imagine. An interesting proposal has been put forth by Schimmel and coworkers (Schimmel and Henderson 1994), involving an intermediate stage of side-by-side interactions between adjacent amino acid-specific proto-tRNAs, whose identity elements were contained exclusively in their acceptor stems. This system of noncoded peptide synthesis would then be converted to a system for template-directed synthesis.

How did the present-day ribosome evolve? The early existence of an all-RNA ribosome of such a level of structural complexity is difficult to imagine. More likely, smaller functional units capable of carrying out the different translational steps such as peptidyl transferase, decoding and so on evolved. These small functional RNA units then merged to form larger structures, which were incrementally refined by incorporation of additional RNA structural elements. Two features of RNA make such a process plausible. First is the fact that small RNAs, unlike most peptides, tend to retain their local structures when excised from larger RNA structures. For example, the anticodon stem-loop of tRNA interacts efficiently and accurately with the ribosome in the absence of the rest of the tRNA structure, and is even capable of accurate codon recognition and translocation (Rose et al. 1983; Joseph and Noller 1998; Ogle et al. 2001). This may be because RNAs do not have hydrophobic cores that are essential for their three-dimensional folding. Thus, building a large functional RNA out of pre-existing small ones can be accomplished while preserving structural and functional properties of the latter. Second, RNAs can readily interact with each other to form complexes, not only by base pairing, but by robust tertiary interactions such as the A-minor interactions described previously. RNAs that have associated noncovalently in this way can then become ligated together to form covalently stable RNAs of increasing size, by well-known ribozyme-catalyzed mechanisms. An explicit proposal for the hierarchical evolution of 23S rRNA was recently described by (Bokov and Steinberg 2009), who analyzed the distribution of donor adenosines and acceptor helices for the A-minor interactions of the 50S ribosomal subunit. They observed a striking assymetry in domain V of 23S rRNA, which contains the elements of the peptidyl transferase center: in all but one case, domain V contributed the acceptor helices for adenosines from other domains. This suggests that 23S rRNA evolved by gradual addition of ancillary RNA domains to the catalytic core in domain V, exploiting A-minor interactions in positioning the added elements. Importantly, in an RNA World, a newly constructed RNA becomes its own gene, whose replication directly provides multiple copies of new functional RNAs. This cycle of noncovalent complex formation, followed by ligation, can then be repeated, while selecting for improvement of ribosome-like function. Finally, the influence of the peptides produced by a primitive ribosome on its own structure and assembly would also begin to impact its own evolution. Ribosome-binding peptides may thus have played an early role in shaping the ultimate form (and function) of the ribosome.

ACKNOWLEDGMENTS

I thank Bill Scott, Melissa Moore, Andrei Korostelev, Gerry Joyce, Jim Dahlberg, Carl Woese, Leslie Orgel and the members of my laboratory for many stimulating discussions. This work was supported by grants from the NIH and NSF, and by a grant from the W.M. Keck Foundation to the Center for Molecular Biology of RNA at UCSC.

REFERENCES

Ban N, Nissen P, Hansen J, Moore PB, Steitz TA. 2000. The complete atomic structure of the large ribosomal subunit at 2.4 Å resolution. *Science* **289:** 905–920.

Barta A, Steiner G, Brosius J, Noller HF, Kuechler E. 1984. Identification of a site on 23S ribosomal RNA located at the peptidyl transferase center. *Proc Natl Acad Sci* **81:** 3607–3611.

Battiste JL, Mao H, Rao NS, Tan R, Muhandiram DR, Kay LE, Frankel AD, Williamson JR. 1996. α helix-RNA major groove recognition in an HIV-1 rev peptide-RRE RNA complex. *Science* **273:** 1547–1551.

Bokov K, Steinberg SV. 2009. A hierarchical model for evolution of 23S ribosomal RNA. *Nature* **457:** 977–980.

Bowman CM, Dahlberg JE, Ikemura T, Konisky J, Nomura M. 1971. Specific inactivation of 16S ribosomal RNA induced by colicin E3 in vivo. *Proc Natl Acad Sci* **68:** 964–968.

Breaker RR. 2010. Riboswitches and the RNA world. *Cold Spring Harb Perspect Biol* **2:** a003566.

Carter AP, Clemons WM, Brodersen DE, Morgan-Warren RJ, Wimberly BT, Ramakrishnan V. 2000. Functional insights from the structure of the 30S ribosomal subunit and its interactions with antibiotics [see comments]. *Nature* **407:** 340–348.

Caskey CT, Beaudet AL, Scolnick EM, Rosman M. 1971. Hydrolysis of fMet-tRNA by peptidyl transferase. *Proc Natl Acad Sci* **68:** 3163–3167.

Cornish PV, Ermolenko DN, Noller HF, Ha T. 2008. Spontaneous intersubunit rotation in single ribosomes. *Mol Cell* **30:** 578–588.

Cornish PV, Ermolenko DN, Staple DW, Hoang L, Hickerson RP, Noller HF, Ha T. 2009. Following movement of the L1 stalk between three functional states in single ribosomes. *Proc Natl Acad Sci* **106:** 2571–2576.

Crick FHC. 1968. The origin of the genetic code. *J Mol Biol* **38:** 367–379.

Dabbs ER. 1986. Mutant studies on the prokaryotic ribosome. in *Structure, function and genetics of ribosomes* (ed. B. Hardesty, G. Kramer), pp. 733–748. Springer-Verlag, N.Y.

Davies J, Nomura M. 1972. The genetics of bacterial ribosomes. *Ann Rev Genet* **6:** 203–234.

Dieckmann T, Suzuki E, Nakamura GK, Feigon J. 1996. Solution structure of an ATP-binding RNA aptamer reveals a novel fold. *RNA* **2:** 628–640.

Endo Y, Wool IG. 1982. The site of action of α-sarcin on eukaryotic ribosomes. The sequence at the α-sarcin cleavage site in 28 S ribosomal ribonucleic acid. *J Biol Chem* **257:** 9054–9060.

Frank J, Agrawal RK. 2000. A ratchet-like inter-subunit reorganization of the ribosome during translocation. *Nature* **406:** 318–322.

Frank J, Gao H, Sengupta J, Gao N, Taylor DJ. 2007. The process of mRNA-tRNA translocation. *Proc Natl Acad Sci* **104:** 19671–19678.

Fredrick K, Noller HF. 2003. Catalysis of ribosomal translocation by sparsomycin. *Science* **300:** 1159–1162.

Gavrilova LP, Kostiashkina OE, Koteliansky VE, Rutkevitch NM, Spirin AS. 1976. Factor-free ("non-enzymic") and factor-dependent systems of translation of polyuridylic acid by *Escherichia coli* ribosomes. *J Mol Biol* **101:** 537–552.

Guerrier-Takada C, Gardiner K, Marsh T, Pace N, Altman S. 1983. The RNA moiety of ribonuclease P is the catalytic subunit of the enzyme. *Cell* **35:** 849–857.

Helser TL, Davies JE, Dahlberg JE. 1972. Mechanism of kasugamycin resistance in *Escherichia coli*. *Nat New Biol* **235:** 6–9.

Hoang L, Fredrick K, Noller HF. 2004. Creating ribosomes with an all-RNA 30S subunit P site. *Proc Natl Acad Sci* **101:** 12439–12443.

Horan LH, Noller HF. 2007. Intersubunit movement is required for ribosomal translocation. *Proc Natl Acad Sci* **104:** 4881–4885.

Illangasekare M, Yarus M. 1999. Specific, rapid synthesis of Phe-RNA by RNA. *Proc Natl Acad Sci* **96:** 5470–5475.

Jiang F, Gorin A, Hu W, Majumdar A, Baskerville S, Xu W, Ellington A, Patel DJ. 1999. Anchoring an extended HTLV-1 Rex peptide within an RNA major groove containing junctional base triples. *Structure Fold Des* **7:** 1461–1472.

Joseph S, Noller HF. 1998. EF-G-catalyzed translocation of anticodon stem-loop analogs of transfer RNA in the ribosome. *EMBO J* **17:** 3478–3483.

Korostelev A, Asahara H, Lancaster L, Laurberg M, Hirschi A, Zhu J, Trakhanov S, Scott WG, Noller HF. 2008. Crystal structure of a translation termination complex formed with release factor RF2. *Proc Natl Acad Sci* **105:** 19684–19689.

Kumar RK, Yarus M. 2001. RNA-catalyzed amino acid activation. *Biochemistry* **40:** 6998–7004.

Kurland CG, Jorgensen F, Richter AA, Ehrenberg M, Bilgin N, Rojas AM. 1990. Through the Accuracy Window. in *The Ribosome: Structure, function, and evolution* (ed. WE. Hill, A. Dahlberg, RA. Garrett, PB. Moore, D. Schlessinger, JR. Warner), pp. 513–526. American Society of Microbiology, Washington, D.C.

Laurberg M, Asahara H, Korostelev A, Zhu J, Trakhanov S, Noller HF. 2008. Structural basis for translation termination on the 70S ribosome. *Nature* **454:** 852–857.

Lee N, Bessho Y, Wei K, Szostak JW, Suga H. 2000. Ribozyme-catalyzed tRNA aminoacylation. *Nat Struct Biol* **7:** 28–33.

Lill R, Robertson JM, Wintermeyer W. 1986. Affinities of tRNA binding sites of ribosomes from *Escherichia coli*. *Biochemistry* **25:** 3245–3255.

Maizels N, Weiner AM. 1993. The genomic tag hypothesis: Modern viruses as molecular fossils of ancient strategies for genomic replication. in *The RNA world* (ed. RF. Gesteland JF. Atkins), pp. 577–602. Cold Spring Harbor Laboratory Press, Cold Spring Harbor, NY.

Mandal M, Breaker RR. 2004. Gene regulation by riboswitches. *Nat Rev Mol Cell Biol* **5:** 451–463.

Moazed D, Noller HF. 1986. Transfer RNA shields specific nucleotides in 16S ribosomal RNA from attack by chemical probes. *Cell* **47:** 985–994.

Moazed D, Noller HF. 1987. Chloramphenicol, erythromycin, carbomycin and vernamycin B protect overlapping sites in the peptidyl transferase region of 23S ribosomal RNA. *Biochimie* **69:** 879–884.

Moazed D, Noller HF. 1989. Interaction of tRNA with 23S rRNA in the ribosomal A, P, and E sites. *Cell* **57:** 585–597.

Moazed D, Noller HF. 1990. Binding of tRNA to the ribosomal A and P sites protects two distinct sets of nucleotides in 16S rRNA. *J Mol Biol* **211:** 135–145.

Moazed D, Noller HF. 1991. Sites of interaction of the CCA end of peptidyl-tRNA with 23S rRNA. *Proc Natl Acad Sci* **88:** 3725–3728.

Monro RE. 1967. Catalysis of peptide bond formation by 50 S ribosomal subunits from *Escherichia coli*. *J Mol Biol* **26:** 147–151.

Moore PB, Steitz TA. 2010. The roles of RNA in the synthesis of protein. *Cold Spring Harb Perspect Biol* **2:** a003780.

Moore VG, Atchison RE, Thomas G, Moran M, Noller HF. 1975. Identification of a ribosomal protein essential for peptidyl transferase activity. *Proc Natl Acad Sci* **72:** 844–848.

Nirenberg M, Leder P. 1964. RNA code words and protein synthesis. *Science* **145:** 1399–1407.

Nissen P, Hansen J, Ban N, Moore PB, Steitz TA. 2000. The structural basis of ribosome activity in peptide bond synthesis. *Science* **289:** 920–930.

Nissen P, Ippolito JA, Ban N, Moore PB, Steitz TA. 2001. RNA tertiary interaction in the large ribosomal subunit: The A-minor motif. *Proc Natl Acad Sci* **98:** 4899–4903.

Noller HF. 1993. On the origin of the ribosome: Coevolution of subdomains of tRNA and rRNA. in *The RNA world* (ed. RF. Gesteland, TR. Cech, JF. Atkins), pp. 137–156. Cold Spring Harbor Laboratory Press, Cold Spring Harbor, NY.

Noller HF. 2004. The driving force for molecular evolution of translation. *RNA* **10:** 1833–1837.

Noller HF, Chaires JB. 1972. Functional modification of 16S ribosomal RNA by kethoxal. *Proc Natl Acad Sci* **69:** 3113–3118.

Noller HF, Hoffarth V, Zimniak L. 1992. Unusual resistance of peptidyl transferase to protein extraction procedures. *Science* **256:** 1416–1419.

Noller HF, Moazed D, Stern S, Powers T, Allen PN, Robertson JM, Weiser B, Triman K. 1990. Structure of rRNA and its functional interactions in translation. in *The ribosome: structure, function, and evolution* (ed. WE. Hill, A. Dahlberg, RA. Garrett, PB. Moore, D. Schlessinger, JR. Warner), pp. 73–92. American Society of Microbiology, Washington, D.C.

Nomura M, Mizushima S, Ozaki M, Traub P, Lowry CV. 1969. Structure and function of ribosomes and their molecular components. *Cold Spring Harbor Symp Quant Biol* **34:** 49–61.

Ogle JM, Brodersen DE, Clemons WM, Tarry MJ, Carter AP, Ramakrishnan V. 2001. Recognition of cognate transfer RNA by the 30S ribosomal subunit. *Science* **292:** 897–902.

Orgel LE. 1968. Evolution of the genetic apparatus. *J Mol Biol* **38:** 381–393.

Orr JW, Hagerman PJ, Williamson JR. 1998. Protein and $Mg(2+)$-induced conformational changes in the S15 binding site of 16 S ribosomal RNA. *J Mol Biol* **275:** 453–464.

Pestka S. 1968. Studies on the formation of transfer ribonucleic acid-ribosome complexes. 3. The formation of peptide bonds by ribosomes in the absence of supernatant enzymes. *J Biol Chem* **243:** 2810–2820.

Poole AM, Jeffares DC, Penny D. 1998. The path from the RNA world. *J Mol Evol* **46:** 1–17.

Powers T, Noller HF. 1990. Dominant lethal mutations in a conserved loop in 16S rRNA. *Proc Natl Acad Sci* **87:** 1042–1046.

Prince JB, Taylor BH, Thurlow DL, Ofengand J, Zimmermann RA. 1982. Covalent crosslinking of tRNA1Val to 16S RNA at the ribosomal P site: Identification of crosslinked residues. *Proc Natl Acad Sci* **79:** 5450–5454.

Puglisi JD, Chen L, Blanchard S, Frankel AD. 1995. Solution structure of a bovine immunodeficiency virus Tat-TAR peptide-RNA complex. *Science* **270:** 1200–1203.

Puglisi JD, Tan R, Calnan BJ, Frankel AD, Williamson JR. 1992. Conformation of the TAR RNA-arginine complex by NMR spectroscopy. *Science* **257:** 76–80.

Reich C, Olsen GJ, Pace B, Pace NR. 1988. Role of the protein moiety of ribonuclease P, a ribonucleoprotein enzyme. *Science* **239:** 178–181.

Rose SJ 3rd, Lowary PT, Uhlenbeck OC. 1983. Binding of yeast tRNAPhe anticodon arm to Escherichia coli 30 S ribosomes. *J Mol Biol* **167:** 103–117.

Saenger W. 1984. *Principles of nucleic acid structure* Springer-Verlag, New York.

Schimmel P, Giege R, Moras D, Yokoyama S. 1993. An operational RNA code for amino acids and possible relationship to genetic code. *Proc Natl Acad Sci* **90:** 8763–8768.

Schimmel P, Henderson B. 1994. Possible role of aminoacyl-RNA complexes in noncoded peptide synthesis and origin of coded synthesis. *Proc Natl Acad Sci* **91:** 11283–11286.

Schluenzen F, Tocilj A, Zarivach R, Harms J, Gluehmann M, Janell D, Bashan A, Bartels H, Agmon I, Franceschi F, et al. 2000. Structure of functionally activated small ribosomal subunit at 3.3 angstroms resolution. *Cell* **102:** 615–623.

Schmeing TM, Huang KS, Strobel SA, Steitz TA. 2005. An induced-fit mechanism to promote peptide bond formation and exclude hydrolysis of peptidyl-tRNA. *Nature* **438:** 520–524.

Schulze H, Nierhaus KH. 1982. Minimal set of ribosomal components for reconstitution of the peptidyltransferase activity. *EMBO J* **1:** 609–613.

Selmer M, Dunham CM, Murphy FVt, Weixlbaumer A, Petry S, Kelley AC, Weir JR, Ramakrishnan V. 2006. Structure of the 70S ribosome complexed with mRNA and tRNA. *Science* **313:** 1935–1942.

Senior BW, Holland IB. 1971. Effect of colicin E3 upon the 30S ribosomal subunit of *Escherichia coli*. *Proc Nat Acad Sci* **69:** 959–964.

Serganov A, Yuan YR, Pikovskaya O, Polonskaia A, Malinina L, Phan AT, Hobartner C, Micura R, Breaker RR, Patel DJ. 2004. Structural basis for discriminative regulation of gene expression by adenine- and guanine-sensing mRNAs. *Chem Biol* **11:** 1729–1741.

Sharp PA. 1991. "Five easy pieces". *Science* **254:** 663.

Shine J, Dalgarno L. 1974. The 3'-terminal sequence of *E coli* 16S ribosomal RNA complementarity to nonsense triplets and ribosome binding sites. *Proc Nat Acad Sci* **71:** 1342–1346.

Southworth DR, Brunelle JL, Green R. 2002. EFG-independent translocation of the mRNA:tRNA complex is promoted by modification of the ribosome with thiol-specific reagents. *J Mol Biol* **324:** 611–623.

Spirin AS. 1985. Ribosomal translocation: Facts and models. *Prog Nucleic Acid Res Mol Biol* **32:** 75–114.

Steitz JA, Jakes K. 1975. How ribosomes select initiator regions in mRNA: Base pair formation between the 3' terminus of 16S rRNA and the mRNA during initiation of protein synthesis in *Escherichia coli*. *Proc Natl Acad Sci* **72:** 4734–4738.

Stern S, Powers T, Changchien LM, Noller HF. 1989. RNA-protein interactions in 30S ribosomal subunits: Folding and function of 16S rRNA. *Science* **244:** 783–790.

von Ahsen U, Noller HF. 1995. Identification of bases in 16S rRNA essential for tRNA binding at the 30S ribosomal P site. *Science* **267:** 234–237.

Weixlbaumer A, Jin H, Neubauer C, Voorhees RM, Petry S, Kelley AC, Ramakrishnan V. 2008. Insights into translational termination from the structure of RF2 bound to the ribosome. *Science* **322:** 953–956.

Wimberly BT, Brodersen DE, Clemons WM Jr, Morgan-Warren RJ, Carter AP, Vonrhein C, Hartsch T, Ramakrishnan V. 2000. Structure of the 30S ribosomal subunit. *Nature* **407:** 327–339.

Woese CR. 1967. *The genetic code. The molecular basis for genetic expression* Harper & Row, New York.

Woese CR, Fox GE, Zablen L, Uchida T, Bonen L, Pechman K, Lewis BJ, Stahl D. 1975. Conservation of primary structure in 16S ribosomal RNA. *Nature* **254:** 83–86.

Wower J, Sylvers LA, Rosen KV, Hixson SS, Zimmermann RA. 1993. A model of the tRNA binding sites on the *Escherichia coli* ribosome. in *The translational apparatus: Structure, function, regulation, evolution* (ed. KH. Nierhaus), pp. 455–464. Plenum Press, New York.

Yoshizawa S, Fourmy D, Puglisi JD. 1999. Recognition of the codon-anticodon helix by ribosomal RNA. *Science* **285:** 1722–1725.

Yusupov M, Yusupova G, Baucom A, Lieberman K, Earnest TN, Cate JH, Noller HF. 2001. Crystal structure of the ribosome at 5.5 Å resolution. *Science* **292:** 883–896.

Zhang B, Cech TR. 1997. Peptide bond formation by in vitro selected ribozymes. *Nature* **390:** 96–100.

The Ribosome: Some Hard Facts about Its Structure and Hot Air about Its Evolution*

V. Ramakrishnan

MRC Laboratory of Molecular Biology, Cambridge CB2 0QH, United Kingdom

Correspondence: ramak@mrc-lmb.cam.ac.uk

SUMMARY

By translating genetically encoded information to synthesize proteins, the ribosome has a central and fundamental role in the molecular biology of the cell. Virtually every molecule made in every cell was made either directly by the ribosome or by enzymes made by the ribosome. Although the ribosome was discovered half a century ago, progress in the field of translation has been revolutionized by the atomic structures of the ribosomal subunits determined in 2000. These structures paved the way not only for more sophisticated biochemical and genetic experiments, but also for the phasing and/or molecular interpretation of all subsequent structures of the ribosome by crystallography or cryoEM (cryo-electron microscopy). In addition to facilitating our understanding of ribosome function, these structures also shed light on the evolution of the ribosome.

Outline

1. The ribosome as an RNA-based machine
2. Decoding by tRNA
3. Peptidyl transfer
4. Peptide release
5. Translocation
6. How did the ribosome evolve?

References

---

*The title is a paraphrase of one by the late eminent crystallographer David M. Blow: "Hard facts on structure: Hot air about mobility" *(Nature* [1982] **297:** 454–455).

Ribosomes from all species consist of approximately two-thirds RNA and one-third protein. Ribosomes from mammalian mitochondria are an exception, with the ratio of protein and RNA reversed (see Sharma et al. 2003). All ribosomes consist of two subunits, termed 50S and 30S in bacteria or 60S and 40S in eukaryotes. Together, they comprise the 70S ribosome in bacteria or the 80S ribosome in eukaryotes (Fig. 1A). The mRNA containing the genetic template binds in a cleft in the small subunit. The amino acids themselves are brought into the ribosome by aminoacylated tRNA substrates. The ribosome has three binding sites for tRNA: the A (aminoacyl) site that brings the new aminoacyl tRNA, the P (peptidyl) site that holds the nascent peptide chain, and the E (exit) site to which the deacylated P-site tRNA moves after peptide bond formation (Fig. 1B).

Translation in all species can be divided into three stages (Fig. 2) (for review, see Schmeing and Ramakrishnan 2009). During initiation, the small subunit of the ribosome binds mRNA at the start site of the coding sequence, in a precise manner that puts the start codon in the P site. This requires three initiation factors and a special initiator tRNA that binds to the P site.

Initiation is followed by the elongation cycle, which consists of three important steps: decoding, peptidyl transfer, and translocation. During decoding, the correct aminoacyl tRNA, which is delivered to the A site of the ribosome as a ternary complex with elongation factor Tu (EF-Tu) and GTP, is selected based on the codon on the mRNA in the A site. Selection of the tRNA leads to hydrolysis of GTP by EF-Tu and release of the factor from the ribosome. The aminoacyl end of the selected tRNA then swings into the peptidyl transferase center (PTC) in the 50S subunit of the ribosome, where peptide bond formation occurs rapidly and spontaneously. Peptidyl transfer leaves the P-site tRNA deacylated, with the A-site tRNA now containing a nascent peptide chain that has been extended by one residue. The 3′ ends of the A- and P-site tRNAs then move first with respect to the 50S subunit to form an intermediate or hybrid state of the ribosome, followed by movement of the mRNA and tRNAs with respect to the 30S subunit, which requires the action of EF-G, another GTPase factor. This leaves the ribosome with an empty A site with a new mRNA codon ready to accept the next aminoacyl tRNA.

The elongation cycle continues until a stop codon is reached in the A site. The so-called class I release factors (RF1 or RF2 in bacteria, eRF1 in eukaryotes) recognize the stop codon and catalyze the cleavage of the polypeptide chain from the P-site tRNA. Finally, a factor known as ribosome recycling factor (RRF), with the help of EF-G,

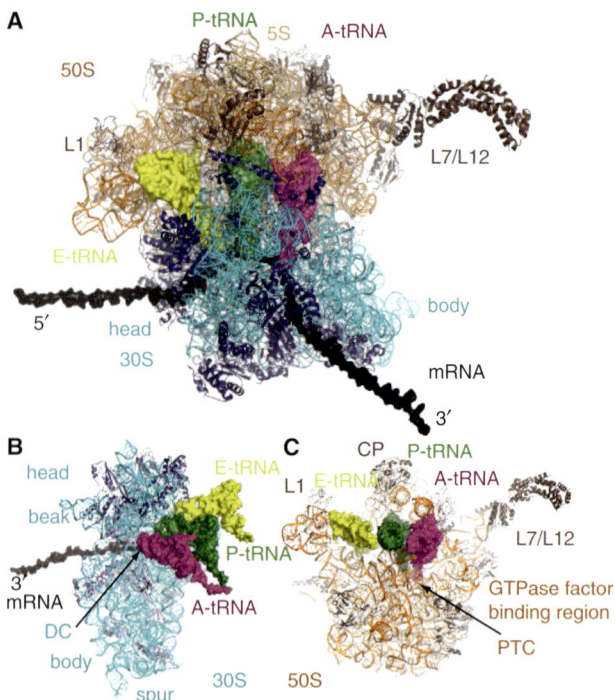

Figure 1. Structure of the ribosome. (*A*) Overview of the bacterial 70S ribosome with the 50S subunit on top and the 30S subunit on the bottom. The mRNA (black) is shown wrapped round the neck of the 30S subunit. (Magenta) A-site, (green) P-site, (yellow) E-site tRNAs. (*B*) The 30S subunit showing the decoding center (DC) where codon-anticodon interactions are monitored during tRNA selection. (*C*) The 50S subunit showing the GTPase-factor-binding region and the PTC where peptide bond formation is catalyzed. (Reprinted, with permission, from Schmeing and Ramakrishnan 2009 [©Nature Publishing Group].)

disassembles the ribosome so that a new round of protein synthesis can begin.

Most of these aspects of translation are common to all kingdoms of life. In eukaryotes, initiation is far more complex and involves a specifically modified mRNA with a 5′ cap and a poly(A) tail at the 3′ end, as well as almost a dozen factors, many of which are large multi-subunit complexes themselves (Kapp and Lorsch 2004).

Recent structural and biochemical work has shed light on many aspects of translation. In particular, the high-resolution structures of the ribosomal subunits (Ban et al. 2000; Wimberly et al. 2000) were useful in the molecular interpretation and/or phasing of all subsequent structures, including a lower-resolution crystal structure of the 70S ribosome with mRNA and tRNA ligands at 5.5-Å resolution (Yusupov et al. 2001), more recent higher-resolution structures of the empty 70S ribosome from *Escherichia coli* (Schuwirth et al. 2005), and the 70S ribosome with mRNA and tRNAs from *Thermus thermophilus* (Selmer et al. 2006). These basic structures have been followed by

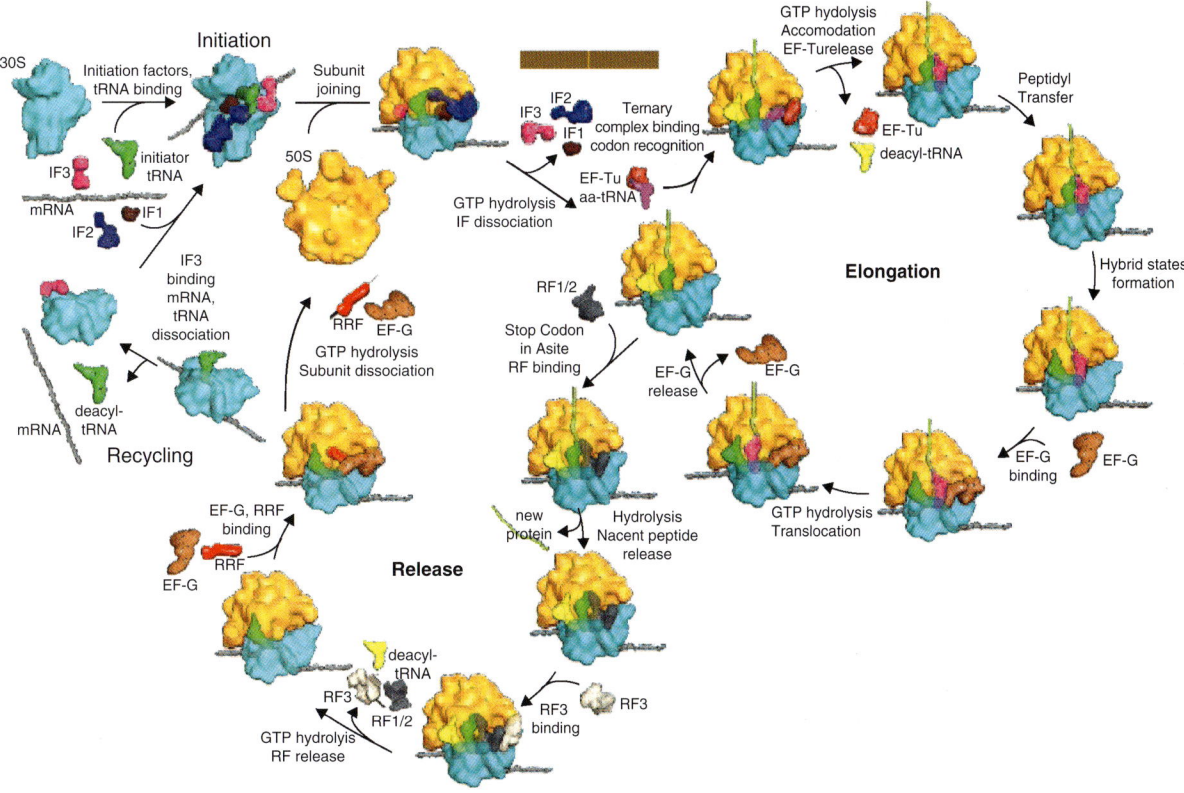

Figure 2. Overview of the translational pathway showing the phases of initiation, elongation, release, and recycling. (Reprinted, with permission, from Schmeing and Ramakrishnan 2009 [©Nature Publishing Group].)

high-resolution structures of the ribosome with protein factors, most notably with release factors (Laurberg et al. 2008; Weixlbaumer et al. 2008) and more recently with elongation factors EF-Tu and EF-G (Gao et al. 2009; Schmeing et al. 2009). In addition, many cryoEM structures of the ribosome represent different functional states at varying resolutions.

1 THE RIBOSOME AS AN RNA-BASED MACHINE

The ribosome itself is a large and complex assembly of RNA and more than 50 proteins. In addition, translation requires a host of protein factors and aminoacyl tRNA substrates. Thus, understanding the evolution of the ribosome poses a difficult challenge. To begin with, the system poses the standard "chicken or egg" question: If the ribosome consists of both RNA and protein, and is needed to make protein, how did it come about? The first attempt to address this was Crick, who presciently wrote, "It is tempting to wonder if the primitive ribosome could have been made *entirely* of RNA" (original italics) (Crick 1968). To my knowledge, this was the first idea that RNA could be both an information carrier and able to perform catalysis, and can be thought of as the origin of the "RNA world hypothesis," which postulates a primordial world consisting of replicating RNA molecules before the advent of proteins. However, in the absence of any known examples of catalysis by RNA, not even Crick could imagine that catalysis in the current ribosome would be RNA based.

It is clear that protein factors could have evolved later to make translation more efficient, because even today it is possible to get inefficient and limited translation without them. For instance, factor-free protein synthesis in vitro was demonstrated by Spirin and coworkers (Gavrilova and Spirin 1974; Gavrilova et al. 1976). But what about the ribosome itself? What are the relative roles of protein and RNA?

The earliest work on ribosome function focused on proteins for two reasons. Partly, proteins were thought to be the molecules responsible for catalytic function. Second, because the standard laboratory organism *E. coli* contains seven genes for rRNA, it was difficult to isolate RNA mutants, and many of the early mutations, such as those for antibiotic resistance, mapped to ribosomal proteins. However, there were hints from quite early on that RNA had a more important role than just providing a scaffolding for functional proteins. For instance, it was shown that chemical modification of rRNA but not proteins would abolish binding of tRNA to 30S subunits (Noller and Chaires 1972). In the absence of any prior evidence for the catalytic properties of RNA, the results were taken to suggest that tRNA-binding sites must therefore consist of both protein and RNA. Subsequent work on the ribosome, notably by Noller and coworkers, continued to provide evidence for the importance of rRNA, but in the absence of an intellectual framework in which RNA catalysis was a real possibility, it was hard to make definite progress.

This situation changed dramatically when catalysis by RNA was discovered in the context of the group I intron (Zaug et al. 1983) or RNase P (Guerrier-Takada et al. 1983). With evidence that RNA could in principle perform catalysis, the ribosome community was far readier to accept that rRNA might have crucial functions, and this prospect renewed interest in the field (Moore 1988). Subsequently, an experiment showed that 50S subunits from *Thermus aquaticus* treated extensively with proteinase K in the presence of SDS nevertheless preserved their peptidyl transferase activity (Noller et al. 1992). This experiment was a major step forward, but not conclusive proof for a variety of reasons. *E. coli* 50S subunits did not maintain activity with such protease treatment, nor did in-vitro-transcribed 23S RNA show activity. Moreover, in the *Thermus* 50S subunit, a large number of protein fragments remained bound after protease treatment, and indeed, subsequent work showed that the treatment left three proteins essentially intact (Khaitovich et al. 1999). A subsequent effort to provide conclusive proof of the role of RNA in peptidyl transfer using in vitro transcription of 23S RNA and its individual domains appeared to have narrowed down the activity to the RNA domain containing the peptidyl transferase center (Nitta et al. 1998), but this work was retracted a year later (Nitta et al. 1999). Thus, although work on the group I intron and RNase P showed that catalysis by RNA was certainly possible, conclusive evidence for a similar role in the ribosome proved to be difficult to obtain by purely biochemical means. It is striking that this limitation was recognized very early on by Crick, who said, "Without a detailed knowledge of the structure of present-day ribosomes it is difficult to make an informed guess" (Crick 1968).

High-resolution structures of ribosomal subunits and the whole ribosome have revealed in stunning detail the environment of the PTC-, tRNA-, and mRNA-binding sites, the intersubunit interface, and many other functionally important regions of the ribosome. These structures at long last allow us to provide conclusive insights into many aspects of ribosome function and, in particular, show unambiguously how widespread the role of RNA is in the contemporary ribosome.

2 DECODING BY tRNA

A crucial event in protein synthesis is the selection of the correct tRNA corresponding to the codon on mRNA. At a fundamental level, this involves base pairing between the codon and the anticodon on tRNA. However, the free energy of base pairing between codon and anticodon is not sufficient to explain the relatively low error rate of the ribosome (Ogle and Ramakrishnan 2005). Kinetic experiments show that binding of the correct tRNA leads to induced conformational changes in the ribosome that accelerate GTP hydrolysis by EF-Tu and tRNA selection (see Rodnina and Wintermeyer 2001).

2.1 Minor Groove Recognition by RNA

Discrimination against incorrect tRNA ultimately depends on recognizing mismatched base pairs. Because base pairing alone is insufficient to explain the accuracy of decoding, and the ribosome appears to have an active role, what is the ultimate nature of this discrimination? Experiments showed that the binding of a cognate tRNA induced three universally conserved bases in the decoding center of the 30S subunit to line the minor groove between codon and anticodon in a manner that would distinguish between canonical Watson–Crick and noncanonical base pairs at the first two positions but did not monitor the geometry of base pairing at the wobble position (Fig. 3) (Ogle et al. 2001). This finding at once explained the additional discrimination provided by the ribosome and the long-standing nature of the genetic code, in which mismatches were allowed at the wobble position but not at the first two positions (Crick 1966). The additional binding energy of cognate tRNA from the induced changes at the decoding center resulted in large-scale movements of the shoulder domain of the 30S subunit, whereas near-cognate tRNA failed to produce such movements (Ogle et al. 2002). This finding led to a model in which the conformational change from an open form to a closed form was required for tRNA selection; such a model could help to rationalize disparate biochemical genetic data on fidelity (Ogle et al. 2002). The rationale for why such a conformational change is needed for tRNA selection has recently been further clarified by the structure of the ribosome bound to elongation factor Tu and tRNA (Schmeing et al. 2009).

2.2 The Possible Role of Minor-groove Recognition in Evolution

DNA and RNA polymerases have exactly the same problem of discriminating against noncanonical base pairs. They are even more accurate than the ribosome. It is striking that both DNA and RNA polymerases use conserved amino acids to monitor the minor groove between the template and transcript strands. Indeed, it is possible for the polymerases to choose base pairs that form no hydrogen bonds between them, as long as they have the same shape as Watson–Crick base pairs at the minor groove (for review, see Kool 2000). Thus, minor-groove recognition is an important feature of ensuring proper base pairing complementarity.

The ribosome shows that such minor-groove recognition can be done by RNA alone. This is significant for the evolution of complexity from a primitive self-replicating RNA system. Such a system initially would have had a very high error rate, because base pairing alone would not have a sufficiently high free-energy difference to allow for substantial accuracy. However, if a primitive replicase also evolved to take advantage of minor-groove recognition of complementary base pairs, the error rate would be reduced by at least two orders of magnitude, allowing for the much more accurate replication required for complex systems to evolve.

2.3 Distortions in tRNA during Decoding

Structures of the ribosome complexed with EF-Tu and tRNA by cryoEM (Valle et al. 2002; Villa et al. 2009; Schuette et al. 2009) or more recently by crystallography (Schmeing et al. 2009) show that when tRNA is delivered to the ribosome by EF-Tu, it is distorted in the anticodon stem (Fig. 4). This suggests that the tRNA molecule has within itself considerable conformational variability. It is possible that this distortion could occur transiently in the absence of EF-Tu. Thus, tRNA in the absence of factors could have still performed decoding in a very similar way, with a transient bend during initial recognition followed by accommodation of the aminoacyl end into the PTC. However, GTP hydrolysis by EF-Tu, an important step in tRNA selection, would have been absent, so that the process would have been both slower and less accurate.

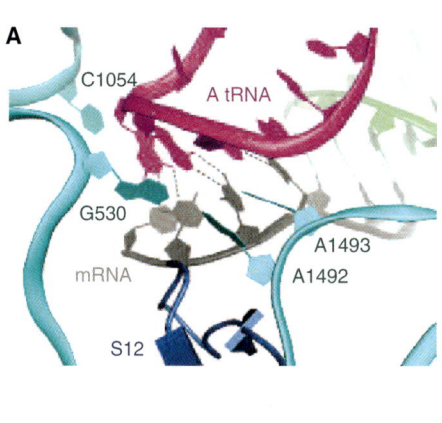
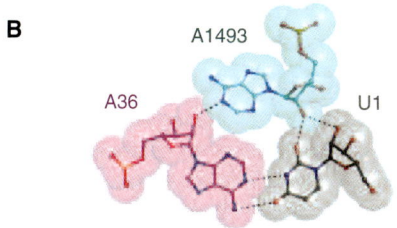
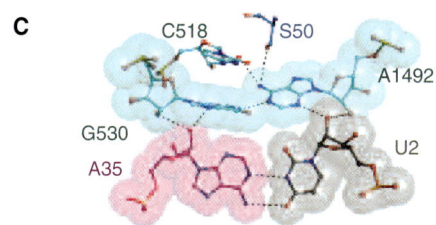
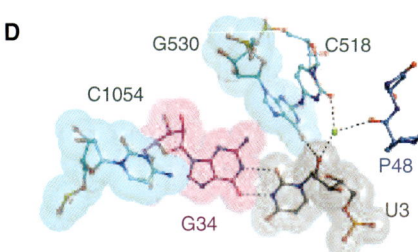

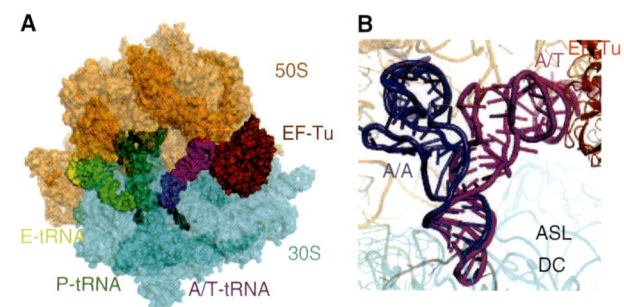

Figure 4. Structure of the complex of EF-Tu and aminoacyl tRNA bound to the ribosome. (A) Overview showing EF-Tu and aminoacyl tRNA (purple) bound to the ribosome. (B) Comparison of the conformation of the tRNA in the distorted state when it is bound to EF-Tu (purple) with the more canonical form after it has swung into the peptidyl transferase center (dark blue). (A, Reprinted, with permission from Schmeing and Ramakrishnan 2009 [©Nature Publishing Group]; B, reprinted, with permission from Schmeing et al. 2009 [©AAAS].)

Figure 3. (A) Decoding center in the 30S subunit of the ribosome, showing how ribosomal bases interact with the minor groove of the codon–anticodon minihelix. (B–D) Interaction of ribosomal bases with the minor groove of the first, second, and third base pairs between the codon (pink) and anticodon (gray) (Ogle et al. 2001). (Reprinted, with permission, from Schmeing and Ramakrishnan 2009 [©Nature Publishing Group].)

3 PEPTIDYL TRANSFER

The central chemical event in translation is the formation of the peptide bond between the nascent polypeptide chain on P-site tRNA and the incoming amino acid on A-site tRNA. This catalysis occurs in the PTC of the ribosome that is located in the 50S subunit. As we have discussed above, the question of whether catalysis is RNA-based or whether proteins are involved could not be settled by purely biochemical experiments.

It was therefore striking that the structure of the 50S subunit in complex with a transition-state analog that defined the catalytic site showed that there were no proteins within 18 Å of the active site (Nissen et al. 2000). Although an acid-base catalytic mechanism involving specific ribosomal bases was proposed, this was disproved by subsequent biochemical work (for review, see Rodnina et al. 2006). Interestingly, none of the ribosomal bases appear to have a catalytic role in the chemical sense of contributing or accepting electrons or protons. Rather, the ribosome's contribution appears to be primarily entropic, by holding the substrates in the proper orientation (Sievers et al. 2004). The one moiety that appears to have a chemical role is the 2′-OH of the terminal adenosine of P-site tRNA itself, showing that the ribosome is an example of substrate-assisted catalysis (Weinger et al. 2004).

One interesting role for the ribosome is that of exposing the ester bond on the P-site tRNA to nucleophilic attack by an induced conformational change in the PTC on A-site tRNA binding (Schmeing et al. 2005b). This is an elegant way of protecting the nascent chain from hydrolysis by water except when the proper A-site substrate is bound, when certain conserved bases move to expose the ester bond that links the nascent chain to P-site tRNA. As in the case of decoding, the induced conformational changes all involve the RNA component of the ribosome.

Although no proteins were found near the active site of the archaeal PTC (Nissen et al. 2000), studies in bacteria have long implicated specific proteins in peptidyl transfer. In particular, two proteins, L16 (Moore et al. 1975) and L27 (Wower et al. 1998), were shown to aid peptidyl transfer. Of these, L27 was known to cross-link to the 3′ ends of both A- and P-site tRNAs, thus placing it right at the PTC (Wower et al. 1998), and just deletion of the first few amino-terminal residues reduced both the cross-linking yield and the rate of peptidyl transfer (Maguire et al. 2005).

The role of these proteins has recently been clarified by crystallography, in which it was shown that L16 becomes ordered and helps to stabilize A-site tRNA, whereas L27 has a long extension that places its amino terminus right at the PTC where it interacts with both A- and P-site tRNAs (Fig. 5) (Voorhees et al. 2009). Thus, at least in the bacterial ribosome, the PTC does have a protein component that has some role in facilitating peptidyl transfer by stabilizing tRNA substrates. In so doing, its role is not fundamentally different from that of rRNA. However, one should bear in mind that it is possible to delete L27 without affecting viability, but the deletion of many conserved RNA residues near the PTC is lethal. So the notion that the ribosome is fundamentally an RNA-based machine is unchanged, but clearly in viewing the contemporary ribosome, we are seeing a snapshot in evolution in which proteins are beginning to play a supporting role.

4 PEPTIDE RELEASE

The process of termination has analogies with both decoding and peptidyl transfer because the stop codon must be recognized and catalysis of the release of the peptide chain must take place at the PTC.

A significant advance has been made recently, owing to the crystal structures of both RF1 and RF2 bound to the ribosome (Korostelev et al. 2008; Laurberg et al. 2008; Weixlbaumer et al. 2008). These structures shed light on both the mechanism of codon recognition by these factors and the role of a conserved GGQ motif in catalysis. In particular, an induced fit of the same three nucleotides involved in decoding by tRNA is required for proper recognition of the stop codon by release factors. Moreover, a similar induced fit on the binding of the GGQ motif in the PTC is seen on release factor binding as was seen for tRNA binding, except in this case, instead of a nucleophilic attack by the amine on A-site tRNA, there is presumably an attack by a water molecule that leads to hydrolysis of the nascent chain.

It is striking that bacterial and eukaryotic release factors have no sequence or structural homology. This suggests that despite their common GGQ motif at the catalytic site, they evolved independently after the divergence of the three kingdoms. If this is true, it is likely that the role of termination was originally played by a tRNA. Presumably, such a tRNA had anticodons complementary to a stop codon so that decoding could occur, but no synthetase was associated with them, so that they bound in the deacylated form by the more inefficient factor-free route, rather than as a complex associated with EF-Tu. They could then still induce a change in the PTC that would expose the ester bond to nucleophilic attack by water. This hypothesis is supported by the fact that even in the contemporary ribosome, deacylated tRNA promotes peptide release but not as efficiently as release factors (Zavialov et al. 2002). Presumably, release factors, in particular their properly positioned GGQ motif, more

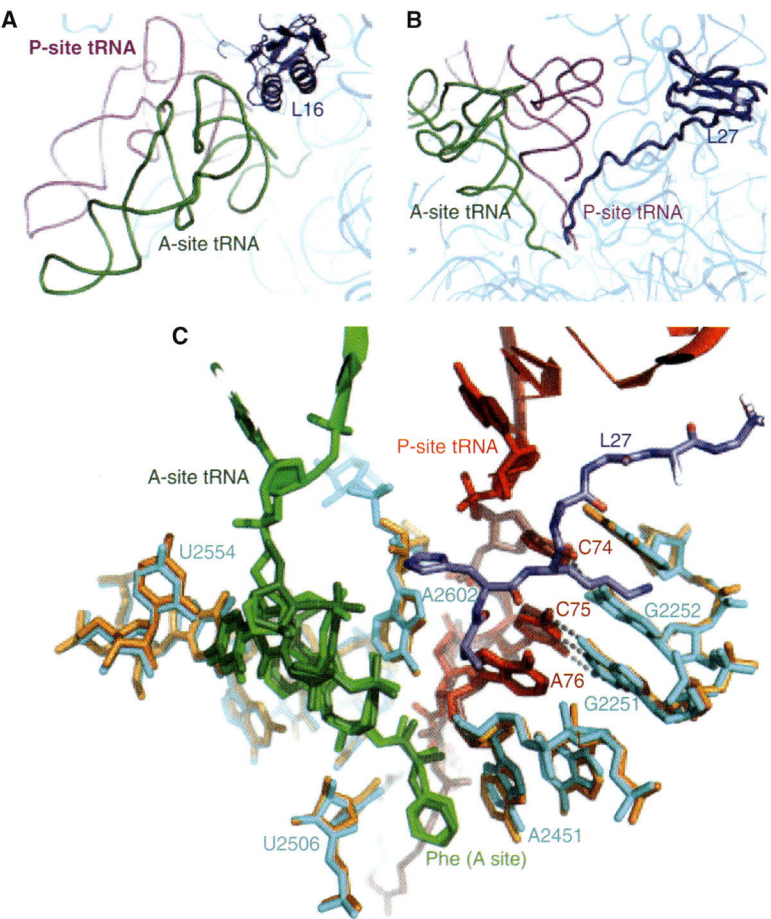

Figure 5. The peptidyl transferase center of the ribosome. (*A,B*) Location of two proteins that aid peptidyl transfer activity by stabilizing tRNA substrates. (*C*) Details of the bacterial PTC with A-site tRNA (green), P-site tRNA (red), and 23S RNA (cyan). A superposition with the archaeal peptidyl transferase center (1VQN, from Schmeing et al. 2005a) with the tRNAs in a slightly darker shade and the 23S RNA (orange) shows that bacterial and archaeal PTC are virtually identical. However, in bacteria, the amino-terminal tail of protein L27 is right at the PTC. (*A* and *B*, Reprinted, with permission, from Voorhees et al. 2009 [©Nature Publishing Group].)

optimally coordinate a water for hydrolysis of the nascent peptide chain. They may also be more efficient at stop codon discrimination, because they have a very low error rate without the proofreading present in normal decoding (Freistroffer et al. 2000). Nevertheless, the structural and biochemical data clearly suggest how a protein factor has taken over a role once likely performed by tRNA.

5 TRANSLOCATION

The sequential nature of protein synthesis requires that the ribosome be able to move relative to mRNA and tRNA after each round of addition of an amino acid to the growing protein chain. This process, translocation, is highly complex and involves large-scale movements that must result in the precise movement by one codon to preserve the reading frame.

The idea that all ribosomes have two subunits because they need to move relative to one another was proposed a long time ago (Bretscher 1968; Spirin 1968). One of these proposed that the tRNAs move first relative to one subunit and only then with respect to the other to generate hybrid states (Bretscher 1968), an idea that was borne out almost two decades later in a landmark experiment using chemical footprinting of rRNA (Moazed and Noller 1989). More recent cryoEM experiments have shown that the ribosomal subunits "ratchet" or rotate relative to one another during translocation (Frank and Agrawal 2000; Valle et al. 2003), and the formation of hybrid states is indeed directly related to this ratcheting movement (Ermolenko et al. 2007).

Strikingly, the interface between the two subunits consists mainly of RNA. This suggests that the features required to ratchet as part of translocation may have existed even in a primordial protein-free ribosome. The finding that factor-free translation, however inefficient, can occur under certain conditions is in keeping with this idea (Gavrilova and Spirin 1974).

5.1 Energy Stored in tRNA

If GTP hydrolysis by EF-G is not strictly required for translocation, what determines the directionality of the movement? The progression of tRNA from A to P to E sites involves a progression of changes in its chemical state, from aminoacyl to peptidyl to deacylated. As has been pointed out previously (Spirin 1985; Noller 2005), the affinity of the various sites has evolved so that changes in the chemical state of tRNA would allow it to progress to the next site on thermodynamic grounds alone. Moreover, Noller has pointed out that the energy from peptide bond formation could be used to drive the process even in the absence of GTP hydrolysis by translational factors (Noller 2005).

Interestingly, it has been observed that the P-site tRNA is distorted relative to free tRNA in solution (Selmer et al. 2006). If tRNA is allowed to relax, e.g., after peptide bond formation when it becomes deacylated, the direction of relaxation would be such as to move it toward the E site. Therefore, some of the energy required may be stored in the distortion in P-site tRNA itself.

5.2 E Site: Conservation and Role

A particular role of the E site in this process could be to trap the intermediate state of translocation by binding the 3′ end of the tRNA that has moved from the P site of the 50S subunit, resulting in a hybrid P/E tRNA. By stably trapping this intermediate, the E site would facilitate translocation.

A comparison of the structures of the 3′ end of tRNA in a bacterial 50S subunit (Korostelev et al. 2006; Selmer et al. 2006) with that in an archaeal 50S subunit (Schmeing et al. 2003) reveals some interesting similarities and differences. The E site can only bind a deacylated tRNA, and the terminal adenine with its 2′-OH is required (Lill et al. 1989; Feinberg and Joseph 2001). This requirement is consistent with its role in trapping deacylated tRNA. Interestingly, both bacterial and archaeal E sites bind the terminal adenine in exactly the same way (Fig. 6). In both cases, the adenine base is intercalated between two conserved purines of 23S RNA and makes identical contacts with a conserved cytidine. This strongly suggests that the E site evolved even before the split among the three kingdoms. Outside the vicinity of the terminal adenine, E-site interactions are quite different. This divergence suggests that those interactions are less essential.

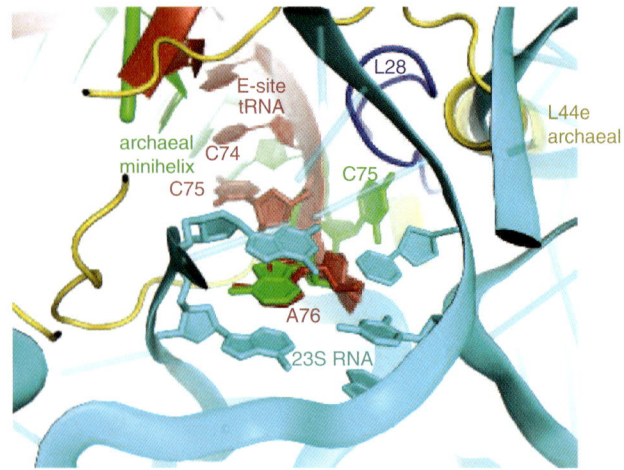

Figure 6. tRNA E site in the 50S subunit of the ribosome. A comparison of the bacterial E-site tRNA (reddish brown) (Selmer et al. 2006) with a minihelix representing the 3′ acceptor arm of tRNA in the archaeal ribosome (green) (Schmeing et al. 2003) shows that the terminal A76 is in an identical conformation making identical interactions with 23S RNA (cyan), suggesting that the E site evolved before the divergence of archaea and bacteria. However, the distinct conformations of C75 as well as the differences in the proteins (L28 in bacteria; L44e in archaea) suggest that other features of the E site have diverged significantly.

6 HOW DID THE RIBOSOME EVOLVE?

Recent structures of the ribosome have shown unambiguously that the essential functions of the ribosome such as decoding, peptidyl transfer, and translocation all appear to be mediated by RNA. The evolution of the ribosome has been much discussed (see, e.g., reviews by Moore 1993; Noller 2005), but there is little detailed understanding of how the process might have occurred. One interesting observation from the structure of the 50S subunit is that the PTC itself has a twofold symmetry that extends beyond the binding sites for A- and P-site tRNAs (Bashan et al. 2003). This suggests that an independently folded domain of RNA may have been duplicated to create the precursor of the PTC. It has been noticed that independent modules that are duplicated and held by tertiary contacts involving precisely the same type of minor-groove interactions as found in decoding could be the basis for the evolution of RNA in the contemporary 50S subunit (Bokov and Steinberg 2009). The evolution of the 30S subunit and coded synthesis involving tRNA and mRNA is even less well understood despite decades of speculation. What is clear is that although the contemporary ribosome appears to be a highly complex assembly of RNA and protein, and additionally involves many different protein factors, the high-resolution structures of the ribosome provide strong support for the idea that the essential functions of the ribosome are mediated by RNA and that the ribosome evolved from a primordial RNA world. In so doing, it appears to have been a Trojan horse that accelerated the transformation of that world into the protein world that we know today.

ACKNOWLEDGMENTS

Work in the author's laboratory is supported by the Medical Research Council (U.K.), the Wellcome Trust, the Agouron Institute, and the Louis-Jeantet Foundation. I thank T. Martin Schmeing for making Figures 1–4 in connection with other publications.

REFERENCES

Ban N, Nissen P, Hansen J, Moore PB, Steitz TA. 2000. The complete atomic structure of the large ribosomal subunit at 2.4 Å resolution. *Science* **289:** 905–920.

Bashan A, Agmon I, Zarivach R, Schluenzen F, Harms J, Berisio R, Bartels H, Franceschi F, Auerbach T, Hansen HA, et al. 2003. Structural basis of the ribosomal machinery for peptide bond formation, translocation, and nascent chain progression. *Mol Cell* **11:** 91–102.

Bokov K, Steinberg SV. 2009. A hierarchical model for evolution of 23S ribosomal RNA. *Nature* **457:** 977–980.

Bretscher MS. 1968. Translocation in protein synthesis: A hybrid structure model. *Nature* **218:** 675–677.

Crick FHC. 1966. Codon-anticodon pairing: The wobble hypothesis. *J Mol Biol* **19:** 548–555.

Crick FHC. 1968. The origin of the genetic code. *J Mol Biol* **38:** 367–379.

Ermolenko DN, Spiegel PC, Majumdar ZK, Hickerson RP, Clegg RM, Noller HF. 2007. The antibiotic viomycin traps the ribosome in an intermediate state of translocation. *Nat Struct Mol Biol* **14:** 493–497.

Feinberg JS, Joseph S. 2001. Identification of molecular interactions between P-site tRNA and the ribosome essential for translocation. *Proc Natl Acad Sci* **98:** 11120–11125.

Frank J, Agrawal RK. 2000. A ratchet-like inter-subunit reorganization of the ribosome during translocation. *Nature* **406:** 318–322.

Freistroffer DV, Kwiatkowski M, Buckingham RH, Ehrenberg M. 2000. The accuracy of codon recognition by polypeptide release factors. *Proc Natl Acad Sci* **97:** 2046–2051.

Gao Y-G, Selmer M, Dunham CM, Weixlbaumer A, Kelley AC, Ramakrishnan V. 2009. The structure of the ribosome with elongation factor G trapped in the post-translocational state. *Science* **326:** 694–699.

Gavrilova LP, Spirin AS. 1974. "Nonenzymatic" translation. *Methods Enzymol* **30:** 452–462.

Gavrilova LP, Kostiashkina OE, Koteliansky VE, Rutkevitch NM, Spirin AS. 1976. Factor-free ("non-enzymic") and factor-dependent systems of translation of polyuridylic acid by *Escherichia coli* ribosomes. *J Mol Biol* **101:** 537–552.

Guerrier-Takada C, Gardiner K, Marsh T, Pace N, Altman S. 1983. The RNA moiety of ribonuclease P is the catalytic subunit of the enzyme. *Cell* **35:** 849–857.

Kapp LD, Lorsch JR. 2004. The molecular mechanics of eukaryotic translation. *Annu Rev Biochem* **73:** 657–704.

Khaitovich P, Mankin AS, Green R, Lancaster L, Noller HF. 1999. Characterization of functionally active subribosomal particles from *Thermus aquaticus*. *Proc Natl Acad Sci* **96:** 85–90.

Kool ET. 2000. Synthetically modified DNAs as substrates for polymerases. *Curr Opin Chem Biol* **4:** 602–608.

Korostelev A, Trakhanov S, Laurberg M, Noller HF. 2006. Crystal structure of a 70S ribosome-tRNA complex reveals functional interactions and rearrangements. *Cell* **126:** 1065–1077.

Korostelev A, Asahara H, Lancaster L, Laurberg M, Hirschi A, Zhu J, Trakhanov S, Scott WG, Noller HF. 2008. Crystal structure of a translation termination complex formed with release factor RF2. *Proc Natl Acad Sci* **105:** 19684–19689.

Laurberg M, Asahara H, Korostelev A, Zhu J, Trakhanov S, Noller HF. 2008. Structural basis for translation termination on the 70S ribosome. *Nature* **454:** 852–857.

Lill R, Robertson JM, Wintermeyer W. 1989. Binding of the 3′ terminus of tRNA to 23S rRNA in the ribosomal exit site actively promotes translocation. *EMBO J* **8:** 3933–3938.

Maguire BA, Beniaminov AD, Ramu H, Mankin AS, Zimmermann RA. 2005. A protein component at the heart of an RNA machine: The importance of protein L27 for the function of the bacterial ribosome. *Mol Cell* **20:** 427–435.

Moazed D, Noller HF. 1989. Intermediate states in the movement of transfer RNA in the ribosome. *Nature* **342:** 142–148.

Moore PB. 1988. The ribosome returns. *Nature* **331:** 223–227.

Moore PB. 1993. Ribosomes and the RNA world. In *The RNA world* (ed. RF Gesteland, JF Atkins), pp. 119–136. Cold Spring Harbor Laboratory Press, Cold Spring Harbor, NY.

Moore VG, Atchison RE, Thomas G, Moran M, Noller HF. 1975. Identification of a ribosomal protein essential for peptidyl transferase activity. *Proc Natl Acad Sci* **72:** 844–848.

Nissen P, Hansen J, Ban N, Moore PB, Steitz TA. 2000. The structural basis of ribosome activity in peptide bond synthesis. *Science* **289:** 920–930.

Nitta I, Ueda T, Watanabe K. 1998. Possible involvement of *Escherichia coli* 23S ribosomal RNA in peptide bond formation. *RNA* **4:** 257–267.

Nitta I, Kamada Y, Noda H, Ueda T, Watanabe K. 1999. Peptide bond formation: Retraction. *Science* **283:** 2019–2020.

Noller HF. 2005. Evolution of ribosomes and translation from an RNA world. In *The RNA world*, 3rd ed. (ed. RF Gesteland et al.), pp. 287–307. Cold Spring Harbor Laboratory Press, Cold Spring Harbor, NY.

Noller HF, Chaires JB. 1972. Functional modification of 16S ribosomal RNA by kethoxal. *Proc Natl Acad Sci* **69:** 3115–3118.

Noller HF, Hoffarth V, Zimniak L. 1992. Unusual resistance of peptidyl transferase to protein extraction procedures. *Science* **256:** 1416–1419.

Ogle JM, Ramakrishnan V. 2005. Structural insights into translational fidelity. *Annu Rev Biochem* **74:** 129–177.

Ogle JM, Brodersen DE, Clemons WMJr, Tarry MJ, Carter AP, Ramakrishnan V. 2001. Recognition of cognate transfer RNA by the 30S ribosomal subunit. *Science* **292:** 897–902.

Ogle JM, Murphy FV, Tarry MJ, Ramakrishnan V. 2002. Selection of tRNA by the ribosome requires a transition from an open to a closed form. *Cell* **111:** 721–732.

Rodnina MV, Wintermeyer W. 2001. Fidelity of aminoacyl-tRNA selection on the ribosome: Kinetic and structural mechanisms. *Annu Rev Biochem* **70:** 415–435.

Rodnina MV, Beringer M, Wintermeyer W. 2006. Mechanism of peptide bond formation on the ribosome. *Q Rev Biophys* **39:** 203–225.

Schmeing TM, Ramakrishnan V. 2009. What recent ribosome structures have revealed about the mechanism of translation. *Nature* **461:** 1234–1242.

Schmeing TM, Moore PB, Steitz TA. 2003. Structures of deacylated tRNA mimics bound to the E site of the large ribosomal subunit. *RNA* **9:** 1345–1352.

Schmeing TM, Huang KS, Kitchen DE, Strobel SA, Steitz TA. 2005a. Structural insights into the roles of water and the 2′ hydroxyl of the P site tRNA in the peptidyl transferase reaction. *Mol Cell* **20:** 437–448.

Schmeing TM, Huang KS, Strobel SA, Steitz TA. 2005b. An induced-fit mechanism to promote peptide bond formation and exclude hydrolysis of peptidyl-tRNA. *Nature* **438:** 520–524.

Schmeing TM, Voorhees RM, Kelley AC, Gao Y-G, Murphy FV IV Weir JR, Ramakrishnan V. 2009. The crystal structure of the ribosome bound to EF-Tu and aminoacyl-tRNA. *Science* **326:** 688–694.

Schuette JC, Murphy FV IV, Kelley AC, Weir JR, Giesebrecht J, Connell SR, Loerke J, Mielke T, Zhang W, Penczek PA, et al. 2009. GTPase activation of elongation factor EF-Tu by the ribosome during decoding. *EMBO J* **28:** 755–765.

Schuwirth BS, Borovinskaya MA, Hau CW, Zhang W, Vila-Sanjurjo A, Holton JM, Cate JH. 2005. Structures of the bacterial ribosome at 3.5 Å resolution. *Science* **310:** 827–834.

Selmer M, Dunham CM, Murphy FV IV Weixlbaumer A, Petry S, Kelley AC, Weir JR, Ramakrishnan V. 2006. Structure of the 70S ribosome complexed with mRNA and tRNA. *Science* **313:** 1935–1942.

Sharma MR, Koc EC, Datta PP, Booth TM, Spremulli LL, Agrawal RK. 2003. Structure of the mammalian mitochondrial ribosome reveals an expanded functional role for its component proteins. *Cell* **115:** 97–108.

Sievers A, Beringer M, Rodnina MV Wolfenden R. 2004. The ribosome as an entropy trap. *Proc Natl Acad Sci* **101:** 7897–7901.

Spirin AS. 1968. How does the ribosome work? A hypothesis based on the two subunit construction of the ribosome. *Curr Mod Biol* **2:** 115–127.

Spirin AS. 1985. Ribosomal translocation: Facts and models. *Prog Nucleic Acid Res Mol Biol* **32:** 75–114.

Valle M, Sengupta J, Swami NK, Grassucci RA, Burkhardt N, Nierhaus KH, Agrawal RK, Frank J. 2002. Cryo-EM reveals an active role for aminoacyl-tRNA in the accommodation process. *EMBO J* **21:** 3557–3567.

Valle M, Zavialov A, Sengupta J, Rawat U, Ehrenberg M, Frank J. 2003. Locking and unlocking of ribosomal motions. *Cell* **114:** 123–134.

Villa E, Sengupta J, Trabuco LG, LeBarron J, Baxter WT, Shaikh TR, Grassucci RA, Nissen P, Ehrenberg M, Schulten K, Frank J. 2009. Ribosome-induced changes in elongation factor Tu conformation control GTP hydrolysis. *Proc Natl Acad Sci* **106:** 1063–1068.

Voorhees RM, Weixlbaumer A, Loakes D, Kelley AC, Ramakrishnan V. 2009. Insights into substrate stabilization from snapshots of the peptidyl transferase center of the intact 70S ribosome. *Nat Struct Mol Biol* **16:** 528–533.

Weinger JS, Parnell KM, Dorner S, Green R, Strobel SA. 2004. Substrate-assisted catalysis of peptide bond formation by the ribosome. *Nat Struct Mol Biol* **11:** 1101–1106.

Weixlbaumer A, Jin H, Neubauer C, Voorhees RM, Petry S, Kelley AC, Ramakrishnan V. 2008. Insights into translational termination from the structure of RF2 bound to the ribosome. *Science* **322:** 953–956.

Wimberly BT, Brodersen DE, Clemons WM Jr, Morgan-Warren RJ, Carter AP, Vonrhein C, Hartsch T, Ramakrishnan V. 2000. Structure of the 30S ribosomal subunit. *Nature* **407:** 327–339.

Wower IK, Wower J, Zimmermann RA. 1998. Ribosomal protein L27 participates in both 50 S subunit assembly and the peptidyl transferase reaction. *J Biol Chem* **273:** 19847–19852.

Yusupov MM, Yusupova GZ, Baucom A, Lieberman K, Earnest TN, Cate JH, Noller HF. 2001. Crystal structure of the ribosome at 5.5 Å resolution. *Science* **292:** 883–896.

Zaug AJ, Grabowski PJ, Cech TR. 1983. Autocatalytic cyclization of an excised intervening sequence RNA is a cleavage-ligation reaction. *Nature* **301:** 578–583.

Zavialov AV, Mora L, Buckingham RH, Ehrenberg M. 2002. Release of peptide promoted by the GGQ motif of class 1 release factors regulates the GTPase activity of RF3. *Mol Cell* **10:** 789–798.

Noncoding RNPs of Viral Origin

Joan Steitz, Sumit Borah, Demian Cazalla, Victor Fok, Robin Lytle, Rachel Mitton-Fry, Kasandra Riley, and Tasleem Samji

Department of Molecular Biophysics and Biochemistry, Howard Hughes Medical Institute, Yale University School of Medicine, New Haven, Connecticut 06536-0812

Correspondence: joan.steitz@yale.edu

SUMMARY

Like their host cells, many viruses produce noncoding (nc)RNAs. These show diversity with respect to time of expression during viral infection, length and structure, protein-binding partners and relative abundance compared with their host-cell counterparts. Viruses, with their limited genomic capacity, presumably evolve or acquire ncRNAs only if they selectively enhance the viral life cycle or assist the virus in combating the host's response to infection. Despite much effort, identifying the functions of viral ncRNAs has been extremely challenging. Recent technical advances and enhanced understanding of host-cell ncRNAs promise accelerated insights into the RNA warfare mounted by this fascinating class of RNPs.

Outline

1. VA RNAs: Multifunctional manipulators of adenovirus host-cell functions
2. EBERs: Abundant but enigmatic contributors to EBV latency
3. HSURs: Managing host microRNA functions during HVS latency
4. PAN RNA: A nuclear sink during the KSHV lytic cycle?
5. Viral microRNAs: Altering host-cell or fine-tuning viral gene expression?
6. Noncoding RNAs of other flavors
7. Prospects

References

Viruses are the ultimate parasites, capable of redirecting a cell's gene expression apparatus to their own ends. Takeover can occur during both lytic growth and latency, whereby the virus retreats into the infected cell in a state that is dormant but can be re-activated to replicate in some distant cellular descendant. Some—but not all—viruses encode noncoding (nc) RNAs that become viral ribonucleoprotein (RNP) complexes, usually by assembling with host rather than viral RNA-binding proteins. These polypeptides often are well-characterized components of cellular RNPs but can appear in unexpected subcellular locations or function in novel ways when recruited into a viral RNP. Because RNAs are less immunogenic than proteins, a viral ncRNP has the potential to slip under the radar of the host's immune system. Thus, viral ncRNAs provide unique perspectives into cellular function, illustrating the many devious ways in which host-cell metabolism can be manipulated.

The viruses that harbor the best-characterized ncRNAs have double-stranded DNA genomes, infect vertebrates and belong to the adenovirus, herpesvirus, or polyomavirus classes. The herpesviruses exist in two states—latency, sometimes accompanied by cellular transformation, and lytic infection, which can be induced from the latent state. Assigning functions to cellular ncRNAs is not straightforward (in contrast to mRNAs), and viral ncRNAs present even more challenging problems. Here, we review current insights into the roles of viral ncRNPs (Tables 1 and 2) and how they relate to the viral takeover agenda.

1 VA RNAs: MULTIFUNCTIONAL MANIPULATORS OF ADENOVIRUS HOST-CELL FUNCTIONS

Adenoviruses infect a wide range of vertebrate hosts, and more than 50 human serotypes have been identified. Human adenoviruses cause a variety of infections including respiratory illness, conjunctivitis (pinkeye), gastrointestinal infections, and urinary tract infections. The study of adenovirus has led to many important discoveries, including pre-mRNA splicing. Adenovirus was the first human virus shown to be oncogenic (although it has never been found in human cancers). Compromised adenoviruses are also used as delivery agents for gene therapy.

The adenovirus VA (virus-associated) RNAs are not only the first viral ncRNAs to be identified (Reich et al. 1966) but also the best understood in molecular terms. VAI ($\sim$160 nt) is a structured RNA transcribed by RNA polymerase III (Pol III), which accumulates up to 10^8 copies/cell in late-stage infection of HeLa cells (about the same level as ribosomes) (Fig. 1A)! VAI RNA is required for high titer viral growth and efficient viral mRNA translation (Thimmappaya et al. 1982). It is localized in the cytoplasm and has long been known to combat the interferon response by formation of a stable RNP with cellular PKR (protein kinase R) (Mathews and Shenk 1991). PKR synthesis is induced by interferon, a cell-signaling molecule produced by the host in response to infection; PKR's kinase activity is turned on by subsequent binding of double-stranded RNA (arising from transcription of the viral genome). Activated PKR phosphorylates translation initiation factor eIF2, leading to inhibition of protein synthesis. However, in adenovirus-infected cells, VAI RNA, with its high abundance and double-stranded upper stem, avidly binds PKR (Fig. 1A) in a way that prevents PKR activation and thus enables the virus to elude this aspect of the cellular antiviral response.

Further research has implicated VAI RNA in manipulating other facets of host-cell function, such as the microRNA (miRNA) and RNA interference (RNAi) pathways. VAI has been shown to competitively inhibit export of miRNA precursors from the nucleus by exportin-5 (Lu and Cullen 2004). Moreover, VAI RNA, as well as the structurally similar VAII RNA (Fig. 1A) encoded by many adenovirus strains, competitively inhibits Dicer activity (see Sharp 2010) (Lu and Cullen 2004; Andersson et al. 2005). Binding of Dicer to the VA RNA terminal stem (Fig. 1A) yields cleavage products that are the size of miRNAs ($\sim$22 nt), albeit very inefficiently (Andersson et al. 2005; Aparicio et al. 2006; Sano et al. 2006). But because of the high abundance of the VA transcripts, the majority of RISCs (RNA-induced silencing complexes) in late-stage infected cells become loaded with VA RNA-derived sequences (Xu et al. 2007). No viral or cellular targets have yet been experimentally shown for the putative adenovirus miRNAs, leaving the in vivo relevance of these observations as yet unresolved.

2 EBERS: ABUNDANT BUT ENIGMATIC CONTRIBUTORS TO EBV LATENCY

Epstein-Barr virus (EBV), a γ herpesvirus, is detected in more than 90% of the human population and persists in most individuals as a lifelong, asymptomatic infection of B lymphocytes. Unfortunately for some, it is also the causative agent of infectious mononucleosis and is associated with malignancies such as Burkitt's lymphoma and nasopharyngeal carcinoma.

The two most abundant viral transcripts in EBV-infected cells are the EBV-encoded RNAs, EBERs1 and 2 (Lerner et al. 1981a) (Fig. 1B). These ncRNAs are so numerous, at approximately 10^6 copies each per infected B cell, that their presence is used in the clinic as diagnostic for EBV's presence (Gulley and Tang 2008). They are also among the few viral gene products expressed in all types of EBV latency (Kieff and Rickinson 2007). EBER1 (167 nts) and EBER2 (172 nts) are highly conserved among

Table 1. Viral ncRNAs excluding miRNAs

Viral Family (subgroup)	Virus	RNA	Abundance (copies/cell)	Length (nt)	RNA Polymerase	Bound Proteins	References
Adenoviridae	Human Adenovirus	VAI	10^8	~160	III	La, PKR, Dicer, Ago2	[1–5]
		VAII	5×10^6	~160	III	La, Dicer, Ago2	[1, 4, 5]
Herpesviridae (α herpesvirus)	HSV-1	LAT	?	~6300	II	?	[6]
		sRNA1	?	65	?	?	[7]
		sRNA2	?	36	?	?	[7]
Herpesviridae (β herpesvirus)	HCMV	β2.7	?	2700	II	GRIM-19	[8–10]
Herpesviridae (γ herpesvirus)	EBV/HHV4	EBER1	5×10^6	167	III	La, L22, (PKR)*	[11–15]
		EBER2	5×10^6	172	III	La, nucleolin[†], (PKR)*	[11,12,14]
		v-snoRNA1	?	65	?	Fibrillarin, Nop56, Nop58	[16]
		v-snoRNA1^{24pp}	?	24	?	?	[16]
	HVS	HSUR1,2,5	10^3–10^4	114–143	II	Sm (HuR, hnRNP D)[‡]	[17–19]
		HSUR3,4,6,7	10^3–10^4	75–106	II	Sm	[18]
	KSHV/HHV8	PAN	5×10^5	1060[#]	II	hnRNP C1; PABPC1	[20, 21] Borah et al., unpub.
	MHV68	tRNA1,2,3,4,5,6,7	?	72–84	III	?	[22]
Tombusviridae	RCNMV	SR1f	?	400	?	?	[23]
Flaviviridae	WNV	sgRNA	?	300–500	?	?	[24]

*In vivo binding has not been shown.
[†]Victor Fok unpublished data.
[‡]Shown in vivo only for HSUR1 (Cook et al. 2004).
[#]This value excludes the poly(A) tail. HSV (herpes simplex virus), HCMV (human cytomegalavirus), EBV (Epstein-Barr virus), HHV (human herpes virus), HVS (Herpesvirus saimiri), KSHV (Kaposi's sarcoma-associated herpesvirus), MHV (murine herpesvirus), RCNMV (Red clover necrotic mosaic virus), WNV (West Nile virus).

1. Lerner et al. 1981b; 2. Mathews and Shenk 1991; 3. Lu and Cullen 2004; 4. Andersson et al. 2005; 5. Aparicio et al. 2006; 6. Stevens et al. 1987; 7. Shen et al. 2009; 8. Greenaway and Wilkinson 1987; 9. Reeves et al. 2007; 10. Spector 1996; 11. Lerner et al. 1981a; 12. Toczyski et al. 1994; 13. Clarke et al. 1990; 14. Sharp et al. 1993; 15. Fok et al. 2006a; 16. Hutzinger et al. 2009; 17. Cook et al. 2004; 18. Lee et al. 1988; 19. Myer et al. 1992; 20. Sun et al. 1996; 21. Conrad 2008; 22. Bowden et al. 1997; 23. Iwakawa et al. 2008; 24. Pijlman et al. 2008.

Table 2. Validated viral microRNAs

Viral Family (subgroup)	Virus	pre-miRNAs	References
Herpesviridae (α herpesvirus)	GHV1/ILTV	7	[1]
	GHV2/MDV type 1	14	[1–7]
	GHV3/MDV type 2	18	[1–7]
	Meleagrid (turkey) HVT	17	[1, 8]
	HSV1/HHV1	5+2*	[9–13]
	HSV2/HHV2	3	[14, 15]
	B virus/macacine HV1	3	[16]
Herpesviridae (β herpesvirus)	HCMV/HHV5	11	[17–21]
	MCMV	12	[22–24]
Herpesviridae (γ herpesvirus)	EBV/HHV4	25	[25–28]
	rLCV	35	[26, 42, 43]
	MHV-68	9	[18]
	KSHV/HHV8	13	[18, 26, 29, 30]
	rRV	7	[31]
Polyomaviridae	BK virus	1	[32]
	JC virus	1	[32]
	murine polyomavirus	1	[23]
	SV40	1	[33]
	merkel cell polyomavirus	1	[34]
Adenoviridae	human adenovirus	1*	[35–37]
Retroviridae	HIV1	3	[38–40]
Ascoviridae	HvAV	1*	[41]

Numbers of miRNAs reflect validated precursor miRNAs recorded in miRBASE as of October 2009. GHV (Gallid herpesvirus), ILTV (infectious laryngotracheitis virus), MDV (Marek's Disease virus), HVT (herpesvirus of turkeys), HSV (herpes simplex virus), HHV (human herpes virus), HCMV (human cytomegalovirus), MCMV (murine cytomegalovirus), EBV (Epstein-Barr virus), rLCV (rhesus lymphocryptovirus), MHV-68 (murine herpesvirus 68), KSHV (Kaposi's sarcoma-associated herpesvirus), rRV (rhesus rhadinovirus), SV40 (simian virus 40), HIV1 (human immunodeficiency type virus 1), HvAV (Heliothis virescens ascovirus).

*Indicates no miRBASE record.

1. Waidner et al. 2009; 2. Burnside et al. 2006; 3. Yao et al. 2007; 4. Morgan et al. 2008; 5. Xu et al. 2008; 6. Yao et al. 2008; 7. Burnside et al. 2008; 8. Yao et al. 2009; 9. Cui et al. 2006; 10. Peng et al. 2008; 11. Umbach et al. 2008; 12. Wu et al. 2009; 13. Umbach and Cullen 2009; 14. Tang et al. 2008; 15. Tang et al. 2009; 16. Besecker et al. 2009; 17. Dunn et al. 2005; 18. Pfeffer et al. 2005; 19. Fannin Rider et al. 2008; 20. Grey and Nelson 2008; 21. Dolken et al. 2009; 22. Dolken et al. 2007; 23. Sullivan et al. 2009; 24. Buck et al. 2007; 25. Pfeffer et al. 2004; 26. Cai et al. 2006b; 27. Grundhoff et al. 2006; 28. Zhu et al. 2009; 29. Cai and Cullen 2006; 30. Samols et al. 2005; 31. Schafer et al. 2007; 32. Seo et al. 2008; 33. Sullivan et al. 2005; 34. Seo et al. 2009; 35. Andersson et al. 2005; 36. Aparicio et al. 2006; 37. Sano et al. 2006; 38. Lin and Cullen 2007; 39. Ouellet et al. 2008; 40. Omoto and Fujii 2006; 41. Hussain et al. 2008; 42. Walz et al. 2010; 43. Riley et al. 2010.

the various EBV strains that infect primates (Arrand et al. 1989). They reside in and remain confined to the nucleoplasm, without shuttling between the nucleus and the cytoplasm (Howe and Steitz 1986; Fok et al. 2006b). The EBERs are transcribed by RNA Pol III and thus possess a tract of uridylate residues at their 3′ ends. Although EBERs1 and 2 are not similar in sequence, both show a high degree of secondary structure, with multiple stem-loop structures that serve as protein-binding sites (see Fig. 1B). The abundance and conservation of the EBERs strongly argue that they are important for EBV infection and latency. Yet, cells infected with mutant EBV lacking the EBER genes are indistinguishable from those infected with wild-type EBV in their growth rate, gene expression or sensitivity to interferon (Swaminathan et al. 1991; Swaminathan et al. 1992).

The earliest notions about the molecular function of the EBERs were fueled by the finding that EBERs appear to substitute for the VA RNAs in rescuing adenovirus-infected cells from the shutdown of protein synthesis caused by activated PKR (Bhat and Thimmappaya 1983; Bhat and Thimmappaya 1985). In vitro, EBERs can bind to PKR directly and inhibit its activity (Clarke et al. 1990; Sharp et al. 1993). However, the distinct subcellular locations of the EBERs and PKR, in the nucleus (Howe and Steitz 1986; Fok et al. 2006b) and cytoplasm (Takizawa et al. 2000), respectively, as well as other data, raise doubts as to whether the EBERs actually modulate PKR activity in infected cells.

Besides PKR, several other proteins are known to interact and form RNP complexes with the EBERs. The La protein, which assists in the correct folding of RNA Pol III transcripts (Wolin and Cedervall 2002), stably binds the 3′-U tails of both EBERs (Lerner et al. 1981a). But because the cellular abundance of La exceeds even that of EBERs, sequestration of this protein is unlikely to be the sole function of EBERs.

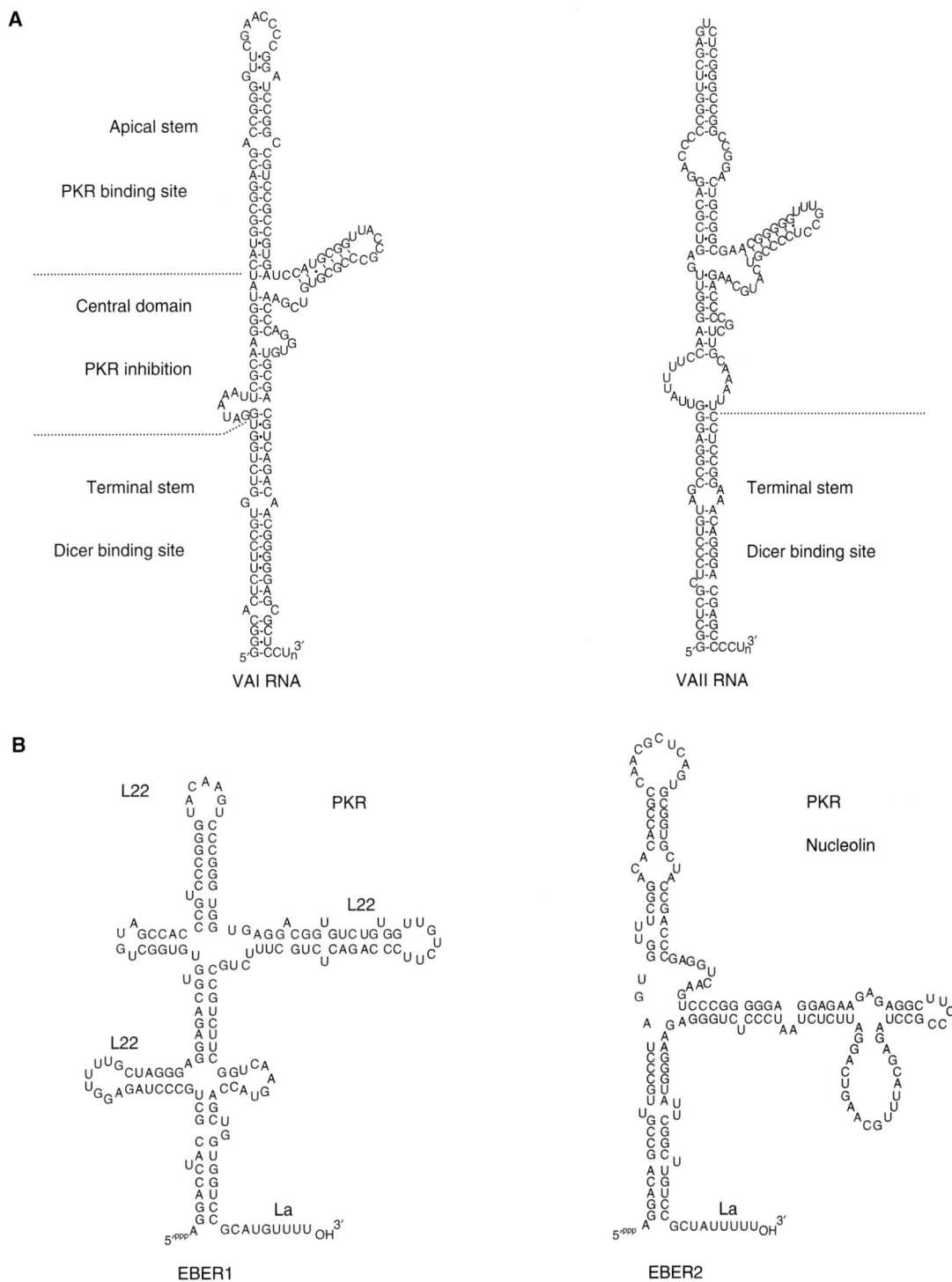

Figure 1. (A) VAI and II RNAs. PKR and Dicer interact with nonoverlapping sites on VAI RNA. The apical stem and central domain of VAI RNA bind to and inhibit PKR, respectively, whereas the terminal stems of VAI and VAII RNA interact with Dicer. Cleavage of VAI to produce putative miRNAs has been proposed to leave an upper fragment that remains a potent inhibitor of PKR activity (Wahid et al. 2008). Although there is variation in sequence and length of VA RNAs from different viral strains (RNAs shown are from the much-studied adenovirus type 2 (Ad2)), their general secondary structure is conserved (Ma and Mathews 1996). (B) EBERs1 and 2. EBERs1 and 2 share a high degree of secondary structure and are highly conserved among viruses related to EBV. L22 has been shown to bind to three of EBER1's stem-loops (Fok et al. 2006a), whereas the La protein binds the polyU tract at the 3' end of the EBERs (Lerner et al. 1981a). The exact sites of binding of PKR to the EBERs or nucleolin to EBER2 have not been determined.

EBER1 also forms an RNP with ribosomal protein L22; indeed, three molecules of ribosomal protein L22 can bind to EBER1 through three of its stem-loops (Fok et al. 2006a) (Fig. 1B). The binding of L22 to EBER1 causes L22 to appear in the nucleoplasm, as well as its normal location in the nucleolus and cytoplasm (Toczyski et al. 1994). Yet, sequestration of L22 may not have much of an effect on the host cell because mice null for the L22 gene have normal B-cell function (Anderson et al. 2007).

A second binding partner for EBER2 is nucleolin (V Fok, unpubl.), an abundant, multifunctional RNA-binding protein involved in diverse cellular processes (Mongelard and Bouvet 2007). This association is observed only after formaldehyde crosslinking of cells, so it may be weak or transient, but nucleolin binding is specific for EBER2 and—as it is also observed after in vivo UV crosslinking—is a direct RNA-protein contact in EBV-containing cells. In contrast to L22, nucleolin does not appear to become relocalized from nucleoli to the nucleoplasm in the presence of EBERs.

Although a molecular mechanism is lacking, EBERs have been reported to be involved in the up-regulation of a variety of cytokines and growth factors and in the modulation of programmed cell death in EBV-transformed cells. EBERs1 and 2 promote the expression of interleukin-10 (IL-10) in B lymphocytes (Kitagawa et al. 2000), IL-9 in T cells (Yang et al. 2004), and insulin-like growth factor 1 (IGFI) in epithelial cells (Iwakiri et al. 2003; Iwakiri et al. 2005). In addition, expression of EBER2, but not EBER1, leads to the up-regulation of IL-6 in lymphoblastoid cell lines (Wu et al. 2007). Expression of the EBERs in EBV-negative Akata cells, a Burkitt's lymphoma cell line, restored resistance to spontaneous apoptosis (Komano et al. 1999). Similarly, the EBERs inhibit apoptosis induced by α-interferon (Nanbo et al. 2002; Ruf et al. 2005).

The EBERs play a role in oncogenesis as well. Akata cells that spontaneously lose their EBV genome lose their tumorigenic potential (Shimizu et al. 1994). Reintroduction of EBERs alone into these EBV-negative cells restores tumor induction in immunodeficient SCID mice (Ruf et al. 2000). EBERs enhance the immortalization of Akata cells and contribute to their growth rate (Yajima et al. 2005). Also, expressing EBERs1 and 2 in EBV-negative Akata cells increases the size and number of colonies in a soft agar growth assay (Houmani et al. 2009). EBER1's association with L22 is important because EBV-negative Akata cells expressing wild-type EBER2 and an EBER1 in which the three stem-loops responsible for L22 binding had been mutated produced virtually no colonies in the soft agar assay (Houmani et al. 2009). In contrast, EBV from which EBER1 had been deleted was as efficient at transforming cord blood lymphocytes as EBV containing both EBERs; in this situation, EBER2 alone appeared to be responsible for efficient B-cell growth transformation (Wu et al. 2007). All these tantalizing phenotypes suggest a well-calculated alteration of host gene expression mechanistically linked to the EBERs, but the molecular details remain unclear.

3 HSURs: MANAGING HOST microRNA FUNCTIONS DURING HVS LATENCY

Herpesvirus saimiri (HVS) is an oncogenic γ herpesvirus that targets the T cells of New World monkeys. Infection with HVS results in either lytic infection or malignant transformation, resulting in aggressive leukemias and lymphomas.

HVS encodes seven ncRNAs, called Herpesvirus saimiri U RNAs (HSURs), that range in size from 75 to 143 nts and are the most abundantly expressed transcripts in latently infected cells (Murthy et al. 1986; Lee et al. 1988; Wassarman et al. 1989; Lee and Steitz 1990; Albrecht and Fleckenstein 1992). Soon after their discovery, HSURs were classified as Sm-class RNAs (Lee et al. 1988) (for more about Sm-class RNAs, see Luhrmann 2010). Besides binding to Sm proteins, HSURs share other structural similarities with cellular Sm snRNPs, including a hypermethylated 5′ cap and a terminal 3′ stem-loop, but they show no significant sequence similarity to any known cellular snRNA. HSURs are well conserved among different HVS subgroups, with HSURs1 and 2 (Fig. 2A) being the most highly conserved and the only snRNAs expressed in the closely related *Herpesvirus ateles* (Albrecht 2000). This conservation of HSURs1 and 2 suggests that they are important to the virus and do not require the presence of the other HSURs to function.

HSURs1 and 2 contain highly conserved sequences at their 5′ ends that mimic AU-rich elements (AREs) (Lee et al. 1988; Fan et al. 1997). AREs are typically found in the 3′-untranslated region (3′UTR) of cellular mRNAs, where they usually confer instability by inducing rapid and tightly regulated mRNA decay (Garneau et al. 2007). Like their cellular counterparts, the AREs present in HSURs bind ARE-binding proteins (Myer et al. 1992; Cook et al. 2004) and, in the case of HSUR1, regulate its stability (Fan et al. 1997). Nonetheless, the association of HSURs1 and 2 with host ARE-binding proteins does not alter the abundance of host mRNAs containing AREs in HVS-transformed cells (Cook et al. 2004). Instead, up-regulation of a handful of proteins that are hallmarks of T-cell activation was observed in cells transformed by the wild-type virus compared with those lacking HSURs1 and 2 (Cook et al. 2005), establishing a phenotype for these viral ncRNAs.

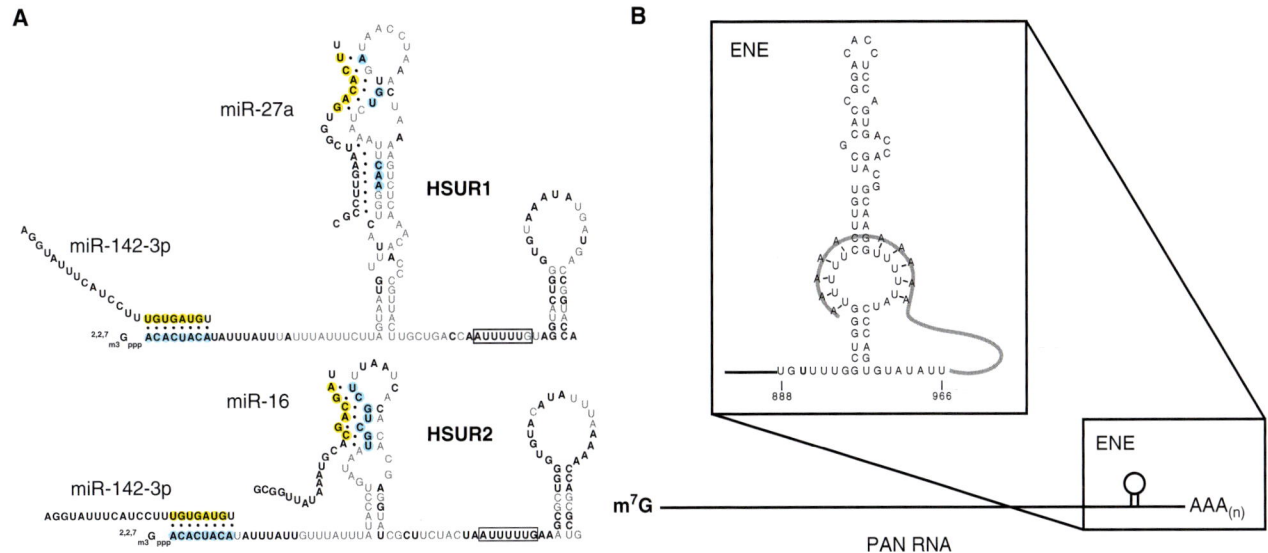

Figure 2. (A) HSURs1 and 2. Predicted base-pairing interactions with miR-16, miR-27a and miR-142-3p are shown. Sm binding sites are boxed. Gray and black residues in the HSURs indicate nonconserved and perfectly conserved nucleotides, respectively. Residues in blue indicate perfectly conserved nucleotides involved in interactions with miRNAs. Yellow residues in the miRNAs indicate seed sequences. (B) PAN RNA and the ENE. Putative base-pairing interactions between the internal U-rich loop of the ENE and the poly(A) tail are shown.

In addition to their AREs, HSURs1 and 2 possess other perfectly conserved sequences that resemble elements found in the 3′UTRs of cellular mRNAs, namely binding sites for host miRNAs. (See Sharp 2010 for more on miRNAs.) HSURs1 and 2 both contain two such sites: their 5′-end sequences are complementary to miR-142-3p; downstream of the ARE sequence, HSUR1 contains a binding site for a second miRNA, miR-27, whereas HSUR2 contains a binding site for the miR-16 family of miRNAs (see Fig. 2A). Strikingly, for all three families of miRNAs, complementarity to HSURs involves the critical seed region of the miRNA (Lewis et al. 2005). Although the binding of miR-142-3p and miR-16 to HSURs1 and 2 does not appear to affect the expression of these miRNAs in latently infected cells, the interaction between HSUR1 and miR-27 results in the degradation of mature miR-27 (D Cazalla, unpubl.). Thus, by expressing HSUR1, HVS selectively lowers the abundance of miR-27, consequently altering the expression of genes that are natural targets of miR-27 in transformed T cells. Interestingly, miR-27 targets RUNX1 and FOXO1 (Ben-Ami et al. 2009; Feng et al. 2009; Guttila and White 2009), transcription factors with roles in tumorigenesis (Blyth et al. 2005; Maiese et al. 2008); their expression is up-regulated in virally transformed cells. This is the first example of a virus using the strategy of miRNA down-regulation to control host gene expression and establishes a function for HSUR1.

The observation that HSURs1 and 2 bind miRNAs in latently infected cells opens new avenues in the search for HSUR functions. Why do HSURs mimic cellular mRNA 3′UTRs in binding miRNAs and ARE-binding proteins? Even though HSURs1 and 2 do not perturb the cellular population of ARE-containing mRNAs (Cook et al. 2004), they may effectively compete with the natural targets of the miR-142-3p and miR-16 families of miRNAs by binding to their seed regions. An alternative possibility is that the 5′ ends of HSURs1 and 2 provide a platform for deposition of miRNAs and ARE-binding proteins onto specific mRNAs through sequence-specific interactions involving other regions of these HSURs, thereby affecting mRNA translation. Further novel functions seem likely to be identified in studies of this unusual class of viral ncRNAs.

4 PAN RNA: A NUCLEAR SINK DURING THE KSHV LYTIC CYCLE?

Kaposi's sarcoma-associated herpesvirus (KSHV) is a γ herpesvirus that is the causative agent of several human cancers and lymphoproliferative disorders, including Kaposi's sarcoma (KS), Multicentric Castleman's Disease (MCD) and Primary Effusion Lymphoma (PEL) (Ganem 2006). The virus infects primarily endothelial cells, in the case of KS, and B cells, in the cases of MCD and PEL (Chang et al. 1994; Cesarman et al. 1995; Soulier et al.

1995; Dupin et al. 1999; Bechtel et al. 2003). Infection with KSHV is particularly serious in patients with compromised or suppressed immune systems, such as AIDS or transplant patients.

KSHV establishes latent infection in the majority of infected cells, and it is unknown which cues stimulate entry of the virus into lytic phase in vivo (Jenner and Boshoff 2002). However, cell cultures harboring the virus can be induced by treatment with sodium butyrate, trichostatin A or valproic acid, which are thought to lead to nonspecific activation of transcription via modification of histones (Countryman et al. 2008). Expression of the viral Orf50 protein is necessary and sufficient to stimulate lytic reactivation, leading it to be referred to as the "master transactivator" switch (Sun et al. 1998).

KSHV expresses an unusual ncRNA called polyadenylated nuclear (PAN) RNA (Sun et al. 1996; Zhong et al. 1996; Zhong and Ganem 1997) on entry into the lytic phase. PAN RNA resembles an mRNA in that it is transcribed by Pol II, is capped at its 5′ end, and ends with a 3′ polyadenylate tail (Fig. 2B). However, unlike mRNAs, PAN RNA is neither spliced nor exported to the cytoplasm. Instead, it remains in the nucleus and accumulates to remarkably high levels, upward of 500,000 copies per lytically infected cell. Ultimately, PAN RNA can account for 80% of the total polyadenylated RNA in the cell and thus greatly outnumbers all other human or viral mRNAs (Sun et al. 1996).

Why does the virus express so many copies of this one transcript? Moreover, why does PAN RNA look like an mRNA, but not behave like one? Unfortunately, answers to these questions remain elusive, in part because of technical hurdles. A direct approach would be to create a mutant version of the virus incapable of expressing PAN RNA: a PAN "knockout virus." Such a virus could then be compared with wild-type KSHV to discern what contribution PAN RNA makes to the lytic phase. This approach, for instance, was useful for studying the HVS HSURs (see earlier discussion) (Cook et al. 2005). Making a PAN knockout KSHV is problematic because the region of the viral genome from which PAN is transcribed overlaps another viral gene called K7 (Feng et al. 2002; Wang et al. 2002), which functions as an anti-apoptotic factor. Any deletion of the PAN region therefore also disrupts the K7 gene, complicating the interpretation of a potential PAN RNA knockout phenotype. An alternative is to use antisense approaches to lower the abundance of PAN RNA during the KSHV lytic phase. Although these experiments are not straightforward because of the nuclear localization of PAN RNA, as well as its extremely high levels, new methodologies (Ideue et al. 2009) to overcome such difficulties are extremely promising and may lead to important insights into the functioning of this unusual ncRNA.

Although we still do not know the function of PAN RNA, we have gained some understanding of why it accumulates to such high levels during the KSHV lytic phase. PAN RNA's nuclear abundance can be traced to a highly structured RNA element located just upstream of its 3′ poly(A) tail, called the ENE (Conrad and Steitz 2005) (Fig. 2B). Deletion of the ENE dramatically lowers PAN levels by increasing the decay rate of the RNA (Conrad et al. 2006). Stabilization appears to depend on the ENE's ability to engage PAN's poly(A) tail by base-pairing to an internal loop structure, thereby inhibiting exonucleolytic trimming of the RNA, the first step in decay. One protein that has been shown to physically interact with the ENE sequence is hnRNP C1, although the significance of this interaction is not known (Conrad 2008).

Recently, Glaunsinger and colleagues have reported nuclear accumulation of the cytoplasmic poly(A)-binding protein C1 (PABPC1) during the lytic phase of KSHV infection (Lee and Glaunsinger 2009). Relocalization appears to be related to the host shutoff effect of the virus, whereby the majority of host mRNAs are selectively degraded while viral mRNAs persist to be translated (Glaunsinger and Ganem 2004). Because PABPC1 has extremely high affinity for poly(A) sequences (Gorlach et al. 1994) and because PAN RNA is an abundant, nuclear polyadenylated transcript, it is likely that PAN RNA serves as a sink for the relocalized PABPC1. How this interaction impacts the host shutoff effect of the virus and the contribution of PAN RNA to the later stages of KSHV lytic growth awaits further investigation. Several other viruses target PABPC1 for nuclear relocalization (rotavirus and herpes simplex virus) or degradation (picornaviruses and caliciviruses), as a component of the shutoff effect for each of these viruses (Dobrikova et al. 2010; Harb et al. 2008; Kuyumcu-Martinez et al. 2002; Kuyumcu-Martinez et al. 2004a; Kuyumcu-Martinez et al. 2004b).

5 VIRAL microRNAs: ALTERING HOST-CELL OR FINE-TUNING VIRAL GENE EXPRESSION?

A novel class of viral ncRNAs was uncovered in 2004 with the discovery of five miRNAs in human B cells latently infected with EBV (Pfeffer et al. 2004). As miRNAs are implicated in regulating expression of the majority of transcripts in mammalian cells (see Sharp 2010), this insight prompted a flurry of investigations that identified miRNAs encoded by a variety of viruses (Table 2). Virtually all herpesviruses express relatively large numbers (7–35) of viral miRNAs; indeed, only one herpesvirus tested, varicella zoster virus, lacks detectable miRNA expression during latency (Umbach and Cullen 2009). Other viruses with DNA genomes express fewer or no miRNAs. Retroviruses, flaviviruses and other viruses with RNA genomes do not generally encode miRNAs (Pfeffer et al. 2005; Cai et al. 2006a; Lin and Cullen 2007). However, the existence of miRNAs processed by Dicer from the RNA genomes of the human immunodeficiency viruses types 1 (Bennasser et al. 2004; Couturier and Root-Bernstein 2005; Omoto and Fujii 2006; Weinberg and Morris 2006; Klase et al. 2007; Ouellet et al. 2008; Kaul et al. 2009) and 2 (Purzycka and Adamiak 2008) remains controversial (Lin and Cullen 2007). With the advent of high-throughput sequencing technology and its application to different types of cells in various stages of viral infection, our current list of viral miRNAs will likely continue to expand in the near future.

As is the case for cellular miRNAs, understanding the function of viral miRNAs requires identification and validation of their target mRNAs. This has proven to be no small task, as miRNA-mRNA interactions are characterized by imperfect base-pairing in a very loose pattern (Lewis et al. 2005; Grimson et al. 2007; Bartel 2009), varying for different miRNA-mRNA combinations (Didiano and Hobert 2008; Nahvi et al. 2009). Although the list of new viral miRNAs continues to expand, only a few target mRNAs have been validated for an even smaller number of viral miRNAs. These targets can be either host or viral mRNA transcripts. Validation typically includes three steps (see also Sharp 2010): (1) a putative target sequence is cloned into the 3′UTR of a reporter construct (e.g., luciferase) and cotransfected into mammalian cells with the miRNA; (2) if repression of reporter activity is observed, then the target must be mutated to abrogate the miRNA-mRNA interaction with predicted loss of repression and ideally, subsequent introduction of compensatory mutations into the miRNA to rescue repression; and (3) miRNA inhibitors (such as antisense) must be shown to increase levels of the putative target protein by relieving miRNA repression. A modest decrease in the abundance of a miRNA-targeted mRNA is often observed in addition to translational downregulation (see Sharp 2010), providing another approach to validation (Grundhoff et al. 2006).

Alteration of as little as a single nucleotide can significantly affect the specificity of a miRNA (Bartel 2004). Whereas a few viral miRNAs are conserved in genomic location, most are not completely conserved in sequence even between very closely related viruses. In notable exception here are the miRNAs of the γ herpesviruses EBV and the related monkey virus rLCV (Cai et al. 2006b; Gottwein et al. 2007; Walz et al. 2010; Riley et al. 2010). The majority of EBV and rLCV miRNAs are identical in both sequence and genomic clustering, but each virus intriguingly encodes several sequence-unique miRNAs located within these clusters (Cai et al. 2006b; Walz et al. 2010; Riley et al. 2010). Whether the EBV and rLCV miRNAs are functionally conserved remains to be explored.

In an instance of functional conservation, three different herpesviruses—human cytomegalovirus (HCMV), KSHV, and EBV—each target a stress-induced human immune ligand (MICB), which is responsible for activating natural killer cell function, with a totally different miRNA (Nachmani et al. 2009). In this way, the three viruses circumvent a key regulator of human immune function. Interestingly, one of these viral miRNAs, KSHV-miR-K12-11, is the only fully validated orthologue of a human miRNA (hsa-miR-155) (Gottwein et al. 2007; Skalsky et al. 2007). Both viral and human miR-155 miRNAs regulate BACH-1, a transcriptional repressor involved in B-cell development (Gottwein et al. 2007). Thus, KSHV uses a miRNA to stall infected B cells at a stage advantageous to the virus.

Although the sequences of their viral miRNAs are not conserved among polyomaviruses, it appears that those expressed by simian virus 40 (SV40) and by the human BK and JC polyomaviruses target homologous viral transcripts (Sullivan et al. 2005; Seo et al. 2008). This functional conservation is interesting for two additional reasons. First, the miRNAs of these polyomaviruses act as small interfering (si) RNAs, as they are perfectly complementary to and direct mRNA cleavage of their target mRNAs, which are early lytic viral transcripts (Sullivan et al. 2005; Seo et al. 2008). They therefore serve as autoregulators to reduce the levels of the transcripts from which they arise at a later time in infection. Second, these miRNAs are an example of a pair of mature miRNAs processed from the opposite strands of the same precursor miRNA (see Sharp 2010), an efficient use of viral genomic material.

Many targets of viral miRNAs are viral mRNAs, as suggested by bioinformatic studies (Murphy et al. 2008). On cloning the first five viral miRNAs, Tuschl and colleagues hypothesized that the EBV miRNA BART2, whose precursor corresponds to the antisense strand of the EBV DNA polymerase, BALF5, acts as an siRNA to reduce levels of BALF5 (Pfeffer et al. 2004). BART2 down-regulation of BALF5 was later documented in EBV-infected cells, where the BALF5 transcript was shown to undergo cleavage at a specific site directed by BART2 (Barth et al. 2008). Interestingly, the insect virus Heliothis virescens ascovirus (HvAV) also encodes a miRNA that down-regulates its own DNA polymerase, even though the HvAV miRNA does not perfectly base-pair with the polymerase mRNA (Hussain et al. 2008). In some cases, the target viral mRNA is regulated by multiple miRNAs. For example, several of the EBV BART miRNAs from a single genomic cluster (BART1-5p, BART3-5p, BART16, and BART17-5p) work together to down-regulate production of the EBV latency membrane protein 1 (LMP1). Of the ten predicted miRNA target sites in the LMP1 3′UTR for four different EBV miRNAs, two tested as nonfunctional, four weakly regulated LMP1, and four appeared to significantly destabilize the luciferase reporter transcript. Thus, it seems that each miRNA contributes differentially to the down-regulation of LMP1 (Lo et al. 2007).

Oncogenic viruses mount multiple strategies to alter the expression of human tumor suppressor and oncogenes, and miRNAs can now be added to the list of mechanisms they use. For instance, gene expression profiling of human transcripts showed repression of the tumor suppressor THBS1, a regulator of cell adhesion, migration, and angiogenesis, in the presence of KSHV miRNAs (Samols et al. 2007). The EBV miRNA BART5 targets PUMA, an apoptotic host protein that is also controlled at the transcriptional level by p53 (Choy et al. 2008), apparently to ensure survival of latently infected cells.

The expression of viral miRNAs can further be highly regulated during the course of infection, which suits the complex life cycle of viruses. Cloning and quantitative PCR have assessed the range of viral miRNA expression levels under various conditions for EBV, the best studied of the viral miRNA-encoding genomes (Cai et al. 2006b; Xing and Kieff 2007; Cosmopoulos et al. 2009). For example, Daudi cells (EBV-infected Burkitt's lymphoma) express basal levels of BARTs1, 3, 7, 10, BHRF1-1, and BHRF1-2 during latency, but on induction of lytic replication, the levels of all these miRNAs dramatically increase, except for BHRF1-1, which decreases in abundance (Cai et al. 2006b). Two of the EBV miRNAs, BARTs21 and 22, were only recently discovered because they are expressed almost exclusively in some nasopharyngeal carcinomas (Zhu et al. 2009). Because some of the miRNA-containing viruses persist latently in host cells, such as EBV in resting memory B cells, it is conceivable that the viral miRNAs expressed under these conditions up-regulate rather than repress translation of their target mRNAs, as has been observed during the G_0 stage of the cell cycle (Vasudevan et al. 2007). Because viral miRNA levels can be regulated by a number of mechanisms, including transcription, processing, and stability, the functional effects of their tight regulation have yet to be fully investigated.

Our increased understanding of the key roles viral miRNAs play in host-virus interactions suggests that viral miRNAs will be suitable targets for antiviral therapy (Kurzynska-Kokorniak et al. 2009). Efficacious therapies may be closer than we might believe. For instance, the polyomaviral miRNAs help maintain evasion of the immune system in infected patients because they down-regulate the expression of viral antigens that trigger an immune response leading to clearance of infected cells (Seo et al. 2008). Thus, agents that down-regulate the polyomaviral miRNAs have significant therapeutic potential.

6 NONCODING RNAs OF OTHER FLAVORS

Herpes simplex virus-1 (HSV-1) is a member of the α subgroup of herpesviruses that infects epithelial and neuronal cells. HSV-1 generally establishes latent infection, but it enters the lytic stage under stress conditions. It encodes an abundant 6.3kb latent transcript called latency-associated transcript (LAT) (Bloom 2004; Stevens et al. 1987). Although LAT is highly unstable, its ~2 kb intron accumulates to extremely high levels (Farrell et al. 1991; Kang et al. 2006). The function of LAT remained enigmatic for years, but it has recently been shown to be a miRNA precursor transcript (Umbach et al. 2008; Umbach and Cullen 2009). Recent data suggest that the first 1.5kb of LAT also give rise to two other small ncRNAs: sRNA1 (65 nt) and sRNA2 (36 nt). These sRNAs co-operate in inhibiting the apoptosis of infected neuronal cells. Furthermore, transfection of rabbit skin cells with the sRNAs prevented viral production. Specifically, sRNA2 inhibits the translation of infected cell protein 4 (ICP4), which is important for the activation and repression of viral transcription during infection (Smith et al. 1993), as ICP4 mRNA levels are not altered. The exact mechanism is unknown (Shen et al. 2009). These two functions imply that the sRNAs are required to prevent reactivation of latent HSV-1 (Shen et al. 2009).

HCMV, a β herpesvirus, is the most frequent congenital viral infection in humans. Approximately 5%–10% of infected infants have symptomatic disease associated with serious physical and mental birth defects and mortality (Dollard et al. 2007); in immunocompromised HIV patients or immunosuppressed transplant patients, HCMV can lead to mental retardation, intellectual impairment and deafness (Revello and Gerna 2002). During the early stages of infection, two major unspliced transcripts, which are 1.2 kb and 2.7 kb long, are expressed at very high levels (Spector 1996; Greenaway and Wilkinson 1987). Recently, the 2.7 kb ncRNA has been found to interact with the GRIM-19 (genes associated with retinoid/interferon-induced mortality-19) protein, targeting it to complex I of the electron transport chain in the inner mitochondrial membrane. This interaction is important for maintaining the electrochemical gradient across mitochondrial membranes and thus aids in the production of sufficient ATP to keep the cell alive during infection (Reeves et al. 2007).

EBV, as discussed earlier, encodes both the EBERs and multiple miRNAs in latently infected B cells. By using a method called subtractive hybridization of RNA transcripts (SHORT), Huttenhofer's group identified a number of ncRNAs that are not miRNAs (Mrazek et al. 2007). One of these is 65 nt long and contains sequence motifs that assign it to the family of C/D box small nucleolar RNAs; v-snoRNA1 accordingly binds three canonical snoRNA proteins: fibrillarin, Nop56, and Nop58. It does not appear to have any known rRNA or snRNA target and is thus categorized as one of many "orphan snoRNAs." V-snoRNA1 potentially also acts as a miRNA precursor as it is further processed to a 24-nt RNA called v-snoRNA1[24pp]. This miRNA appears to target cleavage of the 3′UTR of the viral BALF5 mRNA, which encodes the viral DNA polymerase (Hutzinger et al. 2009), not far from the site where cleavage by EBV-encoded mir-BART2 occurs (see previous). Several cases in which snoRNAs efficiently generate miRNAs have recently been reported (Ender et al. 2008; Scott et al. 2009; Taft et al. 2009).

Murine herpesvirus 68 (MHV-68) belongs to the γ subgroup of herpesviruses and has been shown to encode 8 novel tRNA-like genes interspersed with ORFs 1–3. Like tRNAs, these possess Pol III promoter elements and are predicted to assume a tRNA-like cloverleaf secondary structure. Only MHV-68 tRNA7 contains a short tRNA-type intron. These transcripts are expressed at very high levels during both latent and lytic infection. They are processed into mature tRNAs with posttranscriptional addition of 3′ CCA termini, but they are not aminoacylated. Their function is unknown (Bowden et al. 1997).

The Red clover necrotic mosaic virus (RCNMV), which is a positive strand RNA virus that infects plants, encodes a 0.4 kb ncRNA that is packaged into virions, called SR1f. Two RNAs make up the RCNMV genome, RNA1 and RNA2. SR1f is generated from the 3′UTR of RNA1 and was found to contain a 58-nt sequence (Seq1f58) that protects SR1f from 5′→3′ decay. SR1f appears to inhibit the production of negative-strand RCNMV genomic RNAs via the repression of viral replicase protein production (Iwakawa et al. 2008).

Another ncRNA is found in some members of the Flavivirus family, in particular West Nile virus. Flaviviruses are positive strand RNA viruses that produce a highly structured subgenomic RNA (sgRNA) that is 0.3–0.5 kb long and is derived from incomplete degradation of the 3′UTR. The production of sgRNA plays a role in viral replication as well as determining viral pathogenicity as mice infected with the wild-type virus experience severe symptoms, whereas those infected with a mutant virus that does not express the ncRNA show no signs of infection (Pijlman et al. 2008).

7 PROSPECTS

The roster of viral-encoded ncRNAs has multiplied in recent years, reflecting the variety of distinct classes of ncRNAs produced by cellular genomes. Counterparts of snRNAs, snoRNAs, miRNAs, tRNAs, other RNA pol III-transcribed regulatory RNAs, and large ncRNAs synthesized by RNA pol II are now included in the viral armamentarium. This dazzling diversity is even apparent in the relatively few mechanisms already elucidated whereby viral ncRNAs manipulate gene expression in their host cells. These include sequestering host RNA-binding proteins away from their normal cellular roles and evolving a ncRNA to regulate the activity of a cellular ncRNA.

But many challenges remain, as the functional mechanisms of some viral ncRNAs have remained enigmatic for decades (see Table 1). Clearly, every step of gene expression is a potential target for regulation by a viral ncRNA. Further surprises will undoubtedly emerge as we continue to investigate the molecular basis of the RNA warfare mounted by viruses against their cellular hosts.

ACKNOWLEDGMENTS

We thank all members of the Steitz lab for stimulating discussions. This work was supported by National Institutes of Health grant CA16038. J.A.S. is an investigator of the Howard Hughes Medical Institute. S.B. was funded by an Anna Fuller Predoctoral Fellowship; R.M-F. by a Postdoctoral Fellowship from the Jane Coffin Childs Fund; and K.R. by an American Cancer Society, New England Chapter–Beatrice Cuneo Postdoctoral Fellowship.

REFERENCES

Albrecht JC. 2000. Primary structure of the *Herpesvirus ateles* genome. *J Virol* **74:** 1033–1037.

Albrecht JC, Fleckenstein B. 1992. Nucleotide sequence of HSUR 6 and HSUR 7, two small RNAs of herpesvirus saimiri. *Nucleic Acids Res* **20:** 1810.

Anderson SJ, Lauritsen JP, Hartman MG, Foushee AM, Lefebvre JM, Shinton SA, Gerhardt B, Hardy RR, Oravecz T, Wiest DL. 2007. Ablation of ribosomal protein L22 selectively impairs αβ T cell development by activation of a p53-dependent checkpoint. *Immunity* **26:** 759–772.

Andersson MG, Haasnoot PCJ, Xu N, Berenjian S, Berkhout B, Akusjärvi G. 2005. Suppression of RNA interference by adenovirus virus-associated RNA. *J Virol* **79:** 9556–9565.

Aparicio O, Razquin N, Zaratiegui M, Narvaiza I, Fortes P. 2006. Adenovirus virus-associated RNA is processed to functional interfering RNAs involved in virus production. *J Virol* **80:** 1376–1384.

Arrand JR, Young LS, Tugwood JD. 1989. Two families of sequences in the small RNA-encoding region of Epstein-Barr virus (EBV) correlate with EBV types A and B. *J Virol* **63:** 983–986.

Bartel DP. 2004. MicroRNAs: Genomics, biogenesis, mechanism, and function. *Cell* **116:** 281–297.

Bartel DP. 2009. MicroRNAs: Target recognition and regulatory functions. *Cell* **136:** 215–233.

Barth S, Pfuhl T, Mamiani A, Ehses C, Roemer K, Kremmer E, Jaker C, Hock J, Meister G, Grasser FA. 2008. Epstein-Barr virus-encoded microRNA miR-BART2 down-regulates the viral DNA polymerase BALF5. *Nucleic Acids Res* **36:** 666–675.

Bechtel JT, Liang Y, Hvidding J, Ganem D. 2003. Host range of Kaposi's sarcoma-associated herpesvirus in cultured cells. *J Virol* **77:** 6474–6481.

Ben-Ami O, Pencovich N, Lotem J, Levanon D, Groner Y. 2009. A regulatory interplay between miR-27a and Runx1 during megakaryopoiesis. *Proc Natl Acad Sci* **106:** 238–243.

Bennasser Y, Le SY, Yeung ML, Jeang KT. 2004. HIV-1 encoded candidate micro-RNAs and their cellular targets. *Retrovirology* **1:** 43.

Besecker MI, Harden ME, Li G, Wang XJ, Griffiths A. 2009. Discovery of herpes B virus-encoded microRNAs. *J Virol* **83:** 3413–3416.

Bhat RA, Thimmappaya B. 1983. Two small RNAs encoded by Epstein-Barr virus can functionally substitute for the virus-associated RNAs in the lytic growth of adenovirus 5. *Proc Natl Acad Sci* **80:** 4789–4793.

Bhat RA, Thimmappaya B. 1985. Construction and analysis of additional adenovirus substitution mutants confirm the complementation of VAI RNA function by two small RNAs encoded by Epstein-Barr virus. *J Virol* **56:** 750–756.

Bloom DC. 2004. HSV LAT and neuronal survival. *Int Rev Immunol* **23:** 187–198.

Blyth K, Cameron ER, Neil JC. 2005. The RUNX genes: Gain or loss of function in cancer. *Nat Rev Cancer* **5:** 376–387.

Bowden RJ, Simas JP, Davis AJ, Efstathiou S. 1997. Murine γherpesvirus 68 encodes tRNA-like sequences which are expressed during latency. *J Gen Virol* **78:** 1675–1687.

Buck AH, Santoyo-Lopez J, Robertson KA, Kumar DS, Reczko M, Ghazal P. 2007. Discrete clusters of virus-encoded microRNAs are associated with complementary strands of the genome and the 7.2-kilobase stable intron in murine cytomegalovirus. *J Virol* **81:** 13761–13770.

Burnside J, Bernberg E, Anderson A, Lu C, Meyers BC, Green PJ, Jain N, Isaacs G, Morgan RW. 2006. Marek's disease virus encodes MicroRNAs that map to meq and the latency-associated transcript. *J Virol* **80:** 8778–8786.

Burnside J, Ouyang M, Anderson A, Bernberg E, Lu C, Meyers BC, Green PJ, Markis M, Isaacs G, Huang E, et al. 2008. Deep sequencing of chicken microRNAs. *BMC Genomics* **9:** 185.

Cai X, Cullen BR. 2006. Transcriptional origin of Kaposi's sarcoma-associated herpesvirus microRNAs. *J Virol* **80:** 2234–2242.

Cai X, Li G, Laimins LA, Cullen BR. 2006a. Human papillomavirus genotype 31 does not express detectable microRNA levels during latent or productive virus replication. *J Virol* **80:** 10890–10893.

Cai X, Schafer A, Lu S, Bilello JP, Desrosiers RC, Edwards R, Raab-Traub N, Cullen BR. 2006b. Epstein-Barr virus microRNAs are evolutionarily conserved and differentially expressed. *PLoS Pathog* **2:** e23.

Cesarman E, Chang Y, Moore PS, Said JW, Knowles DM. 1995. Kaposi's sarcoma-associated herpesvirus-like DNA sequences in AIDS-related body-cavity-based lymphomas. *N Engl J Med* **332:** 1186–1191.

Chang Y, Cesarman E, Pessin MS, Lee F, Culpepper J, Knowles DM, Moore PS. 1994. Identification of herpesvirus-like DNA sequences in AIDS-associated Kaposi's sarcoma. *Science* **266:** 1865–1869.

Chang K, Elledge SJ, Hannon GJ. 2006. Lessons from Nature: microRNA-based shRNA libraries. *Nat Methods* **3:** 707–714.

Choy EY, Siu KL, Kok KH, Lung RW, Tsang CM, To KF, Kwong DL, Tsao SW, Jin DY. 2008. An Epstein-Barr virus-encoded microRNA targets PUMA to promote host cell survival. *J Exp Med* **205:** 2551–2560.

Clarke PA, Sharp NA, Clemens MJ. 1990. Translational control by the Epstein-Barr virus small RNA EBER-1. Reversal of the double-stranded RNA-induced inhibition of protein synthesis in reticulocyte lysates. *Eur J Biochem* **193:** 635–641.

Conrad NK. 2008. Chapter 15. Co-immunoprecipitation techniques for assessing RNA-protein interactions in vivo. *Methods Enzymol* **449:** 317–342.

Conrad NK, Steitz JA. 2005. A Kaposi's sarcoma virus RNA element that increases the nuclear abundance of intronless transcripts. *EMBO J* **24:** 1831–1841.

Conrad NK, Mili S, Marshall EL, Shu MD, Steitz JA. 2006. Identification of a rapid mammalian deadenylation-dependent decay pathway and its inhibition by a viral RNA element. *Mol Cell* **24:** 943–953.

Cook HL, Mischo HE, Steitz JA. 2004. The *Herpesvirus saimiri* small nuclear RNAs recruit AU-rich element-binding proteins but do not alter host AU-rich element-containing mRNA levels in virally transformed T cells. *Mol Cell Biol* **24:** 4522–4533.

Cook HL, Lytle JR, Mischo HE, Li MJ, Rossi JJ, Silva DP, Desrosiers RC, Steitz JA. 2005. Small nuclear RNAs encoded by *Herpesvirus saimiri* upregulate the expression of genes linked to T cell activation in virally transformed T cells. *Curr Biol* **15:** 974–979.

Cosmopoulos K, Pegtel M, Hawkins J, Moffett H, Novina C, Middeldorp J, Thorley-Lawson DA. 2009. Comprehensive profiling of EBV microRNAs in nasopharyngeal carcinoma. *J Virol* **83:** 2357–2367.

Countryman JK, Gradoville L, Miller G. 2008. Histone hyperacetylation occurs on promoters of lytic cycle regulatory genes in Epstein-Barr virus-infected cell lines which are refractory to disruption of latency by histone deacetylase inhibitors. *J Virol* **82:** 4706–4719.

Couturier JP, Root-Bernstein RS. 2005. HIV may produce inhibitory microRNAs (miRNAs) that block production of CD28, CD4 and some interleukins. *J Theor Biol* **235:** 169–184.

Cui C, Griffiths A, Li G, Silva LM, Kramer MF, Gaasterland T, Wang XJ, Coen DM. 2006. Prediction and identification of herpes simplex virus 1-encoded microRNAs. *J Virol* **80:** 5499–5508.

Didiano D, Hobert O. 2008. Molecular architecture of a miRNA-regulated 3′ UTR. *RNA* **14:** 1297–1317.

Dobrikova E, Shveygert M, Walters R, Gromeier M. 2010. Herpes simplex virus proteins ICP27 and UL47 associate with polyadenylate-binding protein and control its sub-cellular distribution. *J Virol* **84:** 270–279.

Dolken L, Pfeffer S, Koszinowski UH. 2009. Cytomegalovirus microRNAs. *Virus Genes* **38:** 355–364.

Dolken L, Perot J, Cognat V, Alioua A, John M, Soutschek J, Ruzsics Z, Koszinowski U, Voinnet O, Pfeffer S. 2007. Mouse cytomegalovirus microRNAs dominate the cellular small RNA profile during lytic infection and show features of posttranscriptional regulation. *J Virol* **81:** 13771–13782.

Dollard SC, Grosse SD, Ross DS. 2007. New estimates of the prevalence of neurological and sensory sequelae and mortality associated with congenital cytomegalovirus infection. *Rev Med Virol* **17:** 355–363.

Dunn EF, Hammell CM, Hodge CA, Cole CN. 2005. Yeast poly(A)-binding protein, Pab1, and PAN, a poly(A) nuclease complex recruited by Pab1, connect mRNA biogenesis to export. *Genes Dev* **19:** 90–103.

Dupin N, Fisher C, Kellam P, Ariad S, Tulliez M, Franck N, van Marck E, Salmon D, Gorin I, Escande JP, et al. 1999. Distribution of human herpesvirus-8 latently infected cells in Kaposi's sarcoma, multicentric Castleman's disease, and primary effusion lymphoma. *Proc Natl Acad Sci* **96:** 4546–4551.

Ender C, Krek A, Friedlander MR, Beitzinger M, Weinmann L, Chen W, Pfeffer S, Rajewsky N, Meister G. 2008. A human snoRNA with microRNA-like functions. *Mol Cell* **32:** 519–528.

Fan XC, Myer VE, Steitz JA. 1997. AU-rich elements target small nuclear RNAs as well as mRNAs for rapid degradation. *Genes Dev* **11:** 2557–2568.

Fannin Rider PJ, Dunn W, Yang E, Liu F. 2008. Human cytomegalovirus microRNAs. *Curr Top Microbiol Immunol* **325:** 21–39.

Farrell MJ, Dobson AT, Feldman LT. 1991. Herpes simplex virus latency-associated transcript is a stable intron. *Proc Natl Acad Sci* **88:** 790–794.

Feng J, Iwama A, Satake M, Kohu K. 2009. MicroRNA-27 enhances differentiation of myeloblasts into granulocytes by post-transcriptionally downregulating Runx1. *Br J Haematol* **145:** 412–423.

Feng P, Park J, Lee BS, Lee SH, Bram RJ, Jung JU. 2002. Kaposi's sarcoma-associated herpesvirus mitochondrial K7 protein targets a cellular calcium-modulating cyclophilin ligand to modulate intracellular calcium concentration and inhibit apoptosis. *J Virol* **76:** 11491–11504.

Fok V, Friend K, Steitz JA. 2006b. Epstein-Barr virus noncoding RNAs are confined to the nucleus, whereas their partner, the human La protein, undergoes nucleocytoplasmic shuttling. *J Cell Biol* **173:** 319–325.

Fok V, Mitton-Fry RM, Grech A, Steitz JA. 2006a. Multiple domains of EBER 1, an Epstein-Barr virus non-coding RNA, recruit human ribosomal protein L22. *RNA* **12:** 872–882.

Ganem D. 2006. KSHV infection and the pathogenesis of Kaposi's sarcoma. *Annu Rev Pathol* **1:** 273–296.

Garneau NL, Wilusz J, Wilusz CJ. 2007. The highways and byways of mRNA decay. *Nat Rev Mol Cell Biol* **8:** 113–126.

Glaunsinger B, Ganem D. 2004. Lytic KSHV infection inhibits host gene expression by accelerating global mRNA turnover. *Mol Cell* **13:** 713–723.

Gorlach M, Burd CG, Dreyfuss G. 1994. The mRNA poly(A)-binding protein: Localization, abundance, and RNA-binding specificity. *Exp Cell Res* **211:** 400–407.

Gottwein E, Mukherjee N, Sachse C, Frenzel C, Majoros WH, Chi JT, Braich R, Manoharan M, Soutschek J, Ohler U, et al. 2007. A viral microRNA functions as an orthologue of cellular miR-155. *Nature* **450:** 1096–1099.

Greenaway PJ, Wilkinson GW. 1987. Nucleotide sequence of the most abundantly transcribed early gene of human cytomegalovirus strain AD169. *Virus Res* **7:** 17–31.

Grey F, Nelson J. 2008. Identification and function of human cytomegalovirus microRNAs. *J Clin Virol* **41:** 186–191.

Grimson A, Farh KK, Johnston WK, Garrett-Engele P, Lim LP, Bartel DP. 2007. MicroRNA targeting specificity in mammals: determinants beyond seed pairing. *Mol Cell* **27:** 91–105.

Grundhoff A, Sullivan CS, Ganem D. 2006. A combined computational and microarray-based approach identifies novel microRNAs encoded by human γ-herpesviruses. *RNA* **12:** 733–750.

Gulley ML, Tang W. 2008. Laboratory assays for Epstein-Barr virus-related disease. *J Mol Diagn* **10:** 279–292.

Guttilla IK, White BA. 2009. Coordinate Regulation of FOXO1 by miR-27a, miR-96, and miR-182 in Breast Cancer Cells. *J Biol Chem* **284:** 23204–23216.

Harb M, Becker MM, Vitour D, Baron CH, Vende P, Brown SC, Bolte S, Arold ST, Poncet D. 2008. Nuclear localization of cytoplasmic poly(A)-binding protein upon rotavirus infection involves the interaction of NSP3 with eIF4G and RoXaN. *J Virol* **82:** 11283–11293.

Houmani JL, Davis CI, Ruf IK. 2009. Growth promoting properties of Epstein-Barr virus EBER-1 RNA correlate with ribosomal protein L22 binding. *J Virol* **83:** 9844–9853.

Howe JG, Steitz JA. 1986. Localization of Epstein-Barr virus-encoded small RNAs by *in situ* hybridization. *Proc Natl Acad Sci* **83:** 9006–9010.

Hussain M, Taft RJ, Asgari S. 2008. An insect virus-encoded microRNA regulates viral replication. *J Virol* **82:** 9164–9170.

Hutzinger R, Feederle R, Mrazek J, Schiefermeier N, Balwierz PJ, Zavolan M, Polacek N, Delecluse HJ, Huttenhofer A. 2009. Expression and processing of a small nucleolar RNA from the Epstein-Barr virus genome. *PLoS Pathog* **5:** e1000547.

Ideue T, Hino K, Kitao S, Yokoi T, Hirose T. 2009. Efficient oligonucleotide-mediated degradation of nuclear noncoding RNAs in mammalian cultured cells. *RNA* **15:** 1578–1587.

Iwakawa HO, Mizumoto H, Nagano H, Imoto Y, Takigawa K, Sarawaneeyaruk S, Kaido M, Mise K, Okuno T. 2008. A viral noncoding RNA generated by cis-element-mediated protection against 5′→3′ RNA decay represses both cap-independent and cap-dependent translation. *J Virol* **82:** 10162–10174.

Iwakiri D, Eizuru Y, Tokunaga M, Takada K. 2003. Autocrine growth of Epstein-Barr virus-positive gastric carcinoma cells mediated by an Epstein-Barr virus-encoded small RNA. *Cancer Res* **63:** 7062–7067.

Iwakiri D, Sheen TS, Chen JY, Huang DP, Takada K. 2005. Epstein-Barr virus-encoded small RNA induces insulin-like growth factor 1 and supports growth of nasopharyngeal carcinoma-derived cell lines. *Oncogene* **24:** 1767–1773.

Jenner RG, Boshoff C. 2002. The molecular pathology of Kaposi's sarcoma-associated herpesvirus. *Biochim Biophys Acta* **1602:** 1–22.

Kang W, Mukerjee R, Gartner JJ, Hatzigeorgiou AG, Sandri-Goldin RM, Fraser NW. 2006. Characterization of a spliced exon product of herpes simplex type-1 latency-associated transcript in productively infected cells. *Virology* **356:** 106–114.

Kaul D, Ahlawat A, Gupta SD. 2009. HIV-1 genome-encoded hiv1-mir-H1 impairs cellular responses to infection. *Mol Cell Biochem* **323:** 143–148.

Kieff E, Rickinson AB. 2007. Epstein-Barr virus and its replication. in *Fields Virology* (ed. Knipe D.M., Howley P.M.), pp. 2603–2654. Lippincott Williams & Wilkins, Philadelphia.

Kitagawa N, Goto M, Kurozumi K, Maruo S, Fukayama M, Naoe T, Yasukawa M, Hino K, Suzuki T, Todo S, et al. 2000. Epstein-Barr virus-encoded poly(A)(-) RNA supports Burkitt's lymphoma growth through interleukin-10 induction. *EMBO J* **19:** 6742–6750.

Klase Z, Kale P, Winograd R, Gupta MV, Heydarian M, Berro R, McCaffrey T, Kashanchi F. 2007. HIV-1 TAR element is processed by Dicer to yield a viral micro-RNA involved in chromatin remodeling of the viral LTR. *BMC Mol Biol* **8:** 63.

Komano J, Maruo S, Kurozumi K, Oda T, Takada K. 1999. Oncogenic role of Epstein-Barr virus-encoded RNAs in Burkitt's lymphoma cell line Akata. *J Virol* **73:** 9827–9831.

Kurzynska-Kokorniak A, Jackowiak P, Figlerowicz M. 2009. Human- and virus-encoded microRNAs as potential targets of antiviral therapy. *Mini Rev Med Chem* **9:** 927–937.

Kuyumcu-Martinez M, Belliot G, Sosnovtsev SV, Chang KO, Green KY, Lloyd RE. 2004a. Calicivirus 3C-like proteinase inhibits cellular translation by cleavage of poly(A)-binding protein. *J Virol* **78:** 8172–8182.

Kuyumcu-Martinez NM, Joachims M, Lloyd RE. 2002. Efficient cleavage of ribosome-associated poly(A)-binding protein by enterovirus 3C protease. *J Virol* **76:** 2062–2074.

Kuyumcu-Martinez NM, Van Eden ME, Younan P, Lloyd RE. 2004b. Cleavage of poly(A)-binding protein by poliovirus 3C protease inhibits host cell translation: A novel mechanism for host translation shutoff. *Mol Cell Biol* **24:** 1779–1790.

Lee SI, Steitz JA. 1990. *Herpesvirus saimiri* U RNAs are expressed and assembled into ribonucleoprotein particles in the absence of other viral genes. *J Virol* **64:** 3905–3915.

Lee SI, Murthy SC, Trimble JJ, Desrosiers RC, Steitz JA. 1988. Four novel U RNAs are encoded by a herpesvirus. *Cell* **54:** 599–607.

Lee YJ, Glaunsinger BA. 2009. Aberrant herpesvirus-induced polyadenylation correlates with cellular messenger RNA destruction. *PLoS Biol* **7:** e1000107.

Lerner MR, Andrews NC, Miller G, Steitz JA. 1981a. Two small RNAs encoded by Epstein-Barr virus and complexed with protein are precipitated by antibodies from patients with systemic *Lupus Erythematosus*. *Proc Natl Acad Sci* **78:** 805–809.

Lerner MR, Boyle JA, Hardin JA, Steitz JA. 1981b. Two novel classes of small ribonucleoproteins detected by antibodies associated with *lupus erythematosus*. *Science* **211:** 400–402.

Lewis BP, Burge CB, Bartel DP. 2005. Conserved seed pairing, often flanked by adenosines, indicates that thousands of human genes are microRNA targets. *Cell* **120:** 15–20.

Lin J, Cullen BR. 2007. Analysis of the interaction of primate retroviruses with the human RNA interference machinery. *J Virol* **81:** 12218–12226.

Lo AK, To KF, Lo KW, Lung RW, Hui JW, Liao G, Hayward SD. 2007. Modulation of LMP1 protein expression by EBV-encoded microRNAs. *Proc Natl Acad Sci* **104:** 16164–16169.

Lu S, Cullen BR. 2004. Adenovirus VA1 noncoding RNA can inhibit small interfering RNA and microRNA biogenesis. *J Virol* **78:** 12868–12876.

Luhrmann R. 2010. Spliceosome structure and function. *Cold Spring Harb Perspect Biol* **2:** a003707.

Ma Y, Mathews MB. 1996. Structure, function, and evolution of adenovirus-associated RNA: A phylogenetic approach. *J Virol* **70:** 5083–5099.

Maiese K, Chong ZZ, Shang YC, Hou J. 2008. Clever cancer strategies with FoxO transcription factors. *Cell Cycle* **7:** 3829–3839.

Mathews MB, Shenk T. 1991. Adenovirus virus-associated RNA and translation control. *J Virol* **65:** 5657–5662.

Mongelard F, Bouvet P. 2007. Nucleolin: A multiFACeTed protein. *Trends Cell Biol* **17:** 80–86.

Morgan R, Anderson A, Bernberg E, Kamboj S, Huang E, Lagasse G, Isaacs G, Parcells M, Meyers BC, Green PJ, Burnside J. 2008. Sequence conservation and differential expression of Marek's disease virus microRNAs. *J Virol* **82:** 12213–12220.

Mrazek J, Kreutmayer SB, Grasser FA, Polacek N, Huttenhofer A. 2007. Subtractive hybridization identifies novel differentially expressed ncRNA species in EBV-infected human B cells. *Nucleic Acids Res* **35:** e73.

Murphy E, Vanicek J, Robins H, Shenk T, Levine AJ. 2008. Suppression of immediate-early viral gene expression by herpesvirus-coded microRNAs: Implications for latency. *Proc Natl Acad Sci* **105:** 5453–5458.

Murthy S, Kamine J, Desrosiers RC. 1986. Viral-encoded small RNAs in herpes virus saimiri induced tumors. *EMBO J* **5:** 1625–1632.

Myer VE, Lee SI, Steitz JA. 1992. Viral small nuclear ribonucleoproteins bind a protein implicated in messenger RNA destabilization. *Proc Natl Acad Sci* **89:** 1296–1300.

Nachmani D, Stern-Ginossar N, Sarid R, Mandelboim O. 2009. Diverse herpesviral microRNAs target the stress-induced immune ligand MICB to escape recognition by natural killer cells. *Cell Host Microbe* **5:** 376–385.

Nahvi A, Shoemaker CJ, Green R. 2009. An expanded seed sequence definition accounts for full regulation of the hid 3′ UTR by bantam miRNA. *RNA* **15:** 814–822.

Nanbo A, Inoue K, Adachi-Takasawa K, Takada K. 2002. Epstein-Barr virus RNA confers resistance to interferon-α-induced apoptosis in Burkitt's lymphoma. *EMBO J* **21:** 954–965.

Omoto S, Fujii YR. 2006. Cloning and detection of HIV-1-encoded microRNA. *Methods Mol Biol* **342:** 255–265.

Ouellet DL, Plante I, Landry P, Barat C, Janelle ME, Flamand L, Tremblay MJ, Provost P. 2008. Identification of functional microRNAs released through asymmetrical processing of HIV-1 TAR element. *Nucleic Acids Res* **36:** 2353–2365.

Pawlicki JM, Steitz JA. 2008. Primary microRNA transcript retention at sites of transcription leads to enhanced microRNA production. *J Cell Biol* **182:** 61–76.

Peng W, Vitvitskaia O, Carpenter D, Wechsler SL, Jones C. 2008. Identification of two small RNAs within the first 1.5-kb of the herpes simplex virus type 1-encoded latency-associated transcript. *J Neurovirol* **14:** 41–52.

Pfeffer S, Sewer A, Lagos-Quintana M, Sheridan R, Sander C, Grasser FA, van Dyk LF, Ho CK, Shuman S, Chien M, et al. 2005. Identification of microRNAs of the herpesvirus family. *Nat Methods* **2:** 269–276.

Pfeffer S, Zavolan M, Grasser FA, Chien M, Russo JJ, Ju J, John B, Enright AJ, Marks D, Sander C, et al. 2004. Identification of virus-encoded microRNAs. *Science* **304:** 734–736.

Pijlman GP, Funk A, Kondratieva N, Leung J, Torres S, van der Aa L, Liu WJ, Palmenberg AC, Shi PY, Hall RA, et al. 2008. A highly structured, nuclease-resistant, noncoding RNA produced by flaviviruses is required for pathogenicity. *Cell Host Microbe* **4:** 579–591.

Purzycka KJ, Adamiak RW. 2008. The HIV-2 TAR RNA domain as a potential source of viral-encoded miRNA. A reconnaissance study. *Nucleic Acids Symp Ser (Oxf)* 511–512.

Reeves MB, Davies AA, McSharry BP, Wilkinson GW, Sinclair JH. 2007. Complex I binding by a virally encoded RNA regulates mitochondria-induced cell death. *Science* **316:** 1345–1348.

Reich PR, Forget BG, Weissman SM. 1966. RNA of low molecular weight in KB cells infected with adenovirus type 2. *J Mol Biol* **17:** 428–439.

Revello MG, Gerna G. 2002. Diagnosis and management of human cytomegalovirus infection in the mother, fetus, and newborn infant. *Clin Microbiol Rev* **15:** 680–715.

Riley KJ, Rabinowitz GS, Steitz JA. 2010. Comprehensive analysis of rhesus lymphocryptovirus microRNA expression. *J Virol [AOP]*.

Ruf IK, Rhyne PW, Yang C, Cleveland JL, Sample JT. 2000. Epstein-Barr virus small RNAs potentiate tumorigenicity of Burkitt lymphoma cells independently of an effect on apoptosis. *J Virol* **74:** 10223–10228.

Ruf IK, Lackey KA, Warudkar S, Sample JT. 2005. Protection from interferon-induced apoptosis by Epstein-Barr virus small RNAs is not mediated by inhibition of PKR. *J Virol* **79:** 14562–14569.

Samols MA, Hu J, Skalsky RL, Renne R. 2005. Cloning and identification of a microRNA cluster within the latency-associated region of Kaposi's sarcoma-associated herpesvirus. *J Virol* **79:** 9301–9305.

Samols MA, Skalsky RL, Maldonado AM, Riva A, Lopez MC, Baker HV, Renne R. 2007. Identification of cellular genes targeted by KSHV-encoded microRNAs. *PLoS Pathog* **3:** e65.

Sano M, Kato Y, Taira K. 2006. Sequence-specific interference by small RNAs derived from adenovirus VAI RNA. *FEBS Lett* **580:** 1553–1564.

Schafer A, Cai X, Bilello JP, Desrosiers RC, Cullen BR. 2007. Cloning and analysis of microRNAs encoded by the primate γ-herpesvirus rhesus monkey rhadinovirus. *Virology* **364:** 21–27.

Scott MS, Avolio F, Ono M, Lamond AI, Barton GJ. 2009. Human miRNA precursors with box H/ACA snoRNA features. *PLoS Comput Biol* **5:** e1000507.

Seo GJ, Chen CJ, Sullivan CS. 2009. Merkel cell polyomavirus encodes a microRNA with the ability to autoregulate viral gene expression. *Virology* **383:** 183–187.

Seo GJ, Fink LH, O'Hara B, Atwood WJ, Sullivan CS. 2008. Evolutionarily conserved function of a viral microRNA. *J Virol* **82:** 9823–9828.

Sharp PA. 2010. Gene regulation by RNA. *Cold Spring Harb Perspect Biol* **2:** a003715.

Sharp TV, Schwemmle M, Jeffrey I, Laing K, Mellor H, Proud CG, Hilse K, Clemens MJ. 1993. Comparative analysis of the regulation of the interferon-inducible protein kinase PKR by Epstein-Barr virus RNAs EBER-1 and EBER-2 and adenovirus VA1 RNA. *Nucleic Acids Res* **21:** 4483–4490.

Shen W, Sa e Silva M, Jaber T, Vitvitskaia O, Li S, Henderson G, Jones C. 2009. Two small RNAs encoded within the first 1.5 kilobases of the herpes simplex virus type 1 latency-associated transcript can inhibit productive infection and cooperate to inhibit apoptosis. *J Virol* **83:** 9131–9139.

Shimizu N, Tanabe-Tochikura A, Kuroiwa Y, Takada K. 1994. Isolation of Epstein-Barr virus (EBV)-negative cell clones from the EBV-positive Burkitt's lymphoma (BL) line Akata: Malignant phenotypes of BL cells are dependent on EBV. *J Virol* **68:** 6069–6073.

Skalsky RL, Samols MA, Plaisance KB, Boss IW, Riva A, Lopez MC, Baker HV, Renne R. 2007. Kaposi's sarcoma-associated herpesvirus encodes an ortholog of miR-155. *J Virol* **81:** 12836–12845.

Smith CA, Bates P, Rivera-Gonzalez R, Gu B, DeLuca NA. 1993. ICP4, the major transcriptional regulatory protein of herpes simplex virus type 1, forms a tripartite complex with TATA-binding protein and TFIIB. *J Virol* **67:** 4676–4687.

Soulier J, Grollet L, Oksenhendler E, Cacoub P, Cazals-Hatem D, Babinet P, d'Agay MF, Clauvel JP, Raphael M, Degos L, et al. 1995. Kaposi's sarcoma-associated herpesvirus-like DNA sequences in multicentric Castleman's disease. *Blood* **86:** 1276–1280.

Spector DH. 1996. Activation and regulation of human cytomegalovirus early genes. *Intervirology* **39:** 361–377.

Stevens JG, Wagner EK, Devi-Rao GB, Cook ML, Feldman LT. 1987. RNA complementary to a herpesvirus α gene mRNA is prominent in latently infected neurons. *Science* **235:** 1056–1059.

Sullivan CS, Grundhoff AT, Tevethia S, Pipas JM, Ganem D. 2005. SV40-encoded microRNAs regulate viral gene expression and reduce susceptibility to cytotoxic T cells. *Nature* **435:** 682–686.

Sullivan CS, Sung CK, Pack CD, Grundhoff A, Lukacher AE, Benjamin TL, Ganem D. 2009. Murine Polyomavirus encodes a microRNA that cleaves early RNA transcripts but is not essential for experimental infection. *Virology* **387:** 157–167.

Sun R, Lin SF, Gradoville L, Miller G. 1996. Polyadenylylated nuclear RNA encoded by Kaposi sarcoma-associated herpesvirus. *Proc Natl Acad Sci* **93:** 11883–11888.

Sun R, Lin SF, Gradoville L, Yuan Y, Zhu F, Miller G. 1998. A viral gene that activates lytic cycle expression of Kaposi's sarcoma-associated herpesvirus. *Proc Natl Acad Sci* **95**: 10866–10871.

Swaminathan S, Tomkinson B, Kieff E. 1991. Recombinant Epstein-Barr virus with small RNA (EBER) genes deleted transforms lymphocytes and replicates *in vitro*. *Proc Natl Acad Sci* **88**: 1546–1550.

Swaminathan S, Huneycutt BS, Reiss CS, Kieff E. 1992. Epstein-Barr virus-encoded small RNAs (EBERs) do not modulate interferon effects in infected lymphocytes. *J Virol* **66**: 5133–5136.

Taft RJ, Glazov EA, Lassmann T, Hayashizaki Y, Carninci P, Mattick JS. 2009. Small RNAs derived from snoRNAs. *RNA* **15**: 1233–1240.

Takizawa T, Tatematsu C, Watanabe M, Yoshida M, Nakajima K. 2000. Three leucine-rich sequences and the N-terminal region of double-stranded RNA-activated protein kinase (PKR) are responsible for its cytoplasmic localization. *J Biochem* **128**: 471–476.

Tang S, Patel A, Krause PR. 2009. Novel less-abundant viral microRNAs encoded by herpes simplex virus 2 latency-associated transcript and their roles in regulating ICP34.5 and ICP0 mRNAs. *J Virol* **83**: 1433–1442.

Tang S, Bertke AS, Patel A, Wang K, Cohen JI, Krause PR. 2008. An acutely and latently expressed herpes simplex virus 2 viral microRNA inhibits expression of ICP34.5, a viral neurovirulence factor. *Proc Natl Acad Sci* **105**: 10931–10936.

Thimmappaya B, Weinberger C, Schneider RJ, Shenk T. 1982. Adenovirus VAI RNA is required for efficient translation of viral mRNAs at late times after infection. *Cell* **31**: 543–551.

Toczyski DP, Matera AG, Ward DC, Steitz JA. 1994. The Epstein-Barr virus (EBV) small RNA EBER1 binds and relocalizes ribosomal protein L22 in EBV-infected human B lymphocytes. *Proc Natl Acad Sci* **91**: 3463–3467.

Umbach JL, Cullen BR. 2009. The role of RNAi and microRNAs in animal virus replication and antiviral immunity. *Genes Dev* **23**: 1151–1164.

Umbach JL, Kramer MF, Jurak I, Karnowski HW, Coen DM, Cullen BR. 2008. MicroRNAs expressed by herpes simplex virus 1 during latent infection regulate viral mRNAs. *Nature* **454**: 780–783.

Vasudevan S, Tong Y, Steitz JA. 2007. Switching from repression to activation: microRNAs can up-regulate translation. *Science* **318**: 1931–1934.

Wahid AM, Coventry VK, Conn GL. 2008. Systematic deletion of the adenovirus-associated RNA$_I$ terminal stem reveals a surprisingly active RNA inhibitor of double-stranded RNA-activated protein kinase. *J Biol Chem* **283**: 17485–17493.

Waidner LA, Morgan RW, Anderson AS, Bernberg EL, Kamboj S, Garcia M, Riblet SM, Ouyang M, Isaacs GK, Markis M, et al. 2009. MicroRNAs of Gallid and Meleagrid herpesviruses show generally conserved genomic locations and are virus-specific. *Virology* **388**: 128–136.

Walz N, Christalla T, Tessmer U, Grundhoff A. 2010. A global analysis of evolutionary conservation among known and predicted γherpesvirus miRNAs. *J Virol* **84**: 716–728.

Wang HW, Sharp TV, Koumi A, Koentges G, Boshoff C. 2002. Characterization of an anti-apoptotic glycoprotein encoded by Kaposi's sarcoma-associated herpesvirus which resembles a spliced variant of human survivin. *EMBO J* **21**: 2602–2615.

Wassarman DA, Lee SI, Steitz JA. 1989. Nucleotide sequence of HSUR 5 RNA from *herpesvirus saimiri*. *Nucleic Acids Res* **17**: 1258.

Weinberg MS, Morris KV. 2006. Are viral-encoded microRNAs mediating latent HIV-1 infection? *DNA Cell Biol* **25**: 223–231.

Wolin SL, Cedervall T. 2002. The La protein. *Annu Rev Biochem* **71**: 375–403.

Wu Y, Maruo S, Yajima M, Kanda T, Takada K. 2007. Epstein-Barr virus (EBV)-encoded RNA 2 (EBER2) but not EBER1 plays a critical role in EBV-induced B-cell growth transformation. *J Virol* **81**: 11236–11245.

Wu Z, Zhu Y, Bisaro DM, Parris DS. 2009. Herpes simplex virus type 1 suppresses RNA-induced gene silencing in mammalian cells. *J Virol* **83**: 6652–6663.

Xing L, Kieff E. 2007. Epstein-Barr Virus BHRF1 micro and stable RNAs in Latency III and after induction of replication. *J Virol* **18**: 9967–9975.

Xu H, Yao Y, Zhao Y, Smith LP, Baigent SJ, Nair V. 2008. Analysis of the expression profiles of Marek's disease virus-encoded microRNAs by real-time quantitative PCR. *J Virol Methods* **149**: 201–208.

Xu N, Segerman B, Zhou X, Akusjärvi G. 2007. Adenovirus virus-associated RNAII-derived small RNAs are efficiently incorporated into the RNA-induced silencing complex and associate with polyribosomes. *J Virol* **81**: 10540–10549.

Yajima M, Kanda T, Takada K. 2005. Critical role of Epstein-Barr Virus (EBV)-encoded RNA in efficient EBV-induced B-lymphocyte growth transformation. *J Virol* **79**: 4298–4307.

Yang L, Aozasa K, Oshimi K, Takada K. 2004. Epstein-Barr virus (EBV)-encoded RNA promotes growth of EBV-infected T cells through interleukin-9 induction. *Cancer Res* **64**: 5332–5337.

Yao Y, Zhao Y, Smith LP, Watson M, Nair V. 2009. Novel microRNAs (miRNAs) encoded by herpesvirus of Turkeys: Evidence of miRNA evolution by duplication. *J Virol* **83**: 6969–6973.

Yao Y, Zhao Y, Xu H, Smith LP, Lawrie CH, Sewer A, Zavolan M, Nair V. 2007. Marek's disease virus type 2 (MDV-2)-encoded microRNAs show no sequence conservation with those encoded by MDV-1. *J Virol* **81**: 7164–7170.

Yao Y, Zhao Y, Xu H, Smith LP, Lawrie CH, Watson M, Nair V. 2008. MicroRNA profile of Marek's disease virus-transformed T-cell line MSB-1: Predominance of virus-encoded microRNAs. *J Virol* **82**: 4007–4015.

Zhong W, Ganem D. 1997. Characterization of ribonucleoprotein complexes containing an abundant polyadenylated nuclear RNA encoded by Kaposi's sarcoma-associated herpesvirus (human herpesvirus 8). *J Virol* **71**: 1207–1212.

Zhong W, Wang H, Herndier B, Ganem D. 1996. Restricted expression of Kaposi sarcoma-associated herpesvirus (human herpesvirus 8) genes in Kaposi sarcoma. *Proc Natl Acad Sci* **93**: 6641–6646.

Zhu JY, Pfuhl T, Motsch N, Barth S, Nicholls J, Grasser F, Meister G. 2009. Identification of novel Epstein-Barr Virus miRNA genes from nasopharyngeal carcinomas. *J Virol* **83**: 3333–3341.

Spliceosome Structure and Function

Cindy L. Will and Reinhard Lührmann

Max Planck Institute for Biophysical Chemistry, Department of Cellular Biochemistry, Am Fassberg 11, 37077 Göttingen, Germany

Correspondence: Reinhard.Luehrmann@mpi-bpc.mpg.de

SUMMARY

Pre-mRNA splicing is catalyzed by the spliceosome, a multimegadalton ribonucleoprotein (RNP) complex comprised of five snRNPs and numerous proteins. Intricate RNA-RNA and RNP networks, which serve to align the reactive groups of the pre-mRNA for catalysis, are formed and repeatedly rearranged during spliceosome assembly and catalysis. Both the conformation and composition of the spliceosome are highly dynamic, affording the splicing machinery its accuracy and flexibility, and these remarkable dynamics are largely conserved between yeast and metazoans. Because of its dynamic and complex nature, obtaining structural information about the spliceosome represents a major challenge. Electron microscopy has revealed the general morphology of several spliceosomal complexes and their snRNP subunits, and also the spatial arrangement of some of their components. X-ray and NMR studies have provided high resolution structure information about spliceosomal proteins alone or complexed with one or more binding partners. The extensive interplay of RNA and proteins in aligning the pre-mRNA's reactive groups, and the presence of both RNA and protein at the core of the splicing machinery, suggest that the spliceosome is an RNP enzyme. However, elucidation of the precise nature of the spliceosome's active site, awaits the generation of a high-resolution structure of its RNP core.

Outline

1. Introduction
2. Cis-acting pre-mRNA elements and the catalytic steps of splicing
3. Intron- and exon-defined spliceosome assembly pathways
4. A dynamic network of RNA-RNA interactions in the spliceosome and the catalytic role of RNA
5. A conformational two-state model for the spliceosome's catalytic center
6. The spliceosome possesses a complex and dynamic protein composition
7. A rich protein composition affords flexibility to the metazoan spliceosome
8. Splice site recognition involves the coordinated action of RNA and protein

9 Proteins facilitate structural rearrangements in the spliceosome

10 Posttranslational protein modifications also contribute to splicing dynamics

11 Structure of the spliceosomal snRNPs and non-snRNP splicing factors

12 High resolution structures of snRNP and/or spliceosome components

13 Crystal structure of the human U1 snRNP

14 Elucidation of the spatial organization of spliceosomal complexes using biochemical probes

15 Electron microscopy of spliceosomal complexes

16 Localization of regions of the pre-mRNA in the spliceosome via EM

17 3D structures of the spliceosome obtained by EM

18 Summary

References

1 INTRODUCTION

Most eukaryotic genes are expressed as precursor mRNAs (pre-mRNAs) that are converted to mRNA by splicing, an essential step of gene expression in which noncoding sequences (introns) are removed and coding sequences (exons) are ligated together. Whereas some exons are constitutively spliced—that is, they are present in every mRNA produced from a given pre-mRNA—many are alternatively spliced to generate variable forms of mRNA from a single pre-mRNA species. Alternative splicing is prevalent in higher eukaryotes and it enhances their complexity by increasing the number of unique proteins expressed from a single gene (Nilsen and Graveley 2010). Unraveling splicing at the molecular level is not only important for understanding gene expression, but it is also of medical relevance, as aberrant pre-mRNA splicing is the basis of many human diseases or contributes to their severity (Novoyatleva et al. 2006; Ward and Cooper 2010).

Nuclear pre-mRNA splicing is catalyzed by the spliceosome, a multi-megadalton ribonucleoprotein (RNP) complex. Both the conformation and composition of the spliceosome are highly dynamic, affording the splicing machinery its accuracy and at the same time flexibility. Two unique spliceosomes coexist in most eukaryotes: the U2-dependent spliceosome, which catalyzes the removal of U2-type introns, and the less abundant U12-dependent spliceosome, which is present in only a subset of eukaryotes and splices the rare U12-type class of introns (reviewed by Patel and Steitz 2003). This article focuses on recent advances in our understanding of the structure and function of the U2-dependent spliceosome, with emphasis on constitutive as opposed to alternative pre-mRNA splicing. For more in-depth reviews of alternative splicing and its regulation the reader is referred to several recent reviews (Chen and Manley 2009; Matlin et al. 2005; Black 2003; Smith and Valcárcel 2000).

2 CIS-ACTING PRE-mRNA ELEMENTS AND THE CATALYTIC STEPS OF SPLICING

Information provided by a pre-mRNA that contributes to defining an intron is limited to short, conserved sequences at the 5′ splice site (ss), 3′ss and branch site (BS) (Fig. 1)(see also Burge et al. 1999). The BS is typically located 18-40 nucleotides upstream from the 3′ss and in higher eukaryotes is followed by a polypyrimidine tract (PPT) (Fig. 1B). Different splice site and branch site sequences are found in U2- versus U12-type introns (Burge et al. 1999). The U2-type consensus sequences found in the budding yeast *Saccharomyces cerevisiae* exhibit a higher level of conservation than those in metazoans (Fig. 1B). Additional, cis-acting pre-mRNA elements include exonic and intronic splicing enhancers (ESEs and ISEs) or silencers (ESSs and ISSs). They are typically short and diverse in sequence and modulate both constitutive and alternative splicing by binding regulatory proteins that either stimulate or repress the assembly of spliceosomal complexes at an adjacent splice site (reviewed by Smith and Valcárcel 2000; Wang and Burge 2008).

Nuclear pre-mRNA introns are removed by two consecutive transesterification reactions (reviewed by Moore et al. 1993). First, the 2′ OH group of the branch adenosine of the intron carries out a nucleophilic attack on the 5′ss. This results in cleavage at this site and ligation of the 5′ end of the intron to the branch adenosine, forming a lariat structure (Fig. 1A). Second, the 3′ss is attacked by the 3′OH group of the 5′ exon, leading to the ligation of the 5′ and 3′ exons (forming the mRNA), and release of the intron

(Fig. 1A). The intermediates and products of pre-mRNA splicing are similar to those generated during the removal of group II self-splicing introns (see Jacquier 1990). This similarity led to the hypothesis that catalysis of pre-mRNA splicing is also RNA-based. Likewise similar to group II self-splicing introns, it was recently shown that, under the appropriate conditions, the chemical steps of nuclear pre-mRNA splicing are reversible (Tseng and Cheng 2008), which further emphasizes the mechanistic similarities between the two systems. These studies also underscore the dynamic nature of the spliceosome during the catalytic phase of splicing, as discussed later.

3 INTRON- AND EXON-DEFINED SPLICEOSOME ASSEMBLY PATHWAYS

To compensate for the limited information contained in the splicing substrate itself, a large number of *trans*-acting factors interact with the pre-mRNA to form the spliceosome, in which the reactive groups of the pre-mRNA are spatially positioned for catalysis. The U2-dependent spliceosome is assembled from the U1, U2, U5, and U4/U6 snRNPs and numerous non-snRNP proteins. The main subunits of the U12-dependent spliceosome, in contrast, are the U11, U12, U5, and U4atac/U6atac snRNPs (reviewed by Patel and Steitz 2003). Each snRNP consists of an snRNA (or two in the case of U4/U6), a common set of seven Sm proteins (B/B′, D3, D2, D1, E, F, and G) and a variable number of particle-specific proteins (Will and Lührmann 2006) (Fig. 2). The most probable secondary structure of the human spliceosomal U1, U2, U4/U6, and U5 snRNAs are shown in Figure 2; note that several of the snRNAs undergo structural rearrangements during splicing, as described in detail later. In contrast to ribosomal subunits, none of the spliceosomal snRNPs possesses a pre-formed active site and several of them are substantially remodeled in the course of splicing.

Spliceosome assembly occurs by the ordered interaction of the spliceosomal snRNPs and numerous other splicing factors (reviewed by Brow 2002; Matlin and Moore 2007; Staley and Woolford 2009). In the event that an intron does not exceed ∼200–250 nts, the spliceosome initially assembles across the intron (Fox-Walsh et al. 2005) (Fig. 1C). In the earliest cross-intron spliceosomal complex (i.e., the E complex), the U1 snRNP is recruited to the 5′ss and non-snRNP factors such as SF1/mBBP and U2AF interact with the BS and PPT, respectively. In a subsequent step, the U2 snRNP stably associates with the BS, forming the A complex (also denoted prespliceosome). The U4/U6.U5 tri-snRNP, which is pre-assembled from the U5 and U4/U6 snRNPs, is then recruited, generating the pre-catalytic B complex. Major rearrangements in RNA–RNA and RNA–protein interactions, leading to the destabilization of the U1 and U4 snRNPs, give rise to the activated spliceosome (i.e., the B^{act} complex). Subsequent catalytic activation by the DEAH-box RNA helicase Prp2, generates the B* complex, which catalyzes the first of the two steps of splicing. This yields the C complex, which in turn catalyzes the second step. The spliceosome then dissociates and, after additional remodeling, the released snRNPs take part in additional rounds of splicing.

In addition to this canonical cross-intron assembly pathway, alternative cross-intron assembly pathways leading to a catalytically active spliceosome likely also exist. For example, a complex containing all five snRNPs (the penta-snRNP) but lacking pre-mRNA has been isolated (Stevens et al. 2002). In the presence of pre-mRNA and additional splicing factors, the penta-snRNP can be chased into an active spliceosome without first undergoing disassembly and then subsequent reassembly. Thus, the generation of an active spliceosome does not necessarily require multiple assembly steps before its activation.

Alternative assembly pathways also exist at the earliest stages of spliceosome assembly, at least in metazoans. Most mammalian pre-mRNAs contain multiple introns whose sizes vary from several hundred to several thousand nucleotides (Deutsch and Long 1999), whereas their exons have a rather fixed length of only ∼120 nt on average (Ast 2004). When intron length exceeds ∼200–250 nt (which is the case for many introns in higher eukaryotes), splicing complexes first form across an exon (Fox-Walsh et al. 2005), a process called exon definition (Berget 1995). Recent analysis of the coevolution of the 5′ss and 3′ss support the idea that exon definition is prominent in mammals, but less so in most other metazoans (Xiao et al. 2007). During exon definition, the U1 snRNP binds to the 5′ss downstream of an exon and promotes the association of U2AF with the polypyrimidine tract/3′ss upstream of it (Fig. 1D). This in turn leads to the recruitment of the U2 snRNP to the BS upstream of the exon. Splicing enhancer sequences within the exon (ESEs) recruit proteins of the SR protein family, which establish a network of protein–protein interactions across the exon that stabilize the exon-defined complex (Hoffman and Grabowski 1992; Reed 2000).

As the chemical steps of splicing occur across an intron, subsequent to exon definition the 3′ss must be paired across the adjacent intron with an upstream 5′ss. This switch from an exon-defined to intron-defined splicing complex is currently poorly understood. It is thought that cross-exon interactions are first disrupted and the cross-exon complex is then converted into a cross-intron A complex, where a molecular bridge now forms between U2 and U1 bound to an upstream 5′ss (Reed 2000; Smith and Valcárcel 2000). This

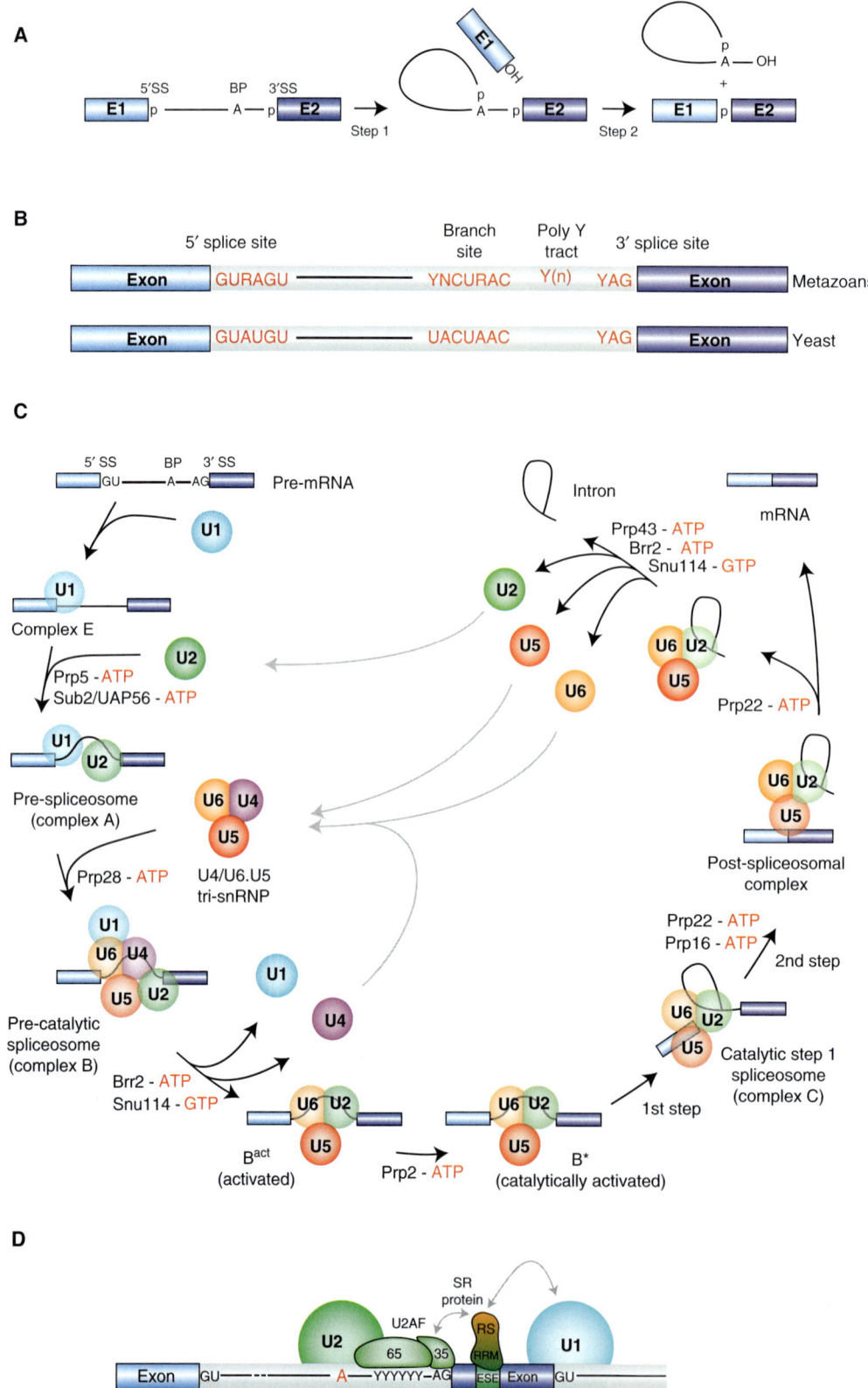

Figure 1. Pre-mRNA splicing by the U2-type spliceosome. (*A*) Schematic representation of the two-step mechanism of pre-mRNA splicing. Boxes and solid lines represent the exons (E1, E2) and the intron, respectively. The branch site adenosine is indicated by the letter A and the phosphate groups (p) at the 5′ and 3′ splice sites, which are conserved in the splicing products, are also shown. (*See facing page for legend.*)

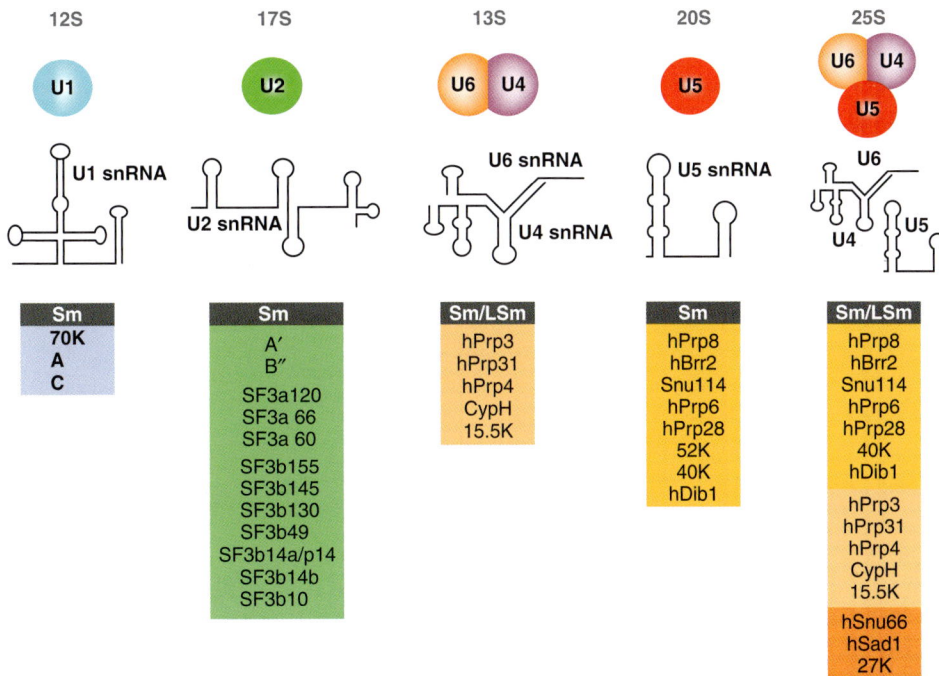

Figure 2. Protein composition and snRNA secondary structures of the major human spliceosomal snRNPs. All seven Sm proteins (B/B', D3, D2, D1, E, F, and G) or LSm proteins (Lsm2-8) are indicated by "Sm" or "LSm" at the top of the boxes showing the proteins associated with each snRNP. The U4/U6.U5 tri-snRNP contains two sets of Sm proteins and one set of LSm proteins.

step is decisive in determining which 5' and 3' exon will ultimately be spliced together; recent data indicate that regulation of exon inclusion or skipping during several alternative splicing events occurs during the switch from a cross-exon to cross-intron complex (House and Lynch 2006; Bonnal et al. 2008; Sharma et al. 2008). Recently, exon-defined complexes were shown to contain not only U1 and U2, but also the U4/U5.U6 tri-snRNP and evidence was provided that it is possible for an exon-defined complex to be converted directly into a cross-intron B complex (Schneider et al. 2010b). These data suggest that multiple pathways leading from an exon-defined complex to an intron-defined spliceosome likely exist and that, although splice site pairing generally occurs during A complex formation (Lim and Hertel 2004), in some instances it can potentially occur at an even later stage of spliceosome assembly.

4 A DYNAMIC NETWORK OF RNA-RNA INTERACTIONS IN THE SPLICEOSOME AND THE CATALYTIC ROLE OF RNA

During spliceosome assembly, an intricate RNA–RNA interaction network is formed that is extensively rearranged during catalytic activation of the spliceosome and the catalytic steps of splicing (reviewed by Nilsen 1998). Whereas RNA-RNA secondary interactions in the spliceosome are, for the most part, well-characterized, information about the nature and dynamics of RNA tertiary interactions is scarce. Thus, conformational rearrangements in the RNA

Figure 1. (*Continued*) (*B*) Conserved sequences found at the 5' and 3' splice sites and branch site of U2-type pre-mRNA introns in metazoans and budding yeast (*S. cerevisiae*). Y = pyrimidine and R = purine. The polypyrimidine tract is indicated by (Yn). (*C*) Canonical cross-intron assembly and disassembly pathway of the U2-dependent spliceosome. For simplicity, the ordered interactions of the snRNPs (indicated by circles), but not those of non-snRNP proteins, are shown. The various spliceosomal complexes are named according to the metazoan nomenclature. Exon and intron sequences are indicated by boxes and lines, respectively. The stages at which the evolutionarily conserved DExH/D-box RNA ATPases/helicases Prp5, Sub2/UAP56, Prp28, Brr2, Prp2, Prp16, Prp22 and Prp43, or the GTPase Snu114, act to facilitate conformational changes are indicated. (*D*) Model of interactions occurring during exon definition.

network of the spliceosome are likely even more complex than current models would suggest.

At the earliest stages of spliceosome assembly, U1 snRNA base pairs with the 5′ss. U2 snRNA then base pairs with the BS, forming a short U2-BS duplex in which the branch adenosine is bulged out, specifying its 2′ OH as the nucleophile for the first catalytic step of splicing. Within the U4/U6.U5 tri-snRNP, the U6 and U4 snRNAs are extensively base paired with each other. After association of the tri-snRNP with the A complex, the U4/U6 interaction is disrupted, and the 5′ end of U6 snRNA base pairs with the 5′ss, displacing the U1 snRNA in the process (Fig. 3). In addition, an extensive base pairing network is formed between U6 and U2, which juxtaposes the 5′ss and BS for the first step of splicing. Furthermore, a central region of the U6 snRNA forms an intramolecular stem-loop structure (U6-ISL) that appears to play a crucial role in splicing catalysis (Fig. 3). Tri-snRNP integration also leads to U5 snRNA interactions with exon nucleotides near the 5′ss.

The precise nature of the U6 and U2 snRNA interaction network is the subject of some debate, with two different models currently proposed. In the first, U2 and U6 form three helices (Ia, Ib, and II) (Fig. 3), with the conserved U6 triad AGC forming three base pairs with U2 (corresponding to helix Ib) (Madhani and Guthrie 1992). In an alternative model the AGC triad no longer base pairs with U2 but rather with other U6 nucleotides, extending the U6-ISL and allowing for an intramolecular U2 stem-loop, thereby generating a U2–U6 four-way junction (Sun and Manley 1995; Sashital et al. 2004). Recent data have revealed a role for helix I in both steps of splicing, and suggest helix I is disrupted after step 1, but reforms before step 2 (Mefford and Staley 2009). Single molecule analyses of U2 and U6 RNA duplexes suggest that multiple conformations of these RNAs exist (Guo et al. 2009). Thus, U2–U6 interactions appear to be highly dynamic, and these snRNAs likely adopt different conformations at different stages of splicing.

Like U6, the U2 snRNA also appears to undergo intramolecular rearrangements during splicing. Two mutually exclusive stem structures (stem IIa and stem IIc) form within the yeast U2 snRNA in the spliceosome, with recent evidence indicating that U2 toggles iteratively between these two conformations (Hilliker et al. 2007; Perriman and Ares 2007). Formation of stem IIa promotes the U2/BS interaction during prespliceosome formation. Subsequent formation of stem IIc promotes the first catalytic step of splicing, and a switch back to the stem IIa conformation is required for the second step (i.e., exon ligation). These iterative conformational changes once again underscore the highly dynamic nature of the RNA network within the spliceosome.

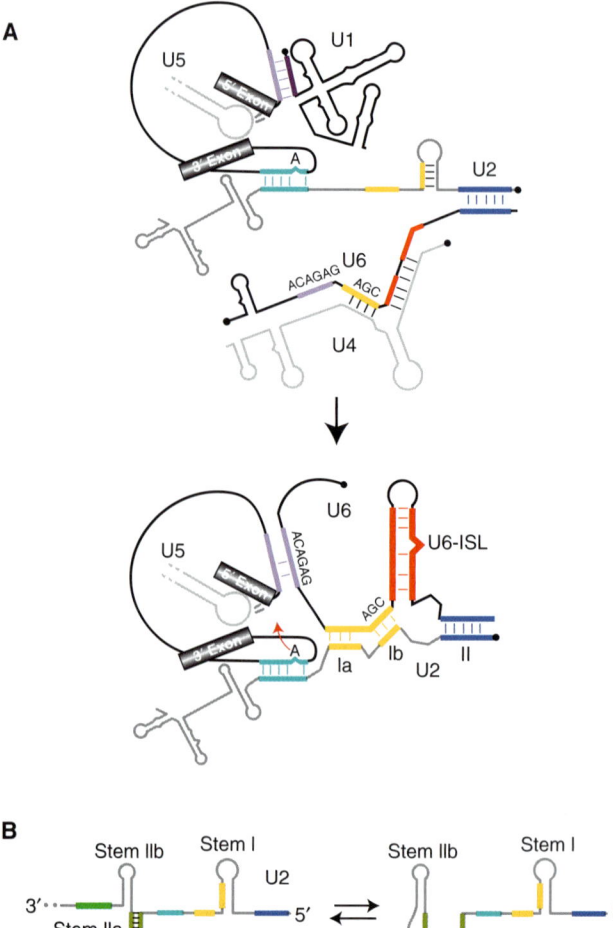

Figure 3. Dynamic network of RNA-RNA interactions in the spliceosome. (A) Exon sequences are indicated by grey boxes and intron sequences by a thin black line. snRNAs are shown schematically (secondary structure as observed in mammals) in grey or black, with those regions engaging in base pairing interactions (indicated by short lines) highlighted in color (not drawn to scale). The 5′ end of the snRNAs is indicated by a black dot. Solely loop 1 of the U5 snRNA is shown. During the transition from a precatalytic spliceosome (upper diagram) to a catalytically activated spliceosome (lower diagram) U1 and U4 are displaced, and U6 and U2 engage in novel base pairing interactions. (B) Conformational toggling of the yeast U2 snRNA. Two mutually exclusive stem structures (stem IIa and stem IIc) are thought to form within the U2 snRNA at different stages of splicing. Solely the 5′ end of the U2 snRNA is shown schematically.

Rearrangements are required after step 1 of splicing to reposition the splicing intermediates for the second step of splicing and allow nucleophilic attack of the 5′ exon at the 3′ss (reviewed by Smith et al. 2008). The precise timing of these changes and the conformation of the RNA–RNA interaction network at this stage is not clear. Recent data have shown that the U6/5′ss interaction must be disrupted before step 2 (Konarska et al. 2006), The U2/BS interaction is not strictly required for the second step and thus it has been proposed that this interaction is also disrupted between the first and second step of splicing (Smith et al. 2007). The conformation of the U2/U6 interaction is also not entirely clear, but likely helix Ia, Ib, and II are formed at this stage (see previous discussion). Before step 2, U5 also contacts exon nucleotides just downstream of the 3'ss, and not only tethers the 5' exon to the spliceosome after step 1, but also aligns both exons for the second catalytic step (reviewed by Turner et al. 2004).

A large body of evidence supports the idea that catalysis of pre-mRNA splicing is at least partially RNA-based with U2 and U6 playing key roles (reviewed by Valadkhan 2005; Wachtel and Manley 2009). Several intermolecular structures formed by the pre-mRNA and the U2, U5, and U6 snRNAs are similar to intramolecular structures formed by self-splicing group II introns (Keating et al. 2010), supporting the idea that pre-mRNA splicing is catalyzed by RNA. Reactions resembling splicing can be catalyzed by short U2 and U6 RNAs in the absence of protein (Valadkhan and Manley 2001; Valadkhan et al. 2007, 2009). However, the kinetics and efficiency of these reactions are slow, suggesting that important cofactors are missing. Furthermore, recent data suggest that RNA generally has a high intrinsic reactivity (Smith and Konarska 2009a), which asks for more profound evidence for the validity of using such snRNA only systems as evidence for RNA-based splicing catalysis (Smith and Konarska 2009b). Perhaps the most compelling evidence that the spliceosome's active site is composed (at least partially) of RNA comes from the crystal structure of a self-splicing group II intron (Toor et al. 2008). These studies revealed that RNA motifs shared by U6 snRNA and group II introns, namely the ACAGAG box and AGC triad, form the basis of the group II intron active site (see also discussion by Keating et al. 2010). On the other hand, data supporting the idea that protein comprises part of the spliceosome's active site also continues to mount (reviewed by Abelson et al. 2008; Wachtel and Manley 2009). Foremost is the recent discovery that the spliceosomal Prp8 protein contains an RNAse H-like domain (see later), suggesting it participates directly in catalysis.

5 A CONFORMATIONAL TWO-STATE MODEL FOR THE SPLICEOSOME'S CATALYTIC CENTER

The spliceosome appears to use a single active site for both catalytic steps, and thus the lariat intermediate formed in the first step is thought to be displaced to allow positioning of the 3'ss for the second step (repositioning of the 5' exon, in contrast, would theoretically not be required). Thus, the spliceosome would need to exist in two distinct conformational states during the catalytic phase of splicing, binding the substrates differently for the two steps. Based on elegant genetic studies in yeast, demonstrating the existence of two opposing classes of suppressor alleles, an equilibrium between two distinct spliceosome conformations, one that promotes the first step and another conformation that promotes the second step was proposed (Query and Konarska 2004; Konarska and Query 2005). By analogy to the ribosome, where tRNA decoding involves transitions between open and closed conformations at the A site of the 30S subunit, the catalytic center of the spliceosome may thus likewise toggle between open and closed states. Indeed, a number of interactions within the spliceosome appear to toggle (i.e., they are disrupted and then reform at a later stage), such as observed with stem II of the U2 snRNA (see earlier; Hilliker et al. 2007; Perriman and Ares 2007). The demonstration that the catalytic steps of splicing are reversible also support this model (Tseng and Cheng 2008). More recent data suggest that the transition between these two states has multiple phases, including a repositioning step (Liu et al. 2007). Thus, the transition from step 1 to step 2 of splicing likely involves several remodeling events (Smith et al. 2008).

6 THE SPLICEOSOME POSSESSES A COMPLEX AND DYNAMIC PROTEIN COMPOSITION

Unlike group II introns, nuclear pre-mRNA introns and the spliceosomal snRNAs do not self-assemble into a catalytically active structure in the absence of spliceosomal proteins. Proteins play critical roles in the recognition and pairing of splice sites, facilitate the dynamics of the RNA-RNA, RNA-protein, and protein-protein interaction networks of the spliceosome, and ensure that the reactive sites of the pre-mRNA are properly positioned for catalysis as discussed in detail later. Proteomic analyses of purified human spliceosomal complexes indicate that over 170 proteins associate with the metazoan spliceosome at some point during the splicing process, with individual assembly intermediates (e.g., B and C complexes) containing significantly fewer (~110) proteins (reviewed by Jurica and Moore 2003; Wahl et al. 2009). Thus, the spliceosome is a particularly protein-rich RNP, with proteins comprising more than two-thirds of its mass in humans in the case of short pre-mRNA introns. Protein–protein, as well as protein–RNA interactions should therefore be prevalent and play functionally important roles in the spliceosome. As a consequence of its complexity, assembly of the spliceosome represents a kinetic challenge that is met, in part, by prepackaging many spliceosomal proteins in the form of snRNPs or in stable pre-formed heteromeric complexes. Indeed, ~45 proteins are recruited to the human spliceosome as part of the spliceosomal snRNPs, whereas non-snRNP proteins comprise the remainder. The composition of the spliceosome is highly dynamic with a remarkable exchange of proteins from one stage of splicing to the next. These changes are also accompanied by extensive remodeling of the snRNPs within the spliceosome.

In addition to human and *Drosophila melanogaster* spliceosomes (Herold et al. 2009), the protein composition of affinity-purified, in vitro assembled *S. cerevisiae* spliceosomal complexes has now been determined by mass spectrometry (Fabrizio et al. 2009). Analysis of yeast B, B^{act} and C complexes revealed that the number of proteins identified in each complex is much lower than that in the corresponding metazoan complex. For example, yeast precatalytic B complexes contained only ~60 proteins (compared to ~110 in humans and flies), including essentially all U1, U2, and tri-snRNP proteins plus proteins of the nineteen complex (NTC) and mRNA retention and splicing (RES) complex (Fig. 4). Likewise, yeast C complexes contained only ~50 proteins compared to ~110 in metazoan C complexes. Altogether ~90 proteins were identified in yeast spliceosomes, nearly all of which have homologs in higher eukaryotes (Fabrizio et al. 2009). Thus, the yeast splicing machinery likely contains the evolutionarily conserved, core set of spliceosomal proteins required for constitutive splicing. Indeed, most of the remaining ~80 proteins found in human and *D. melanogaster* spliceosomes have no counterparts in yeast, with many playing a role in alternative splicing, a process that is essentially absent in yeast (Fabrizio et al. 2009).

A dramatic exchange of proteins occurs during spliceosome assembly and activation (Wahl et al. 2009). The same homologous proteins are subject to dissociation/recruitment events during the transition from the B to C complex in both metazoans and yeast (Deckert et al. 2006; Bessonov et al. 2008; Fabrizio et al. 2009; Herold et al. 2009), indicating that these compositional changes are an evolutionarily conserved design principle of the spliceosome. In yeast, the most extensive compositional exchange occurs during the transition from the precatalytic B complex to the activated B^{act} complex (Fabrizio et al. 2009). During this transition, ~35 proteins dissociate, including among others all U1 and U4/U6 associated proteins, whereas 12 others are recruited (Fig. 4). Thus, the U4/U6.U5 tri-snRNP undergoes massive remodeling during activation, with all U4/U6 associated proteins, the U4 snRNA and some U5 proteins released. It is presently not clear whether these proteins are destabilized/released concomitantly or in separate steps. Indeed it is not known how many discrete RNP remodeling events occur during splicing. Numerous structurally distinct spliceosome intermediates likely exist, with each RNP rearrangement potentially subjected to regulation. As the U6 snRNA appears to have lost most of its preactivation binding partners, it not only engages in novel base pairing interactions with U2 but also new protein–RNA interactions are thought to be established (Wahl et al. 2009). Proteins of the NTC complex (or Prp19/CDC5 complex in humans) and related proteins, as well as SR proteins in humans (Shen and Green 2007), likely are involved in tethering the U6 snRNA at this stage (Chan et al. 2003). In humans there is evidence that the U5 snRNP is also remodeled during activation; ~15 proteins—including those comprising the human Prp19/CDC5 complex associate stably with U5 at this stage, yielding a remodeled 35S form of the U5 snRNP (Makarov et al. 2002).

The transition from B^{act} to C complex is also accompanied by compositional changes, but to a much lesser extent. In yeast, only two proteins are lost and nine proteins, including step two factors and the trimeric NTR spliceosome disassembly complex, are recruited at this stage (Fig. 4). Because of the low number of proteins recruited at this stage, it has been possible to investigate the role of some of these factors in the catalytic steps of splicing using affinity-purified yeast B^{act} complexes of defined composition and adding back recombinant splicing factors (Warkocki et al. 2009; see also below). The ability to restore both steps of *S. cerevisiae* splicing from purified components should allow a fine dissection of the role of RNA helicases in RNP remodeling events accompanying the catalytic steps of splicing. Interestingly, the U2 snRNP also appears to be substantially remodeled just before, or during, C complex formation both in yeast and humans, with an apparent destabilization/loss of the U2-associated SF3a and SF3b proteins (Bessonov et al. 2008; Fabrizio et al. 2009). This suggests that, although required for the U2/BS interaction during the early stages of splicing, SF3a/b are not required after step 1. This is also consistent with the proposed disruption of the U2/BS interaction after step 1 (Smith et al. 2007).

Recently it was possible to purify human spliceosomal C complexes that were capable of catalyzing exon ligation on their own (Bessonov et al. 2008). High salt treatment of these C complexes yielded an RNP core consisting of only ~35 proteins, in which the catalytic RNA–RNA network appeared to be intact. Main components of this spliceosomal RNP core are Prp19/CDC5 proteins and Prp19-related factors, plus U5 proteins including Prp8. These data provide a first glimpse into the RNP core of the step 1 spliceosome and indicate that the aforementioned proteins play a central role in sustaining its catalytically active structure.

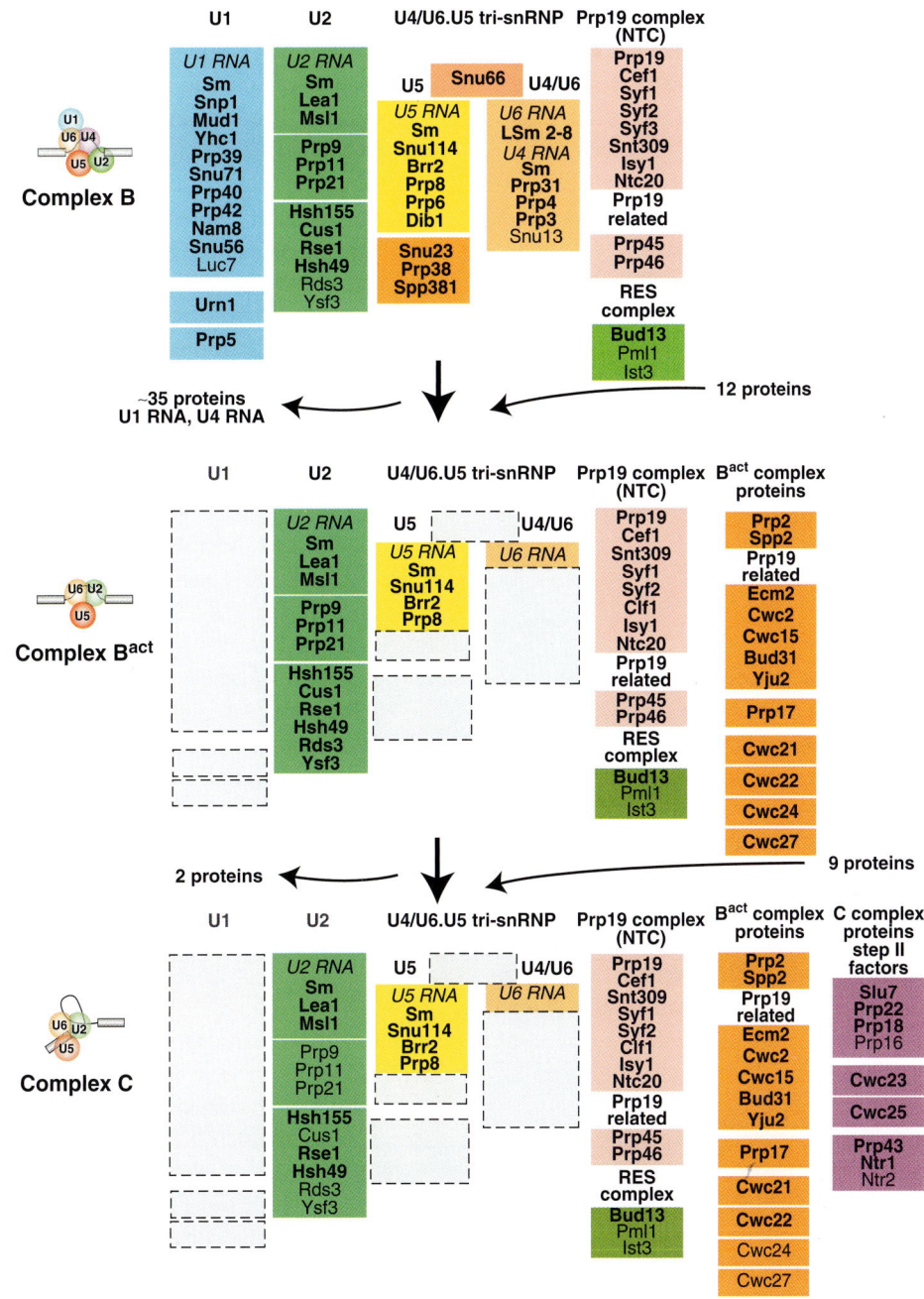

Figure 4. Compositional dynamics of the yeast spliceosome. Proteins identified by mass spectrometry in S. cerevisiae B, B^{act}, and C spliceosomal complexes are shown. Proteins are grouped according to their function or association with an snRNP, protein complex or spliceosomal complex. The relative abundance of the indicated proteins is indicated by light (substoichiometric) or dark (stoichiometric) lettering. (Reprinetd, with permission, from Fabrizio et al. 2009 [© Elsevier].)

7 A RICH PROTEIN COMPOSITION AFFORDS FLEXIBILITY TO THE METAZOAN SPLICEOSOME

The highly complex nature of the metazaon spliceosome in terms of its protein composition is puzzling at first glance. However, as mentioned earlier, a large number of human spliceosome-associated proteins play largely regulatory roles in splicing, whereas others are thought to couple splicing to other molecular machines in the nucleus, including those involved in transcription and polyadenylation (reviewed by Maniatis and Reed 2002). In addition, several proteins that co-purify with human spliceosomes are loosely associated and/or have redundant functions (e.g., different SR-proteins) and thus are likely not required to splice every pre-mRNA substrate (Wahl et al. 2009). Indeed, recent observations suggest that pre-mRNA substrates differ even in their requirements for core components of the spliceosome, presumably because of different affinities between components of the spliceosome and the various pre-mRNAs they encounter or the redundant nature of splicing signals (Clark et al. 2002; Park et al. 2004; Pleiss et al. 2007). Thus, the compositional complexity of the metazoan spliceosome is likely in part a reflection of the wide variety of pre-mRNA substrates that it must engage and the widespread occurrence of regulated splicing events. The expanded repertoire of spliceosome-associated proteins also affords flexibility to the splicing machinery to quickly respond to changes in the cellular environment.

8 SPLICE SITE RECOGNITION INVOLVES THE COORDINATED ACTION OF RNA AND PROTEIN

One of the earliest tasks of the spliceosome is to distinguish bona fide splice sites from the numerous nonauthentic sites found in a pre-mRNA. Proteins contribute to splice site recognition both indirectly by stabilizing snRNA-pre-mRNA base pairing interactions or directly by contacting reactive nucleotides of the pre-mRNA. During splicing, splice sites are recognized multiple times by both protein and RNA, ensuring the remarkable precision of the splicing reaction. Many functionally important binary interactions within the spliceosome are weak. However, the combination of multiple weak interactions enhances the overall stability of the RNP complexes formed, while at the same time affording the spliceosome the plasticity needed for regulated splicing events.

The 5′ss is initially recognized via base pairing of the U1 snRNA. In metazoans, this interaction is stabilized by the U1-70K and U1-C protein (which directly contacts the 5′ss), as well as members of the SR protein family (reviewed by Will and Lührmann 2006). Recent data indicate that SR proteins, via their RS domain, help to stabilize not only the U1/5′ss base pairing interaction, but also subsequent U6 and U5 contacts with the 5′ss (Shen and Green 2004, 2006, 2007). In addition to engaging in protein-protein interactions, positively-charged RS domains are thought to selectively contact the pre-mRNA's splicing signals (i.e., the 5′ss and BS) during the formation of transient snRNA-pre-mRNA duplexes and subsequently promote these interactions by neutralizing the positively charged phosphates of the two RNA strands (Valcárcel et al. 1996; Shen and Green 2006). A dissection of SR protein interactions with the 5′ss revealed that the RS domain contacts the site of the U6/5′ss interaction during step 1 of splicing, and then shifts to the site of the U5/5′ss interaction before step 2, mirroring the remodeling of the spliceosome between the two steps of splicing (Shen and Green 2007). The U5 protein 220K/Prp8 also contacts the 5′ss (as well as the 3′ss, BS and PPT) (reviewed by Grainger and Beggs 2005). Prp8 has thus been implicated in 5′ss recognition, and also in aiding to position the 5′ss within the spliceosome for the first step of splicing. After step 1, 220K/Prp8 is thought, together with the U5 snRNA, to help align the 5′ss and 3′ss for the second step.

Multiple interactions also contribute to the recognition of the BS (reviewed by Will and Lührmann 2006; Wahl et al. 2009). In metazoans, the branch adenosine is initially recognized by SF1/mBBP, which binds cooperatively with the heteromeric dimer U2AF. The 65-kDa subunit of U2AF binds the neighboring polypyrimidine tract, whereas its 35 kDa subunit contacts the 3′ss and plays a role in 3′ss recognition. In a subsequent step, the U2/BS base pairing interaction is formed, SF1/mBBP is displaced and the branch adenosine is now contacted by a subunit of the U2-associated SF3b complex, namely p14/SF3b14a. The U2/BS duplex is stabilized by U2AF and subunits of the heteromeric splicing factors SF3a and SF3b, which contact the pre-mRNA in the vicinity of the branch site. During the catalytic steps of splicing, there are likely multiple remodeling events at or near the BS, but these are presently not well-characterized.

9 PROTEINS FACILITATE STRUCTURAL REARRANGEMENTS IN THE SPLICEOSOME

The sequential rearrangements in the spliceosome's RNA–RNA and RNA–protein networks are driven in most cases by members of the DExD/H-box family of RNA unwindases/RNPases which include Prp5, Sub2/UAP56, Prp28, U5-200K/Brr2, Prp2, Prp16, Prp22, and Prp43 (reviewed by Staley and Guthrie 1998). The energy of ATP hydrolysis by these enzymes is coupled to structural/compositional rearrangements at one or more steps of the splicing cycle (Fig. 1C). Although the precise targets of many of these proteins remain largely unknown, in several cases the mechanisms of their action and regulation are beginning to emerge. The activity of these enzymes must be highly coordinated, a task carried out in part by other spliceosomal components and also modulated in some cases by posttranslational modifications. Several DExD/H-box proteins also play a key role in ensuring the fidelity of the splicing process by facilitating the discard of aberrant/nonproductive splicing intermediates/products (reviewed by Smith et al. 2008).

The DEAD-box proteins Prp5 and Sub2/UAP56 are required for prespliceosome formation. UAP56/Sub2p is thought to facilitate U2 addition to the spliceosome by displacing Mud2p (the yeast homolog of U2AF) and/or BBP from the BS (Kistler and Guthrie 2001). Thus, some spliceosomal DExD/H-box proteins appear to directly catalyze RNA–protein rearrangements in the spliceosome rather than unwind RNA duplexes, as initially thought. Studies in humans suggest that Sub2/UAP56 also acts at a later stage of spliceosome assembly, contacting the U4 and U6 snRNAs and potentially contributing to their unwinding (Shen et al. 2008). Prp5p is thought to use ATP hydrolysis to remodel the U2 snRNP, thereby facilitating U2 snRNA binding to the BS. More recent data indicate that Prp5 also plays a role in proofreading the stability of the U2/BS duplex (Xu and Query 2007).

The DEAD-box protein Prp28 catalyzes the exchange of U1 with U6 at the 5′ss (Staley and Guthrie 1999) during the transition from complex B to B^{act}. Prp28 appears to function by actively disrupting an RNA-protein interaction that stabilizes the U1/5′ss base pairing interaction, namely the interaction between U1-C and the 5′ss (Chen et al. 2001). In humans, phosphorylation of Prp28 is required for its association with the tri-snRNP and for the subsequent stable integration of the tri-snRNP during B-complex formation (Mathew et al. 2008). Whether phosphorylation (or dephosphorylation) of Prp28 modulates its function in releasing U1 from the 5′ss is presently not known.

The DExD/H-box protein Brr2 catalyzes a crucial step in spliceosome activation, namely the unwinding of U4/U6. Most RNA helicases join the spliceosome transiently at the stage at which their activity is required. Brr2, in contrast, is an integral spliceosome component whose activity must be tightly regulated to prevent premature unwinding of U4/U6. Recent evidence indicates that at least two proteins, Prp8 and the GTPase Snu114, regulate Brr2 activity during spliceosome activation. A carboxy-terminal fragment of Prp8 has been shown to interact with Brr2 and stimulate its helicase activity (Maeder et al. 2009). As this region of Prp8 binds ubiquitin (Bellare et al. 2006) and Prp8 is ubiquinated in the tri-snRNP, it has been suggested that this posttranslational modification ultimately plays a key role in regulating Brr2 activity (Bellare et al. 2008; see also later discussion). U4/U6 unwinding during catalytic activation also requires the U5-associated GTPase Snu114 (Bartels et al. 2002; Brenner and Guthrie 2005), which is homologous to the ribosomal elongation factor EF-2 that catalyzes structural rearrangements in the ribosome during translocation. Recent data indicate that Brr2 activity is blocked by Snu114 when it is bound to GDP, but not by its GTP bound form (Small et al. 2006). Snu114 interacts with the same region of Prp8 as Brr2 does, namely its carboxyl terminus, suggesting this region of Prp8 may additionally coordinate Snu114 control of Brr2. Other proteins playing a role in the rearrangements accompanying catalytic activation include members of the yeast NTC or human Prp19/CDC5 complex. During spliceosome activation, the NTC complex acts subsequent to U4 dissociation, apparently by stabilizing the association of U5 and U6 with the activated spliceosome (Chan et al. 2003). More recent data indicate that the NTC plays a key role in specifying the proper interaction of U5 and U6 with the pre-mRNA substrate before step 1 (Chan and Cheng 2005).

The DExH/D-box protein Prp2 is required before step 1 of splicing and it promotes a poorly understood remodeling event that converts B^{act} into the catalytically active B* complex. Studies with purified yeast spliceosomes containing a heat-inactivated Prp2 mutant (prp2-1) revealed a major conformational rearrangement, as evidenced by gradient sedimentation analysis and electron microscopy, upon complementation with purified Prp2 in the presence of ATP (Warkocki et al. 2009). This conformational change leads to a destabilization of SF3a and SF3b proteins, that likely exposes the BS before step 1 (Warkocki et al. 2009). In yeast, Spp2, which recruits Prp2 to the spliceosome (Silverman et al. 2004), Yju2 (Lui et al. 2007), the NTC component Isy1 (Villa and Guthrie 2005) and Cwc25 (Chui et al. 2009; Warkocki et al. 2009) also have been shown to promote step 1 of splicing.

The DExD/H-box ATPase Prp16 promotes a conformational rearrangement in the spliceosome required for step 2 of splicing, the precise nature of which is unclear. In yeast, Prp16 interacts genetically with Prp8 (Query and Konarska 2004), Isy1 (Villa and Guthrie 2005) and U6 snRNA at this stage (Madhani and Guthrie 1994), suggesting it acts on a structure containing one or more of these components. Prp16 also regulates the fidelity of branch site recognition, promoting the discard of aberrant lariat intermediates (Burgess and Guthrie 1993a). These early studies led to a kinetic proof reading model (based on a model initially proposed for proofreading during translation) in which the rate of ATP hydrolysis by Prp16 (or in general by any other ATPase) is proposed to act as a timer to regulate the outcome of two competing events, in this case either discard of an aberrant intermediate or its participation in the next step of splicing (Burgess and Guthrie 1993b). Prp18, Slu7, and Prp22, which act after Prp16, also promote the second step, apparently also by aiding the alignment of the reactive groups responsible for step 2 (reviewed by Umen and Guthrie 1995; Smith et al. 2008). For example, recent data suggest that the interaction of U5 loop 1 with both exons is stabilized by Prp18 at this stage (Crotti et al. 2007).

The DExD/H-box protein Prp22 not only functions subsequent to Prp16 during step 2, but it is also required for the release of the mRNA product from the spliceosome. Prp22 is deposited on the mRNA downstream of the exon–exon junction, concomitant with an RNP rearrangement occurring during step 2 of splicing (Schwer 2008). It is then thought to displace U5 from the mRNA by disrupting Prp8 and U5 snRNA interactions with exon nucleotides, leading to mRNA release (Aronova et al. 2007; Schwer 2008). Prp22 also plays a role in ensuring the fidelity of exon ligation by repressing the splicing of aberrant splicing intermediates (Mayas et al. 2006).

Spliceosome components must function in multiple rounds of splicing, and thus sequestration of splicing factors in postcatalytic or defective splicing complexes would be detrimental. Release of the excised intron from the postsplicing complex, which is accompanied by release of U2, U5, and U6, is catalyzed by Prp43. In yeast, Ntr1 and Ntr2 (which form a stable complex) are also required for spliceosome disassembly (Tsai et al. 2005) and recruit Prp43 to the spliceosome (Tsai et al. 2007). Ntr1 acts as an accessory factor that on binding to Prp43, stimulates its helicase activity (Tanaka et al. 2007). Both proteins have also been implicated in a turnover pathway for defective spliceosomes (Pandit et al. 2006). Brr2 and Snu114 also are required at this stage, where they are thought to facilitate unwinding of U2/U6 duplexes in the postsplicing complex (Small et al. 2006).

10 POSTTRANSLATIONAL PROTEIN MODIFICATIONS ALSO CONTRIBUTE TO SPLICING DYNAMICS

Posttranslational modifications also promote critical RNP rearrangements essential for splicing. The role of reversible protein phosphorylation in splicing continues to grow, and evidence has now been provided that other forms of posttranslational modification also affect the splicing process. The essential role of SR protein phosphorylation/dephosphorylation in splicing is well documented (reviewed by Misteli 1999; Soret and Tazi 2003), with de/phosphorylation shown to modulate protein–protein and protein–RNA interactions involving RS domains (Xiao and Manley 1997; Shin et al. 2004). Indeed, protein modifications likely play a critical role in splicing dynamics by influencing the stability of protein-protein interactions. Dephosphorylation of several spliceosomal phosphoproteins is required for the catalytic steps of splicing, with PP1/PP2A phosphatases playing key roles at this stage. For example, dephosphorylation of the U1-70K protein (Tazi et al. 1993) and SR protein ASF/SF2 (Cao et al. 1997) is required for step 1 in mammals. The U2-associated SF3b155 protein is hyperphosphorylated just before, or during, step 1 of splicing (Wang et al. 1998). SF3b155 and U5-116K (human Snu114) are dephosphorylated by PP1/PP2A phosphatases concomitant with step 2 of splicing (Shi et al. 2006). It was thus proposed that dephosphorylation of these proteins facilitates essential structural rearrangements in the spliceosome during the transition from the first to the second step of splicing (Shi et al. 2006).

Several human tri-snRNP proteins were recently shown to be phosphorylated and their phosphorylation was linked to the stable integration of the U4/U6.U5 tri-snRNP during B-complex formation. Specifically, the human tri-snRNP protein hPrp28 is phosphorylated by the kinase SRPK2 and in the absence of Prp28 phosphorylation, B-complex formation is blocked (Mathew et al. 2008). In addition, human Prp6 and Prp31, both tri-snRNP proteins, are phosphorylated during B-complex formation by Prp4 kinase (Schneider et al. 2010a). The latter kinase is required for stable association of the tri-snRNP during B-complex formation, suggesting that phosphorylation of Prp6 and Prp31 may contribute to this step. Thus, in higher eukaryotes, numerous phosphorylation events contribute to spliceosome assembly, which could therefore potentially be modulated at multiple regulatory checkpoints. Both Prp4 kinase and SRPK2 are absent from *S. cerevisiae*. The latter kinase phosphorylates RS domains that, in higher eukayotes, are typically found in SR proteins and also other spliceosomal proteins (including hPrp28), but are for the most part absent in *S. cerevisiae*. Indeed, there appear to be generally fewer phosphorylation events during splicing in *S. cervisiae* and thus fewer regulatory targets/switches. This is consistent with the paucity of alternative splicing events in yeast, and also with the more flexible nature of interactions among spliceosomal components in higher eukaryotes which thus may be more susceptible to fine tuning by posttranslational modifications.

Aside from phosphorylation, a number of other posttranslational modifications have been implicated in pre-mRNA splicing. Proteomic analyses have revealed that numerous spliceosomal proteins are acetylated (Choudhary et al. 2009), and small-molecule inhibitors of acetylation block spliceosome assembly in vitro at distinct stages before activation (Kuhn et al. 2009), suggesting that acetylation plays a role in splicing. U2AF65 was shown to undergo lysyl-hydroxylation and the responsible enzyme, Jmjd6, was shown to play a role in splicing regulation (Webby et al. 2009). Finally, evidence was provided that ubiquitination plays an important role in splicing (Bellare et al. 2008). Specifically, it was shown to be required to maintain tri-snRNP levels by apparently inhibiting the premature unwinding of U4/U6 (Bellare et al. 2007). These studies also revealed that Prp8 is ubiquitinated within the tri-snRNP. Given Prp8's known role in modulating Brr2 activity (see earlier discussion), it was postulated that ubiquitination/deubiquitination of Prp8 likely plays an indirect role in regulating both U4/U6 unwinding during catalytic activation, and U2/U6 unwinding during spliceosome disassembly (Bellare et al. 2008).

11 STRUCTURE OF THE SPLICEOSOMAL snRNPs AND NON-snRNP SPLICING FACTORS

Insight into the spliceosome's structural organization initially came from the characterization of its main subunits, namely the snRNPs. Much has been learned about protein–protein and protein–RNA interactions within the spliceosomal snRNPs (reviewed by Kambach et al. 1999a; Will and Lührmann 2006) and a complete picture of the spatial arrangement of snRNP components is slowly emerging via ultrastructural analyses of the snRNPs and their components.

Single-particle electron cryomicroscopy has been instrumental in elucidating the morphology and architecture of the spliceosomal snRNPs. Low to moderate resolution ($\sim$10–30 Å) 3D structures of the U1 snRNP, the heteromeric protein complex SF3b (a major subunit of the U2 snRNP), the U11/U12 di-snRNP, and the U4/U6.U5 tri-snRNP and its subunits U5 and U4/U6, have been obtained by EM (reviewed by Stark and Lührmann 2006). The resolution of the U1 snRNP ($\sim$10 Å), SF3b ($\sim$10 Å), and the U11/U12 di-snRNP ($\sim$12 Å) was sufficient to allow the localization of a subset of their protein components by fitting known structures or subdomains of these proteins into the 3D EM map. For example, the RRMs of SF3b49 and SF3b14a/p14 and the carboxy-terminal HEAT repeats of SF3b155, could be localized in the 3D reconstructions of the isolated SF3b complex and also in the U11/U12 di-snRNP (which contains SF3b) (Golas et al. 2003, 2005). The large body of biochemical and structural data available for the U1 snRNP, coupled with its relatively simple composition, allowed the generation of a 3D model of the human U1 snRNP, in which all of its components could be localized (Stark et al. 2001).

EM studies have recently shed light on the molecular architecture of the U4/U6.U5 tri-snRNP. 3D structures of the individual subunits of the human tri-snRNP, namely the U5 and U4/U6 snRNPs, together with the 3D structure of the U4/U6.U5 tri-snRNP have now been obtained by performing cryo-negative stain electron microscopy (Fig. 5A)(Sander et al. 2006). The tri-snRNP possesses an elongated, tetrahedral shape with dimensions of 305 Å $\times$ 200 Å $\times$ 175 Å and its 3D structure could be determined at a resolution of $\sim$25 Å. The position of the U5 and U4/U6 snRNPs within the tri-snRNP could be localized by fitting their 3D structures into the 3D map of the tri-snRNP (Fig. 5A). The positions of the m3G cap and loop 1 of the U5 snRNA were mapped by immunolabeling followed by EM. Because of the low resolution of the EM map, it was not possible to localize individual tri-snRNP proteins using information about their structure obtained by NMR or X-ray crystallography.

By analysing purified yeast U4/U6.U5 tri-snRNPs containing proteins with a genetically introduced tag, the structural arrangement of U5- and U4/U6-specific proteins in the S. cerevisiae tri-snRNP was determined by EM (Häcker et al. 2008). 2D EM images of the "wildtype" yeast tri-snRNP—which has a morphology very similar to the human tri-snRNP—or those containing a tagged tri-snRNP protein were generated and compared. In this way, the carboxyl terminus of the U5 proteins Prp8, the DExH/D-box helicase Brr2, and GTPase Snu114 could be localized in the main body of the tri-snRNP, demonstrating

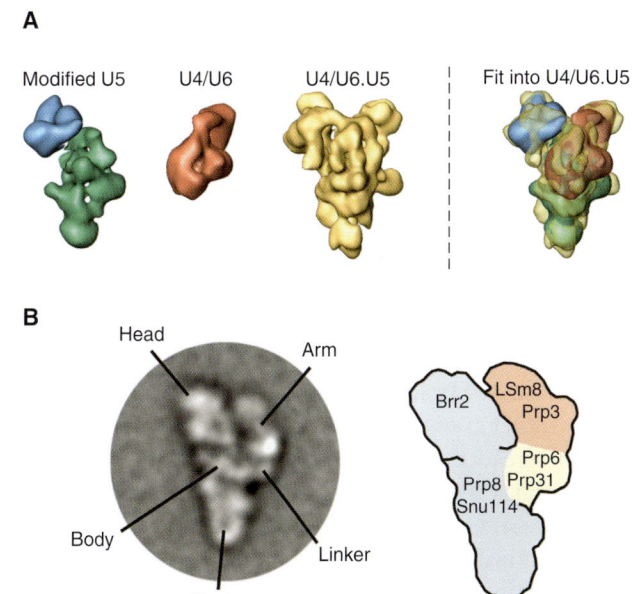

Figure 5. Three dimensional EM structure of the U4/U6.U5 tri-snRNP and localization of functionally important tri-snRNP proteins. (A) 3D reconstructions of the human U5 and U4/U6 snRNPs and tri-snRNP, and fitting of U5 and U4/U6 into the tri-snRNP 3D map (adapted, with permission, from Sander et al. 2006 [© Elsevier]). The head domain of U5 (highlighted blue) appears to be flexible and it is positioned in the U5 snRNP 3D reconstruction shown in a manner favorable for fitting into the tri-snRNP 3D map. (B) Left, representative 2D class average of the affinity purified S. cerevisiae U4/U6.U5 tri-snRNP as visualized by negative-stain electron microscopy after mild fixation using the Grafix protocol. The main structural domains are indicated. Right, cartoon model of the yeast tri-snRNP. Area corresponding to the U5 and U4/U6 snRNPs and the linker region are shaded grey, orange, or yellow, respectively. The position of the carboxyl terminus of several tri-snRNP proteins is indicated (adapted from Häcker et al. 2008).

that the U5 snRNP is located in this region (Fig. 5B). In addition, the U4/U6-proteins LSm8 and Prp3 were mapped to the so-called "arm"—indicating the U4/U6 snRNP is located in this region—and the tri-snRNP-specific bridging proteins Prp6 and Prp31 were detected in the linker region of the tri-snRNP (Fig. 5B). The spatial organization of Brr2 relative to the U4/U6 snRNP has important implications for the mechanism whereby Brr2 facilitates U4/U6 snRNA unwinding during catalytic activation (Häcker et al. 2008). These data thus provide a structural framework for potential mechanisms whereby the molecular motor proteins (Prp8, Brr2, and Snu114) of the tri-snRNP act. A similar genetic tagging approach was used to map the position of the LSm 2-8 proteins in the yeast U6 snRNP, supporting the previously proposed order of the Lsm proteins within the Lsm ring and elucidating their spatial organization relative to the U6-associated Prp24 protein (Karaduman et al. 2008). In the future, the introduction of genetic tags can potentially be used to map via EM the location of these and other proteins in spliceosomal complexes.

12 HIGH RESOLUTION STRUCTURES OF snRNP AND/OR SPLICEOSOME COMPONENTS

Because of its highly dynamic and complex nature, as well as the limited amounts that can be purified, obtaining structural information about the spliceosome at the atomic level represents a major challenge. X-ray crystallography and NMR analyses have been limited (with one notable exception; see later) to individual snRNP and spliceosomal proteins alone or in complex with RNA or a protein binding partner. Crystal structures of a fragment of the U1-A protein bound to stem-loop II of U1 snRNA (Oubridge et al. 1994), the U2-A'/B'' heterodimer bound to U2 snRNA stem-loop IV (Price et al. 1998), the 15.5K protein complexed with the 5' SL of U4 snRNA (Vidovic et al. 2000), and two heteromeric Sm protein dimers (Kambach et al. 1999b), yielded the first insights into the atomic structure of the spliceosomal snRNPs.

More recently, progress has been made in determining the atomic structure of RNA and/or protein fragments of the spliceosome that play important roles in splice site recognition (reviewed by Ritchie et al. 2009). For example, the structural basis for several binary interactions important during early recognition of the branch site or polypyrimidine tract has been elucidated. These include SF1 bound to a branch site RNA (Liu et al. 2001), U2AF65 bound to a polypyrimidine tract (Sickmier et al. 2006), U2AF65's RRM3 bound to an amino-terminal SF1 peptide (Selenko et al. 2003), a heterodimer of U2AF65 and U2AF35 (Kielkopf et al. 2001), and SF3b14a/p14 bound to a SF3b155 peptide (Schellenberg et al. 2006; Spadaccini et al. 2006). Although it cannot be ruled out that the structure of these complexes differs in the context of the spliceosome, the high resolution information obtained can potentially be used to map their position in EM structures of early spliceosomal complexes, provided the resolution of the latter is sufficient (i.e., ~10 Å or less) for integrating such structural information. Indeed, using its atomic structure information, the p14/SF3b155 peptide complex could be fitted into the 3D EM map of the SF3b complex (Schellenberg et al. 2006).

High resolution structures of functionally important proteins of the U4/U6.U5 tri-snRNP have also now been reported. The crystal structure of a portion of the U4 snRNP, consisting of the U4 snRNA 5' stem-loop complexed with the 15.5K protein and part of the hPrp31 protein, was solved (Fig. 6A) (Liu et al. 2007). This study revealed the molecular mechanism underlying the ordered assembly pathway of the U4 snRNP, demonstrating that the 15.5K protein and U4 snRNA form a composite RNP binding site for hPrp31, with 15.5K additionally stabilizing the RNA in a conformation favourable for hPrp31 binding. Induced fit interactions are common for most RNA–protein complexes and the binding of Prp31 leads to pronounced structuring of the pentaloop of the U4 snRNA 5' stem-loop (Fig. 6A).

Insights into the overall structure and mechanism of action the U5-associated DExD/H-box helicase Brr2 have also now been obtained. Brr2 contains two helicase domains (the second of which does not appear to be essential for its helicase activity), each comprised of two RecA-like modules followed by a Sec63-like domain of unknown function. Two groups have now solved the crystal structure of the C-terminal Sec63 domain of S. cerevisiae Brr2 and discovered similarities between two of its three subdomains and two structural modules of archaeal Hel308, a processive 3' to 5' DNA helicase involved in DNA repair (Pena et al. 2009; Zhang et al. 2009). Based on this information, the conservation of sequence throughout the remainder of these helicases, and additional functional data, both groups proposed a structural model for the amino-terminal, functional helicase domain of Brr2 that is analogous to the structure of Hel308. In the latter the Sec63-like domain is an integral component of the active site and is functionally connected to the two RecA-like domains by a winged helix module. The functional implication of the proposed structural organization of Brr2 is that, unlike other spliceosome associated DExH/D-box proteins, Brr2 may act in a more processive manner, which would likely be required to unwind the long stem regions of the U4/U6 duplex during the catalytic activation of the spliceosome.

As discussed earlier, Prp8 is thought to play a key role in organizing the catalytic core of the spliceosome, and it contacts all of the reactive groups (i.e., the 5′ss, 3′ss, and BS) of the pre-mRNA substrate during splicing. Prp8 contains a centrally located RRM, thought to mediate Prp8-RNA interactions. The crystal structure of the carboxyl terminus of the *S. cerevisiae* (Pena et al. 2007) and *Caenorhabditis elegans* (Zhang et al. 2007) Prp8 protein revealed a Jab1/MPN core domain, found in enzymes that remove ubiquitin from ubiquitinated proteins. However, in Prp8 the Jab1/MPN domain is interrupted by insertions and the metal binding site is impaired. As the carboxyl terminus of Prp8 binds both Brr2 and Snu114, it was proposed that this domain of Prp8 represents a pseudoenzyme converted into a protein–protein interaction platform (Pena et al. 2007).

The most intriguing structure to be elucidated recently is that of another domain of the Prp8 protein located just upstream of the Jab1/MPN domain. Three groups independently solved the structure of a ~250 amino acid domain of human and/or yeast Prp8 near its carboxyl terminus, which encompasses amino acids that contact the 5′ss of the pre-mRNA (Pena et al. 2008; Ritchie et al. 2008; Yang et al. 2008). The structure revealed an amino-terminal subdomain with an RNase H-like fold consisting of a five-stranded mixed β-sheet flanked by two α-helices (Fig. 6B), but with a truncated RNase H active center. The RNase H-like domain was interrupted by the insertion of a β-hairpin (atypical for RNase H-like enzymes) and was juxtaposed by a carboxy-terminal cluster of five helices (Fig. 6B). The RNaseH-like domain was shown to interact with a model RNA comprising portions of U2, U6, and the 5′ss (Ritchie et al. 2008), which may mimic part of the activated snRNA/pre-mRNA network. Furthermore, point mutations targeting the apparent RNase H-like active site residues had deleterious effects on cell viability (Pena et al. 2008; Ritchie et al. 2008; Yang et al. 2008). Taken together, these results suggest a role for Prp8's RNase-H-like domain in the assembly and/or maintenance of the spliceosome's catalytic core, and raise the interesting possibility that Prp8 may even directly participate in the catalysis of splicing, for example by coordinating a Mg ion important for catalysis.

13 CRYSTAL STRUCTURE OF THE HUMAN U1 snRNP

The recently reported crystal structure of an in vitro assembled U1 snRNP at 5.5 Å resolution represents a milestone in understanding the 3D structure of a spliceosomal snRNP (Pomeranz-Krummel et al. 2009). The crystallized U1 snRNP contained the U1 RNA, the seven Sm proteins (B, D1, D2, D3, E, F, and G), and part of the U1-C and U1-70K proteins, but lacked the U1-A protein. Because of the relatively low resolution (for a crystal structure), site-specific labeling of individual proteins was performed to unambiguously map their positions, and the generation of an atomic model was aided by prior knowledge of the structure of several of the U1-associated proteins and a solution structure model of the U1 snRNA. The U1 structure was completed by modeling the known structure of U1-A bound to stem-loop II of the U1 snRNA.

The U1 snRNA consists of four stem-loops (I-IV) and a short helix (H). It forms a four-helix junction with two coaxially stacked helices (stem-loop 1/stem-loop 2 and stem loop 3/Helix H) followed by the single-stranded Sm site which separates stem loop 4 from stem loops 1-3 and helix H (Pomeranz-Krummel et al. 2009) (Fig. 6C). This is consistent with the previously proposed structure of free U1 snRNA that was based on biochemical studies (Krol et al. 1990). The crystal structure also confirmed the heptameric ring model of the Sm core, which proposed that the Sm proteins form a seven-membered ring containing one copy of each Sm protein, in the order E, G, D3, B, D1, D2 and F, with the Sm site RNA-Sm protein contacts occurring on the inner surface of the proposed ring (Kambach et al. 1999b) (Fig. 6C). In addition, the crystal structure indicates that each Sm protein contacts one nucleotide of the U1 snRNA with the seven nucleotides of the Sm site (5-AUUUGUG-3′) likely to interact with the Sm proteins E, G, D3, B, D1, D2 and F, respectively (Pomeranz-Krummel et al. 2009). This general arrangement of the Sm core is likely to be found in all other snRNPs that contain the Sm proteins. The U1-70K protein contacts stem loop 1 of the U1 snRNA, and its amino-terminus wraps around the Sm core and contacts the U1-C protein, nearly 180 Å away from its RNA binding site (Fig. 6C). The amino-terminus of the U1-70K protein together with the carboxyl terminus of SmD3, create a binding pocket for the U1-C protein, which provides a structural basis for the previously observed dependence of U1-C binding on the amino terminus of the U1-70K protein. Previous biochemical studies revealed a role for U1-C in stabilizing the base pairing interaction of the U1 snRNA with the 5′ss. In the U1 crystal structure, the 5′-end of the U1 RNA was partially base paired to an adjacent U1 RNA, mimicking the U1/5′ss duplex and thereby providing clues for how the U1 snRNA and U1-C might recognize the 5′ss of the intron during early spliceosome assembly (Fig. 6C). A loop and helix of the zinc finger domain of the U1-C protein binds across the minor groove of this duplex, consistent with a role for U1-C in stabilizing the U1/5′ss duplex in early spliceosomal complexes.

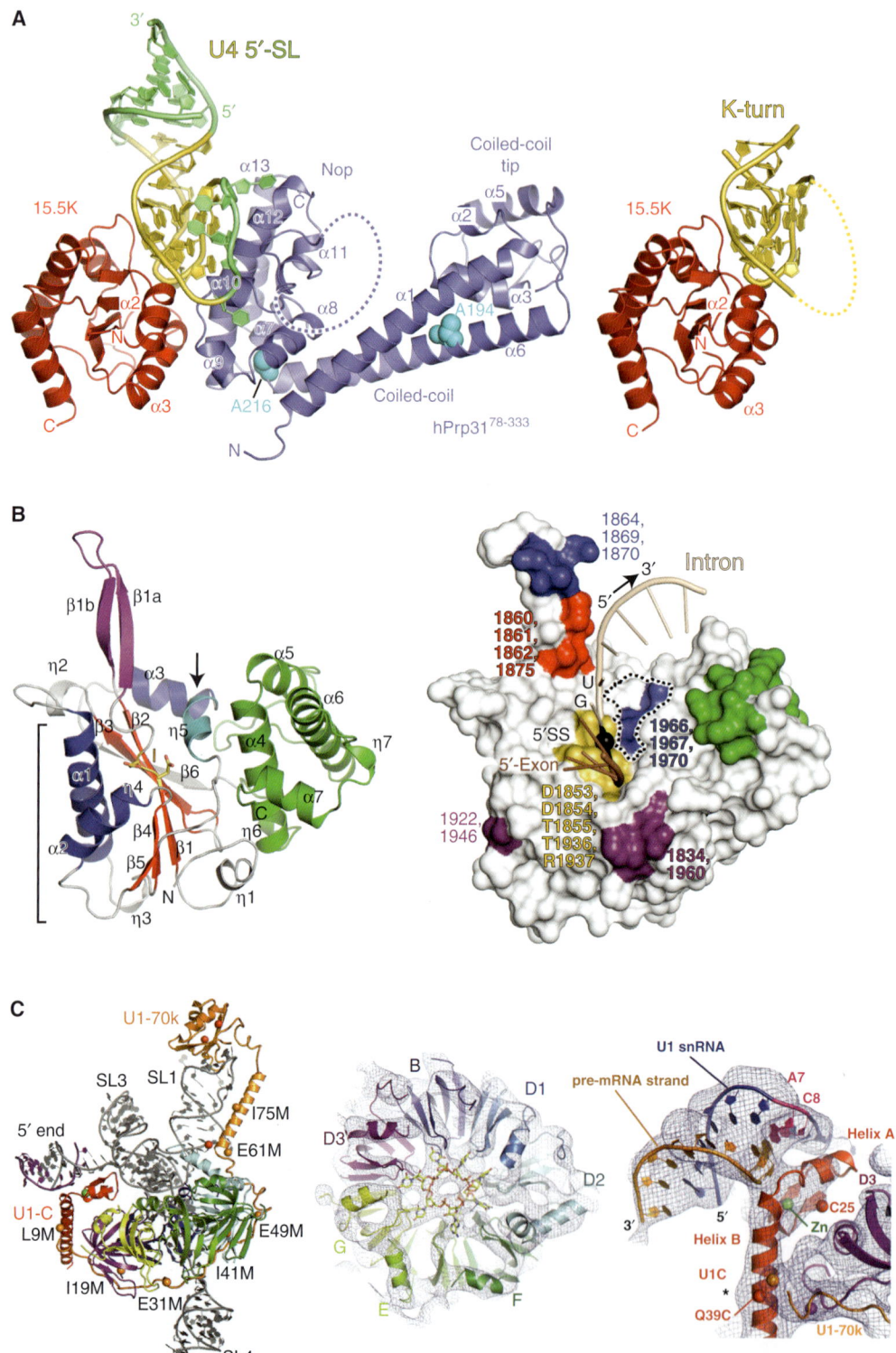

Figure 6. Structures of the U4 snRNA 5′ stem-loop complexed with Prp31 and the 15.5K protein, the RNAse H-like domain of Prp8, and the U1 snRNP. (A) Left, ribbon diagram of hPrp31 (residues 78-333) (purple) and the 15.5K protein (red) complexed with the human U4 snRNA 5′stem loop (nts 20-52) (gold and green). RNA elements absent in the 15.5K-U4 RNA binary complex (because of the shorter RNA used) (shown at right) are highlighted in green. A disordered loop of hPrp31 in the ternary complex and an unstructured region of the U4 pentaloop in the binary complex are indicated by a dashed lines. Mutations at positions A194 and A216 (shown as cyan-colored space-filling models) are linked to retinitis pigmentosa. (*See facing page for legend*).

14 ELUCIDATION OF THE SPATIAL ORGANIZATION OF SPLICEOSOMAL COMPLEXES USING BIOCHEMICAL PROBES

Studies employing site-specifically tethered biochemical probes have demonstrated the power of this method for elucidating the spatial arrangement of spliceosomal components at different stages of the splicing process. For example, hydroxyl radical probing with Fe-BABE-tethered pre-mRNA revealed that the branchpoint and the 3′ss are located within 20 Å of the 5′ss already in the E complex (Kent and MacMillan 2002). These data thus indicate that the pre-mRNA's 5′ss, BS, and 3′ss are spatially preorganized in the E complex. In a more recent study, the organization of U2 relative to U1 and pre-mRNA in spliceosomal complexes was analyzed via site-directed hydroxyl radical probing with Fe-BABE tethered to the 5′ end of U2 snRNA (Dönmez et al. 2007). These studies revealed that functional regions (i.e., the 5′ss, BS, and 3′ss) of the pre-mRNA are located close to the 5′ end of U2 in both the E and A complexes. Furthermore, they showed that U2 is also positioned close to U1 in a defined orientation already in the E complex, and that their relative spatial organization remains largely unchanged during the E to A transition. Finally, hydroxyl radical probing with Fe-BABE site-specifically tethered to the pre-mRNA 10 nts downstream of the 5′ss revealed that the asymmetric bulge of the U6 ISL is in close proximity to the 5′ss/U6 ACAGAG box helix in the activated spliceosome (Rhode et al. 2006). These studies represent important first steps towards understanding the tertiary structure of the spliceosome's RNA–RNA and RNP networks.

In addition to the use of site-specifically introduced biochemical probes, ground work has now been laid to track structural rearrangements and the dynamics of protein–protein and RNA-protein interactions within the spliceosome by introducing fluorescent probes into spliceosomal components for FRET analysis (Rino et al. 2008; Ellis et al. 2009), or by using single molecule approaches (Crawford et al. 2008; Guo et al. 2009). Most of these studies have been carried out in living cells where intermolecular interactions involving spliceosomal proteins are not limited to the spliceosome. In vitro, splicing reconstitution systems—for example as described above for *S. cerevisiae* (Warkocki et al. 2009)—are highly amenable for introducing fluorescently labelled proteins into *in vitro* assembled spliceosomes and tracking intermolecular interactions during activation and the catalytic steps of splicing via FRET or other fluorescence-based assays. Thus, in the near future fluorescence-based techniques may be more commonly used to elucidate the structural dynamics of the spliceosome.

15 ELECTRON MICROSCOPY OF SPLICEOSOMAL COMPLEXES

Because of its highly dynamic and complex nature, single-particle cryo-electron microscopy is currently the method of choice to study the higher-order structure of the spliceosome. 2D EM views of various spliceosomal complexes isolated under physiological conditions from human, *D. melanogaster* and *S. cerevesiae* are now available (Deckert et al. 2006; Fabrizio et al. 2009; Herold et al. 2009). These studies reveal conservation of the overall size and shape of the pre-catalytic B complex (Fig. 7) and the catalytic C complex from various organisms, indicating that higher

Figure 6. (*Continued*) hPrp31 binds to one region of the composite binding platform formed by 15.5K and the U4 snRNA by a lock-and-key type mechanism, and another region of the RNA via an induced fit mechanism. (Reprinted, with permission, from Liu et al. 2007 [© AAAS].) (*B*) Ribbon diagram (*left*) and space filling model (*right*) of the *S. cerevisiae* Prp8 protein (residues 1827-2092). *Left*, the mixed β-sheet and two α-helices typical of RNAse H domains are highlighted, red and purple, respectively. The Prp8-specific β-hairpin and α-helices are colored magenta and green, respectively. Residues comprising the active site in RNAse H (corresponding to Asp1853 and Asp1854 in yeast Prp8) are indicated by sticks and the 3_{10} helix that is crosslinked to the 5′ splice site is highlighted cyan blue. *Right*, modeling of the pre-mRNA (exon and intron nucleotides, brown and beige, respectively and 5′ss phosphate, black) into the Prp8 RNAse-like domain space filling model. Site of Prp8 crosslinks to the 5′ss is encompassed by a dashed line and the predicted active site, gold. The Brr 2 interacting region is shown in green. The sites of Prp8 mutations (amino acid residue indicated) suppressing 5′ss (blue), 3′ss (green), polypyrimidine tract (purple) and U4 cs1 (magenta) mutations are indicated. (*C*) *Left*, ribbon diagram of the U1 snRNP containing the Sm proteins and the 70K and C proteins. The U1 snRNA, with stem-loop (SL) 1, 3 and 4, and the 5′ end indicated, is shaded grey. Orange spheres indicate anomalous peaks from SeMet (introduced at the indicated amino acid position) in U1-70K. *Middle*, ribbon diagram of the Sm proteins (E, F, G, D1, D2, D3, B) and seven nucleotide Sm site RNA, with the experimental electron density map (contoured at 1σ). *Right*, Ribbon diagram with experimental electron density map (contoured at 1σ) of the interaction of the 5′ end of U1 snRNA with a neighboring complex (orange) which mimics the 5′ss of the pre-mRNA. (Reprinted, with permission, from Newman and Nagai 2010 [© Elsevier]; originally Pomeranz-Krummel et al. 2009 [© Macmillan].)

order interactions and the general spatial organization of each of these spliceosomal complexes is conserved between higher and lower eukaryotes.

The structures of the precatalytic B, activated B, and catalytically active C complexes appear to differ dramatically (at least at the 2D level). In a very recent study, the yeast B, B^{act}, and C complexes were affinity purified and their structure determined by single-particle electron cryomicroscopy (Fabrizio et al. 2009). A comparison of 2D class averages of these complexes revealed a maximum dimension of ~400 Å in each case (Fig. 7B). The morphology of the main projection images of B differed clearly from those of B^{act}, which in turn differed from those of the C complex. The most pronounced differences were seen when comparing the B complex, which exhibits a triangular or rhombic shape in most class averages, with B^{act}, whose main body is clearly more compact (Fig. 7B). The structural dynamics revealed by these studies are consistent with the compositional changes and RNP rearrangements occurring during the B to B^{act} and B^{act} to C transitions.

A structural change was also uncovered by EM studies of yeast spliceosomes during the conversion of B^{act} to B*, the catalytically active spliceosome formed after the ATP-dependent action of Prp2 (Warkocki et al. 2009). Taken together, these results underscore the myriad of conformational changes that the spliceosome undergoes during its assembly, activation, and catalytic activity. They further suggest that, unlike the ribosome, a characteristic spliceosome structure does not exist, but rather because of its highly dynamic nature, its structure varies greatly throughout the splicing cycle.

16 LOCALIZATION OF REGIONS OF THE PRE-mRNA IN THE SPLICEOSOME VIA EM

To localize functionally important components of the spliceosome, two groups have now introduced tags into the pre-mRNA and mapped their position in the 2D EM structure of the spliceosomal C or B complex. In the first study, an RNA hairpin that binds the coliphage coat protein PP7

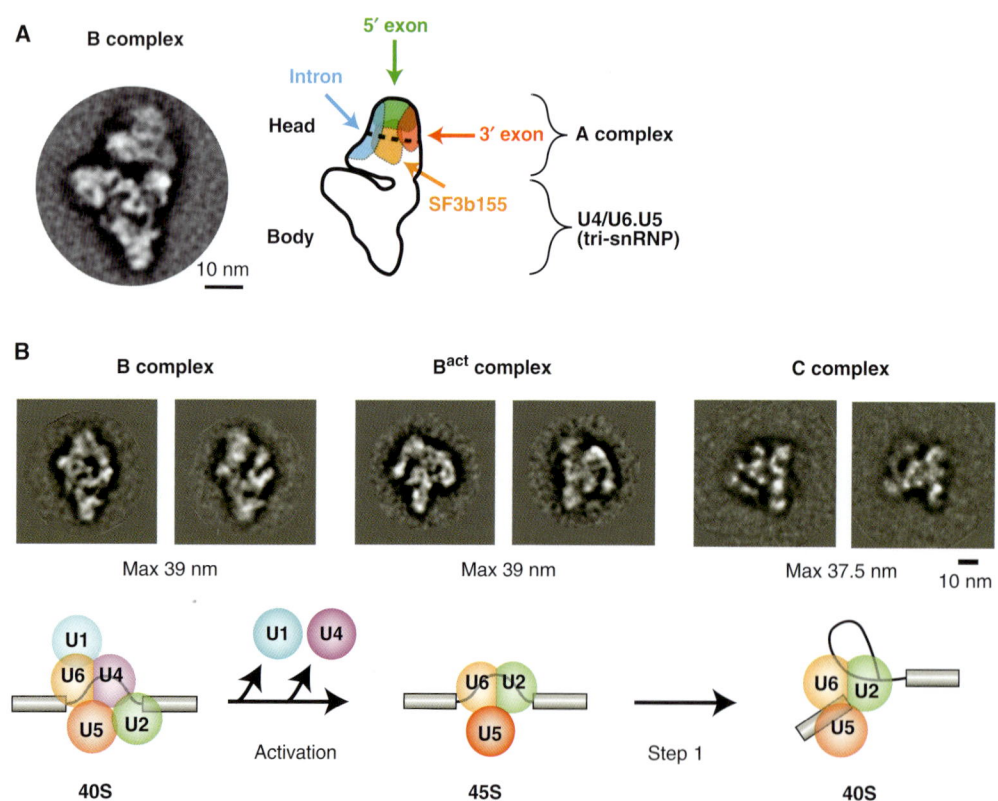

Figure 7. Structural dynamics of the yeast spliceosome as visualized by EM and localization of the pre-mRNA in the human B complex. (A) Class average of electron microscopy images of negatively stained, affinity-purified human B complexes (right). Sketch of the B complex showing regions where the 5' exon, 3' exon, intron and SF3b155 protein were mapped by immuno-EM, and the likely location of components of the A complex and tri-snRNP. (Adapted from Wolf et al. 2009 [© Nature Publishing Group].) (B) Electron microscopy of negatively-stained, affinity-purified S. cerevisiae B, B^{act}, and C complexes. Two prominent class averages of each complex are shown, with the maximum dimension indicated later. (Adapted, with permission, from Fabrizio et al. 2009 [© Elsevier].)

was introduced into either the 5′ or 3′ exon of a pre-mRNA substrate (Alcid and Jurica 2008). After binding a PP7/dnaN fusion protein, C complexes were allowed to form in nuclear extract, and were then affinity-selected and analyzed by single particle EM. Localization of the 5′ versus 3′ tags in 2D EM images revealed that the 5′ and 3′ exons are in close proximity in the C complex. In a second study, MS2 binding sites were introduced into either the 5′ or 3′ exon or the intron of the substrate, and were then bound by an MS2-MBP fusion protein. B complexes assembled in vitro on the pre-bound substrate were subsequently affinity selected and analyzed by EM (Wolf et al. 2009). The positions of the 5′ and 3′ exon and intron, were mapped after binding anti-MBP antibodies and labeling them with colloidal gold. These studies revealed that both exons and the intron are located near each other in the head domain of the B complex (Fig. 7A). Additional immuno-labeling with antibodies against SF3b155, indicated that the latter protein (and as a consequence the U2 snRNP) is also localized in the head region of the B complex (Fig. 7A). These studies represent important first steps toward mapping functionally important sites in the spliceosome.

17 3D STRUCTURES OF THE SPLICEOSOME OBTAINED BY EM

EM analyses of the 3D structure of the spliceosome have been hampered, primarily by the highly dynamic and labile nature of the spliceosome, which has resulted in difficulties in purifying defined, biochemically homogeneous, spliceosomal complexes that are structurally stable. By performing single-particle electron cryomicroscopy, 3D structures at a resolution of 30–40 Å were reported for the human spliceosomal B and C complex, both affinity-purified from splicing extracts under stringent conditions in the presence of heparin (reviewed by Stark and Lührmann 2006). BΔU1 exhibited a triangular main body with a globular head domain and a maximum dimension of 370 Å (Bohringer et al. 2004), whereas the C complex possessed an asymmetric shape with three main subdomains and a maximum dimension of 270 Å (Jurica et al. 2004). At the current level of resolution, it is difficult to conclusively map the relative positions of the snRNP complexes within the spliceosome. Nonetheless, a comparison of the structure of the human U4/U6.U5 tri-snRNP with that of BΔU1, strongly suggested that the tri-snRNP is located in the lower triangular domain of BΔU1 and, as a consequence, that the U2 snRNP must be localized in the upper globular domain (Boehringer et al. 2004; Sander et al. 2006). The 3D structure of the spliceosomal A complex, purified under physiological conditions, has also been elucidated (Behzadnia et al. 2007). It possesses a main globular body ∼205 Å × 195 Å × 150 Å in size, from which several smaller elements protrude. Due to the low resolution of the 3D structure (∼40–50 Å), it was not possible to localize A complex components, such as SF3b, whose 3D structure is known. In addition to these in vitro assembled spliceosomal complexes, the 3D structure of in vivo assembled human spliceosomes (whose assembly/functional stage is presently not clear) (Azubel et al. 2004), and in vivo assembled *Schizosaccharomyces pombe* spliceosomal complexes containing U2, U5, and U6 (Ohi et al. 2007), have also been reported.

Higher resolution 3D EM structures of the spliceosome await improvements in sample preparation and also in image analysis (discussed by Lührmann and Stark 2009). A recent step in this direction was the development of a mild chemical fixation method (termed Grafix), that has considerably improved sample quality and contrast, such that EM images suitable for a reliable 3D structure reconstruction could be obtained by cryo-EM, even in the absence of stain (Kastner et al. 2008). Reducing the compositional heterogeneity of a given complex and obtaining samples where the vast majority are stalled at a specific conformational stage should also lead to improvements in resolution. In this respect, yeast spliceosomes—which are less complex and contain fewer substoichiometrically associated proteins—and which can be stalled at more precise steps of the splicing reaction using temperature-sensitive (*ts*) mutants of the various DExH/D-box spliceosomal proteins, appear to be ideal candidates for obtaining higher resolution images.

18 SUMMARY

Much progress has been made in recent years towards understanding the structure and function of the spliceosome. Both its conformation and composition have proven to be highly dynamic. Despite the hurdles generated by its dynamic nature, a clearer picture of the order and nature of the intricate rearrangements within the spliceosome and their contribution to its function, is slowly emerging. However, answers to a number of questions, foremost the precise nature of its active site, await the generation of a high-resolution structure of the spliceosome.

ACKNOWLEDGMENTS

We are grateful to Berthold Kastner, Patrizia Fabrizio, Holger Stark, Markus Wahl and Elmar Wolf for providing figures. We thank Patrizia Fabrizio, Klaus Hartmuth, and Markus Wahl for constructive comments. We apologize to those of our colleagues who were not cited due to space limitations. This work was funded by grants from the DFG, the European Commission (EURASNET-518238), Fonds der Chemischen Industrie and the Ernst Jung Stiftung to Reinhard Lührmann.

REFERENCES

Abelson J. 2008. Is the spliceosome a ribonucleoprotein enzyme? *Nat Struct Mol Biol* **15**: 1235–1237.

Alcid EA, Jurica MS. 2008. A protein-based EM label for RNA identifies the location of exons in spliceosomes. *Nat Struct Mol Biol* **15**: 213–215.

Aronova A, Bacíková D, Crotti LB, Horowitz DS, Schwer B. 2007. Functional interactions between Prp8, Prp18, Slu7, and U5 snRNA during the second step of pre-mRNA splicing. *RNA* **13**: 1437–1444.

Ast G. 2004. How did alternative splicing evolve? *Nat Rev* **5**: 773–782.

Azubel M, Wolf SG, Sperling J, Sperling R. 2004. Three-dimensional structure of the native spliceosome by cryo-electron microscopy. *Mol Cell* **15**: 833–839.

Bartels C, Klatt C, Lührmann R, Fabrizio P. 2002. The ribosomal translocase homologue Snu114p is involved in unwinding U4/U6 RNA during activation of the spliceosome. *EMBO Rep* **3**: 875–880.

Behzadnia N, Golas M, Hartmuth K, Sander B, Kastner B, Deckert J, Dube P, Will C, Urlaub H, Stark H, Lührmann R. 2007. Composition and three-dimensional EM structure of double affinity-purified, human prespliceosomal A complexes. *EMBO J* **26**: 1737–1748.

Bellare P, Kutach AK, Rines AK, Guthrie C, Sontheimer EJ. 2006. Ubiquitin binding by a variant Jab1/MPN domain in the essential pre-mRNA splicing factor Prp8p. *RNA* **12**: 292–302.

Bellare P, Small EC, Huang X, Wohlschlegel JA, Staley JP, Sontheimer EJ. 2008. A role for ubiquitin in the spliceosome assembly pathway. *Nat Struct Mol Biol* **15**: 444–451.

Berget SM. 1995. Exon recognition in vertebrate splicing. *Mol Cell Biol* **270**: 2411–2414.

Bessonov S, Anokhina M, Will CL, Urlaub H, Lührmann R. 2008. Isolation of an active step 1 spliceosome and composition of its RNP core. *Nature* **452**: 846–850.

Black DL. 2003. Mechanisms of alternative pre-messenger RNA splicing. *Annu Rev Biochem* **72**: 291–336.

Böhringer D, Makarov EM, Sander B, Makarova OV, Kastner B, Lührmann R, Stark H. 2004. Three-dimensional structure of a precatalytic human spliceosomal complex B. *Nat Struct Mol Biol* **11**: 463–468.

Bonnal S, Martinez C, Forch P, Bachi A, Wilm M, Valcárcel J. 2008. RBM5/Luca-15/H37 regulates Fas alternative splice site pairing after exon definition. *Mol Cell* **32**: 81–95.

Brenner TJ, Guthrie C. 2005. Genetic analysis reveals a role for the C terminus of the Saccharomyces cerevisiae GTPase Snu114 during spliceosome activation. *Genetics* **170**: 1063–1080.

Brow DA. 2002. Allosteric cascade of spliceosome activation. *Annu Rev Genet* **36**: 333–360.

Burge CB, Tuschl T, Sharp PA. 1999. Splicing of Precursors to mRNAs by the Spliceosomes. In *The RNA world Second edition* (ed. R.F. Gesteland et al.), pp. 525–560. Cold Spring Harbor Laboratory Press, Cold Spring Harbor, New York.

Burgess SM, Guthrie C. 1993a. A mechanism to enhance mRNA splicing fidelity: The RNA-dependent ATPase Prp16 governs usage of a discard pathway for aberrant lariat intermediates. *Cell* **73**: 1377–1391.

Burgess SM, Guthrie C. 1993b. Beat the clock: Paradigms for NTPases in the maintenance of biological fidelity. *Trends Biochem Sci* **18**: 381–384.

Cao W, Jamison SF, Garcia-Blanco MA. 1997. Both phosphorylation and dephosphorylation of ASF/SF2 are required for pre-mRNA splicing *in vitro*. *RNA* **3**: 1456–1467.

Chan SP, Cheng SC. 2005. The Prp19-associated complex is required for specifying interactions of U5 and U6 with pre-mRNA during spliceosome activation. *J Biol Chem* **280**: 31190–31199.

Chan SP, Kao DI, Tsai WY, Cheng SC. 2003. The Prp19p-associated complex in spliceosome activation. *Science* **302**: 279–282.

Chen M, Manley JL. 2009. Mechanisms of alternative splicing regulation: Insights from molecular and genomics approaches. *Nat Rev Mol Cell Biol* **10**: 741–754.

Chen JY, Stands L, Staley JP, Jackups RR Jr, Latus LJ, Chang TH. 2001. Specific alterations of U1-C protein or U1 small nuclear RNA can eliminate the requirement of Prp28p, an essential DEAD box splicing factor. *Mol Cell* **7**: 227–232.

Chiu YF, Liu YC, Chiang TW, Yeh TC, Tseng CK, Wu NY, Cheng SC. 2009. Cwc25 is a novel splicing factor required after Prp2 and Yju2 to facilitate the first catalytic reaction. *Mol Cell Biol* **29**: 5671–5678.

Choudhary C, Kumar C, Gnad F, Nielsen ML, Rehman M, Walther TC, Olsen JV, Mann M. 2009. Lysine acetylation targets protein complexes and co-regulates major cellular functions. *Science* **325**: 834–840.

Clark TA, Sugnet CW, Ares M Jr. 2002. Genomewide analysis of mRNA processing in yeast using splicing-specific microarrays. *Science* **296**: 907–910.

Crawford DJ, Hoskins AA, Friedman LJ, Gelles J, Moore MJ. 2008. Visualizing the splicing of single pre-mRNA molecules in whole cell extract. *RNA* **14**: 170–179.

Crotti LB, Bacíková D, Horowitz DS. 2007. The Prp18 protein stabilizes the interaction of both exons with the U5 snRNA during the second step of pre-mRNA splicing. *Genes Dev* **21**: 1204–1216.

Deckert J, Hartmuth K, Böhringer D, Behzadnia N, Will CL, Kastner B, Stark H, Urlaub H, Lührmann R. 2006. Protein composition and electron microscopy structure of affinity-purified human spliceosomal B complexes isolated under physiological conditions. *Mol Cell Biol* **26**: 5528–5543.

Dönmez G, Hartmuth K, Kastner B, Will CL, Lührmann R. 2007. The 5′ end of U2 snRNA is in close proximity to U1 and functional sites of the pre-mRNA in early spliceosomal complexes. *Mol Cell* **25**: 399–411.

Deutsch M, Long M. 1999. Intron-exon structures of eukaryotic model organisms. *Nucleic Acids Res* **27**: 3219–3228.

Ellis JD, Llères D, Denegri M, Lamond AI, Cáceres JF. 2008. Spatial mapping of splicing factor complexes involved in exon and intron definition. *J Cell Biol* **181**: 921–934.

Fabrizio P, Dannenberg J, Dube P, Kastner B, Stark H, Urlaub H, Lührmann R. 2009. The evolutionary conserved core design of the catalytic activation step of the yeast spliceosome. *Mol Cell* **36**: 593–608.

Fox-Walsh KL, Dou Y, Lam BJ, Hung SP, Baldi PF, Hertel KJ. 2005. The architecture of pre-mRNAs affects mechanisms of splice-site pairing. *Proc Natl Acad Sci* **102**: 16176–16181.

Golas MM, Sander B, Will CL, Lührmann R, Stark H. 2003. Molecular architecture of the multiprotein splicing factor SF3b. *Science* **300**: 980–984.

Golas MM, Sander B, Will CL, Lührmann R, Stark H. 2005. Major conformational change in the protein complex SF3b upon integration into the spliceosomal U11/U12 di-snRNP as revealed by electron cryomicroscopy. *Mol Cell* **17**: 869–883.

Grainger RJ, Beggs JD. 2005. Prp8 protein: At the heart of the spliceosome. *RNA* **11**: 533–557.

Guo Z, Karunatilaka KS, Rueda D. 2009. Single-molecule analysis of protein-free U2-U6 snRNAs. *Nat Struct Mol Biol* **16**: 1154–1159.

Häcker I, Sander B, Golas MM, Wolf E, Karagoez E, Kastner B, Stark H, Fabrizio P, Lührmann R. 2008. Localization of Prp8, Brr2, Snu114 and U4/U6 proteins in the yeast tri-snRNP by electron microscopy. *Nat Struct Mol Biol* **15**: 1206–1212.

Herold N, Will CL, Wolf E, Kastner B, Urlaub H, Lührmann R. 2009. Conservation of the protein composition and electron microscopy structure of Drosophila melanogaster and human spliceosomal complexes. *Mol Cell Biol* **29**: 281–301.

Hilliker AK, Mefford MA, Staley JP. 2007. U2 toggles iteratively between the stem IIa and stem IIc conformations to promote pre-mRNA splicing. *Genes Dev* **21:** 821–834.

Hoffman BE, Grabowski PJ. 1992. U1 targets an essential splicing factor, U2AF65, to the 3′ splice site by a network of interactions spanning the exon. *Genes Dev* **6:** 2554–2568.

House AE, Lynch KW. 2006. An exonic splicing silencer represses spliceosome assembly after ATP-dependent exon recognition. *Nat Struct Mol Biol* **13:** 937–944.

Jacquier A. 1990. Self-splicing group II and nuclear pre-mRNA introns: how similar are they? *Trends Biochem Sci* **15:** 351–354.

Jurica MS, Moore MJ. 2003. Pre-mRNA splicing: Awash in a sea of proteins. *Mol Cell* **12:** 5–14.

Jurica MS, Sousa D, Moore MJ, Grigorieff N. 2004. Three-dimensional structure of C complex spliceosomes by electron microscopy. *Nat Struct Mol Biol* **11:** 265–269.

Kambach C, Walke S, Nagai K. 1999a. Structure and assembly of the spliceosomal small nuclear ribonucleoprotein particles. *Curr Opin Struct Biol* **9:** 222–230.

Kambach C, Walke S, Young R, Avis JM, de la Fortelle E, Raker VA, Lührmann R, Li J, Nagai K. 1999b. Crystal structures of two Sm protein complexes and their implications for the assembly of the spliceosomal snRNPs. *Cell* **96:** 375–387.

Karaduman R, Dube P, Stark H, Fabrizio P, Kastner B, Lührmann R. 2008. Structure of yeast U6 snRNPs: Arrangement of Prp24p and the LSm complex as revealed by electron microscopy. *RNA* **14:** 2528–2537.

Kastner B, Fischer N, Golas M, Sander B, Dube P, Boehringer D, Hartmuth K, Deckert J, Hauer F, Wolf E, et al. 2008. GraFix: Sample preparation for single-particle electron cryomicroscopy. *Nat Methods* **5:** 53–55.

Keating KS, Toor N, Perlman PS, Pyle AM. 2010. A structural analysis of the group II intron active site and implications for the spliceosome. *RNA* **16:** 1–9.

Kent OA, MacMillan AM. 2002. Early organization of pre-mRNA during spliceosome assembly. *Nat Struct Biol* **9:** 576–581.

Kielkopf CL, Rodionova NA, Green MR, Burley SK. 2001. A novel peptide recognition mode revealed by the X-ray structure of a core U2AF35/U2AF65 heterodimer. *Cell* **106:** 595–605.

Kistler AL, Guthrie C. 2001. Deletion of MUD2, the yeast homolog of U2AF65, can bypass the requirement for Sub2, an essential spliceosomal ATPase. *Genes Dev* **15:** 42–49.

Konarska MM, Query CC. 2005. Insights into the mechanisms of splicing: More lessons from the ribosome. *Genes Dev* **19:** 2255–2260.

Konarska MM, Vilardell J, Query CC. 2006. Repositioning of the reaction intermediate within the catalytic center of the spliceosome. *Mol Cell* **21:** 543–553.

Krol A, Westhof E, Bach M, Lührmann R, Ebel JP, Carbon P. 1990. Solution structure of human U1 snRNA. Derivation of a possible three-dimensional model. *Nucleic Acids Res* **18:** 3803–3811.

Kuhn AN, Van Santen MA, Schwienhorst A, Urlaub H, Lührmann R. 2009. Stalling of spliceosome assembly at distinct stages by small-molecule inhibitors of protein acetylation and deacetylation. *RNA* **15:** 153–175.

Lim SR, Hertel KJ. 2004. Commitment to splice site pairing coincides with A complex formation. *Mol Cell* **15:** 477–483.

Liu L, Query CC, Konarska MM. 2007. Opposing classes of prp8 alleles modulate the transition between the catalytic steps of pre-mRNA splicing. *Nat Struct Mol Biol* **14:** 519–526.

Liu YC, Chen HC, Wu NY, Cheng SC. 2007. A novel splicing factor, Yju2, is associated with NTC and acts after Prp2 in promoting the first catalytic reaction of pre-mRNA splicing. *Mol Cell Biol* **27:** 403–5413.

Liu S, Li P, Dybkov O, Nottrott S, Hartmuth K, Lührmann R, Carlomagno T, Wahl MC. 2007. Binding of the human Prp31 nop domain to a composite RNA-protein platform in U4 snRNP. *Science* **316:** 115–120.

Liu Z, Luyten I, Bottomley MJ, Messias AC, Houngninou-Molango S, Sprangers R, Zanier K, Krämer A, Sattler M. 2001. Structural basis for recognition of the intron branch site RNA by splicing factor 1. *Science* **294:** 1098–1102.

Lührmann R, Stark H. 2009. Structural mapping of spliceosomes by electron microscopy. *Curr Opin Struct Biol* **19:** 96–102.

Madhani HD, Guthrie C. 1992. A novel base-pairing interaction between U2 and U6 snRNAs suggests a mechanism for the catalytic activation of the spliceosome. *Cell* **71:** 803–817.

Madhani HD, Guthrie C. 1994. Genetic interactions between the yeast RNA helicase homolog Prp16 and spliceosomal snRNAs identify candidate ligands for the Prp16 RNA-dependent ATPase. *Genetics* **137:** 677–687.

Maeder C, Kutach AK, Guthrie C. 2009. ATP-dependent unwinding of U4/U6 snRNAs by the Brr2 helicase requires the C terminus of Prp8. *Nat Struct Mol Biol* **16:** 42–48.

Maniatis T, Reed R. 2002. An extensive network of coupling among gene expression machines. *Nature* **416:** 499–506.

Mathew R, Hartmuth K, Moehlmann S, Urlaub H, Ficner R, Lührmann R. 2008. Phosphorylation of human PRP28 by SRPK2 is required for integration of the U4/U6-U5 tri-snRNP into the spliceosome. *Nat Struct Mol Biol* **15:** 435–443.

Matlin AJ, Moore MJ. 2007. Spliceosome assembly and composition. *Adv Exp Med Biol* **623:** 14–35.

Matlin AJ, Clark F, Smith CW. 2005. Understanding alternative splicing: Towards a cellular code. *Nat Rev Mol Cell Biol* **6:** 386–398.

Mayas RM, Maita H, Staley JP. 2006. Exon ligation is proofread by the DExD/H-box ATPase Prp22p. *Nat Struct Mol Biol* **13:** 482–490.

Mefford MA, Staley JP. 2009. Evidence that U2/U6 helix I promotes both catalytic steps of pre-mRNA splicing and rearranges in between these steps. *RNA* **15:** 1386–1397.

Misteli T. 1999. RNA splicing: What has phosphorylation got to do with it? *Curr Biol* **9:** R198–200.

Moore MJ, Query CC, Sharp PA. 1993. Splicing of precursors to mRNA by the spliceosome. In *RNA World* (ed. R.F. Gesteland, J.F. Atkins), pp. 303–357. Cold Spring Harbor Laboratory Press, Cold Spring Harbor, New York.

Nilsen TW. 1998. RNA-RNA interactions in nuclear pre-mRNA splicing. In *RNA Structure and Function* (ed. R.W. Simons, M. Grunberg-Manago), pp. 279–308. Cold Spring Harbor Press, Cold Spring Harbor.

Nilsen TW, Graveley BR. 2010. Expansion of the eukaryotic proteome by alternative splicing. *Nature* **463:** 457–463.

Novoyatleva T., Tang Y, Rafalska I, Stamm S. 2006. Pre-mRNA missplicing as a cause of human disease. *Prog Mol Subcell Biol* **44:** 27–46.

Ohi MD, Ren L, Wall JS, Gould KL, Walz T. 2007. Structural characterization of the fission yeast U5.U2/U6 spliceosome complex. *Proc Natl Acad Sci* **104:** 3195–3200.

Oubridge C, Ito N, Evans PR, Teo CH, Nagai K. 1994. Crystal structure at 1.92 Å resolution of the RNA-binding domain of the U1A spliceosomal protein complexed with an RNA hairpin. *Nature* **372:** 432–438.

Pandit S, Lynn B, Rymond BC. 2006. Inhibition of a spliceosome turnover pathway suppresses splicing defects. *Proc Natl Acad Sci* **103:** 13700–13705.

Park JW, Parisky K, Celotto AM, Reenan RA, Graveley BR. 2004. Identification of alternative splicing regulators by RNA interference in Drosophila. *Proc Natl Acad Sci* **101:** 15974–15979.

Patel AA, Steitz JA. 2003. Splicing double: Insights from the second spliceosome. *Nat Rev Mol Cell Biol* **4:** 960–970.

Pena V, Liu S, Bujnicki JM, Lührmann R, Wahl MC. 2007. Structure of a multipartite protein-protein interaction domain in splicing factor Prp8 and its link to retinitis pigmentosa. *Mol Cell* **25:** 615–624.

Pena V, Rozov A, Fabrizio P, Lührmann R, Wahl MC. 2008. Structure and function of an RNase H domain at the heart of the spliceosome. *EMBO J* **27:** 2929–2940.

Pena V, Mozaffari SJ, Fabrizio P, Orlowski J, Bujnicki JM, Lührmann R, Wahl MC. 2009. Common design principles in the spliceosomal RNA helicase Brr2 and in the Hel308 DNA helicase. *Mol Cell* **35:** 454–466.

Perriman RJ, Ares M Jr. 2007. Rearrangement of competing U2 RNA helices within the spliceosome promotes multiple steps in splicing. *Genes Dev* **21**: 811–820.

Pleiss JA, Whitworth GB, Bergkessel M, Guthrie C. 2007. Transcript specificity in yeast pre-mRNA splicing revealed by mutations in core spliceosomal components. *PLoS Biol* **5**: e90.

Pomeranz-Krummel DA, Oubridge C, Leung AK, Li J, Nagai K. 2009. Crystal structure of human spliceosomal U1 snRNP at 5.5 A resolution. *Nature* **458**: 475–480.

Price SR, Evans PR, Nagai K. 1998. Crystal structure of the spliceosomal U2B"-U2A' protein complex bound to a fragment of U2 small nuclear RNA. *Nature* **394**: 645–650.

Query CC, Konarska MM. 2004. Suppression of multiple substrate mutations by spliceosomal prp8 alleles suggests functional correlations with ribosomal ambiguity mutants. *Mol Cell* **14**: 343–354.

Reed R 2000. Mechanisms of fidelity in pre-mRNA splicing. *Curr Opin Cell Biol* **12**: 340–345.

Rhode BM, Hartmuth K, Westhof E, Lührmann R. 2006. Proximity of conserved U6 and U2 snRNA elements to the 5′ splice site region in activated spliceosomes. *EMBO J* **25**: 2475–2486.

Rino J, Desterro JM, Pacheco TR, Gadella TW Jr, Carmo-Fonseca M. 2008. Splicing factors SF1 and U2AF associate in extraspliceosomal complexes. *Mol Cell Biol* **28**: 3045–3057.

Ritchie DB, Schellenberg MJ, Macmillan AM. 2009. Spliceosome structure: Piece by piece. *Biochim Biophys Act* **1789**: 624–633.

Ritchie DB, Schellenberg MJ, Gesner EM, Raithatha SA, Stuart DT, Macmillan AM. 2008. Structural elucidation of a PRP8 core domain from the heart of the spliceosome. *Nat Struct Mol Biol* **15**: 1199–1205.

Sashital DG, Cornilescu G, McManus CJ, Brow DA, Butcher SE. 2004. U2-U6 RNA folding reveals a group II intron-like domain and a four-helix junction. *Nat Struct Mol Biol* **11**: 1237–1242.

Sander B, Golas MM, Makarov EM, Brahms H, Kastner B, Lührmann R, Stark H. 2006. Organization of the core spliceosomal components U5 snRNA loop I and U4/U6 di-snRNP within the U4/U6.U5 tri-snRNP as revealed by 3D electron microscopy. *Mol Cell* **24**: 267–278.

Schellenberg MJ, Edwards RA, Ritchie DB, Kent OA, Golas MM, Stark H, Lührmann R, Glover JNM, MacMilan AM. 2006. Crystal structure of a core spliceosomal protein interface. *Proc Natl Acad Sci* **103**: 1266–1271.

Schneider M, Hsiao HH, Will CL, Giet R, Urlaub H, Lührmann R. 2010a. Human PRP4 kinase is required for stable tri-snRNP association during spliceosomal B complex formation. *Nat Struct Mol Biol* **17**: 216–221.

Schneider M, Will CL, Anokhina M, Tazi J, Urlaub U, Lührmann R. 2010b. Exon definition complexes contain the tri-snRNP and can be directly converted into B-like pre-catalytic splicing complexes. *Mol Cell* **38**: 223–235.

Schwer B. 2008. A conformational rearrangement in the spliceosome sets the stage for Prp22-dependent mRNA release. *Mol Cell* **30**: 743–754.

Selenko P, Gregorovic G, Sprangers R, Stier G, Rhani Z, Krämer A, Sattler M. 2003. Structural basis for the molecular recognition between human splicing factors U2AF65 and SF1/mBBP. *Mol Cell* **11**: 965–976.

Sharma S, Kohlstaedt LA, Damianov A, Rio DC, Black DL. 2008. Polypyrimidine tract binding protein controls the transition from exon definition to an intron defined spliceosome. *Nat Struct Mol Biol* **15**: 183–191.

Shen H, Green MR. 2004. A pathway of sequential arginine-serine-rich domain-splicing signal interactions during mammalian spliceosome assembly. *Mol Cell* **16**: 363–373.

Shen H, Green MR. 2006. RS domains contact splicing signals and promote splicing by a common mechanism in yeast through humans. *Genes Dev* **20**: 1755–1765.

Shen H, Green MR. 2007. RS domain-splicing signal interactions in splicing of U12-type and U2-type introns. *Nat Struct Mol Biol* **14**: 597–603.

Shen H, Zheng X, Shen J, Zhang L, Zhao R, Green MR. 2008. Distinct activities of the DExD/H-box splicing factor hUAP56 facilitate stepwise assembly of the spliceosome. *Genes Dev* **22**: 1796–1803.

Shi Y, Reddy B, Manley JL. 2006. PP1/PP2A phosphatases are required for the second step of pre-mRNA splicing and target specific snRNP proteins. *Mol Cell* **23**: 819–829.

Sickmier EA, Frato KE, Shen H, Paranawithana SR, Green MR, Kielkopf CL. 2006. Structural basis for polypyrimidine tract recognition by the essential pre-mRNA splicing factor U2AF65. *Mol Cell* **23**: 49–59.

Silverman EJ, Maeda A, Wei J, Smith P, Beggs JD, Lin RJ. 2004. Interaction between a G-patch protein and a spliceosomal DEXD/H-box ATPase that is critical for splicing. *Mol Cell Biol* **24**: 10101–10110.

Small EC, Leggett SR, Winans AA, Staley JP. 2006. The EF-G-like GTPase Snu116p regulates spliceosome dynamics mediated by Brr2p, a DExD/H box ATPase. *Mol Cell* **23**: 389–399.

Smith DJ, Konarska MM. 2009a. Identification and characterization of a short 2′;-3′ bond-forming ribozyme. *RNA* **15**: 8–13.

Smith DJ, Konarska MM. 2009b. A critical assessment of the utility of protein-free splicing systems. *RNA* **15**: 1–3.

Smith CW, Valcárcel J. 2000. Alternative pre-mRNA splicing: The logic of combinatorial control. *Trends Biochem Sci* **25**: 381–388.

Smith DJ, Query CC, Konarska MM. 2007. trans-splicing to spliceosomal U2 snRNA suggests disruption of branch site-U2 pairing during pre-mRNA splicing. *Mol Cell* **26**: 883–890.

Smith DJ, Query CC, Konarska MM. 2008. "Nought may endure but mutability": Spliceosome dynamics and the regulation of splicing. *Mol Cell* **30**: 657–666.

Soret J, Tazi J. 2003. Phosphorylation-dependent control of the pre-mRNA splicing machinery. *Prog Mol Subcell Biol* **31**: 89–126.

Spadaccini R, Reidt U, Dybkov O, Will C, Frank R, Stier G, Corsini L, Wahl MC, Lührmann R, Sattler M. 2006. Biochemical and NMR analyses of a SF3b155-p14-U2AF-RNA interaction network involved in branch point definition during pre-mRNA splicing. *RNA* **12**: 410–425.

Stevens SW, Ryan DE, Ge HY, Moore RE, Young MK, Lee TD, Abelson J. 2002. Composition and functional characterization of the yeast spliceosomal penta-snRNP. *Mol Cell* **9**: 31–44.

Staley JP, Guthrie C. 1998. Mechanical devices of the spliceosome: Motors, clocks, springs, and things. *Cell* **92**: 315–326.

Staley JP, Guthrie C. 1999. An RNA switch at the 5′ splice site requires ATP and the DEAD box protein Prp28p. *Mol Cell* **3**: 55–64.

Staley JP, Woolford JL Jr. 2009. Assembly of ribosomes and spliceosomes: Complex ribonucleoprotein machines. *Curr Opin Cell Biol* **21**: 109–118.

Stark H, Lührmann R. 2006. Electron cryomicroscopy of spliceosomal components. *Ann Rev Biophy Biomol Struct* **35**: 435–457.

Stark H, Dube P, Luhrmann R, Kastner B. 2001. Arrangement of RNA and proteins in the spliceosomal U1 small nuclear ribonucleoprotein particle. *Nature* **409**: 539–542.

Sun JS, Manley JL. 1995. A novel U2-U6 snRNA structure is necessary for mammalian mRNA splicing. *Genes Dev* **9**: 843–854.

Tanaka N, Aronova A, Schwer B. 2007. Ntr1 activates the Prp43 helicase to trigger release of lariat-intron from the spliceosome. *Genes Dev* **21**: 2312–2325.

Tazi J, Kornstädt U, Rossi F, Jeanteur P, Cathala G, Brunel C, Lührmann R. 1993. Thiophosphorylation of U1-70K protein inhibits pre-mRNA splicing. *Nature* **363**: 283–286.

Toor N, Keating KS, Taylor SD, Pyle AM. 2008. Crystal structure of a self-spliced group II intron. *Science* **320**: 77–82.

Tsai RT, Fu RH, Yeh FL, Tseng CK, Lin YC, Huang YH, Cheng SC. 2005. Spliceosome disassembly catalyzed by Prp43 and its associated components Ntr1 and Ntr2. *Genes Dev* **19**: 2991–3003.

Tsai RT, Tseng CK, Lee PJ, Chen HC, Fu RH, Chang KJ, Yeh FL, Cheng SC. 2007. Dynamic interactions of Ntr1-Ntr2 with Prp43 and with U5 govern the recruitment of Prp43 to mediate spliceosome disassembly. *Mol Cell Biol* **27**: 8027–8037.

Tseng CK, Cheng SC. 2008. Both catalytic steps of nuclear pre-mRNA splicing are reversible. *Science* **320**: 1782–1784.

Turner IA, Norman CM, Churcher MJ, Newman AJ. 2004. Roles of the U5 snRNP in spliceosome dynamics and catalysis. *Biochem Soc Trans* **32**: 928–931.

Umen JG, Guthrie C. 1995. The second catalytic step of pre-mRNA splicing. *RNA* **1**: 869–885.

Valadkhan S. 2005. snRNAs as the catalysts of pre-mRNA splicing. *Curr Opin Chem Biol* **9**: 603–608.

Valadkhan S, Manley JL. 2001. Splicing-related catalysis by protein-free snRNAs. *Nature* **413**: 701–707.

Valadkhan S, Mohammadi A, Jaladat Y, Geisler S. 2009. Protein-free small nuclear RNAs catalyze a two-step splicing reaction. *Proc Natl Acad Sci* **106**: 11901–11906.

Valadkhan S, Mohammadi A, Wachtel C, Manley JL. 2007. Protein-free spliceosomal snRNAs catalyze a reaction that resembles the first step of splicing. *RNA* **13**: 2300–2311.

Valcárcel J, Gaur RK, Singh R, Green MR. 1996. Interaction of U2AF65 RS region with pre-mRNA branch point and promotion of base pairing with U2 snRNA. *Science* **273**: 1706–1709.

Vidovic I, Nottrott S, Hartmuth K, Lührmann R, Ficner R. 2000. Crystal structure of the spliceosomal 15.5kD protein bound to a U4 snRNA fragment. *Mol Cell* **6**: 1331–1342.

Wachtel C, Manley JL. 2009. Splicing of mRNA precursors: The role of RNAs and proteins in catalysis. *Mol Biosyst* **5**: 311–316.

Wahl MC, Will CL, Lührmann R. 2009. The spliceosome: Design principles of a dynamic RNP machine. *Cell* **136**: 701–718.

Wang Z, Burge CB. 2008. Splicing regulation: From a parts list of regulatory elements to an integrated splicing code. *RNA* **14**: 802–813.

Wang C, Chua K, Seghezzi W, Lees E, Gozani O, Reed R. 1998. Phosphorylation of spliceosomal protein SAP 155 coupled with splicing catalysis. *Genes Dev* **12**: 1409–1414.

Ward AJ, Cooper TA. 2010. The pathobiology of splicing. *J Pathol* **220**: 152–163.

Warkocki Z, Odenwälder P, Schmitzová J, Platzmann F, Stark H, Urlaub H, Ficner R, Fabrizio P, Lührmann R. 2009. Reconstitution of both steps of Saccharomyces cerevisiae splicing with purified spliceosomal components. *Nat Struct Mol Biol* **16**: 1237–1243.

Webby CJ, Wolf A, Gromak N, Dreger M, Kramer H, Kessler B, Nielsen ML, Schmitz C, Butler DS, Yates JR, et al. 2009. Jmjd6 catalyses lysyl-hydroxylation of U2AF65, a protein associated with RNA splicing. *Science* **325**: 90–93.

Will CL, Lührmann R. 2006. Spliceosome structure and function. In *The RNA world*, 3rd ed. (ed. R.F. Gesteland et al.), pp. 369–400. Cold Spring Harbor Laboratory Press, Cold Spring Harbor, NY.

Wolf E, Kastner B, Deckert J, Merz C, Stark H, Lührmann R. 2009. Exon, intron and splice site locations in the spliceosomal B complex. *EMBO J* **28**: 2283–2292.

Xiao SH, Manley JL. 1997. Phosphorylation of the ASF/SF2 RS domain affects both protein-protein and protein-RNA interactions and is necessary for splicing. *Genes Dev* **11**: 334–344.

Xiao X, Wang Z, Jang M, Burge CB. 2007. Coevolutionary networks of splicing cis-regulatory elements. *Proc Natl Acad Sci* **104**: 18583–18588.

Xu YZ, Query CC. 2007. Competition between the ATPase Prp5 and branch region-U2 snRNA pairing modulates the fidelity of spliceosome assembly. *Mol Cell* **28**: 838–849.

Yang K, Zhang L, Xu T, Heroux A, Zhao R. 2008. Crystal structure of the β-finger domain of Prp8 reveals analogy to ribosomal proteins. *Proc Natl Acad Sci* **105**: 13817–13822.

Zhang L, Shen J, Guarnieri MT, Heroux A, Yang K, Zhao R. 2007. Crystal structure of the C-terminal domain of splicing factor Prp8 carrying retinitis pigmentosa mutants. *Protein Sci* **16**: 1024–1031.

Zhang L, Xu T, Maeder C, Bud LO, Shanks J, Nix J, Guthrie C, Pleiss JA, Zhao R. 2009. Structural evidence for consecutive Hel308-like modules in the spliceosomal ATPase Brr2. *Nat Struct Mol Biol* **16**: 731–739.

Telomerase: An RNP Enzyme Synthesizes DNA

Elizabeth H. Blackburn[1] and Kathleen Collins[2]

[1]Morris Herztein Endowed Professor in Biology and Physiology, University of California, San Francisco, California 94158-2517

[2]Professor of Biochemistry and Molecular Biology, University of California, Berkeley, Department of Molecular & Cell Biology, Berkeley, California 94720-3200

Correspondence: elizabeth.blackburn@ucsf.edu

SUMMARY

Telomerase is a eukaryotic ribonucleoprotein (RNP) whose specialized reverse transcriptase action performs de novo synthesis of one strand of telomeric DNA. The resulting telomerase-mediated elongation of telomeres, which are the protective end-caps for eukaryotic chromosomes, counterbalances the inevitable attrition from incomplete DNA replication and nuclease action. The telomerase strategy to maintain telomeres is deeply conserved among eukaryotes, yet the RNA component of telomerase, which carries the built-in template for telomeric DNA repeat synthesis, has evolutionarily diverse size and sequence. Telomerase shows a distribution of labor between RNA and protein in aspects of the polymerization reaction. This article first describes the underlying conservation of a core set of structural features of telomerase RNAs important for the fundamental polymerase activity of telomerase. These include a pseudoknot-plus-template domain and at least one other RNA structural motif separate from the template-containing domain. The principles driving the diversity of telomerase RNAs are then explored. Much of the diversification of telomerase RNAs has come from apparent gain-of-function elaborations, through inferred evolutionary acquisitions of various RNA motifs used for telomerase RNP biogenesis, cellular trafficking of enzyme components, and regulation of telomerase action at telomeres. Telomerase offers broadly applicable insights into the interplay of protein and RNA functions in the context of an RNP enzyme.

Outline

1. An endogenous reverse transcriptase for telomeric repeat synthesis
2. Interplay of RNA and protein function in a mixed-substance polymerase
3. General features of telomerase RNAs
4. Refinement of the catalytic cycle, including roles for phylogenetically diverse TER motifs
5. Bells and whistles: TER motifs for holoenzyme protein interactions
6. Telomerase as a window to an evolutionary RNP renaissance

References

Copyright © 2011 Cold Spring Harbor Laboratory Press; all rights reserved
Cite this article as *Cold Spring Harb Perspect Biol* doi: 10.1101/cshperspect.a003558

1 AN ENDOGENOUS REVERSE TRANSCRIPTASE FOR TELOMERIC REPEAT SYNTHESIS

DNA and protein assemblies at the ends of eukaryotic chromosomes constitute the end-portions, or "telomeres," essential for genomic stability. Jeopardy is inherent in the terminal location of these chromosome domains: Terminal DNA sequences are under-replicated and even actively eroded in each cell cycle, because of incomplete DNA-templated DNA replication and the nuclease processing required to generate a 3′ overhang for end-capping proteins. The widespread eukaryotic strategy to counter-balance telomeric repeat erosion involves telomerase, a specialized reverse transcriptase capable of de novo DNA synthesis. Telomerase action coupled with additional DNA replication and end-repair activities allows telomere length homeostasis in single-celled organisms and generational maintenance of telomere length in the germline of multicellular organisms (Gilson and Geli 2007; Verdun and Karlseder 2007). Somatic tissues of long-lived organisms repress telomerase, adding a hurdle to tumor progression but limiting tissue renewal (Collins and Mitchell 2002; Garcia et al. 2007). Cancer cells typically activate telomerase and telomere maintenance, with clinical implications described in detail elsewhere (Harley 2008).

The telomerase mechanism is deeply conserved throughout eukaryotes, from flagellated protozoans including *Giardia* and trypanosomes to higher plants and metazoans. Even the genome of a virus that infects chickens has been discovered to encode a telomerase RNA component, likely acquired by lateral transfer from its metazoan host. Eukaryotic species lacking telomerase-dependent telomere maintenance are the exception rather than the rule, with indications that alternative strategies can be independently derived. Soon after its discovery in the pond ciliate *Tetrahymena thermophila* (Greider and Blackburn 1985), the enzymatic activity of telomerase was shown to be sensitive to ribonuclease and protease treatment, implicating an RNA as well as protein contribution to enzymatic activity (Greider and Blackburn, 1987). A region within the integral RNA subunit of the enzyme provides the template for telomeric-repeat DNA synthesis (Greider and Blackburn 1989), defining telomerase as a reverse transcriptase (RT). Many telomerases can reiteratively copy the internal template (Fig. 1) without completely releasing product DNA, achieving both nucleotide addition processivity across the template and repeat addition processivity in the synthesis of a product ladder.

2 INTERPLAY OF RNA AND PROTEIN FUNCTION IN A MIXED-SUBSTANCE POLYMERASE

The demonstration that an RNA moiety provides the template for polymerization of telomeric DNA repeats established a new paradigm for RNA function in a cellular RNP. But how do the various catalytic steps required of a DNA polymerase—binding of primer and nucleotide substrates, conformational change(s) required for substrate discrimination and alignment, catalysis itself, product and template translocation, and eventual product release—distribute as a division of labor between the RNA and protein components of the telomerase RNP enzyme? Together the telomerase RNA (TER) and telomerase reverse transcriptase protein (TERT) can reconstitute telomeric repeat synthesis in vitro, without the additional proteins that coassemble in vivo to form telomerase holoenzymes (Autexier and Lue 2006; Collins 2006). TERT active site motifs shared with other protein-only RTs play an essential role (Lingner et al. 1997), consistent with their expected function in binding the magnesium ions that activate the deoxynucleoside triphosphate for incorporation. In addition to this general polymerase chemistry, there must be enzyme determinants that adapt telomerase for its unique specificity of template and primer use. This article aims to highlight how the catalytic specialization of telomerase occurred in large part through gains of RNA motif function. Coevolution of coordinated roles for protein and RNA subunits of telomerase presents a challenge to deciphering the earliest stages of its evolution (discussed in section 6.1).

In addition to the gains of function necessary for its biochemical and biological specialization, telomerase is notably lacking in some features typically associated with protein RTs. For example, to reuse its internal template, telomerase must eschew the RT-associated RNase H activity important for degradation of the RNA strand of a retroviral RT RNA-cDNA duplex. Although specializations unique to telomerase are described later, we note that the human LINE-1 retroelement RT lacks RNase H activity and like telomerase shows strong preference for copying an associated RNA (Kulpa and Moran 2006; Piskareva and Schmatchenko 2006).

3 GENERAL FEATURES OF TELOMERASE RNAS

3.1 Uncovering the Core Beneath Sequence Divergence and Evolutionary Innovation

Despite broad evolutionary conservation of the telomerase mechanism, the size and sequence of TERs show great diversity. The TERs of smallest size are the group cloned from ciliated protozoa, only 147–209 nt in length (Ye and Romero 2002). Many TERs have been cloned from vertebrate species as well, with lengths ranging from 312 to 559 nt (Xie et al. 2008). On the long side of the spectrum are the sequenced group of budding yeast TERs, with lengths of 779–1817 nt (Gunisova et al. 2009). The current length record is held by the malarial parasite *Plasmodium*

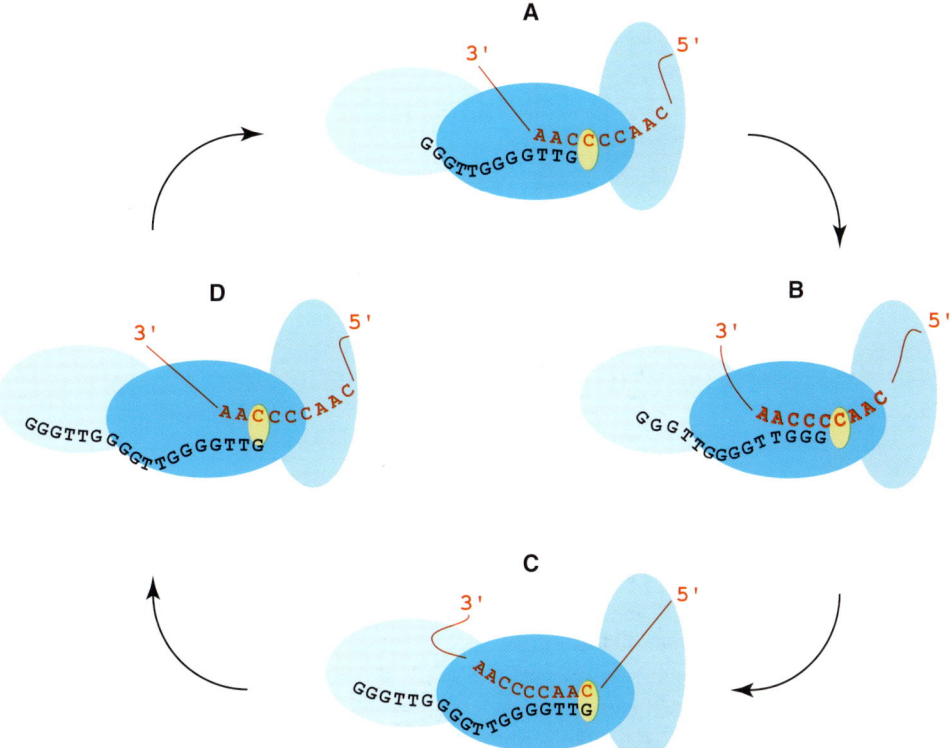

Figure 1. The telomerase catalytic cycle. Stages in the catalytic cycle are illustrated using *T. thermophila* sequences of the TER template (red) and telomeric-repeat DNA (black). The active site (yellow) is within the TERT RT domain (filled with the darkest shade of blue). Some TERT-TER contacts that influence template use are stable across all states of the catalytic cycle (shown for *T. thermophila* as the template 5′-flanking region bound to the TERT TRBD domain in an intermediate shade of blue). Other TERT-TER and TERT-DNA contacts may be specific to particular configurations of template and product relative to the active site (shown here as changing contact between the template 3′-flanking region and the TERT TEN domain in the lightest shade of blue, which also contacts single-stranded DNA). Copying across the template with nucleotide addition processivity is accompanied by changes in the length of hybrid between template RNA and product DNA, depicted in states (A–C). Product released from the template can be held in association with the active enzyme by other interactions, as depicted in state (D). Realignment of the product 3′ end at the template 3′ end, as depicted in the transition from state (D) to state (A), allows for repeat addition processivity.

falciparum, which has a TER of 2.2 kb (Chakrabarti et al. 2007).

What drives this evolutionary diversification, when TERT length and sequence have remained generally more consistent (Fig. 2)? TER phylogenetic diversity could be merely a reflection of freedom to drift. Alternatively, RNA may offer more opportunities than protein for the occurrence, selection, or fixation of functional gains. RNA sequence expansion, gain of interaction, and functional adaptation may be sampled readily over evolution because of inherent properties of RNA structure, leading to advantageous RNP enzyme properties.

It is difficult to chart the evolutionary steps of TER divergence between the phylogenetic groups of ciliates, vertebrates, and budding yeasts using only modern-day TER secondary structure models. Despite this limitation, we will draw within-group and in some cases between-group comparisons of TER structure and function in the sections to follow, beginning in this section with a description of three TER motifs arguably shared by all known TERs with significance for the catalytic cycle: the template, the pseudoknot, and a stem-loop or bulged stem-junction with conserved paired and unpaired nucleotides (Fig. 3).

3.2 The Template

The template region in all TERs includes a 5′ portion that is frequently copied and a 3′ portion that is typically used for the alignment of a primer (a synthetic DNA oligonucleotide in vitro or a single-stranded chromosome 3′ end in vivo). Hybridization-based primer alignment allows the synthesis of a precise repeat sequence. Also, if product released from the end of the template is retained by other enzyme associations, the alignment region facilitates repeat addition

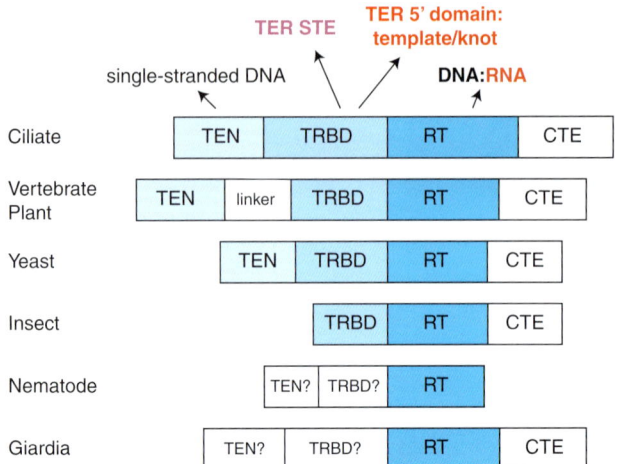

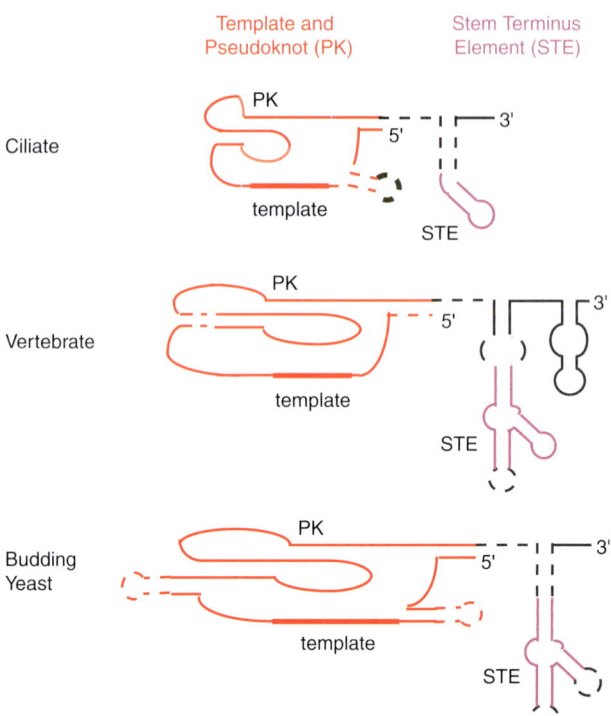

Figure 2. Conserved TERT domains. TERT domain evolutionary conservation and protein-nucleic acid interactions are summarized, with the same color scheme of TERT and TER domains used in other figures.

processivity (Fig. 1). The template-product hybrid must reach a length of up to four to five base-pairs to uniquely match a primer 3' end with its proper template position. However, at least for the telomerases assayed to date, the maximal possible length of template-product hybrid does not form (Wang et al. 1998; Förstemann and Lingner 2005). Instead, as polymerization progresses along the template, unpairing occurs at the template 3' end, even if this requires partial dissociation of formerly annealed primer (Fig. 1). Polymerization along the template halts at a typically, but not always, fixed template 5' boundary. Copying generally results in synthesis of the template complement, but again this is not always true: there are a few examples of deviant nucleotide selection that are biologically used in the normal course of synthesis of mixed or degenerate telomeric repeats, as discussed in detail elsewhere (Collins 2009). Also, some telomerases can extend a primer without hybridization, using a specific template position as the default for initiation (Yu and Blackburn 1991; Wang and Blackburn 1997).

Mutational analyses show that the template is not just a passive carrier of sequence information. Substitutions of one or a few template bases can have large, mutation-specific impacts on rates of dNTP misincorporation, propensity for template slippage or misalignment, or premature product dissociation (Gilley and Blackburn 1996; Lin et al. 2004). Substrate dGTP also affects enzyme function independently of its incorporation, for example altering repeat addition processivity through a presumed allosteric influence on the active site (Hammond and Cech 1997; Hardy et al. 2001). Also, single-stranded DNA can interact with active RNP beyond 3' base-pairing with the template, affecting several aspects of the polymerization

Figure 3. Conserved TER motifs within phylogenetically divergent secondary structures. The thickest region of line represents the template, whereas dashed lines represent sites of sequence variability within the phylogenetic group. PK indicates the pseudoknot. The color scheme of TER motifs matches that in the other figures.

cycle (Collins and Greider 1993; Lee and Blackburn 1993). These and additional results point to a well-tuned codependence of template, nucleotide, and product sequence identity in determining the fidelity and overall activity of internal template use.

3.3 A Shared Pseudoknot

The likelihood of strongly conserved nontemplate TER structures emerged from comparison of the phylogenetically derived TER secondary structures of ciliates, vertebrates, and budding yeasts (Legassie and Jarstfer 2006; Theimer and Feigon 2006). All contain a similarly folded pseudoknot. The template and pseudoknot together are enclosed within a TER domain formed by long-distance pairing of the 5' end of TER (Fig. 3). Ciliate TERs show thermodynamically weak pseudoknot folding within their compact overall structure, whereas yeast TERs tolerate long stem-loop insertions in the template/pseudoknot domain. Curiously, rodent TERs have lost the long-distance 5' end pairing.

One set of open questions concerns the function(s) of the pseudoknot and pseudoknot structural dynamics. The pseudoknot is one of the last elements of ciliate TERs to adopt a stable fold during assembly of the minimal

RNP in vitro, which it does only if TERT contacts are established with flanking TER motifs (Sperger and Cech 2001). On the other hand, the vertebrate TER pseudoknot folds independently into a rigid structure with highly determined triple-helix U-A-U base-pairing architecture (Theimer et al. 2005). Modeling and mutagenesis indicate that triple-helix structure is conserved and essential in budding yeasts as well (Shefer et al. 2007). The thermodynamic stability of human TER pseudoknot folding would seem to argue against catalytic cycle dynamics. However, tertiary structure dynamics could be coupled to distortions of RNA and protein conformation that occur during a cycle of repeat synthesis, as the template transits the active site.

RNA motif transplantation assays have suggested that a particular TER pseudoknot can be replaced by another structure with retention of TER function. Also, circular permutation and trans complementation assays of ciliate and vertebrate TERs indicate that the pseudoknot can stimulate the activity of recombinant TERT even if it is disrupted in backbone continuity or covalently unlinked from adjacent sequence. However, these chimeric, fragmented, and mutated pseudoknot success stories may not have detected important roles of the pseudoknot in the catalytic cycle. In vivo expression of some but not all *T. thermophila* TER pseudoknot-disruption variants reduce cell viability and enzyme activity (Gilley and Blackburn 1999; Cunningham and Collins 2005). Some *Kluyveromyces lactis* pseudoknot mutations that do not abrogate enzyme activity reduce the extent of template copying, leading to variably truncated repeat synthesis in vivo (Tzfati et al. 2003). Also, in vitro assays with a minimized *Saccharomyces cerevisiae* TER suggest that pseudoknot 2′-OH–mediated interactions contribute to primer positioning in the active site (Qiao and Cech 2008).

3.4 A Stem Terminus Element

Beyond the template and pseudoknot, TERs have at least one additional motif that may have been present in the ancestral enzyme. In all groups of TERs characterized to date, a stem-loop or bulged stem-junction is important for activity, with unpaired loop or bulge residues adjacent to a stem terminus being particularly critical. In the context of modern-day TER secondary structures (Fig. 3), this stem terminus element (STE) can occur as a terminal stem-loop (loop IV of ciliate TER), a stem-loop emerging from a stem junction with bulged nucleotides (P6.1 of vertebrate TER), or a stem junction with bulged nucleotides (the three-way junction of budding yeast TER). It is reasonable to surmise that the function of the STE may be conserved among ciliate, vertebrate, and yeast telomerases, but unrelated functions are also possible. The ciliate and vertebrate STE motifs both bind TERT in a manner separable from TERT interaction with the template/pseudoknot domain (discussed in section 4.2) and greatly stimulate telomerase catalytic activity. Robust, highly processive activity can be reconstituted using ciliate or vertebrate TER fragments separately carrying the template/pseudoknot domain and STE, indicating that STE function does not require covalent linkage with the template (Mitchell and Collins 2000; Lai et al. 2003; Mason et al. 2003). Curiously, although *K. lactis* telomerase holoenzyme activity is severely inhibited by STE mutations, the STE is missing from a minimal TER that reconstitutes core *S. cerevisiae* telomerase enzyme activity in vitro (Zappulla et al. 2005; Brown et al. 2007). In contrast, whereas STE substitutions cripple the activity of the *T. thermophila* minimal active RNP reconstituted in vitro, a small fraction of STE-mutant holoenzyme reconstituted in vivo that is stable to purification has normal catalytic activity (Robart et al. 2009).

One plausible model for STE function is in the allosteric modulation of interactions among independently folded domains of TERT (Fig. 2). STE interaction with the TERT high-affinity RNA binding domain (TRBD) could order the adjacent RT domain for a correct disposition of template docking, perhaps holding off active site closure by the TERT amino-terminal domain (TEN) and carboxy-terminal extension (CTE) until the template has been properly placed. The STE could also play a more direct role in the catalytic cycle. Sequence substitutions of certain residues of the STE cripple catalytic activity without severely impacting TERT binding. These residues could map an RNA surface involved in TERT conformational rearrangement, or they could instead map a surface that is oriented by TERT to face the active site. The STE lies far from the template in TER secondary structure, but purified human TER crosslinking and single-molecule FRET studies of ciliate telomerase RNP assembly suggest a tertiary structure proximity of the STE and template (Ueda and Roberts 2004; Stone et al. 2007). Possibly like the pseudoknot, the STE could have a role in shaping the active site that is obscured in the modern-day RNP enzyme by subsequent gains of function.

4 REFINEMENT OF THE CATALYTIC CYCLE, INCLUDING ROLES FOR PHYLOGENETICALLY DIVERSE TER MOTIFS

4.1 Fine-Tuning Repeat Synthesis Activity

Beyond the potentially universal TER motifs described earlier, TER motifs specific to individual phylogenetic groups also contribute to the functional elaboration of telomerase enzymes. Indeed, TERs appear to have finessed the art of telomerase evolutionary adaptation, exploiting it both to fine-tune the catalytic core (discussed in this section) and to recruit holoenzyme proteins important for RNP biogenesis and regulation (discussed in section 5.1).

Importantly, stable association of TERT and TER is not mediated by the RT domain and template. TERT by itself cannot recruit an RNA oligonucleotide to the active site, even an RNA comprised of the native template and its flanking sequence, and even if this RNA is pre-annealed to a telomeric DNA primer (Miller and Collins 2002). If telomerase evolved from a protein-only RT (discussed in section 6.1), a loss-of-function in active site binding to template may have been evolutionarily advantageous as a mechanism for limiting cDNA insertion into the genome. By this model, telomerase would have had to acquire gain-of function mechanisms that productively position the TER internal template and a chromosome 3′ end substrate in the TERT active site.

4.2 RNA Motifs for Binding TERT

One obvious mechanism for telomerase gain-of-function was the physical partnership of protein and RNA, creating the first telomerase RNP. A TERT domain adjacent to the RT motifs, the TRBD (Fig. 2), is necessary and sufficient for the high biological specificity of TERT-TER interaction in ciliate and vertebrate TERTs assayed to date (Lai et al. 2001; O'Connor et al. 2005). Curiously, the TRBD binds to both the template/pseudoknot domain and the STE (Figs. 2 and 3). The logic of two separate TERT-TER interactions appears not to be to increase specificity per se, as one or other TER motif still allows TERT-specific interaction. Rather, these TRBD-TER contacts may be part of a much more complex set of RNP folding cues required to create an active site around an internal template and accommodate the changes in RNA topology that occur during a cycle of repeat synthesis.

4.3 TER Roles in Enforcing the Template 5′ Boundary

TER also functions in template 5′ boundary demarcation. An accurate halt in synthesis at the template 5′ end is an inherent requirement for precise repeat synthesis. In budding and fission yeasts, a stem just 5′ of the template limits template copying (Tzfati et al. 2000; Box et al. 2008). TER substitutions that alter base-pairing of this stem suggest that it acts as a simple steric block. In ciliates, the terminal base pair of a template-adjacent helix and its surrounding single-stranded residues bind TERT to establish the 5′ limit of the template. TER or TERT substitutions that weaken the interaction also weaken the boundary (Lai et al. 2002).

Human telomerase template boundary definition also requires a template-flanking helix, but it is separated from the template 5′ end by intervening single-stranded residues (Chen and Greider 2003). Albeit set in different RNP structural contexts, all known template 5′ boundary determination mechanisms apparently share the feature of steric block or strain against progression of template through the active site. In rodent species in which the TER 5′ end is just a few nucleotides beyond the template 5′ boundary, the biochemical mechanism of template boundary definition remains unknown. The theme of steric occlusion could still hold true, potentially involving the vertebrate TER 5′ trimethylguanosine cap (Chen and Greider 2003; Jády et al. 2004).

4.4 Exchange of Protein-Nucleic Acid Associations across the Catalytic Cycle

Instead of a constant grip on the duplex of RNA template and DNA product, the telomerase active site must recognize a variable length of duplex and also allow for its complete dissociation (Fig. 1). TERT and TER both play important roles in this template handling dynamic, as described in detail elsewhere (Collins 2009). Activity defects imposed by TERT mutagenesis suggest that template residues within and nearby the active site are positioned in part by amino-acid side-chains shared among all RTs (Miller et al. 2000). On the other hand, protein-only RTs would not experience single-stranded RNA at the 3′ end of the template, such as the template residues liberated from the hybrid (Fig. 1). Thus, instead of RT domain interactions, the 3′ end of the template and its adjacent flanking region are likely to make dynamic contact with a TERT-specific protein domain. Transient TER-TERT contacts that fine-tune template use in a dynamic manner over the catalytic cycle appear distinct from stable TER-TRBD interactions that maintain RNP integrity. In addition to several types of TER interactions, TERT also mediates interactions with single-stranded DNA involving the TEN domain (Jacobs et al. 2006; Romi et al. 2007).

5 BELLS AND WHISTLES: TER MOTIFS FOR HOLOENZYME PROTEIN INTERACTIONS

5.1 TER Motifs for Holoenzyme Biogenesis and Regulation

Every phylogenetic group of TERs characterized to date contains group-specific motifs dispensable for catalytic activity reconstitution in vitro yet critical for holoenzyme assembly and function in vivo (Legassie and Jarstfer 2006; Theimer and Feigon 2006). Even the compact ciliate

TERs include a ciliate-specific motif for binding of a holoenzyme protein essential for RNP biogenesis. Vertebrate TER elaboration likewise provides for stable RNP biogenesis and gives the RNP its own nuclear address code, perhaps made necessary by nuclear envelope breakdown with each cell cycle. Fungal TER elaborations are the most extensive, with variable arms extending from the core that offer binding sites for the Sm protein complex (required for RNP biogenesis), the telomerase holoenzyme protein Est1 (required for telomere elongation), and in some species the DNA end-binding heterodimer Ku (required for efficient RNP nuclear localization and recruitment to chromosome ends). The geometry of these binding sites has a degree of positional flexibility for TER function in vivo, leading to the notion of protein beads on an RNA string (Zappulla and Cech 2006). It remains uncertain whether modules are actually flexible relative to one another or rigid but distance-independent in function and whether they are alternately, independently, or cooperatively engaged in protein-RNA interactions.

5.2 Vertebrate TER: Nuclear Assembly and Addressing

The vertebrate-specific motifs of TER create a hairpin-Hinge-hairpin-ACA (H/ACA) motif shared by a large family of RNPs (snoRNPs) that catalyze the site-specific isomerization of uridine to pseudouridine. Each H/ACA-motif stem assembles in highly chaperoned fashion with a set of four proteins: the pseudouridine synthase Cbf5p/NAP57/dyskerin, NHP2, NOP10, and GAR1 (Collins 2008). Bulged regions in each snoRNA stem occur at fixed spacing relative to the Hinge or ACA; unpaired residues of these stem "pockets" hybridize to a target ribosomal or small nuclear RNA. The pocket sequences of vertebrate TERs are not conserved, suggesting that TER does not function as a guide for pseudouridine modification. Why then do vertebrate TERs harbor an H/ACA motif? This motif directs RNP assembly and precursor RNA processing, providing a mechanism for nonpolyadenylated RNA 3′ end formation. In addition, the H/ACA motif and its associated proteins provide vertebrate TERs with nuclear address codes.

Cajal bodies are nuclear domains of concentrated RNP biogenesis and recycling. Subsets of H/ACA and other RNPs, including the human telomerase RNP, are enriched in Cajal bodies. A short RNA element called the CAB box is found in both loops of the small Cajal body H/ACA RNAs that modify small nuclear rather than ribosomal RNA targets and is also found as a single copy in the 3′ loop of many but not all vertebrate TERs (Jády et al. 2004; Xie et al. 2008). CAB box mutation shifts the predominant concentration of human TER from Cajal bodies to nucleoli and slows telomerase-mediated telomere over-elongation in transformed cells but not telomere elongation in primary fibroblasts. The CAB box and H/ACA protein dyskerin bind to a conserved, Cajal body-concentrated WD40-domain protein, WDR79/TCAB1 (Tycowski et al. 2009; Venteicher et al. 2009). Long-term WDR79 depletion in transformed cells shortens telomeres, suggesting that in these cells either TER RNP concentration in Cajal bodies or its lack of concentration in nucleoli improves telomerase function. Cajal bodies are not detectable in all cell types and are disrupted by some forms of cell stress, raising the question of whether TER distribution and its access to telomeres also vary with the state of the cell.

5.3 Yeast TER: Finding and Elongating the Telomere

Beyond base-pairing of the template with single-stranded chromosome-terminal telomeric repeats, TER can play other roles in bringing telomerase to telomeres. One of the *S. cerevisiae* TER "beads on a string" modules binds to Ku80, the larger subunit of a Ku heterodimer that binds double-stranded DNA ends. Telomere maintenance by *S. cerevisiae* telomerase is partially dependent on Ku-TER interaction (Stellwagen et al. 2003). Also, nuclear localization of *S. cerevisiae* telomerase RNP is enhanced by TER association with Ku (Gallardo and Chartrand 2008). At least some budding yeast TERs also harbor a transplantable motif for interaction with the telomerase holoenzyme protein Est1 (Zappulla and Cech 2004). Est1 contributes to telomerase-telomere association in part by bridging TER and Cdc13, a single-stranded telomeric DNA binding protein and component of the Cdc13-Stn1-Ten1 telomeric 3′ overhang capping complex (Pennock et al 2001; Gao et al. 2007). The *S. cerevisiae* Ku-TER and Cdc13-Est1-TER mechanisms of telomerase RNP recruitment to the telomere are distinguished by differential timing: Ku-mediated recruitment of telomerase to telomeres is evident over most of the cell cycle, whereas Cdc13-mediated recruitment of telomerase to telomeres is largely restricted to the DNA replication phase (Chan et al. 2008).

6 TELOMERASE AS A WINDOW TO AN EVOLUTIONARY RNP RENAISSANCE

6.1 Evolutionary Origins of TERT and TER

Even deeply branching eukaryotes harbor a TERT gene with active-site motifs common not just to TERTs but also mobile Group II intron RTs, non-LTR retroelement RTs, and retroviral RTs (Nakamura and Cech 1998). Did the earliest version of a chromosome-end extending telomerase activity begin with the active site of a protein RT that subsequently gained an RNA cofactor, or did this activity begin as a catalytic RNA that later transferred metal-binding active-site duty to a protein? Telomerase likely arose in the last common ancestor of all eukaryotes after the occurrence of fragmented chromosomes, which would have set the stage for RT active-site exaption from a mobile element (as required by the protein-first model). On the other hand, the presence of conserved TER motifs such as the pseudoknot raises the prospect that much of the ancestral telomerase enzyme function could have been RNA-mediated (as required by the RNA-first model). The reshapings of TER over relatively short evolutionary time frames obscure its early origin, but also open a window of opportunity to appreciate the evolving interplay of protein and RNA function in RNP enzyme context.

6.2 RNA Motif Gain-Of-Function in RNP Context

The high thermodynamic stability of even short segments of random RNA sequence creates a barrier to homogeneous folding of large RNAs (Herschlag 1995). Transition from the RNA World to an RNP World of better RNA catalysts exploited peptide chaperones and stably RNA-bound cofactors of hierarchical RNP biogenesis, which gave specificity to the assembly of ribosomes and spliceosomes (described in other articles of this collection). This same machinery has been exploited by telomerase: ciliate, yeast, and vertebrate TERs all require RNP biogenesis chaperones and/or holoenzyme protein cofactors to direct successful maturation and folding of TER in vivo (Collins 2006; Collins 2008; Gallardo and Chartrand 2008). This suggests that the same principles of RNP biogenesis allowed expansion of RNA motif function whether the RNP harbors a catalytic RNA or protein active site. The example of telomerase, among various recently evolving RNPs, provides evidence of an increase in noncoding RNA complexity occurring in RNP context (Hogg and Collins 2008). Future studies of TER motifs and their roles will uncover new insights about telomerase mechanism and also illuminate how an RNP enzyme can accomplish gain-of-function through RNA, providing a window to the evolutionary RNP Renaissance

REFERENCES

Autexier C, Lue NF. 2006. The structure and function of telomerase reverse transcriptase. *Annu Rev Biochem* **75:** 493–517.

Box JA, Bunch JT, Zappulla DC, Glynn EF, Baumann P. 2008. A flexible template boundary element in the RNA subunit of fission yeast telomerase. *J Biol Chem* **283:** 24224–24233.

Brown Y, Abraham M, Pearl S, Kabaha MM, Elboher E, Tzfati Y. 2007. A critical three-way junction is conserved in budding yeast and vertebrate telomerase RNAs. *Nucleic Acids Res* **35:** 6280–6289.

Chakrabarti K, Pearson M, Grate L, Sterne-Weiler T, Deans J, Donohue JP, Ares MJ. 2007. Structural RNAs of known and unknown function identified in malaria parasites by comparative genomics and RNA analysis. *RNA* **13:** 1923–1939.

Chan A, Boulé JB, Zakian VA. 2008. Two pathways recruit telomerase to *Saccharomyces cerevisiae* telomeres. *PLoS Genet* **4:** e1000236.

Chen JL, Greider CW. 2003. Template boundary definition in mammalian telomerase. *Genes Dev* **17:** 2747–2752.

Collins K. 2006. The biogenesis and regulation of telomerase holoenzymes. *Nat Rev Mol Cell Biol* **7:** 484–494.

Collins K. 2008. Physiological assembly and activity of human telomerase complexes. *Mech Ageing Dev* **129:** 91–98.

Collins K. 2009. Forms and functions of telomerase RNA. In *Non-protein coding RNAs* (ed. Walter N.G., Woodson S.A., Batey R.T.), pp. 285–301. Springer-Verlag, Berlin.

Collins K, Greider CW. 1993. Nucleolytic cleavage and non-processive elongation catalyzed by *Tetrahymena* telomerase. *Genes Dev* **7:** 1364–1376.

Collins K, Mitchell JR. 2002. Telomerase in the human organism. *Oncogene* **21:** 564–579.

Cunningham DD, Collins K. 2005. Biological and biochemical functions of RNA in the *Tetrahymena* telomerase holoenzyme. *Mol Cell Biol* **25:** 4442–4454.

Förstemann K, Lingner J. 2005. Telomerase limits the extent of base pairing between template RNA and telomeric DNA. *EMBO Rep* **6:** 361–366.

Gallardo F, Chartrand P. 2008. Telomerase biogenesis: The long road before getting to the end. *RNA Biol* **5:** 212–215.

Gao H, Cervantes RB, Mandell EK, Otero JH, Lundblad V. 2007. RPA-like proteins mediate yeast telomere function. *Nat Struct Mol Biol* **14:** 208–214.

Garcia CK, Wright WE, Shay JW. 2007. Human diseases of telomerase dysfunction: insights into tissue aging. *Nucleic Acids Res* **35:** 7406–7416.

Gilley D, Blackburn EH. 1996. Specific RNA residue interactions required for enzymatic functions of *Tetrahymena* telomerase. *Mol Cell Biol* **16:** 66–75.

Gilley D, Blackburn EH. 1999. The telomerase RNA pseudoknot is critical for the stable assembly of a catalytically active ribonucleoprotein. *Proc Nat Acad Sci* **96:** 6621–6625.

Gilson E, Geli V. 2007. How telomeres are replicated. *Nat Rev Mol Cell Biol* **8:** 825–838.

Greider CW, Blackburn EH. 1985. Identification of a specific telomere terminal transferase activity in *Tetrahymena* extracts. *Cell* **43:** 405–413.

Greider CW, Blackburn EH. 1987. The telomere terminal transferase of *Tetrahymena* is a ribonucleoprotein enzyme with two kinds of primer specificity. *Cell* **51:** 887–898.

Greider CW, Blackburn EH. 1989. A telomeric sequence in the RNA of *Tetrahymena* telomerase required for telomere repeat synthesis. *Nature* **337:** 331–337.

Gunisova S, Elboher E, Nosek J, Gorkovoy V, Brown Y, Lucier JF, Laterreur N, Wellinger RJ, Tzfati Y, Tomaska L. 2009. Identification and comparative analysis of telomerase RNAs from *Candida* species reveal conservation of functional elements. *RNA* **15:** 546–559.

Hammond PW, Cech TR. 1997. dGTP-dependent processivity and possible template switching of *Euplotes* telomerase. *Nucleic Acids Res* **25:** 3698–3704.

Hardy CD, Schultz CS, Collins K. 2001. Requirements for the dGTP-dependent repeat addition processivity of recombinant *Tetrahymena* telomerase. *J Biol Chem* **276:** 4863–4871.

Harley CB. 2008. Telomerase and cancer therapeutics. *Nat Rev Cancer* **8:** 167–179.

Herschlag D. 1995. RNA chaperones and the RNA folding problem. *J Biol Chem* **270:** 20871–20874.

Hogg JR, Collins K. 2008. Structured non-coding RNAs and the RNP Renaissance. *Curr Opin Chem Biol* **12**: 684–689.

Jacobs SA, Podell ER, Cech TR. 2006. Crystal structure of the essential N-terminal domain of telomerase reverse transcriptase. *Nat Struct Mol Biol* **13**: 218–225.

Jády BE, Bertrand E, Kiss T. 2004. Human telomerase RNA and box H/ACA scaRNAs share a common Cajal body-specific localization signal. *J Cell Biol* **164**: 647–652.

Kulpa DA, Moran JV. 2006. Cis-preferential LINE-1 reverse transcriptase activity in ribonucleoprotein particles. *Nat Struct Mol Biol* **13**: 655–660.

Lai CK, Miller MC, Collins K. 2002. Template boundary definition in *Tetrahymena* telomerase. *Genes Dev* **16**: 415–420.

Lai CK, Miller MC, Collins K. 2003. Roles for RNA in telomerase nucleotide and repeat addition processivity. *Mol Cell* **11**: 1673–1683.

Lai CK, Mitchell JR, Collins K. 2001. RNA binding domain of telomerase reverse transcriptase. *Mol Cell Biol* **21**: 990–1000.

Lee MS, Blackburn EH. 1993. Sequence-specific DNA primer effects on telomerase polymerization activity. *Mol Cell Biol* **13**: 6586–6599.

Legassie JD, Jarstfer MB. 2006. The unmasking of telomerase. *Structure* **14**: 1603–1609.

Lin J, Smith DL, Blackburn EH. 2004. Mutant telomere sequences lead to impaired chromosome separation and a unique checkpoint response. *Mol Biol Cell* **15**: 1623–1634.

Lingner J, Hughes TR, Shevchenko A, Mann M, Lundblad V, Cech TR. 1997. Reverse transcriptase motifs in the catalytic subunit of telomerase. *Science* **276**: 561–567.

Mason DX, Goneska E, Greider CW. 2003. Stem-loop IV of *tetrahymena* telomerase RNA stimulates processivity in trans. *Mol Cell Biol* **23**: 5606–5613.

Miller MC, Collins K. 2002. Telomerase recognizes its template by using an adjacent RNA motif. *Proc Natl Acad Sci* **99**: 6585–6590.

Miller MC, Liu JK, Collins K. 2000. Template definition by *Tetrahymena* telomerase reverse transcriptase. *EMBO J* **19**: 4412–4422.

Mitchell JR, Collins K. 2000. Human telomerase activation requires two independent interactions between telomerase RNA and telomerase reverse transcriptase in vivo and in vitro. *Mol Cell* **6**: 361–371.

Nakamura TM, Cech TR. 1998. Reversing time: Origin of telomerase. *Cell* **92**: 587–590.

O'Connor CM, Lai CK, Collins K. 2005. Two purified domains of telomerase reverse transcriptase reconstitute sequence-specific interactions with RNA. *J Biol Chem* **280**: 17533–17539.

Pennock E, Buckley K, Lundblad V. 2001. Cdc13 delivers separate complexes to the telomere for end protection and replication. *Cell* **104**: 387–396.

Piskareva O, Schmatchenko V. 2006. DNA polymerization by the reverse transcriptase of the human L1 retrotransposon on its own template in vitro. *FEBS Lett* **580**: 661–668.

Qiao F, Cech TR. 2008. Triple-helix structure in telomerase RNA contributes to catalysis. *Nat Struct Mol Biol* **15**: 634–640.

Robart AR, O'Connor CM, Collins K. 2010. Ciliate telomerase RNA loop IV nucleotides promote hierarchical RNP assembly and holoenzyme stability. *RNA* **16**: 563–571.

Romi E, Baran N, Gantman M, Shmoish M, Min B, Collins K, Manor H. 2007. High-resolution physical and functional mapping of the template adjacent DNA binding site in catalytically active telomerase. *Proc Natl Acad Sci* **104**: 8791–8796.

Shefer K, Brown Y, Gorkovoy V, Nussbaum T, Ulyanov NB, Tzfati Y. 2007. A triple helix within a pseudoknot is a conserved and essential element of telomerase RNA. *Mol Cell Biol* **27**: 2130–2143.

Sperger JM, Cech TR. 2001. A stem-loop of *Tetrahymena* telomerase RNA distant from the template potentiates RNA folding and telomerase activity. *Biochemistry* **40**: 7005–7016.

Stellwagen AE, Haimberger ZW, Veatch JR, Gottschling DE. 2003. Ku interacts with telomerase RNA to promote telomere addition at native and broken chromosome ends. *Genes Dev* **17**: 2384–2395.

Stone MS, Mihalusova M, O'Connor CM, Prathapam R, Collins K, Zhuang X. 2007. Stepwise protein-mediated RNA folding directs assembly of telomerase ribonucleoprotein. *Nature* **446**: 458–461.

Theimer CA, Feigon J. 2006. Structure and function of telomerase RNA. *Curr Opin Struct Biol* **16**: 307–318.

Theimer CA, Blois CA, Feigon J. 2005. Structure of the human telomerase RNA pseudoknot reveals conserved tertiary interactions essential for function. *Mol Cell* **17**: 671–682.

Tycowski KT, Shu MD, Kukoyi A, Steitz JA. 2009. A conserved WD40 protein binds the Cajal body localization signal of scaRNP particles. *Mol Cell* **34**: 47–57.

Tzfati Y, Fulton TB, Roy J, Blackburn EH. 2000. Template boundary in a yeast telomerase specified by RNA structure. *Science* **288**: 863–867.

Tzfati Y, Knight Z, Roy J, Blackburn EH. 2003. A novel pseudoknot element is essential for the action of a yeast telomerase. *Genes Dev* **17**: 1779–1788.

Ueda CT, Roberts RW. 2004. Analysis of a long-range interaction between conserved domains of human telomerase RNA. *RNA* **10**: 139–147.

Venteicher AS, Abreu EB, Meng Z, McCann KE, Terns RM, Veenstra TD, Terns MP, Artandi SE. 2009. A human telomerase holoenzyme protein required for Cajal body localization and telomere synthesis. *Science* **323**: 644–648.

Verdun RE, Karlseder J. 2007. Replication and protection of telomeres. *Nature* **447**: 924–931.

Wang H, Blackburn EH. 1997. De novo telomere addition by *Tetrahymena* telomerase in vitro. *EMBO J* **16**: 866–879.

Wang H, Gilley D, Blackburn EH. 1998. A novel specificity for the primer-template pairing requirement in *Tetrahymena* telomerase. *EMBO J* **17**: 1152–1160.

Xie M, Mosig A, Qi X, Li Y, Stadler PF, Chen JJ. 2008. Structure and function of the smallest vertebrate telomerase RNA from teleost fish. *J Biol Chem* **283**: 2049–2059.

Ye AJ, Romero DP. 2002. Phylogenetic relationships amongst tetrahymenine ciliates inferred by a comparison of telomerase RNAs. *Int J Syst Evol Microbiol* **52**: 2297–2302.

Yu G, Blackburn EH. 1991. Developmentally programmed healing of chromosomes by telomerase in *Tetrahymena*. *Cell* **67**: 823–832.

Zappulla DC, Cech TR. 2004. Yeast telomerase RNA: A flexible scaffold for protein subunits. *Proc Natl Acad Sci* **101**: 10024–10029.

Zappulla DC, Cech TR. 2006. RNA as a flexible scaffold for proteins: Yeast telomerase and beyond. *Cold Spring Harb Symp Quant Biol* **71**: 217–224.

Zappulla DC, Goodrich K, Cech TR. 2005. A miniature yeast telomerase RNA functions in vivo and reconstitutes activity in vitro. *Nat Struct Mol Biol* **12**: 1072–1077.

Bacterial Small RNA Regulators: Versatile Roles and Rapidly Evolving Variations

Susan Gottesman[1] and Gisela Storz[2]

[1]Laboratory of Molecular Biology, National Cancer Institute, Bethesda, Maryland 20892

[2]Cell Biology and Metabolism Program, Eunice Kennedy Shriver National Institute of Child Health and Human Development, Bethesda, Maryland 20892

Correspondence: susang@helix.nih.gov and storz@helix.nih.gov

SUMMARY

Small RNA regulators (sRNAs) have been identified in a wide range of bacteria and found to play critical regulatory roles in many processes. The major families of sRNAs include true antisense RNAs, synthesized from the strand complementary to the mRNA they regulate, sRNAs that also act by pairing but have limited complementarity with their targets, and sRNAs that regulate proteins by binding to and affecting protein activity. The sRNAs with limited complementarity are akin to eukaryotic microRNAs in their ability to modulate the activity and stability of multiple mRNAs. In many bacterial species, the RNA chaperone Hfq is required to promote pairing between these sRNAs and their target mRNAs. Understanding the evolution of regulatory sRNAs remains a challenge; sRNA genes show evidence of duplication and horizontal transfer but also could be evolved from tRNAs, mRNAs or random transcription.

Outline

1 Introduction

2 How many sRNAs are there?

3 True antisense sRNAs

4 Base pairing sRNAs with limited complementarity

5 sRNAs that modify protein activity

6 sRNAs with intrinsic activities

7 Evolutionary considerations

8 Perspectives

References

Copyright © 2011 Cold Spring Harbor Laboratory Press; all rights reserved
Cite this article as *Cold Spring Harb Perspect Biol* doi: 10.1101/cshperspect.a003798

1 INTRODUCTION

In the third edition of The RNA World book, we described how many small RNA regulators (sRNAs), ranging in length from approximately 50 to 500 nucleotides, were found in bacteria, and summarized what was then known about the functions of the sRNAs. As has been seen for regulatory RNAs in eukaryotes, many bacterial sRNAs act by base pairing, having either extensive or more limited complementarity with their target mRNAs. However, others modulate the activity of proteins by mimicking secondary structures of other nucleic acids. The pace of sRNA discovery in a wide range of bacteria has continued to accelerate and the functions of increasing numbers of sRNAs are being elucidated. Here we cite reviews that discuss older work and emphasize what has been found in the past 5 yr. We will focus on the sRNAs encoded on bacterial chromosomes, particularly in the model organisms *Escherichia coli*, *Salmonella enterica* and *Staphylococcus aureus*, but will point out parallels to phage and plasmid-encoded sRNAs. We also will focus on RNA regulators that act as independent transcripts as opposed to riboswitches, which are part of the mRNA they regulate and are described in (Breaker, *Riboswitches and the RNA World*). CRISPR RNAs, which are central to a defense against foreign DNA in many bacteria and archaea, are yet another class of sRNAs that will not be covered here because they are the topic of (Jone, Brouns and van der Oost, *RNA in Defense: CRISPRS protect prokaryotes against alien nucleic acids*).

2 HOW MANY sRNAs ARE THERE?

The initial screens for sRNA genes in bacteria relied primarily on computational searches of intergenic regions for conserved sequences or orphan promoter and terminator sequences (reviewed in Livny and Waldor 2007). However, with the advances in whole genome expression profiling using tiling arrays or deep sequencing, approaches that rely on direct detection are superseding the computational approaches (reviewed in Sharma and Vogel 2009). Nevertheless, despite the many searches that have now been conducted, the exact number of sRNAs present is still not known for any bacterium. A number of 80–100 is generally cited for *E. coli* (compared with around 4300 proteins); whereas numbers two or three times higher have been reported for other bacteria. Unless a large class of sRNAs has been missed (for instance, sRNAs derived from mRNAs), it seems likely that bacteria will have on the order of a few hundred rather than thousands of sRNAs.

Part of the challenge in establishing the true number of regulatory sRNAs is that many short transcripts have only been detected by one approach and have not been shown to have functions. For example, there are reports of tens or hundreds of antisense RNAs encoded opposite annotated protein-coding genes, based on detection by microarray analysis or deep sequencing (Georg et al. 2009; Guell et al. 2009; Toledo-Arana et al. 2009; Sharma et al. 2010). Chromatin immunoprecipitation experiments similarly suggest that *E. coli* RNA polymerase is transcribing the strands opposite a number of protein-coding genes (Peters et al. 2009). However, few of these signals—some of which potentially could be caused by cross hybridization or spurious cDNA synthesis by reverse transcriptase—have been verified in independent experiments, particularly in experiments with control strains in which the putative gene or promoter has been deleted. In addition, only a limited number of antisense sRNAs have been assigned functions. On the other hand, some sRNAs with true regulatory functions might still be missed because they are only expressed under very specific conditions, their structures make them recalcitrant to the predominant methods of detection, or they are processed from mRNAs and hard to distinguish from 5′ or 3′ UTRs.

With increased sRNA detection and characterization in a wide range of bacteria, we will undoubtedly be able to obtain more definitive answers to questions about the number of sRNAs across bacterial species and whether, as we would predict, all bacteria have sRNA regulators.

3 TRUE ANTISENSE sRNAs

As mentioned earlier, and as has also been noted for eukaryotic organisms, large numbers of antisense RNAs, also referred to as *cis*-encoded RNAs, are being reported to be transcribed opposite annotated genes and thus share extensive complementarity with the corresponding transcripts (Georg et al. 2009; Guell et al. 2009; Toledo-Arana et al. 2009; Sharma et al. 2010). However, clear physiological roles have only been established for a small number of antisense sRNAs. Nevertheless, these examples serve as models for what a *cis*-encoded antisense sRNA is likely to do.

Thus far, the most prevalent role for antisense sRNAs in bacteria has been the repression of genes that encode potentially toxic proteins (reviewed in Gerdes and Wagner 2007; Fozo et al. 2008). This was one of the first functions described for the plasmid-encoded sRNAs, where antisense sRNAs control the expression of small hydrophobic proteins, such as the 50-amino-acid Hok protein encoded on the *E. coli* R1, R100, and F plasmids and the 33 amino acid Fst protein encoded on the *Enterococcus faecalis* plasmid pAD1. Bacterial chromosomes have also been found to encode homologs of these proteins and their corresponding antisense sRNAs. In addition, an increasing number of other small hydrophobic protein-antisense sRNA pairs are being discovered in a variety of organisms, often tandemly duplicated in the same region of the chromosome (Fozo et al. 2010; Sharma et al. 2010). The physiological roles of these gene pairs are not yet known, but in all cases that have been examined, the antisense sRNA represses synthesis of the corresponding protein, which is toxic at high levels. The mechanism by which repression occurs has only been characterized for a limited number of pairs and generally appears to involve a block in translation. However, the mechanism may vary between toxin-antitoxin pairs because the antisense sRNA can overlap the 5′ end of the mRNA, the entire mRNA, or the 3′ end of the mRNA. In many cases both the toxin mRNA and antisense sRNA are predicted to have extensive secondary structures (reviewed in Gerdes and Wagner 2007; Weaver 2007) raising questions about the mechanism of base pairing. The antisense sRNAs appear to be critical for keeping the basal level expression of the toxins low, and where it has been examined, the levels of the antisense sRNA have been reported to be in excess of the mRNA. Thus, another open question is under what conditions antisense regulation is overcome to allow toxin synthesis.

Another role of characterized antisense sRNAs is the directed cleavage of the mRNA encoded on the opposite strand. The first such sRNA to be characterized was the OOP RNA of the bacteriophage λ, where a 77-nucleotide sRNA encoded opposite the *cII-O* mRNA of λ represses cII expression by directing cleavage of this transcript in a mechanism that is partially dependent on RNase III (Krinke and Wulff 1987; Krinke and Wulff 1990). The *E. coli* GadY RNA encoded opposite the *gadX-gadW* mRNA may similarly be directing cleavage of the mRNA, in this case leading to increased levels of the *gadX* transcript (Opdyke et al. 2004; Tramonti et al. 2008). In contrast, the antisense IsrR RNA of *Synechocystis* negatively impacts the levels of the oppositely encoded *isiA* mRNA (Duhring et al. 2006). IsrR synthesis is repressed by iron stress, allowing *isiA* accumulation and translation under these conditions. In addition to blocking translation and directing mRNA processing, one can imagine that simply the act of transcribing the antisense RNA might impact transcription of the mRNA encoded opposite.

4 BASE PAIRING sRNAs WITH LIMITED COMPLEMENTARITY

The most well studied bacterial sRNA regulators are those that act by base pairing but only have limited complementarity with their target RNAs because they are encoded at a different genomic location. They have sometimes been referred to as *trans*-encoded sRNAs, to distinguish them from the antisense, *cis*-encoded sRNAs. Almost all of the sRNAs that fall into this class are expressed under specific growth conditions ranging from limiting iron, oxidative stress, and anaerobic conditions to cell envelope stress, low magnesium, or high glucose-6-phosphate concentrations and glucose starvation. These observations have led to the expectation that base-pairing sRNAs with limited complementarity with their targets will be found to be associated with almost every global response in bacteria. In the past 5 yr, there has been significant progress in characterizing the mechanisms of action and physiological roles of this class of sRNA.

4.1 Mechanisms

These base-pairing sRNAs are reminiscent of eukaryotic microRNAs (miRNAs) and small interfering RNAs (siRNAs) in their ability to modulate mRNA stability and translation. However, they differ in the details of how they are generated and where they pair with their target mRNAs. The bacterial base pairing sRNAs are generally transcribed as single entities around 100 nucleotides in length. Unlike miRNAs and siRNAs, they usually are not processed, although for a few sRNAs, cleavage to a shorter form does occur. In addition, bacterial sRNAs frequently base pair with the 5′ end, rather than with the 3′ end, of target mRNAs. In most cases, the bacterial sRNAs also appear to act stoichiometrically, being degraded along with the mRNA after pairing.

Outcomes of productive base pairing. Pairing between sRNA and mRNA generally involves at least a seed region of 6–8 contiguous base pairs, although significantly longer pairing regions have been predicted in some cases (but rarely tested). Within the sRNA, it is common for the region involved in base pairing, often with multiple targets, to be highly conserved. For instance, the *Salmonella* and *E. coli* GcvB RNAs have a conserved G/U rich, single-stranded region, which is required for base pairing with C/A-rich sequences in the 5′ untranslated regions of the target mRNAs (Sharma et al. 2007).

Base pairing between sRNAs and mRNAs can have a number of regulatory outcomes. Many of the sRNAs base pair at the ribosome binding site, thus blocking translation by preventing ribosome binding. However, systematic shifting of the region involved in *Salmonella* RybB RNA pairing with the *ompN* mRNA showed that RybB RNA can block translation even when the region of pairing is as far as the fifth codon into the open reading frame (ORF) (Bouvier et al. 2008). Translation can also be blocked when the region of pairing is 50 or more nucleotides upstream of the ribosome binding site (Sharma et al. 2007). In most cases where ribosome binding is blocked, an associated decrease in the stability of the mRNA is also observed, possibly as an indirect result of blocking ribosome entry. There are, however, target mRNAs in which base pairing occurs downstream of the Shine-Dalgarno sequence such that ribosome binding would not be blocked. For example, base pairing between the *Salmonella* MicC and the *ompD* mRNA within codons 23–26 accelerates RNase E-dependent decay of the mRNA without blocking ribosome binding (Pfeiffer et al. 2009). In another example, base pairing between *E. coli* RyhB and the *iscRSUA* mRNA in the *iscR-iscS* intergenic region leads to degradation of the downstream region of the polycistronic mRNA (Desnoyers et al. 2009).

In addition to repressing translation, sRNAs can also activate translation, in many cases by preventing or overcoming the formation of an inhibitory secondary structure (Morfeldt et al. 1995; Majdalani et al. 2005; Prévost et al. 2007). Binding of the sRNA leads to remodeling of the mRNA structure, uncovering the ribosome binding site and allowing translation. In some of these cases, binding of the sRNA can be quite distant from the start of translation, making it more difficult to identify targets computationally. In addition, because many targets are found by changes in mRNA abundance in microarrays, targets for positive regulation, which may only show a modest increase in mRNA stability, may be missed. Nonetheless, a growing number of sRNAs have been found to have positive effects on some targets while negatively regulating others (Morfeldt et al. 1995; Lease et al. 1998; Huntzinger et al. 2005; Prévost et al. 2007).

The interactions among and competition between sRNAs, ribosomes, ribonucleases, and the RNA chaperone Hfq and the target mRNAs, as well as the structures of both RNAs are all likely to influence how effective regulation is going to be, but are not well understood. It is interesting to note that effective SgrS repression of the *ptsG* mRNA requires either membrane localization of the target mRNA or a reduction in translation, suggesting that associated changes in mRNA structure or the slower binding of ribosomes is needed for SgrS accessibility to the region of pairing (Kawamoto et al. 2005). It is also intriguing that a surprising number of mRNAs are the targets of multiple sRNAs; perhaps only a subset of mRNAs have features (such as an Hfq binding site) that allow them to be regulated by base pairing sRNAs.

Role of Hfq. Although complex protein machineries are required for miRNA and siRNA processing and to effect the functions of these eukaryotic regulators, the majority of the sRNAs that act via limited base pairing in enteric bacteria have only been found to require the RNA chaperone Hfq, a close relative of the Sm/Lsm family of proteins involved in splicing and mRNA decay (reviewed in Brennan and Link 2007). Hfq binds both sRNAs and mRNAs, and, in vitro, stimulates their pairing. However, a number of questions remain about how this doughnut-shaped hexamer facilitates interactions between RNAs.

When and where is Hfq bound? The sites of Hfq binding on both sRNAs and target mRNAs have only been mapped for a limited number of transcripts, but thus far all are AU-rich single stranded regions. One binding study of the *rpoS* mRNA leader showed that Hfq binds to two sites, one near the region of base pairing with the DsrA RNA as well as another site significantly further upstream (Soper and Woodson 2008). The high affinity upstream Hfq binding site was found to significantly increase the rate of DsrA binding to *rpoS*. This has led to the model that once base-pairing between DsrA and *rpoS* mRNA is established, the second binding site stabilizes the base-paired complex by titrating Hfq away from the sRNA. Several mutational studies indicate that both the proximal and distal face of the Hfq ring bind to RNA (reviewed in Brennan and Link 2007). This view is now supported by structural studies. A previous structure of *S. aureus* Hfq bound to AU_5G shows the RNA bound to the proximal face encircling the pore (Schumacher et al. 2002), whereas the RNA is bound to the distal face in a recent structure of *E. coli* Hfq bound to A_{15} (Link et al. 2009).

How does Hfq act to facilitate base pairing? Hfq binding has been shown to affect RNA secondary structure and has also been proposed to increase the local concentration of mRNA and sRNA by binding both transcripts (reviewed in Brennan and Link 2007). It is not yet known which mechanism facilitates sRNA-mRNA pairing or whether both are used together or independently. Hfq has also been reported to interact with ribosomes and with the RNase E endonuclease responsible for degrading the regulated mRNAs; to what extent these interactions are critical for Hfq action are not clear. It seems likely that the coming years will lead to a much better understanding of the mechanism of Hfq-stimulated pairing and regulation.

Hfq-independent base pairing. Unlike in enteric bacteria, several of the base pairing sRNAs characterized in Gram-positive bacteria such as *S. aureus* and *Bacillus subtilis* do not require Hfq for function, even when Hfq is present in the organism (Silvaggi et al. 2005; Boisset et al. 2007; Heidrich et al. 2007). However, Hfq does bind sRNAs in *Listeria monocytogenes* and *hfq* mutants do affect gene expression in both *B. subtilis* and *L. monocytogenes*, suggesting a possible role in RNA folding or pairing (Christiansen et al. 2004; Christiansen et al. 2006; Heidrich et al. 2006; Nielsen et al. 2009). It is not yet completely clear why Hfq is required in some cases of limited base pairing and not others. Possibly, more extended pairing and a higher proportion of G:C base pairs obviate the need for Hfq for a subset of sRNAs. Alternatively, another RNA chaperone may be playing a parallel role in these cases, as has been proposed for three small basic proteins in *B. subtilis* (Gaballa et al. 2008). This may well be the case in other bacterial species in which sRNAs, but not Hfq, have been found.

4.2 Physiological Roles

The anticipation that base pairing sRNAs would have a variety of physiological roles has been met; both the roles in previously known responses have expanded and new roles have been found.

Repression of outer membrane protein synthesis. In the third edition of *The RNA World*, we noted that the expression of several membrane proteins in *E. coli* is controlled by sRNAs. The list of sRNAs that modulate the expression of membrane proteins, particularly outer membrane β-barrel proteins, many of which function as porins, has continued to expand (Table 1). In *E. coli* and *Salmonella*, all the major outer membrane porins have been found to be down-regulated by one or more sRNAs (reviewed in Guillier et al. 2006; Vogel and Papenfort 2006), and new studies of sRNAs continue to uncover additional examples of regulation of outer membrane porins (Johansen et al. 2008; Papenfort et al. 2008; De Lay and Gottesman 2009; Figueroa-Bossi et al. 2009; Overgaard et al. 2009). Why outer membrane proteins are such predominant targets of sRNA regulation is still not completely clear, especially because many porin proteins are thought to be stable. However, given the abundance of examples, there must be a regulatory advantage to this mode of regulation.

The synthesis of the porin-regulating sRNAs is induced by various growth conditions and regulators (Table 1). For example the *micA* and *rybB* genes are transcribed by σ^E, a sigma factor that regulates genes involved in both the trafficking of outer membrane proteins through the periplasm and quality control when the periplasm accumulates unfolded proteins (Johansen et al. 2006; Papenfort et al. 2006; Thompson et al. 2007; Udekwu and Wagner 2007). These findings have led to the satisfying model that disruptions to the integrity of the outer membrane lead to the induction of sRNAs that are capable of down-regulating the expression of porins, thereby easing the strain on the trafficking machinery and the membrane. The regulatory circuits for some of the other sRNAs in this group, such as CyaR, whose levels increase on glucose starvation, have less obvious connections to membrane stress, but nevertheless may help to down-regulate the σ^E response by keeping synthesis of outer membrane porins low.

Remodeling metabolism. Another theme that is becoming more and more prevalent among the targets of base pairing sRNAs is metabolic remodeling upon environmental shifts. One of the clearest examples of this comes from the Fur-regulated RyhB RNA in *E. coli*, whose levels increase in response to low iron and which represses the synthesis of many nonessential iron-containing enzymes such as aconitase B and succinate dehydrogenase. This allows the limited iron in the cell to be used by the critical enzymes (reviewed in Massé et al. 2007). Other sRNAs have also been found to impact what metabolic enzymes are synthesized. For example, the CRP-regulated Spot42 RNA, whose levels are repressed during glucose starvation, is responsible for the discordant regulation of genes in the galactose operon (Møller et al. 2002). Similarly, the FNR-regulated FnrS RNA, whose levels are induced by low oxygen, represses the expression of enzymes such as lactate dehydrogenase that are not needed during anaerobic growth (Boysen et al. 2010; Durand and Storz 2010).

Modulating the synthesis of key transcription factors. The *rpoS* gene encoding the stationary phase sigma factor σ^S in *E. coli* was one of the first targets of base pairing sRNA targets to be characterized, being positively regulated by DsrA and RprA and negatively regulated by OxyS (reviewed in Majdalani et al. 2005). Yet another sRNA, denoted ArcZ, has recently been found to increase the levels of σ^S (Mandin and Gottesman, submitted). DsrA and RprA are induced in response to low temperature and cell surface stress, respectively, whereas OxyS is induced by oxidative stress and ArcZ is repressed under anaerobic conditions. Thus, the sRNAs can finely tune σ^S synthesis in response to a range of environmental signals (Fig. 1). Another transcription factor whose synthesis is under the control of multiple sRNAs is the LuxR regulator of quorum sensing in *Vibrio harveyi* and its homolog HapR in *V. cholerae*. In this case the Qrr RNAs (five in *V. harveyi* and four in *V. cholerae*), which repress translation, are all homologous, but again the redundancy allows for nuanced regulation (Lenz et al. 2004; Tu and Bassler 2007; Svenningsen et al. 2008; Tu et al. 2008; Svenningsen et al. 2009). In *S. aureus*, RNA III has a significant impact on virulence by repressing the synthesis of Rot, a repressor of exotoxin genes (Boisset et al. 2007). sRNAs also directly negatively regulate a number of two-component systems; OmrA and OmrB repress *ompR* expression as part of a negative feed-back loop (Guillier and Gottesman 2008) and the σ^E- regulated MicA RNA connects one regulatory input (outer membrane stress) to another (the PhoPQ regulon) (Cornaert et al. 2010) (Fig. 1). We predict that the interconnections between sRNAs and transcription regulators will only increase as more sRNA targets are identified.

Table 1. Physiological roles of Hfq-binding RNAs in E. coli and S. typhimurium

RNA name	Length	Regulation	Physiological Response*	References
MicF	93	Induced by high osmolarity; Activated by OmpR, SoxS, MarA	Repression of porin synthesis (ompF)	(Andersen and Delihas 1990; Coyer et al. 1990)
MicC	109	Increased at low temperature; Repressed by OmpR	Repression of porin synthesis (ompC, ompD)	(Chen et al. 2004)
MicA	72	Cell envelope stress; σ^E-regulated	Repression of porin synthesis and PhoPQ (ompA, ompX, phoP)	(Rasmussen et al. 2005; Udekwu et al. 2005; Cornaert et al. 2010)
RybB	80	Cell envelope stress; σ^E-regulated	Repression of porin synthesis (ompC, ompW)	(Johansen et al. 2006; Papenfort et al. 2006; Thompson et al. 2007)
RseX	91	Not reported	Repression of porin synthesis (ompC, ompA)	(Douchin et al. 2006)
MicM/ChiX	85	Repressed by chitosugars; chbBCARFG decoy mRNA stimulates degradation	Repression of chitoporin synthesis (chiP) (also dpiBA)	(Figueroa-Bossi et al. 2009; Overgaard et al. 2009)
CyaR	87	Induced by low glucose; Activated by CRP (σ^E-regulated)	Repression of porin synthesis and group behavior (ompX, luxS, nadE)	(Johansen et al. 2008; Papenfort et al. 2008; De Lay and Gottesman 2009)
OmrA	88	Induced by high osmolarity; Activated by OmpR	Repression of outer membrane protein synthesis (cirA, fecA, fepA, ompT, gntP, ompR)	(Guillier and Gottesman 2006; Guillier and Gottesman 2008)
OmrB	82	Induced by high osmolarity; Activated by OmpR	Repression of outer membrane protein synthesis (cirA, fecA, fepA, ompT, gntP, ompR)	(Guillier and Gottesman 2006; Guillier and Gottesman 2008)
GcvB	205	Induced by high glycine levels; Activated by GcvB	Repression of peptide transport (oppA, dppA, gltI, livK, livJ, argT, cycA, sstT)	(Urbanowski et al. 2000; Sharma et al. 2007; Pulvermacher et al. 2009a; Pulvermacher et al. 2009b)
RydC	64	Not reported	Repression of putative ABC transporter (yejABEF)	(Antal et al. 2005)
MgrR	99	Induced by low magnesium; Activated by PhoP	Repression of LPS modification gene (eptB, ygdQ)	(Moon and Gottesman 2009)
SgrS	227	Induced by glucose-phosphate or analogs; Activated by dedicated SgrR Bifunctional RNA	Protection against glucose-phosphate stress (ptsG)	(Vanderpool and Gottesman 2004; Wadler and Vanderpool 2007)
GlmZ	207	Repressed by high nitrogen; GlmY decoy sRNA blocks degradation	Induction of GlcN-6-P synthase (discoordinate regulation of glmUS operon)	(Kalamorz et al. 2007; Reichenbach et al. 2008; Urban and Vogel 2008)
Spot42	109	Repressed by low glucose; Repressed by CRP	Repression of galactokinase blocks degradation (discoordinate regulation of galETKM operon)	(Møller et al. 2002)
GadY	105, 90, 59	Induced in stationary phase; σ^S-regulated	Activation of acid response (gadX)	(Opdyke et al. 2004; Tramonti et al. 2008)
RyhB	90	Induced by limiting iron; Repressed by Fur	Iron-sparing (sodB, sdhC, frdA, activates shiA)	(Massé and Gottesman 2002; Massé et al. 2005; Prévost et al. 2007)
FnrS	113	Induced under anaerobic conditions; Activated by FNR, ArcA and CRP	Repression of unneeded enzymes (sodB, maeA, gpmA, folE, folX)	(Durand and Storz 2010; Boysen et al. 2010)
OxyS	109	Induced by oxidative stress; Activated by OxyR	Repression of unneeded activities (fhlA, yobF-cspC, ybaY wrbA, rpoS)	(Altuvia et al. 1997; Altuvia et al. 1998)
ArcZ	120, 50	Induced under aerobic conditions; Repressed by ArcA	Activation of σ^S; repression of alternative activities (rpoS, sdaCB, tpx)	(Papenfort et al. 2009; Mandin and Gottesman 2010)
DsrA	87	Increased at low temperature	Activation of σ^S; repression of hns (rpoS, hns)	(Sledjeski and Gottesman 1995; Sledjeski et al. 1996; Majdalani et al. 1998; Lease and Belfort 2000)
RprA	105	Induced by cell surface stress; Activated by RcsB	Activation of σ^S (rpoS)	(Majdalani et al. 2001; Majdalani et al. 2002)
DicF	53	Not reported (cryptic prophage gene)	Inhibition of cell division (ftsZ)	(Bouché and Bouché 1989; Faubladier et al. 1990; Tetart and Bouché 1992)

*Some, but not all, published targets of these sRNAs are listed in parentheses.

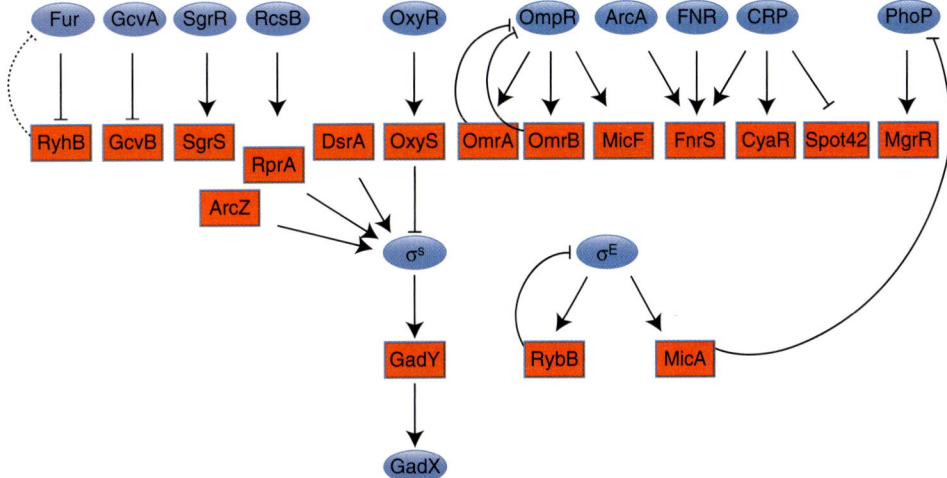

Figure 1. Regulatory circuits for base pairing sRNAs. sRNAs are shown as red boxes, and the transcription regulators known to regulate their synthesis as blue ovals. Both positive and negative regulators are shown. Feedback regulation can be direct (as for OmrA and OmrB regulation of OmpR) or indirect (as for RyhB regulation of Fur).

4.3 Variations

With the characterization of an increasing number of base pairing sRNAs, more and more regulatory variations are being discovered; here three are described in more detail.

Dual functions as mRNA and sRNA: SgrS and RNAIII. First, not all base pairing sRNAs are "noncoding." The *E. coli* SgrS RNA was initially identified as an Hfq binding sRNA whose expression was induced by the presence of sugar phosphates and which base paired with and blocked translation of *ptsG* mRNA, encoding a sugar phosphate transporter (reviewed in Vanderpool 2007). However, it was noted that the sRNA is unusually long for a base pairing sRNA and further inspection of the sequence revealed that SgrS also encodes a 43 amino acid protein denoted SgrT (Wadler and Vanderpool 2007). Because expression of just the small protein has some of the same effects as expression of the noncoding region of SgrS involved in base pairing (such as decreased glucose uptake), it was hypothesized that SgrT reinforces the regulation brought about by the sRNA; SgrS inhibits PtsG synthesis by base pairing with the *ptsG* mRNA whereas SgrT inhibits the activity of the pre-existing transporter. Interestingly, a phylogenetic analysis of SgrS from a variety of related bacteria showed that the most conserved feature of the sRNA is the region of base pairing whereas the region encoding the small protein is much more variable (Horler and Vanderpool 2009).

The *S. aureus* RNA III is one of the longest base pairing sRNAs and also encodes a small protein, in this case the exotoxin hemolysin δ (reviewed in Romby and Charpentier 2009). Here again, the base pairing function and mRNA function of the sRNA are complementary, with pairing indirectly leading to increased synthesis of exotoxins by repressing expression of Rot (Boisset et al. 2007). These two examples open up the likelihood that RNAs now considered to be only mRNAs might in fact be bifunctional, also having a role as regulatory RNAs. This possibility significantly complicates the search strategies for regulatory RNAs, because we do not currently know how to distinguish a bifunctional RNA from an mRNA.

Base-pairing riboswitch: SreA, SreB. Bifunctional RNAs exist in other forms as well. As described in (Breaker, *Riboswitches and the RNA World*), riboswitches are leader sequences that control expression of the downstream genes through metabolite (ranging from flavin mononucleotide to molybdenum cofactor) induced changes in secondary structures. The changes in mRNA structure can alter ribosome binding or transcription termination. The short transcripts generated by riboswitch promoted transcription termination have been detected in a number of sRNA screens, raising the possibility that they also could act in *trans* (Vogel et al. 2003; Kawano et al. 2005a). Recently S-adenosylmethionine riboswitches of *Listeria monocytogenes*, denoted SreA and SreB, have been reported to base pair with the mRNA encoding the PrfA virulence regulator, leading to an approximately twofold reduction in the levels of this regulator (Loh et al. 2009). This example raises the possibility that other riboswitches have additional functions.

Regulating sRNA levels by mimics: chbBC and GlmY. Although the synthesis of many base-pairing sRNAs is regulated by specific transcription factors, this is not the only means by which the levels of the sRNAs can be controlled. The *E. coli* and *S. enterica* sRNA, alternatively called ChiX and MicM, was shown to strongly repress expression of the mRNA encoding the ChiP chitoporin (also denoted YbfM) (Rasmussen et al. 2009). Surprisingly, genetic screens for regulators of ChiX did not result in the identification of transcription factors. Instead an intergenic region in the *chbBCARFG* (chitobiose use) operon was found to modulate ChiX levels (Figueroa-Bossi et al. 2009; Overgaard et al. 2009). Further characterization of this region revealed that ChiX can base pair with the mRNA in the *chbBC* intergenic region and that this base pairing triggers decay of ChiX. Thus ChiX levels are regulated by an "mRNA target mimic" that promotes degradation of the sRNA. In a slightly different twist, the levels of the GlmZ RNA, which base pairs with the *glmUS* mRNA to promote translation of the *glmS*-encoded GlcN-6-P synthase, are modulated by the GlmY "sRNA mimic" (Kalamorz et al. 2007; Urban and Vogel 2008). In this case, high levels of the GlmY RNA appear to titrate a factor responsible for GlmZ degradation (Reichenbach et al. 2008; Urban and Vogel 2008). In both examples, the transcription of the mimics, *chbBCARFG* and *glmY*, is controlled by environmental signals sensed by transcription regulators (Figueroa-Bossi et al. 2009; Reichenbach et al. 2009).

Regulation of sRNAs solely by target mimics may be rare, but if the sRNAs are generally used stoichiometrically (Massé et al. 2003), all mRNA targets will act as competitors with one another, unless the sRNA is in excess. Thus, it seems quite possible that hierarchies of mRNA regulation exist, and are not fully appreciated under the usual laboratory test conditions of sRNA overproduction.

5 sRNAs THAT MODIFY PROTEIN ACTIVITY

In addition to acting by base pairing, sRNAs can interact with proteins, modifying their activities by mimicking and thus competing with RNA or DNA targets. In bacteria, two families of regulatory sRNAs that act by mimicking other nucleic acids have been characterized most extensively and will be discussed here. The first, exemplified by the *E. coli* 6S RNA, mimics a DNA promoter open complex and interacts with RNA polymerase. The second, the sRNAs that regulate the CsrA/RsmA family of translation regulatory proteins by competing with mRNA targets, will be referred to here as the CsrB family of sRNAs.

5.1 6S sRNA

Although the *E. coli* 6S RNA was first identified as a distinct RNA species in 1967, the function of this sRNA was not understood until more than 30 years later, when 6S was found to bind tightly to the RNA polymerase holoenzyme containing σ^{70} but not the related σ^S holoenzyme (reviewed in Wassarman 2007). Binding of 6S inhibits RNA polymerase activity in vitro. In vivo, it preferentially inhibits a subset of promoters (those with weak -35 regions) (Cavanagh et al. 2008). The 6S structure, determined both by probing experiments and by comparing the 6S RNA in a variety of bacteria (see the following), is a double-stranded RNA hairpin with a critical bubble that mimics the DNA in an open complex promoter, binding at the active site of RNA polymerase. Evidence that this sRNA can mimic DNA is provided by the finding that, both in vitro and in vivo, RNA polymerase can synthesize a short transcript encoded by the 6S RNA, starting within the bubble (Wassarman and Saecker 2006; Gildehaus et al. 2007; Sharma et al. 2010). Interaction of the 6S RNA with RNA polymerase requires region 4.2 of σ^{70} (Cavanagh et al. 2008). The same region of σ^{70} is involved in binding to the -35 region of promoter DNA, although the amino acids required for the interactions with the promoter DNA and 6S RNA are somewhat distinct (Klocko and Wassarman 2009).

Unlike most other *E. coli* regulatory sRNAs, 6S is processed from a longer mRNA, although the function of the downstream gene (*ygfA*) is unknown. The RNA accumulates in stationary phase and acts as one of the multiple regulatory inputs for down-regulating vegetative genes and allowing preferential expression of stationary phase genes (transcribed by RNA polymerase containing σ^S). Recovery from 6S inhibition appears to be caused by the release of 6S from RNA polymerase on transcription from the inhibited complex; it may be that this is the purpose of this transcription. Alternatively or in addition, the short 6S templated RNAs, termed pRNAs, may have independent functions. In vitro, transcription from 6S requires high levels of nucleotides and therefore may only occur during the transition from stationary phase into exponential growth when nucleotide levels rise. Once released, 6S becomes sensitive to degradation (Wassarman and Saecker 2006; Wassarman 2007; Wurm et al. 2010).

6S RNAs are found fairly broadly in other bacterial species, including Gram-positive as well as Gram-negative bacteria (Barrick et al. 2005; Trotochaud and Wassarman 2005; Shi et al. 2009). However, primary sequence conservation is very limited. Thus computational searches for 6S homologs have depended on both secondary structure conservation and the linkage to the downstream *ygfA* ORF, found in some but not all organisms. Because 6S RNA is generally 180–200 nucleotides, not overlapping in size with tRNAs, it has also been possible to find 6S as an abundant and stable RNA (as it was first found in *E. coli*) or by its ability to bind to RNA polymerase (Watanabe et al. 1997; Trotochaud and Wassarman 2005;

Wilkomm et al. 2005). Some organisms have multiple 6S RNAs (Barrick et al. 2005). For example, two 6S-like RNAs are present in B. subtilis, both able to bind the vegetative RNA polymerase (Trotochaud and Wassarman 2005), whereas Clostridium has three 6S-like RNAs (Barrick et al. 2005). The purpose of multiple RNAs is not yet known, but it is certainly possible that 6S-like RNAs in other organisms will bind different forms of RNA polymerase, opening up the possibility of a broader role for 6S-like RNAs.

5.2 CsrB Family of RNAs

The CsrB family of sRNAs counteracts the activities of the small CsrA protein and its homologs (such as RsmA) by titrating them away from their mRNA binding sites. In E. coli, CsrA was first identified as a critical posttranscriptional regulator of the switch between gluconeogenesis and glycolytic growth, inhibiting glycogen synthesis (reviewed in Babitzke et al. 2009). Most of the studied regulation by CsrA is negative; the protein inhibits translation, usually by binding near and thus blocking the binding site. There are reported cases of positive regulation that appear to be direct because they occur in a coupled transcription/translation system (Wei et al. 2001), but the mechanism for positive regulation has not been reported. Binding of CsrA may block other negative translational regulators from binding or may remodel the mRNA. The CsrA protein has a preference for binding GGA in the loop of an RNA hairpin (Dubey et al. 2005; Gutierrez et al. 2005; Heeb et al. 2005; Schubert et al. 2007), and target mRNAs frequently contain multiple binding motifs.

Although synthesis of the family of CsrA proteins may change with environmental conditions (Cui et al. 2005), the major regulation appears to be via inhibition of CsrA activity by the CsrB family of RNAs. These RNAs contain as many as 18 CsrA binding motifs. When they are present at high levels, CsrA binds to them and therefore is not available for interacting with the target mRNAs. Thus, mutations in csrB and csrC, the genes for the two E. coli sRNAs in this family, lead to stringent repression of CsrA targets, and overexpression of CsrB (the more abundant of the two sRNAs) mimics a csrA mutation, leading to increased glycogen synthesis (reviewed in Babitzke and Romeo 2007).

Synthesis of the CsrB and CsrC RNAs is dependent on the two-component system BarA/UvrY, and this regulatory cascade is conserved in many bacteria, although the signals feeding into this cascade are not well defined. Additionally, the stability of CsrB and CsrC is regulated. The turnover of the sRNAs is generally fairly rapid (half-life of 2 minutes for CsrB, around 4 min for CsrC in E. coli), but these sRNAs are fully stabilized in cells mutant for either RNase E, the essential endonuclease involved in degradation of many mRNAs, or CsrD, a protein with homology to, but lacking the activity of, proteins associated with synthesis and degradation of the signaling molecule cyclic diGMP (Suzuki et al. 2009). The effects of CsrD in destabilizing these RNAs appear to be fairly specific, raising a number of questions about how this occurs, as well as the question of whether or not this is a regulatory point and if so, what regulates CsrD (Suzuki et al. 2009).

CsrA-like proteins are fairly widespread, found in both Gram-negative and Gram-positive bacteria, sometimes in more than one copy, although the majority of studies have been with the Gram-negative proteins (White et al. 1996). The CsrB/Rsm RNAs have been found in Enterobacteria, Pseudomonads, and Legionella (Kulkarni et al. 2006; Babitzke and Romeo 2007; Sahr et al. 2009); possibly they are more widespread but have diverged sufficiently to make identification of the equivalent RNAs difficult in more distantly related species. The network of CsrA-like proteins, the BarA/UvrY family of two-component regulators, and downstream sRNA regulators seem to play central roles in the selection of bacterial life-styles between swarming, free-swimming, and biofilm formation, as well as modulating virulence, although the precise roles vary between bacterial species. Much remains to be understood about this network, including how environmental signals impinge on it to up-regulate or down-regulate the sRNAs by transcription or RNA turnover.

5.3 More sRNAs that Regulate Proteins?

As additional sRNA regulators are characterized, it seems likely that more will be found to interact with and regulate proteins, as has been found with the 6S and CsrB RNAs. Certainly there are a large number of known RNA-binding proteins involved in translation, RNA splicing, and trafficking, and generally affecting mRNA fates; many more are likely to exist. It is easy to imagine that sRNAs will be found that regulate these proteins. Other sRNAs may well mimic DNA, as 6S does, and therefore may act as regulators of a variety of DNA binding proteins. In principle, regulatory sRNAs could also fold into a form that can interact with proteins in other ways, such as acting as a scaffold or tether to bring proteins together.

6 sRNAs WITH INTRINSIC ACTIVITIES

Bacteria also have a small number of sRNAs with known activities that do not involve pairing or the regulation of proteins. The ribozyme RNase P has been well characterized (reviewed in Gopalan and Altman 2006; Altman 2007).

A detailed description of 4.5S, the RNA component of bacterial signal recognition particle can also be found elsewhere (reviewed in Egea et al. 2005). Here we will discuss tmRNA (also denoted 10Sa and SsrA RNA), which is widespread within bacteria, is also found in some eukaryotic organelles, and carries out both quality control and regulatory roles.

tmRNA was named for its hybrid characteristics; it has both a tRNA-like domain that is charged by alanine synthetase and an mRNA segment encoding a short ORF that is used as an extension for cotranslational tagging of proteins (reviewed in Moore and Sauer 2007). The mechanism by which tmRNA is recruited to a ribosome when translation stalls and/or the mRNA end lacks a translation termination codon is well studied. On recognizing a stalled ribosome, tmRNA acts first like a tRNA, adding an alanine to the stalled polypeptide chain, and then like a mRNA by encoding a short amino acid tag (10 amino acids in *E. coli*). Because the short ORF ends in a termination codon, it allows release of the polypeptide and re-use of the stalled ribosome. The amino acid tag on the released polypeptide is efficiently recognized by a number of bacterial ATP-dependent proteases, particularly ClpXP. This leads to rapid degradation of the tagged protein and therefore rescue from protein fragments that might otherwise accumulate (reviewed in Moore and Sauer 2007; Keiler 2008). One of the tmRNA roles is reducing stress associated with ribosome stalling/sequestration. This is reflected in the induction of the heat-shock response in cells mutant for tmRNA and in an increased need for tmRNA in response to antibiotics or genetic changes that lead to more stalled ribosomes (reviewed in Keiler 2008). In *Neisseria* and *Helicobacter*, tmRNA is essential (Huang et al. 2000; Thibonnier et al. 2008).

However, recent findings in a number of systems suggest that tmRNAs also may play specific regulatory roles. For example, in *Yersinia*, the lack of tmRNA is accompanied by avirulence and a defect in the transcription of effector proteins (Okan et al. 2006). In *Caulobacter crescentus*, a bacterium with a complex developmental cycle, tmRNA expression is regulated with the cell cycle and tmRNA mutants have aberrant timing of DNA replication (Keiler and Shapiro 2003). This disruption seems to be because of misregulation of *dnaA* transcription, encoding a key regulator of DNA replication; it may be that an unidentified transcription factor is inappropriately synthesized in the tmRNA mutants (Cheng and Keiler 2009).

There are several possibilities for how tmRNAs might play specific roles in regulation. It is known that some ORFs contain sequences that promote tmRNA-dependent tagging and therefore degradation (Roche and Sauer 2001; Sunohara et al. 2002). Conceivably, the absence of tagging leads to overproduction of the untagged protein under inappropriate conditions. Another route to tagging and therefore tmRNA-dependent degradation has been studied in the *lacI* gene encoding the Lac repressor. In this case, the Lac repressor binds near and within the *lacI* gene, blocking full transcription and therefore translation to the end of the gene; this then leads to tagging and degradation of the partially translated protein (Abo et al. 2000). These regulatory mechanisms do not require changes in the levels of tmRNA action, but rather that tmRNA maintains low levels of specific proteins. If tmRNA is transmitting regulatory information, we would guess that this is at the level of general translational stress (stalled ribosomes), possibly leading to changing efficiencies of tagging of a given protein.

7 EVOLUTIONARY CONSIDERATIONS

One interesting question for bacterial sRNAs is whether they have their origins in the RNA world or are newly evolving regulators. Although one can imagine that base pairing between RNAs was one of the earliest forms of regulation, the question of whether any of the bacterial sRNAs date from the RNA world cannot yet be answered. However, the rapid accumulation of genome sequence information together with the interest in uncovering sRNAs in a wide range of bacteria have provided glimpses into the evolution of these RNAs that may help to discern their history.

The protein-binding regulatory 6S and CsrB RNAs, as well as the functional RNase P RNA, SRP/4.5S, and SsrA/tmRNAs, are more broadly conserved than those that act by base pairing. These RNAs are a bit larger, and, in the case of 6S RNA and tmRNA, there are constraints on structure which have facilitated the tracking of these sRNAs in different species. The ability to find 6S-like RNAs over a wide range of bacteria both supports an early evolutionary origin for this regulatory RNA and points out the difficulties in tracing the evolution of sRNAs. Although the structure of 6S is reasonably well conserved, the sequence is not (only 23% of positions conserved at >80% throughout eubacteria) (Barrick et al. 2005). The linkage of this sRNA to a protein-coding gene in many organisms, more easily identified by homology, allowed the robust development of a structural model, therefore facilitating further searches for RNAs unlinked to this ORF.

7.1 Recent Evolution

For the Hfq-binding sRNAs, evolution of the regulatory RNAs appears to be rapid. As a result, neither sequence similarities nor structural similarities are sufficient to provide a clear picture of their evolution, even between relatively closely related species (such as in Gram-negative γ-Proteobacteria). Nevertheless some general constraints on, or contributions to, their evolution can be noted.

Constraints imposed by the targets. The Hfq-binding sRNAs have to pair with specific mRNA targets, and so the evolution of the sRNA and the targets must be linked. The observation that many, but not all, of these sRNAs contain a highly conserved core region, and that this region is frequently involved in pairing with targets (Sharma et al. 2007), supports the idea that pairing helps constrain evolution. For sRNAs with many targets, it may be that some particular target or targets are particularly critical and therefore may correlate better in terms of pairing with the sRNA. In this context, it is intriguing that a number of mRNAs are the targets of multiple sRNAs. This overlap complicates the possibilities for evolution and suggests that these multiple sRNAs may have a common ancestor.

Conserved regulation. The regulators controlling the expression of the sRNAs are other factors that can be used to trace the evolution of sRNAs. The most broadly studied class of potentially related base-pairing sRNAs are those regulated by iron availability via the Fur repressor. This repressor and the sites to which it binds are strikingly well conserved from Gram-negative to Gram-positive bacteria (reviewed in Carpenter et al. 2009). It is becoming increasingly apparent that one or more sRNA is part of the Fur regulon in many organisms. The *E. coli* RyhB RNA was the first example of a Fur-regulated RNA found. Other Enterobacterial species have sRNAs clearly related to RyhB, sharing the Fur binding site in the promoter and a conserved core region involved in pairing to many but not all of the same targets (reviewed in Gottesman et al. 2007). Second copies of RyhB-like RNAs are found in some of these species. In organisms such as *Pseudomonas*, the Fur-regulated sRNAs, called PrrF, have Fur binding sites in the promoter and regulate similar types of genes to those regulated in *E. coli*, but have no obvious sequence similarity to the other RyhB RNAs (Wilderman et al. 2004). Either these sRNAs evolved independently, certainly a possibility, or they have diverged significantly. More recently, Fur-regulated sRNAs have been found in *Neisseria* and *B. subtilis*; again, these sRNAs bear no sequence similarity to RyhB beyond the Fur binding sites (Mellin et al. 2007; Gaballa et al. 2008), also suggesting either independent evolution or rapid divergence.

Gene duplication. Similar to eukaryotic miRNAs, where proliferation of related families has occurred, there is evidence of sRNA gene duplication in bacteria. As mentioned earlier, two Fur-regulated sRNAs with significant overlap in sequence are found in some bacteria. In most Pseudomonads, the two *prrF* genes are located at two different chromosomal locations. In *P. aeruginosa*, however, the two highly similar genes are located adjacent to each other. One evolutionary scenario is that the original *prrF* gene duplicated and then moved to a new site. In an ancestor of *P. aeruginosa*, one of these copies was lost or did not survive the move, and the original gene then duplicated more recently (Gottesman et al. 2007). There are other duplicated sRNAs as well. The highly similar *omrA* and *omrB* genes are adjacent and commonly regulated in *E. coli*. *Vibrio* species have four to five copies of the Qrr genes, at separate chromosomal locations, with varying degrees of divergence (Lenz et al. 2004). Thus duplication of genes has played a role in the distribution of the several of the sRNAs. Having more than one copy of an sRNA gene can have regulatory implications, allowing for increased induction or differential regulation of the copies.

Conserved gene neighborhood. Given the lack of sequence similarity between similarly regulated sRNAs with related functions, the use of gene context is proving to be another means of investigating the functional conservation of sRNAs. For example, tracking of the Hfq-binding GcvB RNA has been simplified because it is always found divergently transcribed from its transcription regulator, GcvA. GcvB is found throughout γ-Proteobacteria, except for Pseudomonads (Sharma et al. 2007). Spot42, a conserved sRNA found by homology only in Enterobacterial species, occupies an intergenic region between two highly conserved protein-coding genes; in *Pseudomonas*, this same intergenic neighborhood encodes an sRNA of unknown function, with no sequence similarity to Spot42 (Gottesman et al. 2007). One explanation is that Spot42 and this sRNA are evolutionarily related but have diverged rapidly. Alternatively, this is a location into which sRNA genes are inserted from different (and unknown) sources. The characterization of more mRNA targets of both of these sRNAs may give insights into whether they are indeed related.

Horizontal transfer. Are sRNAs more or less likely than other genes to be part of mobile genetic elements, and might they be subject to horizontal transfer? In *E. coli*, DicF, an Hfq-binding sRNA, is found within a cryptic prophage, but regulates the host *ftsZ* gene (Bouché and Bouché 1989). IpeX, an Hfq-independent sRNA that regulates porin genes, was also found to be encoded in a cryptic *E. coli* prophage (Castillo-Keller et al. 2006). In specific searches for sRNAs in pathogenicity islands in *S. enterica*, a number of such regulatory RNAs were found, including a second copy of a Fur-regulated RyhB-like sRNA

(Padalon-Brauch et al. 2008). Both the RyhB-related sRNA and at least one other pathogenicity island sRNA, InvR, regulate genes encoded in the core genome (Pfeiffer et al. 2007). Pathogenicity island sRNAs have also been discovered for *S. aureus* (Pichon and Felden 2005) and are likely to be present in other pathogens. The reciprocal case where a host-encoded sRNA regulates the expression of an island-encoded gene has been found as well (Papenfort and Vogel, unpublished). A number of the chromosomally located sRNAs lie adjacent to regions into which phage or transposons have been shown to insert, suggesting that they may either be picked up by these elements when the elements excise, or that they were first brought into the genome with these elements (De Lay and Gottesman 2009).

It is worth remembering that antisense sRNAs are intimately involved in regulating both plasmid copy number and plasmid conjugation (reviewed in Wagner et al. 2002; Brantl 2007). As in eukaryotes, where RNA interference plays a critical role in limiting the spread of transposable elements, transposons in bacteria are also limited in their activity by antisense RNAs. Thus, the elements for horizontal transfer carry within them regulatory sRNAs that could evolve to have complementarity with targets other than the mobile elements.

7.2 Speculation on Deeper Evolution

The examples described in the previous section provide a glimpse at the recent evolution of the base-pairing sRNAs in bacteria, and suggest that the constraints on their evolution are loose enough so that they change sequence and structure relatively rapidly. Given a rapid rate of change, it is not yet possible to develop hard evidence for original sources of these sRNAs, leaving us free to speculate. There is no reason to assume a single evolutionary pathway for the rather divergent sRNAs already identified in bacteria, so all (or none) of the possibilities may turn out to be correct.

Capture of random transcription for the purpose of regulation. As more deep sequencing and tiling array analysis is performed, a certain level of what might be considered transcriptional "noise," low-level transcription antisense to genes and within spacer regions, with no assigned function thus far, is apparent. The hypothesis that there is low level transcription that does not lead to productive transcripts is supported by the finding of antisense promoter activity without the detection of stable transcripts (Kawano et al. 2005b). It is relatively easy to imagine that if some low-level promoter activity were sufficient to give advantageous regulation under some condition, both the promoter and the RNA might evolve into a highly expressed regulatory sRNA.

Capture of transcripts of other function: sRNA-tRNA connections. Another possible source of regulatory sRNAs is tRNAs, or the ancestors of tRNAs. These are folded, stable RNAs of about the same size as many of the regulatory RNAs. Several characteristics are suggestive of a relationship between regulatory sRNAs and tRNAs. As for tRNAs, both tmRNA (clearly tRNA-related) and *E. coli* CyaR, an Hfq-dependent sRNA, act as attachment sites for phage integration (Kirby et al. 1994; De Lay and Gottesman 2009). Because it is not really clear why tRNA genes are used this way, the implications of this are not known (Campbell 1992). In eukaryotic cells, retrotransposons make use of tRNA primers. A number of studies have suggested that specific tRNAs also interact with Hfq, possibly because Hfq modulates their processing (Zhang et al. 2003; Lee and Feig 2008). This ability of tRNAs to bind Hfq could be considered support for a common ancestry. One prediction is that tRNAs (or more specifically tRNA fragments) will be found to have regulatory roles. This prediction is supported by recent studies in eukaryotic cells (Cole et al. 2009; Lee et al. 2009; Thompson and Parker 2009); investigations of functions for tRNA fragments in bacteria have not yet been reported. A second prediction is that genes which appear to be similar to, but not quite like, tRNAs will be shown to be regulatory sRNAs rather than functional tRNAs.

Capture of transcripts of other function: sRNA-mRNA connections. A third possible source for regulatory sRNAs are mRNAs. The sRNAs that also encode proteins and the bifunctional riboswitches recently described in *L. monocytogenes* (see earlier) provide examples for what precursors for free-standing regulatory sRNAs might be. In *E. coli*, mRNAs that are targets for Hfq-dependent sRNA regulation all appear to have Hfq-binding sites; it is not difficult to imagine that an Hfq-binding UTR or intergenic region could be both the target of regulation and, under other conditions, the regulator. An example of an mRNA regulator has already been described (see earlier discussion, section on RNA mimics). Separation from the mRNA or loss of the downstream ORF might then lead to what we now detect as a free-standing sRNA. This model might predict that further sequencing will reveal cases where sequences similar to sRNAs are found as UTRs or within mRNAs. Certainly, in the eukaryotic RNA world, the appearance of miRNAs encoded within introns is consistent with the possibility that pieces of evolving mRNAs may become regulatory sRNAs, or vice versa.

8 PERSPECTIVES

Although regulatory RNAs have been a major topic of interest in eukaryotic cells, studies of bacterial sRNAs have been equally exciting. Exactly why the cell regulates a given gene or set of genes via regulatory RNAs, rather than at the level of transcription initiation, is not yet understood, but the availability of multiple sRNAs, identification of their upstream regulators and of their targets, is allowing investigators to begin to test the advantages of sRNA as regulators in mathematical models (Levine and Hwa 2008; Mehta et al. 2008; Mitarai et al. 2009). In many instances, it may not be a question of one type of regulation over another, but that the requirement for many bacteria to respond rapidly to changing environments has led to multiple levels of regulation, among them the very versatile and easily adaptable regulatory sRNAs.

ACKNOWLEDGMENTS

The writing of this review was supported by the Intramural Research Program of the National Institutes of Health, National Cancer Institute, Center for Cancer Research and Eunice Kennedy Shriver National Institute of Child Health and Human Development. We thank members of our laboratories, C. Vanderpool and J. Vogel for comments on the manuscript, and thank J. Vogel for sharing unpublished results.

REFERENCES

Abo T, Inada T, Ogawa K, Aiba H. 2000. SsrA-mediated tagging and proteolysis of LacI and its role in the regulation of *lac* operon. *EMBO J* **19:** 3762–3769.

Altman S. 2007. A view of RNase P. *Mol Biosyst* **3:** 604–607.

Altuvia S, Weinstein-Fischer D, Zhang A, Postow L, Storz G. 1997. A small stable RNA induced by oxidative stress: Role as a pleiotropic regulator and antimutator. *Cell* **90:** 43–53.

Altuvia S, Zhang A, Argaman L, Tiwari A, Storz G. 1998. The *Escherichia coli oxyS* regulatory RNA represses *fhlA* translation by blocking ribosome binding. *EMBO J* **17:** 6069–6075.

Andersen J, Delihas N. 1990. *micF* RNA binds to the 5′ end of *ompF* mRNA and to a protein from *Escherichia coli*. *Biochemistry* **29:** 9249–9256.

Antal M, Bordeau V, Douchin V, Felden B. 2005. A small bacterial RNA regulates a putative ABC transporter. *J Biol Chem* **280:** 7901–7908.

Babitzke P, Baker CS, Romeo T. 2009. Regulation of translation initiation by RNA binding proteins. *Annu Rev Microbiol* **63:** 27–44.

Babitzke P, Romeo T. 2007. CsrB sRNA family: Sequestration of RNA-binding regulatory proteins. *Curr Opin Microbiol* **10:** 156–163.

Barrick JE, Sudarsan N, Weinberg Z, Ruzzo WL, Breaker RR. 2005. 6S RNA is a widespread regulator of eubacterial RNA polymerase that resembles an open promoter. *RNA* **11:** 774–784.

Boisset S, Geissmann T, Huntzinger E, Fechter P, Bendridi N, Possedko M, Chevalier C, Helfer AC, Benito Y, Jacquier A, et al. 2007. *Staphylococcus aureus* RNAIII coordinately represses the synthesis of virulence factors and the transcription regulator Rot by an antisense mechanism. *Genes Dev* **21:** 1353–1366.

Bouché F, Bouché JP. 1989. Genetic evidence that DicF, a second division inhibitor encoded by the *Escherichia coli dicB* operon, is probably RNA. *Mol Microbiol* **3:** 991–994.

Bouvier M, Sharma CM, Mika F, Nierhaus KH, Vogel J. 2008. Small RNA binding to 5′ mRNA coding region inhibits translational initiation. *Mol Cell* **32:** 827–837.

Boysen A, Moller-Jensen J, Kallipolitis B, Valentin-Hansen P, Overgaard M. 2010. Translational regulation of gene expression by an anaerobically induced small non-coding RNA in *Escherichia coli*. *J Biol Chem* **285:** 10690–10702.

Brantl S. 2007. Regulatory mechanisms employed by cis-encoded antisense RNAs. *Curr Opin Microbiol* **10:** 102–109.

Brennan RG, Link TM. 2007. Hfq structure, function and ligand binding. *Curr Opin Microbiol* **10:** 125–133.

Campbell AM. 1992. Chromosomal insertion sites for phages and plasmids. *J Bacteriol* **174:** 7495–7499.

Carpenter BM, Whitmire JM, Merrell DS. 2009. This is not your mother's repressor: the complex role of Fur in pathogenesis. *Infect Immun* **77:** 2590–2601.

Castillo-Keller M, Vuong P, Misra R. 2006. Novel mechanism of *Escherichia coli* porin regulation. *J Bacteriol* **188:** 576–586.

Cavanagh AT, Klocko AD, Liu X, Wassarman KM. 2008. Promoter specificity for 6S RNA regulation of transcription is determined by core promoter sequences and competition for region 4.2 of σ^{70}. *Mol Microbiol* **67:** 1242–1256.

Chen S, Zhang A, Blyn LB, Storz G. 2004. MicC, a second small-RNA regulator of Omp protein expression in *Escherichia coli*. *J Bacteriol* **186:** 6689–6697.

Cheng L, Keiler KC. 2009. Correct timing of *dnaA* transcription and initiation of DNA replication requires *trans* translation. *J Bacteriol* **191:** 4268–4275.

Christiansen JK, Larsen MH, Ingmer H, Sogaard-Andersen L, Kallipolitis BH. 2004. The RNA-binding protein Hfq of *Listeria monocytogenes*: Role in stress tolerance and virulence. *J Bacteriol* **186:** 3355–3362.

Christiansen JK, Nielsen JS, Ebersbach T, Valentin-Hansen P, Sogaard-Andersen L, Kallipolitis BH. 2006. Identification of small Hfq-binding RNAs in *Listeria monocytogenes*. *RNA* **12:** 1383–1396.

Cole C, Sobala A, Lu C, Thatcher SR, Bowman A, Brown JWS, Green PJ, Barton GJ, Hutvagner G. 2009. Filtering of deep sequencing data reveals the existence of abundant Dicer-dependent small RNAs derived from tRNAs. *RNA* **15:** 2147–2160.

Coornaert A, Lu A, Mandin P, Springer M, Gottesman S, Guillier M. 2010. MicA sRNA links the PhoP regulon to cell envelope stress. *Mol Microbiol* **76:** 467–479.

Coyer J, Andersen J, Forst SA, Inouye M, Delihas N. 1990. *micF* RNA in *ompB* mutants of *Escherichia coli*: different pathways regulate *micF* RNA levels in response to osmolarity and temperature change. *J Bacteriol* **172:** 4143–4150.

Cui Y, Chatterjee A, Hasegawa H, Dixit V, Leigh N, Chatterjee AK. 2005. ExpR, a LuxR homolog of *Erwinia carotovora* subsp. *carotovora*, activates transcription of *rsmA*, which specifies a global regulatory RNA-binding protein. *J Bacteriol* **187:** 4792–4803.

De Lay N, Gottesman S. 2009. The Crp-activated small noncoding regulatory RNA CyaR (RyeE) links nutritional status to group behavior. *J Bacteriol* **191:** 461–476.

Desnoyers G, Morissette A, Prevost K, Massé E. 2009. Small RNA-induced differential degradation of the polycistronic mRNA *iscRSUA*. *EMBO J* **28:** 1551–1561.

Douchin V, Bohn C, Bouloc P. 2006. Down-regulation by porins by a small RNA bypasses the essentiality of the regulated intramembrane proteolysis protease RseP in *Escherichia coli*. *J Biol Chem* **281:** 12253–12259.

Dubey AK, Baker CS, Romeo T, Babitzke P. 2005. RNA sequence and secondary structure participate in high-affinity CsrA-RNA interaction. *RNA* **11:** 1579–1587.

Duhring U, Axmann IM, Hess WR, Wilde A. 2006. An internal antisense RNA regulates expression of the photosynthesis gene *isiA*. *Proc Natl Acad Sci* **103:** 7054–7058.

Durand S, Storz G. 2010. Reprogramming of anaerobic metabolism by the FnrS small RNA. *Mol Microbiol* **75:** 1215–1231.

Egea PF, Stroud RM, Walter P. 2005. Targeting proteins to membranes: Structure of the signal recognition particle. *Curr Opin Struct Biol* **15:** 213–220.

Faubladier M, Cam K, Bouché J-P. 1990. *Escherichia coli* cell division inhibitor DicF-RNA of the *dicB* operon. Evidence for its generation *in vivo* by transcription termination and by RNase III and RNase E-dependent processing. *J Mol Biol* **212:** 461–471.

Figueroa-Bossi N, Valentini M, Malleret L, Bossi L. 2009. Caught at its own game: Regulatory small RNA inactivated by an inducible transcript mimicking its target. *Genes Dev* **23:** 1981–1985.

Fozo EM, Hemm MR, Storz G. 2008. Small toxic proteins and the antisense RNAs that repress them. *Microbiol Mol Biol Rev* **72:** 579–589.

Fozo EM, Makarova KS, Shabalina SA, Yutin N, Koonin EV, Storz G. 2010. Abundance of type I toxin-antitoxin systems in bacteria: searches for new candidates and discovery of novel families. *Nucleic Acids Res* (in press).

Gaballa A, Antelmann H, Aguilar C, Khakh SK, Song K-B, Smaldone GT, Helmann JD. 2008. The *Bacillus subtilis* iron-sparing response is mediated by a Fur-regulated small RNA and three small, basic proteins. *Proc Natl Acad Sci* **105:** 11927–11932.

Georg J, Vos B, Scholz I, Mitschke J, Wilde A, Hess WR. 2009. Evidence for a major role of antisense RNAs in cyanobacterial gene regulation. *Mol Sys Biol* **5:** 305.

Gerdes K, Wagner EG. 2007. RNA antitoxins. *Curr Opin Microbiol* **10:** 117–124.

Gildehaus N, Neusser T, Wurm R, Wagner R. 2007. Studies on the function of the riboregulator 6S RNA from *E. coli*: RNA polymerase binding, inhibition of *in vitro* transcription and synthesis of RNA-directed *de novo* transcripts. *Nucleic Acids Res* **35:** 1885–1896.

Gopalan V, Altman S. 2006. Ribonuclease P: Structure and catalysis. In *The RNA world* (ed. Gesteland R.F., Cech T.R., Atkins J.F.). Cold Spring Harbor Laboratory Press, Cold Spring Harbor, NY.

Gottesman S, McCullen CA, Guillier M, Vanderpool CK, Majdalani N, Benhammou J, Thompson KM, FitzGerald PC, Sowa NA, FitzGerald DJ. 2007. Small RNA regulators and the bacterial response to stress. *Cold Spring Harbor Symp Quant Biol* **71:** 1–11.

Guell M, van Noort V, Yus E, Chen W-H, Leigh-Bell J, Michalodimitrakis K, Yamada T, Arumugam M, Doerks T, Kuhner S, et al. 2009. Transcriptome complexity in a genome-reduced bacterium. *Science* **326:** 1268–1271.

Guillier M, Gottesman S. 2006. Remodelling of the *Escherichia coli* outer membrane by two small regulatory RNAs. *Mol Microbiol* **59:** 231–247.

Guillier M, Gottesman S. 2008. The 5′ end of two redundant sRNAs is involved in the regulation of multiple targets, including their own regulator. *Nucleic Acids Res* **36:** 6781–6794.

Guillier M, Gottesman S, Storz G. 2006. Modulating the outer membrane with small RNAs. *Genes Dev* **20:** 2338–2348.

Gutierrez P, Li Y, Osborne MJ, Pomerantseva E, Liu Q, Gehring K. 2005. Solution structure of the carbon storage regulator protein CsrA from *Escherichia coli*. *J Bacteriol* **187:** 3496–3501.

Heeb S, Kuehne SA, Bycroft M, Crivii S, Allen MD, Haas D, Camara M, Williams P. 2005. Functional analysis of the post-transcriptional regulator RsmA reveals a novel RNA-binding site. *J Mol Biol* **355:** 1026–1036.

Heidrich N, Moll I, Brantl S. 2007. *In vitro* analysis of the interaction between the small RNA SR1 and its primary target *ahrC* mRNA. *Nucleic Acids Res* **35:** 4331–4346.

Heidrich N, Chinali A, Gerth U, Brantl S. 2006. The small untranslated RNA SR1 from the *Bacillus subtilis* genome is involved in the regulation of arginine catabolism. *Mol Microbiol* **62:** 520–536.

Horler RS, Vanderpool CK. 2009. Homologs of the small RNA SgrS are broadly distributed in enteric bacteria but have diverged in size and sequence. *Nucleic Acids Res* **37:** 5465–5476.

Huang C, Wolfgang MC, Withey J, Koomey M, Friedman DI. 2000. Charged tmRNA but not tmRNA-mediated proteolysis is essential for *Neisseria gonorrhoeae* viability. *EMBO J* **19:** 1098–1107.

Huntzinger E, Boisset S, Saveneau C, Benito Y, Geissmann T, Namane A, Lina G, Etienne J, Ehresmann B, Ehresmann C, et al. 2005. *Staphylococcus aureus* RNAIII and the endoribonuclease III coordinately regulate *spa* gene expression. *EMBO J* **24:** 824–835.

Johansen J, Eriksen M, Kallipolitis B, Valentin-Hansen P. 2008. Down-regulation of outer membrane proteins by noncoding RNAs: Unraveling the cAMP-CRP- and σ^E-dependent CyaR-*ompX* regulatory case. *J Mol Biol* **383:** 1–9.

Johansen J, Rasmussen AA, Overgaard M, Valentin-Hansen P. 2006. Conserved small non-coding RNAs that belong to the σ^E regulon: role in down-regulation of outer membrane proteins. *J Mol Biol* **364:** 1–8.

Kalamorz F, Reichenbach B, Marz W, Rak B, Gorke B. 2007. Feedback control of glucosamine-6-phosphate synthase GlmS expression depends on the small RNA GlmZ and involves the novel protein YhbJ in *Escherichia coli*. *Mol Microbiol* **65:** 1518–1533.

Kawamoto H, Morita T, Shimizu A, Inada T, Aiba H. 2005. Implication of membrane localization of target mRNA in the action of a small RNA: mechanism of post-transcriptional regulation of glucose transporter in *Escherichia coli*. *Genes Dev* **19:** 328–338.

Kawano M, Reynolds AA, Miranda-Rios J, Storz G. 2005a. Detection of 5′- and 3′-UTR-derived small RNAs and cis-encoded antisense RNAs in *Escherichia coli*. *Nucleic Acids Res* **33:** 1040–1050.

Kawano M, Storz G, Rao BS, Rosner JL, Martin RG. 2005b. Detection of low-level promoter activity within open reading frame sequences of *Escherichia coli*. *Nucleic Acids Res* **33:** 6268–6276.

Keiler KC. 2008. Biology of trans-translation. *Annu Rev Microbiol* **62:** 133–151.

Keiler KC, Shapiro L. 2003. tmRNA is required for correct timing of DNA replication in *Caulobacter crescentus*. *J Bacteriol* **185:** 573–580.

Kirby JE, Trempy JE, Gottesman S. 1994. Excision of a P4-like cryptic prophage leads to Alp protease expression in *Escherichia coli*. *J Bacteriol* **176:** 2068–2081.

Klocko AD, Wassarman KM. 2009. 6S RNA binding to Eσ^{70} requires a positively charged surface of σ^{70} region 4.2. *Mol Microbiol* **73:** 152–164.

Krinke L, Wulff DL. 1987. OOP RNA, produced from multicopy plasmids, inhibits λ *cII* gene expression through an RNase III-dependent mechanism. *Genes Dev* **1:** 1005–1013.

Krinke L, Wulff DL. 1990. RNase III-dependent hydrolysis of λ *cII-O* gene mRNA mediated by λ OOP antisense RNA. *Genes Dev* **4:** 2223–2233.

Kulkarni PR, Cui X, Williams JW, Stevens AM, Kulkarni RV. 2006. Prediction of CsrA-regulating small RNAs in bacteria and their experimental verification in *Vibrio fischeri*. *Nucleic Acids Res* **34:** 3361–3369.

Lease RA, Belfort M. 2000. Riboregulation by DsrA RNA: Trans-actions for global economy. *Mol Microbiol* **38:** 667–672.

Lease RA, Cusick M, Belfort M. 1998. Riboregulation in *Escherichia coli*: DsrA RNA acts by RNA:RNA interactions at multiple loci. *Proc Natl Acad Sci USA* **95:** 12456–12461.

Lee T, Feig AL. 2008. The RNA binding protein Hfq interacts specifically with tRNAs. *RNA* **14:** 514–523.

Lee YS, Shibata Y, Malhotra A, Dutta A. 2009. A novel class of small RNAs: tRNA-derived RNA fragments (tRFs). *Genes Dev* **23:** 2639–2649.

Lenz DH, Mok KC, Lilley BN, Kulkarni RV, Wingreen NS, Bassler BL. 2004. The small RNA chaperone Hfq and multiple small RNAs control quorum sensing in *Vibrio harveyi* and *Vibrio cholerae*. *Cell* **118:** 69–82.

Levine E, Hwa T. 2008. Small RNAs establish gene expression thresholds. *Curr Opin Microbiol* **11:** 574–579.

Link TM, Valentin-Hansen P, Brennan RG. 2009. Structure of *Escherichia coli* Hfq bound to polyriboadenylate RNA. *Proc Natl Acad Sci USA* **106:** 19292–19297.

Livny J, Waldor MK. 2007. Identification of small RNAs in diverse bacterial species. *Curr Opin Microbiol* **10:** 96–101.

Loh E, Dussurget O, Gripenland J, Vaitkevicius K, Tiensuu T, Mandin P, Repoila F, Buchrieser C, Cossart P, Johansson J. 2009. A *trans*-acting riboswitch controls expression of the virulence regulator PrfA in *Listeria monocytogenes*. *Cell* **139:** 770–779.

Majdalani N, Hernandez D, Gottesman S. 2002. Regulation and mode of action of the second small RNA activator of RpoS translation, RprA. *Mol Microbiol* **46:** 813–826.

Majdalani N, Vanderpool CK, Gottesman S. 2005. Bacterial small RNA regulators. *CRC Crit Rev Biochem* **40:** 93–113.

Majdalani N, Chen S, Murrow J, St John K, Gottesman S. 2001. Regulation of RpoS by a novel small RNA: The characterization of RprA. *Mol Microbiol* **39:** 1382–1394.

Majdalani N, Cunning C, Sledjeski D, Elliott T, Gottesman S. 1998. DsrA RNA regulates translation of RpoS message by an anti-antisense mechanism, independent of its action as an antisilencer of transcription. *Proc Natl Acad Sci USA* **95:** 12462–12467.

Mandin P, Gottesman S. 2010. submitted.

Massé E, Gottesman S. 2002. A small RNA regulates the expression of genes involved in iron metabolism in *Escherichia coli*. *Proc Natl Acad Sci USA* **99:** 4620–4625.

Massé E, Escorcia FE, Gottesman S. 2003. Coupled degradation of a small regulatory RNA and its mRNA targets in *Escherichia coli*. *Genes Dev* **17:** 2374–2383.

Massé E, Vanderpool CK, Gottesman S. 2005. Effect of RyhB small RNA on global iron use in *Escherichia coli*. *J Bacteriol* **187:** 6962–6972.

Massé E, Salvail H, Desnoyers G, Arguin M. 2007. Small RNAs controlling iron metabolism. *Curr Opin Microbiol* **10:** 140–145.

Mehta P, Goyal S, Wingreen NS. 2008. A quantitative comparison of sRNA-based and protein-based gene regulation. *Mol Syst Biol* **4:** 221.

Mellin JR, Goswami S, Grogan S, Tjaden B, Genco CA. 2007. A novel Fur- and Iron-regulated small RNA, NrrF, is required for indirect Fur-mediated regulation of the *sdhA* and *sdhC* genes in *Neisseria meningitidis*. *J Bacteriol* **189:** 3686–3694.

Mitarai N, Benjamin J-AM, Krishna S, Semsey S, Csiszovszki Z, Massé E, Sneppen K. 2009. Dynamic features of gene expression control by small regulatory RNAs. *Proc Natl Acad Sci USA* **106:** 10655–10659.

Møller T, Franch T, Udesen C, Gerdes K, Valentin-Hansen P. 2002. Spot 42 RNA mediates discoordinate expression of the *E. coli* galactose operon. *Genes Dev* **16:** 1696–1706.

Moon K, Gottesman S. 2009. A PhoQ/P-regulated small RNA regulates sensitivity of *Escherichia coli* to antimicrobial peptides. *Molec Microbiol* **74:** 1314–1330.

Moore SD, Sauer RT. 2007. The tmRNA system for translational surveillance and ribosome rescue. *Annu Rev Biochem* **76:** 101–124.

Morfeldt E, Taylor D, von Gabain A, Arvidson S. 1995. Activation of alpha-toxin translation in *Staphylococcus aureus* by the *trans*-encoded antisense RNA, RNAIII. *EMBO J* **14:** 4569–4577.

Nielsen JS, Lei LK, Ebersbach T, Olsen AS, Klitgaard JK, Valentin-Hansen P, Kallipolitis BH. 2009. Defining a role for Hfq in gram-positive bacteria: Evidence for Hfq-dependent antisense regulation in *Listeria monocytogenes*. *Nucleic Acids Res* **38:** 907–919.

Okan NA, Bliska JB, Karzai AW. 2006. A role for the SmpB-SsrA system in *Yersinia pseudotuberculosis* pathogenesis. *PLoS Pathog* **2:** e6.

Opdyke JA, Kang JG, Storz G. 2004. GadY, a small-RNA regulator of acid response genes in *Escherichia coli*. *J Bacteriol* **186:** 6698–6705.

Overgaard M, Johansen J, Moller-Jensen J, Valentin-Hansen P. 2009. Switching off small RNA regulation with trap-mRNA. *Mol Microbiol* **73:** 790–800.

Padalon-Brauch G, Hershberg R, Elgrably-Weiss M, Baruch K, Rosenshine I, Margalit H, Altuvia S. 2008. Small RNAs encoded within genetic islands of *Salmonella typhimurium* show host-induced expression and role in virulence. *Nucleic Acids Res* **36:** 1913–1927.

Papenfort K, Pfeiffer V, Lucchini S, Sonawane A, Hinton JC, Vogel J. 2008. Systematic deletion of *Salmonella* small RNA genes identifies CyaR, a conserved CRP-dependent riboregulator of OmpX synthesis. *Mol Microbiol* **68:** 890–906.

Papenfort K, Pfeiffer V, Mika F, Lucchini S, Hinton JCD, Vogel J. 2006. σ^{E}-dependent small RNAs of *Salmonella* respond to membrane stress by accelerating global *omp* mRNA decay. *Mol Microbiol* **62:** 1674–1688.

Papenfort K, Said N, Welsink T, Lucchini S, Hinton JC, Vogel J. 2009. Specific and pleiotropic patterns of mRNA regulation by ArcZ, a conserved, Hfq-dependent small RNA. *Mol Microbiol* **74:** 139–158.

Peters JM, Mooney RA, Kuan PF, Rowland JL, Keles S, Landick R. 2009. Rho directs widespread termination of intragenic and stable RNA transcription. *Proc Natl Acad Sci USA* **106:** 15406–15411.

Pfeiffer V, Papenfort K, Lucchini S, Hinton JCD, Vogel J. 2009. Coding sequence targeting by MicC RNA reveals bacterial mRNA silencing downstream of translational initiation. *Nat Struct Mol Biol* **16:** 840–846.

Pfeiffer V, Sittka A, Tomer R, Tedin K, Brinkmann V, Vogel J. 2007. A small non-coding RNA of the invasion gene island (SPI-1) represses outer membrane protein synthesis from the *Salmonella* core genome. *Mol Microbiol* **66:** 1174–1191.

Pichon C, Felden B. 2005. Small RNA genes expressed from *Staphylococcus aureus* genomic and pathogenicity islands with specific expression among pathogenic strains. *Proc Natl Acad Sci USA* **102:** 14249–14254.

Prévost K, Salvail H, Desnoyers G, Jacques JF, Phaneuf E, Massé E. 2007. The small RNA RyhB activates the translation of *shiA* mRNA encoding a permease of shikimate, a compound involved in siderophore synthesis *Mol Microbiol* **64:** 1260–1273.

Pulvermacher SC, Stauffer LT, Stauffer GV. 2009a. Role of the sRNA GcvB in regulation of *cycA* in *Escherichia coli*. *Microbiology* **155:** 106–114.

Pulvermacher SC, Stauffer LT, Stauffer GV. 2009b. The small RNA GcvB regulates *sstT* mRNA expression in *Escherichia coli*. *J Bacteriol* **191:** 238–248.

Rasmussen AA, Eriksen M, Gilany K, Udesen C, Franch T, Peterson C, Valentin-Hansen P. 2005. Regulation of *ompA* mRNA stability: The role of a small regulatory RNA in growth phase-dependent control. *Mol Microbiol* **58:** 1421–1429.

Rasmussen AA, Johansen J, Nielsen JS, Overgaard M, Kallipolitis B, Valentin-Hansen P. 2009. A conserved small RNA promotes silencing of the outer membrane protein YbfM. *Mol Microbiol* **72:** 566–577.

Reichenbach B, Gopel Y, Gorke B. 2009. Dual control by perfectly overlapping σ^{54}- and σ^{70}-promoters adjusts small RNA GlmY expression to different environmental signals. *Mol Microbiol* **74:** 1054–1070.

Reichenbach B, Maes A, Kalamorz F, Hajnsdorf E, Gorke B. 2008. The small RNA GlmY acts upstream of the sRNA GlmZ in the activation of *glmS* expression and is subject to regulation by polyadenylation in *Escherichia coli*. *Nucleic Acids Res* **36:** 2570–2580.

Roche ED, Sauer RT. 2001. Identification of endogenous SsrA-tagged proteins reveals tagging at positions corresponding to stop codons. *J Biol Chem* **276:** 28509–28515.

Romby P, Charpentier E. 2009. An overviews of RNAs with regulatory functions in gram-positive bacteria. *Cell Mol Life Sci* **67:** 217–237.

Sahr T, Bruggemann H, Jules M, Lomma M, Albert-Weissenberger C, Cazalet C, Buchrieser C. 2009. Two small ncRNAs jointly govern virulence and transmission in *Legionella pneumophila*. *Mol Microbiol* **72:** 741–762.

Schubert M, Lapouge K, Duss O, Oberstrass FC, Jelesarov I, Haas D, Alain FH. 2007. Molecular basis of messenger RNA recognition by the specific bacterial repressing clamp RsmA/CsrA. *Nat Struct Mol Biol* **14:** 807–813.

Schumacher MA, Pearson RF, Møller T, Valentin-Hansen P, Brennan RG. 2002. Structures of the pleiotropic translational regulator Hfq and an Hfq-RNA complex: A bacterial Sm-like protein. *EMBO J* **21:** 3546–3556.

Sharma CM, Vogel J. 2009. Experimental approaches for the discovery and characterization of regulatory small RNAs. *Curr Opin Microbiol* **12:** 536–546.

Sharma CM, Darfeuille F, Plantinga TH, Vogel J. 2007. A small RNA regulates multiple ABC transporter mRNAs by targeting C/A-rich elements inside and upstream of ribosome-binding sites. *Genes Dev* **21:** 2804–2817.

Sharma CM, Hoffman S, Darfeuille F, Reignier J, Findeiss S, Sittka A, Chabas S, Reiche K, Hackermuller J, Reinhardt R, et al. 2010. The primary transcriptome of the major human pathogen *Helicobacter pylori*. *Nature* **464:** 250–255.

Shi Y, Tyson GW, DeLong EF. 2009. Metatranscriptomics reveals unique microbial small RNAs in the ocean's water column. *Nature* **459:** 266–269.

Silvaggi JM, Perkins JB, Losick R. 2005. Small untranslated RNA antitoxin in *Bacillus subtilis*. *J Bacteriol* **187:** 6641–6650.

Sledjeski D, Gottesman S. 1995. A small RNA acts as an antisilencer of the H-NS-silenced rcsA gene of *Escherichia coli*. *Proc Natl Acad Sci* **92:** 2003–2007.

Sledjeski DD, Gupta A, Gottesman S. 1996. The small RNA, DsrA, is essential for the low temperature expression of RpoS during exponential growth in *Escherichia coli*. *EMBO J* **15:** 3993–4000.

Soper TJ, Woodson SA. 2008. The rpoS mRNA leader recruits Hfq to facilitate annealing with DsrA sRNA. *RNA* **14:** 1907–1917.

Sunohara T, Abo T, Inada T, Aiba H. 2002. The C-terminal amino acid sequence of nascent peptide is a major determinant of SsrA tagging at all three stop codons. *RNA* **8:** 1416–1427.

Suzuki K, Babitzke P, Kushner SR, Romeo T. 2009. Identification of a novel regulatory protein (CsrD) that targets the global regulatory RNAs CsrB and CsrC for degradation by RNase E. *Genes Dev* **20:** 2605–2617.

Svenningsen SL, Tu KC, Bassler BL. 2009. Gene dosage compensation calibrates four regulatory RNAs to control *Vibrio cholerae* quorum sensing. *EMBO J* **28:** 429–439.

Svenningsen SL, Waters CM, Bassler BL. 2008. A negative feedback loop involving small RNAs accelerates *Vibrio cholerae*'s transition out of quorum-sensing mode. *Genes Dev* **22:** 226–238.

Tetart F, Bouché JP. 1992. Regulation of the expression of the cell-cycle gene *ftsZ* by DicF antisense RNA. Division does not require a fixed number of FtsZ molecules. *Mol Microbiol* **6:** 615–620.

Thibonnier M, Thiberge J-M, De Reuse H. 2008. *Trans*-translation in *Helicobacter pylori*: Essentiality of ribosome rescue and requirement of protein tagging for stress resistance and competence. *PLoS One* **3:** e3810.

Thompson DM, Parker R. 2009. Stressing out over tRNA cleavage. *Cell* **138:** 215–219.

Thompson KM, Rhodius VA, Gottesman S. 2007. σ^E regulates and is regulated by a small RNA in *Escherichia coli*. *J Bacteriol* **189:** 4243–4256.

Toledo-Arana A, Dussurget O, Nikitas G, Sesto N, Guet-Revillet H, Balestrino D, Loh E, Gripenland J, Tensuu T, Vaitkevicius K, et al. 2009. The *Listeria* transcriptional landscape from saprophytism to virulence. *Nature* **459:** 950–956.

Tramonti A, De Canio M, De Biase D. 2008. GadX/GadW-dependent regulation of the *Escherichia coli* acid fitness island: Transcriptional control at the gadY-gadW divergent promoters and identification of four novel 42 bp GadX/GadW-specific binding sites. *Mol Microbiol* **70:** 965–982.

Trotochaud AE, Wassarman KM. 2005. A highly conserved 6S RNA structure is required for regulation of transcription. *Nat Struct Mol Biol* **12:** 313–319.

Tu KC, Bassler BL. 2007. Multiple small RNAs act additively to integrate sensory information and control quorum sensing in *Vibrio harveyi*. *Genes Dev* **21:** 221–233.

Tu KC, Waters CM, Svenningsen SL, Bassler BL. 2008. A small-RNA-mediated negative feedback loop controls quorum-sensing dynamics in *Vibrio harveyi*. *Mol Microbiol* **70:** 896–907.

Udekwu KI, Wagner EG. 2007. Sigma E controls biogenesis of the antisense RNA MicA. *Nucleic Acids Res* **35:** 1279–1288.

Udekwu KI, Darfeuille F, Vogel J, Reimegard J, Holmqvist E, Wagner EG. 2005. Hfq-dependent regulation of OmpA synthesis is mediated by an antisense RNA. *Genes Dev* **19:** 2355–2366.

Urban JH, Vogel J. 2008. Two seemingly homologous noncoding RNAs act hierarchically to activate *glmS* mRNA translation. *PLoS Biol* **6:** e64.

Urbanowski ML, Stauffer LT, Stauffer GV. 2000. The *gcvB* gene encodes a small untranslated RNA involved in expression of the dipeptide and oligopeptide transport systems in *Escherichia coli*. *Mol Microbiol* **37:** 856–868.

Vanderpool CK, Gottesman S. 2004. Involvement of a novel transcriptional activator and small RNA in post-transcriptional regulation of the glucose phosphoenolpyruvate phosphotransferase system. *Mol Microbiol* **54:** 1076.

Vanderpool CK. 2007. Physiological consequences of small RNA-mediated regulation of glucose-phosphate stress. *Curr Opin Microbiol* **10:** 146–151.

Vogel J, Papenfort K. 2006. Small non-coding RNAs and the bacterial outer membrane. *Curr Opin Microbiol* **9:** 605–611.

Vogel J, Bartels V, Tang HH, Churakov G, Slagter-Jager JG, Huttenhofer A, Wagner EGH. 2003. RNomics in *Escherichia coli* detects new sRNA species and indicates parallel transcriptional output in bacteria. *Nucleic Acids Res* **31:** 6435–6443.

Wadler CS, Vanderpool CK. 2007. A dual function for a bacterial small RNA: SgrS performs base pairing-dependent regulation and encodes a functional polypeptide. *Proc Natl Acad Sci USA* **104:** 20454–20459.

Wagner EG, Altuvia S, Romby P. 2002. Antisense RNAs in bacteria and their genetic elements. *Adv Genet* **46:** 361–398.

Wassarman KS. 2007. 6S RNA: A regulator of transcription. *Mol Microbiol* **65:** 1425–1431.

Wassarman KM, Saecker RM. 2006. Synthesis-mediated release of a small RNA inhibitor of RNA polymerase. *Science* **314:** 1601–1613.

Watanabe T, Sugiura M, Sugita M. 1997. A novel small stable RNA, 6Sa RNA, from the cyanobacterium *Synechococcus* sp. strain PCC3601. *FEBS Lett* **416:** 302–306.

Weaver KE. 2007. Emerging plasmid-encoded antisense RNA regulated systems. *Curr Opin Microbiol* **10:** 110–116.

Wei BL, Brun-Zinkernagel AM, Simecka JW, Pruss BM, Babitzke P, Romeo T. 2001. Positive regulation of motility and *flhDC* expression by the RNA-binding protein CsrA of *Escherichia coli*. *Mol Microbiol* **40:** 245–256.

White D, Hart ME, Romeo T. 1996. Phylogenetic distribution of the global regulatory gene *csrA* among eubacteria. *Gene* **182:** 221–223.

Wilderman PJ, Sowa NA, FitzGerald DJ, FitzGerald PC, Gottesman S, Ochsner UA, Vasil ML. 2004. Identification of tandem duplicate regulatory small RNAs in *Pseudomonas aeruginosa* involved in iron homeostasis. *Proc Natl Acad Sci* **101:** 9792–9797.

Wilkomm DK, Minnerup J, Huttenhofer A, Hartmann RK. 2005. Experimental RNomics in *Aquifex aeolicus*: Identification of small non-coding RNAs and the putative 6S RNA homolog. *Nucleic Acids Res* **33:** 1949–1960.

Wurm R, Neusser T, Wagner R. 2010. 6S RNA-dependent inhibition of RNA polymerase is released by RNA-dependent synthesis of small de novo products. *Biol Chem* **391:** 187–196.

Zhang A, Wassarman KM, Rosenow C, Tjaden BC, Storz G, Gottesman S. 2003. Global analysis of small RNA and mRNA targets of Hfq. *Mol Microbiol* **50:** 1111–1124.

RNA in Defense: CRISPRs Protect Prokaryotes against Mobile Genetic Elements

Matthijs M. Jore, Stan J.J. Brouns, and John van der Oost

Laboratory of Microbiology, Wageningen University, Netherlands

Correspondence: John.vanderoost@wur.nl

SUMMARY

The CRISPR/Cas system in prokaryotes provides resistance against invading viruses and plasmids. Three distinct stages in the mechanism can be recognized. Initially, fragments of invader DNA are integrated as new spacers into the repetitive CRISPR locus. Subsequently, the CRISPR is transcribed and the transcript is cleaved by a Cas protein within the repeats, generating short RNAs (crRNAs) that contain the spacer sequence. Finally, crRNAs guide the Cas protein machinery to a complementary invader target, either DNA or RNA, resulting in inhibition of virus or plasmid proliferation. In this article, we discuss our current understanding of this fascinating adaptive and heritable defense system, and describe functional similarities and differences with RNAi in eukaryotes.

Outline

1 Introduction
2 CRISPR loci and Cas genes
3 Mode of action
4 Analogy with RNAi in eukaryotes
5 Concluding remarks
References

1 INTRODUCTION

The evolution of micro-organisms is significantly influenced both qualitatively and quantitatively by the continuous exchange of genomic material with mobile genetic elements: viruses and plasmids. Viruses are among the most abundant entities on earth (Bergh et al. 1989; Wommack and Colwell 2000) and they proliferate by a series of events: adsorption of the virion to the host's cell wall, injection of the viral genome (DNA, RNA) through the cell membrane(s), expression of viral genes, replication of the viral genome and assembly of viral protein capsids, and finally release of progeny virions (Sturino and Klaenhammer 2004). Plasmids are another main class of selfish mobile elements. After entry, plasmid DNA resides in a host, either free in the cytoplasm or as integrated sequence in the host genome. Plasmids can be transferred from donor to recipient via conjugation, making use of dedicated transfer systems (Llosa et al. 2002).

Despite the occasional gain of function as a result of horizontal gene transfer, recombination with mobile elements can also cause severe damage (disruption of either structural or regulatory regions on the host genome leads to loss of function). Additionally, phage infections can eventually lead to host cell lysis. To avoid these detrimental effects, sophisticated mechanisms have evolved to defend host organisms against nucleic acids of invading mobile elements. Several defense systems have been recognized in prokaryotes that are very different from the eukaryotic immune systems. A passive defense mechanism may act at the level of virion adsorption and/or injection of its genomic material. Spontaneous mutations in virus receptor proteins of the host can perturb virus attachment and genome injection, not affecting host fitness, for example, the maltoporin of *Escherichia coli* used by phage lambda (Hofnung et al. 1976). A well known active defense mechanism is the restriction-modification (R-M) system. Dedicated methyltransferases modify potential cleavage sites of the host DNA, preventing strand cleavage by restriction enzymes. Incoming invader DNA lacks these modifications, and is therefore a target for digestion by these endonucleases (Review, Tock and Dryden 2005). An additional mechanism that appears functionally analogous to eukaryotic apoptosis is the prokaryotic Abortive infection mechanism (Abi). This mechanism inhibits phage multiplication either by blocking the phage replication machinery, or by inhibiting host translation. This results in death of both host and virus, a sacrifice that will save the rest of the population (Chopin et al. 2005).

Recently, another defense mechanism has been discovered that is based on clusters of regularly interspersed short palindromic repeats (CRISPRs) and CRISPR-associated genes (*cas* genes). The CRISPR/Cas system can integrate nucleic acid fragments from invading mobile elements into the CRISPR locus. The CRISPR is transcribed and cleaved into short mature RNAs (crRNAs). These crRNAs specifically guide the Cas protein machinery to their complementary targets: either DNA or RNA from invading viruses or plasmids. Thus, the CRISPR/Cas system can provide the host with acquired and heritable resistance (reviewed in Sorek et al. 2008; van der Oost et al. 2009; Horvath and Barrangou 2010; Karginov and Hannon 2010; Marraffini and Sontheimer 2010a). In this article, we describe mechanistic features of the CRISPR/Cas system, and we discuss the similarities and differences with RNA interference in eukaryotes.

2 CRISPR LOCI AND CAS GENES

CRISPRs were first discovered in 1987 when a chromosomal fragment from *E. coli* K12 was sequenced (Ishino et al. 1987). Since then, many CRISPR sequences have been identified in prokaryotic genomes (for overview, see: http://crispr.u-psud.fr/crispr/CRISPRdatabase.php). CRISPRs have been detected in 48% of the sequenced bacterial genomes and in 95% of the sequenced archaeal genomes. CRISPRs are composed of a cluster of identical repetitive sequences that are separated by nonidentical spacer sequences of similar length (see a later discussion). The CRISPR array is often preceded by an AT-rich leader sequence of up to 500 base pairs (bp) (Jansen et al. 2002). The number of CRISPR loci per genome ranges from one to 20, varying in length from a few to hundreds of repeat-spacer pairs. The present record holder is a CRISPR of *Chloroflexus* sp. with 374 repeats and spacers. Twelve major types of CRISPR have been proposed, based on sequence similarity of the repeats (Kunin et al. 2007). The repeat size varies from 24 to 47 bp, whereas the spacer size ranges from 24 to 72 bp. The size of repeats and spacers are typically around 30 bp. Some repeats have palindromic sequences that encode CRISPR RNAs with potentially strong secondary structures, whereas other sequences appear to lack such structures (Fig. 1B). Each cluster correlates mainly with one Cas subtype, as discussed later. In 2005, three different research groups independently observed that at least a subset of the spacer sequences are identical to phage and plasmid DNA sequences (Bolotin et al. 2005; Mojica et al. 2005; Pourcel et al. 2005). The virus or plasmid fragment matching the spacer sequence is called the proto-spacer (Deveau et al. 2008). The observation that spacer sequences were derived from viral sequences has led to the hypothesis that the CRISPR/Cas system might be involved in prokaryotic resistance to alien nucleic acids (Reviewed, Makarova et al. 2006). The composition of

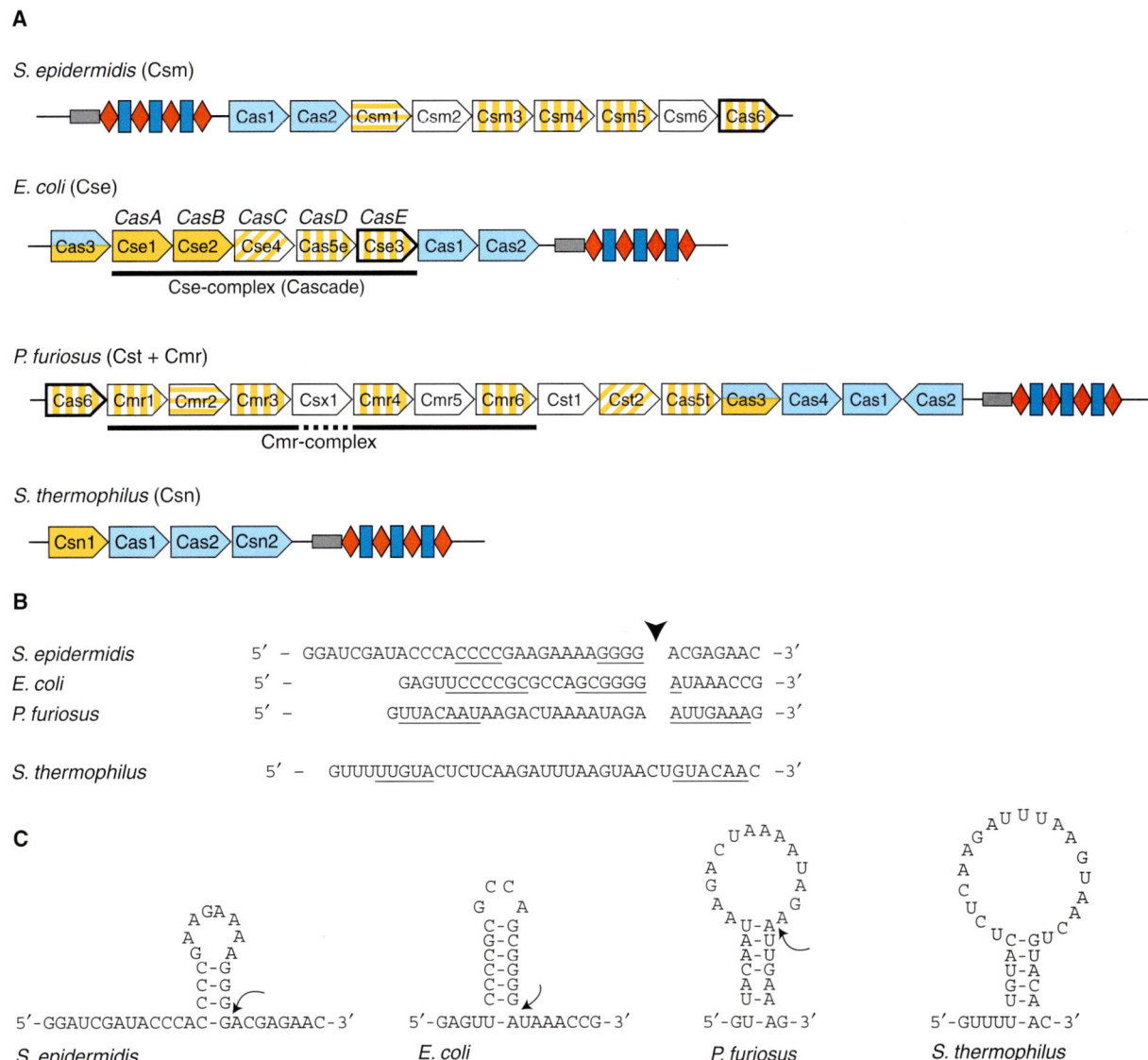

Figure 1. Overview of the four CRISPR/Cas subtypes that are described in this article. For an overview of all eight CRISPR/Cas subtypes, see Haft et al. 2005 and van der Oost et al. 2009. (A) cas gene neighborhoods in four experimentally studied organisms, each representing a different subtype indicated between brackets. CRISPRs consist of a leader (grey box), repeats (red diamonds), and spacers (blue boxes). Only a fragment of the CRISPR is shown. Genes are indicated as arrows. Blue arrows indicate genes that are (possibly) involved in spacer acquisition. Yellow arrows indicate genes that are involved in CRISPR transcription and processing and target interference. The endonucleases that cleave pre-crRNA generating crRNA are highlighted as bold arrows. Hatching patterns indicate gene similarity: RAMP genes have vertical lines, polymerase genes have horizontal lines, CasC homologs have diagonal lines, and other genes that are not related to each other are filled. Genes that encode proteins from isolated complexes (Cse-complex from *E. coli* and Cmr-complex from *Pyrococcus furiosus*) are underlined. (B) CRISPR RNA repeat sequences from each organism are given. The cleavage site is indicated by a triangle. Although the repeat sequences are different, all CRISPR RNA cleavage events generate an eight-nucleotide 5′ handle. Please note that the cleavage site in *Streptococcus thermophilus* CRISPR RNA has not been determined. Palindromic sequences are underlined. (C) Predicted secondary structures of the different CRISPR RNA repeats. Cleavage sites are indicated with an arrow. As described previously by Kunin et al., the repeat of *P. furiosus* is not likely to form a stem loop (Kunin et al. 2007).

the CRISPR is hypervariable and is rapidly shaped by extrachromosomal elements in the host's environment (Lillestol et al. 2006; Andersson and Banfield 2008; Tyson and Banfield 2008; Banfield and Young 2009; Held and Whitaker 2009; Lillestol et al. 2009). Extrachromosomal elements in turn respond by extensive gene shuffling (Andersson and Banfield 2008) or mutations (Deveau et al. 2008; Heidelberg et al. 2009; Semenova et al. 2009; van der Ploeg 2009) to escape the CRISPR defense mechanism, illustrating the ongoing battle between hosts and their predators.

A set of conserved *cas* genes can be found in close proximity of the CRISPR array. The encoded proteins were initially thought to be involved in DNA repair, because they had predicted nucleic acid related functions (Makarova et al. 2002). The link between *cas* genes and CRISPRs was made shortly thereafter (Jansen et al. 2002) and the four most conserved *cas* genes were identified. The *cas* gene products were further classified into ~45 distinct families (Haft et al. 2005). That number was later reduced to ~25 families (Makarova et al. 2006). The set of Cas proteins is composed of core proteins (Cas1-6), a diverse group of Repeat-Associated Mysterious Proteins (RAMPs), and more loosely associated Cas proteins, such as a polymerase. Based on the composition of the *cas* operons, eight Cas subtypes (Csa, Csd, Cse, Csh, Csm, Csn, Cst, and Csy) by Haft et al. (2005), or seven Cas systems (CASS1–7) by Makarova et al. (2006), have been proposed that each contain a certain set of core proteins and a subtype specific module that in most cases contains at least one RAMP (Fig. 1A). An additional subtype (Cas module RAMP, Cmr) includes many RAMP proteins and a polymerase/nuclease. This system seems to share core Cas proteins with another subtype that resides on the same genome. As discussed later, the Cmr cluster at least to some extent, resembles the Csm-subtype. The distribution of the closely related subtypes in phylogenetically distant organisms, suggests that the CRISPR/Cas system has frequently been exchanged by horizontal gene transfer between distant micro-organisms (Makarova et al. 2006; Horvath et al. 2009). This hypothesis is supported by the observation that the CRISPR/Cas can be located on plasmids (e.g., the megaplasmids from *Thermus thermophilus*). Other plasmids have been reported to contain a CRISPR locus without associated *cas* genes (e.g., the pNOB8 conjugative plasmid from *Sulfolobus solfataricus*) (Godde and Bickerton 2006).

The best conserved *cas* genes are *cas1* and *cas2* that are present in all subtypes (Haft et al. 2005). Therefore, they are suitable markers for the presence of CRISPR/Cas. The putative nuclease/integrase Cas1 (Makarova et al. 2006) has been demonstrated to be a metal dependent nuclease that cleaves ssDNA and dsDNA, generating ~80 bp DNA fragments from dsDNA. The Cas1 structure reveals a novel fold with a two-domain architecture (Wiedenheft et al. 2009). The small Cas2 protein cleaves ssRNAs in U-rich regions. Crystal structures of Cas2 from several species have been solved, revealing a ferredoxin fold, which is not common for endoribonucleases (Beloglazova et al. 2008). Cas1 has been proposed to be involved in spacer integration (Makarova et al. 2006), a prediction that is in agreement with the observation that Cas1 and Cas2 in *E. coli* are not involved in the antiviral defense stage of the mechanism when a spacer is already present in the CRISPR array (Brouns et al. 2008; Hale et al. 2009). Fusion of *cas1* and *cas4* genes in several genomes, including that of *Geobacter sulfurreducens*, suggests that Cas4, a putative RecB-like nuclease (Makarova et al. 2006), might also be involved in spacer acquisition (van der Oost et al. 2009). Cas3 is a special case, typically being a single polypeptide composed of two domains: an HD domain that has metal-dependent nuclease activity on double-stranded oligonucleotides (Aravind and Koonin 1998; Han and Krauss 2009) and a DEAD/H box helicase domain (Makarova et al. 2006). Interestingly, in the Csa-subtype the domains are separated, and in the Csy-subtype Cas3 is fused to Cas2 (Makarova et al. 2006). Cas5 and Cas6, previously annotated as core Cas proteins as well, represent a group of distantly related Cas proteins referred to as RAMPs. They appear to have similar 3D structures, and share at least a carboxy-terminal glycine-rich loop (Makarova et al. 2002). Two RAMP proteins (CasE and Cas6) have recently been demonstrated to be metal-independent endonucleases involved in the processing of CRISPR RNA (pre-crRNA), as described later (Brouns et al. 2008; Carte et al. 2008). Additionally, two types of multisubunit Cas complexes have recently been characterized. In *E. coli*, a complex is encoded by five clustered genes *cse1–4* and *cas5e* (Cas5e and Cse3 are RAMPs) and the gene products form a Cse-complex termed Cascade (<u>C</u>RISPR-<u>a</u>ssociated <u>c</u>omplex for <u>a</u>ntiviral <u>de</u>fense) (Brouns et al. 2008). A crRNA-binding Cmr-complex comprising Cmr1-6 has been isolated from *Pyrococcus furiosus* (Hale et al. 2009). An overview of experimentally determined and putative activities and structures of core Cas proteins and Cas complexes is provided in Table 1.

3 MODE OF ACTION

The CRISPR/Cas mechanism can be divided into three distinct stages. The first stage concerns the integration of nucleic acid fragments of invading mobile genetic elements as new spacers into the CRISPR locus. In the second stage, the CRISPR is transcribed as a precursor (pre-crRNA), which is subsequently cleaved by a dedicated endoribonuclease, resulting in mature crRNAs that remain associated

Table 1. Cas proteins involved in different stages.*

Stage involved	Name	Activity	Remarks
Acquisition	Cas1	DNA endonuclease (Wiedenheft et al. 2009), RNA and DNA binding (Han et al. 2009)	Crystal structures (Wiedenheft et al. 2009) (3GOD and 2YZS)
	Cas2	Ribonuclease activity (Beloglazova et al. 2008)	Crystal structures (Beloglazova et al. 2008) (2I8E, 1ZPW, 2I0X, and 2IVY)
	Cas4		RecB-like nuclease
Processing	CasE (part of Cascade), Cas6	pre-crRNA cleavage (Brouns et al. 2008; Carte et al. 2008)	Crystal structures of CasE (Ebihara et al. 2006) (1WJ9) and Cas6 (Carte et al. 2008) (3I4H)
Interference	Cse-complex (Cascade)		Comprises CasA-E. Crystal structures solved of CasB (Agari et al. 2008) (2ZCA) and CasE (see earlier discussion)
	Cas3	Nuclease activity of HD domain (Han and Krauss 2009)	Helicase, often fused to HD-domain (Makarova et al. 2006). Possibly also involved in acquisition, according to fusion to Cas2 in Csy-subtype (van der Oost et al. 2009)
	Cmr-complex	Cleavage of RNA complementary to crRNA (Hale et al. 2008)	Comprises Cmr1-6. Two structures of Cmr5 are available (Sakamoto et al. 2009) (2OEB and 2ZOP)

*Experimentally determined activities and PDB ID codes are indicated.

with a Cas protein complex. During the third and final stage, the crRNA guides the Cas complex to known, invading nucleic acids to neutralize the invader, most likely by cleavage.

3.1 Integration of New Spacers

The first experimental evidence that the CRISPR/Cas system is indeed an antiviral defense system was obtained from phage infection experiments of the lactic acid bacterium *Streptococcus thermophilus* that has a CRISPR/Cas locus of the Csn-subtype (Fig. 1A) (Barrangou et al. 2007; Deveau et al. 2008; Horvath et al. 2008). Screening for adaptation of the CRISPR locus in the surviving bacteria revealed that a subpopulation of survivors had acquired new phage-specific spacer sequences (Fig. 2). Subsequent deletion of these new spacer sequences resulted in loss of the acquired resistance, demonstrating the correlation of spacer presence and phage resistance (Barrangou et al. 2007). Comparative analysis of the spacer-targeted region of the viral genome revealed a sequence motif called CRISPR motif or proto-spacer adjacent motif (PAM) (i.e., NNAGAAW) downstream of the proto-spacer. The phages responded to the resistance of the host by mutations in the proto-spacer, but also by mutations in the motif (Deveau et al. 2008), illustrating the constant battle between phages and bacteria. Spacers that did not have a perfect PAM were integrated as well, but these were always accompanied by spacers with a perfect PAM (Deveau et al. 2008), the latter probably being essential for resistance. The data suggests that the PAM is crucial for target interference, but not for integration of new spacers. Genomic analysis of proto-spacers revealed the presence of slightly different PAMs in many extrachromosomal elements, and the conservation of PAM sequences correlates well with the CRISPR repeat types, and thus cas gene subtypes (Mojica et al. 2009).

The spacers in *S. thermophilus* were integrated at the leader proximal end of the CRISPR locus, suggesting that a CRISPR array is a chronological record of past phage infections. The role of the leader sequence is not exactly known, but it is possibly required for repeat duplication and/or spacer integration. Polarity of the CRISPR integration had been previously predicted by Lillestol and coworkers after comparing the CRISPRs of two *Sulfolobus solfataricus* strains with major variation at the leader-side

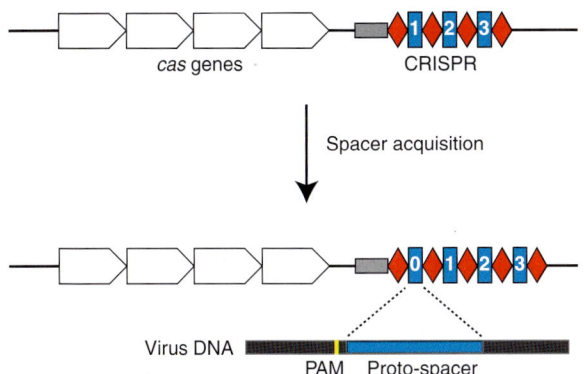

Figure 2. Integration of a new spacer. A new spacer is acquired at the leader proximal side of the CRISPR during virus infection, resulting in resistance. The CRISPR consist of a leader (grey box), repeats (red diamonds), and spacers (blue boxes). The newly acquired spacer is numbered 0 and matches the sequence of the virus (proto-spacer). The proto-spacer adjacent motif (PAM) is located downstream or upstream the proto-spacer.

of the CRISPR (Lillestol et al. 2006). Next to acquired resistance by spacer integration in *S. thermophilus*, the above mentioned study also showed that *cas* genes are involved in CRISPR-mediated defense. Disruption of the *csn1* gene resulted in loss of viral resistance. An interrupted *csn2* gene did not lead to loss of resistance, but rather to a disrupted ability to integrate new spacers. This does not only show that both *cas* genes are involved in resistance, but also that they play a role at different stages. *Csn2* is apparently needed for integration of new spacers, but it is restricted to the Csn-subtype. In *Wollinella succinogenes*, which carries the same subtype, *csn2* is absent and seems to be replaced by *cas4* (van der Oost et al. 2009). Thus, despite the unknown role of Csn2 in spacer integration, it is likely that it is replaced by Cas proteins with analogous functions in other Cas subtypes, Cas4 being a likely candidate. Csn1 may have a function analogous to Cascade/Cas3 (as discussed later, Cas3 is also involved in target interference) (van der Oost et al. 2009).

3.2 CRISPR Transcription and Processing

Before it was hypothesized that the CRISPR/Cas system protects the host against invading nucleic acids, studies on short RNAs in the archaea *Archaeoglobus fulgidus* and *S. solfataricus* had already indicated that CRISPRs are actively transcribed and the pre-crRNA processed (Tang et al. 2002; Tang et al. 2005). Although studies on transcription and processing in *S. solfataricus* showed that transcription can be bidirectional (Lillestol et al. 2006; Lillestol et al. 2009), most other studies have reported unidirectional transcription from the leader proximal side (Brouns et al. 2008; Hale et al. 2008; Marraffini and Sontheimer 2008; Semenova et al. 2009). It is therefore anticipated that in most cases a single promoter at the leader side controls transcription of a CRISPR locus. It has recently been shown that in the case of *E. coli* K12 the promoter indeed resides in the leader region (Pul et al. 2010). CRISPR transcription and processing has been studied in more detail for this *E. coli* K12 system, which consists of 8 *cas* genes upstream of a CRISPR (Fig. 1A). The Cas proteins were overexpressed in *E. coli* BL21, a strain that lacks endogenous *cas* genes (Studier et al. 2009). Pull down analysis revealed the presence of Cascade, a protein complex that contains five different subunits, CasABCDE. Northern blot analysis of crRNAs in *E.coli* K12 revealed short RNAs, the size of which corresponded to approximately one spacer and one repeat. Omitting the *cas* genes one-by-one identified CasE, a RAMP protein, as a potential candidate for pre-crRNA processing. This finding was confirmed by in vitro activity assays, showing that CasE is a metal-independent endoribonuclease that specifically cleaves a precursor (pre-crRNA) into mature crRNAs. After processing, these crRNAs remain tightly bound to the Cascade complex. Cloning and subsequent sequence analysis revealed more crRNA products derived from the leader proximal end of the CRISPR (Brouns et al. 2008). Independent analysis of crRNA from *P. furiosus* also uncovered more products from the leader proximal side of the CRISPR, and fewer from the distal end. This can be explained by premature termination of CRISPR transcription (Hale et al. 2008). Sequence analysis of cloned *E. coli* crRNA products revealed that the product contains eight nucleotides of the repeat termed the 5′ handle, the spacer, and a large part of the next repeat including a part of the stem-loop termed the 3′ handle (Brouns et al. 2008). It has been proposed that these handles, the conserved parts of the crRNAs, are bound by subunits of Cascade (Brouns et al. 2008). Based on sequence conservation in CasE and the crystal structure of a CasE homolog from *Thermus thermophilus* (Ebihara et al. 2006), the histidine residue at position 20 has been predicted to be a residue involved in endonuclease activity. Indeed, the activity was lost when the histidine was substituted by an alanine. In vivo analysis showed that the mutation resulted in loss of resistance, thus showing that pre-crRNA cleavage is a mechanistic requirement.

Despite the fact that Cas6 from *P. furiosus* shares low sequence identity (except for the carboxy-terminal glycine-rich loop, a common feature of RAMP proteins), its structure, a duplicated ferredoxin fold, is surprisingly similar to CasE from *T. thermophilus* (Fig. 3) (van der Oost et al. 2009). Like CasE, Cas6 displays metal independent endoribonuclease activity (Carte et al. 2008). Although the folded secondary structure of the repeat RNA of *P. furiosus* is debatable (Kunin et al. 2007), its crRNA cleavage product also contains an eight-nucleotide 5′ handle (psi-tag) (Fig. 1B and C). Unlike the proposed single histidine site in CasE (Brouns et al. 2008), a potential catalytic triad with a histidine, tyrosine, and a lysine residue is present in Cas6 (Fig. 3) (Carte et al. 2008). This predicted catalytic site is structurally similar to that of tRNA splicing enzymes. Analysis of the cleaved pre-crRNA products revealed that cleavage occurs at the 3′ side of the phosphodiester bond, generating a 5′ end hydroxyl group and a 2′, 3′ end cyclic phosphate group, analogous to tRNA splicing enzymes (Calvin and Li 2008; Carte et al. 2008). Both CasE and Cas6 probably cleave the pre-crRNA following a general acid-base hydrolysis mechanism (Fig. 3c). Besides its endoribonucleolytic activity, Cas6 binds the 3′ handle of the crRNA. However, in *P. furiosus*, the endonucleolytic product is further trimmed to active mature crRNAs lacking a 3′ handle (Hale et al. 2008; Hale et al. 2009), whereas mature crRNAs in *E. coli* do contain a 3′ handle. Another difference is the fact that CasE remains part of the Cascade

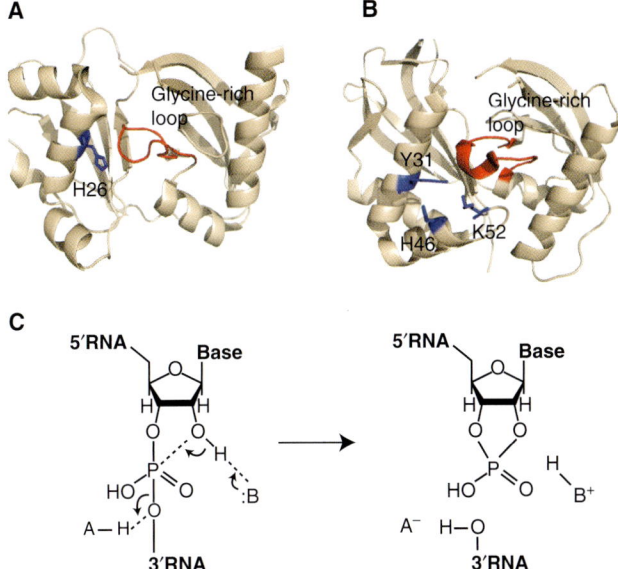

Figure 3. The catalytic sites of CasE and Cas6, and the proposed reaction mechanism of pre-crRNA cleavage. (A) Proposed catalytic site of CasE from *T. thermophilus* showing the conserved histidine residue (H26) and the glycine-rich carboxy-terminal loop. The catalytic site of Cas6 from *P. furiosus* (B) contains a catalytic triad of tyrosine (Y31), histidine (H46) and lysine (K52) and a glycine-rich carboxy-terminal loop. The loop and the overall duplicated ferredoxin fold are conserved among CasE and Cas6. Pre-crRNA cleavage might follow a general acid–base hydrolysis mechanism (C). A base (B) draws a proton from the 2′OH of the ribose ring. A subsequent nucleophilic attack on the phosphorus atom is simultaneously compensated by the acid (A) that donates a proton to the leaving 3′RNA. The tyrosine residue of Cas6 is proposed to be the base and the histidine the acid residue (Carte et al. 2008). In CasE the histidine and a water molecule might be the catalytic residues. Pictures in (A) and (B) are generated with pymol (www.pymol.org), potential catalytic residues are depicted in blue; the glycine-rich loop is depicted in red. Coordinates were obtained from the Protein Data Bank (www.pdb.org).

complex (Brouns et al. 2008), whereas Cas6 has not been identified as part of the pyrococcal Cmr-complex (see the following) (Hale et al. 2009); apparently cleaved crRNAs in *P. furiosus* are transferred to the Cmr-complex. Besides the Cmr-complex, an additional Cst-complex might be present in *P. furiosus*, that is encoded by three *cst* genes downstream of the *cmr* module (Fig. 1A). Overall, it has become clear that despite some mechanistic similarity, also substantial differences exists between the different CRISPR/Cas subtypes.

3.3 Target Interference

Although it was initially hypothesized that the CRISPR/Cas system would target alien RNA, analogous to RNAi (Makarova et al. 2006), several studies have indicated that the target of the CRISPR system rather is invading DNA.

The first observation supporting this hypothesis was made in virus infection studies with *S. thermophilus*, revealing that spacer sequences corresponding to either the coding or noncoding strand were integrated (Barrangou et al. 2007). The CRISPR locus has been demonstrated to be transcribed only from the leader proximal side in *E. coli*, *Staphyloccuus epidermidis* and *P. furiosus* (Brouns et al. 2008; Hale et al. 2008; Marraffini and Sontheimer 2008); to date the only exception appears to be the case of *S. solfataricus* (Lillestol et al. 2006; Lillestol et al. 2009). A consequence of monodirectional CRISPR transcription would be that generated crRNAs have to be complementary to the mRNA of the virus. Only spacers from one strand should be incorporated in case of an antisense RNA mechanism. The observation that this is generally not the case suggests that DNA is the target. Additional evidence that DNA is the target was provided by a study on an engineered *E. coli* Cas system in which artificial spacers were unidirectionally transcribed. Generating crRNAs complementary to both the coding strand and the template strand were successful in inhibiting virus proliferation (Brouns et al. 2008). Furthermore, a study on plasmid conjugation in *S. epidermidis* convincingly proved that DNA is being targeted (Marraffini and Sontheimer 2008). A natural spacer from the CRISPR of *S. epidermidis* has a perfect match with a gene of a conjugative plasmid. This spacer confers resistance and prevents conjugation of this plasmid. A self-splicing intron was inserted into the center of the proto-spacer sequence in the plasmid. After conjugation and subsequent transcription, this intron is spliced, generating a mature mRNA that contains a fully complementary sequence to the crRNA. The plasmid was able to escape the CRISPR/Cas system, showing that mRNA is not being targeted; hence, the target must be DNA that could not be recognized by the crRNA because of the intron DNA sequence interrupting the proto-spacer. This finding is confirmed by an additional experiment in which a fragment of the targeted gene is inserted in a plasmid, in both orientations. The fragment is not essential for propagation of the plasmid. Nevertheless, the inserted fragment, in both orientations, dramatically decreased the transformation efficiencies, again indicating that DNA is being targeted. Recently, an exceptional CRISPR/Cas system has been described, i.e., the Cmr-subtype from *P. furiosus* that appears to target and degrade RNA (discussed later). For this system, however, no natural targets have been found yet.

Most biochemical and mechanistic information on the pre-crRNA and target interference is derived from the earlier mentioned *E. coli* and *P. furiosus* model systems. Requirements for CRISPR-based resistance were determined in *E. coli* BL21. Artificial CRISPRs were designed to target four different genes of phage lambda. Overexpression of

Cascade and CRISPR RNA was sufficient to yield mature crRNAs that were bound and protected by Cascade (Brouns et al. 2008). No resistance was observed when the CRISPRs targeting phage lambda were coexpressed with Cascade only. However, coexpression of this crRNA-loaded Cascade with the Cas3 protein did result in a dramatic increase of resistance towards phage lambda infection. The role of the two-domain protein Cas3 remains to be elucidated, but it is tempting to speculate that the HD nuclease domain cleaves targeted DNA. This Cse-subtype from *E. coli* K12 targets DNA, but how does it discriminate between self DNA (namely the CRISPR that contains a spacer complementary to the crRNA) and nonself DNA (the invader)? A recent study in *S. epidermidis* revealed that the flanking sequence of the proto-spacer is crucial for self versus nonself DNA recognition, and thus for interference (Marraffini and Sontheimer 2010b). It was shown that a targeted conjugative plasmid could bypass host resistance by the CRISPR/Cas system if three bases downstream of the proto-spacer were complementary to the CRISPR repeat sequence (Marraffini and Sontheimer 2010b). In other subtypes, however, the PAM might help prevent autoimmunity by being present in the targeted invading DNA sequence, and absent from the CRISPR DNA sequence. The PAM determines if invading nucleic acids are being targeted in the case of *S. thermophilus*, as can be deduced from phages that have mutated their PAM sequence and thus can escape the CRISPR/Cas system (Deveau et al. 2008). The two mechanisms of self versus non-self DNA recognition appear to be fundamentally different. In *S. epidermidis* the potential of the downstream sequence of the proto-spacer to basepair with the CRISPR repeat determines whether the DNA is being targeted, while in, for example, *S. thermophilus* the PAM determines whether the DNA is being targeted. To bypass immunity in *S. epidermidis*, the invader can only mutate its flanking nucleic acids to a sequence complementary to the repeat DNA, while in *S. thermophilus*, the invader can mutate its PAM to any other sequence.

Interestingly, the Cas-system from *P. furiosus* has recently been reported to be capable of interfering with target RNA rather than DNA (Hale et al. 2009). With native Northern blot analysis, a Cmr-type protein complex was identified that forms a stable interaction of crRNAs (psiRNA) (Hale et al. 2008; Hale et al. 2009). The crRNAs are 39 and 45 nucleotides in length, containing an identical 5′ handle but different 3′ ends. The RNP complex isolated from *Pyrococcus* comprised six distinct Cas proteins: Cmr1–6. Cas6 was not part of the complex (Hale et al. 2009), unlike its functional analog CasE that is a core subunit of the Cascade complex (Cse-complex) in *E. coli* (Brouns et al. 2008). The isolated RNP complex cleaved complementary RNA but not ssDNA. The 39 and 45 nucleotide-long crRNAs resulted in two different cleavage sites in the target mRNA. Both cleaved 14 nucleotides upstream from the 3′ end of the crRNA, suggesting a molecular ruler mechanism for cleavage (Fig. 4B). Reconstitution of the RNP complex from purified subunits revealed that all except Cmr5 are essential for target RNA cleavage (Hale et al. 2008; Hale et al. 2009). The Cmr2 protein that is part of the Cmr-complex contains a PALM (polymerase) domain, fused to a HD-nuclease domain. Whereas the biological function of the polymerase is not known, the HD-nuclease might be responsible for degradation of the target RNA. At least two proteins from the Cmr-complex are related to proteins that are encoded by the *csm* module in *S. epidermidis* (Fig. 1A). The polymerase Cmr2 is related to Csm1 and the RAMP protein Cmr4 is related to Csm3 (Haft et al. 2005). The Csm-type system from *S. epidermidis* has been demonstrated to target DNA in vivo, which is in contrast to the observed in vitro RNA cleavage activity of the Cmr-complex. Future biochemical and in vivo analyses are required to resolve this apparent contradiction.

4 ANALOGY WITH RNAi IN EUKARYOTES

The function and biogenesis of crRNAs and the mechanism of target interference display striking analogies with small regulatory RNAs in eukaryotes. The eukaryotic regulatory RNAs can be divided into three major groups: endogenous miRNAs that generally silence host gene expression, piRNAs that silence transposable elements in animal germ cells, and siRNAs that can be involved in viral RNA silencing. In short, siRNAs are derived from dsRNA that is randomly cleaved by Dicer, generating fragments with a 3′ dinucleotide overhang. Dicer subsequently binds the overhang and cleaves ~20 bases away from the first cleavage site following a molecular ruler mechanism, generating short dsRNAs with 3′ overhangs and 5′ phosphates (Bernstein et al. 2001; Macrae et al. 2006; MacRae and Doudna 2007). These dsRNAs are transferred to the Argonaute protein in the RNA induced silencing complex (RISC) that also consists of Dicer and a dsRNA binding protein; the latter determines in which orientation the dsRNA molecule is loaded onto Argonaute (Tomari et al. 2004). Argonaute recognizes the passenger strand and degrades it, retaining the guide strand (Rand et al. 2005). When RISC encounters a perfectly complementary RNA, it can interact with it by Watson-Crick base pairing. Argonaute subsequently cleaves the complementary target RNA strand, 10 nucleotides away from the 5′ end of the guide, after which the target fragments are released and the RISC complex is recycled for a new target degradation event, eventually resulting in silencing the virus (Baulcombe 2004) (Fig. 4C).

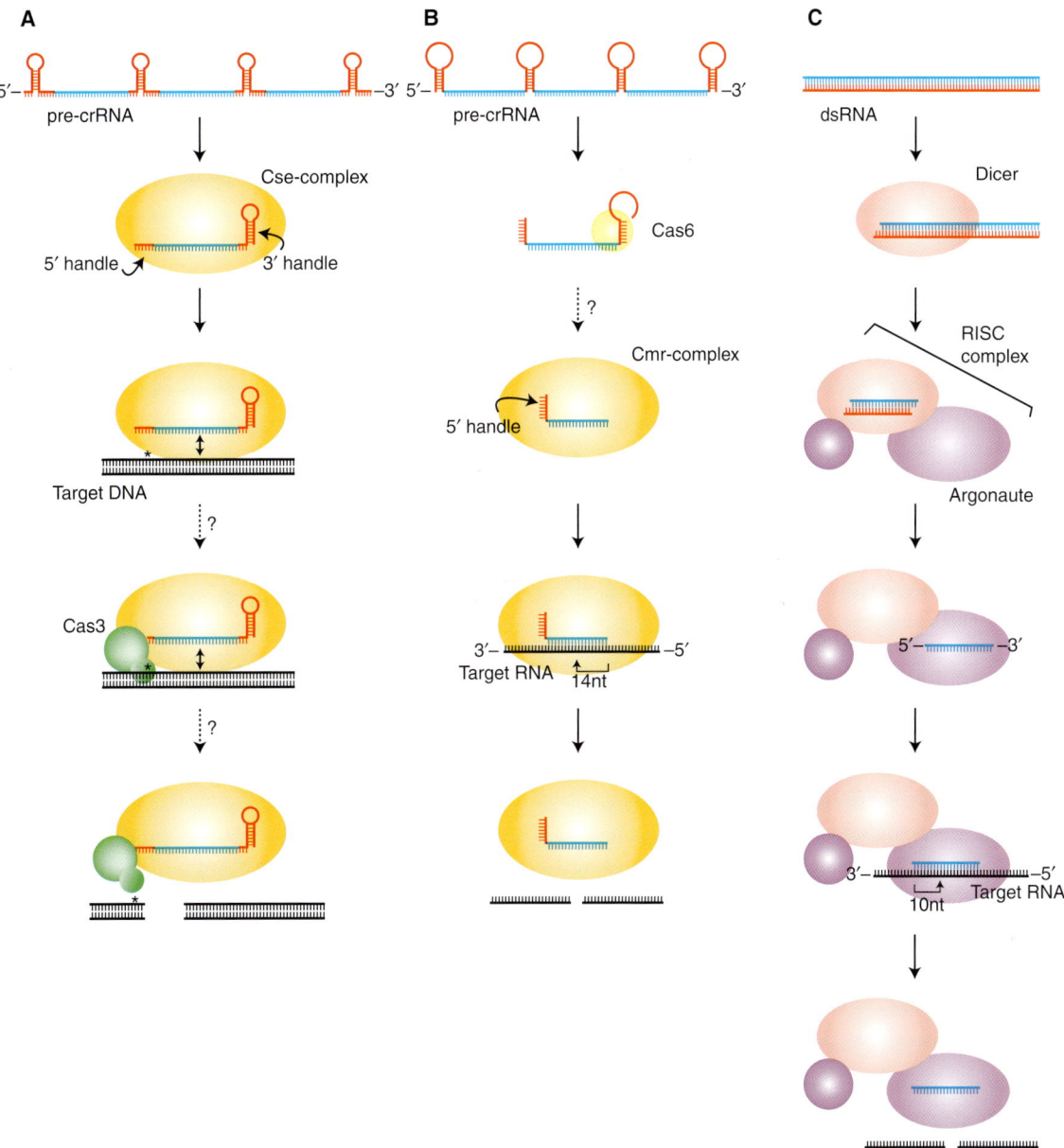

Figure 4. Antiviral DNA and RNA silencing pathways in prokaryotes and eukaryotes. (*A*) crRNA mediated DNA silencing pathway in *E. coli*. pre-crRNA is cleaved by the CasE subunit of Cascade (Cse-complex) and the mature crRNA remains bound to Cascade. How Cascade, assisted by Cas3, recognizes and neutralizes invading DNA remains to be elucidated. (*B*) crRNA mediated RNA silencing pathway in *P. furiosus*. Pre-crRNA is cleaved by Cas6 and then further trimmed to generate crRNAs of two different lengths. These crRNAs are bound by the Cmr-complex. This loaded Cmr-complex specifically binds viral RNA and cleaves the complementary strand 14 nucleotides away from the 3′ end of the crRNA. This pathway shares functional analogies with siRNA mediated antiviral resistance in eukaryotes. (*C*) siRNAs are generated from viral dsRNA by dicer. The first (random) cleavage event by dicer generates dsRNA with a 3′ dinucleotide overhang. The second cleavage by dicer takes place 20–25 bases away from the overhang, generating short dsRNAs. The dsRNA is transferred to the Argonaute protein of the RISC complex and the passenger strand is removed. The retained guide strand can basepair with a complementary viral mRNA molecule, followed by a cleavage of the scissile bond between the 10th and 11th base from the 3′ end of the guide strand. The cleaved target RNA dissociates and the recycled RISC can be used in a second round of RNA binding and cleavage. Please note that dashed arrows indicate processes that are based on hypotheses.

Although sequence comparison of the prokaryotic CRISPR/Cas and the eukaryotic siRNA system indicates they are phylogenetically unrelated (Makarova et al. 2006), they share some functional and mechanistic analogies. Both crRNA and siRNA are derived from large RNA precursors. Furthermore, both the Cmr-complex from *P. furiosus* and RISC complex specifically bind their target RNA by base pairing and degrade it by a molecular ruler mechanism. The major difference between the two systems is that crRNAs are transcribed from the host chromosome and are part of an adaptive immune system, whereas antiviral siRNAs are derived from the invaders and are part of an innate immune system. Moreover, the Cse- and Csm-type CRISPR/Cas systems target DNA rather than RNA. The RNAi pathways in which Dicer and Argonaute are the key players are restricted to eukaryotes. The only conserved component among prokaryotes and eukaryotes is the Argonaute protein. The ones from prokaryotes have been useful models in crystallization studies (Jinek and Doudna 2009; Wang et al. 2009). It has recently been hypothesized that prokaryotic Argonaute proteins are involved in yet another DNA-mediated antiviral defense mechanism (Makarova et al. 2009).

5 CONCLUDING REMARKS

The adaptive and heritable antiviral and antiplasmid CRISPR/Cas system described here is of particular interest because of potential applications in research (silencing of genes in prokaryotes), in medical diagnostics (strain typing by comparison of CRISPR spacer sequences), and industry (development of phage resistant strains) (Sorek et al. 2008). Although the rapidly expanding research field on CRISPR/Cas yields a steady gain of insight, many questions on this fascinating defense system still remain to be answered. One poorly understood aspect is the mechanism of spacer integration. A prime candidate to be involved in this process is the Cas1 protein, because it is very well conserved among different subtypes (Jansen et al. 2002; Haft et al. 2005; Makarova et al. 2006; Marraffini and Sontheimer 2009), it has nuclease activity (Wiedenheft et al. 2009), and it is not necessary for crRNA processing or target interference (Brouns et al. 2008). Lactic acid bacteria such as *S. thermophilus* seem to be the best organisms to study this process because it is the only class of organisms to date where spacer integration has been observed in laboratory experiments (Barrangou et al. 2007). Significant advances have been made as far as CRISPR transcription and processing are concerned. Recently, a possible regulator, Heat-stable Nucleoid Structuring protein (H-NS), has been identified in *E. coli* K12. This regulator possibly represses transcription of the CRISPR and *cascade* genes by binding to their promoter regions (Pul et al. 2010). This poses an interesting lead for future research on how the system is regulated, although this likely differs substantially in other organisms. The biogenesis of crRNAs has been unraveled in more detail but it is unknown how these mature crRNAs guide the Cas protein machinery to their complementary target. Some first light has been shed on how the Cmr-subtype of *P. furiosus* degrades complementary RNA. But how other subtypes, such as the Csm- and Cse-subtype, trace and target complementary incoming DNA remains to be answered. Major breakthroughs with answers to these questions are to be expected in the near future.

ACKNOWLEDGMENTS

The authors are financially supported by a Vici grant to J.v.d.O. and a Veni grant to S.J.J.B. from the Dutch Organisation for Scientific Research (Nederlandse Organisatie voor Wetenschappelijk Onderzoek).

REFERENCES

Agari Y, Yokoyama S, Kuramitsu S, Shinkai A. 2008. X-ray crystal structure of a CRISPR-associated protein, Cse2, from Thermus thermophilus HB8. *Proteins* **73:** 1063–1067.

Andersson AF, Banfield JF. 2008. Virus population dynamics and acquired virus resistance in natural microbial communities. *Science* **320:** 1047–1050.

Aravind L, Koonin EV. 1998. The HD domain defines a new superfamily of metal-dependent phosphohydrolases. *Trends Biochem Sci* **23:** 469–472.

Banfield JF, Young M. 2009. Microbiology. Variety–the splice of life–in microbial communities. *Science* **326:** 1198–1199.

Barrangou R, Fremaux C, Deveau H, Richards M, Boyaval P, Moineau S, Romero DA, Horvath P. 2007. CRISPR provides acquired resistance against viruses in prokaryotes. *Science* **315:** 1709–1712.

Baulcombe D. 2004. RNA silencing in plants. *Nature* **431:** 356–363.

Beloglazova N, Brown G, Zimmerman MD, Proudfoot M, Makarova KS, Kudritska M, Kochinyan S, Wang S, Chruszcz M, Minor W, et al. 2008. A novel family of sequence-specific endoribonucleases associated with the clustered regularly interspaced short palindromic repeats. *J Biol Chem* **283:** 20361–20371.

Bergh O, Borsheim KY, Bratbak G, Heldal M. 1989. High abundance of viruses found in aquatic environments. *Nature* **340:** 467–468.

Bernstein E, Caudy AA, Hammond SM, Hannon GJ. 2001. Role for a bidentate ribonuclease in the initiation step of RNA interference. *Nature* **409:** 363–366.

Bolotin A, Quinquis B, Sorokin A, Ehrlich SD. 2005. Clustered regularly interspaced short palindrome repeats (CRISPRs) have spacers of extrachromosomal origin. *Microbiology* **151:** 2551–2561.

Brouns SJ, Jore MM, Lundgren M, Westra ER, Slijkhuis RJ, Snijders AP, Dickman MJ, Makarova KS, Koonin EV, van der Oost J. 2008. Small CRISPR RNAs guide antiviral defense in prokaryotes. *Science* **321:** 960–964.

Calvin K, Li H. 2008. RNA-splicing endonuclease structure and function. *Cell Mol Life Sci* **65:** 1176–1185.

Carte J, Wang R, Li H, Terns RM, Terns MP. 2008. Cas6 is an endoribonuclease that generates guide RNAs for invader defense in prokaryotes. *Genes Dev* **22:** 3489–3496.

Chopin MC, Chopin A, Bidnenko E. 2005. Phage abortive infection in lactococci: variations on a theme. *Curr Opin Microbiol* **8:** 473–479.

Deveau H, Barrangou R, Garneau JE, Labonte J, Fremaux C, Boyaval P, Romero DA, Horvath P, Moineau S. 2008. Phage response to CRISPR-encoded resistance in *Streptococcus thermophilus*. *J Bacteriol* 190: 1390–1400.

Ebihara A, Yao M, Masui R, Tanaka I, Yokoyama S, Kuramitsu S. 2006. Crystal structure of hypothetical protein TTHB192 from *Thermus thermophilus* HB8 reveals a new protein family with an RNA recognition motif-like domain. *Protein Sci* 15: 1494–1499.

Godde JS, Bickerton A. 2006. The repetitive DNA elements called CRISPRs and their associated genes: evidence of horizontal transfer among prokaryotes. *J Mol Evol* 62: 718–729.

Haft DH, Selengut J, Mongodin EF, Nelson KE. 2005. A guild of 45 CRISPR-associated (Cas) protein families and multiple CRISPR/Cas subtypes exist in prokaryotic genomes. *PLoS Comput Biol* 1: e60.

Hale C, Kleppe K, Terns RM, Terns MP. 2008. Prokaryotic silencing (psi)RNAs in *Pyrococcus furiosus*. *RNA* 14: 2572–2579.

Hale CR, Zhao P, Olson S, Duff MO, Graveley BR, Wells L, Terns RM, Terns MP. 2009. RNA-guided RNA cleavage by a CRISPR RNA-Cas protein complex. *Cell* 139: 945–956.

Han D, Krauss G. 2009. Characterization of the endonuclease SSO2001 from *Sulfolobus solfataricus* P2. *FEBS Lett* 583: 771–776.

Han D, Lehmann K, Krauss G. 2009. SSO1450–a CAS1 protein from *Sulfolobus solfataricus* P2 with high affinity for RNA and DNA. *FEBS Lett* 583: 1928–1932.

Heidelberg JF, Nelson WC, Schoenfeld T, Bhaya D. 2009. Germ warfare in a microbial mat community: CRISPRs provide insights into the co-evolution of host and viral genomes. *PLoS One* 4: e4169.

Held NL, Whitaker RJ. 2009. Viral biogeography revealed by signatures in *Sulfolobus islandicus* genomes. *Environ Microbiol* 11: 457–466.

Hofnung M, Jezierska A, Braun-Breton C. 1976. lamB mutations in *E. coli* K12: growth of λ host range mutants and effect of nonsense suppressors. *Mol Gen Genet* 145: 207–213.

Horvath P, Barrangou R. 2010. CRISPR/Cas, the immune system of bacteria and archaea. *Science* 327: 167–170.

Horvath P, Coute-Monvoisin AC, Romero DA, Boyaval P, Fremaux C, Barrangou R. 2009. Comparative analysis of CRISPR loci in lactic acid bacteria genomes. *Int J Food Microbiol* 131: 62–70.

Horvath P, Romero DA, Coute-Monvoisin AC, Richards M, Deveau H, Moineau S, Boyaval P, Fremaux C, Barrangou R. 2008. Diversity, activity, and evolution of CRISPR loci in *Streptococcus thermophilus*. *J Bacteriol* 190: 1401–1412.

Ishino Y, Shinagawa H, Makino K, Amemura M, Nakata A. 1987. Nucleotide sequence of the iap gene, responsible for alkaline phosphatase isozyme conversion in *Escherichia coli*, and identification of the gene product. *J Bacteriol* 169: 5429–5433.

Jansen R, Embden JD, Gaastra W, Schouls LM. 2002. Identification of genes that are associated with DNA repeats in prokaryotes. *Mol Microbiol* 43: 1565–1575.

Jinek M, Doudna JA. 2009. A three-dimensional view of the molecular machinery of RNA interference. *Nature* 457: 405–412.

Karginov FV, Hannon GJ. 2010. The CRISPR system: small RNA-guided defense in bacteria and archaea. *Mol Cell* 37: 7–19.

Kunin V, Sorek R, Hugenholtz P. 2007. Evolutionary conservation of sequence and secondary structures in CRISPR repeats. *Genome Biol* 8: R61.

Lillestol RK, Redder P, Garrett RA, Brugger K. 2006. A putative viral defence mechanism in archaeal cells. *Archaea* 2: 59–72.

Lillestol RK, Shah SA, Brugger K, Redder P, Phan H, Christiansen J, Garrett RA. 2009. CRISPR families of the crenarchaeal genus Sulfolobus: bidirectional transcription and dynamic properties. *Mol Microbiol* 72: 259–272.

Llosa M, Gomis-Ruth FX, Coll M, de la Cruz Fd F. 2002. Bacterial conjugation: a two-step mechanism for DNA transport. *Mol Microbiol* 45: 1–8.

MacRae IJ, Doudna JA. 2007. Ribonuclease revisited: Structural insights into ribonuclease III family enzymes. *Curr Opin Struct Biol* 17: 138–145.

Macrae IJ, Zhou K, Li F, Repic A, Brooks AN, Cande WZ, Adams PD, Doudna JA. 2006. Structural basis for double-stranded RNA processing by Dicer. *Science* 311: 195–198.

Makarova KS, Aravind L, Grishin NV, Rogozin IB, Koonin EV. 2002. A DNA repair system specific for thermophilic Archaea and bacteria predicted by genomic context analysis. *Nucleic Acids Res* 30: 482–496.

Makarova KS, Grishin NV, Shabalina SA, Wolf YI, Koonin EV. 2006. A putative RNA-interference-based immune system in prokaryotes: Computational analysis of the predicted enzymatic machinery, functional analogies with eukaryotic RNAi, and hypothetical mechanisms of action. *Biol Direct* 1: 7.

Makarova KS, Wolf YI, van der Oost J, Koonin EV. 2009. Prokaryotic homologs of Argonaute proteins are predicted to function as key components of a novel system of defense against mobile genetic elements. *Biol Direct* 4: 29.

Marraffini LA, Sontheimer EJ. 2008. CRISPR interference limits horizontal gene transfer in staphylococci by targeting DNA. *Science* 322: 1843–1845.

Marraffini LA, Sontheimer EJ. 2009. Invasive DNA, chopped and in the CRISPR. *Structure* 17: 786–788.

Marraffini LA, Sontheimer EJ. 2010a. CRISPR interference: RNA-directed adaptive immunity in bacteria and archaea. *Nat Rev Genet* 11: 181–190.

Marraffini LA, Sontheimer EJ. 2010b. Self versus non-self discrimination during CRISPR RNA-directed immunity. *Nature* 463: 568–571.

Mojica FJ, Diez-Villasenor C, Garcia-Martinez J, Almendros C. 2009. Short motif sequences determine the targets of the prokaryotic CRISPR defence system. *Microbiology* 155: 733–740.

Mojica FJ, Diez-Villasenor C, Garcia-Martinez J, Soria E. 2005. Intervening sequences of regularly spaced prokaryotic repeats derive from foreign genetic elements. *J Mol Evol* 60: 174–182.

Pourcel C, Salvignol G, Vergnaud G. 2005. CRISPR elements in Yersinia pestis acquire new repeats by preferential uptake of bacteriophage DNA, and provide additional tools for evolutionary studies. *Microbiology* 151: 653–663.

Pul U, Wurm R, Arslan Z, Geissen R, Hofmann N, Wagner R. 2010. Identification and characterization of E. coli CRISPR-cas promoters and their silencing by H-NS. *Mol Microbiol* 75: 1495–1512.

Rand TA, Petersen S, Du F, Wang X. 2005. Argonaute2 cleaves the antiguide strand of siRNA during RISC activation. *Cell* 123: 621–629.

Sakamoto K, Agari Y, Agari K, Yokoyama S, Kuramitsu S, Shinkai A. 2009. X-ray crystal structure of a CRISPR-associated RAMP superfamily protein, Cmr5, from Thermus thermophilus HB8. *Proteins* 75: 528–532.

Semenova E, Nagornykh M, Pyatnitskiy M, Artamonova II, Severinov K. 2009. Analysis of CRISPR system function in plant pathogen *Xanthomonas oryzae*. *FEMS Microbiol Lett* 296: 110–116.

Sorek R, Kunin V, Hugenholtz P. 2008. CRISPR–a widespread system that provides acquired resistance against phages in bacteria and archaea. *Nat Rev Microbiol* 6: 181–186.

Studier FW, Daegelen P, Lenski RE, Maslov S, Kim JF. 2009. Understanding the differences between genome sequences of *Escherichia coli* B strains REL606 and BL21(DE3) and comparison of the *E. coli* B and K-12 genomes. *J Mol Biol* 394: 653–680.

Sturino JM, Klaenhammer TR. 2004. Bacteriophage defense systems and strategies for lactic acid bacteria. *Adv Appl Microbiol* 56: 331–378.

Tang TH, Bachellerie JP, Rozhdestvensky T, Bortolin ML, Huber H, Drungowski M, Elge T, Brosius J, Huttenhofer A. 2002. Identification of 86 candidates for small non-messenger RNAs from the archaeon *Archaeoglobus fulgidus*. *Proc Natl Acad Sci* 99: 7536–7541.

Tang TH, Polacek N, Zywicki M, Huber H, Brugger K, Garrett R, Bachellerie JP, Huttenhofer A. 2005. Identification of novel non-coding RNAs as potential antisense regulators in the archaeon *Sulfolobus solfataricus*. *Mol Microbiol* 55: 469–481.

Tock MR, Dryden DT. 2005. The biology of restriction and anti-restriction. *Curr Opin Microbiol* 8: 466–472.

Tomari Y, Matranga C, Haley B, Martinez N, Zamore PD. 2004. A protein sensor for siRNA asymmetry. *Science* 306: 1377–1380.

Tyson GW, Banfield JF. 2008. Rapidly evolving CRISPRs implicated in acquired resistance of microorganisms to viruses. *Environ Microbiol* **10:** 200–207.

van der Oost J, Jore MM, Westra ER, Lundgren M, Brouns SJ. 2009. CRISPR-based adaptive and heritable immunity in prokaryotes. *Trends Biochem Sci* **34:** 401–407.

van der Ploeg JR. 2009. Analysis of CRISPR in *Streptococcus mutans* suggests frequent occurrence of acquired immunity against infection by M102-like bacteriophages. *Microbiology* **155:** 1966–1976.

Wang Y, Juranek S, Li H, Sheng G, Wardle GS, Tuschl T, Patel DJ. 2009. Nucleation, propagation and cleavage of target RNAs in Ago silencing complexes. *Nature* **461:** 754–761.

Wiedenheft B, Zhou K, Jinek M, Coyle SM, Ma W, Doudna JA. 2009. Structural basis for DNase activity of a conserved protein implicated in CRISPR-mediated genome defense. *Structure* **17:** 904–912.

Wommack KE, Colwell RR. 2000. Virioplankton: Viruses in aquatic ecosystems. *Microbiol Mol Biol Rev* **64:** 69–114.

Ancestral Roles of Small RNAs: An Ago-Centric Perspective

Leemor Joshua-Tor and Gregory J. Hannon

Watson School of Biological Sciences, Howard Hughes Medical Institute, Cold Spring Harbor Laboratory, Cold Spring Harbor, New York 11724

Correspondence: hannon@cshl.edu and leemor@cshl.edu

SUMMARY

RNAi has existed at least since the divergence of prokaryotes and eukaryotes. This collection of pathways responds to a diversity of "abberant" RNAs and generally silences or eliminates genes sharing sequence content with the silencing trigger. In the canonical pathway, double-stranded RNAs are processed into small RNAs, which guide effector complexes to their targets by complementary base pairing. Many alternative routes from silencing trigger to small RNA are continuously being uncovered. Though the triggers of the pathway and the mechanisms of small RNA production are many, all RNAi-related mechanisms share Argonaute proteins as the heart of their effector complexes. These can act as self-contained silencing machines, binding directly to small RNAs, carrying out homology-based target recognition, and in some cases cleaving targets using an endogenous nuclease domain. Here, we discuss the diversity of Argonaute proteins from a structural and functional perspective.

Outline

1 Introduction
2 Argonautes in three clades
3 Ago structure reveals function
4 Multiple modes of RNA binding
5 Small RNA guided cleavage
6 Comparative studies of argonaute biology hint at ancestral function

References

1 INTRODUCTION

The evolution of genomes capable of sustaining their own replication was almost certainly followed by the rapid emergence of parasites capable of exploiting this innovation. Hosts able to control or combat such parasites would have a strong selective advantage. Thus, the deeply rooted conflict between hosts and those seeking to exploit them has driven the development of a panoply of immune strategies that permit discrimination of self from nonself at the genomic level and provide strategies for selectively negating the propagation of pathogens at the expense of their host.

Small regulatory RNAs are pervasive throughout eukaryotes and components of small RNA pathways are also found in select archea and eubacteria (Hock and Meister 2008; Chen 2009; Karginov and Hannon 2010). In multiple kingdoms of life, small RNAs play key roles in responses to both exogenous nucleic acid pathogens, such as viruses, and to resident genomic parasites, namely transposons (Aravin et al. 2007). Though the precise architecture of these small RNA-based immune systems varies, their sharing of core components and of general regulatory strategies strongly suggests genome defense as an ancestral function of small RNAs. Numerous observations suggest later diversification of the biological roles of small RNAs, extending these to gene regulation at nearly every conceivable level and even to control of genomic architecture and content.

The diversity of small RNA families seems to be growing at an alarming rate, catalyzed in large part by the application of increasingly deep next-generation sequencing. At times, it seems as if we will find that nearly every position in every eukaryotic genome will be soon be reflected within at least one small RNA species. However, here we will confine our discussion to the classes of small RNAs that are united under the broad umbrella of RNA interference pathways. Though small RNAs may enter these pathways through a surprising number of different avenues (Kim et al. 2009), all pathways falling under the RNAi rubric share one common and immutable feature. They exert their functions through an effector complex, called RISC or the RNA-induced silencing complex, that has an Argonaute-family protein at its core.

Argonaute proteins were first linked to RNAi through genetic studies in *Caenorhabditis elegans*. A screen for RNAi-resistant mutants uncovered RDE-1 (RNAi-defective-1) as being essential for gene silencing in response to exogenously introduced double-stranded RNA (Tabara et al. 1999). This was a representative of a little-studied family of proteins, which had previously only been noted because of developmental phenotypes that arose upon their inactivation. In fact, the Argonaute designation came from the appearance of an *Arabidopsis* mutant whose morphology was reminiscent of a squid (Bohmert et al. 1998). However, even from the beginning, it was clear that Argonaute proteins were members of a highly conserved and quite large family of proteins, with multiple members in nearly every eukaryotic genome (Cerutti et al. 2000). Later studies placed Ago proteins within the emerging biochemical framework of RNA silencing mechanisms, showing that Argonaute was the component of RISC, which directly interacted with the small RNA species that contribute target specificity to the complex (Hammond et al. 2001).

2 ARGONAUTES IN THREE CLADES

A phylogenic analysis of Argonautes from *Homo sapiens* (Hs), *Arabidopsis thaliana* (At), *C. elegans* (Ce), *C. briggsae* (Cb), *Drosophila melanogaster* (Dm), and *Schizosaccharomyces pombe* (Sp), representatives of the plant, animal, and fungal kingdoms, suggested that these proteins can be divided into three clades (Fig. 1) (Tolia and Joshua-Tor 2007). In many ways, this diversity mirrors the breadth of biological processes, which Argonaute proteins can impact and mechanisms through which they can act. The Ago-like subfamily proteins, defined based on their similarity to AtAgo1, are present in animals, plants, and fission yeast and are involved in transcriptional and posttranscriptional silencing and in genome organization (Joshua-Tor 2006). They show broad expression patterns being present in nearly every tissue and cell type examined in these representative species. They most often interact with what have become known as canonical small RNAs, which based upon their mechanisms of biogenesis average around 22 nucleotides in length and have 5′ phosphate and 3′ hydroxyl groups (Zamore et al. 2000; Bernstein et al. 2001). These include the well-known microRNAs and small interfering RNAs, both derived from double-stranded RNA precursors. The Piwi subfamily is an animal-specific clade that is expressed almost exclusively in gonadal tissues, often only in germline cells, where they form the core of animal transposon silencing pathways (Aravin et al. 2007). Piwi proteins interact with piRNAs (Piwi-interacting) RNAs, which are generated via a pathway distinct from canonical small RNAs. They have 5′ P and 3′ OH ends, suggesting nucleolytic processing from longer precursors, but are often larger than RNAs of the canonical classes (23–29 nt). The Wago clade is constituted entirely of worm Argonautes. It is unusually diverse, contributing 18 of the 27 family members in the *C. elegans* genome (Tolia and Joshua-Tor 2007). Many that have been characterized so

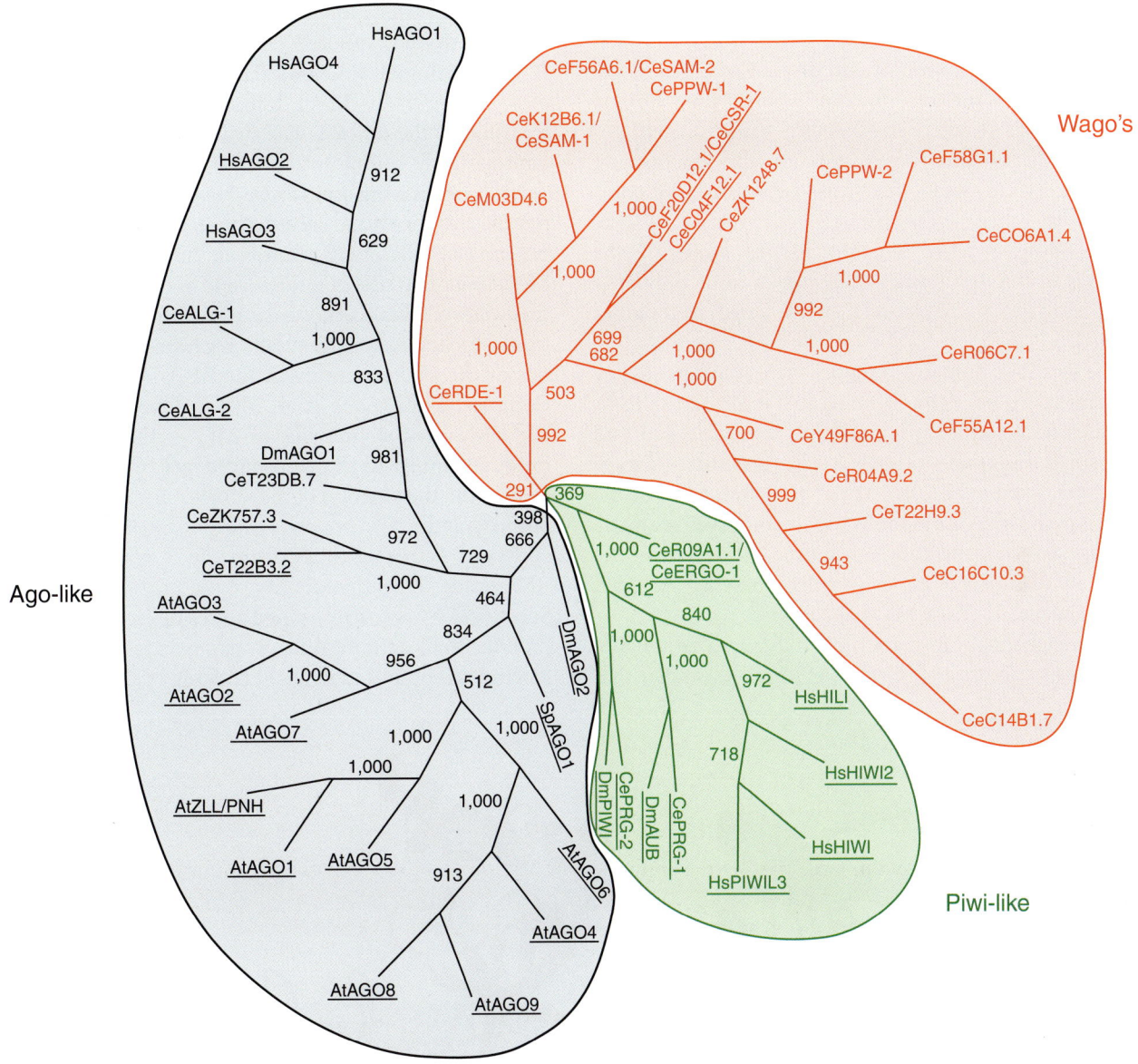

Figure 1. The three clades of the Argonautes. The Ago-like clade, shown in black, is based on similarity to AtAgo1; the Piwi-like clade, shown in green based on similarity to DmPiwi and the Wago's, which is a worm-specific clade is shown in red. Argonautes that contain a complete catalytic motif are underlined (Adapted from Tolia and Joshua-Tor 2007).

far bind small RNAs generated by direct synthesis from RNA templates (Yigit et al. 2006). Thus, Wago proteins are generally occupied by "secondary siRNAs" which bear 5′ triphosphate and 3′ hydroxyl termini. The biological roles of Wago proteins are still emerging; however, it is clear that they are important for silencing responses to exogenously delivered dsRNAs, for small RNA-based regulation of endogenous genes, and chromosome structure and segregation (Tabara et al. 1999; Yigit et al. 2006; Claycomb et al. 2009; Conine et al. 2010; Gu et al. 2009).

3 AGO STRUCTURE REVEALS FUNCTION

A continuous accumulation of genetic and biochemical evidence pointed to central roles for Ago proteins in virtually all RNAi-related processes. Characterization of the RISC, particularly from *Drosophila* and mammals demonstrated that it acted as a multiple turnover enzyme that could cleave RNA substrates in a manner determined by the sequence of its bound RNA, an activity termed "Slicer" (Martinez and Tuschl 2004; Schwarz et al. 2004). How

Argonaute proteins contributed to this activity, however, remained a mystery.

Sequence conservation of Ago proteins pointed to the presence of two discrete domains, which were termed PAZ and PIWI (Cerutti et al. 2000). Unfortunately, sequence alone provided no insight into the functions of these regions nor could their roles be inferred by comparisons to other better-characterized proteins. A true understanding of their function and consequently the functions of the Ago proteins themselves came only after structures of PAZ domains and then full-length Argonautes emerged.

The first PAZ domain structures were accompanied by biochemical studies demonstrating that it formed an RNA binding module, which specifically interacts with the 3' end of the small RNA that guides Ago target specificity (Lingel et al. 2003; Song et al. 2003; Yan et al. 2003). Most of the contacts to the RNA are made by highly conserved aromatic residues—two tyrosines and a histidine—that contact the phosphate that bridges the two terminal bases of the small RNA (Lingel et al. 2004; Ma et al. 2004).

The determination of the first full-length Argonaute protein structure from *P. furiosus* (PfAgo) illustrated that Argonautes also contain three additional domains: the amino-terminal, middle (Mid), and PIWI domains (Fig. 2) (Song et al. 2004). The big surprise came upon examination of the PIWI domain located at the carboxyl terminus of the protein. The structure revealed that Argonautes clearly belong to the RNase H family of enzymes. Apart from having the RNase H fold at its core, it also has the two highly conserved aspartates that are invariably present on adjacent β-strands in this class of enzymes. Modeling of an siRNA guide strand and target mRNA further showed this to be consistent with what was known from biochemical experiments regarding the characteristics of the slicing activity of the RNA-Induced Silencing Complex, RISC (Song et al. 2004). These include the dependence on the presence of Mg^{++}, the generation of products similar to those produced by RNase H family enzymes, and the position of the scissile phosphate, or the endonucleolytic cut, opposite the phosphate between the 10th and 11th nucleotide from the 5' end of the small RNA guide. Thus the identity of the previously unknown Slicer activity was determined to reside in the Argonaute protein itself, defining this protein as a self-contained silencing machine (Liu et al. 2004; Song et al. 2004).

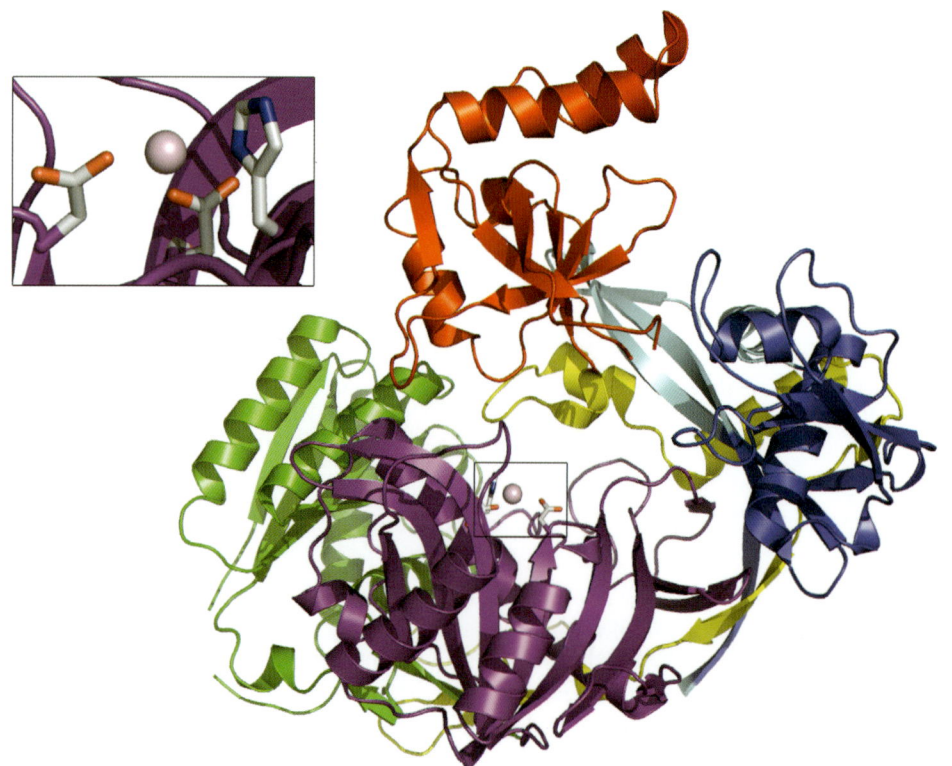

Figure 2. The crystal structure of Argonaute from *P. furiosus* shown as a ribbon diagram. The amino-terminal domain is in blue, the PAZ domain is red, the middle domain is green, and the PIWI domain is in purple. The interdomain connector is shown in yellow. A close-up of the active site residues coordinating a Mn^{++} ion (pink ball) is shown in the inset.

The PAZ domain, which resembles an OB (Oligonucleotide Binding) fold domain with a slightly different topology (Lingel et al. 2003; Song et al. 2003; Yan et al. 2003), appears to be rather mobile as evident from the two structures of *Aquifex aeolicus* Argonaute (Yuan et al. 2005; Rashid et al. 2007). In one case the PAZ domain is in a similar orientation to that of the PAZ domain in the *P. furiosus* structure and in the other there is a shift of approximately 24°. Even though crystal contacts could play a role, this disparity does illustrate the conformational space that the PAZ domain may assume. The hypothesis of a flexible PAZ is underscored by molecular dynamics simulations (Rashid et al. 2007) and normal mode analyses of these structures (Ming et al. 2007) showing low frequency motions, especially implicating movement of the PAZ domain in opening the putative binding groove for RNA.

4 MULTIPLE MODES OF RNA BINDING

Within RISC, Ago proteins must exist in a large number of RNA bound configurations. Initially, the small RNA guide must be loaded to prime the complex. In the canonical pathway, it still remains unclear whether all guides are loaded initially as double-stranded RNAs. However, biochemical studies clearly demonstrate that following cleavage into discrete segments, double-stranded guides can enter RISC (Matranga et al. 2005). These can be converted by cleavage or by simple dissociation of the nonguide strand into an active complex, with a single-stranded guide ready to search for targets. In most instances, Ago clade proteins preferentially use a portion of the guide to identify targets. This region, known as the "seed" encompasses nucleotides two to eight of the small RNA (Bartel 2009). Although the remainder of the small RNA and its degree of pairing to its substrate can certainly contribute to both the efficiency of silencing and the mode of RISC action, structural studies have provided clues to the dominant role of the seed in target selection. They have also suggested that cycles of target cleavage and substrate release are accompanied not only by conformational changes in the Ago proteins themselves but also by changes in the way Ago engages its small RNA guide.

Much of this insight has been gained through a series of structures of the *T.thermophilus* Argonaute bound to various combinations of DNA guide strands and RNA and DNA target strands (Wang et al. 2008a; Wang et al. 2008b; Wang et al. 2009). The *thermophilus* protein actually prefers DNA as both target and guide; however the latter fact became apparent over time, causing initial studies to be done with DNA guide/RNA target combinations.

Bound to a 21-mer DNA guide in the absence of target, Argonaute anchors both ends of the small DNA (Wang et al. 2008b). The 3′ end resides within the PAZ domain as predicted based upon prior structural and biochemical studies of other family members (Lingel et al. 2003; Song et al. 2003; Ma et al. 2004). The 5′ end of the guide lies within a binding site in the mid domain that was identified previously through structures of a PIWI-domain protein with RNA, though the trajectory of the strand differs in the two complexes (Ma et al. 2005; Parker et al. 2005).

The existence of a 5′ end-binding pocket within the Ago protein itself is consistent with the many indications that different family members have different 5′ end preferences. In *Arabidopsis*, the 12 Ago proteins each show a different spectrum of preferences for the 5′ nucleotide of the bound small RNA (Mi et al. 2008; Montgomery et al. 2008). In fact, individual species can be redirected from one family member to another simply by changing the terminal base. In worms, different Ago proteins likely alter this pocket to engage small RNAs with mono versus triphosphate termini, an accommodation that is also likely made in certain fungal Argonautes. Some Piwi proteins bind almost exclusively to small RNAs beginning with U (Aravin et al. 2007), though it is presently unclear whether this preference is established during biogenesis or small RNA selection by members of this clade. Though the *pyrococcus*, *thermophilus*, and *aquifex* structures form a basis from which to build hypotheses regarding small RNA selection by Ago proteins, what is clearly required as a future step is to define binding pockets within family members whose natural small RNA preferences are clear.

A second correlate between structural, biochemical, and genetic studies extends to the way in which the seed is used for target selection. There appears to be no pressure on natural targets to pair with the first base of the small RNA (Lewis et al. 2003). This is in accord with its limited availability within the structure (Wang et al. 2008b). Moreover, the structure reveals that the bases of the first and second nucleotides are destacked and that the edges of the bases for nucleotides 2–6 are exposed for base pairing. The middle part of the guide (nucleotides 12–17) is disordered in the structure lacking substrate. However, one can see an arginine intercalating between nucleotides 10 and 11 resulting in a catalytically incompetent conformation.

The exploitation of active site mutants of the *thermophilus* enzyme, substituting one of the active site aspartates with asparagine, enabled the examination of structures with target RNAs (Wang et al. 2009). In a complex containing a quite short, 12-mer target, the guide remained anchored at both ends as in the binary enzyme (Fig. 3). The 5′-phosphate was bound to several conserved residues and a Mg^{++} ion, underscoring the importance of the

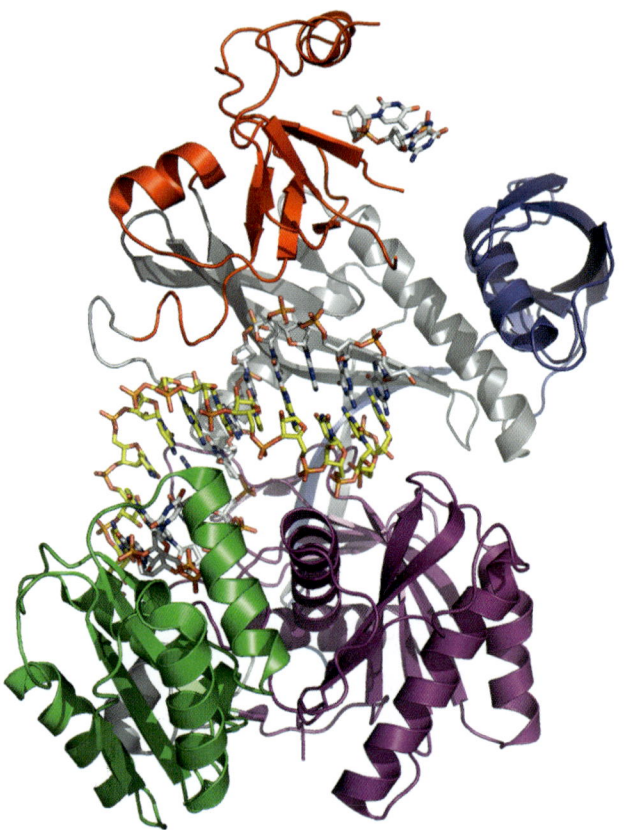

Figure 3. The crystal structure of Argonaute from *Thermus thermophilus* in complex with a 21-nucleotide DNA guide (in stick with gray carbons) and a 12-nucleotide RNA target (in stick with yellow carbons). The amino-terminal domain is in blue, the PAZ domain in red, the mid domain in green and the PIWI domain in purple. The two linkers are shown in gray. The first 12 nucleotides and the two 3′-nucleotides of the DNA guide are observed in the structure. The 3′ nucleotides are bound to the PAZ domain.

affect these conformational changes, however, it does illustrate the range of possible motions it could sample. A different active site mutant with a longer 15-mer target RNA resulted in a smaller movement of the PAZ domain. However, in this case, the guide strand appears to have exited the PAZ domain to pair with the RNA and its 3′ end is now disordered. It is worth noting that guide-strand exiting the PAZ during the catalytic cycle had been previously predicted by careful kinetic analysis of *Drosophila* RISC (Schwarz et al. 2004). The 3′ end pairing between the passenger and guide is accompanied by a widening of the channel between the PIWI and N-domains. An even longer RNA target, 19-nucleotides in length (Fig. 4), helps localize the 16th nucleotide of the target RNA and in this case the N-domain appears to block simple propagation of the DNA-RNA duplex, though additional conformational changes in the protein might occur with an even longer target. Throughout the seed pairing, Argonaute contacts only the guide strand, though contacts to the target are observed around nucleotides 10–13.

The reloading of an RNA target into the complex during the catalytic cycle is a problem worth examining. In some of the aforementioned structures, the DNA guide appears to be blocking easy exit of the RNA. It appears to be nestled well between the guide and the protein. Would two turns of a duplex need to be unwound and rewound for each cycle to allow entrance of a new target RNA, or does the protein "help" distort the duplex so that there is no hindrance for exit and entrance of successive RNA target strands to be cleaved? This question is clearly relevant for achieving multiple catalytic cycles by each RISC but also likely for the conversion of RISC loaded with dsRNA guides to those complexes primed with ssRNA and ready for action.

5 SMALL RNA GUIDED CLEAVAGE

Though Argonaute proteins can impact their targets in many ways, one of the most critical and conserved is endonucleolytic cleavage (Song et al. 2004). The substrate specificity of Argonaute is determined by the sequence of the bound guide RNA, as the target is recognized via hybridization to the guide strand. As described earlier, Argonaute is an RNase H-like enzyme creating a 5′ product with a 3-OH and a 3′ product carrying a 5′ phosphate, with the DNA strand generally being replaced by a guide RNA. The RNAse H protein family consists of well-characterized enzymes such as retroviral integrases and transposases (Nowotny 2009). *Escherichia coli* RNase H1 catalyzes a single reaction resulting in substrate cleavage. Integrases and transposases catalyze two consecutive reactions, donor-end processing and nucleotidyl transfer, resulting in strand

presence of this phosphate for selectivity in guide binding and for establishing fidelity in cleavage sites selection (Ma et al. 2005; Rivas et al. 2005). The first guide-strand base remained destacked from the next as was shown previously. The DNA-RNA duplex that follows at positions 2 through 12 are reported to resemble an A-form conformation and the scissile bond is placed at the active site of the enzyme in a helical and catalytically competent conformation in contrast to their orthogonal arrangement in the binary complex. The following nucleotides of the guide strand are presumably unstructured, because they are not observed. The final two nucleotides again appear anchored within the PAZ domain. The placement of the guide is a bit different in the ternary structure, but most notable is the substantial movement of the PAZ domain with respect to the binary complex. One must keep in mind that the PAZ domain is involved in significant crystal contacts that might

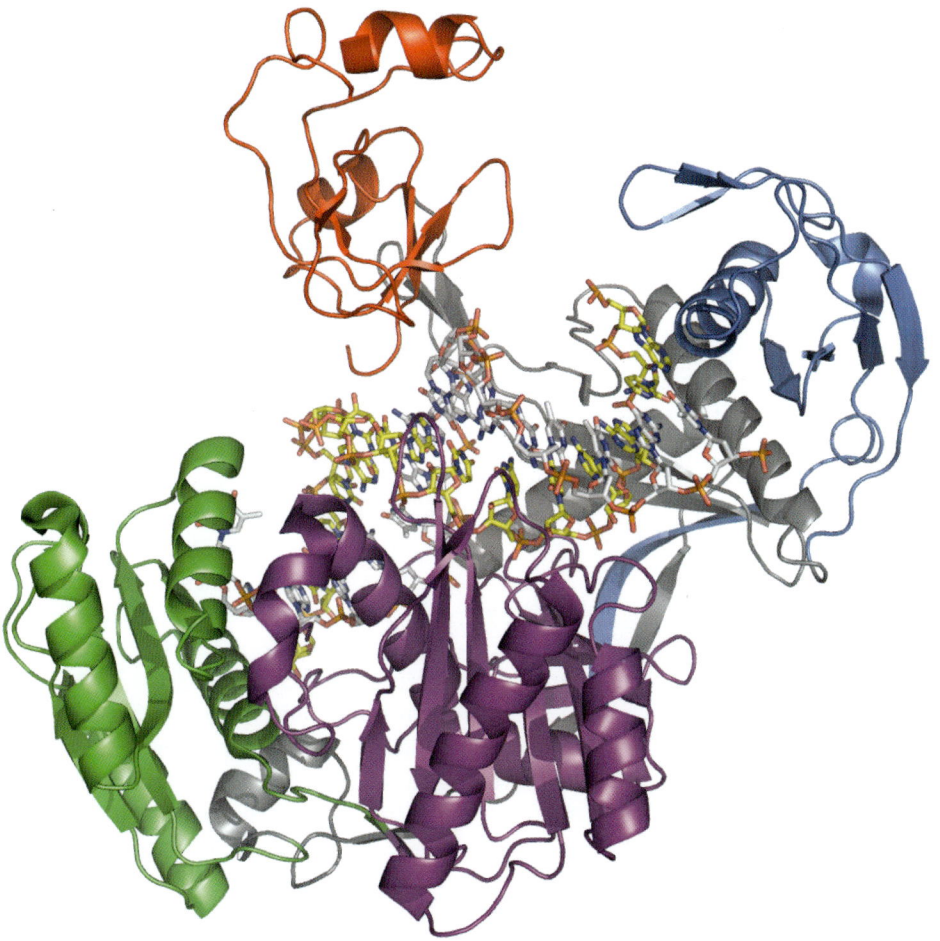

Figure 4. The crystal structure of Argonaute from *T. thermophilus* in complex with a 21-nucleotide DNA guide (in stick with gray carbons) and a 19-nucleotide RNA target (in stick with yellow carbons). Domains are colored as in Figure 3. The first 16 nucleotides of the DNA guide and 15 nucleotides of the RNA target strand are observed in the structure. The 3′-end nucleotides of the guide are no longer bound to the PAZ domain in this structure.

transfer. The nucleophile in these reactions is either a water molecule or a nucleotide 3′-OH. They involve a two-metal ion catalysis mechanism, with one metal activating the nucleophile and the second stabilizing the intermediate (Nowotny et al. 2005). Two metals have been observed in crystal structures of substrate-bound complexes of Tn5 transposase and both human and *B. halodurans* RNase H1 (Lovell et al. 2002; Nowotny et al. 2005; Nowotny et al. 2007). So far, the presence of at least one metal ion is clear in some of the Argonaute structures (Rivas et al. 2005; Yuan et al. 2005; Wang et al. 2008a; Wang et al. 2009). A second metal ion was reported in one of the ternary complexes, largely based on possible coordination with neighboring oxygen atoms (Wang et al. 2009).

A combination of structural and mutational analyses identified the two invariant aspartates characteristic of RNase H enzymes as well as a third residue that in most cases is a histidine rather than another carboxylate, critical for catalysis (Fig. 2) (Joshua-Tor 2006). The presence of all three residues is not sufficient to predict whether a particular Argonaute has catalytic activity. For example, the four human Argonaute clade proteins, HsAgo1, HsAgo2, HsAgo3, and HsAgo4 have been tested for activity, with HsAgo2 alone having slicer activity, even though HsAgo3 has an intact catalytic triad (Liu et al. 2004; Meister et al. 2004). The remaining Piwi clade proteins also conserve catalytic potential, and are presumed active based upon their conserved mechanism of transposon recognition and silencing (Faehnle and Joshua-Tor 2007).

The presence of the Argonaute protein and the guide RNA in a minimal recombinant RISC, is sufficient for target cleavage, but does not show ATP-stimulated product release and turns over rather slowly. However, other factors likely also contribute to the function of a complete RISC. In one report, a complex containing Argonaute, Dicer, and human immunodeficiency virus-1 trans-activating response

(TAR) RNA-binding protein (TRBP) is capable of guide-strand loading and multiple rounds of target cleavage, stimulated by nucleotides (Gregory et al. 2005). Another report indicated that Dicer dissociates once Ago2 is loaded with the guide RNA (Maniataki and Mourelatos 2005). More recent studies support the dissociation of both Dicer and TRBP from Ago2 after loading, and these appear to not participate in target cleavage (MacRae et al. 2008).

Many questions still remain to be answered to fully understand target cleavage by the Argonautes. How is multiple turnover achieved? Does PAZ domain movement assist in substrate engagement and placement, and in product release? What is essential to constitute an active Slicer in addition to the active site requirements?

6 COMPARATIVE STUDIES OF ARGONAUTE BIOLOGY HINT AT ANCESTRAL FUNCTION

In essence, RISC is a programmable homology search engine, which can confer a variety of properties upon a target. Some of these depend upon the catalytic activity of Ago proteins, whereas others do not. This is reflected in a diversification of the Ago family into multiple clades and into groups of family members, which either retain or have lost catalytic potential. Given the staggering breadth of Ago functions, it seems difficult to infer the ancestral roles of the RNAi pathway. In both plants and animals, Argonautes are programmed by microRNAs to regulate gene expression, mainly at the posttranscriptional level (Bartel 2009). However, differences in microRNA biogenesis and effector mechanisms and the lack of any shared microRNAs between plant and animal kingdoms strongly suggests that gene regulatory functions of small RNAs may have evolved several times independently. Plants, animals, and fungi have a variety of small RNA-based pathways which impact chromatin structure and organization and which in some cases are essential for chromosome function or even control chromosome content (Mochizuki et al. 2002; Volpe et al. 2002; Mathieu and Bender 2004; Alexander et al. 2008). Although these pathways play different roles in different species, one shared feature is that all are driven by the recognition of repetitive sequences. In fact, recognition and control of repeated elements, transposons, and viruses—generally foreign nucleic acids—seems the one theme that unifies small RNA pathways wherever they are found.

The first hints of immune roles for RNAi came from plants, where canonical RNAi mechanisms—involving Dicer, siRNAs and Ago-clade proteins—proved a key mediator of antiviral defense (Hamilton and Baulcombe 1999; Mlotshwa et al. 2008). Here, simply the double-stranded nature of replication intermediates seemed sufficient to indicate that an RNA was "foreign" and to tag it for nucleolytic destruction. Similar observations in *Drosophila* and *C. elegans* extended the antiviral paradigm to animals (Lu et al. 2005; Wang et al. 2006). In mammals, where an adaptive immune system has evolved to fight viral infection, the roles of RNAi in antiviral immunity are less clear. However, mammals clearly use small RNA pathways to recognize and control endogenous nucleic acid parasites.

An animal-specific clade of Ago proteins, the Piwis, forms the basis of an elegant, RNA-based innate immune system that selectively silences transposons in the germ cells of animals (Aravin et al. 2007). This incorporates two key elements, a programmable evolutionary record of transposon exposure and control and an adaptive mechanism to optimize the response to ongoing transposon challenge.

Small RNAs that enter the piRNA pathway are produced by at least two distinct biogenesis mechanisms. So-called "primary piRNAs" are generated from specific loci within animal genomes, called piRNA clusters. These are often highly enriched for fragmented, diverged transposon remnants, and in some cases represent the most transposon-rich areas of the host genome. These loci appear to be transcribed as large single-stranded precursors, which are parsed into individual small RNAs to create the equivalent of an immune repertoire. In order to gain control over a new element, it must be incorporated into the piRNA pathway by integration into a piRNA cluster, and some clusters are known to act as hotspots for insertion of the elements that they control.

Once made, primary piRNA/Piwi protein complexes can recognize targets much as in the canonical RNAi pathway, catalyzing small RNA-dependent cleavage, in this case of transposon mRNAs. Here, the canonical and piRNA pathways diverge in the sense that target cleavage by Piwi can create the 5′ end of a new small RNA (Brennecke et al. 2007; Gunawardane et al. 2007). This target-derived RNA can catalyze the production of further small RNAs through the same mechanism. If target RNAs can recognize antisense transposon content synthesized as part of the precursors generated from piRNA clusters, then an amplification loop can be formed where interactions between the immune repertoire and their targets can increase populations of small RNAs to meet a particular transposon challenge. In this way, the adaptive amplification loop, called the ping-pong cycle, is analogous to the expansion of adaptive immune cells triggered by antigen-receptor interactions (Aravin et al. 2007).

piRNA pathways in animals clearly use post-transcriptional mechanisms as part of their regulatory arsenal. However, there is clear evidence that piRNA pathways direct DNA methylation of repeats in mammals (Carmell et al. 2007) and that posttranscriptional controls may also figure prominently in silencing of mobile elements in flies

(Pal-Bhadra et al. 2004; Brower-Toland et al. 2007). In plants, a 24-nucleotide class of small RNAs directs the methylation of repeat elements via binding to an Ago-clade protein (Zilberman et al. 2003). These depend upon a Dicer for their production, implicating dsRNA precursors that are absent from the piRNA pathway. Unlike the piRNA pathway in animals where we are at least starting to gain glimpses into the mechanism of self-non-self recognition, there are no concrete hypotheses regarding how repeat elements are singled out for silencing in plants. Two distinct RNA polymerases, polIV and polV, appear essential for RNA-directed DNA methylation (Herr et al. 2005; Onodera et al. 2005) suggesting at least one construction for a reinforcing loop similar to the one in animals. If transcription by these polymerases triggers small RNA production, which in turn directs methylation, one could close a cycle if that same methylation also helped recruit the polymerase variants. The question, however, would be how the cycle would be initiated in the first place.

In ciliates, small RNAs are critical for repeat management through strategies that show some conceptual similarities to the piRNA pathway but also some remarkable differences (Mochizuki and Gorovsky 2004). *Tetrahymena thermophila* is unusual in many respects. Among these are that somatic and germ nuclei, the macronucleus and micronucleus respectively, are propagated within a single vegetatively growing cell. The nuclei differ not only in their transcriptional activity but also in their genomic content. During the sexual cycle when a new somatic nucleus is produced following fusion of the meiotic products of the micronucleus, a significant fraction of the genome is discarded through process that is guided by small RNAs. The small RNA population is adjusted based on the content of the old macronucleus through mechanisms that are not precisely clear. However the result is that through a process of genome comparison, small RNA populations targeting repeats are enriched, and these direct the physical elimination of sequences from the developing somatic genome. Small RNAs that direct elimination reside in a Piwi family protein but are Dicer-dependent (Malone et al. 2005; Mochizuki and Gorovsky 2005), a strange convergence of pathways that seem distinct in other species.

Examples of small RNA and repeat management can also be drawn from the fungal kingdom. In fission yeast, small RNAs are essential for the functional integrity of centromeres (Folco et al. 2008; White and Allshire 2008). Their effect in this case is on the organization of the centromeric heterochromatin domain, which is in turn essential for proper segregation. In *S. pombe*, a series of centromere-associated repeats becomes periodically transcribed during the S-phase of the cell cycle (Chen et al. 2008; Kloc et al. 2008). These are targeted by the RNAi machinery, in this case via a pathway that requires Dicer, an RdRP, and an Ago-clade protein, resulting in the deposition of histone methyl marks characteristic of heterochromatin (White and Allshire 2008). Here, the conserved theme of targeting repeats seems to have been co-opted to serve a positive role in the host cell. A second example can be seen in the phenomenon of unpaired silencing (MSUD, meiotic silencing of unpaired DNA) in *Neurospora* (Shiu et al. 2001). During MSUD, sequences that differ between two homologs are selectively and heritably silenced during the sexual reproductive cycle. Such a comparison would presumably pick up new transposon insertions in addition to other types of copy number variation between synapsed chromosomes. Though no small RNAs driving this process have yet been identified, it does require the same types of RNAi-related components required for the creation of centromeric heterochromatin in *pombe* (Alexander et al. 2008).

Though RNAi-related pathways are not the universal mechanism by which hosts battle genomic parasites, the conceptual framework that emerged particularly by studies of the piRNA pathway also underlies phage resistance in both archea and eubacteria. Here the CRISPR-CAS system uses small RNAs (crRNAs) to identify and cleave phage mRNAs or genomic DNAs (Jore et al. 2010; Karginov and Hannon 2010). There is also analogy in the mechanism by which invading elements are incorporated into a genetic memory of resistance. Fragments of invading phage become integrated into a locus (CRISPR), which serves as the source of phage-targeting crRNAs. Upon challenge with an unrecognized phage, the clusters can clearly evolve by the acquisition of new sequence information near the 5′ end of the CRISPR unit. Fragments can also be lost from this heritable resistance determinant in the absence of pressure to maintain them.

Thus, through all kingdoms of life, small RNAs underlie the ability of genomes to selectively propagate in the presence of endogenous and exogenous invaders that seek to hijack replicative potential. However, the flexible nature of small RNAs themselves and the protein complexes, which use these to guide target selection, has resulted in an explosion of regulatory function that now pervades almost every aspect of biology.

REFERENCES

Alexander WG, Raju NB, Xiao H, Hammond TM, Perdue TD, Metzenberg RL, Pukkila PJ, Shiu PK. 2008. DCL-1 colocalizes with other components of the MSUD machinery and is required for silencing. *Fungal Genet Biol* **45:** 719–727.

Aravin AA, Hannon GJ, Brennecke J. 2007. The Piwi-piRNA pathway provides an adaptive defense in the transposon arms race. *Science* **318:** 761–764.

Bartel DP. 2009. MicroRNAs: Target recognition and regulatory functions. *Cell* **136:** 215–233.

Bernstein E, Caudy AA, Hammond SM, Hannon GJ. 2001. Role for a bidentate ribonuclease in the initiation step of RNA interference. *Nature* **409**: 363–366.

Bohmert K, Camus I, Bellini C, Bouchez D, Caboche M, Benning C. 1998. AGO1 defines a novel locus of *Arabidopsis* controlling leaf development. *Embo J* **17**: 170–180.

Brennecke J, Aravin AA, Stark A, Dus M, Kellis M, Sachidanandam R, Hannon GJ. 2007. Discrete small RNA-generating loci as master regulators of transposon activity in *Drosophila*. *Cell* **128**: 1089–1103.

Brower-Toland B, Findley SD, Jiang L, Liu L, Yin H, Dus M, Zhou P, Elgin SC, Lin H. 2007. *Drosophila* PIWI associates with chromatin and interacts directly with HP1a. *Genes Dev* **21**: 2300–2311.

Carmell MA, Girard A, van de Kant HJ, Bourc'his D, Bestor TH, de Rooij DG, Hannon GJ. 2007. MIWI2 is essential for spermatogenesis and repression of transposons in the mouse male germline. *Dev Cell* **12**: 503–514.

Cerutti L, Mian N, Bateman A. 2000. Domains in gene silencing and cell differentiation proteins: The novel PAZ domain and redefinition of the Piwi domain. *Trends Biochem Sci* **25**: 481–482.

Chen X. 2009. Small RNAs and their roles in plant development. *Annu Rev Cell Dev Biol* **25**: 21–44.

Chen ES, Zhang K, Nicolas E, Cam HP, Zofall M, Grewal SI. 2008. Cell cycle control of centromeric repeat transcription and heterochromatin assembly. *Nature* **451**: 734–737.

Claycomb JM, Batista PJ, Pang KM, Gu W, Vasale JJ, van Wolfswinkel JC, Chaves DA, Shirayama M, Mitani S, Ketting RF, et al. 2009. The Argonaute CSR-1 and its 22G-RNA cofactors are required for holocentric chromosome segregation. *Cell* **139**: 123–134.

Conine CC, Batista PJ, Gu W, Claycomb JM, Chaves DA, Shirayama M, Mello CC. 2010. Argonautes ALG-3 and ALG-4 are required for spermatogenesis-specific 26G-RNAs and thermotolerant sperm in *Caenorhabditis elegans*. *Proc Natl Acad Sci* **107**: 3588–3593.

Faehnle CR, Joshua-Tor L. 2007. Argonautes confront new small RNAs. *Curr Opin Chem Biol* **11**: 569–577.

Folco HD, Pidoux AL, Urano T, Allshire RC. 2008. Heterochromatin and RNAi are required to establish CENP-A chromatin at centromeres. *Science* **319**: 94–97.

Gregory RI, Chendrimada TP, Cooch N, Shiekhattar R. 2005. Human RISC couples microRNA biogenesis and posttranscriptional gene silencing. *Cell* **123**: 631–640.

Gu W, Shirayama M, Conte D Jr, Vasale J, Batista PJ, Claycomb JM, Moresco JJ, Youngman EM, Keys J, Stoltz MJ, et al. 2009. Distinct argonaute-mediated 22G-RNA pathways direct genome surveillance in the *C. elegans* germline. *Mol Cell* **36**: 231–244.

Gunawardane LS, Saito K, Nishida KM, Miyoshi K, Kawamura Y, Nagami T, Siomi H, Siomi MC. 2007. A slicer-mediated mechanism for repeat-associated siRNA 5′ end formation in *Drosophila*. *Science* **315**: 1587–1590.

Hamilton AJ, Baulcombe DC. 1999. A species of small antisense RNA in posttranscriptional gene silencing in plants. *Science* **286**: 950–952.

Hammond SM, Boettcher S, Caudy AA, Kobayashi R, Hannon GJ. 2001. Argonaute2, a link between genetic and biochemical analyses of RNAi. *Science* **293**: 1146–1150.

Herr AJ, Jensen MB, Dalmay T, Baulcombe DC. 2005. RNA polymerase IV directs silencing of endogenous DNA. *Science* **308**: 118–120.

Hock J, Meister G. 2008. The Argonaute protein family. *Genome Biol* **9**: 210.

Jore MM, Brouns SJJ, van der Oost J. 2010. RNA in defense: CRISPRs protect prokaryotes against alien nucleic acids. *Cold Spring Harb Perspect Biol* doi: 10.1101/cshperspect.a003772.

Joshua-Tor L. 2006. The Argonautes. *Cold Spring Harb Symp Quant Biol* **71**: 67–72.

Karginov FV, Hannon GJ. 2010. The CRISPR system: Small RNA-guided defense in bacteria and archaea. *Mol Cell* **37**: 7–19.

Kim VN, Han J, Siomi MC. 2009. Biogenesis of small RNAs in animals. *Nat Rev Mol Cell Biol* **10**: 126–139.

Kloc A, Zaratiegui M, Nora E, Martienssen R. 2008. RNA interference guides histone modification during the S phase of chromosomal replication. *Curr Biol* **18**: 490–495.

Lewis BP, Shih IH, Jones-Rhoades MW, Bartel DP, Burge CB. 2003. Prediction of mammalian microRNA targets. *Cell* **115**: 787–798.

Lingel A, Simon B, Izaurralde E, Sattler M. 2003. Structure and nucleic-acid binding of the *Drosophila* Argonaute 2 PAZ domain. *Nature* **426**: 465–469.

Lingel A, Simon B, Izaurralde E, Sattler M. 2004. Nucleic acid 3′-end recognition by the Argonaute2 PAZ domain. *Nat Struct Mol Biol* **11**: 576–577.

Liu J, Carmell MA, Rivas FV, Marsden CG, Thomson JM, Song JJ, Hammond SM, Joshua-Tor L, Hannon GJ. 2004. Argonaute2 is the catalytic engine of mammalian RNAi. *Science* **305**: 1437–1441.

Lovell S, Goryshin IY, Reznikoff WR, Rayment I. 2002. Two-metal active site binding of a Tn5 transposase synaptic complex. *Nat Struct Biol* **9**: 278–281.

Lu R, Maduro M, Li F, Li HW, Broitman-Maduro G, Li WX, Ding SW. 2005. Animal virus replication and RNAi-mediated antiviral silencing in *Caenorhabditis elegans*. *Nature* **436**: 1040–1043.

Ma JB, Ye K, Patel DJ. 2004. Structural basis for overhang-specific small interfering RNA recognition by the PAZ domain. *Nature* **429**: 318–322.

Ma JB, Yuan YR, Meister G, Pei Y, Tuschl T, Patel DJ. 2005. Structural basis for 5′-end-specific recognition of guide RNA by the *A. fulgidus* Piwi protein. *Nature* **434**: 666–670.

MacRae IJ, Ma E, Zhou M, Robinson CV, Doudna JA. 2008. In vitro reconstitution of the human RISC-loading complex. *Proc Natl Acad Sci USA* **105**: 512–517.

Malone CD, Anderson AM, Motl JA, Rexer CH, Chalker DL. 2005. Germ line transcripts are processed by a Dicer-like protein that is essential for developmentally programmed genome rearrangements of *Tetrahymena thermophila*. *Mol Cell Biol* **25**: 9151–9164.

Maniataki E, Mourelatos Z. 2005. A human, ATP-independent, RISC assembly machine fueled by pre-miRNA. *Genes Dev* **19**: 2979–2990.

Martinez J, Tuschl T. 2004. RISC is a 5′ phosphomonoester-producing RNA endonuclease. *Genes Dev* **18**: 975–980.

Mathieu O, Bender J. 2004. RNA-directed DNA methylation. *J Cell Sci* **117**: 4881–4888.

Matranga C, Tomari Y, Shin C, Bartel DP, Zamore PD. 2005. Passenger-strand cleavage facilitates assembly of siRNA into Ago2-containing RNAi enzyme complexes. *Cell* **123**: 607–620.

Meister G, Landthaler M, Patkaniowska A, Dorsett Y, Teng G, Tuschl T. 2004. Human Argonaute2 mediates RNA cleavage targeted by miRNAs and siRNAs. *Mol Cell* **15**: 185–197.

Mi S, Cai T, Hu Y, Chen Y, Hodges E, Ni F, Wu L, Li S, Zhou H, Long C, et al. 2008. Sorting of small RNAs into *Arabidopsis* argonaute complexes is directed by the 5′ terminal nucleotide. *Cell* **133**: 116–127.

Ming D, Wall ME, Sanbonmatsu KY. 2007. Domain motions of Argonaute, the catalytic engine of RNA interference. *BMC Bioinformatics* **8**: 470.

Mlotshwa S, Pruss GJ, Vance V. 2008. Small RNAs in viral infection and host defense. *Trends Plant Sci* **13**: 375–382.

Mochizuki K, Gorovsky MA. 2004. Small RNAs in genome rearrangement in *Tetrahymena*. *Curr Opin Genet Dev* **14**: 181–187.

Mochizuki K, Gorovsky MA. 2005. A Dicer-like protein in *Tetrahymena* has distinct functions in genome rearrangement, chromosome segregation, and meiotic prophase. *Genes Dev* **19**: 77–89.

Mochizuki K, Fine NA, Fujisawa T, Gorovsky MA. 2002. Analysis of a piwi-related gene implicates small RNAs in genome rearrangement in *Tetrahymena*. *Cell* **110**: 689–699.

Montgomery TA, Howell MD, Cuperus JT, Li D, Hansen JE, Alexander AL, Chapman EJ, Fahlgren N, Allen E, Carrington JC. 2008. Specificity of ARGONAUTE7-miR390 interaction and dual functionality in TAS3 trans-acting siRNA formation. *Cell* **133**: 128–141.

Nowotny M. 2009. Retroviral integrase superfamily: The structural perspective. *EMBO Rep* **10**: 144–151.

Nowotny M, Gaidamakov SA, Crouch RJ, Yang W. 2005. Crystal structures of RNase H bound to an RNA/DNA hybrid: Substrate specificity and metal-dependent catalysis. *Cell* **121:** 1005–1016.

Nowotny M, Gaidamakov SA, Ghirlando R, Cerritelli SM, Crouch RJ, Yang W. 2007. Structure of human RNase H1 complexed with an RNA/DNA hybrid: Insight into HIV reverse transcription. *Mol Cell* **28:** 264–276.

Onodera Y, Haag JR, Ream T, Nunes PC, Pontes O, Pikaard CS. 2005. Plant nuclear RNA polymerase IV mediates siRNA and DNA methylation-dependent heterochromatin formation. *Cell* **120:** 613–622.

Pal-Bhadra M, Leibovitch BA, Gandhi SG, Rao M, Bhadra U, Birchler JA, Elgin SC. 2004. Heterochromatic silencing and HP1 localization in *Drosophila* are dependent on the RNAi machinery. *Science* **303:** 669–672.

Parker JS, Roe SM, Barford D. 2005. Structural insights into mRNA recognition from a PIWI domain-siRNA guide complex. *Nature* **434:** 663–666.

Rashid UJ, Paterok D, Koglin A, Gohlke H, Piehler J, Chen JC. 2007. Structure of *Aquifex aeolicus* argonaute highlights conformational flexibility of the PAZ domain as a potential regulator of RNA-induced silencing complex function. *J Biol Chem* **282:** 13824–13832.

Rivas FV, Tolia NH, Song JJ, Aragon JP, Liu J, Hannon GJ, Joshua-Tor L. 2005. Purified Argonaute2 and an siRNA form recombinant human RISC. *Nat Struct Mol Biol* **12:** 340–349.

Schwarz DS, Tomari Y, Zamore PD. 2004. The RNA-induced silencing complex is a Mg2+-dependent endonuclease. *Curr Biol* **14:** 787–791.

Shiu PK, Raju NB, Zickler D, Metzenberg RL. 2001. Meiotic silencing by unpaired DNA. *Cell* **107:** 905–916.

Song JJ, Liu J, Tolia NH, Schneiderman J, Smith SK, Martienssen RA, Hannon GJ, Joshua-Tor L. 2003. The crystal structure of the Argonaute2 PAZ domain reveals an RNA binding motif in RNAi effector complexes. *Nat Struct Biol* **10:** 1026–1032.

Song JJ, Smith SK, Hannon GJ, Joshua-Tor L. 2004. Crystal structure of Argonaute and its implications for RISC slicer activity. *Science* **305:** 1434–1437.

Tabara H, Sarkissian M, Kelly WG, Fleenor J, Grishok A, Timmons L, Fire A, Mello CC. 1999. The rde-1 gene, RNA interference, and transposon silencing in *C. elegans. Cell* **99:** 123–132.

Tolia NH, Joshua-Tor L. 2007. Slicer and the argonautes. *Nat Chem Biol* **3:** 36–43.

Volpe TA, Kidner C, Hall IM, Teng G, Grewal SI, Martienssen RA. 2002. Regulation of heterochromatic silencing and histone H3 lysine-9 methylation by RNAi. *Science* **297:** 1833–1837.

Wang XH, Aliyari R, Li WX, Li HW, Kim K, Carthew R, Atkinson P, Ding SW. 2006. RNA interference directs innate immunity against viruses in adult *Drosophila. Science* **312:** 452–454.

Wang Y, Juranek S, Li H, Sheng G, Tuschl T, Patel DJ. 2008a. Structure of an argonaute silencing complex with a seed-containing guide DNA and target RNA duplex. *Nature* **456:** 921–926.

Wang Y, Juranek S, Li H, Sheng G, Wardle GS, Tuschl T, Patel DJ. 2009. Nucleation, propagation and cleavage of target RNAs in Ago silencing complexes. *Nature* **461:** 754–761.

Wang Y, Sheng G, Juranek S, Tuschl T, Patel DJ. 2008b. Structure of the guide-strand-containing argonaute silencing complex. *Nature* **456:** 209–213.

White SA, Allshire RC. 2008. RNAi-mediated chromatin silencing in fission yeast. *Curr Top Microbiol Immunol* **320:** 157–183.

Yan KS, Yan S, Farooq A, Han A, Zeng L, Zhou MM. 2003. Structure and conserved RNA binding of the PAZ domain. *Nature* **426:** 468–474.

Yigit E, Batista PJ, Bei Y, Pang KM, Chen CC, Tolia NH, Joshua-Tor L, Mitani S, Simard MJ, Mello CC. 2006. Analysis of the *C. elegans* Argonaute family reveals that distinct Argonautes act sequentially during RNAi. *Cell* **127:** 747–757.

Yuan YR, Pei Y, Ma JB, Kuryavyi V, Zhadina M, Meister G, Chen HY, Dauter Z, Tuschl T, Patel DJ. 2005. Crystal structure of *A. aeolicus* argonaute, a site-specific DNA-guided endoribonuclease, provides insights into RISC-mediated mRNA cleavage. *Mol Cell* **19:** 405–419.

Zamore PD, Tuschl T, Sharp PA, Bartel DP. 2000. RNAi: double-stranded RNA directs the ATP-dependent cleavage of mRNA at 21 to 23 nucleotide intervals. *Cell* **101:** 25–33.

Zilberman D, Cao X, Jacobsen SE. 2003. ARGONAUTE4 control of locus-specific siRNA accumulation and DNA and histone methylation. *Science* **299:** 716–719.

RNA Interference and Heterochromatin Assembly

Tom Volpe[1] and Robert A. Martienssen[2]

[1]Department of Molecular and Cellular Biology, Northwestern University, Chicago, Illinois 60611
[2]Cold Spring Harbor Laboratory, Cold Spring Harbor, New York 11724

Correspondence: martiens@cshl.edu

SUMMARY

In most eukaryotes, histone and DNA modifications are responsible for the silencing of genes integrated in heterochromatic sequences, as well as the silencing of pericentromeric repeats and transposable elements themselves. But the mechanisms that guide these modifications to heterochromatin during the cell cycle have been elusive. RNA interference takes advantage of heterochromatic transcription to process small RNAs and recruit enzymes required for both histone and DNA modifications, and is one such mechanism that has been identified. The processes are best understood in fission yeast and plants, but recent work in mammalian cells, especially in the germline, suggests these mechanisms may be highly conserved.

Outline

1. Introduction
2. Heterochromatin and heterochromatic modifications
3. RNA interference
4. Heterochromatic RNA interference at fission yeast centromeres
5. Heterochromatic silencing on chromosome arms
6. RNAi-dependent heterochromatic silencing in other eukaryotes
7. Conclusions

References

1 INTRODUCTION

Heterochromatin is defined as chromosomal material that remains condensed during interphase, when euchromatin unravels co-incident with gene expression. Heterochromatin is faithfully inherited from one cell cycle to the next, but (at least in some cases) can be reset in each sexual generation, and so fulfills many of the criteria for epigenetic inheritance (Djupedal and Ekwall 2009). Nonetheless the function of heterochromatin is still incompletely understood. One important function is the ability to silence genes that are closely linked to heterochromatin—leading to the concept of "spreading" of silencing from one chromosomal location to the next. This type of heterochromatic silencing is known as "position effect variegation" (PEV) (Fig. 1), in which variegation refers to the mosaic pattern of silencing visualized by clones of cells in the developing eye of *Drosophila* (Huisinga and Elgin 2009). Since its discovery in the 1930s, more than 100 loci have been identified that either enhance or suppress PEV, which is also acutely sensitive to the position of the gene on a chromosome, to dosage, and to environmental factors such as temperature and nutrition (Ebert et al. 2006).

Importantly, silencing of transposons shares many of the same properties as PEV, and was recognized as a related phenomenon when it was first discovered in maize in the 1950s (Lippman and Martienssen 2004). Since the advent of genome sequencing, it is clear that transposons make up a substantial portion of heterochromatic sequences in most eukaryotes, and heterochromatic silencing has become a paradigm for epigenetic mechanisms of gene regulation. However, although silencing clearly has a role in regulating transposable elements, the biological function of PEV remains unclear. In addition to its role in silencing, heterochromatin also has profound effects on chromosome organization, especially at centromeres in which it plays a major role in establishing the kinetochore (Durand-Dubief and Ekwall 2008).

RNA is an attractive intermediate in heterochromatic silencing and spreading, as base-pairing with target sequences might account for sequence specificity. Furthermore, RNA interference has been established as a potent silencing mechanism, albeit at the posttranscriptional level. Silencing is accomplished by the Argonaute family of RNase H-related endonucleases, which use 20-30nt small RNA to guide cleavage or translational inhibition of target mRNA (see Joshua-Tor and Hannon 2010). In this article we review the role of RNA interference in establishment and maintenance of heterochromatic transcriptional silencing, focusing on studies in the fission yeast *Schizosaccharomyces pombe*.

2 HETEROCHROMATIN AND HETEROCHROMATIC MODIFICATIONS

Heterochromatin is distinguished by the covalent modification of histones and of DNA. Histone modifications specific to heterochromatin are more highly conserved and are found in organisms that lack DNA methylation, such as fission yeast and *Caenorhabditis elegans*, as well as in most animals and plants, in which DNA methylation is widespread. 5-methyl cytosine at symmetric CpG dinucleotides, on the other hand, is found in all plant and vertebrate genomes, but is lacking from some invertebrate and fungal genomes (Feng et al. 2010). CpG methylation is maintained by the methyltransferase Dnmt1, and guided during DNA replication by the hemimethylation binding protein Uhfr1, and their orthologs MET1 and VIM1 in plants, respectively. In plants and filamentous fungi heterochromatic sequences are also enriched in CpNpG as well as asymmetric (CpHpH) methylation, which are targeted by alternative DNA methyltransferases (Henderson and Jacobsen 2007). However, DNA methylation in these contexts is much rarer in animal genomes, in which CG methylation is distributed more widely (Popp et al. 2010).

Histone modifications include methylation of histone H3 lysine-9 (H3K9), methylation of H3K27 (in multicellular animals and plants) and H3 arginine-2, as well as demethylation of H3K4, deubiquitination of H2BK123 and deacetylation of several lysine residues in histones H3 and H4 (Garcia et al. 2007). Importantly, these modifications occur in readily accessible, unstructured histone "tails" that extrude from the nucleosome, and enzymes such as SET domain methyltransferases can modify nucleosomes

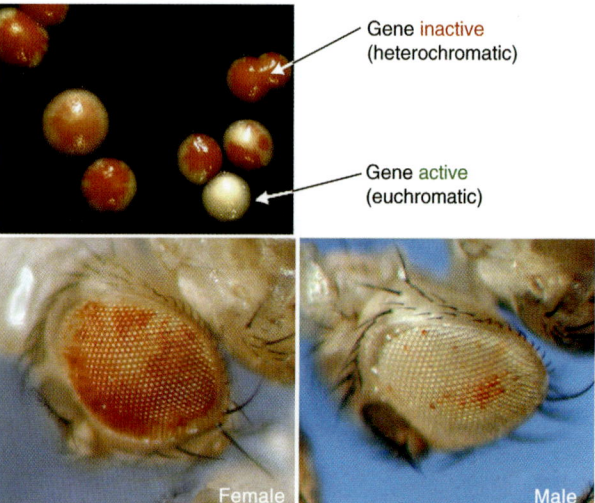

Figure 1. Position effect variegation in *Drosophila* and fission yeast. In *Drosophila*, the white gene w[m4h] is silenced by heterochromatin, causing white sectors on an otherwise red eye. Silencing is enhanced in males. In fission yeast, loss of *ade6* silencing when integrated near heterochromatic sequences gives rise to white sectors on red colonies.

in situ, without having to (completely) unravel individual nucleosomes. Additional covalent modifications are found in internal domains, which presumably can only be efficiently targeted to histone monomers and dimers. Methylation of lysine residues by SET domain methyltransferases can result in mono- (me1), di- (me2), or tri- (me3) methylated derivatives, and different derivatives are associated with heterochromatin in different species. Demethylation is accomplished sequentially by Jumonji C (JmJC)-domain oxidoreductases (from me3 to me2 to me1) and Lysine Specific Demethylase (LSD)-like amine oxidases (from me2 to me1) (Mosammaparast and Shi 2010). Deacetylation is accomplished by type I and type II deacetylases as well as type III, or "sirtuin" deacetylases that use an NADH cofactor(Shahbazian and Grunstein 2007). Importantly, many of these enzymes can also modify nonhistone proteins, such as the DNA methyltransferase Dnmt1, and so impact chromatin modification indirectly as well (Nicholson and Chen 2009).

Modifications of histone tails are specifically recognized by "effector" and "adaptor" proteins that recognize intact modified nucleosomes (Taverna et al. 2007). Adaptor domains include chromodomains, PHD domains, and WD repeat domains (all of which bind methylated lysines), bromodomains and BAH domains (which bind acetylated lysines) and 14-3-3 domains (which bind phosphorylated serines). Recognition can be reduced by nearby modifications, such as S10 phosphorylation (which prevents K9me2/3 recognition by chromodomains) or R2 methylation (which prevents K4me3 recognition by PHD domains), or enhanced by modification of other histone tails or even of the underlying DNA sequence itself (Cheng and Blumenthal 2010). Effector domains include a variety of enzymatic activities such as ATP-dependent chromatin remodeling and further histone and DNA modification. These observations have led to the idea that a combination (or code) of chromatin modifications is required to dictate any given readout (Jenuwein and Allis 2001).

Many of the enzymes that modify histones, as well as proteins that recognize them, were first discovered in genetic screens for mutants that either enhance or suppress position effect variegation in *Drosophila*, and related silencing phenomena in yeast, other animals and plants. For example SuVar (suppressor of variegation) 3–9, responsible for H3K9 methylation, was first discovered in *Drosophila*, but homologs were subsequently found in almost every other eukaryotic genome (Ebert et al. 2006).

3 RNA INTERFERENCE

Double-stranded RNA (dsRNA) can trigger posttranscriptional silencing of cognate genes through sequence specific recognition of endogenous transcripts (see Joshua-Tor and Hannon 2010). The RNase III enzyme, Dicer, specifically cleaves dsRNA into small interfering RNAs (siRNA) 21–24nt in length. These siRNAs can act as guides to target effector complexes, such as the RNA-induced silencing complex (RISC), to endogenous transcripts in a sequence-specific manner resulting in cleavage and subsequent degradation or in translational inhibition. Several components of RISC have been isolated. One of these components, the small RNA binding protein Argonaute, is a key effector and can cleave endogenous messages via its endonuclease domain, which resembles RNaseH. RNA dependent RNA polymerase (RdRP) is thought to be involved in amplification of small interfering RNAs (siRNAs) in some organisms by using endogenous messages as templates to generate additional dsRNA substrates.

RNAi is found in almost all eukaryotes though interesting exceptions exist. The budding yeast *S. cerevisiae* appears to have lost the RNAi machinery (Aravind et al. 2000), although recent studies of other budding yeast species reveal the presence of Argonaute as well as noncanonical Dicer genes and suggest their involvement in transposon regulation (Drinnenberg et al. 2009). Similarly, whereas the function of RdRP has been well studied in plants, worms, and fission yeast, RdRP genes in insects and mammals have been elusive. Recent studies, however, had revealed RdRP activity in *Drosophila* and in humans suggesting that the activity itself is conserved (Lipardi and Paterson 2009; Maida et al. 2009).

Many RNAi mechanisms depend on dsRNA, and aside from exogenous sources (such as viruses and transgenes), dsRNA can also be generated endogenously. One mechanism involves bidirectional transcription followed by annealing of complementary transcripts, although the large amount of antisense transcription observed in eukaryotic genomes raises questions regarding specificity. Another method involves dsRNA synthesis via RdRP activity on single stranded RNA templates. Exogenously generated siRNA, or "primary siRNAs" directing amplification of endogenous "secondary siRNAs" via activity of RdRP has been shown in plants and worms, and occurs in fission yeast (Halic and Moazed 2010).

In higher eukaryotes endogenously expressed single stranded stem-loop RNAs are recognized and processed by Dicer resulting in small RNA molecules known as micro RNAs (miRNAs). miRNAs are incorporated into Argonaute, which then targets RISC to endogenous transcripts. This targeting results in posttranscriptional silencing via inhibition of translation or, in some cases, cleavage of the associated message (Aravin and Hannon 2008). The versatility of miRNAs in regulation of gene expression is vast, and some miRNAs have also been shown to result in transcriptional activation as well as repression (Kim et al. 2008).Yet another class of small RNAs, PIWI interacting RNAs (piRNAs), require neither Dicer nor RdRP for biogenesis and associate with a subset of argonaute homologs called PIWI proteins. These piRNAs have been implicated in germ line development, transposon regulation, and genome organization (Thomson and Lin 2009). The biogenesis and function of miRNA and piRNA are described by Joshua-Tor and Hannon (2010).

4 HETEROCHROMATIC RNA INTERFERENCE AT FISSION YEAST CENTROMERES

S. pombe has proven to be a powerful system for studying the endogenous function of RNAi. One reason is that the fission yeast genome contains only a single homolog each of Dicer, Argonaute, and RdRP thus eliminating much of the redundancy exhibited in other organisms (Aravind et al. 2000; Carmell et al. 2002). One startling discovery from early studies in fission yeast was the dramatic loss of reporter gene silencing near centromeres in the absence of RNAi, which was accompanied by the accumulation of transcripts derived from heterochromatic repeats (Volpe et al. 2002). These observations suggested that heterochromatic silencing was mediated, at least in part, by RNA interference.

Heterochromatin in *S. pombe* exists at centromeres, telomeres, and the silent mating-type locus (Cam et al. 2005). Proper heterochromatin assembly at centromeres and telomeres is required for chromosome stability whereas heterochromatin within regions of the mating-type locus is required for proper cell fate determination. A major function of heterochromatin at centromeres appears to be the recruitment of cohesin, which is required for proper chromosome segregation. This recruitment is mediated by the *S. pombe* HP1 homolog Swi6, which can bind both H3K9me2 and cohesin (Bernard et al. 2001). The centromeres of the three fission yeast chromosomes (Fig. 2) are organized in a similar manner in that each is composed of a central core region (cnt) enriched with the histone variant Cenp-A and the site of kinetochore assembly (Durand-Dubief and Ekwall 2008). The central core is then flanked by large inverted repeats referred to as the innermost repeats (imr). The central and innermost regions are further flanked by tandem alternating copies of dg and dh centromeric repeats (Takahashi et al. 1991); it is from these the outermost repeats (otr) that most centromere transcription originates (Volpe et al. 2002).

Heterochromatic RNAs are expressed bidirectionally from both the top (forward transcripts) and the bottom (reverse transcripts) DNA strands in $rdp1^-$ (RNA-dependent RNA polymerase), $dcr1^-$ (Dicer), and $ago1^-$ (Argonaute) mutant cells. Analysis of nascent RNA by nuclear run-on revealed reverse strand centromeric transcripts are synthesized but rapidly degraded in both wild type and $swi6^-$ mutant cells, suggesting that RNAi is involved in the posttranscriptional degradation of reverse strand centromeric transcripts. In addition, nascent forward strand transcripts were not detected by nuclear run-on in wild type. These transcripts do accumulate in $swi6^-$ mutants, however, suggesting the promoter driving forward transcription is transcriptionally silenced in wild type cells and this silencing is dependent on Swi6. Furthermore, accumulation of forward strand centromeric transcripts

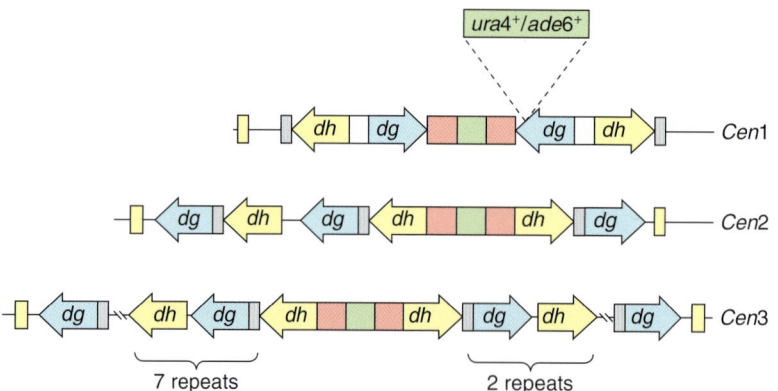

Figure 2. Centromeric heterochromatin in the fission yeast *S. pombe*. The three centromeres of *S. pombe* each include a central region (green rectangles) flanked by large inverted innermost repeats (red rectangles). These are flanked by tandem copies of outermost elements (blue arrows, orange arrows, white rectangles and grey rectangles) that are composed of dg and dh repeats. Centromeres also contain clusters of tRNA genes (yellow rectangles) at the boundaries between heterochromatin and euchromatin. Reporter genes integrated into centromere 1 (ade6 or ura4) are silenced by RNAi-mediated heterochromatic silencing.

in RNAi mutants suggests transcriptional regulation of the forward promoter is also regulated by RNAi (Volpe et al. 2002). Therefore, both transcriptional and posttranscriptional mechanisms are involved in silencing of centromeric repeats in *S. pombe*.

Heterochromatic regions such as the *S. pombe* centromeric repeats are normally enriched in the chromodomain-containing Swi6 protein and H3K9me2 whereas sites of active transcription are associated with H3K4me2 (Cam et al. 2005). However, in RNAi mutants Swi6 and H3K9me2 enrichment is replaced by H3K4me2 (Volpe et al. 2002). This led to a model in which transcripts derived from heterochromatic centromeric repeat sequences are processed by the RNAi machinery resulting in sequence specific targeting of histone modifications to centromeres. This was further supported by the identification of small interfering RNAs (siRNAs) corresponding to *S. pombe* centromeric transcripts and the discovery that at least one component of the RNAi machinery, RdRP, is enriched at centromeric repeat sequences (Reinhart and Bartel 2002; Volpe et al. 2002). Insights into effector complexes involved in RNAi mediated heterochromatin assembly mechanisms would soon follow (summarized in Fig. 3).

Biochemical purification of the fission yeast chromodomain-containing protein, Chp1, identified an RNAi effector complex consisting of Chp1, Ago1, and a novel protein, Tas3 as well as heterochromatic siRNAs (Verdel et al. 2004). This complex, the RNA-induced initiation of transcriptional silencing complex (RITS), is associated with heterochromatin (Verdel et al. 2004; Cam et al. 2005). This enrichment appears to be mediated by base pairing of RITS associated siRNAs with heterochromatic nascent transcripts, although it has also been shown that binding of RITS to heterochromatic regions requires H3K9me2, which is recognized by the Chp1 chromodomain (Schalch et al. 2009). RITS can then recruit several other protein complexes to heterochromatin.

One protein complex recruited by RITS is the RNA-directed RNA polymerase complex (RDRC), which consists of Rdp1, the RNA helicase Hrr1, and the oligoadenylate polymerase Cid12. This complex acts on single-stranded RNA templates to generate dsRNA independently of siRNA primers (Motamedi et al. 2004). RITS association with RDRC is dependent on heterochromatin as well as siRNAs (Motamedi et al. 2004). The endonuclease activity of Ago1 is used first in a siRNA processing step whereby one strand of the siRNA duplex is "sliced" to yield single stranded siRNA. This siRNA processing step occurs in a complex called ARC, which acts to chaperone siRNA duplexes generated by Dicer to the RITS complex (Buker et al. 2007). Once in the RITS complex, the slicer activity of Ago1 on heterochromatin transcripts has a further function in generating 3' ends to act as substrates for RdRP (Irvine et al. 2006; Halic and Moazed 2010).

But most importantly, the catalytic activity of RITS is required for recruitment and spreading of a multifunctional histone modification complex known as the CLRC complex (Irvine et al. 2006). Along with the WD repeat protein Rik1, which resembles DDB1 (DNA damage binding protein), CLRC contains the sole *S. pombe* lysine 9 histone methyltransferase Clr4, as well as the H3K4 demethylase Lid2 and the cullin4 ubiquitin E3 ligase, and two other

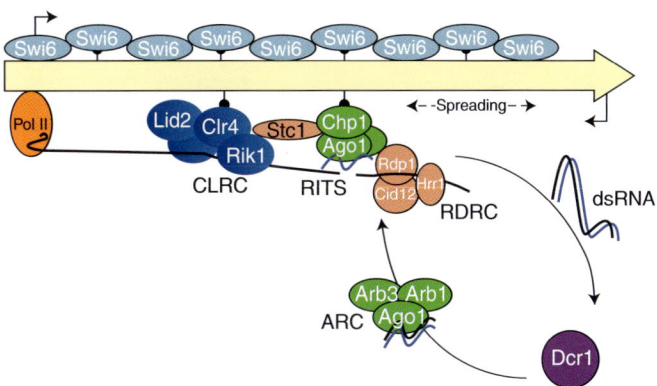

Figure 3. Model for heterochromatin assembly and spreading at *S. pombe* centromeric outer repeats. Heterochromatic centromere sequences (yellow arrow) are transcribed by RNA Polymerase II. These centromere transcripts are targeted by RITS via siRNA loaded Ago1. Association of RITS with centromere heterochromatin is strengthened by binding of Chp1 to H3mK9. RITS activity can recruit both CLRC, via interactions with Stc1, and RDRC resulting in spreading of H3mK9 and amplification of siRNAs, respectively (see text for details). dsRNA generated either by bi-directional transcription from centromere promoters (black arrows) or by RDRC activity is recognized and processed by Dicer (Dcr1). The resulting centromere siRNAs are then loaded onto Ago1 first in the ARC complex and then in RITS.

proteins, Clr7 and Clr8 (Horn et al. 2005; Jia et al. 2005; Li et al. 2005; Li et al. 2008). Therefore, RITS association with heterochromatin acts both to amplify siRNAs and also to modify nucleosomal histones and thus reinforce heterochromatin in a sequence specific manner (Fig. 3). The bridging protein that allows interaction between RITS and CLRC has been identified as the LIM domain protein Stc1, which can interact with both Ago1 and Clr4 (Bayne et al. 2010). Further, Clr4 has the ability to bind H3K9me2 via its chromodomain(Zhang et al. 2008). This suggests a mechanism of heterochromatin spreading whereby recruitment of Clr4 by H3K9me2 allows modification of adjacent nucleosomes.

A key element in the formation of heterochromatic siRNAs, and therefore heterochromatin assembly, is the regulation of transcription within these domains. It seems paradoxical that transcription of heterochromatin triggers transcriptional silencing of these very same sequences. Insight into this paradox was provided by the finding that centromeric transcription is regulated in a cell cycle dependent manner (Kloc et al. 2008). Centromere repeat transcription is limited to the G1 and S phase of the cell cycle coinciding with increases in Pol II occupancy at centromeres (Chen et al. 2008). This is the time when H3K9me2 is at its lowest, and Swi6 occupancy is reduced because of phosphorylation of H3S10, which prevents chromodomain binding (Fig. 4). Repeat transcription in early S-phase is followed by increases in siRNAs and in H3K9me2 that peaks in G2 (Kloc et al. 2008). Thus, centromere repeat expression in early S-phase occurs when pericentromeric heterochromatin is replicated and could provide a timely means for inheritance of the silent state by newly replicated chromatin (Fig. 4).

Mutations in RNA Polymerase II subunits have effects on centromere silencing, indicating that heterochromatin transcripts are Pol II derived (Djupedal et al. 2005) and that pol II plays an important role in the silencing process. For example, viable mutants in Rpb7 prevent accumulation of centromeric reverse transcripts and siRNA, whereas mutants in Rpb2 lose siRNA, but not centromere transcription, indicating that Pol II participates in both precursor (pre-siRNA) transcription and processing (Kato et al. 2005). As might be expected, factors involved in RNA processing also play a role in centromeric silencing. Mutants in several RNA splicing factors, including *prp5*, *prp8*, and *prp10* have strong defects in siRNA accumulation and centromeric reporter gene silencing (Bayne et al. 2010). Mutations in the exosome exonucleases Dis3 and Rrp6 also affect

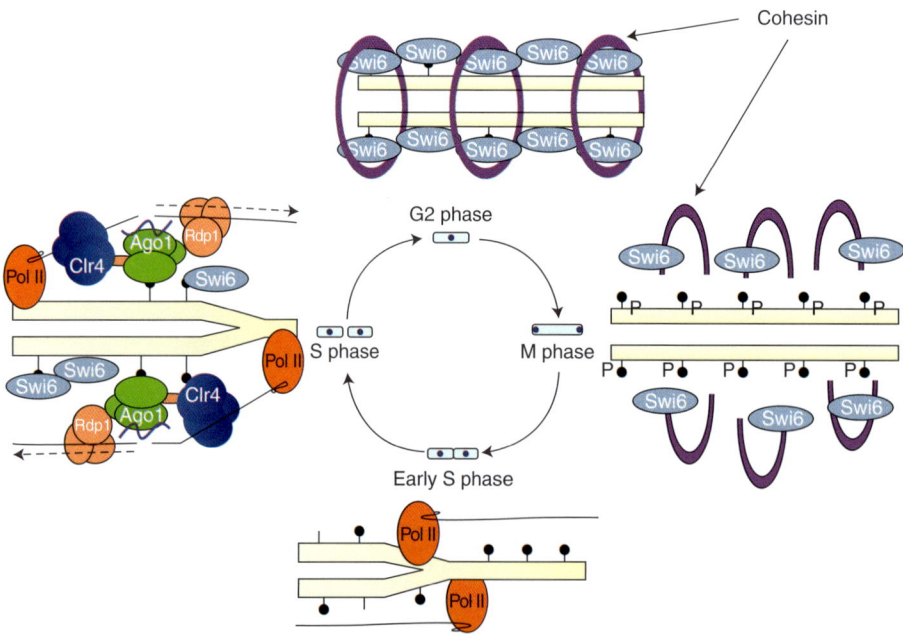

Figure 4. Cell cycle regulation of heterochromatin transcription and assembly. Pericentromeric heterochromatin is silent through most of the cell cycle (G2) but is transcriptionally activated when Swi6 (the HP-1 homolog) is evicted from methylated histone H3 lysine-9 by phosphorylation of histone H3 serine-10. Transcription continues during G1 and S phase, when transcripts are processed by RNAi and converted into siRNA. Replication during early S phase replaces approximately half of the parental nucleosomes with freshly assembled nucleosomes that lack H3K9me2. The RITS complex promotes H3K9 methylation and H3K4 demethylation, by the CLRC complex, which recruits Swi6 and silences heterochromatin in the subsequent G2. Thus transient expression and siRNA production during S phase promotes epigenetic inheritance of heterochromatic modifications.

reporter gene silencing whereas mutants in the TRAMP complex oligoadenylate polymerase Cid14 prevent siRNA accumulation as well (Murakami et al. 2007; Buhler et al. 2008). Most of these mutations, however, have a much lower impact on centromere transcript accumulation and/or heterochromatin assembly relative to RNAi. One possibility is that these effects are mostly posttranscriptional and that transcriptional silencing remains largely intact. For example, pre-siRNA transcripts that are cleaved by Argonaute are stabilized in cells deficient for Rrp6 (Irvine et al. 2006).

RNAi is not the only mechanism that guides histone modification to the centromeric repeats. The histone deacetylase Clr3 is partially redundant with RNA interference, so that double mutants lose H3K9me2 completely, whereas single RNAi mutants lose H3K9me2 predominantly from embedded reporter genes but retain residual levels over centromeric repeats. In *clr3* mutants, the levels of siRNA actually increase, indicating enhanced silencing at the post-transcriptional level (Yamada et al. 2005).

5 HETEROCHROMATIC SILENCING ON CHROMOSOME ARMS

Unlike at centromeres, heterochromatin contained within the silent mating type-locus and sub-telomeric regions is maintained in $rdp1^-$, $dcr1^-$ and $ago1^-$ mutants. Heterochromatin initiation at the mating type locus, however, seems to require RNAi (Hall et al. 2002). Once established, heterochromatin maintenance at the silent mating-type locus is dependent on the stress activated ATF/CREB (activating transcription factor/cAmp response-element binding protein) family members Atf1 and Pcr1 which can recruit Swi6 protein independently of RNAi (Jia et al. 2004). Interestingly, RNAi dependent heterochromatin assembly also occurs at sites of convergent gene transcription resulting in transient recruitment of Swi6 and binding of cohesin along chromosome arms (Gullerova and Proudfoot 2008).

Proper targeting of heterochromatin is extremely important, because misregulation of assembly can lead either to the expression or silencing of the wrong genes. Interestingly, some portions of centromeric repeat sequences integrated within euchromatic sequences are sufficient to recruit heterochromatin assembly resulting in silencing of adjacent reporter genes suggesting that these sequences act as heterochromatin nucleation elements (Ayoub et al. 2000; Partridge et al. 2002). This ectopic silencing requires factors involved in heterochromatin assembly at endogenous centromeres including RNAi (Hall et al. 2002; Volpe et al. 2003).

Whether regions of centromere homology normally nucleate heterochromatin formation in *cis* or in *trans* is not clear. Early studies of the effects of hairpin derived siRNAs in *S. pombe* suggested that only posttranscriptional silencing of target mRNAs occurs with no obvious transcriptional silencing effects found in other organisms (Mette et al. 2000; Morris et al. 2004; Sigova et al. 2004). Further studies addressed *cis* versus *trans* silencing by tethering components of the RITS complex to RNA transcripts of a euchromatic $ura4^+$ reporter gene. Interestingly, this resulted in heterochromatin assembly at the euchromatic $ura4^+$ locus. In addition, siRNAs corresponding to *ura4* sequences were observed in these strains suggesting that *ura4* transcripts were being processed by RNAi (Buhler et al. 2006). These studies also addressed whether *ura4* siRNAs were capable of targeting heterochromatin assembly to a second *ura4* reporter in *trans*. In a wild-type background no *trans* targeting was observed suggesting that the *ura4* siRNAs could only silence in *cis*. However, in certain strain backgrounds, such as an *eri1* mutant, heterochromatin appeared to be assembled in *trans* albeit inefficiently (Buhler et al. 2006).

Eri1 is an RNA nuclease that was first identified in *C. elegans* in a screen for mutants that enhanced RNA interference on injection of exogenous siRNAs. Mutations in *Eri-1* resulted in a more robust silencing of the reporter suggesting that it acts as an inhibitor of RNAi (Kennedy et al. 2004). Consistent with this early report, Eri1 appears to play a negative role in the assembly of heterochromatin in *S. pombe* (Iida et al. 2006). Another study suggests, however, that although *C. elegans* Eri-1 may interfere with effects from exogenous siRNAs, it is actually required for post-transcriptional gene silencing triggered by endogenously expressed siRNAs (Duchaine et al. 2006). Therefore Eri1 could have a role in promoting as well as inhibiting heterochromatin assembly.

Results from two recent studies indicate *trans*-targeting can in fact occur under some circumstances in fission yeast. First, it was found that siRNAs corresponding to reporter gene sequences resulted in heterochromatin targeting to a euchromatic site (Simmer et al. 2010). Interestingly, this *trans* targeting appears to be dependent on the location of the euchromatic target being near established heterochromatin. Second, Dcr1 protein is found in intracellular foci that are distinct from those containing centromere chromatin (Emmerth et al. 2010). This suggests that processing of centromere siRNAs does not occur at the site of transcription. *Trans* targeting of heterochromatin assembly has been observed in other systems (Mette et al. 2000; Morris et al. 2004) and in fission yeast portions of both the silent mating-type locus and telomeres contain regions of centromere homology (Grewal and Klar 1997; Mandell et al. 2005) suggesting these sequences can act as *trans* targets for signals such as siRNAs. Importantly, these regions of centromere homology coincide with the centromeric regions known to be expressed and have been shown to function as heterochromatin nucleation elements suggesting that heterochromatin formation within these regions is mediated by a common mechanism (Volpe et al. 2003).

6 RNAi-DEPENDENT HETEROCHROMATIC SILENCING IN OTHER EUKARYOTES

Although the mechanisms of RNAi mediated heterochromatin assembly discovered in *S. pombe* are highly conserved, their role in higher eukaryotes is still being explored. In part this is because other transcriptional silencing mechanisms, most importantly DNA methylation, are found in mammals as well as in plants and filamentous fungi (such as *Neurospora crassa*). Furthermore, RNAi in animals appears to operate largely in the germline, rather than in cell lines and primary cells in which its impact on heterochromatin would be easier to measure. The extent to which DNA methylation in the mammalian germline is influenced by RNAi is discussed in the article by Joshua-Tor and Hannon (2010). Recently, a similar prominence for RNAi in germline silencing has been discovered in plants, and indeed it is in plants that the role of RNA in gene silencing was first described and is arguably the best understood.

When cDNA copies of plant RNA viruses are artificially integrated into the genome as transgenes, they are susceptible to DNA methylation, which is promoted by subsequent infection with viral RNA (Wassenegger et al. 1994). Viral infection can also lead to posttranscriptional silencing, if copies of chromosomal genes are included in the infectious viral genome (Ruiz et al. 1998). Similarly, when plant transgenes are integrated in the genome, they are often methylated by mechanisms that depend on RNAi especially when they carry sequences that match transposon small RNA (Chan et al. 2006).

The genes required for both posttranscriptional and transcriptional silencing have been painstakingly isolated in genetic screens, and have led to a comprehensive picture of the mechanism by which small RNA leads to transcriptional gene silencing in plants (Vaucheret et al. 2001; Baulcombe 2004; Bender 2004; Henderson and Jacobsen 2007; Matzke et al. 2009). Silent transgenes, and many transposable elements, are transcribed by a plant-specific RNA polymerase related to Pol II, known as Pol V (Wierzbicki et al. 2009). Silencing is promoted by a factor related to SPT5, first described as a protein required for transposon transcription in yeast (He et al. 2009), as well as a chromatin remodeling enzyme (DRD1) and an SMC hinge domain protein (Matzke et al. 2009). These transcripts interact with one of the ten Argonaute proteins found in plants, namely Ago4. Usually (though not always) this interaction depends on cleavage or "slicing" of the target transcripts (Qi et al. 2006). These transcripts are processed by a combination of RNA polymerases PolIV and RDR2 (an RdRP) into dsRNA intermediates that are substrates for one or more of the four classes of Dicer enzymes found in plants (Pikaard et al. 2008). The cycle is complete when dsRNA binding proteins such as SGS3 and its homologs bind and presumably unwind the dsRNA duplexes and load one strand onto Ago4 (Ausin et al. 2009).

As in *S. pombe*, the subsequent steps that connect RNA processing with transcriptional silencing in plants are still unclear. What is clear is that RNAi leads to histone modification, via Suvar 3-9 homologs including SUVH4/KYP, and to DNA methylation, via Dnmt3 homologs DRM1 and DRM2 and the chromomethylase CMT3. In each case, the enzymes responsible for one modification are guided by the other—for example, SUVH4/KYP has a methylcytosine-binding SRA domain with a preference for methylated CpNpG sites (the product of CMT3). On the other hand CMT3 has a chromodomain that binds H3K9me2, which is the product of SUVH4 (Johnson et al. 2007). The precise details of how the 3 CMT and 3 DRM methyltransferase homologs are recruited by RNAi are still being determined, but histone modifications and SUVH enzymes may play at least an intermediate role, reminiscent of the mechanism in fission yeast (Johnson et al. 2008).

Thus DNA and histone methylation are regulated via the 23-24nt siRNA pathway in plants. This pathway closely resembles the RNAi pathway in fission yeast, and as in *S. pombe*, RNAi-mediated heterochromatin formation is partially redundant with RNAi-independent mechanisms of heterochromatic modification, including (in plants) both histone deacetylation and DNA methylation (Lippman et al. 2003). The targets of heterochromatic modifications in plants include pericentromeric satellite sequences, as in fission yeast, but also many transposable elements (Hamilton et al. 2002; Lippman et al. 2003; May et al. 2005). As in *S. pombe*, RNAi can both establish as well as maintain heterochromatic modifications, unlike the methyltransferase MET1 and the chromatin remodeler DDM1, which can only maintain these modifications from one generation to the next (Teixeira et al. 2009).

In the pollen grain, down-regulation of the DDM1/MET1 pathway leaves transposons uniquely sensitive to RNAi, giving rise to a new class of 21nt heterochromatic small RNA that are mobilized into the sperm cells (Slotkin et al. 2009). Transposons are similarly sensitive to RNAi in the egg cell, indicating a similar loss of RNAi-independent silencing on the female side (Olmedo-Monfil et al. 2010). It is thought that this prevalence of transposon small RNA in germ cells may contribute to interspecific "genome incompatibility" when they fail to recognize transposons from the other parent. Very importantly, mutants in RNAi on the male side result in meiotic failure and pollen inviability (Nonomura et al. 2007), whereas those on the female side result in misspecification of germ cells and the production of clonal, diploid eggs (Olmedo-Monfil et al. 2010). These mechanisms therefore have a profound role in plant reproduction.

In *N. crassa*, RNAi also plays a role in histone modification, gene silencing and in DNA methylation, although all three mechanisms do not always contribute to reporter gene silencing at the same time (Cogoni and Macino 1999; Tamaru and Selker 2001; Chicas et al. 2005). In meiosis, genes normally expressed at this time (such as spore color genes) are silenced if they are paired with a partially or completely deleted homolog (Kelly and Aramayo 2007). "Meiotic silencing of unpaired DNA," also occurs in worms, and also depends on RNAi, which is required to silence the unpaired X chromosome as well as transgene arrays in the germline (Kelly and Aramayo 2007). In both *C. elegans* and *N. crassa* H3K9 methylation and silencing depend on Argonaute and RdRP, although a role for Dicer has so far proved elusive (She et al. 2009), and the origin of siRNA precursor transcripts that are presumably involved remains a mystery. In perhaps the most dramatic example of RNAi-guided histone modification, RNAi and H3K9me2 are required for elimination of intervening DNA and transposon related sequences in the protists *Tetrahymena* and *Paramecium* (Chalker 2005). Comparison of genome transcripts and small RNA via "scanning" is thought to result in modification and elimination of sequences from the new somatic macronucleus, guided by the germline micronucleus (Mochizuki et al. 2002). Clearly, RNAi-guided heterochromatin assembly has taken a prominent role in the germline of even the simplest eukaryotes.

7 CONCLUSIONS

The finding that RNA interference can guide epigenetic modification of histones and DNA raises important questions as to the origin and function of heterochromatin. Well known as a key factor in the organization of chromosomes, a developmental role for heterochromatin has been revealed by RNAi in the germlines of plants, protists, and animals. As a repository of transposon and other repeat sequences, heterochromatin is a source of sequence-specific epigenetic information—namely small RNA—that can silence transposons and modify chromosomes during meiosis and during mitosis, where small RNA provide a mechanism for epigenetic inheritance. Conversely, the programmed (or reprogrammed) loss of RNAi-mediated heterochromatin, for example during development, reproduction or senescence, has profound effects on cell fate, especially of germ cells and their descendents. The role of RNAi in heterochromatic modification may yet have more surprises in store.

ACKNOWLEDGMENTS

We thank Mikel Zaratiegui and Michele McDonough for helpful comments on the manuscript, Derek Goto and Jim Birchler for the images in Figure 1, and to our many colleagues and collaborators for their contributions to this field. Work in the authors' laboratories is funded by grants from the NIH R01-GM067014 (to RM) and R01-GM074986 (to TV).

REFERENCES

Aravin AA, Hannon GJ. 2008. Small RNA silencing pathways in germ and stem cells. *Cold Spring Harb Symp Quant Biol* **73:** 283–290.

Aravind L, Watanabe H, Lipman DJ, Koonin EV. 2000. Lineage-specific loss and divergence of functionally linked genes in eukaryotes. *Proc Natl Acad Sci* **97:** 11319–11324.

Ausin I, Mockler TC, Chory J, Jacobsen SE. 2009. IDN1 and IDN2 are required for de novo DNA methylation in *Arabidopsis thaliana*. *Nat Struct Mol Biol* **16:** 1325–1327.

Ayoub N, Goldshmidt I, Lyakhovetsky R, Cohen A. 2000. A fission yeast repression element cooperates with centromere-like sequences and defines a mat silent domain boundary. *Genetics* **156:** 983–994.

Baulcombe D. 2004. RNA silencing in plants. *Nature* **431:** 356–363.

Bayne EH, White SA, Kagansky A, Bijos DA, Sanchez-Pulido L, Hoe KL, Kim DU, Park HO, Ponting CP, Rappsilber J, et al. 2010. Stc1: A critical link between RNAi and chromatin modification required for heterochromatin integrity. *Cell* **140:** 666–677.

Bender J. 2004. Chromatin-based silencing mechanisms. *Curr Opin Plant Biol* **7:** 521–526.

Bernard P, Maure JF, Partridge JF, Genier S, Javerzat JP, Allshire RC. 2001. Requirement of heterochromatin for cohesion at centromeres. *Science* **294:** 2539–2542.

Buhler M, Verdel A, Moazed D. 2006. Tethering RITS to a nascent transcript initiates RNAi- and heterochromatin-dependent gene silencing. *Cell* **125:** 873–886.

Buhler M, Spies N, Bartel DP, Moazed D. 2008. TRAMP-mediated RNA surveillance prevents spurious entry of RNAs into the *Schizosaccharomyces pombe* siRNA pathway. *Nat Struct Mol Biol* **15:** 1015–1023.

Buker SM, Iida T, Buhler M, Villen J, Gygi SP, Nakayama J, Moazed D. 2007. Two different Argonaute complexes are required for siRNA generation and heterochromatin assembly in fission yeast. *Nat Struct Mol Biol* **14:** 200–207.

Cam HP, Sugiyama T, Chen ES, Chen X, FitzGerald PC, Grewal SI. 2005. Comprehensive analysis of heterochromatin- and RNAi-mediated epigenetic control of the fission yeast genome. *Nat Genet* **37:** 809–819.

Carmell MA, Xuan Z, Zhang MQ, Hannon GJ. 2002. The Argonaute family: Tentacles that reach into RNAi, developmental control, stem cell maintenance, and tumorigenesis. *Genes Dev* **16:** 2733–2742.

Chalker DL. 2005. Genome rearrangements: Mother knows best! *Curr Biol* **15:** 827–829.

Chan SW, Zhang X, Bernatavichute YV, Jacobsen SE. 2006. Two-step recruitment of RNA-directed DNA methylation to tandem repeats. *PLoS Biol* **4:** e363.

Chen ES, Zhang K, Nicolas E, Cam HP, Zofall M, Grewal SI. 2008. Cell cycle control of centromeric repeat transcription and heterochromatin assembly. *Nature* **451:** 734–737.

Cheng X, Blumenthal RM. 2010. Coordinated chromatin control: Structural and functional linkage of DNA and histone methylation. *Biochemistry* **49:** 2999–3008.

Chicas A, Forrest EC, Sepich S, Cogoni C, Macino G. 2005. Small interfering RNAs that trigger posttranscriptional gene silencing are not required for the histone H3 Lys9 methylation necessary for transgenic tandem repeat stabilization in *Neurospora crassa*. *Mol Cell Biol* **25:** 3793–3801.

Cogoni C, Macino G. 1999. Gene silencing in *Neurospora crassa* requires a protein homologous to RNA-dependent RNA polymerase. *Nature* **399:** 166–169.

Djupedal I, Ekwall K. 2009. Epigenetics: heterochromatin meets RNAi. *Cell Res* **19:** 282–295.

Djupedal I, Portoso M, Spahr H, Bonilla C, Gustafsson CM, Allshire RC, Ekwall K. 2005. RNA Pol II subunit Rpb7 promotes centromeric transcription and RNAi-directed chromatin silencing. *Genes Dev* **19:** 2301–2306.

Drinnenberg IA, Weinberg DE, Xie KT, Mower JP, Wolfe KH, Fink GR, Bartel DP. 2009. RNAi in budding yeast. *Science* **326:** 544–550.

Duchaine TF, Wohlschlegel JA, Kennedy S, Bei Y, Conte D, Pang K, Brownell DR, Harding S, Mitani S, Ruvkun G et al. 2006. Functional proteomics reveals the biochemical niche of *C. elegans* DCR-1 in multiple small-RNA-mediated pathways. *Cell* **124:** 343–354.

Durand-Dubief M, Ekwall K. 2008. Heterochromatin tells CENP-A where to go. *Bioessays* **30:** 526–529.

Ebert A, Lein S, Schotta G, Reuter G. 2006. Histone modification and the control of heterochromatic gene silencing in *Drosophila*. *Chromosome Res* **14:** 377–392.

Emmerth S, Schober H, Gaidatzis D, Roloff T, Jacobeit K, Buhler M. 2010. Nuclear retention of fission yeast dicer is a prerequisite for RNAi-mediated heterochromatin assembly. *Dev Cell* **18:** 102–113.

Feng S, Cokus SJ, Zhang X, Chen PY, Bostick M, Goll MG, Hetzel J, Jain J, Strauss SH, Halpern ME, et al. 2010. Conservation and divergence of methylation patterning in plants and animals. *Proc Natl Acad Sci* **107:** 8689–8694.

Garcia BA, Hake SB, Diaz RL, Kauer M, Morris SA, Recht J, Shabanowitz J, Mishra N, Strahl BD, Allis CD, et al. 2007. Organismal differences in post-translational modifications in histones H3 and H4. *J Biol Chem* **282:** 7641–7655.

Grewal SI, Klar AJ. 1997. A recombinationally repressed region between mat2 and mat3 loci shares homology to centromeric repeats and regulates directionality of mating-type switching in fission yeast. *Genetics* **146:** 1221–1238.

Gullerova M, Proudfoot NJ. 2008. Cohesin complex promotes transcriptional termination between convergent genes in S. pombe. *Cell* **132:** 983–995.

Halic M, Moazed D. 2010. Dicer-independent primal RNAs trigger RNAi and heterochromatin formation. *Cell* **140:** 504–516.

Hall IM, Shankaranarayana GD, Noma K, Ayoub N, Cohen A, Grewal SI. 2002. Establishment and maintenance of a heterochromatin domain. *Science* **297:** 2232–2237.

Hamilton A, Voinnet O, Chappell L, Baulcombe D. 2002. Two classes of short interfering RNA in RNA silencing. *EMBO J* **21:** 4671–4679.

He XJ, Hsu YF, Zhu S, Wierzbicki AT, Pontes O, Pikaard CS, Liu HL, Wang CS, Jin H, Zhu JK. 2009. An effector of RNA-directed DNA methylation in *Arabidopsis* is an ARGONAUTE 4- and RNA-binding protein. *Cell* **137:** 498–508.

Henderson IR, Jacobsen SE. 2007. Epigenetic inheritance in plants. *Nature* **447:** 418–424.

Horn PJ, Bastie JN, Peterson CL. 2005. A Rik1-associated, cullin-dependent E3 ubiquitin ligase is essential for heterochromatin formation. *Genes Dev* **19:** 1705–1714.

Huisinga KL, Elgin SC. 2009. Small RNA-directed heterochromatin formation in the context of development: What flies might learn from fission yeast. *Biochim Biophys Acta* **1789:** 3–16.

Iida T, Kawaguchi R, Nakayama J. 2006. Conserved ribonuclease, Eri1, negatively regulates heterochromatin assembly in fission yeast. *Curr Biol* **16:** 1459–1464.

Irvine DV, Zaratiegui M, Tolia NH, Goto DB, Chitwood DH, Vaughn MW, Joshua-Tor L, Martienssen RA. 2006. Argonaute slicing is required for heterochromatic silencing and spreading. *Science* **313:** 1134–1137.

Jenuwein T, Allis CD. 2001. Translating the histone code. *Science* **293:** 1074–1080.

Jia S, Kobayashi R, Grewal SI. 2005. Ubiquitin ligase component Cul4 associates with Clr4 histone methyltransferase to assemble heterochromatin. *Nat Cell Biol* **7:** 1007–1013.

Jia S, Noma K, Grewal SI. 2004. RNAi-independent heterochromatin nucleation by the stress-activated ATF/CREB family proteins. *Science* **304:** 1971–1976.

Johnson LM, Bostick M, Zhang X, Kraft E, Henderson I, Callis J, Jacobsen SE. 2007. The SRA methyl-cytosine-binding domain links DNA and histone methylation. *Curr Biol* **17:** 379–384.

Johnson LM, Law JA, Khattar A, Henderson IR, Jacobsen SE. 2008. SRA-domain proteins required for DRM2-mediated de novo DNA methylation. *PLoS Genet* **4:** e1000280.

Joshua-Tor L, Hannon GJ. 2010. Ancestral roles of small RNAs: An agocentric perspective. *Cold Spring Harb Perspect Biol* doi: 10.1101/cshperspect.a003772.

Kato H, Goto DB, Martienssen RA, Urano T, Furukawa K, Murakami Y. 2005. RNA polymerase II is required for RNAi-dependent heterochromatin assembly. *Science* **309:** 467–469.

Kelly WG, Aramayo R. 2007. Meiotic silencing and the epigenetics of sex. *Chromosome Res* **15:** 633–651.

Kennedy S, Wang D, Ruvkun G. 2004. A conserved siRNA-degrading RNase negatively regulates RNA interference in *C. elegans*. *Nature* **427:** 645–649.

Kim DH, Saetrom P, Snove O Jr, Rossi JJ. 2008. MicroRNA-directed transcriptional gene silencing in mammalian cells. *Proc Natl Acad Sci* **105:** 16230–16235.

Kloc A, Zaratiegui M, Nora E, Martienssen R. 2008. RNA interference guides histone modification during the S phase of chromosomal replication. *Curr Biol* **18:** 490–495.

Li F, Goto DB, Zaratiegui M, Tang X, Martienssen R, Cande WZ. 2005. Two novel proteins, dos1 and dos2, interact with rik1 to regulate heterochromatic RNA interference and histone modification. *Curr Biol* **15:** 1448–1457.

Li F, Huarte M, Zaratiegui M, Vaughn MW, Shi Y, Martienssen R, Cande WZ. 2008. Lid2 is required for coordinating H3K4 and H3K9 methylation of heterochromatin and euchromatin. *Cell* **135:** 272–283.

Lipardi C, Paterson BM. 2009. Identification of an RNA-dependent RNA polymerase in Drosophila involved in RNAi and transposon suppression. *Proc Natl Acad Sci* **106:** 15645–15650.

Lippman Z, Martienssen R. 2004. The role of RNA interference in heterochromatic silencing. *Nature* **431:** 364–370.

Lippman Z, May B, Yordan C, Singer T, Martienssen R. 2003. Distinct mechanisms determine transposon inheritance and methylation via small interfering RNA and histone modification. *PLoS Biol* **1:** E367.

Maida Y, Yasukawa M, Furuuchi M, Lassmann T, Possemato R, Okamoto N, Kasim V, Hayashizaki Y, Hahn WC, Masutomi K. 2009. An RNA-dependent RNA polymerase formed by TERT and the RMRP RNA. *Nature* **461:** 230–235.

Mandell JG, Bahler J, Volpe TA, Martienssen RA, Cech TR. 2005. Global expression changes resulting from loss of telomeric DNA in fission yeast. *Genome Biol* **6:** pR1.

Matzke M, Kanno T, Daxinger L, Huettel B, Matzke AJ. 2009. RNA-mediated chromatin-based silencing in plants. *Curr Opin Cell Biol* **21:** 367–376.

May BP, Lippman ZB, Fang Y, Spector DL, Martienssen RA. 2005. Differential regulation of strand-specific transcripts from *Arabidopsis* centromeric satellite repeats. *PLoS Genet* **1**: e79.

Mette MF, Aufsatz W, van der Winden J, Matzke MA, Matzke AJ. 2000. Transcriptional silencing and promoter methylation triggered by double-stranded RNA. *EMBO J* **19**: 5194–5201.

Mochizuki K, Fine NA, Fujisawa T, Gorovsky MA. 2002. Analysis of a piwi-related gene implicates small RNAs in genome rearrangement in tetrahymena. *Cell* **110**: 689–699.

Morris KV, Chan SW, Jacobsen SE, Looney DJ. 2004. Small interfering RNA-induced transcriptional gene silencing in human cells. *Science* **305**: 1289–1292.

Mosammaparast N, Shi Y. 2010. Reversal of histone methylation: Biochemical and molecular mechanisms of histone demethylases. *Annu Rev Biochem* **79**: 155–179.

Motamedi MR, Verdel A, Colmenares SU, Gerber SA, Gygi SP, Moazed D. 2004. Two RNAi complexes, RITS and RDRC, physically interact and localize to noncoding centromeric RNAs. *Cell* **119**: 789–802.

Murakami H, Goto DB, Toda T, Chen ES, Grewal SI, Martienssen RA, Yanagida M. 2007. Ribonuclease activity of Dis3 is required for mitotic progression and provides a possible link between heterochromatin and kinetochore function. *PLoS ONE* **2**: e317.

Nicholson TB, Chen T. 2009. LSD1 demethylates histone and non-histone proteins. *Epigenetics* **4**: 129–132.

Nonomura K, Morohoshi A, Nakano M, Eiguchi M, Miyao A, Hirochika H, Kurata N. 2007. A germ cell specific gene of the ARGONAUTE family is essential for the progression of premeiotic mitosis and meiosis during sporogenesis in rice. *Plant Cell* **19**: 2583–2594.

Olmedo-Monfil V, Duran-Figueroa N, Arteaga-Vazquez M, Demesa-Arevalo E, Autran D, Grimanelli D, Slotkin RK, Martienssen RA, Vielle-Calzada JP. 2010. Control of female gamete formation by a small RNA pathway in *Arabidopsis*. *Nature* **464**: 628–632.

Partridge JF, Scott KS, Bannister AJ, Kouzarides T, Allshire RC. 2002. cis-acting DNA from fission yeast centromeres mediates histone H3 methylation and recruitment of silencing factors and cohesin to an ectopic site. *Curr Biol* **12**: 1652–1660.

Pikaard CS, Haag JR, Ream T, Wierzbicki AT. 2008. Roles of RNA polymerase IV in gene silencing. *Trends Plant Sci* **13**: 390–397.

Popp C, Dean W, Feng S, Cokus SJ, Andrews S, Pellegrini M, Jacobsen SE, Reik W. 2010. Genome-wide erasure of DNA methylation in mouse primordial germ cells is affected by AID deficiency. *Nature* **463**: 1101–1105.

Qi Y, He X, Wang XJ, Kohany O, Jurka J, Hannon GJ. 2006. Distinct catalytic and non-catalytic roles of ARGONAUTE4 in RNA-directed DNA methylation. *Nature* **443**: 1008–1012.

Reinhart BJ, Bartel DP. 2002. Small RNAs correspond to centromere heterochromatic repeats. *Science* **297**: 1831.

Ruiz MT, Voinnet O, Baulcombe DC. 1998. Initiation and maintenance of virus-induced gene silencing. *Plant Cell* **10**: 937–946.

Schalch T, Job G, Noffsinger VJ, Shanker S, Kuscu C, Joshua-Tor L, Partridge JF. 2009. High-affinity binding of Chp1 chromodomain to K9 methylated histone H3 is required to establish centromeric heterochromatin. *Mol Cell* **34**: 36–46.

Shahbazian MD, Grunstein M. 2007. Functions of site-specific histone acetylation and deacetylation. *Annu Rev Biochem* **76**: 75–100.

She X, Xu X, Fedotov A, Kelly WG, Maine EM. 2009. Regulation of heterochromatin assembly on unpaired chromosomes during *Caenorhabditis elegans* meiosis by components of a small RNA-mediated pathway. *PLoS Genet* **5**: e1000624.

Sigova A, Rhind N, Zamore PD. 2004. A single Argonaute protein mediates both transcriptional and posttranscriptional silencing in *Schizosaccharomyces pombe*. *Genes Dev* **18**: 2359–2367.

Simmer F, Buscaino A, Kos-Braun IC, Kagansky A, Boukaba A, Urano T, Kerr AR, Allshire RC. 2010. Hairpin RNA induces secondary small interfering RNA synthesis and silencing in trans in fission yeast. *EMBO Rep* **11**: 112–118.

Slotkin RK, Vaughn M, Borges F, Tanurdzic M, Becker JD, Feijo JA, Martienssen RA. 2009. Epigenetic reprogramming and small RNA silencing of transposable elements in pollen. *Cell* **136**: 461–472.

Takahashi K, Murakami S, Chikashige Y, Niwa O, Yanagida M. 1991. A large number of tRNA genes are symmetrically located in fission yeast centromeres. *J Mol Biol* **218**: 13–17.

Tamaru H, Selker EU. 2001. A histone H3 methyltransferase controls DNA methylation in *Neurospora crassa*. *Nature* **414**: 277–283.

Taverna SD, Li H, Ruthenburg AJ, Allis CD, Patel DJ. 2007. How chromatin-binding modules interpret histone modifications: Lessons from professional pocket pickers. *Nat Struct Mol Biol* **14**: 1025–1040.

Teixeira FK, Heredia F, Sarazin A, Roudier F, Boccara M, Ciaudo C, Cruaud C, Poulain J, Berdasco M, Fraga MF, et al. 2009. A role for RNAi in the selective correction of DNA methylation defects. *Science* **323**: 1600–1604.

Thomson T, Lin H. 2009. The biogenesis and function of PIWI proteins and piRNAs: Progress and prospect. *Annu Rev Cell Dev Biol* **25**: 355–376.

Vaucheret H, Beclin C, Fagard M. 2001. Post-transcriptional gene silencing in plants. *J Cell Sci* **114**: 3083–3091.

Verdel A, Jia S, Gerber S, Sugiyama T, Gygi S, Grewal SI, Moazed D. 2004. RNAi-mediated targeting of heterochromatin by the RITS complex. *Science* **303**: 672–676.

Volpe TA, Kidner C, Hall IM, Teng G, Grewal SI, Martienssen RA. 2002. Regulation of heterochromatic silencing and histone H3 lysine-9 methylation by RNAi. *Science* **297**: 1833–1837.

Volpe T, Schramke V, Hamilton GL, White SA, Teng G, Martienssen RA, Allshire RC. 2003. RNA interference is required for normal centromere function in fission yeast. *Chromosome Res* **11**: 137–146.

Wassenegger M, Heimes S, Riedel L, Sanger HL. 1994. RNA-directed de novo methylation of genomic sequences in plants. *Cell* **76**: 567–576.

Wierzbicki AT, Ream TS, Haag JR, Pikaard CS. 2009. RNA polymerase V transcription guides ARGONAUTE4 to chromatin. *Nat Genet* **41**: 630–634.

Yamada T, Fischle W, Sugiyama T, Allis CD, Grewal SI. 2005. The nucleation and maintenance of heterochromatin by a histone deacetylase in fission yeast. *Mol Cell* **20**: 173–185.

Zhang K, Mosch K, Fischle W, Grewal SI. 2008. Roles of the Clr4 methyltransferase complex in nucleation, spreading and maintenance of heterochromatin. *Nat Struct Mol Biol* **15**: 381–388.

The X as Model for RNA's Niche in Epigenomic Regulation

Jeannie T. Lee

Howard Hughes Medical Institute, Department of Molecular Biology, Massachusetts General Hospital, Department of Genetics, Harvard Medical School, Boston, Massachusetts 02115

Correspondence: lee@molbio.mgh.harvard.edu

SUMMARY

The X-linked region now known as the "X-inactivation center" (*Xic*) was once dominated by protein-coding genes but, with the rise of Eutherian mammals some 150–200 million years ago, became infiltrated by genes that produce long noncoding RNA (ncRNA). Some of the noncoding genes have been shown to play crucial roles during X-chromosome inactivation (XCI), including the targeting of chromatin modifiers to the X. The rapid establishment of ncRNA hints at a possible preference for long transcripts in some aspects of epigenetic regulation. This article discusses the role of RNA in XCI and considers the advantages RNA offers in delivering allelic, *cis*-limited, and locus-specific control. Unlike proteins and small RNAs, long ncRNAs are tethered to the site of transcription and effectively tag the allele of origin. Furthermore, long ncRNAs are drawn from larger sequence space than proteins and can mark a unique region in a complex genome. Thus, like their small RNA cousins, long ncRNAs may emerge as versatile and powerful regulators of the epigenome.

Outline

1 Introduction

2 The unique challenges of XCI

3 XCI's repertoire of NCRNA regulators

4 RNA in pairing, counting, and choice

5 Symmetry break by Tsix RNA

6 Polycomb proteins targeted and regulated by RNA

References

1 INTRODUCTION

It has become very fashionable to study long noncoding RNA (ncRNA). Given that only 1%–2% of the mammalian genome carries protein-coding information, genomicists have oft pondered the relevance of the remaining 98% which, as a whole, bear little phylogenetic conservation among any except for the most closely related species. It is by now clear, however, that the vast intergenic space is anything but quiescent. An astounding 70%–90% of our nucleotides are apparently transcribed (Kapranov et al. 2007b; Mercer et al. 2009). The mouse and human transcriptomes contain 80,000-180,000 RNAs (Carninci et al. 2005; Wheeler et al. 2008), the majority of which lack conserved open reading frames and belong to a rising class of transcripts formerly thought to be "junk," ranging in size from 50 nt to >100 kb. Affectionately dubbed "macroRNA," they are distinct from catalytic and structural RNAs, as well as the well-known small RNAs of the RNA-interference pathway (Cam et al. 2009; Cech 2009; Ghildiyal and Zamore 2009; Moazed 2009; Sharp 2009), and can be antisense, promoter-associated, or intergenic (Claverie 2005; Katayama et al. 2005; Kapranov et al. 2007a; Kapranov et al. 2007b; He et al. 2008; Guttman et al. 2009; Khalil et al. 2009; Mercer et al. 2009). Whether they possess function or are merely incidental biproducts of genome activity remains unknown, but many are clearly developmentally regulated and linked to genes with undisputed function.

Although the occurrence of genome-wide transcription has sparked widespread interest in long ncRNA only recently, such RNAs have captivated investigators in some fields for some time. Epigeneticists have long been drawn to the mystique surrounding RNA behemoths, beginning with the discovery of the 2.6-kb H19 RNA in the field of genomic imprinting (Brannan et al. 1990; Bartolomei et al. 1991) and the 17-kb XIST RNA for X-chromosome inactivation (XCI) (Borsani et al. 1991; Brown et al. 1991a; Brockdorff et al. 1992; Brown et al. 1992). Genes subject to imprinting are expressed from only one of two alleles in a manner dependent on parent of origin, and such genes are almost always clustered within domains and coordinately regulated *in cis* by a single (sometimes two) "imprinting centers" (Edwards and Ferguson-Smith 2007; Wan and Bartolomei 2008). Likewise, XCI—the mechanism of dosage compensation in mammals—is a domain phenomenon that extends to include the entire X-chromosome (Lyon 1961; Wutz 2003; Lucchesi et al. 2005; Masui and Heard 2006; Wutz and Gribnau 2007; Payer and Lee 2008). During XCI, almost all of the ∼1000 protein-coding genes on one of two X-chromosomes become transcriptionally inactivated *in cis* by a single control region known as the "X-inactivation center" (Cattanach and Isaacson 1967; Rastan and Robertson 1985). Because the nature of XCI is also allelic, silencing must be coordinately controlled to take place only on genes of one chromosome. Thus, imprinting and XCI are unified by the fact that a small control region exerts long-range effects on chromatin and by the allelic nature of the regulatory process.

The existence of two macroRNAs for two quintessentially "epigenetic" phenomena brought early suspicion that RNA may be central to the control of epigenetic inheritance. The subsequent discoveries of the regulatory antisense RNAs, *Airn* for the imprinted *Igf2r* locus (Wutz et al. 1997) and *Tsix* for XCI (Lee et al. 1999a; Lee and Lu 1999)—themselves even longer ncRNAs—raised that level of suspicion. The emerging understanding that Tsix RNA may help target DNA methylation by Dnmt3a and other chromatin modifiers, either directly or indirectly, to the *Xist* promoter suggested that RNA may be a recruiting tool for chromatin modifiers on the X (Navarro et al. 2005; Sado et al. 2005; Sun et al. 2006). That RepA RNA directly binds and targets Polycomb proteins to the X then argued that RNA may be a direct mechanism for recruiting specific enzymatic activities *in cis* (Zhao et al. 2008). The antisense Airn and Kcnq1ot1 RNAs have also been hinted to play similar roles at imprinted loci, by direct or indirect means (Nagano et al. 2008; Pandey et al. 2008). Such findings have brought recognition of RNA's potential in regulating our epigenomic landscape *in cis* and led to further ruminations about the possible roles of pervasive noncoding transcription in our genomes.

This article will develop the concept of RNAs as effectors of choice by drawing examples from the XCI paradigm. To date, at least seven distinct noncoding genes have been found within the X-linked region loosely defined as the "X-inactivation center" (*Xic*), a 100–500-kb domain with few, if any, protein-coding genes (Brown et al. 1991b; Lee et al. 1996; Simmler et al. 1996; Willard 1996; Lee et al. 1999b; Chureau et al. 2002). Why have long ncRNAs come to dominate the *Xic* landscape? RNA's supremacy seems to imply that this nucleic acid polymer offers something that proteins do not (Lee 2009). Herein, I will argue that RNA is ideally suited to epigenetic inheritance, particularly in cases that demand allelic regulation (*cis* effects) and action at a single locus (singularity).

2 THE UNIQUE CHALLENGES OF XCI

XCI presents a number of interesting challenges. Two X-chromosomes sharing the same nucleus must be treated in diametrically opposite ways during early development. There is first the problem of chromosome counting—the determination of whether it possesses one (XY) or two

Xs (XX) and whether it should therefore initiate XCI. Genetic analyses have shown that an X:autosome ratio of 1 or greater triggers the XCI cascade (Kay et al. 1994; Lyon 1999; Avner and Heard 2001; Boumil and Lee 2001) and that noncoding elements of the *Xic* influence the X:A ratio (Clerc and Avner 1998; Morey et al. 2004; Lee 2005; Donohoe et al. 2009; Monkhorst et al. 2009).

There is second the problem of X-chromosome choice, which implies the existence of a random selection mechanism that chooses the active X (Xa) and the inactive X (Xi) in a mutually exclusive manner (Lee 2002). The mutually exclusive nature of choice necessitates communication between the Xs *in trans* to ensure that a cell does not adopt the lethal XaXa or XiXi fate. This second step of XCI requires the action of noncoding genes as well.

Finally, silencing factors must be recruited to the future Xi in a colinear fashion, spreading along the chromosome in a strictly *cis*-limited manner and without *trans* effects on homologous loci of the future Xa. It is widely believed that inactivation begins at the *Xic* before spreading outwardly to encompass the rest of the X-chromosome. During XCI, the Xi becomes distinguished from Xa by its enrichment for Polycomb proteins, the SET domain protein HP1, variant histones such as macroH2A, and a number of distinct chromatin modifications (Plath et al. 2002; Wutz 2003; Lucchesi et al. 2005; Payer and Lee 2008). How heterochromatinization and gene silencing are coordinated along 1000 genes of the X remains a largely unsolved problem, but there is widespread belief that Xist RNA is the messenger of the *Xic*, that it spreads along the X, and that this action is directly responsible for reprogramming the chromosome's gene expression state.

3 XCI'S REPERTOIRE OF NCRNA REGULATORS

At least seven noncoding loci have been identified within the 100- to 500-kb region in or around the *Xic* (Brown et al. 1991b; Lee et al. 1996; Simmler et al. 1996; Willard 1996; Lee et al. 1999b; Chureau et al. 2002) (Fig. 1A). Several have been shown to play vital roles during XCI (Fig. 1B). The 17-kb Xist RNA is transcribed only from the Xi and its accumulation on the X is thought to induce chromosome-wide silencing by binding and spreading silencing complexes throughout the X *in cis* (Brockdorff et al. 1992; Brown et al. 1992; Clemson et al. 1996; Penny et al. 1996; Marahrens et al. 1997). Embedded within the 5' end of the *Xist* locus is a repeated motif known as "Repeat A" (Brown et al. 1992), previously shown to be essential for Xist RNA's silencing function (Wutz et al. 2002). It is now known to encode a separate transcription unit, named "*RepA*." RepA RNA directly binds Polycomb proteins and recruits them to the *Xic* in an action that paradoxically activates *Xist* expression and the initiation of XCI (Zhao et al. 2008).

The actions of RepA and Xist RNAs are controlled by Tsix, a 40-kb ncRNA that is antisense to both RNAs (Lee et al. 1999a; Lee and Lu 1999; Lee 2000; Luikenhuis et al.

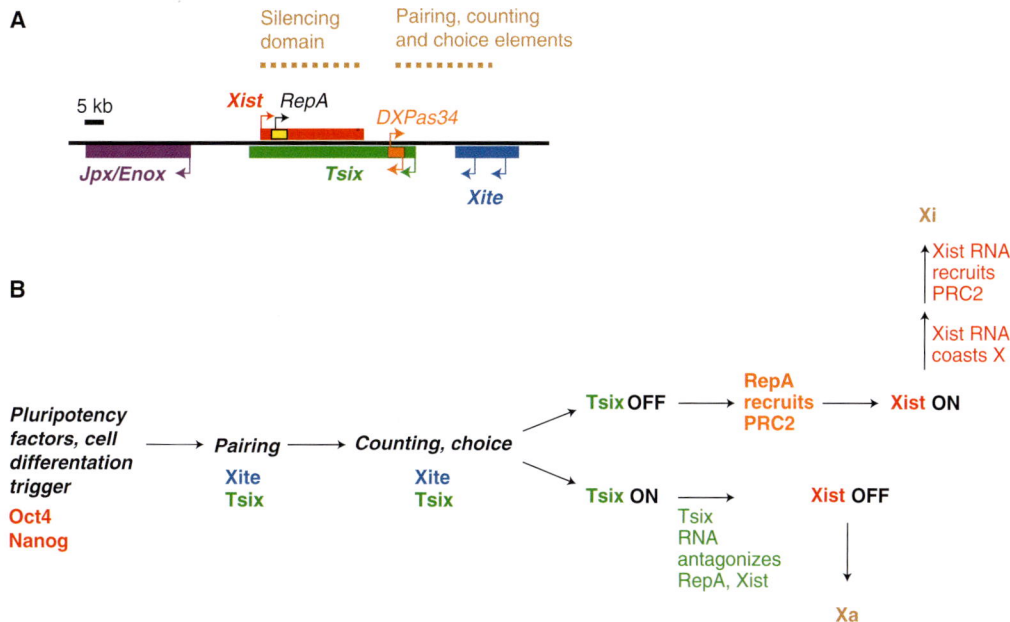

Figure 1. The X-inactivation center, its noncoding genes, and their functions during the XCI cascade. (*A*) More than seven noncoding genes have been found within or around the *Xic*. *Xist, RepA, Tsix,* and *Xite* have ascribed function. The regions responsible for X-chromosome pairing, counting/choice, and spreading/silencing are indicated. (*B*) Steps of XCI and the factors known to regulate them.

2001; Sado et al. 2001). *Tsix* bears a crucial relationship to *Xist* (Fig. 2). In undifferentiated cells (pre-XCI), Tsix is expressed on both Xs. During this stage, Xist RNA is expressed at extremely low "basal" levels. At the onset of cell differentiation, Tsix persists only on the chromosome selected to become Xa. It is the persistence of Tsix RNA that prevents the up-regulation of *Xist* on Xa. Conversely, the down-regulation of Tsix on Xi creates a permissive state for *Xist* transactivation. Whether Tsix RNA will persist during cell differentiation is controlled by the upstream locus, *Xite*. *Xite* harbors a developmentally regulated enhancer that maintains expression of *Tsix* on Xa and prevents inactivation of Xa (Ogawa and Lee 2003; Stavropoulos et al. 2005). Thus, the ncRNAs have a Yin-yang relationship to each other, with RepA and Xist RNAs promoting the initiation of XCI, and Xite and Tsix RNAs blocking that action.

Several other ncRNAs have been identified, including Jpx/Enox (Chureau et al. 2002; Johnston et al. 2002), Ftx (Chureau et al. 2002), and the bidirectional transcripts (Cohen et al. 2007) associated with a repeat element called "DXPas34," located at *Tsix*'s 5′ end (Courtier et al. 1995; Debrand et al. 1999). Small RNAs of 21-42 nucleotides are also observed in this region, presumably cleaved from long Xist/Tsix RNA duplexes (Ogawa et al. 2008). Although some ncRNAs remain uncharacterized, several have well-defined roles and their mechanisms of action have come into sharper focus during the past years. The discrete roles of four noncoding genes—*Xite*, *Tsix*, *RepA*, and *Xist*—will be discussed in the next sections.

4 RNA IN PAIRING, COUNTING, AND CHOICE

In the widely used embryonic stem (ES) cell model, XCI is triggered directly by decreasing Oct4 levels and the onset of cell differentiation (Monk and Harper 1979; Navarro et al. 2008; Donohoe et al. 2009). Evidence suggests that "counting" involves interaction between Oct4 and a 15-kb region encompassing *Xite* and the 5′ end of *Tsix* (Fig. 1A)(Clerc and Avner 1998; Morey et al. 2004; Lee 2005; Donohoe et al. 2009). This same region has also been shown to function during "choice," Its mutually exclusive nature implies interchromosomal communication to ensure that each cell acquires exactly one Xa and one Xi (Marahrens 1999; Lee 2005). Recent studies have proposed that homologous X-chromosome "pairing" may mediate communication *in trans* and, in doing so, help achieve correct counting and choice of Xa and Xi (Bacher et al. 2006; Xu et al. 2006; Xu et al. 2007; Donohoe et al. 2009). Time course analyses have indicated that the Xs make very brief contact ($t_{1/2}$ <30–60 min) shortly after the onset of cell differentiation and just before the up-regulation of Xist RNA (Xu et al. 2007). When pairing is disrupted genetically or pharmacologically, cells adopt either the XaXa or XiXi phenotype and they die or arrest in culture. Thus, pairing achieves counting and the mutually exclusive choice of Xa and Xi by providing a platform for inter-chromosomal communication and by breaking symmetry between two X-chromosomes (Bacher et al. 2006; Xu et al. 2006; Nicodemi and Prisco 2007b).

Unlike homologous pairing during meiosis and in somatic cells of fruitflies, X-X pairing does not involve whole chromosomes—it occurs only at the *Xic*. Not surprisingly, pairing elements coincide with those responsible for counting and choice in the 15-kb region including *Xite* and *Tsix* (Xu et al. 2006)(Fig. 1A). Deleting 3.7 kb of sequence at the 5′ end of *Tsix* on both X-chromosomes abrogates pairing. For *Xite*, deleting even just one allele severely compromises pairing potential in female cells. *Tsix* and *Xite* are not only required but are also sufficient to induce pairing. When

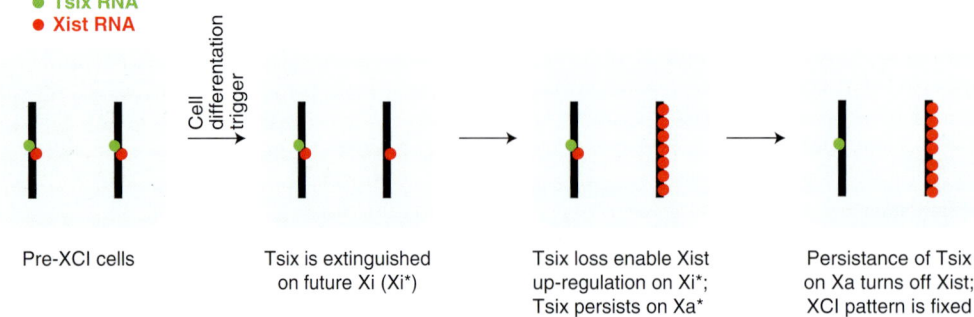

Figure 2. Yin-Yang relationship between Tsix and Xist RNAs. *Tsix* bears a critical relationship to *Xist* expression. In undifferentiated cells (pre-XCI), Tsix is expressed from both Xs at high levels and Xist RNA is expressed at low basal levels (~5 copies). At the onset of cell differentiation, Tsix persists only on the chromosome selected to become Xa. It is the persistence of Tsix RNA that prevents the up-regulation of *Xist* on Xa. Conversely, the down-regulation of Tsix on Xi creates a permissive state for *Xist* transactivation and spread of the RNA along the X to initiate chromosome-wide silencing. Tsix is eventually turned off on Xa once the pattern of XCI is fixed.

inserted at ectopic locations in the genome, *Tsix* and *Xite* direct pairing of the X with the chromosome into which they are inserted (Bacher et al. 2006; Xu et al. 2006). Unexpectedly, pairing can be induced by 1–3-kb fragments that carry very diverse sequences, and even sequences of very low complexity recapitulate pairing. For example, a 1.6-kb element (*DXPas34*) carrying a 34-nucleotide motif repeated 29 times induces X-autosome interactions when multimerized on autosomes (Xu et al. 2007).

How do elements of such diverse nature and low complexity drive two chromosomes together? Two interesting features are shared by most if not all pairing-competent sequences (Xu et al. 2007). First, they bind Ctcf protein, an 11-Zn finger protein involved in many aspects of epigenetic regulation (Lobanenkov et al. 1990; Bell et al. 1999; Ohlsson et al. 2001; Kim et al. 2007). Ctcf has a large number of binding sites within *Tsix* and *Xite* loci and is thought to function as a transcriptional activator of both genes (Chao et al. 2002; Xu et al. 2007). Ctcf interacts with Oct4, a pluripotency factor thought to trigger counting as Oct4 levels fall during cell differentiation, and knocking down either Ctcf or Oct4 leads to an abrogation of pairing in ES cells (Donohoe et al. 2009; Xu et al. 2007).

Sequences sufficient for pairing also share transcriptional competence, a fact that raises the intriguing possibility of an RNA requirement. Despite their diversity of sequence and low sequence complexity, nearly all pairing-competent fragments show Pol-II promoter activity (Sado et al. 2001; Ogawa and Lee 2003; Stavropoulos et al. 2005; Cohen et al. 2007; Xu et al. 2007). Inhibiting Pol-II transcription by a 2.0–4.0 h actinomycin D treatment completely inhibits X-X pairing, indicating that new transcription is necessary for X-chromosome coupling (Xu et al. 2007). In contrast, inhibiting Pol-II for as little as 0.5 h disrupted new pair formation but did not affect cohesion of previously formed pairs. Time course analysis shows that cohesion of the previously formed pairs decayed with a half-life of ∼0.5–1.0 h. Combined, these data show that pairing is very transient and that an RNA, or the act of transcribing it, is crucial for formation of new X-chromosome pairs.

It is therefore proposed that a complex of RNA and protein mediates X-X pairing (Fig. 3). In undifferentiated pre-XCI cells, Oct4 and Ctcf proteins would dimerize and bind *Tsix* and *Xite* to activate expression of the ncRNAs. Falling Oct4 levels during differentiation would then trigger trans-association of the *Xic*'s in an RNA-dependent manner. The *trans*-interactions would then lead to physical differences between two Xs and signal one to become Xa and the other Xi. For example, proximity of *Xic*'s might result in the irreversible shift of protein factors such as Oct4 and Ctcf—which are initially bound to both *Xic*'s—from one *Xic* (future Xi) to the other (future Xa) (Xu et al. 2006; Xu et al. 2007; Donohoe et al. 2009). Although this idea has yet to be tested, computational modeling suggests that, under certain conditions, the shift of factors to one chromosome could create greater thermodynamic stability (Nicodemi and Prisco 2007a; Nicodemi

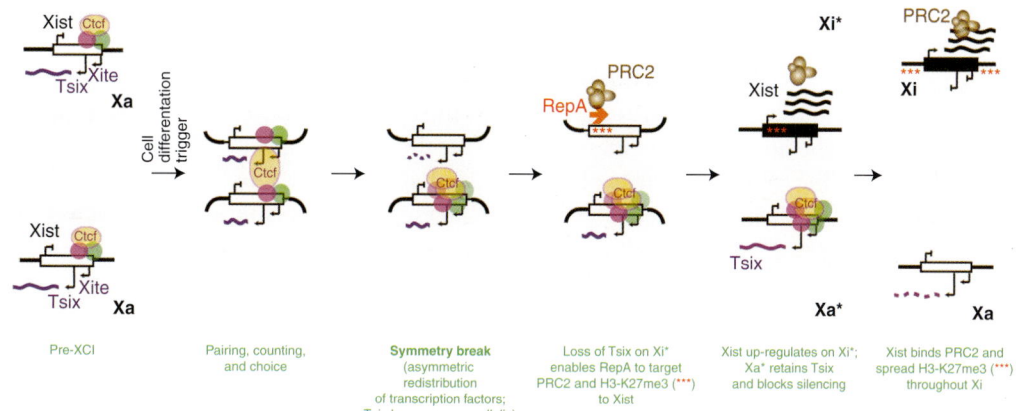

Figure 3. Symmetry break mediated by pairing and *Tsix*. Adapted from (Anguera et al. 2006) and (Lee 2009): In the pre-XCI state, the Xs are epigenetically equivalent, both expressing Tsix at high levels and Xist at low basal levels. Falling levels of Oct4 during cell differentiation triggers X-X pairing. Pairing is mediated by Ctcf, Tsix, and Xite and enables cross-talk to achieve correct counting and mutually exclusive Xa and Xi fates. Transcription factors such as Ctcf, Oct4, and others (red, green circles), which were previously randomly distributed between the two Xs, shift to bind one X, a state that is thermodynamically favored. The chromosome that retains the transcription factors maintains Tsix expression and becomes Xa. Loss of factor binding on the future Xi enables RepA to target PRC2 and H3-K27me3 to the 5′ end of *Xist*, an event that leads to *Xist*'s transcriptional activation. Xist RNA spreads along the future Xi and recruits PRC2 to it to initiate chromosome-wide silencing.

and Prisco 2007b; Nicodemi et al. 2008). This shift of activators would then result in allele-specific persistence of *Tsix* on the future Xa, which in turn would establish the necessary X-chromosome asymmetry. In this way, the pairing of X's makes possible both the counting of X's and the mutually exclusive choice of Xa and Xi at the onset of cell differentiation.

5 SYMMETRY BREAK BY TSIX RNA

Such a model suggests that pairing generates asymmetric Tsix expression and that *Tsix* must therefore function as a molecular switch for XCI. Accordingly, *Tsix* is one of the most heavily mutagenized genes in the history of mouse genetics. Toward understanding its mechanism of action, multiple knockouts, knock-ins, and transgene manipulations have been created to delete, truncate, overexpress, and transplant parts or all of the antisense RNA (Lee et al. 1999a; Lee and Lu 1999; Lee 2000; Luikenhuis et al. 2001; Sado et al. 2001; Stavropoulos et al. 2001; Sado et al. 2002; Morey et al. 2004; Shibata and Lee 2004; Lee 2005; Ohhata et al. 2006; Sado et al. 2006; Xu et al. 2007; Ohhata et al. 2008; Shibata et al. 2008). Abrogating *Tsix* expression results in constitutively elevated Xist expression, whereas its forced expression prevents *Xist* up-regulation. Such analyses have shown that not only does *Tsix* participate in the most upstream steps of XCI (counting, choice, pairing), but it also plays an active role during cell differentiation. Indeed, it is *Tsix*'s binary state during cell differentiation that breaks X-chromosome symmetry (Fig. 3).

In the pre-XCI cells, both alleles of *Tsix* are expressed, with the consequence that both *Xist* alleles are uninduced (very low level, basal expression state). Time course analyses of chromatin structure using chromatin immunoprecipitations (ChIP) have revealed dynamic changes during cell differentiation. Several of the analyses show that *Tsix* regulates *Xist* by modulating the chromatin state of the *Xist* promoter (Navarro et al. 2005; Sado et al. 2005; Navarro et al. 2006; Sun et al. 2006). One study followed allelic differences in female ES cells during XCI and found that *Xist* promoter's configuration mirrored that of *Tsix*'s transcriptional state (Sun et al. 2006). When *Tsix* is expressed in pre-XCI cells, both *Xist* alleles are maintained in an open chromatin state. During cell differentiation, loss of *Tsix* on the future Xi results in a closed chromatin conformation along the 40-kb *Tsix*/*Xist* locus. The repressive chromatin strangely correlates with *Xist* transactivation, as the 5′ end of *Xist* becomes H3-K27 trimethylated (H3-K27me3) just prior to the 100-fold induction of *Xist* transcription (see next section for further discussion). These unexpected results suggested that *Xist* expression may, paradoxically, be enhanced by repressive chromatin.

On the future Xa, cell differentiation leads to continued expression of *Tsix*, which in turn maintains the domain in an open state that is enriched for H3-K4 methylation and H4 acetylation and depleted of H3-K27 trimethylation. This "open" chromatin configuration occurs on the repressed *Xist* allele. RNA ChIP in ES cells suggests that Tsix RNA complexes with the de novo methyltransferase, Dnmt3a, and facilitates CpG methylation and silencing of the *Xist* promoter (Sun et al. 2006), consistent with in vivo analysis in *Tsix*-deficient and Dnmt3a-deficient mice (Sado et al. 2004; Sado et al. 2005). Interestingly, small RNAs can be generated from longer Tsix:Xist RNA duplexes in vivo, a finding that raises the possibility of RNA-directed DNA methylation and transcriptional gene silencing (Ogawa et al. 2008). Consistent with this idea, Dicer-deficient ES cells, which have markedly reduced small RNA production, have reduced *Xist* DNA methylation and significantly increased Xist RNA levels (Nesterova et al. 2008; Ogawa et al. 2008; Kanellopoulou et al. 2009). These results lead to a working hypothesis in which Tsix RNA maintains the activity of the X in two ways—first by directing euchromatic modifications to the *Xist* locus and paradoxically blocking *Xist* activation, and second by recruiting Dnmt3a activity to CpG-methylate and stably silencing the *Xist* promoter (Fig. 4).

Therefore, *Tsix*'s binary state—"off" on Xi and "on" on Xa—underlies the asymmetry of XCI. A crucial aspect of the mechanism may be that the chromatin states of *Tsix* and *Xist* are intimately locked together, such that when *Tsix* is euchromatic, *Xist* is also; when *Tsix* is heterochromatic, *Xist* is likewise. This lock-step relationship elicits the obvious question of why their expression patterns do not also mirror each other—and in fact are even diametrically opposed. It is in this paradox, we believe, that one of the challenges of XCI is met. We suppose that *Tsix* and *Xist* thrive in opposite chromatin environments, with *Tsix* being activated by open chromatin, and *Xist* being repressed by it. This unusual structure-function relationship serves XCI exceptionally well, when considered from the standpoint that *Xist* must escape inactivation on the X that it ultimately silences. A gene that is activated by heterochromatin and remains robust in it would be singularly suited to both initiate and maintain the Xi state.

6 POLYCOMB PROTEINS TARGETED AND REGULATED BY RNA

Here we consider how *Xist* is induced and initiates chromosome-wide silencing. One of the first changes to occur during the silencing step is the recruitment of Polycomb repressive complex 2 (PRC2). PRC2 contains four

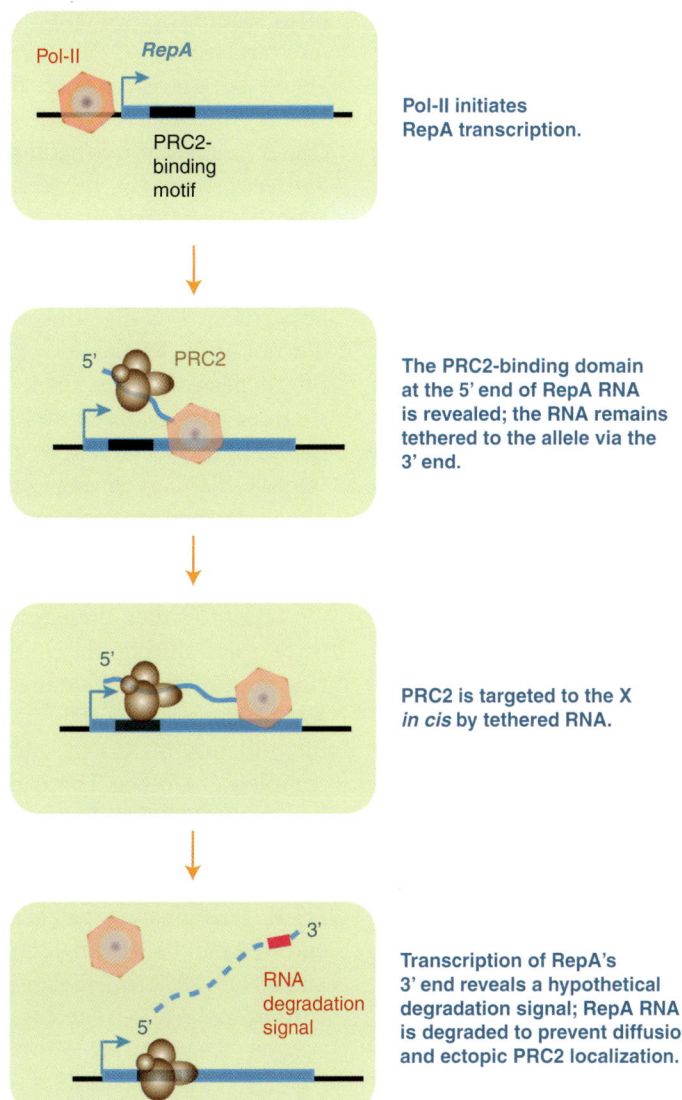

Figure 4. Long ncRNAs recruit chromatin modifiers in a *cis*-limited and locus-specific manner. The example of PRC2 recruitment by RepA RNA is shown.

subunits (Eed, Ezh2, Suz12, and RbAp46/48) and belongs to the Polycomb group (PcG) family of proteins that have genome-wide roles in the maintenance of stem-cell identity, proper developmental progression, and etiology of diseases such as cancer (Ringrose and Paro 2004; Schwartz and Pirrotta 2008). How Polycomb complexes are recruited to specific genomic locations remains unclear. *Drosophila* PcG complexes are known to contain sequence-specific DNA binding proteins such as Zeste, Pipsqueak (PSQ), and Pho, but whether mammalian complexes also contain such DNA-binding proteins remains somewhat controversial (Ringrose and Paro 2004; Schwartz and Pirrotta 2008). The possibility of an RNA cofactor was initially raised by the curious observation that PcG complexes become unstable when treated with RNAse (R. Paro, personal communication). Experimental evidence first came to light with the discovery of HOTAIR for human *HOX* gene regulation (Rinn et al. 2007) and *RepA* for XCI (Zhao et al. 2008).

XCI has provided a convenient model to study Polycomb regulation, as PRC2 is localized *en masse* to the X almost immediately following Xist RNA accumulation (Plath et al. 2003; Silva et al. 2003; Kohlmaier et al. 2004). As discussed in the previous section, one of the first changes to follow Tsix down-regulation is appearance of a PRC2 and H3-K27me3 domain at the 5′ end of *Xist* (Sun et al. 2006; Zhao et al. 2008). PRC2 is brought there *in cis* by RepA RNA, a 1.6 kb ncRNA originating from the 5′ end of *Xist*, precisely in the region showing H3-K27me3 activity. RepA consists of 7.5 tandem repeats of two stem-loop structures (Wutz et al. 2002), and is transcribed in the same orientation as *Xist*. RNA gel shift analysis shows that RepA RNA directly interacts with PRC2 through its 28-nt stem loop structures and binds the catalytic subunit, Ezh2 (Zhao et al. 2008). RepA is apparently sufficient to recruit PRC2, as when an inducible *RepA* transgene is placed at an ectopic location, increased recruitment of PRC2 can be observed. Importantly, this recruitment is observed only when *RepA* is transcribed (transgene expression induced by doxycycline treatment). When a transcribed RepA RNA is knocked down by shRNA, PRC2 recruitment *in cis* is significantly disrupted. Taken together, these experiments show that RepA functions as a ncRNA and directly targets PRC2 to the *Xic*.

RepA-mediated PRC2 recruitment and H3-K27 trimethylation of the *Xist* promoter are believed to create the "heterochromatic state" (Sun et al. 2006)(discussed in the previous section) that is required for transcriptional induction of *Xist*. Indeed, knocking down RepA RNA and PRC2 components (Zhao et al. 2008) or knocking out the Repeat A motif in mice (Hoki et al. 2009) results in the inability to activate *Xist* expression. In undifferentiated ES cells, RepA RNA is biallelically expressed and binds PRC2 without catalyzing H3-K27 trimethylation. During cell differentiation, the down-regulation of Tsix RNA on the future Xi allows the RepA-PRC2 complex to load onto the *Xist* chromatin and induce H3-K27me3, an event that would then lead to activation of the *Xist* promoter, accumulation of Xist RNA, and its spread along the X. RNA immunoprecipitations (RIP) show that Xist RNA can also bind PRC2, presumably via the Repeat A motif, which it shares with RepA. Because Xist RNA greatly exceeds the length of RepA and contains a number of putative chromosome-binding domains (Wutz et al. 2002), Xist RNA can perform what RepA RNA cannot—the spreading of PRC2 along the rest of the X-chromosome.

In the course of characterizing the RepA-PRC2 interaction by RNA EMSA, it was discovered that the antisense strand of RepA—i.e., Tsix—could bind PRC2 (Zhao et al. 2008). Similarly, RIP analysis in ES cells indicated that Tsix RNA occurred in a complex with PRC2. These unexpected findings led to the idea that Tsix RNA might act as a competitive inhibitor for PRC2. Several mechanisms could be envisioned. First, if Tsix and RepA bound PRC2 in a mutually exclusive manner, Tsix RNA could directly compete with RepA RNA for PRC2. Because Tsix is present in vast molar excess over RepA RNA ($\sim$100:1) in undifferentiated ES cells (Shibata and Lee 2003), formation of a Tsix-PRC2 complex would be stoichiometrically favored over a RepA-PRC2 complex. Alternatively, Tsix RNA could block loading of the RepA-PRC2 complex to chromatin, or interfere with PRC2's catalytic activity—both of which would be consistent with the fact that H3-K27me3 on *Xist* is not observed until Tsix RNA is down-regulated (Zhao et al. 2008). Whether PRC2 binds both sense and antisense strands at the same time or mutually exclusively is not known. Pretreating RIP samples with ribonucleases showed that the RNAs were destroyed in the presence of RNAse I and RNAse V1, which digest ssRNA and dsRNA respectively, but digesting with RNAse H and DNAse I did not affect RIP pulldowns. Therefore, RNA bound by PRC2 possess both single-stranded (ss) and double-stranded (ds) character and is not annealed to DNA. The bound transcript(s) may be a single strand of RNA with intramolecular basepairing (secondary structure) or an RNA duplex formed by RepA and Tsix RNAs. The existence of such duplex RNA is supported by in vivo RNAse protection assays in ES cells (Ogawa et al. 2008).

Taken together, these results suggest that ncRNAs play three roles during the initiation of the silencing step (Fig. 3): (1) In undifferentiated ES cells and on the future Xa of differentiating cells, high levels of Tsix RNA prevent the initiation of XCI by interfering with the RepA-PRC2 complex. (2) When Tsix is down-regulated on the future Xi, RepA RNA productively engages PRC2, targets the Polycomb activity to the *Xist* promoter, and triggers the 100-fold up-regulation of Xist RNA by trimethylating H3-K27 to create a patch of heterochromatin at the 5′ end of *Xist*. (3) Xist then binds PRC2 and spreads PRC2 and its H3-K27 trimethylase activity along the X to initiate chromosome-wide silencing.

6.1 The Advantages of Long ncRNA

The X-linked region that is now the X-inactivation center was once dominated by protein-coding genes (Duret et al. 2006). Comparative genomic analysis suggests that ncRNAs infiltrated the region (and transformed the original coding residents) only within the last 120–150 million years with the rise of eutherian mammals and the evolution of random XCI. The marsupial mammal, which uses an imprinted form of XCI (Sharman 1971), completely lacks an *Xic* and its noncoding genes (Duret et al. 2006; Davidow et al. 2007; Hore et al. 2007; Shevchenko et al. 2007). Clearly, ncRNAs of the *Xic* established themselves very quickly and effectively within a short space of time. Does RNA's preeminence here hint at similar strategies elsewhere in the genome, especially considering the degree of noncoding transcription that is now known to occur generally? For epigenetic regulation, there may be two simple reasons to favor RNA over protein (and small molecules). I contend that RNAs deliver superior *cis*-control of allelism and make possible the targeting of a unique genomic location (Lee 2009).

Unlike proteins and small RNAs, large ncRNAs are stably tethered to the site of transcription and therefore tag the allele of origin. Because a long ncRNA remains bound via RNA polymerase to its parent locus during the act of transcription, the large RNA presents itself as a sequence-specific tag for the locus. The proteome cannot serve this function, as memory of allelic origin is always lost once mRNA detaches from the locus and exits the nucleus to be translated to protein. The ncRNA tags must be lengthy, with a 5′ "business end" that binds its target protein(s) once the RNA is synthesized and exposed, and a transcriptionally lagging 3′ "tethering end" that remains associated with the transcription machinery and chromatin (Fig. 4). With the potential to be locus-specific "beacons," long ncRNA may be the missing link between chromatin modifier and genomic target. It is probably no coincidence that the Repeat A motif is at the proximal terminus of both RepA and Xist RNAs. This strategic position enables the RNAs to bind PRC2 co-transcriptionally and tether the proteins to the X *in cis*. If the RNA is then destabilized by degradation signals at the 3′ end that would be revealed by Pol-II as transcription is completed, the RNA could be prevented from diffusing away from the site of synthesis. In this manner, long ncRNAs such as RepA and Xist RNA could function monoallelically and act strictly *in cis*.

Large ncRNAs are drawn from larger sequence space than proteins and can therefore achieve greater sequence specificity. The diversity of sequences specified by RNA is a direct function of its length, composition, and nucleotide permutation. RNA sequence space exceeds that of the proteome, as there is virtually no limit as to how long a macroRNA can be. A single transcription unit can identify a unique position in the genome. In contrast, transcription factors and other DNA-binding proteins recognize DNA motifs of a limited number of basepairs and the number of possible binding sites is therefore generally $\gg 1$, enabling a transcription factor to operate within a large regulatory network. For instance, the transcription factor and chromatin insulator, Ctcf, binds a 20-bp motif that appears thousands of times in the genome, and the pluripotency factor, Oct4, likewise binds an 8-bp motif with many genomic repetitions (Kim et al. 2007; Chen et al. 2008). On the other hand, the Tsix and RepA ncRNAs occur only once, specifically at the *Xic*. Therefore, whereas transcription factors tend to function within a large gene network, long ncRNA can specify a unique address in the genome.

The concept of an "RNA guide" to target chromatin modifiers takes RNA away from its well-recognized role as intermediary between genotype (DNA) and phenotype (protein) and shifts it toward a dynamic role in epigenome control. RNA-mediated targeting need not be restricted to the X. PRC2 has also been associated, either directly or indirectly, with autosomal ncRNAs such as HOTAIR (Rinn et al. 2007; Khalil et al. 2009) and Kcnq1ot1 (Pandey et al. 2008). Other chromatin modifiers, such as the *de novo* DNA methyltransferase Dnmt3a (Sun et al. 2006) and the H3-K9 methylase G9a (Nagano et al. 2008), may also be regulated by long ncRNA, as suggested by RNA ChIP. Further work is required to determine if there is direct RNA-protein interaction or if the proteins are recruited by the underlying chromatin, of which RNA may be one component. An RNA-mediated mechanism may explain why many chromatin modifiers lack a sequence-specific DNA-binding activity but often harbor RNA-binding motifs (Denisenko et al. 1998; Bernstein and Allis 2005; Bernstein et al. 2006).

With an ever-growing number of long ncRNAs being discovered across the genome (Claverie 2005; Kapranov et al. 2007a; Kapranov et al. 2007b; Guttman et al. 2009; Mercer et al. 2009), it is easy to imagine that RNA guidance may become a recurrent strategy in epigenomic regulation. Interchromosomal associations have been similarly proposed to underlie allelic choice at autosomal loci (LaSalle and Lalande 1996; Spilianakis et al. 2005; Lomvardas et al. 2006), begging the question of whether transcription and RNA guide these *trans*-interactions. Furthermore, many genes harbor antisense RNA (Katayama et al. 2005; He et al. 2008), raising the possibility of RNA-mediated recruitment of chromatin modifiers at those loci as well. On the X-chromosome, ncRNA's full range of function has yet to be explored fully, given that many *Xic*-linked ncRNAs remain uncharacterized.

As we discover more examples of regulation by long ncRNA, one might predict that the large transcripts will eventually rival small RNAs and proteins in their versatility as epigenetic regulators. This article has argued that long ncRNAs are particularly well-suited to allelic regulation and to any action that requires locus-specific targeting and communication *in cis*. Thus, the vast intergenic sequence space that was once widely thought to harbor "junk" may actually be primary determinants of phenotypic variation within and among species. Closely related organisms (e.g., humans versus chimps, mice versus rats), and even individuals of a species, are virtually identical in their "ORF-omes" but can vary significantly in their transcriptomes. For example, differences in coat color and disease susceptibility have been attributed to a retrotransposon near the *Agouti* gene in mice (Waterland and Jirtle 2003; Blewitt et al. 2006). Human cancer and metastasis have been associated with loss of regulation by long ncRNAs, including MALAT1 for lung adenocarcinomas,

GTL2 in neuronal tumors, and H19 for Wilms tumor (Lin et al. 2007; Edwards et al. 2008; Wan and Bartolomei 2008; Wilusz et al. 2008). Interestingly, many of these RNAs operate in cis. Before long, we may be looking into the ncRNA world, as a matter of routine, for answers to many questions relating to health and disease.

ACKNOWLEDGMENTS

I am greatly indebted to all members of my lab for their invaluable discussions, insights, and inspiration. Work described in this article has been sponsored by either the National Institutes of Health. (RO1-GM58839) or by the HHMI, where JTL is an investigator.

REFERENCES

Anguera MC, Sun BK, Xu N, Lee JT. 2006. X-chromosome kiss and tell: How the Xs go their separate ways. *Cold Spring Harb Symp Quant Biol* **71:** 429–437.

Avner P, Heard E. 2001. X-chromosome inactivation: Counting, choice and initiation. *Nat Rev Genet* **2:** 59–67.

Bacher CP, Guggiari M, Brors B, Augui S, Clerc P, Avner P, Eils R, Heard E. 2006. Transient colocalization of X-inactivation centres accompanies the initiation of X inactivation. *Nat Cell Biol* **8:** 293–299.

Bartolomei MS, Zemel S, Tilghman SM. 1991. Parental imprinting of the mouse H19 gene. *Nature* **351:** 153–155.

Bell AC, West AG, Felsenfeld G. 1999. The protein CTCF is required for the enhancer blocking activity of vertebrate insulators. *Cell* **98:** 387–396.

Bernstein E, Allis CD. 2005. RNA meets chromatin. *Genes Dev* **19:** 1635–1655.

Bernstein E, Duncan EM, Masui O, Gil J, Heard E, Allis CD. 2006. Mouse polycomb proteins bind differentially to methylated histone H3 and RNA and are enriched in facultative heterochromatin. *Mol Cell Biol* **26:** 2560–2569.

Blewitt ME, Vickaryous NK, Paldi A, Koseki H, Whitelaw E. 2006. Dynamic reprogramming of DNA methylation at an epigenetically sensitive allele in mice. *PLoS Genet* **2:** e49.

Borsani G, Tonlorenzi R, Simmler MC, Dandolo L, Arnaud D, Capra V, Grompe M, Pizzuti A, Muzny D, Lawrence C, et al. 1991. Characterization of a murine gene expressed from the inactive X chromosome. *Nature* **351:** 325–329.

Boumil RM, Lee JT. 2001. Forty years of decoding the silence in X-chromosome inactivation. *Hum Mol Genet* **10:** 2225–2232.

Brannan CI, Dees EC, Ingram RS, Tilghman SM. 1990. The product of the H19 gene may function as an RNA. *Mol Cell Biol* **10:** 28–36.

Brockdorff N, Ashworth A, Kay GF, McCabe VM, Norris DP, Cooper PJ, Swift S, Rastan S. 1992. The product of the mouse Xist gene is a 15 kb inactive X-specific transcript containing no conserved ORF and located in the nucleus. *Cell* **71:** 515–526.

Brown CJ, Ballabio A, Rupert JL, Lafreniere RG, Grompe M, Tonlorenzi R, Willard HF. 1991a. A gene from the region of the human X inactivation centre is expressed exclusively from the inactive X chromosome. *Nature* **349:** 38–44.

Brown CJ, Hendrich BD, Rupert JL, Lafreniere RG, Xing Y, Lawrence J, Willard HF. 1992. The human XIST gene: Analysis of a 17 kb inactive X-specific RNA that contains conserved repeats and is highly localized within the nucleus. *Cell* **71:** 527–542.

Brown CJ, Lafreniere RG, Powers VE, Sebastio G, Ballabio A, Pettigrew AL, Ledbetter DH, Levy E, Craig IW, Willard HF. 1991b. Localization of the X inactivation centre on the human X chromosome in Xq13. *Nature* **349:** 82–84.

Cam HP, Chen ES, Grewal SI. 2009. Transcriptional scaffolds for heterochromatin assembly. *Cell* **136:** 610–614.

Carninci P, Kasukawa T, Katayama S, Gough J, Frith MC, Maeda N, Oyama R, Ravasi T, Lenhard B, Wells C, et al. 2005. The transcriptional landscape of the mammalian genome. *Science* **309:** 1559–1563.

Cattanach BM, Isaacson JH. 1967. Controlling elements in the mouse X chromosome. *Genetics* **57:** 331–346.

Cech TR. 2009. Crawling out of the RNA world. *Cell* **136:** 599–602.

Chao W, Huynh KD, Spencer RJ, Davidow LS, Lee JT. 2002. CTCF, a candidate trans-acting factor for X-inactivation choice. *Science* **295:** 345–347.

Chen X, Xu H, Yuan P, Fang F, Huss M, Vega VB, Wong E, Orlov YL, Zhang W, Jiang J, et al. 2008. Integration of external signaling pathways with the core transcriptional network in embryonic stem cells. *Cell* **133:** 1106–1117.

Chureau C, Prissette M, Bourdet A, Barbe V, Cattolico L, Jones L, Eggen A, Avner P, Duret L. 2002. Comparative sequence analysis of the X-inactivation center region in mouse, human, and bovine. *Genome Res* **12:** 894–908.

Claverie JM. 2005. Fewer genes, more noncoding RNA. *Science* **309:** 1529–1530.

Clemson CM, McNeil JA, Willard HF, Lawrence JB. 1996. XIST RNA paints the inactive X chromosome at interphase: Evidence for a novel RNA involved in nuclear/chromosome structure. *J Cell Biol* **132:** 259–275.

Clerc P, Avner P. 1998. Role of the region 3' to Xist exon 6 in the counting process of X-chromosome inactivation. *Nat Genet* **19:** 249–253.

Cohen DE, Davidow LS, Erwin JA, Xu N, Warshawsky D, Lee JT. 2007. The DXPas34 repeat regulates random and imprinted X inactivation. *Dev Cell* **12:** 57–71.

Courtier B, Heard E, Avner P. 1995. Xce haplotypes show modified methylation in a region of the active X chromosome lying 3' to Xist. *Proc Natl Acad Sci* **92:** 3531–3535.

Davidow LS, Breen M, Duke SE, Samollow PB, McCarrey JR, Lee JT. 2007. The search for a marsupial XIC reveals a break with vertebrate synteny. *Chromosome Res* **15:** 137–146.

Debrand E, Chureau C, Arnaud D, Avner P, Heard E. 1999. Functional analysis of the DXPas34 locus, a 3' regulator of Xist expression. *Mol Cell Biol* **19:** 8513–8525.

Denisenko O, Shnyreva M, Suzuki H, Bomsztyk K. 1998. Point mutations in the WD40 domain of Eed block its interaction with Ezh2. *Mol Cell Biol* **18:** 5634–5642.

Donohoe ME, Silva SS, Pinter SF, Xu N, Lee JT. 2009. The pluripotency factor Oct4 interacts with Ctcf and also controls X-chromosome pairing and counting. *Nature* **460:** 128–132.

Donohoe ME, Zhang LF, Xu N, Shi Y, Lee JT. 2007. Identification of a Ctcf cofactor, Yy1, for the X chromosome binary switch. *Mol Cell* **25:** 43–56.

Duret L, Chureau C, Samain S, Weissenbach J, Avner P. 2006. The Xist RNA gene evolved in eutherians by pseudogenization of a protein-coding gene. *Science* **312:** 1653–1655.

Edwards CA, Ferguson-Smith AC. 2007. Mechanisms regulating imprinted genes in clusters. *Curr Opin Cell Biol* **19:** 281–289.

Edwards CA, Mungall AJ, Matthews L, Ryder E, Gray DJ, Pask AJ, Shaw G, Graves JA, Rogers J, Dunham I, et al. 2008. The evolution of the DLK1-DIO3 imprinted domain in mammals. *PLoS Biol* **6:** e135.

Ghildiyal M, Zamore PD. 2009. Small silencing RNAs: an expanding universe. *Nat Rev Genet* **10:** 94–108.

Guttman M, Amit I, Garber M, French C, Lin MF, Feldser D, Huarte M, Zuk O, Carey BW, Cassady JP, et al. 2009. Chromatin signature reveals over a thousand highly conserved large non-coding RNAs in mammals. *Nature* **458:** 223–227.

He Y, Vogelstein B, Velculescu VE, Papadopoulos N, Kinzler KW. 2008. The antisense transcriptomes of human cells. *Science* **322:** 1855–1857.

Hoki Y, Kimura N, Kanbayashi M, Amakawa Y, Ohhata T, Sasaki H, Sado T. 2009. A proximal conserved repeat in the Xist gene is essential as a genomic element for X-inactivation in mouse. *Development* **136**: 139–146.

Hore TA, Koina E, Wakefield MJ, Marshall Graves JA. 2007. The region homologous to the X-chromosome inactivation centre has been disrupted in marsupial and monotreme mammals. *Chromosome Res* **15**: 147–161.

Johnston CM, Newall AE, Brockdorff N, Nesterova TB. 2002. Enox, a novel gene that maps 10 kb upstream of Xist and partially escapes X inactivation. *Genomics* **80**: 236–244.

Kanellopoulou C, Muljo SA, Dimitrov SD, Chen X, Colin C, Plath K, Livingston DM. 2009. X chromosome inactivation in the absence of Dicer. *Proc Natl Acad Sci* **106**: 1122–1127.

Kapranov P, Willingham AT, Gingeras TR. 2007b. Genome-wide transcription and the implications for genomic organization. *Nat Rev Genet* **8**: 413–423.

Kapranov P, Cheng J, Dike S, Nix DA, Duttagupta R, Willingham AT, Stadler PF, Hertel J, Hackermuller J, Hofacker IL, et al. 2007a. RNA maps reveal new RNA classes and a possible function for pervasive transcription. *Science* **316**: 1484–1488.

Katayama S, Tomaru Y, Kasukawa T, Waki K, Nakanishi M, Nakamura M, Nishida H, Yap CC, Suzuki M, Kawai J, et al. 2005. Antisense transcription in the mammalian transcriptome. *Science* **309**: 1564–1566.

Kay GF, Barton SC, Surani MA, Rastan S. 1994. Imprinting and X chromosome counting mechanisms determine Xist expression in early mouse development. *Cell* **77**: 639–650.

Khalil AM, Guttman M, Huarte M, Garber M, Raj A, Rivea Morales D, Thomas K, Presser A, Bernstein BE, van Oudenaarden A, et al. 2009. Many human large intergenic noncoding RNAs associate with chromatin-modifying complexes and affect gene expression. *Proc Natl Acad Sci* **106**: 11667–11672.

Kim TH, Abdullaev ZK, Smith AD, Ching KA, Loukinov DI, Green RD, Zhang MQ, Lobanenkov VV, Ren B. 2007. Analysis of the vertebrate insulator protein CTCF-binding sites in the human genome. *Cell* **128**: 1231–1245.

Kohlmaier A, Savarese F, Lachner M, Martens J, Jenuwein T, Wutz A. 2004. A chromosomal memory triggered by Xist regulates histone methylation in X inactivation. *PLoS Biol* **2**: E171.

LaSalle JM, Lalande M. 1996. Homologous association of oppositely imprinted chromosomal domains. *Science* **272**: 725–728.

Lee JT. 2000. Disruption of imprinted X inactivation by parent-of-origin effects at Tsix. *Cell* **103**: 17–27.

Lee JT. 2002. Homozygous Tsix mutant mice reveal a sex-ratio distortion and revert to random X-inactivation. *Nat Genet* **32**: 195–200.

Lee JT. 2005. Regulation of X-chromosome counting by Tsix and Xite sequences. *Science* **309**: 768–771.

Lee JT. 2009. Lessons from X-chromosome inactivation: Long ncRNA as guides and tethers to the epigenome. *Genes Dev* **23**: 1831–1842.

Lee JT, Lu N. 1999. Targeted mutagenesis of Tsix leads to nonrandom X inactivation. *Cell* **99**: 47–57.

Lee JT, Davidow LS, Warshawsky D. 1999a. Tsix, a gene antisense to Xist at the X-inactivation centre. *Nat Genet* **21**: 400–404.

Lee JT, Lu N, Han Y. 1999b. Genetic analysis of the mouse X inactivation center defines an 80-kb multifunction domain. *Proc Natl Acad Sci* **96**: 3836–3841.

Lee JT, Strauss WM, Dausman JA, Jaenisch R. 1996. A 450 kb transgene displays properties of the mammalian X-inactivation center. *Cell* **86**: 83–94.

Lin R, Maeda S, Liu C, Karin M, Edgington TS. 2007. A large noncoding RNA is a marker for murine hepatocellular carcinomas and a spectrum of human carcinomas. *Oncogene* **26**: 851–858.

Lobanenkov VV, Nicolas RH, Adler VV, Paterson H, Klenova EM, Polotskaja AV, Goodwin GH. 1990. A novel sequence-specific DNA binding protein which interacts with three regularly spaced direct repeats of the CCCTC-motif in the 5' flaking sequence of the chicken c-myc gene. *Oncogene* **5**: 1743–1753.

Lomvardas S, Barnea G, Pisapia DJ, Mendelsohn M, Kirkland J, Axel R. 2006. Interchromosomal interactions and olfactory receptor choice. *Cell* **126**: 403–413.

Lucchesi JC, Kelly WG, Panning B. 2005. Chromatin remodeling in dosage compensation. *Annu Rev Genet* **39**: 615–651.

Luikenhuis S, Wutz A, Jaenisch R. 2001. Antisense transcription through the Xist locus mediates Tsix function in embryonic stem cells. *Mol Cell Biol* **21**: 8512–8520.

Lyon MF. 1961. Gene action in the X-chromosome of the mouse (*Mus musculus* L.). *Nature* **190**: 372–373.

Lyon MF. 1999. Imprinting and X chromosome inactivation. In *Results and problems in cell differentiation* (ed. R Ohlsson), pp. 73–90. Springer-Verlag, Heidelberg.

Marahrens Y. 1999. X-inactivation by chromosomal pairing events. *Genes Dev* **13**: 2624–2632.

Marahrens Y, Panning B, Dausman J, Strauss W, Jaenisch R. 1997. Xist-deficient mice are defective in dosage compensation but not spermatogenesis. *Genes Dev* **11**: 156–166.

Masui O, Heard E. 2006. RNA and protein actors in X-chromosome inactivation. *Cold Spring Harb Symp Quant Biol* **71**: 419–428.

Mercer TR, Dinger ME, Mattick JS. 2009. Long non-coding RNAs: Insights into functions. *Nat Rev Genet* **10**: 155–159.

Moazed D. 2009. Small RNAs in transcriptional gene silencing and genome defense. *Nature* **457**: 413–420.

Monk M, Harper MI. 1979. Sequential X chromosome inactivation coupled with cellular differentiation in early mouse embryos. *Nature* **281**: 311–313.

Monkhorst K, de Hoon B, Jonkers I, Mulugeta Achame E, Monkhorst W, Hoogerbrugge J, Rentmeester E, Westerhoff HV, Grosveld F, Grootegoed JA, et al. 2009. The probability to initiate X chromosome inactivation is determined by the X to autosomal ratio and X chromosome specific allelic properties. *PLoS ONE* **4**: e5616.

Morey C, Navarro P, Debrand E, Avner P, Rougeulle C, Clerc P. 2004. The region 3' to Xist mediates X chromosome counting and H3 Lys-4 dimethylation within the Xist gene. *Embo J* **23**: 594–604.

Nagano T, Mitchell JA, Sanz LA, Pauler FM, Ferguson-Smith AC, Feil R, Fraser P. 2008. The Air noncoding RNA epigenetically silences transcription by targeting G9a to chromatin. *Science* **322**: 1717–1720.

Navarro P, Chambers I, Karwacki-Neisius V, Chureau C, Morey C, Rougeulle C, Avner P. 2008. Molecular coupling of Xist regulation and pluripotency. *Science* **321**: 1693–1695.

Navarro P, Page DR, Avner P, Rougeulle C. 2006. Tsix-mediated epigenetic switch of a CTCF-flanked region of the Xist promoter determines the Xist transcription program. *Genes Dev* **20**: 2787–2792.

Navarro P, Pichard S, Ciaudo C, Avner P, Rougeulle C. 2005. Tsix transcription across the Xist gene alters chromatin conformation without affecting Xist transcription: Implications for X-chromosome inactivation. *Genes Dev* **19**: 1474–1484.

Nesterova TB, Popova BC, Cobb BS, Norton S, Senner CE, Tang YA, Spruce T, Rodriguez TA, Sado T, Merkenschlager M, et al. 2008. Dicer regulates Xist promoter methylation in ES cells indirectly through transcriptional control of Dnmt3a. *Epigenetics Chromatin* **1**: 2.

Nicodemi M, Prisco A. 2007a. Self-assembly and DNA binding of the blocking factor in x chromosome inactivation. *PLoS Comput Biol* **3**: e210.

Nicodemi M, Prisco A. 2007b. Symmetry-breaking model for X-chromosome inactivation. *Phys Rev Lett* **98**: 108104.

Nicodemi M, Panning B, Prisco A. 2008. A thermodynamic switch for chromosome colocalization. *Genetics* **179**: 717–721.

Ogawa Y, Lee JT. 2003. Xite, X-inactivation intergenic transcription elements that regulate the probability of choice. *Mol Cell* **11**: 731–743.

Ogawa Y, Sun BK, Lee JT. 2008. Intersection of the RNA interference and X-inactivation pathways. *Science* **320**: 1336–1341.

Ohhata T, Hoki Y, Sasaki H, Sado T. 2006. Tsix-deficient X chromosome does not undergo inactivation in the embryonic lineage in males: Implications for Tsix-independent silencing of Xist. *Cytogenet Genome Res* **113**: 345–349.

Ohhata T, Hoki Y, Sasaki H, Sado T. 2008. Crucial role of antisense transcription across the Xist promoter in Tsix-mediated Xist chromatin modification. *Development* **135**: 227–235.

Ohlsson R, Renkawitz R, Lobanenkov VV. 2001. CTCF is a uniquely versatile transcription regulator linked to epigenetics and disease. *Trends Genet* **7**: 520–527.

Pandey RR, Mondal T, Mohammad F, Enroth S, Redrup L, Komorowski J, Nagano T, Mancini-Dinardo D, Kanduri C. 2008. Kcnq1ot1 antisense noncoding RNA mediates lineage-specific transcriptional silencing through chromatin-level regulation. *Mol Cell* **32**: 232–246.

Payer B, Lee JT. 2008. X Chromosome dosage compensation: How mammals keep the balance. *Annu Rev Genet* **42**: 733–772.

Penny GD, Kay GF, Sheardown SA, Rastan S, Brockdorff N. 1996. Requirement for Xist in X chromosome inactivation. *Nature* **379**: 131–137.

Plath K, Fang J, Mlynarczyk-Evans SK, Cao R, Worringer KA, Wang H, de la Cruz CC, Otte AP, Panning B, Zhang Y. 2003. Role of histone H3 lysine 27 methylation in X inactivation. *Science* **300**: 131–135.

Plath K, Mlynarczyk-Evans S, Nusinow DA, Panning B. 2002. Xist RNA and the mechanism of X chromosome inactivation. *Annu Rev Genet* **36**: 233–278.

Rastan S, Robertson EJ. 1985. X-chromosome deletions in embryo-derived (EK) cell lines associated with lack of X-chromosome inactivation. *J Embryol Exp Morphol* **90**: 379–388.

Ringrose L, Paro R. 2004. Epigenetic regulation of cellular memory by the Polycomb and Trithorax group proteins. *Annu Rev Genet* **38**: 413–443.

Rinn JL, Kertesz M, Wang JK, Squazzo SL, Xu X, Brugmann SA, Goodnough LH, Helms JA, Farnham PJ, Segal E, et al. 2007. Functional demarcation of active and silent chromatin domains in human HOX loci by noncoding RNAs. *Cell* **129**: 1311–1323.

Sado T, Hoki Y, Sasaki H. 2005. Tsix silences Xist through modification of chromatin structure. *Dev Cell* **9**: 159–165.

Sado T, Hoki Y, Sasaki H. 2006. Tsix defective in splicing is competent to establish Xist silencing. *Development* **133**: 4925–4931.

Sado T, Li E, Sasaki H. 2002. Effect of TSIX disruption on XIST expression in male ES cells. *Cytogenet Genome Res* **99**: 115–118.

Sado T, Okano M, Li E, Sasaki H. 2004. De novo DNA methylation is dispensable for the initiation and propagation of X chromosome inactivation. *Development* **131**: 975–982.

Sado T, Wang Z, Sasaki H, Li E. 2001. Regulation of imprinted X-chromosome inactivation in mice by Tsix. *Development* **128**: 1275–1286.

Schwartz YB, Pirrotta V. 2008. Polycomb complexes and epigenetic states. *Curr Opin Cell Biol* **20**: 266–273.

Sharman GB. 1971. Late DNA replication in the paternally derived X chromosome of female kangaroos. *Nature* **230**: 231–232.

Sharp PA. 2009. The centrality of RNA. *Cell* **136**: 577–580.

Shevchenko AI, Zakharova IS, Elisaphenko EA, Kolesnikov NN, Whitehead S, Bird C, Ross M, Weidman JR, Jirtle RL, Karamysheva TV, et al. 2007. Genes flanking Xist in mouse and human are separated on the X chromosome in American marsupials. *Chromosome Res* **15**: 127–136.

Shibata S, Lee JT. 2003. Characterization and quantitation of differential Tsix transcripts: implications for Tsix function. *Hum Mol Genet* **12**: 125–136.

Shibata S, Lee JT. 2004. Tsix transcription- versus RNA-based mechanisms in Xist repression and epigenetic choice. *Curr Biol* **14**: 1747–1754.

Shibata S, Yokota T, Wutz A. 2008. Synergy of Eed and Tsix in the repression of Xist gene and X-chromosome inactivation. *Embo J* **27**: 1816–1826.

Silva J, Mak W, Zvetkova I, Appanah R, Nesterova TB, Webster Z, Peters AH, Jenuwein T, Otte AP, Brockdorff N. 2003. Establishment of histone h3 methylation on the inactive X chromosome requires transient recruitment of Eed-Enx1 polycomb group complexes. *Dev Cell* **4**: 481–495.

Simmler MC, Cunningham DB, Clerc P, Vermat T, Caudron B, Cruaud C, Pawlak A, Szpirer C, Weissenbach J, Claverie JM, et al. 1996. A 94 kb genomic sequence 3′ to the murine Xist gene reveals an AT rich region containing a new testis specific gene Tsx. *Hum Mol Genet* **5**: 1713–1726.

Spilianakis CG, Lalioti MD, Town T, Lee GR, Flavell RA. 2005. Interchromosomal associations between alternatively expressed loci. *Nature* **435**: 637–645.

Stavropoulos N, Lu N, Lee JT. 2001. A functional role for Tsix transcription in blocking Xist RNA accumulation but not in X-chromosome choice. *Proc Natl Acad Sci* **98**: 10232–10237.

Stavropoulos N, Rowntree RK, Lee JT. 2005. Identification of developmentally specific enhancers for Tsix in the regulation of X chromosome inactivation. *Mol Cell Biol* **25**: 2757–2769.

Sun BK, Deaton AM, Lee JT. 2006. A transient heterochromatic state in Xist preempts X inactivation choice without RNA stabilization. *Mol Cell* **21**: 617–628.

Wan LB, Bartolomei MS. 2008. Regulation of imprinting in clusters: Noncoding RNAs versus insulators. *Adv Genet* **61**: 207–223.

Waterland RA, Jirtle RL. 2003. Transposable elements: Targets for early nutritional effects on epigenetic gene regulation. *Mol Cell Biol* **23**: 5293–5300.

Wheeler DL, Barrett T, Benson DA, Bryant SH, Canese K, Chetvernin V, Church DM, Dicuccio M, Edgar R, Federhen S, et al. 2008. Database resources of the National Center for Biotechnology Information. *Nucleic Acids Res* **36**: D13–21.

Willard HF. 1996. X chromosome inactivation, XIST, and pursuit of the X-inactivation center. *Cell* **86**: 5–7.

Wilusz JE, Freier SM, Spector DL. 2008. 3′ end processing of a long nuclear-retained noncoding RNA yields a tRNA-like cytoplasmic RNA. *Cell* **135**: 919–932.

Wutz A. 2003. RNAs templating chromatin structure for dosage compensation in animals. *Bioessays* **25**: 434–442.

Wutz A, Gribnau J. 2007. X inactivation Xplained. *Curr Opin Genet Dev* **17**: 387–393.

Wutz A, Rasmussen TP, Jaenisch R. 2002. Chromosomal silencing and localization are mediated by different domains of Xist RNA. *Nat Genet* **30**: 167–174.

Wutz A, Smrzka OW, Schweifer N, Schellander K, Wagner EF, Barlow DP. 1997. Imprinted expression of the Igf2r gene depends on an intronic CpG island. *Nature* **389**: 745–749.

Xu N, Tsai CL, Lee JT. 2006. Transient homologous chromosome pairing marks the onset of X inactivation. *Science* **311**: 1149–1152.

Xu N, Donohoe ME, Silva SS, Lee JT. 2007. Evidence that homologous X-chromosome pairing requires transcription and Ctcf protein. *Nat Genet* **39**: 1390–1396.

Zhao J, Sun BK, Erwin JA, Song JJ, Lee JT. 2008. Polycomb proteins targeted by a short repeat RNA to the mouse X chromosome. *Science* **322**: 750–756.

The Long Arm of Long Noncoding RNAs: Roles as Sensors Regulating Gene Transcriptional Programs

Xiangting Wang[1] and Xiaoyuan Song[1], Christopher K. Glass[2], and Michael G. Rosenfeld[1]

[1]Howard Hughes Medical Institute, School and Department of Medicine, University of California, San Diego School of Medicine, La Jolla, California 92093-0651

[2]Cellular and Molecular Medicine and Department of Medicine, University of California, San Diego School of Medicine, La Jolla, California 92093-0651

Correspondence: mrosenfeld@ucsd.edu

SUMMARY

A major surprise arising from genome-wide analyses has been the observation that the majority of the genome is transcribed, generating noncoding RNAs (ncRNAs). It is still an open question whether some or all of these ncRNAs constitute functional networks regulating gene transcriptional programs. However, in light of recent discoveries and given the diversity and flexibility of long ncRNAs and their abilities to nucleate molecular complexes and to form spatially compact arrays of complexes, it becomes likely that many or most ncRNAs act as sensors and integrators of a wide variety of regulated transcriptional responses and probably epigenetic events. Because many RNA-binding proteins, on binding RNAs, show distinct allosteric conformational alterations, we suggest that a ncRNA/RNA-binding protein-based strategy, perhaps in concert with several other mechanistic strategies, serves to integrate transcriptional, as well as RNA processing, regulatory programs.

Outline

1 Introduction
2 Characterization of long ncRNAs
3 Long ncRNA-dependent recruitment of protein complexes
4 Mechanisms mediating transcriptional regulation by long ncRNAs
5 Biological roles of long ncRNAs
6 Conclusions and perspectives
References

1 INTRODUCTION

Although the roles of cognate elements for DNA binding proteins and assorted coactivators and corepressors in organizing and encoding patterns of transcriptional responses are well established (Reviewed in Rosenfeld et al. 2006), the discovery that a huge percentage of the genome is transcribed into noncoding RNAs (ncRNAs), has presented the challenge of determining whether there is a vast functional network of ncRNAs that might regulate gene transcriptional programs (Bertone et al. 2004; Carninci et al. 2005; Cheng et al. 2005; Johnson et al. 2005; Katayama et al. 2005; Carninci et al. 2006; Gustincich et al. 2006; Willingham and Gingeras 2006).

The critical roles of some classes of small ncRNAs (reviewed in Chu and Rana 2007; Kutter and Svoboda 2008; Choudhuri 2009) are beginning to emerge; however, the broad spatial distribution of long ncRNA expression, often at remarkably low levels, has itself raised the challenge to delineate the mechanisms underlying their transcription, regulation, and potential functional roles. The basic questions of "who, when, what, where, and how" are still incompletely defined. Here, we will attempt to organize some of the rapidly expanding information regarding the long ncRNAs, focusing on their roles in transcriptional regulation of nonimprinted genes.

Based on a overview of the cumulative data in the literature regarding ncRNAs, we propose that ncRNAs use several mechanisms to exert important regulatory functions in the control of transcriptional programs (Fig. 1). This large number of ncRNAs have the inherent flexibility to function in "nucleating" molecules for combinatorial complex assembly, ideal for functioning as "sensors" and integrators for a variety of developmental and physiological signals. The alternative recruitment of "corepressors," "coactivators," and DNA-modifying complexes in various combinations might additionally be dictated by alterations in strandedness, transcriptional length, or splicing of ncRNAs, functionally modulating their biological actions. We suggest that most of the functions of long ncRNAs relate to their interactions with RNA-binding proteins, most of which contain multiple RNA-binding domains that may recognize distinct "targets." This implies that many RNA-binding proteins might, as a consequence, show distinct "allosterically dictated" conformations based on the specific RNA sequences they bind—in a sense RNAs would be analogous to the effect of distant "ligands" on nuclear receptors (Glass and Rosenfeld 2000; McKenna and O'Malley 2002). Given that RNA-binding proteins constitute the largest "transcriptional" regulatory family in the genome (Burd and Dreyfuss 1994; Lunde et al. 2007), such an ncRNA/RNA-binding protein-based strategy to integrate transcriptional programs is very likely a general rule in the transcriptional as well as RNA processing regulation. Taken together, the ncRNA sensor code appears to be a robust and critical strategy underlying a wide variety of gene regulatory programs.

2 CHARACTERIZATION OF LONG NCRNAS

Classification

Compared with small ncRNAs (Montgomery and Fire 1998; Grishok and Mello 2002; Ambros 2001; Aravin et al. 2006; Girard et al. 2006; Grivna et al. 2006; Lau et al. 2006; Vagin et al. 2006; Watanabe et al. 2006; Amaral and Mattick 2008; Lee et al. 2009), long ncRNAs are often arbitrarily considered as >200 nt and include long ncRNAs implicated in dosage compensation and imprinting. The most recently identified long ncRNAs in transcriptional regulation of nonimprinted genes are usually shorter than 10 kb. These long ncRNAs can be grouped into a remarkably diverse set of transcripts. One classification has been based on abundance, i.e., the "high abundance" (such as *NEAT1* and *NEAT2*; Hutchinson et al. 2007) and "low abundance" long ncRNAs (such as *CCND1*ncRNA; Wang et al. 2008a). Another classification is by function, i.e., "*cis*-acting" and "*trans*-acting" long ncRNAs. The conservative estimated number of long ncRNAs is ~17,000 in the human and ~10,000 in the mouse genome (www.Invitrogene.com). The number is likely to be greatly underestimated, especially, because many primary transcripts are often processed into smaller ncRNAs.

Strandedness

Long ncRNAs can be expressed from either or both DNA strands. Sense–antisense pairing between ncRNAs from two DNA strands has been observed in some cases. For example, massive sense–antisense pairing transcripts are found in *Alu* repeats (Wang et al. 2008b) and UTRs of protein coding genes (Kapranov et al. 2007). The significance of forming a double-stranded (ds) ncRNA is not clear. However, a ds ncRNA will lose the ability to pair with its complementary RNA or DNA sequence. Therefore, we hypothesize that the formation of a ds ncRNA may often act to block/regulate at least some functions of single-stranded (ss) ncRNAs. However, this does not exclude the possibility that ds ncRNAs form to change their affinity for RNA-binding proteins.

Subcellular Localization

In contrast to most mRNAs, which ultimately localize to the cytoplasm after processing, most long ncRNAs are

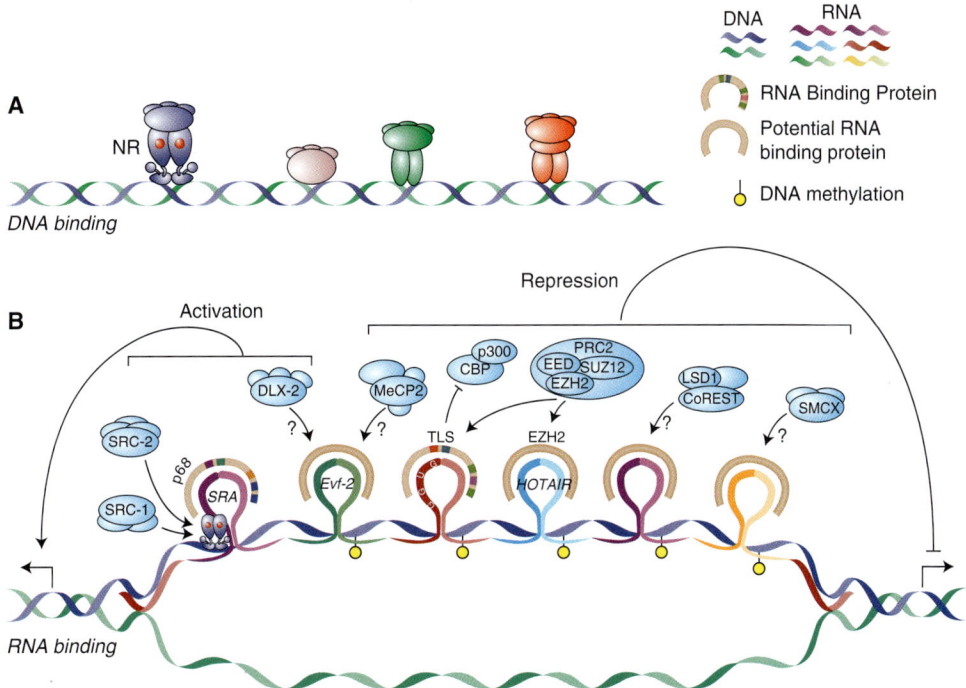

Figure 1. NcRNA-protein network in gene transcription program regulation. Comparing with the well-established DNA-transcriptional factor network (A), the ncRNA-protein network (B) provides a flexible platform capable of acting as sensors for various signals, integrating multiple regulatory complexes. In *cis* and in *trans* ncRNAs may work synergistically to recruit protein complexes involved in transcription regulation via RNA-binding proteins, as a bridge, to serve as a combinatorial DNA and histone modification code, altering chromosomal architecture.

permanently localized in the nucleus, including polyA-negative long ncRNAs that account for a large portion of the total transcribed sequences (Wu et al. 2008) and long ncRNAs transcribed from intronic regions (Cheng et al. 2005). A subset of long ncRNAs is located in both the nucleus and the cytoplasm (Imamura et al. 2004; Cheng et al. 2005; Kapranov et al. 2007; Wu et al. 2008). There are also some long ncRNAs selectively localized in the cytoplasm (Louroa et al. 2009).

Regulation/Processing

The tissue-/organ-specific expression patterns of many long ncRNAs in development, and the distinct subcellular localization of long ncNRAs, strongly suggest that their expression is under precise control (Amaral and Mattick 2008; Dinger et al. 2008; Mercer et al. 2008). Although some long ncRNAs have been reported to be transcribed by RNA polymerase III (Pol III) (Liu et al. 1995; Nguyen et al. 2001; Yang et al. 2001; Dieci et al. 2007), the majority are transcribed by RNA polymerase II (Pol II). However, it remains largely unknown how long ncRNAs are regulated at the level of their transcription and/or processing. For the ~1600 more abundant (conserved mammalian) long ncRNAs (Guttman et al. 2009; Khalil et al. 2009), it is reported that the basic "rules" are analogous to those of conventional Pol II transcription units: having histone H3K4 trimethylation mark at their promoters and H3K36 trimethylation marks within transcription body, and forming the so-called chromatin signature of the K4-K36 domain (Fig. 2A) and having a 5′ CAP site (Fig. 2B). Moreover, several well-known transcription factors for coding mRNA genes are also found to be present on a large number of ncRNA transcription units (Martone et al. 2003; Cawley et al. 2004; Kim et al. 2005). Other reported factors include Cha4, SAGA and Swi/Snf in *Saccharomyces cerevisiae* and REST in mammalian cells (Reviewed in Neurosci Letter 2009. Aug 11, Epub ahead of print). A central question for those ncRNAs that do not show the conventional H3 trimethylation of K4 and K36 marks and the Cap site (Fig. 2C) is whether this reflects processing from very long ncRNA transcripts that do harbor such sites at their transcription origins, or represents an alternative transcriptional regulatory strategy, such as polymerase entry at nicked sites.

Long ncRNAs are often spliced (Fig. 2D) and appear to use mechanisms similar to those used for miRNA processing. For example, we found that knock-down of the miRNA regulator DROSHA, but not DICER, increased the expression level of *CCND1*ncRNA about four-fold (TW, XS and

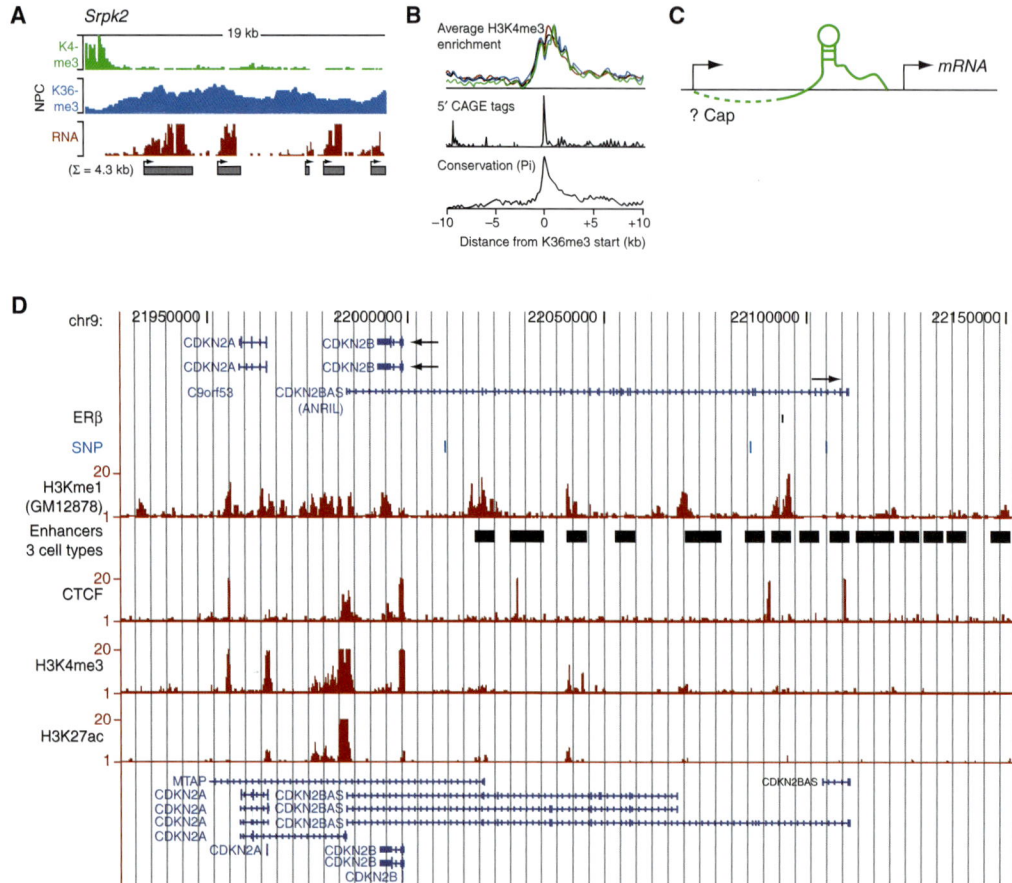

Figure 2. Characterization of long ncRNAs. (*A*) Long ncRNAs can be generated from intergenic K4–K36 domains as exemplified by the *Srpk2* intergenic long ncRNAs, which were identified by hybridizing RNA to DNA tiling arrays (Adopted from Guttman et al. 2009). (*B*) Expression of promoter-associated long ncRNAs can be identified by their correlation to H3K4 me3 and CAGE-tag (Adopted from Guttman et al. 2009). (*C*) Some long ncRNAs have no apparent Cap site been identified (such as the *CCND1*ncRNAs). (*D*) Some long ncRNAs are processed by splicing such as the disease loci in *CDKN2BAS* (*ANRIL*; Pasmant et al. 2007).

MGR, unpublished data), suggesting that *CCND1*ncRNA may be processed in a DROSHA-dependent manner.

3 LONG NCRNA-DEPENDENT RECRUITMENT OF PROTEIN COMPLEXES

One critical function of many, if not all, ncRNAs would appear to reside in their ability to interact with specific regulatory proteins/protein complexes, probably usually involving members of the large families of RNA-binding proteins. Because of RNA sequence and structural flexibility, ncRNAs are well-suited to accommodate binding of multiple complexes. Indeed, many ncRNAs associate with proteins and form complexes that function as a unit, as exemplified by the RNA Pol III-transcribed ribosome, RNA-induced silencing complex (RISC), and signal recognition particle (the conserved SRP, protein-RNA complex; Batey et al. 2000; Storz 2002). As more and more ncRNAs have been discovered, some generalized rules regarding potentially associated protein complexes are beginning to be discerned.

Studies from several groups suggested a critical role of long ncRNAs in epigenetic regulation by orienting chromatin-modifying factors/complexes to specific locations in the genome and in the nucleus (Sanchez-Elsner et al. 2006; Rinn et al. 2007; Chen et al. 2008; Zhao et al. 2008; Khalil et al. 2009). Initially described factors/complexes that are recruited by long ncRNAs include Ash1 by *TRE1-3* (Sanchez-Elsner et al. 2006) and MSL/MSL2 by *roX* RNA (Li et al. 2008; and reviewed in Scott and Li 2008) in *Drosophila*; the SRC1 complex by *SRA* (Lanz et al. 1999), PRC1 and PRC2 complexes by *Xist* (Plath et al. 2002; de Napoles et al. 2004; Schoeftner et al. 2006), PRC2 complex by *HOTAIR* and *RepA* (Rinn et al. 2007; Zhao et al. 2008), G9a by *Air* (Nagano et al. 2008), and TLS/CBP/p300 complex by *CCND1*ncRNA (Wang et al. 2008) in mammalian cells. Recently, a genome-wide ChIP-RNA sequencing analysis found that up to 38% of the ~3300 conserved large intergenic ncRNAs are associated with one of the following four chromatin-modifying factors-EZH2, SUZ12, CoREST, and JARID1C/SMCX (Khalil al. 2009). Besides, the homeodomain protein Dlx-2 has been shown to be recruited to the intergenic enhancer region of *Dlx-5* and *Dlx-6 via* a brain specific ncRNA, *Evf2* (Feng et al. 2006).

It is very interesting that most of the reported long ncRNAs-associated proteins are chromatin-modifying factors. Other chromatin/histone-modifying factors, such as LSD1, a component of the CoREST complex (Shi et al. 2005), are likely to combinatorialy impose strong effects as well. We propose that the complexity and diversity of ncRNAs promote the formation of chromatin-modifying complexes to establish the "epigenetic" memory. Despite the fact that most of the reported ncRNA-associated chromatin-modifying complexes are involved in gene repression, we suggest that an equally large numbers of long ncRNAs can recruit coactivator complexes, including components of the trithorax/COMPASS/MLL complex (Beisel et al. 2007; Schuettengruber et al. 2007; Shilatifard 2008). An immediate question is how regulatory ncRNAs recruit protein complexes. One possibility is that ncRNAs form RNA:RNA or RNA:DNA structures, which provide sequence specificity and serve as platforms to bind proteins that are not strictly sequence-specific, and thus directing these proteins to target sites (Mattick and Gagen 2001; Mattick 2007). It is also possible that ncRNAs can alter their structure on "ligand" binding and function as "riboswitches" (Wickiser et al. 2005; St Laurent and Wahlestedt 2007). It will be of interest to explore more open questions as: How many distinct complexes are recruited to various ncRNAs? What are the combinatorial "programs" of multiple coregulator complexes that might be required for maintaining epigenetic memory?

4 MECHANISMS MEDIATING TRANSCRIPTIONAL REGULATION BY LONG NCRNAS

Given their widespread distribution, long ncRNAs are likely to play roles in gene repression and/or activation, acting as sensors of various regulatory signals. The central strategy is the use of epigenetic regulation, including histone and DNA methylation, and many other post-translational modifications and remodeling complexes. In addition, long ncRNAs can affect the loading of general transcription factors as well as polymerase, or modulate the activities of specific transcription factors.

Serving as Ligands or Cofactors to Mediate Histone Modification

SRA

Nuclear receptors (NRs) comprise a super family of ligand-dependent transcription factors that regulate metabolism, development and reproduction (reviewed in Glass and Rosenfeld 2000; McKenna and O'Malley 2002). It has been well-established that the activities of NRs are mediated by the ligand-dependent exchange of coactivators and corepressors, which were initially thought to all be proteins. The finding that an ncRNA, *SRA*, functions as a coactivator (Lanz et al. 1999) significantly expanded concepts of mechanisms enabling transcriptional coactivation. *SRA* coactivates a range of NRs, including ER, AR, GR, PR, RARα, PPARδ and γ, TR, and VDR (Lanz et al. 1999; Deblois and Giguere 2003; Kawashima et al. 2003; Zhao et al. 2004; Hatchell et al. 2006), and some other classes of transcription factors, such as MyoD (Caretti et al. 2006). *SRA* has multiple stem-loops and a series of mutational studies showed that discrete stem-loops are required for the coactivator activity of *SRA*. Many SRA-associated RNA-binding proteins have been found to either positively or negatively regulate the coactivator activity of *SRA*. For example, the coactivator activity of *SRA* on *Myo D* is augmented in the presence of DEAD box-containing RNA-binding proteins p68 and p72 (Caretti et al. 2006). In the absence of ligand, SRA is sequestered by the transcriptionally silent TRα2 to a repressive protein complex containing RNA-binding proteins SHARP and SLIRP (Shi et al. 2001; Hatchell et al. 2006). SHARP and SLIRP repress the activities of a range of NRs through the recruitment of HDAC and NCoR/SMRT (Shi, Downes et al. 2001; Hatchell, Colley et al. 2006). When the ligand is present, *SRA* is released from TRα2 and switches between binding corepressor complexes and binding coactivator complexes (e.g., SRC1 and SRC2; Xu and Koenig 2004; Xu and Koenig 2005).

Evf2

Another example for a long ncRNA serving as a transcription coactivator/corepressor is *Evf2* in the developing mouse forebrain. *Evf2* is a long, polyadenylated ncRNA transcribed from an ultraconserved intergenic enhancer region associated with the *Dlx-5/6* locus (Feng et al. 2006) (Fig. 3A). The *Dlx* genes are related to the *Drosophila Distalless* gene (*dll*) homeodomain-containing protein family, and play crucial roles in neuronal development. Ei and eii are intergenic enhancers identified from *Dlx-5/6* loci, regulated by homeodomain protein Dlx-2. *Evf2* is transcribed from the Ei region and its expression is highly correlated

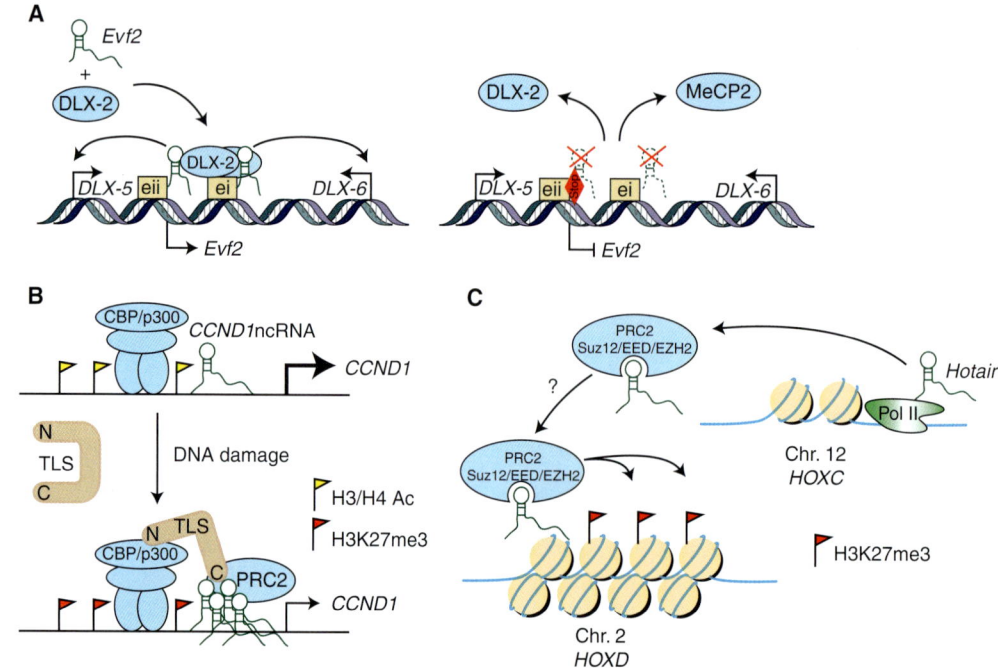

Figure 3. NcRNAs act as co-factors or ligands to regulate transcription. (*A*) A brain-specific ncRNA, *Evf2* is generated from an intergenic enhancer region of *DLX-5/DLX-6* and promotes the enhancer activity *via* recruitment of homeodomain protein DLX-2 (*Left*). Both DLX-2 and MeCP2 are dismissed from the intergenic enhancer region when the expression of *Evf2* is disrupted (*Right*). (*B*) Low-copy numbered *CCND1*ncRNAs are induced on DNA damage and *in cis* recruit an RNA-binding protein, TLS to inhibit the *CCND1* mRNA expression via inhibiting the HAT activities of CBP/p300. PRC2 complex is recruited to the *CCND1* 5′ regulatory region via TLS. (*C*) *HOTAIR* generated from *HOXC* on Chr. 12, acts *in trans* on *HOXD* on Chr. 2 to repress the gene expression via recruitment of PRC2 complex.

with the expression of *Dlx-5* and *-6*. *Evf2* forms a complex with Dlx-2 and recruits Dlx-2 to induce the enhancer activities of ei and eii, resulting in induced expression of both *Dlx-5* and *-6*. These data suggest that the *Evf2* ncRNA functions as a coactivator molecule, analogous to the functions of *SRA*.

Evf2-null mice were generated by inserting transcription stop sites into *Evf* exon 1 (Bond et al. 2009). ChIP analysis on these *Evf2*-null mice revealed that both Dlx and MeCP2, a previous known repressor of *Dlx-5*, are dismissed from ei and eii. MeCP2-mediated repression is suggested to be through the recruitment of HDACs. The binding of HDACs, however, is not changed in the absence of MeCP2 in *Evf2*-null mice. Surprisingly, the levels of *Dlx-5* and *-6* transcripts increased in *Evf2*-null mice. It is unclear how *Dlx-5* and *-6* are transcribed in the absence of Dlx binding. It is possible that Dlx proteins may only be required for initial activation of *Dlx-5/6* in an *Evf2*-independent manner, whereas subsequent regulation of *Dlx-5/6* by Dlx and MeCP2 is *Evf2*-dependent. Alternatively, other Dlx-binding sites compensate in the absence of Dlx-ei/eii interactions; or the major role of Dlx proteins is to prevent repressors such as MeCP2 from binding ei/eii, rather than acting as direct activators.

CCND1 ncRNA

Promoter-associated ncRNAs can function as ligands to mediate histone modifications, exemplified by our study of *CCND1*ncRNA and members of the TET RNA-binding protein family, including TLS, EWS, and TAFII68 (Wang et al. 2008) (Fig. 3B). TLS, translocated in liposarcoma, is an RNA-binding protein with RNA-binding domains at its C-terminus. These RNA-binding domains are frequently deleted in human tumors in which the amino terminus of TLS is fused with other transcriptional factors, suggesting an important role of RNA-binding domains of TLS in human disease. The amino-terminal glutamine-rich domain of TLS, on the other hand, is responsible for the interaction with two well-known histone acetyltransferases, CBP and p300. The interaction between TLS and CBP/p300 results in the substrate-specific inhibition of the HAT activities of CBP/p300 (Wang et al. 2008). We reported that a series of ncRNAs (*CCND1*ncRNAs) are generated from the 5′ regulatory regions of *CCND1*. The *CCND1*ncRNAs are upregulated in response to genotoxic stress, when the *CCND1* mRNA is down-regulated. The induction the *CCND1*ncRNAs recruit TLS and cause a close-to-open conformation change in TLS that licenses its amino-terminal binding of CBP/p300, resulting in substrate-specific inhibition of their HAT enzymatic activities, and thus establishing the hypo-acetylation status of the chromatin and repressing of the *CCND1* mRNA expression. Surprisingly, these *CCND1*ncRNAs are at very low abundance (two to eight copies/cell) and are associated with chromatin, functioning *in cis* as ss RNAs to recruit and modulate the activity of TLS. Interestingly, these ncRNAs appear to facilitate the recruitment of PRC1 and PRC2 complexes as well (Fig. 1) (TW, XS, BS, and MGR, unpublished data). In combination with our preliminary data that TLS binds to the EZH2 complex (BS, TW, XS, and MGR, unpublished data), it further suggested that TLS might be required for the recruitment of EZH2 to *CCND1* promoter during repression.

HOTAIR

HOX genes are a group of genes that control the anterior–posterior axis and segmentation during development. Humans contain four clusters of *HOX* genes, *HOXA* on chromosome 7, *HOXB* on chromosome 17, *HOAC* on chromosome 12, and *HOXD* on chromosome 7 (Wellik 2009). Tiling arrays covering all four human *HOX* clusters identified 231 novel ncRNAs, which are spatially expressed along developmental axes and show distinct histone methylation patterns (Rinn, Kertesz et al. 2007). Among these, a 2.2 kb long ncRNA, *HOTAIR*, transcribed from the boundary of two diametric chromatin domains in the *HOXC* locus was preferentially expressed in posterior and distal sites. *HOTAIR* recruits PRC2 complex *in trans* across 40 kb of the *HOXD* locus on chromosome 2, promotes the histone H3K27 trimethylation and results in transcriptional repression of *HOXD* locus (Fig. 3C). Therefore, *HOTAIR* represents an ncRNA that exerts transcriptional repression, at least in part, through recruitment of the chromatin modifying enzyme complex PRC2.

DNA Methylation

Many protein-coding genes have antisense partners, including tumor suppressor genes (Yu et al. 2008). Misregulation of the associated antisense ncRNAs may subsequently silence the tumor suppressor gene (Fig. 4A), leading to oncogenesis (Yu et al. 2008).

P15AS

A long (~200 bp) antisense ncRNA, *P15AS*, was identified from *P15* gene. *P15* is frequently hypermethylated, thus silenced, in leukaemia (Lubbert 2003). It is well established that small ncRNAs, such as microRNA and piRNA, can mediate DNA methylation in many species (Bao et al. 2004; Matzke et al. 2004; Ronemus and Martienssen 2005). The opposite expression patterns between *P15AS* and *P15* raise the possibility that *P15AS* may trigger epigenetic silencing of the *p15* genes in cancer cell lines. Indeed, the activity of the *P15* promoter could not be reactivated by removal of *p15AS* (Yu et al. 2008). In the presence of exogenous *p15AS*, the *p15* promoter showed a marked increase in dimethylation of H3K9 and a decrease in dimethylation of H3K4 in human cancer cell lines and mouse embryonic stem cells (Yu et al. 2008). DNA hypermethylation was induced by *p15AS* when embryonic stem cells were differentiated into embryoid bodies (Yu, Gius et al. 2008). Antisense transcript-mediated silencing could be a general mechanism for the silencing of many other tumor suppressor genes in cancer, many of which have long antisense transcripts (Fig. 4A).

Khps1

Khps1 is an antisense, long ncRNAs generated from the regulatory region (T-DMR) of the sense transcript *Sphk1* (Imamura et al. 2004). When overexpressed, *Khps1* leads to CpG demethylation and simultaneous non-CG methylation in the T-DMR (Imamura et al. 2004), the methylation status of which correlates with the regulation of *Sphk1* expression (Imamura et al. 2001). This RNA-induced CG demethylation and non-CG methylation suggests an intriguing and important connection between ncRNAs and epigenetic regulation (Imamura et al. 2004).

General Transcription Factors and RNA pol II Loading

It is known that RNA has the ability to regulate bacterial RNA polymerase activity (Wassarman and Storz 2000). There are now quite a few cases in which long ncRNAs have been reported to promote or repress transcription by binding to mammalian RNA pol II and modulating the latter's loading on the promoters of regulated genes (Fig. 4B).

Alu RNA

Massive ncRNAs have been found to be transcribed from the *SINE*s retrotransposon elements, including *Alu* repeats in human cells and *SINE B1* and *SINE B2* in mouse cells (Maraia et al. 1993; Kramerov and Vassetzky 2005). These ncRNAs have been suggested to be involved in gene regulation through roles in transcription, mRNA editing, or even miRNA regulation (Reviewed in Hasler and Strub 2006; Hasler et al. 2007). *Alu* RNA and *SINE B2* RNA have been shown to repress mRNA transcription by preventing RNA pol II loading during heat shock (Allen et al. 2004; Espinoza et al. 2004; Espinoza et al. 2007; Mariner et al. 2008; Yakovchuk et al. 2009). Both *SINE B2* and *Alu* RNA

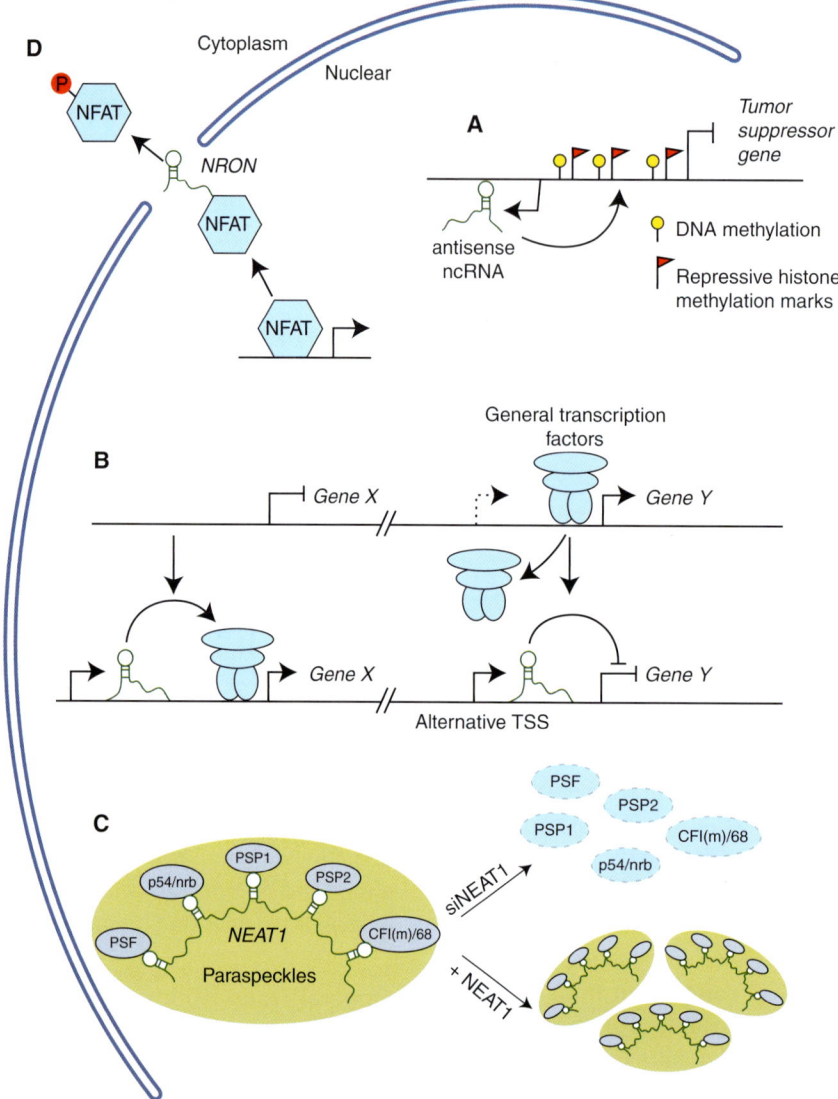

Figure 4. NcRNAs act as transcription regulators *via* diverse mechanism. (*A*) Antisense ncRNA from tumor suppressor gene (TSG) induces DNA and repressive histone methylation on the promoter of TSG. (*B*) NcRNA generated from an alternative transcription start site (TSS) plays either a repressive or activating role by modulating the recruitment of general transcription factors on the promoter of its adjacent gene. (*C*) NcRNA *NEAT1* is specifically expressed in paraspeckles and colocalized with paraspeckle markers (PSF, p54/nrb, PSP1, PSP2 and CFI(m)/68). Knockdown of *NEAT1* by specific siRNA disrupts paraspeckles, whereas overexpression of *NEAT1* increases the number of paraspeckles. (*D*) NcRNA *NRON* prevents nuclear localization of the transcription factor NFAT.

could directly bind to RNA pol II with a high affinity (Allen et al. 2004; Espinoza et al. 2004; Mariner et al. 2008). Surprisingly, *Alu* RNA contains modular domains, a hallmark of protein regulators of RNA pol II transcription (Mariner et al. 2008). The modular domain of *Alu* RNA can be fused to a RNA pol II-binding *SINE* B1 RNA to mediate transcription repression, in a manner similar to trans-acting protein transcription factors (Mariner et al. 2008). Because *SINE B2* RNA is able to block the association of RNA pol II as a DNA/RNA hybrid, it was proposed that the modular repression domains in *Alu* RNA target the DNA-binding channel of RNA pol II and thus prevent the polymerase from forming proper contacts with promoter DNA (Mariner et al. 2008). Indeed, this hypothesis was supported by a series of biochemical assays (Yakovchuk et al. 2009). *Alu* RNA and *SINE B2* RNA apparently prevented the closed complexes between RNA pol II and the promoter DNA, despite the fact that RNA pol II remains associated with the promoter DNA. This effect is reversible in the presence of RNase I, but only so before the formation of the closed complex, suggesting these two ncRNAs block the re-engagement of RNA pol II on the promoter DNA (Yakovchuk et al. 2009). Together, these data suggest that ncRNAs derived from *SINE* elements function as transrepressors of gene expression.

Fbp1+ ncRNA

In the fission yeast *S. pombe*, the expression of *fbp1+* is induced by glucose starvation (Hirota et al. 2008). Glucose starvation also induces at least three large (∼ several kb) and rare sense-stranded ncRNA transcripts before induction of the *fbp1+* transcript. The transcription start sites (TSSs) for these ncRNAs are located at ∼-1.3 kb to -530 bp from the ATG. During the glucose starvation, the RNA pol II binding sites shift from the 5′ to 3′ region in the *fbp1+* promoter (from ncRNA to *fbp1+* transcript), paralleling a corresponding alteration of chromatin structure determined by MNase digestion. Disruption of the ncRNA transcripts by inserting a terminator between the ncRNA transcripts and the TATA box resulted in short and premature ncRNA products, failed RNA pol II recruitment, and chromatin remodeling in the downstream region of the terminator, and eliminated *fbp1+* induction in the absence of glucose (Hirota et al. 2008). These results suggest that the RNA pol II-mediated ncRNA transcripts across the *fbp1+* promoter increased access to the RNA polymerase and transcriptional activators through progressive opening of chromatin structure (Fig. 4B). Although they affect the activity of general transcription factors and/or RNA pol II, these long ncRNAs regulate gene transcription in a gene-specific manner. It will therefore be of interest to determine whether additional *cis* acting ncNRAs are involved in these functions.

Transcription Interference

Alternative TSSs have been found on >80% of the tested genes by the Encyclopedia of DNA Elements (ENCODE) project (Trinklein et al. 2003; Carninci et al. 2005; Cooper et al. 2006; Birney et al. 2007; Kawaji et al. 2009). These newly identified TSSs are located at 5′ distal or internal to the annotated gene boundary. Long ncRNA transcripts generated from the alternative TSS may affect the transcription of the adjacent mRNA genes by interference with the loading of the general transcriptional factor, such as TBP (Fig. 4B). The examples of this mechanism are a long ncRNA generated from an upstream minor promoter that inhibits the *DHFR* gene encoding dihydrofolate in human (Martianov et al. 2007) and a long ncRNA, *SRG1*, transcribed from the upstream alternative TSS of the *SER3* gene in *S. cerevisiae* (Martens et al. 2004; Martens et al. 2005).

Regulating the Activity of Transcriptional Factors

By screening a short hairpin RNA library against 512 evolutionarily conserved putative ncRNAs, an ncRNA – *NRON* (noncoding repressor of NFAT), was found to repress the nuclear factor of activated T cells (NFAT) family (Willingham et al. 2005). *NRON* has three splicing variants, with size ranging from 0.8kb to 3.7kb. *NRON* showed tissue-specific expression and it was particularly enriched in lymphoid tissues. Eleven proteins were found to bind to *NRON*, including importin-β and factors directly mediating the nuclear-cytoplasmic transport of cargoes such as NFAT, suggesting that *NRON* may act as a modulator of NFAT nuclear trafficking (Fig. 4D). This was supported by the evidence that the level of nuclear NFAT was greatly elevated in the presence of *NRON* shRNA (Willingham et al. 2005). It is unknown if the role of *NRON* is to repress nuclear import or to promote the nuclear export of NFAT.

The mechanisms of long ncRNA regulating transcription are diverse, many of which are not included in above discussion. For example, a newly identified ncRNA, *HSR1*, has been shown to be involved in regulating the activity of the heat shock transcriptional factor 1 (*HSF1*) (Shamovsky et al. 2006). Because *HSR1* was coimmunoprecipitated from *HSF1*-eEF1A complex, the interaction of *HSR1* and eEF1A was suggested to be required for the activation of *HSF1* during heat shock.

5 BIOLOGICAL ROLES OF LONG NCRNAS

Although still unresolved, the idea that many or even a majority of the ncRNA transcripts are functional has been suggested by the fact that the percentage of the ncRNAs transcribed in the genome is proportional to the complexity of the organism (Taft et al. 2007), and it has received initial experimental support.

Roles of Promoter-Associated ncRNAs

Transcriptome maps in the entire nonrepetitive portion of the human genome revealed many long ncRNAs (and short ncRNAs) around promoter regions (Kapranov et al. 2007). The number of identified promoter-associated ncRNAs continues to grow in human (Calin et al. 2007; Guenther et al. 2007; Kapranov et al. 2007) as well as other species (Davis and Ares 2006; Guenther et al. 2007). Many of these transcripts are under precise control of diverse signals. Another class of less stable ncRNAs, promoter upstream transcripts, was revealed after siRNA deletion of hRrp40, a core component of the human 3′ to 5′ exoribonucleolytic exosome, which is one of the major RNA degradation complexes (Preker et al. 2008). These transcripts are reminiscent of the cryptic unstable transcripts in S. cerevisiae (Wyers et al. 2005; Neil et al. 2009; Xu et al. 2009), which are located 0.5–2.5 kb upstream of active TSSs and require the downstream promoters. Promoter upstream transcripts partially overlap with promoter-associated ncRNAs, but lack their own known promoters. Both promoter-associated and upstream ncRNAs can be bidirectionally transcribed. These ncRNAs are postulated to have regulatory functions, as suggested by a few ncRNAs that are experimentally supported to regulate transcription of downstream genes, e.g., CCND1ncRNA (Wang et al. 2008), Khps1a (Imamura et al. 2004), and DHFR ncRNA (Martianov et al. 2007), although some events have been suggested to reflect "occlusion" rather than via recruiting RNA-binding proteins or controlling CpG demethylation (Imamura et al. 2004); however, those events are likely to be mechanistically linked. An alternative, nonexclusive function of these long ncRNAs around or upstream of active promoters is to alter chromatin structure through their own transcription.

Roles of ncRNAs as "Transcriptional" Boundary Markers

Almost half of the human protein-coding genes have been suggested to be marked by both promoter- and 3′-associated short ncRNAs (Kapranov et al. 2007; Borel et al. 2008). That 40% of the small ncRNAs could be processed from long ncRNAs as indicated by the genome-wide, high resolution tiling arrays (Kapranov et al. 2007) raises the possibility that at least a portion of the short ncRNAs represent processed products of the long ncRNAs, thus serving as markers of boundaries of mRNA transcription units. In addition, they may function in stabilizing "transcriptional" boundaries (Kapranov et al. 2007), conceptually analogous to the proposed network for boundary elements, such as CTCF (Phillips and Corces 2009).

Roles of ncRNAs as Sensors of Signals

In concert with their ability to recruit specific complexes, long ncRNAs appear to function as sensors of developmental signals and other signaling pathways, such as genotoxic stress. As an example, we have reported that CCND1 ncRNAs, generated in the regulatory region of the CCND1, sense the genotoxic stress signal and up-regulate, followed by in cis recruiting TLS, which leads to its changing conformation to be able to bind and inhibit CBP/p300 and results in CCND1 mRNA repression (Wang et al. 2008).

Although we will not discuss in detail the function of the long ncRNAs in dosage compensation and imprinting in this review, we would like to point out a 1.6-kb long ncRNA (RepA) within another long ncRNA, Xist, functions as a developmental signal sensor, at onset of X-inactivation when the antisense ncRNA Tsix is down-regulated, and recruits the PRC2 complex to Xist, which leads to the initiation and spreading of X-inactivation (Zhao et al. 2008). Likewise, preferentially expressed in posterior and distal sites, HOTAIR in the HOXC cluster targets PRC2 to distant loci in HOXD and regresses its expression in trans, representing an example of long ncRNAs sensing developmental signals that control anterior–posterior axis (Rinn et al. 2007).

Potential Roles in Structural Integrity of Cellular Organellels and Architecture

Analogous to the role of rRNAs in ribosome assembly, long ncRNAs can exert functional roles in specific nuclear organelle assembly, exemplified by the actions of the long ncRNA NEAT1 (MENε), which is functionally essential for structural integrity of nuclear paraspeckles (Fox et al., 2002; Clemson et al. 2009; Sasaki et al. 2009; Sunwoo et al. 2009). NEAT1 (MENε) and NEAT2 are the two (out of three) large, polyadenylated, nuclear enriched ncRNAs identified in two human female cell lines using Affymetrix U133A and U133B expression arrays (Hutchinson et al. 2007). Both genes are located on human chromosome 11, and have no significant homology to each other. However, they both are highly conserved within the mammalian lineage, suggesting a role specific for mammals during evolution. NEAT2 is colocalized with SC-35+ nuclear speckles/inter-chromatin granules, whereas NEAT1 (MENε) is localized predominantly to domains adjacent to nuclear speckles—paraspeckles (Fox et al. 2002; Clemson et al. 2009; Sasaki et al. 2009; Sunwoo et al. 2009). Paraspeckles are marked by paraspeckle-associated proteins—PSP1, PSP2, PSF/PSFQ, and p54/nrb, and possibly CFI(m)/68, whose functions have been suggested in transcription, pre-mRNA splicing and nuclear retention of RNA (Fox et al. 2002; Dettwiler et al. 2004; Fox et al. 2005). Knock-down of NEAT1 (MENε), but not NEAT2, specifically eradicated paraspeckles in many cell lines including 293, HeLa, HT-1080, and Tig1, suggesting that NEAT1 (MENε) RNA is required for paraspeckle formation (Fig. 4C). On the other hand, overexpression of NEAT1 (MENε) induced more paraspeckle staining (Fig. 4C). In addition, the flexibility of RNA to accommodate a series of complexes and to extend across large physical distances, suggests the potential to play key roles in regulation of nuclear architecture.

6 CONCLUSIONS AND PERSPECTIVES

Together, the cumulative data suggest that ncRNAs exert important regulatory functions in the control of transcriptional programs (Fig. 1). Because of the large number of long ncRNAs, and their functions in "nucleating" molecules for combinatorial complex assembly, the resultant ncRNA-protein interactions are likely to serve as "sensors" for a variety of developmental and signaling pathways, nuclear receptor ligands, and genotoxic-stress. The presence of ss ncRNA versus regions of ds ncRNAs is likely to be dynamically regulated in the environment of the cell. This relationship may dictate the alternative recruitment of "corepressors," "coactivators," and DNA-modifying complexes. Thus it is tempting to postulate that the ncRNAs might themselves serve as the critical arbiters of local DNA methylation patterns and even as key arbiters in establishing regions of "epigenetic" memory. Conversely, it is likely that alterations in strandedness/transcriptional length/splicing of ncRNAs functionally modulate their actions. An additional consideration in the actions of long ncRNAs relates to the architecture of most RNA-binding proteins, which contain multiple RNA-binding domains that may recognize distinct "targets." This implies that many RNA-binding proteins might, as a consequence, show distinct "allosterically-dictated" conformations based on the specific RNA sequences they bind that might regulate interactions with other proteins required for distinct functions—for example at promoters versus during transcriptional elongation/splicing. Given that RNA-binding proteins constitute the largest "transcriptional" regulatory family in the genome (Burd and Dreyfuss 1994; Lunde et al. 2007), ncRNA/RNA-binding protein-based strategy to integrate transcriptional programs is very likely a general rule in the transcriptional as well as RNA processing regulation. Considering the broad expression of ncRNAs, and the ambiguity of any sequence specificity of the reported ncRNAs acting *in trans*, it is even possible that they are recruited by *cis*-acting ncRNAs. Together, the ncRNA sensor code appears to be a robust and critical strategy underlying a wide variety of gene regulatory programs and we are certain to see an explosive increment of knowledge and insights into this area in the coming months.

ACKNOWLEDGMENTS

We thank J. Hightower and D. Benson for assistance figure preparation. We thank X. Su for useful discussion. Work cited from our laboratories was supported by National Institutes of Health (NIH) grants to CKG and to MGR. XW is supported by NIH grant 5T32DK007044-28. XS is supported by the Irvington Institute Fellowship Program of the Cancer Research Institute. MGR is an HHMI investigator. X. Wang and X. Song contributed equally to this work.

REFERENCES

Allen TA, Von Kaenel S, Goodrich JA, Kugel JF. 2004. The SINE-encoded mouse B2 RNA represses mRNA transcription in response to heat shock. *Nat Structural Mol Biol* **11**: 816–821.

Amaral PP, Mattick JS. 2008. Noncoding RNA in development. *Mamm Genome* **19**: 454–492.

Ambros V. 2001. microRNAs: tiny regulators with great potential. *Cell* **107**: 823–826.

Aravin A, Gaidatzis D, Pfeffer S, Lagos-Quintana M, Landgraf P, Iovino N, Morris P, Brownstein MJ, Kuramochi-Miyagawa S, Nakano T, et al. 2006. A novel class of small RNAs bind to MILI protein in mouse testes. *Nature* **442**: 203–207.

Bao N, Lye KW, Barton MK. 2004. MicroRNA binding sites in Arabidopsis class III HD-ZIP mRNAs are required for methylation of the template chromosome. *Develop Cell* **7**: 653–662.

Batey RT, Rambo RP, Lucast L, Rha B, Doudna JA. 2000. Crystal structure of the ribonucleoprotein core of the signal recognition particle. *Science* **287**: 1232–1239.

Beisel C, Buness A, Roustan-Espinosa IM, Koch B, Schmitt S, Haas SA, Hild M, Katsuyama T, Paro R. 2007. Comparing active and repressed expression states of genes controlled by the Polycomb/Trithorax group proteins. *Proc Nat Acad Sci* **104**: 16615–16620.

Bernstein E, Allis CD. 2005. RNA meets chromatin. *Genes & Development* **19**: 1635–1655.

Bertone P, Stolc V, Royce TE, Rozowsky JS, Urban AE, Zhu X, Rinn JL, Tongprasit W, Samanta M, Weissman S, et al. 2004. Global identification of human transcribed sequences with genome tiling arrays. *Science* **306**: 2242–2246.

Birney E, Stamatoyannopoulos JA, Dutta A, Guigo R, Gingeras TR, Margulies EH, Weng Z, Snyder M, Dermitzakis ET, Thurman RE, et al. 2007. Identification and analysis of functional elements in 1% of the human genome by the ENCODE pilot project. *Nature* **447**: 799–816.

Bond AM, Vangompel MJ, Sametsky EA, Clark MF, Savage JC, Disterhoft JF, Kohtz JD. 2009. Balanced gene regulation by an embryonic brain ncRNA is critical for adult hippocampal GABA circuitry. *Nat Neurosci* **12**: 1020–1027.

Borel C, Gagnebin M, Gehrig C, Kriventseva EV, Zdobnov EM, Antonarakis SE. 2008. Mapping of small RNAs in the human ENCODE regions. *Am J Human Gen* **82**: 971–981.

Brown CJ, Hendrich BD, Rupert JL, Lafreniere RG, Xing Y, Lawrence J, Willard HF. 1992. The human XIST gene: Analysis of a 17kb inactive X-specific RNA that contains conserved repeats and is highly localized within the nucleus. *Cell* **71**: 527–542.

Bruce AW, Lopez-Contreras AJ, Flicek P, Down TA, Dhami P, Dillon SC, Koch CM, Langford CF, Dunham I, Andrews RM, et al. 2009. Functional diversity for REST (NRSF) is defined by in vivo binding affinity hierarchies at the DNA sequence level. *Genome Research* **19**: 994–1005.

Burd CG, Dreyfuss G. 1994. Conserved structures and diversity of functions of RNA-binding proteins. *Science* **265**: 615–621.

Calin GA, Liu CG, Ferracin M, Hyslop T, Spizzo R, Sevignani C, Fabbri M, Cimmino A, Lee EJ, Wojcik SE, et al. 2007. Ultraconserved regions encoding ncRNAs are altered in human leukemias and carcinomas. *Cancer Cell* **12**: 215–229.

Caretti G, Schiltz RL, Dilworth FJ, Di Padova M, Zhao P, Ogryzko V, Fuller-Pace FV, Hoffman EP, Tapscott SJ, Sartorelli V. 2006. The RNA helicases p68/p72 and the noncoding RNA SRA are coregulators of MyoD and skeletal muscle differentiation. *Develop Cell* **11**: 547–560.

Carninci P, Kasukawa T, Katayama S, Gough J, Frith MC, Maeda N, Oyama R, Ravasi T, Lenhard B, Wells C, et al. 2005. The transcriptional landscape of the mammalian genome. *Science* **309**: 1559–1563.

Carninci P, Sandelin A, Lenhard B, Katayama S, Shimokawa K, Ponjavic J, Semple CA, Taylor MS, Engstrom PG, Frith MC, et al. 2006. Genome-wide analysis of mammalian promoter architecture and evolution. *Nat Gen* **38:** 626–635.

Cawley S, Bekiranov S, Ng HH, Kapranov P, Sekinger EA, Kampa D, Piccolboni A, Sementchenko V, Cheng J, Williams AJ, et al. 2004. Unbiased mapping of transcription factor binding sites along human chromosomes 21 and 22 points to widespread regulation of noncoding RNAs. *Cell* **116:** 499–509.

Chen X, Xu H, Yuan P, Fang F, Huss M, Vega VB, Wong E, Orlov YL, Zhang W, Jiang J, et al. 2008. Integration of external signaling pathways with the core transcriptional network in embryonic stem cells. *Cell* **133:** 1106–1117.

Cheng J, Kapranov P, Drenkow J, Dike S, Brubaker S, Patel S, Long J, Stern D, Tammana H, Helt G, et al. 2005. Transcriptional maps of 10 human chromosomes at 5-nucleotide resolution. *Science* **308:** 1149–1154.

Clemson CM, Hutchinson JN, Sara SA, Ensminger AW, Fox AH, Chess A, Lawrence JB. 2009. An architectural role for a nuclear noncoding RNA: NEAT1 RNA is essential for the structure of paraspeckles. *Mol Cell* **33:** 717–726.

Cooper SJ, Trinklein ND, Anton ED, Nguyen L, Myers RM. 2006. Comprehensive analysis of transcriptional promoter structure and function in 1% of the human genome. *Gen Res* **16:** 1–10.

Costa FF. 2007. Non-coding RNAs: lost in translation? *Gene* **386:** 1–10.

Davis CA, Ares MJr., 2006. Accumulation of unstable promoter-associated transcripts upon loss of the nuclear exosome subunit Rrp6p in Saccharomyces cerevisiae. *Proc Nat Acad Sci* **103:** 3262–3267.

de Napoles M, Mermoud JE, Wakao R, Tang YA, Endoh M, Appanah R, Nesterova TB, Silva J, Otte AP, Vidal M, et al. 2004. Polycomb group proteins Ring1A/B link ubiquitylation of histone H2A to heritable gene silencing and X inactivation. *Develop Cell* **7:** 663–676.

Deblois G, Giguere V. 2003. Ligand-independent coactivation of ERα AF-1 by steroid receptor RNA activator (SRA) via MAPK activation. *J Steroid Biochem Mol Biol* **85:** 123–131.

Dettwiler S, Aringhieri C, Cardinale S, Keller W, Barabino SM. 2004. Distinct sequence motifs within the 68-kDa subunit of cleavage factor Im mediate RNA binding, protein-protein interactions, subcellular localization. *J Biol Chem* **279:** 35788–35797.

Dieci G, Fiorino G, Castelnuovo M, Teichmann M, Pagano A. 2007. The expanding RNA polymerase III transcriptome. *Trends Genet* **23:** 614–622.

Dinger ME, Amaral PP, Mercer TR, Pang KC, Bruce SJ, Gardiner BB, Askarian-Amiri ME, Ru K, Solda G, Simons C, et al. 2008. Long noncoding RNAs in mouse embryonic stem cell pluripotency and differentiation. *Gen Res* **18:** 1433–1445.

Espinoza CA, Allen TA, Hieb AR, Kugel JF, Goodrich JA. 2004. B2 RNA binds directly to RNA polymerase II to repress transcript synthesis. *Nat Structural Mol Biol* **11:** 822–829.

Espinoza CA, Goodrich JA, Kugel JF. 2007. Characterization of the structure, function, mechanism of B2 RNA, an ncRNA repressor of RNA polymerase II transcription. *RNA* **13:** 583–596.

Feng J, Bi C, Clark BS, Mady R, Shah P, Kohtz JD. 2006. The Evf-2 noncoding RNA is transcribed from the Dlx-5/6 ultraconserved region and functions as a Dlx-2 transcriptional coactivator. *Genes Develop* **20:** 1470–1484.

Fox AH, Bond CS, Lamond AI. 2005. P54nrb forms a heterodimer with PSP1 that localizes to paraspeckles in an RNA-dependent manner. *Mol Biol Cell* **16:** 5304–5315.

Fox AH, Lam YW, Leung AK, Lyon CE, Andersen J, Mann M, Lamond AI. 2002. Paraspeckles: A novel nuclear domain. *Curr Biol* **12:** 13–25.

Ginger MR, Shore AN, Contreras A, Rijnkels M, Miller J, Gonzalez-Rimbau MF, Rosen JM. 2006. A noncoding RNA is a potential marker of cell fate during mammary gland development. *Proc Nat Acad Sci* **103:** 5781–5786.

Girard A, Sachidanandam R, Hannon GJ, Carmell MA. 2006. A germline-specific class of small RNAs binds mammalian Piwi proteins. *Nature* **442:** 199–202.

Glass CK, Rosenfeld MG. 2000. The coregulator exchange in transcriptional functions of nuclear receptors. *Genes Develop* **14:** 121–141.

Goodrich JA, Kugel JF. 2006. Non-coding-RNA regulators of RNA polymerase II transcription. *Nat Rev Mol Cell Biol* **7:** 612–616.

Grishok A, Mello CC. 2002. RNAi (Nematodes: *Caenorhabditis elegans*). *Adv Gen* **46:** 339–360.

Grivna ST, Beyret E, Wang Z, Lin H. 2006. A novel class of small RNAs in mouse spermatogenic cells. *Genes Develop* **20:** 1709–1714.

Guenther MG, Levine SS, Boyer LA, Jaenisch R, Young RA. 2007. A chromatin landmark and transcription initiation at most promoters in human cells. *Cell* **130:** 77–88.

Gustincich S, Sandelin A, Plessy C, Katayama S, Simone R, Lazarevic D, Hayashizaki Y, Carninci P. 2006. The complexity of the mammalian transcriptome. *J Physiol* **575:** 321–332.

Guttman M, Amit I, Garber M, French C, Lin MF, Feldser D, Huarte M, Zuk O, Carey BW, Cassady JP, et al. 2009. Chromatin signature reveals over a thousand highly conserved large non-coding RNAs in mammals. *Nature* **458:** 223–227.

Hasler J, Samuelsson T, Strub K. 2007. Useful 'junk': Alu RNAs in the human transcriptome. *Cell Mol Life Sci* **64:** 1793–1800.

Hasler J, Strub K. 2006. Alu elements as regulators of gene expression. *Nuc Acids Res* **34:** 5491–5497.

Hatchell EC, Colley SM, Beveridge DJ, Epis MR, Stuart LM, Giles KM, Redfern AD, Miles LE, Barker A, MacDonald LM, et al. 2006. SLIRP, a small SRA binding protein, is a nuclear receptor corepressor. *Mol Cell* **22:** 657–668.

Hirota K, Miyoshi T, Kugou K, Hoffman CS, Shibata T, Ohta K. 2008. Stepwise chromatin remodelling by a cascade of transcription initiation of non-coding RNAs. *Nature* **456:** 130–134.

Hutchinson JN, Ensminger AW, Clemson CM, Lynch CR, Lawrence JB, Chess A. 2007. A screen for nuclear transcripts identifies two linked noncoding RNAs associated with SC35 splicing domains. *BMC Genomics* **8:** 39.

Imamura T, Ohgane J, Ito S, Ogawa T, Hattori N, Tanaka S, Shiota K. 2001. CpG island of rat sphingosine kinase-1 gene: tissue-dependent DNA methylation status and multiple alternative first exons. *Genomics* **76:** 117–125.

Imamura T, Yamamoto S, Ohgane J, Hattori N, Tanaka S, Shiota K. 2004. Non-coding RNA directed DNA demethylation of Sphk1 CpG island. *Biochem Biophys Res Comm* **322:** 593–600.

Johnson JM, Edwards S, Shoemaker D, Schadt EE. 2005. Dark matter in the genome: evidence of widespread transcription detected by microarray tiling experiments. *Trends Genet* **21:** 93–102.

Johnson R, Teh CH, Jia H, Vanisri RR, Pandey T, Lu ZH, Buckley NJ, Stanton LW, Lipovich L. 2009. Regulation of neural macroRNAs by the transcriptional repressor REST. *RNA* **15:** 85–96.

Kapranov P, Cheng J, Dike S, Nix DA, Duttagupta R, Willingham AT, Stadler PF, Hertel J, Hackermuller J, Hofacker IL, et al. 2007. RNA maps reveal new RNA classes and a possible function for pervasive transcription. *Science* **316:** 1484–1488.

Katayama S, Tomaru Y, Kasukawa T, Waki K, Nakanishi M, Nakamura M, Nishida H, Yap CC, Suzuki M, Kawai J, et al. 2005. Antisense transcription in the mammalian transcriptome. *Science* **309:** 1564–1566.

Kawaji H, Severin J, Lizio M, Waterhouse A, Katayama S, Irvine KM, Hume DA, Forrest AR, Suzuki H, Carninci P, et al. 2009. The FANTOM web resource: from mammalian transcriptional landscape to its dynamic regulation. *Gen Biol* **10:** pR40.

Kawashima H, Takano H, Sugita S, Takahara Y, Sugimura K, Nakatani T. 2003. A novel steroid receptor co-activator protein (SRAP) as an alternative form of steroid receptor RNA-activator gene: Expression in prostate cancer cells and enhancement of androgen receptor activity. *Biochem J* **369:** 163–171.

Khalil AM, Guttman M, Huarte M, Garber M, Raj A, Rivea Morales D, Thomas K, Presser A, Bernstein BE, van Oudenaarden A, et al. 2009. Many human large intergenic noncoding RNAs associate with chromatin-modifying complexes and affect gene expression. *Proc Nat Acad Sci* **106:** 11667–11672.

Kim TH, Barrera LO, Zheng M, Qu C, Singer MA, Richmond TA, Wu Y, Green RD, Ren B. 2005. A high-resolution map of active promoters in the human genome. *Nature* **436:** 876–880.

Koch F, Jourquin F, Ferrier P, Andrau JC. 2008. Genome-wide RNA polymerase II: Not genes only! *Trends in Biochem Sci* **33:** 265–273.

Kramerov DA, Vassetzky NS. 2005. Short retroposons in eukaryotic genomes. *International Review of Cytology* **247:** 165–221.

Kraner SD, Chong JA, Tsay HJ, Mandel G. 1992. Silencing the type II sodium channel gene: A model for neural-specific gene regulation. *Neuron* **9:** 37–44.

Kurokawa R, Rosenfeld MG, Glass CK. 2009. Transcriptional regulation through noncoding RNAs and epigenetic modifications. *RNA Biology* **6:** 233–236.

Kuwabara T, Hsieh J, Nakashima K, Taira K, Gage FH. 2004. A small modulatory dsRNA specifies the fate of adult neural stem cells. *Cell* **116:** 779–793.

Lanz RB, McKenna NJ, Onate SA, Albrecht U, Wong J, Tsai SY, Tsai MJ, O'Malley BW. 1999. A steroid receptor coactivator, SRA, functions as an RNA and is present in an SRC-1 complex. *Cell* **97:** 17–27.

Lau NC, Seto AG, Kim J, Kuramochi-Miyagawa S, Nakano T, Bartel DP, Kingston RE. 2006. Characterization of the piRNA complex from rat testes. *Science* **313:** 363–367.

Lee HC, Chang SS, Choudhary S, Aalto AP, Maiti M, Bamford DH, Liu Y. 2009. qiRNA is a new type of small interfering RNA induced by DNA damage. *Nature* **459:** 274–277.

Lee Y, Kim M, Han J, Yeom KH, Lee S, Baek SH, Kim VN. 2004. MicroRNA genes are transcribed by RNA polymerase II. *EMBO J* **23:** 4051–4060.

Li F, Schiemann AH, Scott MJ. 2008. Incorporation of the noncoding roX RNAs alters the chromatin-binding specificity of the *Drosophila* MSL1/MSL2 complex. *Mol Cell Biol* **28:** 1252–1264.

Liu WM, Chu WM, Choudary PV, Schmid CW. 1995. Cell stress and translational inhibitors transiently increase the abundance of mammalian SINE transcripts. *Nucleic Acids Res* **23:** 1758–1765.

Louro R, Smirnova AS, Verjovski-Almeida S. 2009. Long intronic noncoding RNA transcription: Expression noise or expression choice? *Genomics* **93:** 291–298.

Lubbert M. 2003. Gene silencing of the p15/INK4B cell-cycle inhibitor by hypermethylation: An early or later epigenetic alteration in myelodysplastic syndromes? *Leukemia* **17:** 1762–1764.

Lunde BM, Moore C, Varani G. 2007. RNA-binding proteins: Modular design for efficient function. *Nat Rev Mol Cell Biol* **8:** 479–490.

Maraia RJ, Driscoll CT, Bilyeu T, Hsu K, Darlington GJ. 1993. Multiple dispersed loci produce small cytoplasmic Alu RNA. *Mol Cell Biol* **13:** 4233–4241.

Mariner PD, Walters RD, Espinoza CA, Drullinger LF, Wagner SD, Kugel JF, Goodrich JA. 2008. Human Alu RNA is a modular transacting repressor of mRNA transcription during heat shock. *Mol Cell* **29:** 499–509.

Martens JA, Laprade L, Winston F. 2004. Intergenic transcription is required to repress the *Saccharomyces cerevisiae* SER3 gene. *Nature* **429:** 571–574.

Martens JA, Wu PY, Winston F. 2005. Regulation of an intergenic transcript controls adjacent gene transcription in *Saccharomyces cerevisiae*. *Genes Develop* **19:** 2695–2704.

Martianov I, Ramadass A, Serra Barros A, Chow N, Akoulitchev A. 2007. Repression of the human dihydrofolate reductase gene by a non-coding interfering transcript. *Nature* **445:** 666–670.

Martone R, Euskirchen G, Bertone P, Hartman S, Royce TE, Luscombe NM, Rinn JL, Nelson FK, Miller P, Gerstein M, et al. 2003. Distribution of NFkappaB- binding sites across human chromosome 22. *Proc Nat Acad Sci* **100:** 12247–12252.

Mattick JS. 1994. Introns: evolution and function. *Current Opinion Gen Develop* **4:** 823–831.

Mattick JS. 2007. A new paradigm for developmental biology. *J Exp Biol* **210:** 1526–1547.

Mattick JS, Gagen MJ. 2001. The evolution of controlled multitasked gene networks: the role of introns and other noncoding RNAs in the development of complex organisms. *Mol Biol Evolution* **18:** 1611–1630.

Mattick JS, Makunin IV. 2006. Non-coding RNA. *Human Mol Genet* **15 Spec No 1,** R17–29.

Matzke M, Aufsatz W, Kanno T, Daxinger L, Papp I, Mette MF, Matzke AJ. 2004. Genetic analysis of RNA-mediated transcriptional gene silencing. *Biochim Biophys Acta* **1677:** 129–141.

McKenna NJ, O'Malley BW. 2002. Combinatorial control of gene expression by nuclear receptors and coregulators. *Cell* **108:** 465–474.

Mercer TR, Dinger ME, Mattick JS. 2009. Long non-coding RNAs: Insights into functions. *Nat Rev* **10:** 155–159.

Mercer TR, Dinger ME, Sunkin SM, Mehler MF, Mattick JS. 2008. Specific expression of long noncoding RNAs in the mouse brain. *Proc Nat Acad Sci* **105:** 716–721.

Montgomery MK, Fire A. 1998. Double-stranded RNA as a mediator in sequence-specific genetic silencing and co-suppression. *Trends Genet* **14:** 255–258.

Mori N, Schoenherr C, Vandenbergh DJ, Anderson DJ. 1992. A common silencer element in the SCG10 and type II Na+ channel genes binds a factor present in nonneuronal cells but not in neuronal cells. *Neuron* **9:** 45–54.

Nagano T, Mitchell JA, Sanz LA, Pauler FM, Ferguson-Smith AC, Feil R, Fraser P. 2008. The Air noncoding RNA epigenetically silences transcription by targeting G9a to chromatin. *Science* **322:** 1717–1720.

Neil H, Malabat C, d'Aubenton-Carafa Y, Xu Z, Steinmetz LM, Jacquier A. 2009. Widespread bidirectional promoters are the major source of cryptic transcripts in yeast. *Nature* **457:** 1038–1042.

Nguyen VT, Kiss T, Michels AA, Bensaude O. 2001. 7SK small nuclear RNA binds to and inhibits the activity of CDK9/cyclin T complexes. *Nature* **414:** 322–325.

Nishihara S, Tsuda L, Ogura T. 2003. The canonical Wnt pathway directly regulates NRSF/REST expression in chick spinal cord. *Biochem Biophys Res Commun* **311:** 55–63.

Otto SJ, McCorkle SR, Hover J, Conaco C, Han JJ, Impey S, Yochum GS, Dunn JJ, Goodman RH, Mandel G. 2007. A new binding motif for the 7 transcriptional repressor REST uncovers large gene networks devoted to neuronal functions. *J Neurosci* **27:** 6729–6739.

Pasmant E, Laurendeau I, Heron D, Vidaud M, Vidaud D, Bieche I. 2007. Characterization of a germ-line deletion, including the entire INK4/ARF locus, in a melanoma-neural system tumor family: Identification of ANRIL, an antisense noncoding RNA whose expression coclusters with ARF. *Cancer Res* **67:** 3963–3969.

Phillips JE, Corces VG. 2009. CTCF: Master weaver of the genome. *Cell* **137:** 1194–1211.

Plath K, Mlynarczyk-Evans S, Nusinow DA, Panning B. 2002. Xist RNA and the mechanism of X chromosome inactivation. *Ann Rev Gen* **36:** 233–278.

Ponting CP, Oliver PL, Reik W. 2009. Evolution and functions of long noncoding RNAs. *Cell* **136:** 629–641.

Preker P, Nielsen J, Kammler S, Lykke-Andersen S, Christensen MS, Mapendano CK, Schierup MH, Jensen TH. 2008. RNA exosome depletion reveals transcription upstream of active human promoters. *Science* **322:** 1851–1854.

Rinn JL, Kertesz M, Wang JK, Squazzo SL, Xu X, Brugmann SA, Goodnough LH, Helms JA, Farnham PJ, Segal E, et al. 2007. Functional demarcation of active and silent chromatin domains in human HOX loci by noncoding RNAs. *Cell* **129:** 1311–1323.

Ronemus M, Martienssen R. 2005. RNA interference: methylation mystery. *Nature* **433:** 472–473.

Rosenfeld MG, Glass CK. 2001. Coregulator codes of transcriptional regulation by nuclear receptors. *J Biol Chem* **276:** 36865–36868.

Rosenfeld MG, Lunyak VV, Glass CK. 2006. Sensors and signals: A coactivator/corepressor/epigenetic code for integrating signal-dependent programs of transcriptional response. *Genes Develop* **20:** 1405–1428.

Sanchez-Elsner T, Gou D, Kremmer E, Sauer F. 2006. Noncoding RNAs of trithorax response elements recruit *Drosophila* Ash1 to Ultrabithorax. *Science* **311**: 1118–1123.

Sasaki YT, Ideue T, Sano M, Mituyama T, Hirose T. 2009. MENepsilon/β noncoding RNAs are essential for structural integrity of nuclear paraspeckles. *Proc Nat Acad Sci* **106**: 2525–2530.

Schoeftner S, Sengupta AK, Kubicek S, Mechtler K, Spahn L, Koseki H, Jenuwein T, Wutz A. 2006. Recruitment of PRC1 function at the initiation of X inactivation independent of PRC2 and silencing. *EMBO J* **25**: 3110–3122.

Schuettengruber B, Chourrout D, Vervoort M, Leblanc B, Cavalli G. 2007. Genome regulation by polycomb and trithorax proteins. *Cell* **128**: 735–745.

Scott MJ, Li F. 2008. How do ncRNAs guide chromatin-modifying complexes to specific locations within the nucleus? *RNA Biol* **5**: 13–16.

Shamovsky I, Ivannikov M, Kandel ES, Gershon D, Nudler E. 2006. RNA-mediated response to heat shock in mammalian cells. *Nature* **440**: 556–560.

Shi Y, Downes M, Xie W, Kao HY, Ordentlich P, Tsai CC, Hon M, Evans RM. 2001. Sharp, an inducible cofactor that integrates nuclear receptor repression and activation. *Genes Develop* **15**: 1140–1151.

Shi YJ, Matson C, Lan F, Iwase S, Baba T, Shi Y. 2005. Regulation of LSD1 histone demethylase activity by its associated factors. *Mol Cell* **19**: 857–864.

Shilatifard A. 2008. Molecular implementation and physiological roles for histone H3 lysine 4 (H3K4) methylation. *Current Opinion Cell Biol* **20**: 341–348.

St Laurent G, 3rd, Wahlestedt C. 2007. Noncoding RNAs: couplers of analog and digital information in nervous system function? *Trends Neurosci* **30**: 612–621.

Storz G. 2002. An expanding universe of noncoding RNAs. *Science* **296**: 1260–1263.

Sunwoo H, Dinger ME, Wilusz JE, Amaral PP, Mattick JS, Spector DL. 2009. MEN epsilon/β nuclear-retained non-coding RNAs are up-regulated upon muscle differentiation and are essential components of paraspeckles. *Gen Res* **19**: 347–359.

Taft RJ, Pheasant M, Mattick JS. 2007. The relationship between nonproteincoding DNA and eukaryotic complexity. *Bioessays* **29**: 288–299.

Trinklein ND, Aldred SJ, Saldanha AJ, Myers RM. 2003. Identification and functional analysis of human transcriptional promoters. *Gen Res* **13**: 308–312.

Umlauf D, Fraser P, Nagano T. 2008. The role of long non-coding RNAs in chromatin structure and gene regulation: Variations on a theme. *Biol Chem* **389**: 323–331.

Vagin VV, Sigova A, Li C, Seitz H, Gvozdev V, Zamore PD. 2006. A distinct small RNA pathway silences selfish genetic elements in the germline. *Science* **313**: 320–324.

Walter P, Blobel G. 1982. Signal recognition particle contains a 7S RNA essential for protein translocation across the endoplasmic reticulum. *Nature* **299**: 691–698.

Wang P, Yin S, Zhang Z, Xin D, Hu L, Kong X, Hurst LD. (2008b). Evidence for common short natural trans sense-antisense pairing between transcripts from protein coding genes. *Gen Biol* **9**: R169.

Wang X, Arai S, Song X, Reichart D, Du K, Pascual G, Tempst P, Rosenfeld MG, Glass CK, Kurokawa R. (2008a). Induced ncRNAs allosterically modify RNA-binding proteins in cis to inhibit transcription. *Nature* **454**: 126–130.

Wassarman KM, Storz G. 2000. 6S RNA regulates *E.coli* RNA polymerase activity. *Cell* **101**: 613–623.

Watanabe T, Takeda A, Tsukiyama T, Mise K, Okuno T, Sasaki H, Minami N, Imai H. 2006. Identification and characterization of two novel classes of small RNAs in the mouse germline: Retrotransposon-derived siRNAs in oocytes and germline small RNAs in testes. *Genes Develop* **20**: 1732–1743.

Wellik DM. 2009. Hox genes and vertebrate axial pattern. *Current Topics Develop Biol* **88**: 257–278.

Wickiser JK, Winkler WC, Breaker RR, Crothers DM. 2005. The speed of RNA transcription and metabolite binding kinetics operate an FMN riboswitch. *Mol Cell* **18**: 49–60.

Willingham AT, Gingeras TR. 2006. TUF love for "junk" DNA. *Cell* **125**: 1215–1220.

Willingham AT, Orth AP, Batalov S, Peters EC, Wen BG, Aza-Blanc P, Hogenesch JB, Schultz PG. 2005. A strategy for probing the function of noncoding RNAs finds a repressor of NFAT. *Science* **309**: 1570–1573.

Wilusz JE, Sunwoo H, Spector DL. 2009. Long noncoding RNAs: functional surprises from the RNA world. *Genes Develop* **23**: 1494–1504.

Wu Q, Kim YC, Lu J, Xuan Z, Chen J, Zheng Y, Zhou T, Zhang MQ, Wu CI, Wang SM. 2008. Poly A- transcripts expressed in HeLa cells. *PloS one* **3**: e2803.

Wyers F, Rougemaille M, Badis G, Rousselle JC, Dufour ME, Boulay J, Regnault B, Devaux F, Namane A, Seraphin B, et al. 2005. Cryptic pol II transcripts are degraded by a nuclear quality control pathway involving a new poly(A) polymerase. *Cell* **121**: 725–737.

Xu B, Koenig RJ. 2004. An RNA-binding domain in the thyroid hormone receptor enhances transcriptional activation. *J Biol Chem* **279**: 33051–33056.

Xu B, Koenig RJ. 2005. Regulation of thyroid hormone receptor α2 RNA binding and subcellular localization by phosphorylation. *Mol Cell Endocrinology* **245**: 147–157.

Xu Z, Wei W, Gagneur J, Perocchi F, Clauder-Munster S, Camblong J, Guffanti E, Stutz F, Huber W, Steinmetz LM. 2009. Bidirectional promoters generate pervasive transcription in yeast. *Nature* **457**: 1033–1037.

Yakovchuk P, Goodrich JA, Kugel JF. 2009. B2 RNA and Alu RNA repress transcription by disrupting contacts between RNA polymerase II and promoter DNA within assembled complexes. *Proc Nat Acad Sci* **106**: 5569–5574.

Yang Z, Zhu Q, Luo K, Zhou Q. 2001. The 7SK small nuclear RNA inhibits the CDK9/cyclin T1 kinase to control transcription. *Nature* **414**: 317–322.

Yelin R, Dahary D, Sorek R, Levanon EY, Goldstein O, Shoshan A, Diber A, Biton S, Tamir Y, Khosravi R, et al. 2003. Widespread occurrence of antisense transcription in the human genome. *Nat Biotechnol* **21**: 379–386.

Yu W, Gius D, Onyango P, Muldoon-Jacobs K, Karp J, Feinberg AP, Cui H. 2008. Epigenetic silencing of tumour suppressor gene p15 by its antisense RNA. *Nature* **451**: 202–206.

Zhao J, Sun BK, Erwin JA, Song JJ, Lee JT. 2008. Polycomb proteins targeted by a short repeat RNA to the mouse X chromosome. *Science* **322**: 750–756.

Zhao X, Patton JR, Davis SL, Florence B, Ames SJ, Spanjaard RA. 2004. Regulation of nuclear receptor activity by a pseudouridine synthase through posttranscriptional modification of steroid receptor RNA activator. *Mol Cell* **15**: 549–558.

Folding and Finding RNA Secondary Structure

David H. Mathews[1], Walter N. Moss[2], and Douglas H. Turner[3]

[1]Department of Biochemistry and Biophysics and Center for RNA Biology, University of Rochester School of Medicine and Dentistry, Rochester, New York 14642

[2]Department of Chemistry, University of Rochester, Rochester, New York 14627-0216

[3]Department of Chemistry and Center for RNA Biology, University of Rochester, Rochester, New York 14627-0216

Correspondence: turner@chem.rochester.edu

SUMMARY

Optimal exploitation of the expanding database of sequences requires rapid finding and folding of RNAs. Methods are reviewed that automate folding and discovery of RNAs with algorithms that couple thermodynamics with chemical mapping, NMR, and/or sequence comparison. New functional noncoding RNAs in genome sequences can be found by combining sequence comparison with the assumption that functional noncoding RNAs will have more favorable folding free energies than other RNAs. When a new RNA is discovered, experiments and sequence comparison can restrict folding space so that secondary structure can be rapidly determined with the help of predicted free energies. In turn, secondary structure restricts folding in three dimensions, which allows modeling of three-dimensional structure. An example from a domain of a retrotransposon is described. Discovery of new RNAs and their structures will provide insights into evolution, biology, and design of therapeutics. Applications to studies of evolution are also reviewed.

Outline

1 Folding RNA into secondary structures
2 Restraining folding space
3 Automating comparative sequence analysis
4 Finding Functional RNA
5 Future directions
6 Application to the study of evolution
References

Copyright © 2011 Cold Spring Harbor Laboratory Press; all rights reserved
Cite this article as *Cold Spring Harb Perspect Biol* doi: 10.1101/cshperspect.a003665

Cellular RNA was considered merely an intermediate between DNA and protein for much of the history of molecular biology (except in RNA viruses). The discovery of catalytic RNA showed that this schema had to be revised. The finding that RNA could possess functionality once believed to be the sole domain of protein enzymes, led to the hypothesis that RNA could have preceded protein and DNA in an "RNA World." Echoes of the RNA world remain in that RNA continues to perform functions developed early in evolution, e.g., as the catalyst for protein synthesis (Moore and Steitz 2010). There is recent evidence that an unexpectedly large fraction of DNA in higher eukaryotes, perhaps 90%, is transcribed into RNA (Birney et al. 2007). A positive correlation between an increased proportion of noncoding versus coding RNA and an organism's developmental complexity has been observed (Taft and Mattick 2003). This trend has been suggested to mean that noncoding RNA represents a "new genetics" and may very well be the engine of eukaryotic complexity (Mattick 2004). RNA may have evolved and may continue to evolve many yet unknown functions. Evolution may be slow to eliminate nonfunctional RNAs. RNAs without current function, however, may constitute a pool of molecules that might be adapted to fill novel roles, such as gene regulatory elements or defenses against transposons and viruses. One of the themes of this article is to describe methods for finding functional RNAs and their structures, which in turn can reveal structure-function and evolutionary relationships.

Evolution is restrained by fundamental chemical and physical principles of which thermodynamics is one. Although much of the sequence dependence of RNA thermodynamics is unknown, for RNA sequences of fewer than 700 nucleotides it is possible to correctly predict roughly 70% of secondary structure from thermodynamics alone (Mathews et al. 2004). This success suggests that thermodynamics is a major determinant of secondary structure and thus of evolution of structured RNAs. Perhaps thermodynamics was particularly important in the early stages of evolution when RNA molecules had a high degree of structural plasticity. Once a functional structure was generated in an evolving population of RNAs, there would be a driving force for stabilizing that particular structure over alternative folds in the population. Thus, structures developed early in evolution may be determined more by free energy minimization of the RNA than structures developed later. For example, later structures may depend more on the kinetics of folding and on interaction with proteins. Our understanding of evolution and structure at the molecular level is still evolving and currently somewhat primitive.

A second theme of this article is overcoming current limitations through the combined application of thermodynamics, sequence comparison, and experiment to rapidly model RNA secondary structures. The methods facilitate finding RNA sequences with functions that rely on secondary structure.

1 FOLDING RNA INTO SECONDARY STRUCTURES

1.1 Free Energy Minimization

If folding was determined by thermodynamics alone and if the sequence dependence of thermodynamics was completely understood, then it would be possible to predict secondary structure from sequence alone. For a unimolecular reaction such as the folding of an RNA molecule:

$$U \leftrightarrow F \quad K = [F]/[U] = e^{-\Delta G^\circ/RT}.$$

Here, K is the equilibrium constant giving the ratio of concentrations for folded, F, and unfolded, U, species at equilibrium; ΔG° is the standard free energy difference between F and U; R is the gas constant; and T is the temperature in kelvins. The challenge of predicting secondary structure from thermodynamics is to find the base-pairing that gives the lowest free energy change in going from the unfolded to folded state, and therefore the highest concentration of folded species. Generally this search is accomplished with a dynamic programming algorithm, a type of recursive algorithm that is commonly used to solve optimization problems in biology (e.g., sequence alignment) and elsewhere. Dynamic programming can implicitly search the entire set of possible RNA secondary structures to find the lowest free energy structure without the necessity of generating all structures explicitly. The free energy change is typically approximated with a nearest neighbor model in which the ΔG° is the sum of free energy increments for the various nearest neighbor motifs (e.g., stacked base pairs in an RNA helix) that occur in a structure (Turner 2000; Mathews et al. 2005). Parameters for the nearest neighbor increments have been experimentally determined by optical melting studies (Xia et al. 1998; Turner 2000; Mathews et al. 2004), by relating parameters to the number of occurrences of various motifs in known secondary structures (Do et al. 2006), by optimizing parameters to predict known secondary structures (Ninio 1979; Papanicolaou et al. 1984), or by some combination of the previous (Jaeger et al. 1989; Mathews et al. 1999; Andronescu et al. 2007).

1.2 Partition Functions and Probabilities

Because the accuracy of secondary structure prediction is limited in part by an incomplete knowledge of the folding rules, there is significant interest in determining the quality of predictions. One approach to estimating the quality of prediction is to assign a probability to the prediction using a partition function.

The partition function, Q, contains a description of the ensemble thermodynamic properties of a system and is defined as the sum of the equilibrium constants for all possible secondary structures of a given sequence. The fraction of strands that will fold into a particular structure is the equilibrium constant for that structure, divided by Q. Counter-intuitively, the lowest free energy structure often occurs with a vanishingly small probability. For example, given the calculated partition function for the *Tetrahymena* group I intron (1.8×10^{107}), the probability that a strand folds into its predicted minimum free energy structure is 1 in 760 million. Many base pairs, however, are common to a large number of the low free energy structures. These common pairs are well represented in the structural ensemble and thus can have high pairing probability. In fact, 80 base pairs out of 144 predicted for the *Tetrahymena* group I intron have 90% or higher pairing probability.

Using probabilistic methods, three approaches are taken to enhance the information provided by structure prediction. The first is to predict the lowest free energy structure and then color annotate the structure with base pairing probabilities (Mathews 2004). Pairs of higher probability are more likely to be correctly predicted pairs (Mathews 2004). Therefore, the user can have greater confidence in the highly probable pairs (>90%) being truly informative of native secondary structure.

A second approach is to simply assemble structures of highly probable pairs (Mathews 2004). For example, structures can be assembled of pairs that exceed a given pairing probability threshold. These are valid structures, i.e., a nucleotide will only pair with up to one other nucleotide, if the threshold is set at 50% pairing probability or higher. The quality of the predicted pairs is high, but these structures are generally not saturated with pairs (Mathews 2004). Alternatively, structures can be assembled using the most probable pairs using a dynamic programming algorithm (Do et al. 2006; Hamada et al. 2009; Lu et al. 2009). These structures are called maximum expected accuracy structures.

A third approach is to sample structures from the folding ensemble according to their probability of occurring, using stochastic sampling (Ding and Lawrence 2003). The sampled structures can be analyzed to determine base pairing probabilities. Additionally, predicted structures can be clustered. A representative structure, called a centroid, of the most populated cluster can be a more accurate prediction of the native structure than the predicted lowest free energy structure (Ding et al. 2005). Stochastic sampling and clustering are especially useful for analyzing an RNA that natively populates more than one structure, e.g., a riboswitch sequence, because each structure should appear as a distinct cluster.

Table 1. Secondary structure prediction programs. This table provides a list of software packages that predict secondary structures using thermodynamics

Program:	URL:	Features:
RNAstructure	http://rna.urmc.rochester.edu/RNAstructure.html	JAVA/Windows Graphical User Interface; Command Line Interface; C++ Class Library
Sfold	http://sfold.wadsworth.org/	Web server
UNAFold/mfold	http://mfold.bioinfo.rpi.edu/	Web server; Command Line Interface
Vienna RNA Package	http://www.tbi.univie.ac.at/RNA/	Web server; Command Line Interface; C Function Library

1.3 Available Programs

Table 1 summarizes some of the available computer programs for RNA secondary structure prediction and their features. This list is confined to those that predict structure based on thermodynamics, although alternative approaches based on reproducing structural features in the database of known structures also show promise (Dowell and Eddy 2004; Do et al. 2006).

2 RESTRAINING FOLDING SPACE

Free energy minimization alone typically predicts correctly only about 70% of secondary structure. There are several reasons for the limited accuracy. For example, folding may not be determined only by thermodynamics, the sequence dependence of free energy changes is far from completely known, and the folding space for RNA is enormous; an RNA of n nucleotides has 1.8^n possible secondary structures (Zuker and Sankoff 1984). Finding the correct secondary structure can be compared to the difficulty of using random keystrokes to type correctly a sentence with 28 letters and spaces. This would take 27^{28}, or about 10^{40}, keystrokes. If, however, a letter is fixed whenever it is typed correctly, then it would only take a few thousand keystrokes (Dawkins 1987; Zwanzig et al. 1992). There are only 4 letters in the RNA alphabet and the "words" are helices and loops. In an analogous way, knowing that a given nucleotide is in a base pair or a loop greatly reduces the remaining folding space. Thus, the folding problem becomes tractable if there are ways to deduce when a base pair or loop is correct. Several approaches for this are described below.

2.1 Experiments

Experiments can provide constraints and restraints to reduce folding space. The methods most commonly used employ chemical modification of bases (Inoue and Cech 1985; Moazed et al. 1986; Ehresmann et al. 1987; Mathews et al. 2004) or ribose sugars (Merino et al. 2005; Deigan et al. 2009) to identify nucleotides that are unpaired or in loosely structured regions. Modified sites can be rapidly read out by primer extension using reverse transcriptase, which will stop at the base 3′ of the modified site. By using multiple primers, any length RNA can be interrogated. Chemical agents are useful for restraining unpaired or loosely paired bases. Nuclear magnetic resonance (NMR) spectra can be used to constrain base paired nucleotides (Hart et al. 2008).

Dimethyl sulfate (DMS), 1-cyclohexyl-3-(2-morpholinoethyl)carbodiimide metho-p-toluenesulfanate (CMCT), and kethoxal react with the Watson-Crick faces of A and C, U, and G, respectively. DMS is of special use as it can be applied in living cells for in vivo mapping of truly native RNA structures (Harris et al. 1995; Zaug and Cech 1995). Chemical reactivity has been applied as a constraint in the RNAstructure folding program (Mathews et al. 2004). The constraint applied is that a base that reacts cannot be in a Watson-Crick pair flanked by Watson-Crick pairs (Fig. 1). The same constraint can be applied for reactivity that cleaves the backbone in loops: such as with Pb^{2+} (Lindell et al. 2002) or with hydrolysis (Li and Breaker 1999; Soukup and Breaker 1999).

N-methylisotoic anhydride (NMIA) and related molecules react with flexible ribose groups (Merino et al. 2005; Mortimer and Weeks 2007). Thus, reactive nucleotides are presumably not in strong Watson-Crick pairs or rigid tertiary interactions. Reactivity is less sensitive to local environment than that of DMS, CMCT, and kethoxal (Wilkinson et al. 2009), presumably because the electrostatic environment of the sugars is more uniform than that of the bases. Because this "SHAPE" chemistry interrogates the ribose of every nucleotide, relative reactivity can be assigned to every nucleotide. Readout by capillary gel electrophoresis has provided quantification that allows relative reactivity to be used as restraints. That is, relative reactivity provides a measure of the likelihood of a nucleotide being unpaired or paired, rather than an absolute constraint (Fig. 1). This allows lack of reactivity to be interpreted as a favorable likelihood for Watson-Crick base pairing (Deigan et al. 2009). Although lack of reactivity may also reflect strong tertiary interactions, runs of consecutive

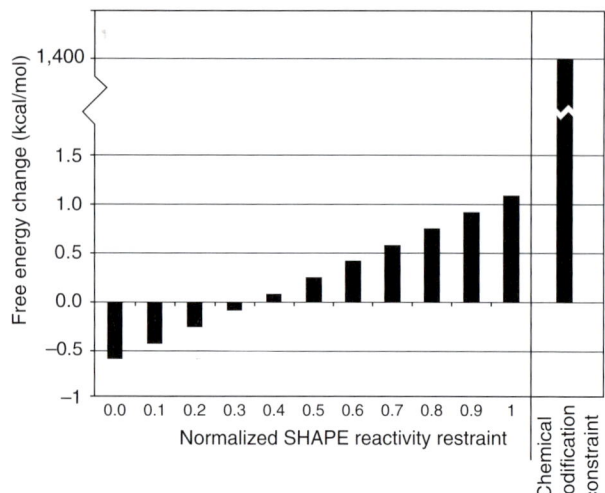

Figure 1. Folding constraints and restraints. Traditional chemical agents that act on bases are applied as folding constraints, i.e., a base accessible to chemical modification cannot be in a base pair flanked by Watson-Crick pairs on each side. In RNAstructure, this is implemented by assigning a large positive free energy to any conformation that violates the constraint. SHAPE reactivity is applied as a folding restraint, i.e., a free energy change bonus or penalty for pairing of a nucleotide. For nucleotides with low SHAPE reactivity, a pairing stabilization is provided and for high reactivity, a pairing penalty is provided. The SHAPE restraint is provided per nucleotide in a base pair stack. Therefore, the free energy change is applied twice per nucleotide buried in a helix and once per nucleotide in a pair at the end of a helix (Deigan et al. 2009).

Watson-Crick base pairs in helixes are usually longer than consecutive tertiary interactions so that helixes are favored more by the enhancement in the probability increment for Watson-Crick base pairing.

NMR interrogates the chemical environment of nuclei. The chemical shifts of an imino hydrogen proton and the attached nitrogen nucleus provide a signature that identifies whether the imino proton is in a Watson-Crick GC, AU, or wobble GU pair. Because the imino protons are close to each other in the middle of a helix, they can exchange energy, which is measurable using two-dimensional NMR spectroscopy. Thus, it is possible to determine that helixes with certain sequences of base pairs must be present in the secondary structure. This provides information complementary to chemical modification, which identifies unpaired nucleotides most definitively. An algorithm for NMR assisted prediction of secondary structure (NAPSS) is available (Hart et al. 2008). Because the method identifies double helixes, it is especially effective for revealing pseudoknots. It also provides a few initial assignments of resonances, which can facilitate determination of three-dimensional structure. It is limited, however, to RNAs that are labeled with ^{15}N and that can be studied by NMR. The latter limitation probably restricts the maximum length of the RNA to somewhere between 100 and 300 nucleotides. Another disadvantage of NMR is that relatively large amounts of RNA are required compared to chemical methods.

2.2 Structure Comparison

RNAs whose biological function depends on their structure (e.g., tRNA, rRNA, etc.) should show structural conservation when comparisons are made between RNAs of related species (Woese and Pace 1993). Thus, another approach for identifying correct base pairs and loops is their occurrence in the same or similar locations in homologous (i.e., evolutionarily related) RNAs. The structure comparison approach has the advantages that it gives the structure in the cell, should work for sequences not governed by thermodynamics, can identify pseudoknots, non-Watson-Crick base pairs and elements of the tertiary structure; it also leverages the exploding database of sequences. Structure comparison may not be applicable to all RNAs, however. To determine secondary structure de novo from sequence analysis, multiple, homologous sequences are required, as are high quality alignments of the sequences. These requirements may be hard to meet for rare RNAs or RNAs with high levels of variability.

The manual determination of a conserved structure from a set of sequences is called comparative sequence analysis and involves aligning available sequences to identify covariant sites: sites that show correlated mutations. Synchronized mutation between sites is interpreted as the manifestation of functional (structural) constraints on the molecular evolution of the RNA. Double point mutations in aligned sites that preserve base pairing (e.g., G-C mutating to A-U, C-G, etc.) are the simplest covariations to identify and interpret. Simply, to model base pairing one searches an alignment of RNA sequences for "structurally silent" mutations. Comparative sequence analysis is phenomenally accurate at predicting a structure (>95% of predicted pairs correct) when significant human effort and skill are applied (Gutell et al. 2002).

2.3 Combined Methods for Determining Secondary Structure: An Example from the 5′ Regions of R2 Retrotransposon RNA

An illustrative example for determining RNA secondary structure comes from the region that occurs toward the 5′ terminus of silk moth R2 retroelements. This roughly 350 nucleotide structured region was discovered as a persistent "contaminant" in preparations of the silk moth, *Bombyx mori*, R2 encoded protein. This RNA is strongly bound by R2 protein and this binding is an essential part of R2 element insertion into the host genome (Christensen et al. 2006). An initial structural model was proposed using free energy minimization with a single sequence, constrained by chemical modification and oligonucleotide binding data (Kierzek et al. 2008). Unusual structural features of this model inspired further probing of a 74 nucleotide fragment using NMR, which revealed a pseudoknot structure for this fragment (Hart et al. 2008).

As additional R2 sequences became available, it was possible to use comparative sequence analysis to interrogate the structure of this RNA (Kierzek et al. 2009). Each of four additional R2 sequences was subjected to the same battery of chemical reagents (i.e., DMS, CMCT, and NMIA) and oligonucleotide binding experiments as *B. mori*. These data were used in constrained free energy minimization and combined with the *B. mori* NAPSS results to provide initial structural hypotheses for manual alignment and comparative analysis. The reasonability of the structural hypotheses was gauged with a partition function calculation and annotation of the base pairing probabilities. The alignment and secondary structures were altered to maximize the conservation of structure and the formation of compensatory base changes.

The results of this modeling are summarized in Figure 2. There are five regions in this RNA that are structurally conserved. These are organized into four hairpin loop structures and a pseudoknot. One of the interesting features revealed by structure alignment was that two of the conserved hairpins, falling within the coding region of this RNA, correspond to conserved protein coding regions. Evidently, evolution acted on two levels: mutations that preserved RNA base pairing could also be synonymous substitutions with respect to amino acid coding. The secondary structures shown in Figure 2 are consistent with chemical mapping and NMR results; they are well supported by consistent and compensatory mutations (single and double point mutations that preserve base pairing). Moreover, for each single sequence fold, these conserved structures are composed of high probability base pairs as determined from partition function calculations. This wealth of data is summarized in Figure 2.

3 AUTOMATING COMPARATIVE SEQUENCE ANALYSIS

The example of comparative analysis given above represents a significant investment in time and a certain degree of artisanal craftsmanship. To date, no completely accurate computational approach exists for automating comparative analysis, but a number of distinct approaches have been applied to the problem. Overall, these methods are

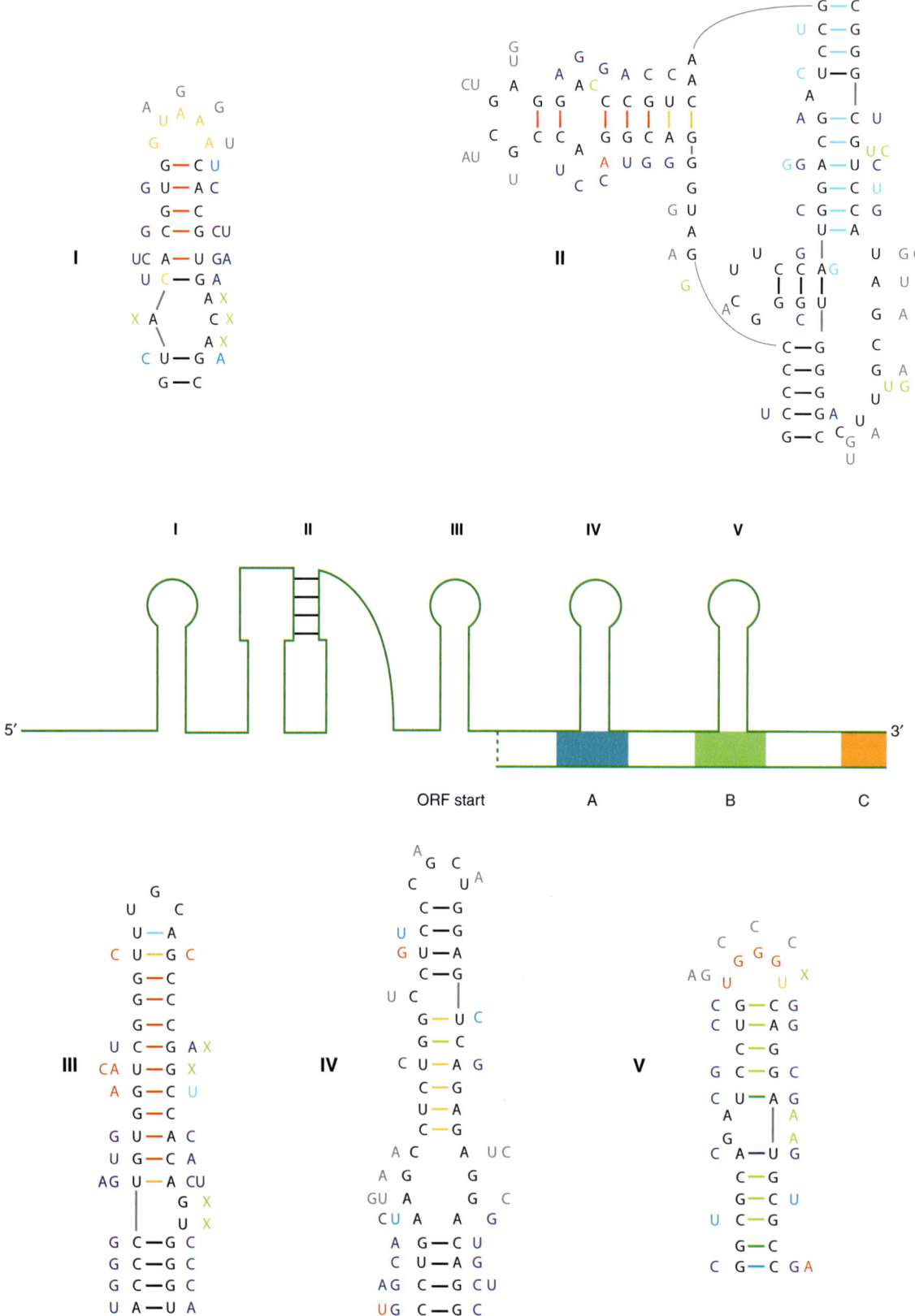

Figure 2. (See facing page for legend)

helpful at generating hypotheses that can be tested with manual comparison or experiments.

3.1 Three General Approaches to Predicting Conserved Secondary Structures

It would be most efficient to deduce correctly secondary structure from sequence alone, and programs are available for attempting this when multiple sequences are available. The problem of predicting the lowest free energy structure common to multiple sequences has been approached from three directions. The first approach is to simultaneously find both the lowest free energy structure and the optimum alignment of sequences. The second approach is to start with the sequences aligned by nucleotide identity and then find the conserved pairs in the given alignment. The final approach is to predict low free energy structures for each sequence separately and then to sort through the predicted structures to find the structures common to all sequences.

3.2 Approach 1: Fold and Align

The concept of using a dynamic programming algorithm to simultaneously fold and align a set of sequences was introduced by Sankoff (Sankoff 1985). This idea was implemented in practical computer programs such as Dynalign and Foldalign to find lowest free energy common structures (Mathews and Turner 2002; Havgaard et al. 2005). To make calculations feasible for long sequences (currently up to about 2000 nucleotides), other data are used to restrict the possible structures or alignments. The current approach in Dynalign, for example, is to rule out base pairs that can only exist in structures with high folding free energies (Uzilov et al. 2006) and to rule out alignments that are extremely unlikely ($<10^{-3}$) based on statistical analysis of aligned sequences (Harmanci et al. 2007). Additional algorithms have been developed to use scoring schemes based on producing structures similar to secondary structures present in databases of known structures (Holmes 2005; Dowell and Eddy 2006; Do et al. 2008).

To determine confidence estimates for predictions of base pairs common to two sequences, a partition function algorithm for common structures was developed, called PARTS (Harmanci et al. 2008). This algorithm calculates equilibrium constants for common structures with a pseudo free energy score derived from base pair probabilities determined for each sequence and sequence alignment probabilities. The shortcut of using base pairing probabilities for scoring saves significant computation time and had been previously proposed (Hofacker et al. 2004; Hofacker and Stadler 2004) and used in the structure prediction programs LocARNA (Will et al. 2007) and Murlet (Kiryu et al. 2007). As with single sequences, base pairs that form with greater probability in common structures are more likely to be correctly predicted than those of low probability (Harmanci et al. 2008). Additionally, the partition function allows for stochastic sampling and clustering of structures conserved between the two sequences (Harmanci et al. 2009). As expected, the additional information contained in two sequences improves the fidelity of structure prediction as evidenced by the fact that the individual clusters contain structures that are much more similar to each other than for single sequence structure sampling (Harmanci et al. 2009).

Because of the computational cost, practical calculations are only performed with two sequences. To determine a structure common to more than two sequences, methods have been developed that use greedy heuristics to find the common structure using multiple pairwise calculations (Bellamy-Royds and Turcotte 2007; Kiryu et al. 2007; Torarinsson et al. 2007; Will et al. 2007). The drawback to this approach is that any calculations performed early in the set that have poor prediction accuracy will make the subsequent predictions poor as well.

Figure 2. Determination of Structured Regions of an RNA. A cartoon of the 5′ region of the silk moth R2 retrotransposon is shown. The conserved structure is organized into four hairpin loops (labeled I and III–V) and a pseudoknot (labeled II). Also shown are three conserved coding regions (A–C) and a putative open reading frame (ORF) start site. The five conserved structures are detailed with data that went into the structural modeling. The sequences shown are for *B. mori* whereas mutations are those that occur in four other moth species. Mutational data appear next to the main sequence and is color annotated: dark blue are double mutations that maintain base pairing (compensatory), light blue are single point mutations that maintain pairing (consistent), gray are mutations in loops, red disrupt canonical base pairs (inconsistent), green are insertions (green X represents a deletion). Experimental mapping is color annotated on the backbone sequence: red are NMIA only modifications and orange are modifications by both traditional mapping agents (DMS or CMCT) and NMIA. Base pairs are indicated with dashes between nucleotides and are color annotated for probability from partition function calculation: Red, probability (P) $\geq 99\%$; Orange, $99\% > P \geq 95\%$; Yellow, $95\% > P \geq 90\%$; Dark Green, $90\% > P \geq 80\%$; Light Green, $80\% > P \geq 70\%$; Light Blue, $70\% > P \geq 60\%$; Dark Blue, $60\% > P \geq 50\%$; Black $<50\%$. Many base pairs in the pseudoknot have low probability because the RNAstructure program does not allow pseudoknots and thus, under-counts them in the partition function.

3.3 Approach 2: Align then Fold

In the second approach, a multiple sequence alignment is constructed based on sequence information alone and then the lowest free energy structure is predicted that is common to all or most sequences (Lück et al. 1996; Hofacker et al. 2002; Bernhart et al. 2008). Calculations are improved by also providing free energy change bonuses for base pair formation at sites of covariation, where structure is conserved, but sequence is not.

The advantage to this approach is speed. It can be applied to almost any number of sequences and takes roughly the same calculation time as structure prediction for a single sequence of the same length as the alignment length. The drawback is that the quality of the structure prediction depends on the quality of the alignment. Alignments based on sequence alone may not properly reflect the structural homology that relates the set of sequences, and it is possible to miss compensating base pair changes that are key to evaluating the quality of the structure prediction. In a Catch 22, without a structurally informed alignment, it is difficult to develop a structural model to refine the alignment. The program ConStruct, however, addresses this limitation by providing a graphical user interface by which the user can manually adjust the alignment to facilitate the testing of structural models and iteratively refine the alignment and structure (Lück et al. 1999).

3.4 Approach 3: Fold then Align

In the third approach, a set of low free energy secondary structures is determined for each of multiple sequences and then the predicted structures are analyzed to find the lowest free energy structure common to all sequences (Reeder and Giegerich 2005). The direct implementation of this would be nearly computationally intractable because the number of low free energy structures for a given sequence is enormous (Wuchty et al. 1999). It is also known that the number of structures for a sequence with a folding free energy change below a threshold increases exponentially as the threshold is raised higher. Therefore, if the structures were explicitly analyzed, then it would be hard or impossible to sort through enough low free energy structures to make this approach feasible.

Instead of directly using this approach on structures, Giegerich and coworkers apply the algorithm on folding topologies, called abstract shapes (Giegerich et al. 2004). For example, one level of shape abstraction is to examine just the branching topology of the structure, without considering the internal or bulge loops. With increasing threshold above the lowest free energy structure, the number of abstract shapes increases much more slowly than the increase in number of structures (Voss et al. 2006).

The Fold then Align approach has the advantages of being faster than Fold and Align and also not being subject to the limited accuracy of sequence alignment as in Align then Fold. It has the drawback that the common abstract shape is found, which does not exactly predict which pairs are homologous, although an estimate can be generated by postprocessing (Höchsmann et al. 2004).

3.5 Available Programs

Table 2 shows a list of the available programs for predicting conserved secondary structures. This table is restricted to those programs that work using thermodynamics as a basis, although these approaches have been explored using alternative scoring methods.

4 FINDING FUNCTIONAL RNA

Given the number of sequenced whole genomes and the fact that much of these genomes are transcribed, there is significant interest in finding genes for noncoding RNA (ncRNA) sequences, i.e., genes that encode RNA sequences that function without being translated to a protein. This work fits into two categories, in which the first is the discovery of RNA sequences of a specific, known type and the second is the discovery of new types of RNA. Predictions of thermodynamic stability play important roles in both types of searches.

Because RNA structure is more highly conserved than sequence, methods to scan for specific ncRNAs test for the formation of a specific secondary structure. The earliest successful methods required training to a specific type of

Table 2. Programs for the prediction of a conserved RNA secondary structure. This table provides a list of programs that predict conserved secondary structures using thermodynamics

Program:	URL:	Type:
ConStruct	http://www.biophys.uni-duesseldorf.de/local/ConStruct/ConStruct.html	Align then Fold
Dynalign	http://rna.urmc.rochester.edu/dynalign.html	Fold and Align
FOLDALIGN	http://foldalign.ku.dk/	Fold and Align
LocARNA	http://www.bioinf.uni-freiburg.de/Software/LocARNA/	Fold and Align
Murlet	http://murlet.ncrna.org/	Fold and Align
PARTS	http://rna.urmc.rochester.edu/parts.html	Fold and Align
RNAalifold	http://rna.tbi.univie.ac.at/	Align then Fold
RNAcast	http://bibiserv.techfak.uni-bielefeld.de/rnacast/	Fold then Align

RNA either by automated training to a sequence alignment (Eddy and Durbin 1994) or by development of scores based on specific knowledge (Fichant and Burks 1991; Lowe and Eddy 1997; Lowe and Eddy 1999). A different program, called RNAmotif, was developed to scan for a user-specified secondary structure or class of structures, in which the user provides a descriptor of the structure (Macke et al. 2001). The drawback to this search method is that it is prone to predicting false positives. For example, a large number of potential cloverleaf structures encoded in genome sequences are not tRNA sequences (Tsui et al. 2003). Fortunately, predicted folding free energy change is an excellent criterion for separating the true positives from false positives (Tsui et al. 2003). In other words, the sequences with the potential to fold as cloverleafs, but are not tRNA sequences, nearly always had less favorable folding free energy change compared to true tRNA sequences folded as cloverleafs.

The problem of finding novel ncRNAs also can rely on predicted folding free energy change. It was hypothesized early that ncRNA sequences have lower folding free energy changes than random sequences (Le et al. 1988; Chen et al. 1990). This hypothesis proved controversial and one reason for the controversy is whether the correct controls for testing this hypothesis are random sequences with the same nucleotide content or dinucleotide content; this is because the stacking nearest neighbor parameters depend on dinucleotides (Seffens and Digby 1999; Workman and Krogh 1999). It is now generally accepted that there is a statistical trend for ncRNAs to have lower folding free energy change than matched control sequences with identical dinucleotide content (Clote et al. 2005; Uzilov et al. 2006). This trend, however, is not large enough to find with high sensitivity and specificity ncRNA sequences in genomes because of a large overlap in the distributions of folding free energy changes for ncRNAs and controls (Rivas and Eddy 2000; Uzilov et al. 2006).

The discovery of conserved ncRNA genes by scanning genome alignments, however, is achievable by evaluating thermodynamic stability. The programs that perform these scans have, as their basis, algorithms that predict conserved secondary structures using either "align then fold" or "fold and align" approaches as described earlier. For example, RNAz uses the align then fold algorithm RNAalifold to identify stable RNA structures in multiple genome alignments (Washietl et al. 2005). The fold and align algorithm, Dynalign, adjusts the original genome alignment to reflect an alignment based on RNA structure and therefore is capable of finding ncRNAs that have diverged farther in sequence than RNAz (Uzilov et al. 2006). The drawback is that it is slower. Foldalign, another fold and align algorithm, has also been used to find conserved, structured RNA in genomes (Torarinsson et al. 2006). It has been applied to compare genome sequences in regions that are not alignable based on sequence alone and it found numerous conserved putative ncRNA genes.

5 FUTURE DIRECTIONS

The progress in rapid determination of secondary structure lays the foundation for accelerating determination of three dimensional structure. There are NMR fingerprints for various loop motifs (Varani et al. 1996; Moore 2001) and there will likely also be chemical mapping fingerprints. Models of three-dimensional (3D) structures can be tested for consistency with chemical mapping and NMR data. Computers are constantly becoming more powerful so that it is possible to envision methods based on physics (e.g., molecular mechanics) or homology or a combination of the two for correctly predicting secondary and even 3D structure (Westhof et al. 2010) quickly on the basis of sequence alone. Physics based methods, however, will require a more fundamental understanding of molecular interactions determining thermodynamics and structure (Yildirim and Turner 2005). Understanding the physics of the molecular interactions should also lead to improved predictions of RNA dynamics, which are likely to be important for many functions.

A hindrance to homology modeling of RNA 3D structure is that, in comparison to protein structures, the collection of high resolution RNA 3D structures is relatively impoverished. The list of high resolution RNA structures has been growing steadily, however, and "information-based" approaches to 3D structure determination show promise. For example, the MC-Sym algorithm (Parisien and Major 2008) decomposes elements of RNA 3D structure into graphical representations, called "cyclic motifs," which can facilitate homology modeling. Resulting 3D models can be constrained by complementary data, such as NMR and chemical mapping to weed out poor models.

The *B. mori* R2 RNA 5′ region again provides an illustrative example of the process of moving from primary to secondary to 3D structure. As indicated in Figure 2, energy minimization guided by chemical mapping, oligonucleotide binding, NMR and comparative analysis was able to determine the base pairing of the R2 pseudoknot. Knowledge of the correct base pairing is a strong restraint on possible 3D folding, and this was used to constrain MC-Sym modeling of the R2 pseudoknot. The resulting 3D models were further screened by searching for helix–helix stacking that was consistent with NMR results: namely the NMR signature that connected the minor hairpin and the longer of the two pseudoknot helices (Hart et al. 2008). The final 3D model for this pseudoknot (Fig. 3) was selected based

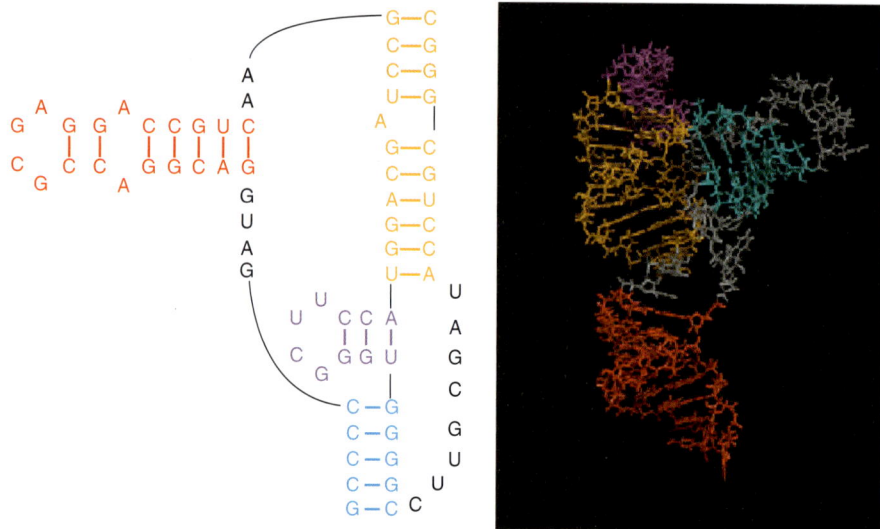

Figure 3. Experiment and Sequence Comparison are Used to Model 3D Structure. Homology with known structures was used to propose 3D folds for the *B. mori* R2 element pseudoknot from MC-Sym (Parisien and Major 2008) which were then screened with respect to experimental data (e.g. solvent accessibility to chemical reagents [Kierzek 2009] and helix stacking from NMR [Hart et al. 2008]). Helical motifs in the 3D model are color coded to match the secondary structural model.

on consistency with chemical mapping data (mainly used to rule out possible tertiary interactions) and similarity to models of the four other silk moth pseudoknots.

The ability to rapidly model RNA structure may facilitate discovery of therapeutics that target RNA. Structure in the target mRNA is an important consideration in designing siRNAs (Lu and Mathews 2007; Shao et al. 2007; Tafer et al. 2008) and determining microRNA targets (Rehmsmeier et al. 2004; Long et al. 2007). Additionally, the Disney group is using small-molecule microarray methods to deduce the basis for matching potential drugs with RNA motifs that bind them strongly (Childs-Disney et al. 2007; Disney and Childs-Disney 2007). Microarray methods based on short, chemically modified oligonucleotides (Kierzek et al. 2008; Kierzek et al. 2009) are also providing insight into the rules that govern oligonucleotide binding to structured RNAs, which should facilitate design of nucleic acid based therapeutics.

6 APPLICATION TO THE STUDY OF EVOLUTION

Structural comparisons may allow discovery of new fundamental principles of evolution and biology. For example, studies of systems biology are revealing intricate regulatory networks in cells and providing hypotheses of their evolution (Feschotte 2008). It is likely that RNA/RNA interactions are important in at least some networks. In turn, new biological principles discovered on the basis of RNA/RNA interactions can then be applied to accelerating discovery of new functional RNAs and of their structures.

The identification of microRNAs (miRNAs) and miRNA targets is an excellent example of how methods for folding and finding RNA have contributed to our knowledge of biology. MicroRNAs are one type of RNA important for regulatory networks. MicroRNAs are involved in a number of important cellular processes, such as development, apoptosis, and cell differentiation/proliferation (Bartel 2004). As well, miRNAs may have roles in the progression of at least 70 human diseases (Lu et al. 2008), including cancer and cardiovascular disease. Identification of putative miRNAs can be accomplished using algorithms based on RNA folding thermodynamics (Lim et al. 2003a; Lim et al. 2003b). Additionally, thermodynamics lay at the heart of many miRNA target prediction software (Kiriakidou et al. 2004; Rehmsmeier et al. 2004; Krek et al. 2005). Applications of such software have revealed complex miRNA networks where single miRNAs have multiple binding sites in target mRNAs.

The evolutionary origins of miRNAs are still being unraveled, but interesting results suggest at least some arise from transposable elements. Fifty five miRNAs, representing 12% of experimentally characterized human miRNAs, apparently originated from transposons (Piriyapongsa et al. 2007). Indeed, certain families of transposable elements appear to be natural fodder for the evolution of miRNAs: miniature inverted-repeat transposable elements (MITES) possess complementary palindromic termini separated by short linkers (Feschotte et al. 2002). When transcribed, these MITES have the ability to fold into hairpins that are structurally similar to precursor miRNA hairpins.

In addition to being able to generate new miRNAs, transposons may be responsible for evolving miRNA networks (Feschotte 2008). In their replication in host genomes, transposons may replicate miRNA genes, or insert new miRNA target sites into host genes. This process is evidenced by the finding that multiple genes may be regulated by the same miRNA. RNA structural constraints act on the evolution of miRNA biogenesis and targeting. RNA thermodynamics are crucial to many miRNA target site prediction programs (Berezikov et al. 2006; Kruger and Rehmsmeier 2006) and have played an important role in predicting and classifying families of miRNAs (Kaczkowski et al. 2009) and miRNA regulatory networks (Rehmsmeier et al. 2004).

Another key conserved regulatory pathway is RNA mediated transcriptional gene silencing. In one mode of action, small RNAs regulate DNA methylation, an important epigenetic mechanism of gene control (Hawkins and Morris 2008). This mode of action silences repetitive "junk" elements: a process vital for maintaining genome health. Again, tandem repeat sequences (associated with repetitive elements) appear to be important for the production of the small double-stranded RNAs needed to stimulate DNA methylation (Chan et al. 2006).

6.1 RNA Structure and Phylogenetic Reconstruction

Prediction and analyses of structured RNAs play fundamentally important roles in elucidating the evolutionary connections that link all organisms. It was the analysis of ribosomal RNA sequences that led Woese to propose the *Archaea* as a distinct major branch on the "Tree of Life" (Woese et al. 1990).

In addition to resolving these deep phylogenetic relationships, structured RNAs are commonly used markers for phylogenetic reconstruction at higher taxonomical levels. Internally transcribed spacer (ITS) regions of ribosomal RNAs are popular targets for phylogenetic analysis (Alvarez and Wendel 2003). These particular RNAs are not under the strict functional constraints of ribosomal RNA, and thus have enough variation to make them appropriate for higher level classifications. Though evolving under less stringent evolutionary constraints, the ITS2 RNA shows a conserved core secondary structure common throughout eukaryotes (Schultz et al. 2005). Presence of secondary structure in phylogenetic markers has important implications for phylogeny reconstruction. Sequence alignments, the basis for phylogenetic comparison, that do not account for structural homology may not reflect true evolutionary relationships. Compensatory mutations can confound phylogenetic analysis, as the nucleotides that constitute the alignments are not independently evolving characters, but are rather linked by higher order constraints, e.g., base pairing (Alvarez and Wendel 2003).

With the ability to generate good structural models for these phylogenetic markers RNA structure can facilitate, rather than confound, phylogeny reconstruction. Knowledge of RNA secondary structure can improve alignment quality (Goertzen et al. 2003). With accurate knowledge of paired sites, patterns of compensatory mutations between aligned species can be used to infer phylogeny (Wolf et al. 2005). Phylogenetic reconstruction methods that rely on models of sequence evolution, such as likelihood-based methods, also benefit from structural knowledge: paired and loop regions of structured RNAs are better accounted for using evolutionary models that account for different mutational rates for changes in paired or unpaired nucleotides (Telford et al. 2005). To facilitate such studies, an ITS2 secondary structure database ($>$100,000 entries) has been created using structural models based on free energy minimization and guided by comparative analysis (Selig et al. 2008).

Elements of RNA secondary structure themselves can be treated as evolving characters and phylogenetic connections may be traced by changes in structural character states (Knudsen and Caetano-Anolles 2008). Deconstructing RNA secondary structure into evolving characters and subjecting them to cladistic analysis allows for the study of the origin of particular substructures, e.g., hairpin loops. Such cladistic analysis of RNA secondary structure has led to insights into the molecular evolution of the tRNA cloverleaf structure (Sun and Caetano-Anolles 2008). This method of using structure as a character has also been applied to the classification of species using rRNA (Caetano-Anolles 2002), ITS RNA (Tippery and Les 2008) and tRNA (Sun and Caetano-Anolles 2008) and to investigate evolutionary trends in the structures of SINE elements (Sun et al. 2007).

6.2 Investigating Evolution: The RNA Model

The prediction of RNA structure is useful for understanding evolution from both in silico and in vitro studies. A number of fundamental evolutionary concepts can be explored using RNA. In RNA, genotype may be considered as the sequence of nucleotides, whereas the phenotype is the structure that may be formed by that sequence. Genetic variation may be simply modeled with mutations in the sequence (e.g., mutations introduced in silico). Selection may be introduced as a constraint on structure or thermodynamic stability in computer modeling. The concept of phenotypic plasticity applies to RNA: a single sequence may have multiple accessible secondary structures; as does the concept of neutrality: A single structure may be accessible to a number of sequences (Fontana 2002).

For computational modeling of evolutionary principles, RNA has a number of qualities that can be exploited to draw generalized conclusions. The thermodynamic model of RNA folding is physically grounded and results in a straight forward mapping of genotype (sequence) to phenotype (secondary structure). From a given sequence, it is possible to explore the entire phenotypic space (the set of accessible structures). Studies of genotype-phenotype mappings have revealed neutral networks connecting phenotypes (common structures) with sequence space (sequences that have the given phenotype). Such neutral networks explain the evolvability of nucleic acids by linking neutral drift and selection (Schuster and Stadler 2003). Neutral drift, the accumulation of structurally silent mutations, allows an evolving RNA to sample sequences with different plastic repertoires (available phenotypes); this is the basis of structural innovation. Such an evolutionary path was simulated for evolving a tRNA structure: long phases of phenotypic stasis were punctuated by structural (evolutionary) innovations along the path to the optimal tRNA structure (Schuster 2001). Neutral network theory found practical application in the discovery of an in vitro evolved RNA sequence at the intersection of two neutral networks that simultaneously performed catalytic activities of two different ribozymes (cleavage ribozyme and RNA ligase) (Schultes and Bartel 2000). Similarly, structure calculations were used to engineer a neutral path between two aptamer variants, whose intermediates were capable of binding to two substrates (FAD and GMP) (Held et al. 2003).

The availability of tools for folding and finding RNA made possible the studies discussed earlier, and many more. These tools will improve with advances in computer science, our understanding of the forces that govern RNA folding, and our understanding of fundamental biology. If even a tiny fraction of the noncoding portions of eukaryotic genomes represents functional RNA, then there is a colossal task to identify and understand these molecules. What other fascinating RNAs and complex RNA-based networks remain to be discovered?

With new RNAs come new opportunities to study evolution. Newly revealed RNAs may provide important markers for refining the phylogenetic relations that map the tree of life. We may also better understand the trajectories of and the evolutionary forces acting on structured RNAs: a critical component to understanding the RNA World.

ACKNOWLEDGMENTS

This work was supported by National Institutes of Health grants GM22939 (DHT) and GM076485 (DHM). We thank Francoise Major for demonstrating MC-Sym, and Ela Kierzek for sharing chemical mapping data for the R2 pseudoknot prior to publication.

REFERENCES

Alvarez I, Wendel JF. 2003. Ribosomal ITS sequences and plant phylogenetic inference. *Mol Phylogenet Evol* **29:** 417–434.

Andronescu M, Condon A, Hoos HH, Mathews DH, Murphy KP. 2007. Efficient parameter estimation for RNA secondary structure prediction. *Bioinformatics* **23:** i19–i28.

Bartel DP. 2004. MicroRNAs: genomics, biogenesis, mechanism, and function. *Cell* **116:** 281–297.

Bellamy-Royds AB, Turcotte M. 2007. Can Clustal-style progressive pairwise alignment of multiple sequences be used in RNA secondary structure prediction? *BMC Bioinformatics* **8:** 190.

Berezikov E, Cuppen E, Plasterk RH. 2006. Approaches to microRNA discovery. *Nat Genet* **38 Suppl:** S2–7.

Bernhart SH, Hofacker IL, Will S, Gruber AR, Stadler PF. 2008. RNAalifold: improved consensus structure prediction for RNA alignments. *BMC Bioinformatics* **9:** 474.

Birney E, Stamatoyannopoulos JA, Dutta A, Guigo R, Gingeras TR, Margulies EH, Weng Z, Snyder M, Dermitzakis ET, Thurman RE, et al. 2007. Identification and analysis of functional elements in 1% of the human genome by the ENCODE pilot project. *Nature* **447:** 799–816.

Caetano-Anolles G. 2002. Tracing the evolution of RNA structure in ribosomes. *Nucleic Acids Res* **30:** 2575–2587.

Chan SW, Zhang X, Bernatavichute YV, Jacobsen SE. 2006. Two-step recruitment of RNA-directed DNA methylation to tandem repeats. *PLoS Biol* **4:** e363.

Chen JH, Le SY, Shapiro B, Currey KM, Maizel JV. 1990. A computational procedure for assessing the significance of RNA secondary structure. *Comput Appl Biosci* **6:** 7–18.

Childs-Disney JL, Wu M, Pushechnikov A, Aminova O, Disney MD. 2007. A small molecule microarray platform to select RNA internal loop-ligand interactions. *ACS Chem Biol* **2:** 745–754.

Christensen SM, Ye J, Eickbush TH. 2006. RNA from the 5' end of the R2 retrotransposon controls R2 protein binding to and cleavage of its DNA target site. *Proc Natl Acad Sci* **103:** 17602–17607.

Clote P, Ferre F, Kranakis E, Krizanc D. 2005. Structural RNA has lower folding energy than random RNA of the same dinucleotide frequency. *RNA* **11:** 578–591.

Dawkins R. 1987. *The blind watchmaker*. Norton, New York.

Deigan KE, Li TW, Mathews DH, Weeks KM. 2009. Accurate SHAPE-directed RNA structure determination. *Proc Natl Acad Sci* **106:** 97–102.

Ding Y, Lawrence CE. 2003. A statistical sampling algorithm for RNA secondary structure prediction. *Nucleic Acids Res* **31:** 7280–7301.

Ding Y, Chan CY, Lawrence CE. 2005. RNA secondary structure prediction by centroids in a Boltzmann weighted ensemble. *RNA* **11:** 1157–1166.

Disney MD, Childs-Disney JL. 2007. Using selection to identify and chemical microarray to study the RNA internal loops recognized by 6'-N-acylated kanamycin A. *Chembiochem* **8:** 649–656.

Do CB, Foo CS, Batzoglou S. 2008. A max-margin model for efficient simultaneous alignment and folding of RNA sequences. *Bioinformatics* **24:** 68–76.

Do CB, Woods DA, Batzoglou S. 2006. CONTRAfold: RNA secondary structure prediction without physics-based models. *Bioinformatics* **22:** e90–98.

Dowell RD, Eddy SR. 2004. Evaluation of several lightweight stochastic context-free grammars for RNA secondary structure prediction. *BMC Bioinformatics* **5:** 71.

Dowell RD, Eddy SR. 2006. Efficient pairwise RNA structure prediction and alignment using sequence alignment constraints. *BMC Bioinformatics* **7:** 400.

Eddy SR, Durbin R. 1994. RNA sequence analysis using covariance models. *Nucleic Acids Res* **22:** 2079–2088.

Ehresmann C, Baudin F, Mougel M, Romby P, Ebel J, Ehresmann B. 1987. Probing the structure of RNAs in solution. *Nucleic Acids Res* **15:** 9109–9128.

Feschotte C. 2008. Transposable elements and the evolution of regulating networks. *Nat Rev Gen* **9:** 397–405.

Feschotte C, Jiang N, Wessler SR. 2002. Plant transposable elements: Where genetics meets genomics. *Nat Rev Genet* **3:** 329–341.

Fichant GA, Burks C. 1991. Identifying potential tRNA genes in genomic DNA sequences. *J Mol Biol* **220:** 659–671.

Fontana W. 2002. Modelling 'evo-devo' with RNA. *Bioessays* **24:** 1164–1177.

Giegerich R, Voss B, Rehmsmeier M. 2004. Abstract shapes of RNA. *Nucleic Acids Res* **32:** 4843–4851.

Goertzen LR, Cannone JJ, Gutell RR, Jansen RK. 2003. ITS secondary structure derived from comparative analysis: Implications for sequence alignment and phylogeny of the Asteraceae. *Mol Phylogenet Evol* **29:** 216–234.

Gutell RR, Lee JC, Cannone JJ. 2002. The accuracy of ribosomal RNA comparative structure models. *Curr Opin Struct Biol* **12:** 301–310.

Hamada M, Kiryu H, Sato K, Mituyama T, Asai K. 2009. Prediction of RNA secondary structure using generalized centroid estimators. *Bioinformatics* **25:** 465–473.

Harmanci AO, Sharma G, Mathews DH. 2007. Efficient pairwise RNA structure prediction using probabilistic alignment constraints in Dynalign. *BMC Bioinformatics* **8:** 130.

Harmanci AO, Sharma G, Mathews DH. 2008. PARTS: Probabilistic alignment for RNA joinT secondary structure prediction. *Nucleic Acids Res* **36:** 2406–2417.

Harmanci AO, Sharma G, Mathews DH. 2009. Stochastic sampling of the RNA structural alignment space. *Nucleic Acids Res* **37:** 4063–4075.

Harris KA Jr, Crothers DM, Ullu E. 1995. In vivo structural analysis of spliced leader RNAs in *Trypanosoma brucei* and *Leptomonas collosoma*: A flexible structure that is independent of cap4 methylations. *RNA* **1:** 351–362.

Hart JM, Kennedy SD, Mathews DH, Turner DH. 2008. NMR-assisted prediction of RNA secondary structure: Identification of a probable pseudoknot in the coding region of an R2 retrotransposon. *J Am Chem Soc* **130:** 10233–10239.

Havgaard JH, Lyngso RB, Stormo GD, Gorodkin J. 2005. Pairwise local structural alignment of RNA sequences with sequence similarity less than 40%. *Bioinformatics* **21:** 1815–1824.

Hawkins PG, Morris KV. 2008. RNA and transcriptional modulation of gene expression. *Cell Cycle* **7:** 602–607.

Held DM, Greathouse ST, Agrawal A, Burke DH. 2003. Evolutionary landscapes for the acquisition of new ligand recognition by RNA aptamers. *J Mol Evol* **57:** 299–308.

Höchsmann M, Voss B, Giegerich R. 2004. Pure multiple RNA secondary structure alignments: A progressive profile approach. *IEEE Transactions on Computational Biology and Bioinformatics* **1:** 1–10.

Hofacker IL, Stadler PF. 2004. The partition function variant of Sankoff's algorithm. in *Computational Science—ICCS 2004, volume 3039 of Lecture Notes in Computer Science* (ed. Marian Bubak G.D.v.A., Sloot Peter M. A., Dongarra Jack J.), p. 728–735, Kraków.

Hofacker IL, Bernhart SH, Stadler PF. 2004. Alignment of RNA base pairing probability matrices. *Bioinformatics* **20:** 2222–2227.

Hofacker IL, Fekete M, Stadler PF. 2002. Secondary structure prediction for aligned RNA sequences. *J Mol Biol* **319:** 1059–1066.

Holmes I. 2005. Accelerated probabilistic inference of RNA structure evolution. *BMC Bioinformatics* **6:** 73.

Inoue T, Cech TR. 1985. Secondary structure of the circular form of the *Tetrahymena* rRNA intervening sequence: A technique for RNA structure analysis using chemical probes and reverse transcriptase. *Proc Natl Acad Sci* **82:** 648–652.

Jaeger JA, Turner DH, Zuker M. 1989. Improved predictions of secondary structures for RNA. *Proc Natl Acad Sci* **86:** 7706–7710.

Kaczkowski B, Torarinsson E, Reiche K, Havgaard JH, Stadler PF, Gorodkin J. 2009. Structural profiles of human miRNA families from pairwise clustering. *Bioinformatics* **25:** 291–294.

Kierzek K. 2009. Binding of Short Oligonucleotides to RNA: Studies of the binding of common RNA structural motifs to isoenergetic microarrays. *Biochemistry* **48:** 11344–11356.

Kierzek E, Christensen SM, Eickbush TH, Kierzek R, Turner DH, Moss WN. 2009. Secondary structures for 5' regions of R2 retrotransposon RNAs reveal a novel conserved pseudoknot and regions that evolve under different constraints. *J Mol Biol* **390:** 428–442.

Kierzek E, Kierzek R, Moss WN, Christensen SM, Eickbush TH, Turner DH. 2008. Isoenergetic penta- and hexanucleotide microarray probing and chemical mapping provide a secondary structure model for an RNA element orchestrating R2 retrotransposon protein function. *Nucleic Acids Res* **36:** 1770–1782.

Kiriakidou M, Nelson PT, Kouranov A, Fitziev P, Bouyioukos C, Mourelatos Z, Hatzigeorgiou A. 2004. A combined computational-experimental approach predicts human microRNA targets. *Genes Dev* **18:** 1165–1178.

Kiryu H, Kin T, Asai K. 2007. Robust prediction of consensus secondary structures using averaged base pairing probability matrices. *Bioinformatics* **23:** 434–441.

Knudsen V, Caetano-Anolles G. 2008. NOBAI: A web server for character coding of geometrical and statistical features in RNA structure. *Nucleic Acids Res* **36:** W85–90.

Krek A, Grun D, Poy MN, Wolf R, Rosenberg L, Epstein EJ, MacMenamin P, da Piedade I, Gunsalus KC, Stoffel M, et al. 2005. Combinatorial microRNA target predictions. *Nat Genet* **37:** 495–500.

Kruger J, Rehmsmeier M. 2006. RNAhybrid: microRNA target prediction easy, fast and flexible. *Nucleic Acids Res* **34:** W451–454.

Le SV, Chen JH, Currey KM, Maizel JV Jr. 1988. A program for predicting significant RNA secondary structures. *Comput Appl Biosci* **4:** 153–159.

Li Y, Breaker RR. 1999. Kinetics of RNA degradation by specific base catalysis of transesterification involving the 2'-hydroxyl group. *J Am Chem Soc* **121:** 5364–5372.

Lim LP, Glasner ME, Yekta S, Burge CB, Bartel DP. 2003a. Vertebrate microRNA genes. *Science* **299:** 1540.

Lim LP, Lau NC, Weinstein EG, Abdelhakim A, Yekta S, Rhoades MW, Burge CB, Bartel DP. 2003b. The microRNAs of *Caenorhabditis elegans*. *Genes Dev* **17:** 991–1008.

Lindell M, Romby P, Wagner EG. 2002. Lead(II) as a probe for investigating RNA structure in vivo. *RNA* **8:** 534–541.

Long D, Lee R, Williams P, Chan CY, Ambros V, Ding Y. 2007. Potent effect of target structure on microRNA function. *Nat Struct Mol Biol* **14:** 287–294.

Lowe TM, Eddy SR. 1997. tRNAscan-SE: A Program for improved detection of transfer RNA genes in genomic sequence. *Nucleic Acids Res* **25:** 955–964.

Lowe TM, Eddy SR. 1999. A computational screen for methylation guide snoRNAs in yeast. *Science* **283:** 1168–1171.

Lu ZJ, Mathews DH. 2007. Efficient siRNA selection using hybridization thermodynamics. *Nucleic Acids Res* **36:** 640–647.

Lu ZJ, Gloor JW, Mathews DH. 2009. Improved RNA secondary structure prediction by maximizing expected pair accuracy. *RNA* **15:** 1805–1813.

Lu M, Zhang Q, Deng M, Miao J, Guo Y, Gao W, Cui Q. 2008. An analysis of human microRNA and disease associations. *PLoS One* **3:** e3420.

Lück R, Gräf S, Steger G. 1999. ConStruct: A tool for thermodynamic controlled prediction of conserved secondary structure. *Nucleic Acids Res* **27:** 4208–4217.

Lück R, Steger G, Riesner D. 1996. Thermodynamic prediction of conserved secondary structure: Application to the RRE element of HIV, the tRNA-like element of CMV and the mRNA of prion protein. *J Mol Biol* **258**: 813–826.

Macke T, Ecker D, Gutell R, Gautheret D, Case DA, Sampath R. 2001. RNAMotif: A new RNA secondary structure definition and discovery algorithm. *Nucl Acids Res* **29**: 4724–4735.

Mathews DH. 2004. Using an RNA secondary structure partition function to determine confidence in base pairs predicted by free energy minimization. *RNA* **10**: 1178–1190.

Mathews DH, Disney MD, Childs JL, Schroeder SJ, Zuker M, Turner DH. 2004. Incorporating chemical modification constraints into a dynamic programming algorithm for prediction of RNA secondary structure. *Proc Natl Acad Sci* **101**: 7287–7292.

Mathews DH, Sabina J, Zuker M, Turner DH. 1999. Expanded sequence dependence of thermodynamic parameters provides improved prediction of RNA secondary structure. *J Mol Biol* **288**: 911–940.

Mathews DH, Schroeder SJ, Turner DH, Zuker M. 2005. Predicting RNA secondary structure. In *The RNA world, third edition* (ed. Gesteland R.F., Cech T.R., Atkins J.F.), p. 631–657. Cold Spring Harbor Laboratory Press, Cold Spring Harbor.

Mathews DH, Turner DH. 2002. Dynalign: An algorithm for finding the secondary structure common to two RNA sequences. *J Mol Biol* **317**: 191–203.

Mattick JS. 2004. RNA regulation: A new genetics? *Nat Rev Genet* **5**: 316–323.

Merino EJ, Wilkinson KA, Coughlan JL, Weeks KM. 2005. RNA structure analysis at single nucleotide resolution by selective 2'-hydroxyl acylation and primer extension (SHAPE). *J Am Chem Soc* **127**: 4223–4231.

Moazed D, Stern S, Noller HF. 1986. Rapid chemical probing of conformation in 16S ribosomal RNA and 30S ribosomal subunits using primer extension. *J Mol Biol* **187**: 399–416.

Moore PB. 2001. A spectroscopist's view of RNA conformation: RNA structural motifs. In *RNA* (ed. Soll D., Nishimura S., Moore P.B.), p. 1–19. Elsevier, Oxford.

Moore PB, Steitz TA. 2010. The roles of RNA in the synthesis of protein. *Cold Spring Harb Perspect Biol* **2**: a003780.

Mortimer SA, Weeks KM. 2007. A fast-acting reagent for accurate analysis of RNA secondary and tertiary structure by SHAPE chemistry. *J Am Chem Soc* **129**: 4144–4145.

Ninio J. 1979. Prediction of pairing schemes in RNA molecules—loop contributions and energy of wobble and non-wobble pairs. *Biochimie* **61**: 1133–1150.

Papanicolaou C, Gouy M, Ninio J. 1984. An energy model that predicts the correct folding of both the tRNA and the 5S RNA molecules. *Nucleic Acids Res* **12**: 31–44.

Parisien M, Major F. 2008. The MC-Fold and MC-Sym pipeline infers RNA structure from sequence data. *Nature* **452**: 51–55.

Piriyapongsa J, Marino-Ramirez L, Jordan IK. 2007. Origin and evolution of human microRNAs from transposable elements. *Genetics* **176**: 1323–1337.

Reeder J, Giegerich R. 2005. Consensus shapes: An alternative to the Sankoff algorithm for RNA consensus structure prediction. *Bioinformatics* **21**: 3516–3523.

Rehmsmeier M, Steffen P, Hochsmann M, Giegerich R. 2004. Fast and effective prediction of microRNA/target duplexes. *RNA* **10**: 1507–1517.

Rivas E, Eddy SR. 2000. Secondary structure alone is not statistically significant for the detection of noncoding RNAs. *Bioinformatics* **16**: 583–605.

Sankoff D. 1985. Simultaneous solution of the RNA folding, alignment and protosequence problems. *Siam J Appl Math* **45**: 810–825.

Schultes EA, Bartel DP. 2000. One sequence, two ribozymes: implications for the emergence of new ribozyme folds. *Science* **289**: 448–452.

Schultz J, Maisel S, Gerlach D, Muller T, Wolf M. 2005. A common core of secondary structure of the internal transcribed spacer 2 (ITS2) throughout the Eukaryota. *RNA* **11**: 361–364.

Schuster P. 2001. Evolution in silico and in vitro: the RNA model. *Biol Chem* **382**: 1301–1314.

Schuster P, Stadler PF. 2003. Networks in molecular evolution. *Complexity* **8**: 34–42.

Seffens W, Digby D. 1999. mRNAs have greater negative folding free energies than shuffled or codon choice randomized sequences. *Nucleic Acids Res* **27**: 1578–1584.

Selig C, Wolf M, Muller T, Dandekar T, Schultz J. 2008. The ITS2 Database II: homology modelling RNA structure for molecular systematics. *Nucleic Acids Res* **36**: D377–380.

Shao Y, Chan CY, Maliyekkel A, Lawrence CE, Roninson IB, Ding Y. 2007. Effect of target secondary structure on RNAi efficiency. *RNA* **13**: 1631–1640.

Soukup GA, Breaker RR. 1999. Relationship between internucleotide linkage geometry and the stability of RNA. *RNA* **5**: 1308–1325.

Sun FJ, Caetano-Anolles G. 2008. The origin and evolution of tRNA inferred from phylogenetic analysis of structure. *J Mol Evol* **66**: 21–35.

Sun FJ, Fleurdepine S, Bousquet-Antonelli C, Caetano-Anolles G, Deragon JM. 2007. Common evolutionary trends for SINE RNA structures. *Trends Genet* **23**: 26–33.

Tafer H, Ameres SL, Obernosterer G, Gebeshuber CA, Schroeder R, Martinez J, Hofacker IL. 2008. The impact of target site accessibility on the design of effective siRNAs. *Nat Biotechnol* **26**: 578–583.

Taft RJ, Mattick JS. 2003. Increasing biological complexity is positively correlated with the relative genome-wide expansion of non-protein-coding DNA sequences. *Genome Res* **5**: P1.

Telford MJ, Wise MJ, Gowri-Shankar V. 2005. Consideration of RNA secondary structure significantly improves likelihood-based estimates of phylogeny: Examples from the bilateria. *Mol Biol Evol* **22**: 1129–1136.

Tippery NP, Les DH. 2008. Phylogenetic analysis of the internal transcribed spacer (ITS) region in Menyanthaceae using predicted secondary structure. *Mol Phylogenet Evol* **49**: 526–537.

Torarinsson E, Havgaard JH, Gorodkin J. 2007. Multiple structural alignment and clustering of RNA sequences. *Bioinformatics* **23**: 926–932.

Torarinsson E, Sawera M, Havgaard JH, Fredholm M, Gorodkin J. 2006. Thousands of corresponding human and mouse genomic regions unalignable in primary sequence contain common RNA structure. *Genome Res* **16**: 885–889.

Tsui V, Macke T, Case DA. 2003. A novel method for finding tRNA genes. *RNA* **9**: 507–517.

Turner DH. 2000. Conformational changes. In *Nucleic acids* (ed. Bloomfield V., Crothers D., Tinoco I. Jr), pp. 259–334. University Science Books, Sausalito, CA.

Uzilov AV, Keegan JM, Mathews DH. 2006. Detection of non-coding RNAs on the basis of predicted secondary structure formation free energy change. *BMC Bioinformatics* **7**: 173.

Varani G, Aboul-ela F, Allain F. 1996. NMR investigation of RNA structure. *Prog Nucl Magn Reson Spectrosc* **29**: 51–127.

Voss B, Giegerich R, Rehmsmeier M. 2006. Complete probabilistic analysis of RNA shapes. *BMC Biol* **4**: 5.

Washietl S, Hofacker IL, Stadler PF. 2005. Fast and reliable prediction of noncoding RNAs. *Proc Natl Acad Sci* **102**: 2454–2459.

Westhof E, Masquida B, Jossinet F. 2010. Predicting and modeling RNA architecture. *Cold Spring Harb Perspect Biol* **2**: a003632.

Wilkinson KA, Vasa SM, Deigan KE, Mortimer SA, Giddings MC, Weeks KM. 2009. Influence of nucleotide identity on ribose 2'-hydroxyl reactivity in RNA. *RNA* **15**: 1314–1321.

Will S, Reiche K, Hofacker IL, Stadler PF, Backofen R. 2007. Inferring noncoding RNA families and classes by means of genome-scale structure-based clustering. *PLoS Comput Biol* **3**: e65.

Woese CR, Pace NR. 1993. Probing RNA structure, function, and history by comparative analysis. In *The RNA world* (ed. Gesteland R.F., Atkins J.F.), p. 91–117. Cold Spring Harbor Laboratory Press, Cold Spring Harbor.

Woese CR, Kandler O, Wheelis ML. 1990. Towards a natural system of organisms: proposal for the domains Archaea, Bacteria, and Eucarya. *Proc Natl Acad Sci* **87**: 4576–4579.

Wolf M, Friedrich J, Dandekar T, Muller T. 2005. CBCAnalyzer: Inferring phylogenies based on compensatory base changes in RNA secondary structures. *In Silico Biol* **5**: 291–294.

Workman C, Krogh A. 1999. No evidence that mRNAs have lower folding free energies than random sequences with the same dinucleotide distribution. *Nucleic Acids Res* **27**: 4816–4822.

Wuchty S, Fontana W, Hofacker IL, Schuster P. 1999. Complete suboptimal folding of RNA and the stability of secondary structures. *Biopolymers* **49**: 145–165.

Xia T, SantaLucia J Jr, Burkard ME, Kierzek R, Schroeder SJ, Jiao X, Cox C, Turner DH. 1998. Thermodynamic parameters for an expanded nearest-neighbor model for formation of RNA duplexes with Watson-Crick pairs. *Biochemistry* **37**: 14719–14735.

Yildirim I, Turner DH. 2005. RNA challenges for computational chemists. *Biochemistry* **44**: 13225–13234.

Zaug AJ, Cech TR. 1995. Analysis of the structure of *Tetrahymena* nuclear RNAs in vivo: Telomerase RNA, the self-splicing rRNA Intron, and U2 snRNA. *RNA* **1**: 363–374.

Zuker M, Sankoff D. 1984. RNA secondary structures and their prediction. *Bull Math Biol* **46**: 591–621.

Zwanzig R, Szabo A, Bagchi B. 1992. Levinthal's paradox. *Proc Natl Acad Sci* **89**: 20–22.

Predicting and Modeling RNA Architecture

Eric Westhof, Benoît Masquida, and Fabrice Jossinet

Architecture et réactivité de l'ARN, Université de Strasbourg, Institut de biologie moléculaire et cellulaire du CNRS, 15 rue René Descartes, 67084 Strasbourg, France

Correspondence: e.westhof@ibmc-cnrs.unistra.fr

SUMMARY

A general approach for modeling the architecture of large and structured RNA molecules is described. The method exploits the modularity and the hierarchical folding of RNA architecture that is viewed as the assembly of preformed double-stranded helices defined by Watson-Crick base pairs and RNA modules maintained by non-Watson-Crick base pairs. Despite the extensive molecular neutrality observed in RNA structures, specificity in RNA folding is achieved through global constraints like lengths of helices, coaxiality of helical stacks, and structures adopted at the junctions of helices. The Assemble integrated suite of computer tools allows for sequence and structure analysis as well as interactive modeling by homology or *ab initio* assembly with possibilities for fitting within electronic density maps. The local key role of non-Watson-Crick pairs guides RNA architecture formation and offers metrics for assessing the accuracy of three-dimensional models in a more useful way than usual root mean square deviation (RMSD) values.

Outline

1. Introduction
2. The analysis of the secondary structure and the RNA modules
3. Comparative sequence analysis
4. The extraction of tertiary structure constraints
5. How to search for RNA modules
6. Modeling large RNA assemblies
7. Modeling RNA–protein complexes
8. Modeling and fitting into medium to low resolution electron density maps
9. Interactive molecular modeling with *Assemble*
10. Comparisons between RNA models and crystal structures
11. Conclusions

References

1 INTRODUCTION

In this article, a general approach for predicting three-dimensional (3D) contacts and modeling the architecture of large and structured RNA molecules on the basis of sequence analysis and sequence alignments is described. The method assumes that the folding is sequential, with modular units being incorporated hierarchically in the final architecture. The aim is the architecture of the assembly and not all the fine atomic details, although the large RNA molecules are assembled using all-atom components. Recently, following the increase in the number of RNA crystal structures, new approaches have been proposed, most of which can be coupled or used in parallel.

The main driving force for RNA architecture is the packing of RNA helices and modules through stacking between terminal base pairs and specific molecular recognition contacts between RNA segments. Structured RNA molecules are able to self-assemble into complex architectural folds because they contain, beyond the Watson-Crick base pairs that maintain the secondary structure, additional tertiary base pairs, often non-Watson-Crick, as well as various types of contacts between segments of the polynucleotide chain. This hierarchical assembly of 3D RNA structures is coupled with the binding of cations (Misra and Draper 2002; Rangan et al. 2003). A reduction in the net charge of the molecule with the initial association of cations induces a collapse of the RNA chain into compact structures (Fang et al. 1999; Heilman-Miller et al. 2001; Sosnick and Pan 2003) that then favor and promote the formation of tertiary interactions. Experimentally, the initial processes that lead to compaction of the RNA are clearly distinguishable from those leading to the formation of the native tertiary structure (Pan and Woodson 1999; Thirumalai et al. 2001). In the following, we consider only the final assembled structure. However, in the modeling approach, it is worthwhile to take into account the possible folding pathways leading to the modeled architecture.

RNA modeling started with attempts at transfer RNAs, first the anticodon loop (Fuller and Hodgson 1967) and later the full tRNA (Levitt 1969; Ninio et al. 1969). Crystallographic data, although reduced to regular RNA helices and tRNA structures, led to a revival in RNA modeling some 20 years after those early attempts (Dock-Bregeon et al. 1989; Krol et al. 1990; Romby et al. 1988; Westhof et al. 1989). Since then, striking progress has occurred in RNA crystallography and the large number of available RNA crystal structures has considerably strengthened our knowledge of RNA structure and folding. It is now apparent that large RNA structures can be parsed into various structural elements: regular RNA helices, junctions between helices, hairpin loops, and RNA modules (Westhof et al. 1996). Several of those structural elements are recurrent and occur in structured RNAs of very diverse origins or functions (Costa and Michel 1995; Leontis and Westhof 2003; Moore 1999). The set of experimental structures forms the basis of all knowledge-based modeling approaches.

2 THE ANALYSIS OF THE SECONDARY STRUCTURE AND THE RNA MODULES

Almost all of the secondary structure information and some of the tertiary structure information can be deduced from comparative sequence analysis (Michel et al. 2000; Pace et al. 1986; Pace et al. 1989). RNA complexity here is linked with topology, which means the content of the RNA sequence that forms branched junctions, terminal and internal loops, pseudoknots, and non-Watson-Crick tertiary contacts. This topological complexity can be translated into internal constraints that can significantly help find the relative spatial positions of the secondary structure elements.

A second critical point in RNA modeling is that, despite six torsion angles along the main chain, only two contribute strongly to the overall folding pathway of the polynucleotide backbone: the two contiguous torsion angles around the phosphate group. They mainly populate three domains (*gauche -*, *gauche +*, and *trans*). The other four torsion angles oscillate around an invariant value most of the time except in particular cases. In addition, the torsion angle between the base and the sugar adopts two conformational domains, *syn* and *anti*, with the *anti* conformation overwhelmingly more frequent. These general rules are still apparent in recent crystal structures (Richardson et al. 2008; Westhof and Fritsch 2000). This ensemble of observations contributed to the prediction power of the RNA modeling method.

This article is intended to describe an overall process of RNA modeling, mainly based on phylogenetic analysis and illustrated with various examples from small to large size assemblies compared, whenever possible, to the available crystal structures. Other approaches relying on automatic folding predictions and at different levels of granulometry have been developed but will not be extensively addressed here (Shapiro et al. 2007).

Energetically, the secondary structure is the main component of an RNA architecture, whereas tertiary structure contributes only minimally to the stability of the native state in terms of Gibbs free energy (Brion and Westhof 1997; Tinoco and Bustamante 1999). Quite naturally, the construction of the tertiary structure of an RNA molecule always assumes and starts from a given secondary structure. Therefore, the determination of the secondary structure is an essential step in the study of the structure–function relationships of an RNA molecule. Among the recent modeling tools based on conformational space searching, three of them (Das and Baker 2007; Ding et al. 2008; Parisien and Major 2008) attempt to predict simultaneously the secondary and tertiary structure, whereas a fourth coarse-grained approach requires the knowledge of the secondary structure (Jonikas et al. 2009). Two main situations need to be considered. In the first case, several sequences of homologous RNAs are known and sequence comparisons can be applied. In the second case, only one sequence is known. Additional information, usually based on chemical and enzymatic probing, is then needed. In this latter case, in which no or few homologs have been reported, the secondary structure can be explored using folding algorithms (Zuker 1989, 2003).

The term "secondary structure" may carry some ambiguity because it includes not only all segments that can build helices formed by any combination of the isosteric Watson-Crick pairings but also, in variable proportions, Watson-Crick as well as non-Watson-Crick pairs involved in tertiary structure (Westhof and Michel 1994). A secondary structure can be broken down into recognizable elementary modules such as the helical regions (stems and pseudoknots) and nonhelical linking elements (hairpin and internal loops, bulges, and multiple junctions). In the secondary structure, a pseudoknot is a specific RNA module that results from standard Watson-Crick pairs involving a single-stranded stretch, located between paired strands, and a distal single-stranded region (Dam et al. 1992; Westhof and Jaeger 1992). The single-stranded regions may belong to a hairpin loop, an internal loop, or a 3′ (or 5′) dangling end; but at least one of them must occur between base-paired helical strands. When both single-stranded regions are hairpin loops in a single RNA molecule, they are said to form a loop–loop module (Brunel et al. 2002; Lehnert et al. 1996), which is formally equivalent to a pseudoknot. Intermolecular loop–loop interactions between two RNA molecules occur in dimer formation (Bourassa and Major 2002; Ferrandon et al. 1997; Wagner et al. 2004). It is worth noticing that in intramolecular or intermolecular loop–loop motifs, the interactions are not always of the standard Watson-Crick pairing types (Khvorova et al. 2003; Oubridge et al. 2002; Weichenrieder et al. 2000). Formally, the two-dimensional (2D) structure reduces the secondary structure to the set of Watson-Crick base pairs that form a planar graph (i.e., without crossing edges) when the sequence of bases is arranged along a circle and the base pairs are connected by edges. Thus, pseudoknots or loop–loop motifs, which occur in the folding process once one hairpin at least has been formed, belong to the 3D structure and not to the 2D structure.

Although the 2D structure is dominated by Watson-Crick pairs (which represent 60%–70% in a structured RNA), non-Watson-Crick pairs underlie most of the 3D structure. This point was clear already from the structure of tRNAs, where among the tertiary pairs (at least seven), only one is of the standard Watson-Crick type. The x-ray structures that appeared since then have strengthened this conclusion. At the level of the active tertiary structural organization, an architectural module is an arrangement containing a few secondary structure elements associated with a specific geometry and topology. The combination of such substructures leads to compact domains (Batey et al. 1999), which often fold autonomously and independently of the rest of the RNA architecture. An ensemble of observations bears out a view of RNA folding whereby the architecture results from the cooperative compaction of separate and stable substructures, which might undergo only minor and local rearrangement during the process. The introduction of modular units, hierarchically organized and folded, circumvents most of the numerical nightmares inherent to the Levinthal's paradox of a purely mathematically based prediction of RNA structure at the atomic level even with coarse-grained approximations or *ad hoc* potentials.

3 COMPARATIVE SEQUENCE ANALYSIS

The comparative approach is based on the assumption that the function has been conserved by the folding architecture during evolution and, consequently, that a consensus secondary structure should be derivable by an alignment of RNA sequences based on a maximization of Watson-Crick covariations (Michel et al. 1982; Pace et al. 1986). An alignment of RNA sequences is, thus, formally equivalent to a secondary structure common to the set of sequences. This approach is the method of choice when a set of RNA sequences with identical biological function is available. The sequences should be arranged in groups and subgroups (ideally of similar size), either according to the phylogenetic classification (Pace et al. 1989) or following a phenotypic parsing (Michel and Westhof 1990). The first step in an alignment consists in the establishment of the paired regions along each sequence and those should be arranged horizontally so that the lengths of the Watson-Crick paired regions juxtapose vertically. In a second step, conservation or semiconservation of bases can be highlighted by a vertical alignment with inclusion of blanks or gaps in a fashion similar to the alignments of protein sequences. Alignments of residues do not necessarily imply structural superimposition but instead a structural correspondence (Brown et al. 2009).

The overall robustness of the approach increases with the diversity of the sequences and the evolutionary distances between them, whereas the accuracy of each prediction depends on the number of covariation events in each group and subgroup (Michel et al. 2000). Within such a scenario, conserved residues, potentially forming Watson-Crick base-pairs, do not display any covariation and, thus, should be regarded as nonproven or with extreme caution. Indeed, conservation of Watson-Crick pairs can reveal either tertiary contacts or an alternative pairing geometry. This can be the case for A-U pairs when they form Hoogsteen/Watson-Crick interactions; for example, the U80A14 pair in tRNAs or the pair U135-A187 in the crystal structure of the P4-P6 domain in the Tetrahymena group I intron (Cate et al. 1996a). One advantage of comparative analysis is that near-Watson-Crick pairs, like the wobble GoU pair or GoA pairs of the Watson-Crick type at the ends of helices, can be noticed and properly assessed. Critically, sequence comparisons allow us to also delimit rather precisely the helical regions and do not intrude into those segments that should form non-Watson-Crick pairs. Another important and nonnegligeable advantage of comparative analysis is that pseudoknotted regions are easily recognized by visual inspection.

4 THE EXTRACTION OF TERTIARY STRUCTURE CONSTRAINTS

The efficiency of the comparative approach stems from the fact that molecular 3D architectures evolve much more slowly than sequences which sample sequence space on a given 3D fold. Global architecture changes extremely slowly as it relies on conserved long-range tertiary interactions. However, phylogenetic methods are fraught with problems related to statistical relevance. With only four bases to choose among, purely coincidental compensatory base changes (or covariations between positions) are bound to occur, leading to ambiguities. Phylogenetically, the level of ambiguity is reduced by new sequences presenting additional covariations. Thus, if the function is identical and the sequences are sufficiently diverse, the noise level (or covariations resulting from historical contingencies) will be decreased by comparisons. The more compensatory base change events there are in the sequences, the more firmly the secondary structure will be established. But the extraction of 3D content from sequences is difficult (Gautheret et al. 1995; Gautheret and Gutell 1997; Michel and Westhof 1990) first because the rules of tertiary interactions are not as well-defined as those contained in the complementarity of the Watson-Crick pairs and, second, because several contacts involve sequence-independent properties like phosphate or ribose hydroxyl groups hydrogen bonding to other chemical groups (Cate et al. 1996a; Zirbel et al. 2009).

5 HOW TO SEARCH FOR RNA MODULES

Here, search is not meant through genomic sequences (Lambert et al. 2004; Macke et al. 2001; Nawrocki et al. 2009) but instead through a common secondary structure deduced from a set of aligned sequences. After alignment and the derivation of a common secondary structure, the RNA parts that do not display Watson-Crick covariations can be suspected to form specific tertiary modules that should be scrutinized for sequence similarities and characteristics to known modules seen in x-ray structures. The compatibility between a new sequence and a given interaction scheme, as provided by a crystal structure, can be checked using isostericity matrices or known variations observed in aligned sequences. The isostericity matrices have been verified for several RNA motifs using structural alignments anchored by crystallographic structures (Leontis et al. 2002b; Lescoute et al. 2005). Still, it is worth noting that sequence analysis of non-Watson-Crick interactions is difficult to perform without a crystallographic structure because many base–base geometries can be accommodated by any base combination (Stombaugh et al. 2009).

RNA–RNA or RNA–protein interactions are mediated by RNA modules, defined recurrent ensembles of ordered non-Watson-Crick base pairs (Leontis et al. 2002a; Leontis and Westhof 2003). A single RNA module comprises a family of sequences all of which can fold into the same 3D structure and can mediate the same types of interaction(s) (Fig. 1). The chemistry and geometry of base pairing constrain the evolution of modules in such a way that random mutations that occur within them are accepted or rejected insofar as they can mediate a similar ordered array of interactions.

RNA modules (Fig. 1) can be either architectural, forming a bend, e.g., the kink-turn (Klein et al. 2001), or a reorientation within a helix or between helices, or anchors for association as in the tetraloop–tetraloop receptor interaction (Cate et al. 1996b; Costa and Michel 1995, 1997). Up to now, it appears that modules are limited in number and are recurrent, as well as some of the rules of association between them (Leontis and Westhof 2003). Thus, the modules can show sequence variability without impairing their ability to adopt a structure close to the archetype. Surprisingly, the most common RNA–RNA interaction motif, the A-minor motif (Nissen et al. 2001), is also the least specific in its local requirements (Fig. 2). A-minor motifs are mediated by adenines binding into the

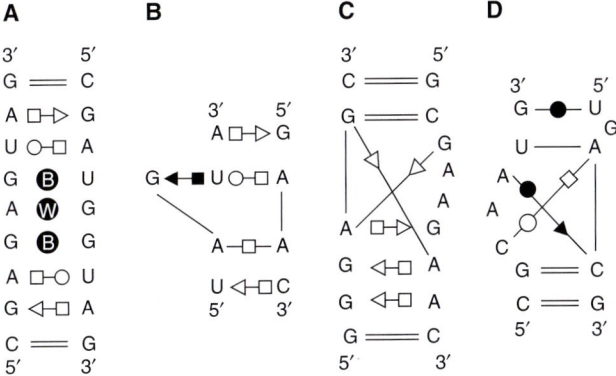

Figure 1. Some examples of annotated common RNA modules are represented. The annotations use the geometric nomenclature where a circle indicates the Watson-Crick sites, the square the Hoogsteen sites, and the triangle the sugar edge sites (filled black, the bases interact in *cis*; empty symbols, they interact in *trans*) (Leontis and Westhof 2001). Using this nomenclature, all base–base pairwise interactions present in nucleic acids have been classified in 12 families in which each family is a 4 × 4 matrix of the bases A, G, C, and U. This classification allows to deduce the isotericity matrices that yield all the possible and geometrically equivalent base pairs in a given family. (*A*) The double loop E motif of bacterial 5S rRNA. The B within the circle indicates a bifurcated pair and a W a water-mediated interaction. The simple motif occurs in the 16S as well as the 23S rRNAs. (*B*) The S or bulged G motif in which a base triple occurs. This motif is typical of the eukaryotic loop E of rRNA and of the sarcin/ricin loop in bacterial 23S rRNA. The adenines often interact through their Watson-Crick or sugar-edge sites with other nucleotides (Leontis et al. 2002a). (*C*) The kink-turn motif (Klein et al. 2001). (*D*) The C-motif (Leontis and Westhof 2003).

shallow/minor groove of any combination of stacked and helical Watson-Crick base pairs. Thus, A-minor motifs are mutationally robust and can accommodate many combinations of neutral mutations. RNA nano-objects could successfully be built following some of the above rules (Chworos et al. 2004; Liao and Seeman 2004). New tools are appearing for automatically extracting recurrent modules from databases of x-ray structures (Djelloul and Denise 2008; Sarver et al. 2008).

6 MODELING LARGE RNA ASSEMBLIES

There are three main categories of tertiary structure interactions, those between two double-stranded helices, those between a helix and a single strand, and those between two single-stranded regions. Sequence analyses together with the growing number of RNA crystal structures have shown that RNA architectures are assembled from modules in a hierarchical manner. The modeling process developed in the laboratory is based on this principle of natural folding processes (Westhof et al. 1996; Westhof and Michel 1994). The process is highly iterative in most cases. The secondary structure is first parsed into modules based on elementary structural elements, the 3D coordinates of which can be generated using appropriate programs (see later). These modules are afterwards assembled to form the RNA architecture. One starts with some module identifications and proceeds to assemble interactively the fragments. During this process, new potential contacts can be identified or suspected. These are then assessed in the set of available aligned sequences. Finally, the geometry of the model is regularized using least-square refinement. In this process, standard hydrogen bonds between nucleotides are used as explicit constraints.

One of the most difficult tasks is the arrangement of the multiple-way junctions between helices. The main problem is the proper choice of helices that stack on each other (Duckett et al. 1995; Hohng et al. 2004; Krol et al. 1990). The natural tendency for right-handedness in RNA strands helps often in the decision process. Thus, the Hoogsteen edge (a purine N7 atom) is more accessible when the purine base is 3′ terminal than when it is either internal or 5′ terminal (Westhof et al. 1989). The avoidance of knot formation should be kept in mind and, in large structures with several long-range loop–loop contacts, like the sense–antisense complex between CopA and CopT, topological criteria have to be considered carefully to avoid knot formation (Kolb et al. 2001). Three-way junctions with two helices approximately coaxially stacked can be divided into three main families depending on the relative lengths of the segments linking the three Watson-Crick helices. Each family has topological characteristics with some conservation in the non-Watson-Crick pairs within the linking segments as well as in the types of contacts between the segments and the helices (Lescoute and Westhof 2006b). The determination of such three-way junctions is only applicable in case of coaxial stacking of two of the three helices, a rather frequent situation.

7 MODELING RNA–PROTEIN COMPLEXES

The intricacies of RNA–protein complexes render the modeling of RNA complexes particularly difficult and challenging. Together with extensive chemical and enzymatic probing, information about some crucial contacts (Romby et al. 1990) and the relative positioning of the interacting surfaces (Caprara et al. 1996; Webb et al. 2001) can be gained, even in the absence of the structures of the binding proteins. However, with the knowledge of the protein crystal structures and additional biochemical evidence, for example cross-linking data, actual docking of protein to the RNA model or directed probing can be attempted (Tsai et al. 2003). Modeling of RNA–protein complexes will constitute a major challenge for the next years.

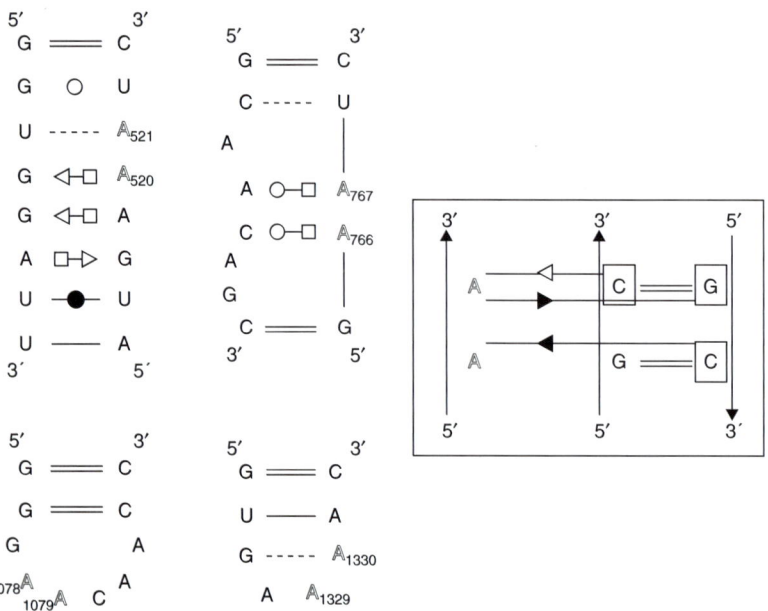

Figure 2. The figure shows four examples of A-minor interactions (the A residues involved are shown in gray tone). Notice how, despite the similar types of contacts, the two consecutive adenines can belong to very different local environments. At the *left*, four different types of motifs in which two consecutive adenines form similar types of A-minor interactions (Nissen et al. 2001) with two consecutive base pairs, as shown at the *right* (first contact at the *top*, type I A-minor, the adenine H-bonds to both bases of the Watson-Crick pair; second contact *below*, type II A-minor, the adenine H-bonds to only one base of the Watson-Crick pair). It has been established (Battle and Doudna 2002; Doherty et al. 2001): (1) consecutive adenines recognize without any strong bias any two stacked Watson-Crick pairs; (2) consecutive adenines have a strong preference for complementary Watson-Crick pairs compared to noncomplementary pairs.

8 MODELING AND FITTING INTO MEDIUM TO LOW RESOLUTION ELECTRON DENSITY MAPS

The progress in cryoelectron microscopy techniques has led to density maps of large functional objects at resolution around or below 7 Å (Schuler et al. 2006). This allows the fitting of atomic models into the density map (Mitra and Frank 2006). Such models can be assembled from homology models derived from solved crystallographic structures but generally de novo construction based on various RNA modules is required. This approach has been applied to the cricket paralysis virus IRES RNA bound to ribosomes in cryoelectron densities at 7.3 Å (Schuler et al. 2006). The resulting model, obtained before crystallography (Pfingsten et al. 2006), agrees very well with the structure of the IRES RNA structure alone (Pfingsten and Kieft 2008).

9 INTERACTIVE MOLECULAR MODELING WITH *ASSEMBLE*

Assemble contains a complete set of interactively connected computer tools with web service capabilities dedicated to the analysis of RNA structures, the structural alignments of RNA sequences with or without a known 3D structure, and the modeling with assembly of RNA modules into an RNA architecture. *Assemble* extends considerably our previous *Manip* program (Massire and Westhof 1998). Importantly, it can be linked to the S2S application dedicated to RNA alignments (Jossinet and Westhof 2005). As discussed previously, the construction of an RNA 3D model with *Assemble* starts with the secondary structure. RNA algorithms (Mfold [Zuker 2003] or the RNAVienna package [Hofacker 2009]) or online repositories (CRW site [Cannone et al. 2002] or RNASTRAND [Andronescu et al. 2008]) provide precomputed secondary structures in CT or BPSEQ files. The secondary structure can also be computed directly from an RNA sequence stored in a FASTA file. Assemble outsources this prediction task to a series of RNA algorithms available as web services (developed in the lab and by third parties). Finally, *Assemble* can also start the 3D modeling from a solved tertiary structure described in a PDB file. In this latter case, the secondary structure is the result of a 3D annotation process done by RNA algorithms like RNAVIEW (Yang et al. 2003) or MCAnnotate (Lemieux and Major 2002).

The RNA secondary structure is displayed in a 2D panel connected to a 3D one. This 2D panel plays several roles to assist the modeling process. It allows the selection of residues with their visualization in the 3D panel and the definition or edition of 2D structure elements like helices, single-strands, and secondary and tertiary interactions. Helices or single stranded regions defined in the RNA secondary structure can be exported as 3D building blocks with a default regular A-form helical fold. Several options are then available in the 3D panel to adapt the model to the RNA peculiarities. First, the default folding can be altered by modifying the six torsion angles along the sugar-phosphate backbone for any single residue within the 3D structure. The folding of an RNA motif stored in a local repository can be applied to the current 3D selection. The building blocks can be reorganized in the 3D scene to produce the overall shape of the 3D architectural model. If available, electron density maps can be simultaneously displayed for fitting models into density.

Finally, stereochemical and geometrical errors introduced during the modeling process can be fixed using a refinement algorithm embedded into Assemble. The set of base–base interactions defined in the 2D panel are used to deduce the structural and geometrical constraints needed to improve the 3D model. The refinement is achieved by geometrical least-squares using the Konnert-Hendrickson algorithm (Konnert and Hendrickson 1980) implemented in the program *Nuclin/Nuclsq* (Westhof et al. 1985). The algorithm takes into account bond lengths, valence angles, dihedrals, and has antibump restraints. The resulting function is minimized against a dictionary of distances that have been observed in high-resolution crystal structures of nucleotides and oligonucleotides. Because the refinement program uses the steepest descent algorithm, the conformation of the starting model should not present extremely distorted regions to avoid refinement failure. The refined model can then be collated to the data and the process of interactive modeling and least-square refinement can be looped until the model is satisfactory. A subset of the data should be used as a blind test during the building of the model so as to help validate the model. Further steps of interactive modeling may then be required until a satisfactory solution is reached.

10 COMPARISONS BETWEEN RNA MODELS AND CRYSTAL STRUCTURES

Several RNAs were predicted, sometimes several years, before the x-ray structures became available. This situation is rather unique to the RNA world; this is not the case for example in the protein field where special competitions had to be installed *ad hoc* (Critical Assessment of Techniques for Protein Structure Prediction; http://predictioncenter.org/). Root mean square deviations between a chosen set of atoms in the structures to be assessed (RMSDs) are usually the main measure for benchmarking prediction tools and comparing predicted structures with experimentally derived structures. For example, Lsqman (Kleywegt 1996) calculates normalized RMSD values according to (Carugo and Pongor 2001). Because RMSDs increase with the size of the molecule, normalization is necessary to allow the comparisons between models of different size. As can be expected, the values for the RMSDs tend to improve with the size of the x-ray crystallographic database on which the modeling is based. However, RMSDs, being based on a least-squares approach, spread errors over the whole molecule so as to minimize the final value. Thus, low values can be obtained for compact structures that may be missing most of the key intramolecular contacts, whereas high values can be observed when the relationship of two domains is wrongly deduced despite the fact that each domain is correctly predicted. An accurate model should present most of the internal interaction contacts maintaining the overall architecture (Lescoute and Westhof 2006a). New metrics have been introduced to calibrate the interaction network fidelity (Parisien et al. 2009). The results show that RMSDs do not provide information about the quality and precision of the base–base interaction networks.

Globally, the interactively modeled architectures with all of the long-range contacts (loop–helix, loop–loop, pseudoknots, . . .) are in excellent agreement with crystallographic structures, especially for group I introns (Jaeger et al. 1993; Lehnert et al. 1996; Michel and Westhof 1990) and the RNA component of ribonuclease P (Brown et al. 1996; Chen et al. 1998; Haas et al. 1991; Massire et al. 1998). Thus, for the specificity domain of the RNA of ribonuclease P (Fig. 3), the overall RMSD is 11.2 Å but the normalized one is 4.8 Å between the modeled (Massire et al. 1998) and the 3.15 Å resolution x-ray structure (Krasilnikov et al. 2003). Interestingly, the RMSD between the crystal structures of the specificity domains of the two main families of the ribonuclease P RNAs (Krasilnikov et al. 2004) is 1.6 Å. In addition, locally, several motifs were properly identified and inserted into large structures: the ribose zipper in the hairpin ribozyme (Earnshaw et al. 1997; Rupert and Ferre-D'Amare 2001) based on experimental data (Chowrira et al. 1993); the presence of a C-motif in the 5′-UTR of the Thr-aminoacylsynthetase messenger (Caillet et al. 2003; Torres-larios et al. 2002); the presence of a loop-E like structure in domain P7 of subgroup IA2 in group I introns involved in tertiary contacts (Golden et al. 2005; Leontis and Westhof 1998a; Waldsich et al. 2002), or other tertiary contacts like that between P3 and J6/6a in the *Azoarcus* group I intron (Adams

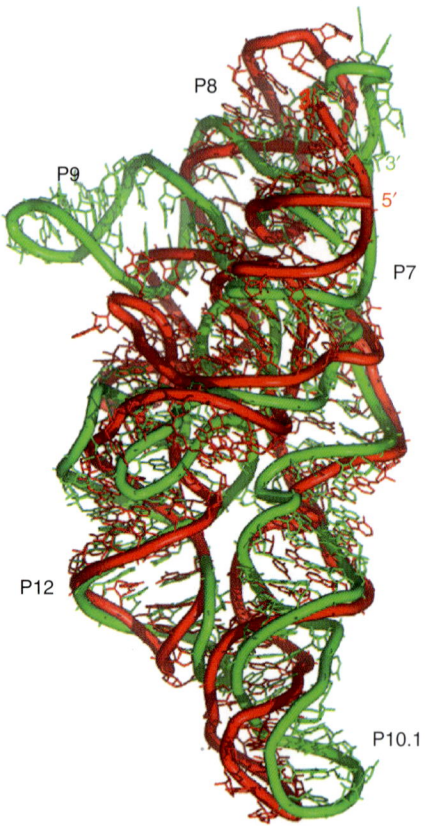

Figure 3. Models can be compared with x-ray structures to assess their accuracy. One measure, the root mean square deviations (RMSD), is obtained after minimizing the sum of the distances between corresponding atoms. Superimposition of the ribonuclease P specificity domain of *Bacillus subtilis* the model (in red) (Massire et al. 1998) was assembled 5 years before the x-ray structure (in green) (Krasilnikov et al. 2003).

et al. 2004; Rangan et al. 2003). A systematic comparison between phylogenetic (Leontis and Westhof 1998b), chemical probing (Romby et al. 1988), and x-ray data (Correll et al. 1997) led to a refined model of the 5S loop E in spinach chorosplast, which was later strongly supported by NMR data (Vallurupalli and Moore 2003). However, despite a correct architecture with the presence of the proper long-range contacts and the ensuing rather good normalized RMSD between modeled and x-ray structures of the *Azoarcus* group I intron, 3.85 Å, at the atomic level many contacts are off, sometimes one nucleotide away. Further, a comparison with the crystal structures of group I introns (Adams et al. 2004; Golden et al. 2005; Guo et al. 2004) shows that, although the binding mode of the guanine was correctly predicted (Michel et al. 1989), the adjacent invariant nucleotides were not (Michel and Westhof 1990). A recent analysis between models and crystallographic results of group II introns has been presented (Michel et al. 2009).

Up to now, the RNA architectures, assembled using all-atom components, agree very well with the determined x-ray structures despite a rather poor congruence at the atomic level. The interactive 3D modeling of RNA is still the most efficient and reliable method for assembling large structures. The ultimate goal is to model RNA folds with atomic precision as automatically as possible. Despite great progress, new tools based on conformational space searches have not yet produced a novel fold before x-ray determination of major biological impact. Some of them do reproduce known structures with excellent accuracy (Parisien et al. 2009). In the near future, through systematic comparisons between crystal structures and sets of aligned sequences, refined and new key molecular rules will be unraveled, which should lead to an improved accuracy in automatic model prediction.

11 CONCLUSIONS

The modeling approach described here is based on a corpus of observations leading to the paradigm that RNA architecture results from the hierarchical assembly of preformed double-stranded helices defined by Watson-Crick base pairs and RNA modules maintained by non-Watson-Crick base pairs (Fig. 4). Thus, metrics for assessing the accuracy of RNA models should include checks on the number and correctness of non-Watson-Crick pairs in the predicted models (Parisien et al. 2009). The most common long-range RNA–RNA contacts are the A-minor interactions that are mutationally robust and can accommodate many combinations of neutral mutations. This characteristic dilutes the links between RNA sequence and structure. To achieve specificity in RNA folding, global, positional, and orientational, constraints on the native fold must occur upstream in the folding process. Critical parameters are the lengths of the helices, the coaxiality of the helical stacks, and the structure adopted at the junctions of helices (Lescoute and Westhof 2006b). Thus, the molecular neutrality present in the local interactions is partially compensated by global topological criteria, much less accessible to sequence analysis because they are attached to the 3D architecture (Cruz and Westhof 2009).

Because modeling requires the integration of a vast amount of data at various levels of complexity, the quality of the modeling reflects, in the end, the current understanding of the modeled systems and the quality and usefulness in the integration of knowledge. This present understanding of RNA architecture and modeling bears on the RNA World hypothesis. RNA architecture can be parsed into recurrent modules, limited in number, with defined borders. Modules interact through defined protocols of interaction applicable in many topological

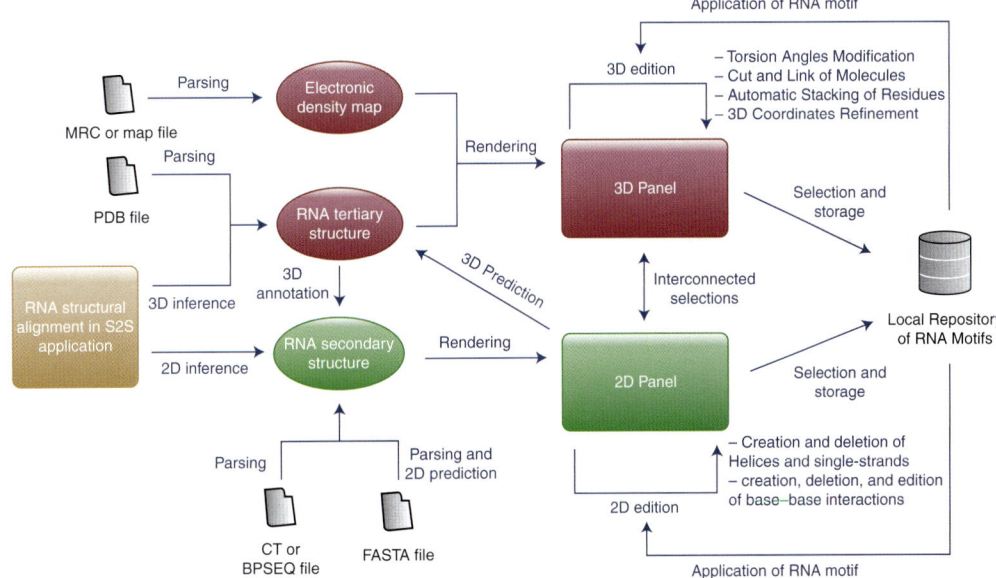

Figure 4. Integrated sets of computer programs are a necessity for analyzing RNA sequences and structures as well as for modeling RNA structures by the assembly of identified modules. Details of the workflow of the *Assemble* package (http://bioinformatics.org/assemble). This application is divided into two main components: (1) a 3D panel rendering the 3D model and (2) a 2D panel displaying the secondary structure scaffold of this model. Because of their interconnections, the secondary structure is used as a guide during the modeling process. The secondary structure can be loaded or computed from several file formats or can be inferred from a dedicated tool like the *S2S* application (Jossinet and Westhof 2005).

situations. Modules can accommodate sequence variations that are neutral within defined geometrical limits. Such hierarchical networks with embedded modularity are typical of self-organizing networks (Ravasz et al. 2002). The physicochemical and structural characteristics of RNA molecules are ideally suited to the evolution of such complex networks. They lead to the appearance of similar modules capable of adaptation. The neutrality of the contacts between the modules allows for a great diversity of architectures despite the use of a limited number of recurrent modular building blocks. For example (Lescoute and Westhof 2006a), it is striking that a central piece of the peptidyl transferase center is made of the most frequent three-way junctions, observed in riboswitches for binding activating ligands or at the active site of the hammerhead ribozymes.

ACKNOWLEDGMENTS

Some of this work was made possible by a grant from the Human Frontier Science Program (RGP0032/2005-C to E.W.).

REFERENCES

Adams PL, Stahley MR, Gill ML, Kosek AB, Wang J, Strobel SA. 2004. Crystal structure of a group I intron splicing intermediate. *RNA* **10**: 1867–1887.

Andronescu M, Bereg V, Hoos HH, Condon A. 2008. RNA STRAND: the RNA secondary structure and statistical analysis database. *BMC Bioinformatics* **9**: 340.

Batey RT, Rambo RP, Doudna JA. 1999. Tertiary motifs in RNA structure and folding. *Angew Chem Int Ed Engl* **38**: 2326–2343.

Battle DJ, Doudna JA. 2002. Specificity of RNA-RNA helix recognition. *Proc Natl Acad Sci* **99**: 11676–11681.

Bourassa N, Major F. 2002. Implication of the prohead RNA in phage phi29 DNA packaging. *Biochimie* **84**: 945–951.

Brion P, Westhof E. 1997. Hierarchy and dynamics of RNA folding. *Annu Rev Biophys Biomol Struct* **26**: 113–137.

Brown JW, Birmingham A, Griffiths PE, Jossinet F, Kachouri-Lafond R, Knight R, Lang BF, Leontis N, Steger G, Stombaugh J, et al. 2009. The RNA structure alignment ontology. *RNA* **15**: 1623–1631.

Brown JW, Nolan JM, Haas ES, Rubio MA, Major F, Pace NR. 1996. Comparative analysis of ribonuclease P RNA using gene sequences from natural microbial populations reveals tertiary structural elements. *Proc Natl Acad Sci* **93**: 3001–3006.

Brunel C, Marquet R, Romby P, Ehresmann C. 2002. RNA loop-loop interactions as dynamic functional motifs. *Biochimie* **84**: 925–944.

Caillet J, Nogueira T, Masquida B, Winter F, Graffe M, Dock-Bregeon AC, Torres-Larios A, Sankaranarayanan R, Westhof E, Ehresmann B, et al. 2003. The modular structure of *Escherichia coli* threonyl-tRNA synthetase as both an enzyme and a regulator of gene expression. *Mol Microbiol* **47**: 961–974.

Cannone JJ, Subramanian S, Schnare MN, Collett JR, D'Souza LM, Du Y, Feng B, Lin N, Madabusi LV, Muller KM, et al. 2002. The comparative RNA web (CRW) site: An online database of comparative sequence and structure information for ribosomal, intron, and other RNAs. *BMC bioinformatics* **3**: 2.

Caprara MG, Lehnert V, Lambowitz AM, Westhof E. 1996. A tyrosyl-tRNA synthetase recognizes a conserved tRNA-like structural motif in the group I intron catalytic core. *Cell* **87**: 1135–1145.

Carugo O, Pongor S. 2001. A normalized root-mean-square distance for comparing protein three-dimensional structures. *Protein Sci* **10**: 1470–1473.

Cate JH, Gooding AR, Podell E, Zhou K, Golden BL, Kundrot CE, Cech TR, Doudna JA. 1996a. Crystal structure of a group I ribozyme domain: Principles of RNA packing. *Science* **273**: 1678–1684.

Cate JH, Gooding AR, Podell E, Zhou K, Golden BL, Szewczak AA, Kundrot CE, Cech TR, Doudna JA. 1996b. RNA tertiary structure mediation by adenosine platforms. *Science* **273**: 1696–1699.

Chen JL, Nolan JM, Harris ME, Pace NR. 1998. Comparative photocrosslinking analysis of the tertiary structures of *Escherichia coli* and *Bacillus subtilis* RNase P RNAs. *Embo J* **17**: 1515–1525.

Chowrira BM, Berzal-Herranz A, Keller CF, Burke JM. 1993. Four ribose 2'-hydroxyl groups essential for catalytic function of the hairpin ribozyme. *J Biol Chem* **268**: 19458–19462.

Chworos A, Severcan I, Koyfman AY, Weinkam P, Oroudjev E, Hansma HG, Jaeger L. 2004. Building programmable jigsaw puzzles with RNA. *Science* **306**: 2068–2072.

Correll CC, Freeborn B, Moore PB, Steitz TA. 1997. Metals, motifs, and recognition in the crystal structure of a 5S rRNA domain. *Cell* **28**: 705–712.

Costa M, Michel F. 1995. Frequent use of the same tertiary motif by self-folding RNAs. *EMBO J* **14**: 1276–1285.

Costa M, Michel F. 1997. Rules for RNA recognition of GNRA tetraloops deduced by in vitro selection: Comparison with in vivo evolution. *EMBO J* **16**: 3289–3302.

Cruz JA, Westhof E. 2009. The dynamic landscapes of RNA architecture. *Cell* **136**: 604–609.

Dam E, Pleij K, Draper D. 1992. Structural and functional aspects of RNA pseudoknots. *Biochemistry* **31**: 11665–11676.

Das R, Baker D. 2007. Automated de novo prediction of native-like RNA tertiary structures. *Proc Natl Acad Sci* **104**: 14664–14669.

Ding F, Sharma S, Chalasani P, Demidov VV, Broude NE, Dokholyan NV. 2008. Ab initio RNA folding by discrete molecular dynamics: From structure prediction to folding mechanisms. *RNA* **14**: 1164–1173.

Djelloul M, Denise A. 2008. Automated motif extraction and classification in RNA tertiary structures. *RNA* **14**: 2489–2497.

Dock-Bregeon AC, Westhof E, Giege R, Moras D. 1989. Solution structure of a tRNA with a large variable region: Yeast tRNASer. *J Mol Biol* **206**: 707–722.

Doherty EA, Batey RT, Masquida B, Doudna JA. 2001. A universal mode of helix packing in RNA. *Nat Struct Biol* **8**: 339–343.

Duckett DR, Murchie AI, Lilley DM. 1995. The global folding of four-way helical junctions in RNA, including that in U1 snRNA. *Cell* **83**: 1027–1036.

Earnshaw DJ, Masquida B, Muller S, Sigurdsson ST, Eckstein F, Westhof E, Gait MJ. 1997. Inter-domain cross-linking and molecular modelling of the hairpin ribozyme. *J Mol Biol* **274**: 197–212.

Fang X, Pan T, Sosnick TR. 1999. A thermodynamic framework and cooperativity in the tertiary folding of a Mg2+-dependent ribozyme. *Biochemistry* **38**: 16840–16846.

Ferrandon D, Koch I, Westhof E, Nusslein-Volhard C. 1997. RNA-RNA interaction is required for the formation of specific bicoid mRNA 3' UTR-STAUFEN ribonucleoprotein particles. *Embo J* **16**: 1751–1758.

Fuller W, Hodgson A. 1967. Conformation of the anticodon loop intRNA. *Nature* **215**: 817–821.

Gautheret D, Gutell RR. 1997. Inferring the conformation of RNA base pairs and triples from patterns of sequence variation. *Nucleic Acids Res* **25**: 1559–1564.

Gautheret D, Damberger SH, Gutell RR. 1995. Identification of base-triples in RNA using comparative sequence analysis. *J Mol Biol* **248**: 27–43.

Golden BL, Kim H, Chase E. 2005. Crystal structure of a phage Twort group I ribozyme-product complex. *Nat Struct Mol Biol* **12**: 82–89.

Guo F, Gooding AR, Cech TR. 2004. Structure of the *Tetrahymena* ribozyme: Base triple sandwich and metal ion at the active site. *Mol Cell* **16**: 351–362.

Haas ES, Morse DP, Brown JW, Schmidt FJ, Pace NR. 1991. Long-range structure in ribonuclease P RNA. *Science* **254**: 853–856.

Heilman-Miller SL, Pan J, Thirumalai D, Woodson SA. 2001. Role of counterion condensation in folding of the *Tetrahymena* ribozyme. II. Counterion-dependence of folding kinetics. *J Mol Biol* **309**: 57–68.

Hofacker IL. 2009. RNA secondary structure analysis using the Vienna RNA package. *Current protocols in bioinformatics/editoral board, Andreas D Baxevanis* et al Chapter 12, Unit12 12.

Hohng S, Wilson TJ, Tan E, Clegg RM, Lilley DM, Ha T. 2004. Conformational flexibility of four-way junctions in RNA. *J Mol Biol* **336**: 69–79.

Jaeger L, Westhof E, Michel F. 1993. Monitoring of the cooperative unfolding of the sunY group I intron of bacteriophage T4. The active form of the sunY ribozyme is stabilized by multiple interactions with 3' terminal intron components. *J Mol Biol* **234**: 331–346.

Jonikas MA, Radmer RJ, Laederach A, Das R, Pearlman S, Herschlag D, Altman RB. 2009. Coarse-grained modeling of large RNA molecules with knowledge-based potentials and structural filters. *RNA* **15**: 189–199.

Jossinet F, Westhof E. 2005. Sequence to Structure (S2S): display, manipulate and interconnect RNA data from sequence to structure. *Bioinformatics* **21**: 3320–3332.

Khvorova A, Lescoute A, Westhof E, Jayasena SD. 2003. Sequence elements outside the hammerhead ribozyme catalytic core enable intracellular activity. *Nat Struct Biol* **10**: 708–712.

Klein DJ, Schmeing TM, Moore PB, Steitz TA. 2001. The kink-turn: A new RNA secondary structure motif. *Embo J* **20**: 4214–4221.

Kleywegt GJ. 1996. Use of non-crystallographic symmetry in protein structure refinement. *Acta Cryst* **D52**: 842–857.

Kolb FA, Westhof E, Ehresmann B, Ehresmann C, Wagner EG, Romby P. 2001. Four-way junctions in antisense RNA-mRNA complexes involved in plasmid replication control: A common theme? *J Mol Biol* **309**: 605–614.

Konnert JH, Hendrickson WA. 1980. Restrained parameters thermal factors refinement procedures. *Acta Crystallographica* **A36**: 344–349.

Krasilnikov AS, Xiao Y, Pan T, Mondragon A. 2004. Basis for structural diversity in homologous RNAs. *Science* **306**: 104–107.

Krasilnikov AS, Yang X, Pan T, Mondragon A. 2003. Crystal structure of the specificity domain of ribonuclease P. *Nature* **421**: 760–764.

Krol A, Westhof E, Bach M, Luhrmann R, Ebel JP, Carbon P. 1990. Solution structure of human U1 snRNA. Derivation of a possible three-dimensional model. *Nucleic Acids Res* **18**: 3803–3811.

Lambert A, Fontaine JF, Legendre M, Leclerc F, Permal E, Major F, Putzer H, Delfour O, Michot B, Gautheret D. 2004. The ERPIN server: An interface to profile-based RNA motif identification. *Nucleic Acids Res* **32**: W160–165.

Lehnert V, Jaeger L, Michel F, Westhof E. 1996. New loop-loop tertiary interactions in self-splicing introns of subgroup IC and ID: A complete 3D model of the *Tetrahymena thermophila* ribozyme. *Chem Biol* **3**: 993–1009.

Lemieux S, Major F. 2002. RNA canonical and non-canonical base pairing types: A recognition method and complete repertoire. *Nucleic Acids Res* **30**: 4250–4263.

Leontis NB, Stombaugh J, Westhof E. 2002a. Motif prediction in ribosomal RNAs Lessons and prospects for automated motif prediction in homologous RNA molecules. *Biochimie* **84**: 961–973.

Leontis NB, Stombaugh J, Westhof E. 2002b. The non-Watson-Crick base pairs and their associated isostericity matrices. *Nucleic Acids Res* **30**: 3497–3531.

Leontis NB, Westhof E. 1998a. A common motif organizes the structure of multi-helix loops in 16S and 23S ribosomal RNAs. *J Mol Biol* **283**: 571–583.

Leontis NB, Westhof E. 1998b. The 5S rRNA loop E: Chemical probing and phylogenetic data versus crystal structure. *RNA* **4:** 1134–1153.

Leontis NB, Westhof E. 2001. Geometric nomenclature and classification of RNA base pairs. *RNA* **7:** 499–512.

Leontis NB, Westhof E. 2003. Analysis of RNA motifs. *Curr Opin Struct Biol* **13:** 300–308.

Lescoute A, Westhof E. 2006a. The interaction networks of structured RNAs. *Nucleic Acids Res* **34:** 6587–6604.

Lescoute A, Westhof E. 2006b. Topology of three-way junctions in folded RNAs. *RNA* **12:** 83–93.

Lescoute A, Leontis NB, Massire C, Westhof E. 2005. Recurrent structural RNA motifs, isostericity matrices and sequence alignments. *Nucleic Acids Res* **33:** 2395–2409.

Levitt M. 1969. Detailed molecular model for transfer ribonucleic acid. *Nature* **224:** 759–763.

Liao S, Seeman NC. 2004. Translation of DNA signals into polymer assembly instructions. *Science* **306:** 2072–2074.

Macke TJ, Ecker DJ, Gutell RR, Gautheret D, Case DA, Sampath R. 2001. RNAMotif, an RNA secondary structure definition and search algorithm. *Nucleic Acids Res* **29:** 4724–4735.

Massire C, Westhof E. 1998. MANIP: An interactive tool for modelling RNA. *J Mol Graph Model* **16:** 197–205, 255–197.

Massire C, Jaeger L, Westhof E. 1998. Derivation of the three-dimensional architecture of bacterial ribonuclease P RNAs from comparative sequence analysis. *J Mol Biol* **279:** 773–793.

Michel F, Westhof E. 1990. Modelling of the three-dimensionnal architecture of group-I catalytic introns based on comparative sequence analysis. *J Mol Biol* **216:** 585–610.

Michel F, Costa M, Westhof E. 2009. The ribozyme core of group II introns: A structure in want of partners. *Trends Biochem Sci* **34:** 189–199.

Michel F, Jaquier A, Dujon B. 1982. Comparison of fungal mitochondrial introns reveals extensive homologies in RNA secondary structure. *Biochimie* **64:** 867–881.

Michel F, Costa M, Massire C, Westhof E. 2000. Modeling RNA tertiary structure from patterns of sequence variation. *Methods Enzymol* **317:** 491–510.

Michel F, Hanna M, Green R, Bartel DP, Szostak JW. 1989. The guanosine binding site of the *Tetrahymena* ribozyme. *Nature* **342:** 391–395.

Misra VK, Draper DE. 2002. The linkage between magnesium binding and RNA folding. *J Mol Biol* **317:** 507–521.

Mitra K, Frank J. 2006. Ribosome dynamics: Insights from atomic structure modeling into cryo-electron microscopy maps. *Annu Rev Biophys Biomol Struct* **35:** 299–317.

Moore PB. 1999. Structural motifs in RNA. *Annu Rev Biochem* **68:** 287–300.

Nawrocki EP, Kolbe DL, Eddy SR. 2009. Infernal 1.0: Inference of RNA alignments. *Bioinformatics* (Oxford, England) **25:** 1335–1337.

Ninio J, Favre A, Yaniv M. 1969. Molecular model for transfer RNA. *Nature* **223:** 1333–1335.

Nissen P, Ippolito JA, Ban N, Moore PB, Steitz TA. 2001. RNA tertiary interactions in the large ribosomal subunit: The A-minor motif. *Proc Natl Acad Sci* **98:** 4899–4903.

Oubridge C, Kuglstatter A, Jovine L, Nagai K. 2002. Crystal structure of SRP19 in complex with the S domain of SRP RNA and its implication for the assembly of the signal recognition particle. *Mol Cell* **9:** 1251–1261.

Pace NR, Olsen GJ, Woese CR. 1986. Ribosomal RNA phylogeny and the primary lines of evolutionary descent. *Cell* **45:** 325–326.

Pace NR, Smith DK, Olsen GJ, James BD. 1989. Phylogenetic comparative analysis and the secondary structure of ribonuclease P RNA–a review. *Gene* **82:** 65–75.

Pan J, Woodson SA. 1999. The effect of long-range loop-loop interactions on folding of the *Tetrahymena* self-splicing RNA. *J Mol Biol* **294:** 955–965.

Parisien M, Major F. 2008. The MC-Fold and MC-Sym pipeline infers RNA structure from sequence data. *Nature* **452:** 51–55.

Parisien M, Cruz JA, Westhof E, Major F. 2009. New metrics for comparing and assessing discrepancies between RNA 3D structures and models. *RNA* **15:** 1875–1885.

Pfingsten JS, Kieft JS. 2008. RNA structure-based ribosome recruitment: Lessons from the Dicistroviridae intergenic region IRESes. *RNA* **14:** 1255–1263.

Pfingsten JS, Costantino DA, Kieft JS. 2006. Structural basis for ribosome recruitment and manipulation by a viral IRES RNA. *Science* **314:** 1450–1454.

Rangan P, Masquida B, Westhof E, Woodson SA. 2003. Assembly of core helices and rapid tertiary folding of a small bacterial group I ribozyme. *Proc Natl Acad Sci* **100:** 1574–1579.

Ravasz E, Somera AL, Mongru DA, Oltvai ZN, Barabasi AL. 2002. Hierarchical organization of modularity in metabolic networks. *Science* **297:** 1551–1555.

Richardson JS, Schneider B, Murray LW, Kapral GJ, Immormino RM, Headd JJ, Richardson DC, Ham D, Hershkovits E, Williams LD, et al. 2008. RNA backbone: Consensus all-angle conformers and modular string nomenclature (an RNA Ontology Consortium contribution). *RNA* **14:** 465–481.

Romby P, Baudin F, Brunel C, Leal de Stevenson I, Westhof E, Romaniuk PJ, Ehresmann C, Ehresmann B. 1990. Ribosomal 5S RNA from *Xenopus laevis* oocytes: Conformation and interaction with transcription factor IIIA. *Biochimie* **72:** 437–452.

Romby P, Westhof E, Toukifimpa R, Mache R, Ebel JP, Ehresmann C, Ehresmann B. 1988. Higher order structure of chloroplastic 5S ribosomal RNA from spinach. *Biochemistry* **27:** 4721–4730.

Rupert PB, Ferre-D'Amare AR. 2001. Crystal structure of a hairpin ribozyme-inhibitor complex with implications for catalysis. *Nature* **410:** 780–786.

Sarver M, Zirbel CL, Stombaugh J, Mokdad A, Leontis NB. 2008. FR3D: Finding local and composite recurrent structural motifs in RNA 3D structures. *J Math Biol* **56:** 215–252.

Schuler M, Connell SR, Lescoute A, Giesebrecht J, Dabrowski M, Schroeer B, Mielke T, Penczek PA, Westhof E, Spahn CM. 2006. Structure of the ribosome-bound cricket paralysis virus IRES RNA. *Nat Struct Mol Biol* **13:** 1092–1096.

Shapiro BA, Yingling YG, Kasprzak W, Bindewald E. 2007. Bridging the gap in RNA structure prediction. *Curr Opin Struct Biol* **17:** 157–165.

Sosnick TR, Pan T. 2003. RNA folding: Models and perspectives. *Curr Opin Struct Biol* **13:** 309–316.

Stombaugh J, Zirbel CL, Westhof E, Leontis NB. 2009. Frequency and isostericity of RNA base pairs. *Nucleic Acids Res* **37:** 2294–2312.

Thirumalai D, Lee N, Woodson SA, Klimov D. 2001. Early events in RNA folding. *Annu Rev Phys Chem* **52:** 751–762.

Tinoco I Jr, Bustamante C. 1999. How RNA folds. *J Mol Biol* **293:** 271–281.

Torres-Larios A, Dock-Bregeon AC, Romby P, Rees B, Sankaranarayanan R, Caillet J, Springer M, Ehresmann C, Ehresmann B, Moras D. 2002. Structural basis of translational control by Escherichia coli threonyl tRNA synthetase. *Nat Struct Biol* **9:** 343–347.

Tsai HY, Masquida B, Biswas R, Westhof E, Gopalan V. 2003. Molecular modeling of the three-dimensional structure of the bacterial RNase P holoenzyme. *J Mol Biol* **325:** 661–675.

Vallurupalli P, Moore PB. 2003. The solution structure of the loop E region of the 5S rRNA from spinach chloroplasts. *J Mol Biol* **325:** 843–856.

Wagner C, Ehresmann C, Ehresmann B, Brunel C. 2004. Mechanism of dimerization of bicoid mRNA: Initiation and stabilization. *J Biol Chem* **279:** 4560–4569.

Waldsich C, Masquida B, Westhof E, Schroeder R. 2002. Monitoring intermediate folding states of the td group I intron in vivo. *Embo J* **21:** 5281–5291.

Webb AE, Rose MA, Westhof E, Weeks KM. 2001. Protein-dependent transition states for ribonucleoprotein assembly. *J Mol Biol* **309:** 1087–1100.

Weichenrieder O, Wild K, Strub K, Cusack S. 2000. Structure and assembly of the Alu domain of the mammalian signal recognition particle. *Nature* **408:** 167–173.

Westhof E, Jaeger L. 1992. RNA Pseudoknots: Structural and functional aspects. *Curr Op Struct Biol* **2:** 327–333.

Westhof E, Fritsch V. 2000. RNA folding: Beyond Watson-Crick pairs. *Structure Fold Des* **8:** R55–65.

Westhof E, Michel F. 1994. Prediction and experimental investigation of RNA secondary and tertiary foldings. In *RNA-Proteins interactions: Frontiers in molecular biology* IRL Press at Oxford University Press, pp. 25–51.

Westhof E, Dumas P, Moras D. 1985. Crystallographic refinement of yeast aspartic acid transfer RNA. *J Mol Biol* **184:** 119–145.

Westhof E, Masquida B, Jaeger L. 1996. RNA tectonics: Towards RNA design. *Fold Des* **1:** R78–88.

Westhof E, Romby P, Romaniuk PJ, Ebel JP, Ehresmann C, Ehresmann B. 1989. Computer modeling from solution data of spinach chloroplast and of *Xenopus laevis* somatic and oocyte 5 S rRNAs. *J Mol Biol* **207:** 417–431.

Yang H, Jossinet F, Leontis N, Chen L, Westbrook J, Berman H, Westhof E. 2003. Tools for the automatic identification and classification of RNA base pairs. *Nucleic Acids Res* **31:** 3450–3460.

Zirbel CL, Sponer JE, Sponer J, Stombaugh J, Leontis NB. 2009. Classification and energetics of the base-phosphate interactions in RNA. *Nucleic Acids Res* **37:** 4898–4918.

Zuker M. 1989. On finding all suboptimal foldings of an RNA molecule. *Science* **244:** 48–52.

Zuker M. 2003. Mfold web server for nucleic acid folding and hybridization prediction. *Nucleic Acids Res* **31:** 3406–3415.

RNA Reactions One Molecule at a Time

Ignacio Tinoco, Gang Chen, and Xiaohui Qu

Department of Chemistry, University of California, Berkeley, California 94720-1460

Correspondence: intinoco@lbl.gov

SUMMARY

Much of the dynamics information is lost in bulk measurements because of the population averaging. Single-molecule methods measure one molecule at a time; they provide knowledge not obtainable by other means. In this article, we review the application of the two most widely used single-molecule methods—fluorescence resonance energy transfer (FRET) and force versus extension measurements—to several RNA reactions. First, we discuss folding/unfolding studies on a hairpin ribozyme that revealed multiple conformations of the RNA with distinct kinetics, and on a series of RNA pseudoknots, whose mechanical stabilities were found to show a strong correlation with their frameshifting efficiency during translation. We also discuss several RNA-related molecular motors. Single-molecule experiments revealed detailed mechanisms for the interaction of HIV reverse transcriptase and nucleic acid helicases (NS3 and RIG-1) with their substrates. Optical tweezers studies showed that translation of a single messenger RNA by a ribosome occurs by successive translocation-and-pause cycles. Single-molecule FRET experiments yielded important information on ribosome conformational changes and tRNA dynamics during translation. Overall, single-molecule experiments have been very valuable for understanding RNA reactions.

Outline

1 Introduction
2 Methods
3 Folding, unfolding, and function
4 Molecular motors
5 Conclusions
References

1 INTRODUCTION

In the usual bulk, or ensemble, measurements, the properties of all the molecules in the sample, contribute to the signal. For example, a fluorescence signal from the solution is a consequence of the different absorbance and emission properties of all the molecules with their varied dynamics. In ensemble measurements, the averaging over all conformations and species obscures the effects of minor contributions to the signal; the major contributors dominate the spectrum. However, single-molecule methods allow the measurement of the properties and reactions of one molecule at a time. By observing the conformational changes of each single molecule over time, one can learn about the distributions of properties and their dynamics, not just averages.

Because of the stochastic nature of kinetics, single-molecule methods are especially advantageous for studying reactions. Each molecule has a probability of reacting, but when it will actually react is not predictable. Thus, in bulk, reactions do not remain synchronized, and the resulting population average hides many kinetic details; intermediates may be difficult to detect. However, following the progression of a single molecule from reactant to product can reveal each intermediate; a detailed mechanism can be obtained. The kinetics and thermodynamics of the reaction can be obtained from the lifetimes of each conformation.

Any property that can be measured for one molecule at a time can be used to characterize molecules and their reactions. The two most widely used single-molecule methods applied to RNA structures and functions are fluorescence resonance energy transfer (FRET), and force versus extension measurements. A unique capability of studying one molecule at a time is to apply force to the molecule while not perturbing the rest of the solution. In such experiments, the molecule is attached to two beads controlled by optical tweezers, or it is attached to a surface and an atomic force microscope cantilever; the force on the molecule and the distance between the attached points are measured. Force becomes a thermodynamic variable, like temperature or pressure, that can influence a reaction. Force affects the equilibrium if there is a change in the length of the molecule during the reaction. Similarly, force affects rates of reactions depending on the distances to the transition states. Thus, force can be used to study thermodynamics and kinetics of reactions—such as unfolding of RNA—that would otherwise only occur at high temperatures, or in the presence of a denaturant. Unfolding and refolding of an RNA can be studied in the presence of proteins, other RNAs, ligands, and even mixtures approximating the contents of biological cells. Although the process is not the same as what occurs for RNA molecules in cells, it should be a better approximation than the conventional unfolding and folding studies in high concentrations of urea, or by thermal melting curves. The reversible mechanical work (force times distance) for unfolding an RNA is equal to the Gibbs free energy of unfolding. The temperature dependence of this work gives the enthalpy, and thus the entropy of the process.

In this article, we briefly describe methods used in studying single molecules, and review the applications of these methods to RNA reactions, including mechanical unfolding and folding of RNA secondary and tertiary structures, interactions with HIV reverse transcriptase, unfolding by helicases, and translation of messenger RNAs.

2 METHODS

2.1 Force

A standard method to apply force to a molecule using optical tweezers is shown in Figure 1 (Liphardt et al. 2001; Tinoco et al. 2006; Li et al. 2008). There, we see an RNA flanked by RNA•DNA-handles attached to beads by biotin-streptavidin or digoxigenin-antidigoxigenin. The motion of a helicase or a ribosome on the RNA is monitored by measuring the change of distance between the beads while the force is kept constant. As the helicase unwinds the hairpin (Dumont et al. 2006), or the ribosome translates the mRNA (Wen et al. 2008), the distance between the beads increases as base pairs become single stranded. Alternatively, the folding and unfolding of an RNA species (hairpin [Li et al. 2007], pseudoknot [Chen et al. 2007], ribozyme [Onoa et al. 2003], kissing complex [Li and Tinoco 2009], riboswitch [Greenleaf et al. 2008], and so forth) is studied by increasing, then decreasing, the force on the molecule. As each element of the structure unfolds or refolds, there is an abrupt increase or decrease in the end-to-end distance.

The kinetics of reactions are studied by measuring the lifetimes of each species in the reaction; for example, the time a ribosome spends at each codon, or the lifetimes of intermediates in an unfolding reaction. Every time the reaction is repeated for a single molecule, the lifetime of each species will be different because of the stochastic nature of kinetics. Several steps in a reaction can occur without changing the measured property of the species. The distribution of lifetimes for a species tells you about the number of such kinetic steps that occur during the lifetime. For a single kinetic step, the distribution of lifetimes is exponential. The mean lifetime $<\tau>$ is the reciprocal of the single-step rate constant, k.

$$<\tau> = \frac{1}{k} \qquad (1)$$

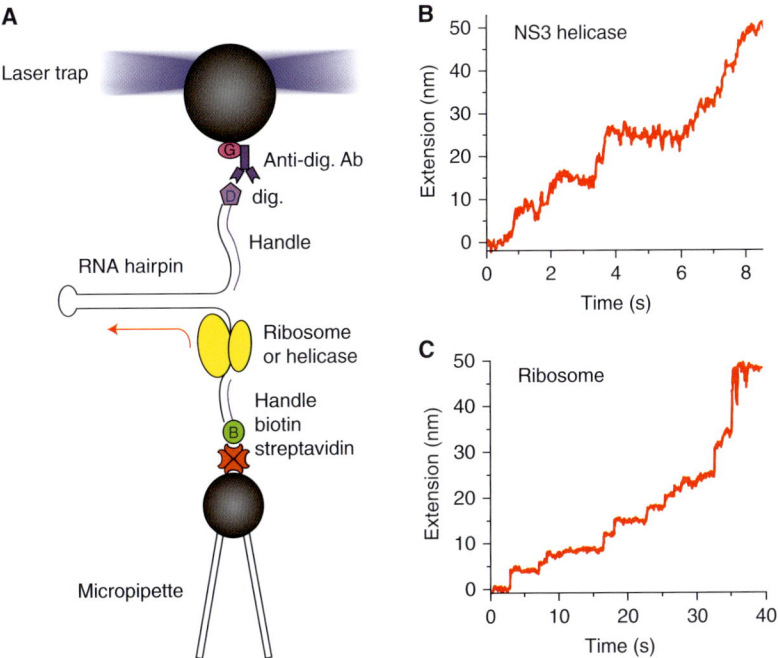

Figure 1. Optical tweezers assay for studying NS3 helicase unwinding or ribosome translation. (A) The RNA has a short single-stranded region for loading NS3 helicase or ribosome, followed by a hairpin region to be unwound or translated. Note that NS3 translocates 3′ to 5′, whereas the ribosome translates mRNA 5′ to 3′. The two ends of the RNA are attached to long handles for separating the RNA/enzyme construct from the beads. The ends are then held between two micron-size polystyrene beads coated with antidigoxigenin antibody and streptavidin. Drawings are schematic and not to scale. (B, C) Representative traces of extension versus time for NS3 unwinding (B; force is maintained at 15 pN) and ribosome translation (C; force is maintained at 20 pN). Adapted from Wen et al. 2008 and Dumont et al. 2006.

For more than one step contributing to the lifetime, the distribution has a maximum that does not occur at zero time and the mean lifetime depends on all the rate constants involved. The force dependence of the kinetics characterizes the position of the transition state along the reaction coordinate.

The Gibbs free energy change, ΔG, of a reaction is obtained from the *reversible* mechanical work done. Mechanical work is the integral of force times distance; if the force remains constant during the process, it is just the force times the change in distance.

$$w = \int F dx = F \Delta x \qquad (2)$$

An unfolding/folding reaction is reversible if it occurs at the same force in the forward process and the reverse process; there is no hysteresis. However, even for a nonreversible reaction, the reversible work can be obtained from the distribution of nonreversible work values. There are two ways this can be done. The intersection of the distribution of work values for an unfolding reaction and its refolding reaction corresponds to the reversible work—the Gibbs free energy (Crooks 1999). The unfolding of a three-helix junction from *Escherichia coli* ribosomal RNA that binds S15 ribosomal protein was analyzed this way (Collin et al. 2005). If the distribution of nonreversible work values for unfolding, or for refolding, but not both, is measured, the free energy is obtained from the exponential average of the distribution of work, w, values (Jarzynski 1997)

$$e^{-\Delta G/kT} = <e^{-w/kT}> \qquad (3)$$

with k the Boltzmann constant and T the absolute temperature. Equation (3) is exact in the limit of averaging over infinite repetitions of the experiment. This method was tested on the unfolding of the P5abc three-helix junction domain from the *Tetrahymena thermophila* group I intron (Liphardt et al. 2002), where it was shown that a few hundred measurements of nonreversible unfolding accurately provided the free energy of unfolding.

The unfolding of single RNA molecules can also be studied using atomic force microscopy (AFM). However, up to now, nearly all AFM unfolding studies have been

2.2 Single-molecule Fluorescence Resonance Energy Transfer (smFRET)

Single-molecule fluorescence techniques allow the dissection of biochemical reactions and their molecular mechanisms in great detail (Selvin and Ha 2008; Roy et al. 2008). In FRET experiments, molecules are labeled with a donor fluorophore and an acceptor fluorophore at two specific positions. The distance between the donor and acceptor is monitored by a technique based on the 1948 theory of Förster, who showed that electronic excitation energy can be efficiently transferred through transition dipole–dipole interactions. The efficiency of energy transfer depends on the reciprocal of the distance to the sixth power, $(1/R)^6$, the optical properties of the fluorophores, and on an orientation factor. The distance, R_o, at which energy transfer is 50% efficient, is specific to a given set of donor and acceptor molecules. The efficiency of energy transfer is given by

$$\% \ FRET = 100 \frac{1}{1 + (R^6/R_o^6)}, \quad (4)$$

in which R is the distance between donor and acceptor. The largest variation in FRET signal occurs when R and R_o are equal, therefore donor-acceptor pairs of fluorophores are chosen to maximize the sensitivity of the experiment for the distance of interest. Usually, distances in the range of 2 nm to 10 nm are measurable by FRET.

FRET values can be obtained from donor and acceptor fluorescence intensities, and then used to calculate the distance between donor and acceptor using Equation (4). Measuring accurate FRET values, thus accurately determining R, requires knowledge of quantum yields of donor and acceptor. In practice, FRET values are often approximated as:

$$\% \ FRET = 100 \frac{I_A}{I_D + I_A}, \quad (5)$$

in which I_D and I_A are fluorescence intensities of donor and acceptor, respectively. Equation (5) suffices for identifying changes in the donor–acceptor distance. As a protein or RNA unfolds, or a molecular motor moves, the distance between donor and acceptor changes and FRET changes; in principle, FRET can also change if R_o changes, for example, if the orientation factor changes (Iqbal et al. 2008).

The usual single-molecule FRET (smFRET) method involves excitation of a donor by the evanescent wave in a total internal reflection fluorescence (TIRF) microscope, as shown in Figure 3C. Only molecules very close to the surface (usually within 100 nm) are excited by the evanescent laser light; data can be collected from hundreds of individual molecules simultaneously. Each molecule can be studied during multiple transitions or reactions, until one of the fluorophores undergoes photobleaching. Single-molecule FRET has been applied to translation of mRNA (Blanchard et al. 2004b; Fei et al. 2008; Cornish et al. 2009), unfolding by helicases (Rasnik et al. 2006), protein-mediated folding (Stone et al. 2007), ribozyme dynamics (Pereira et al. 2008), etc.

3 FOLDING, UNFOLDING, AND FUNCTION

RNA function depends on its folded structure and on dynamic fluctuations between functional states. Single-molecule methods are ideal for following the dynamics of folding and unfolding of the elements of RNA secondary and tertiary structure. Knowledge of the stabilities and dynamics of the folded structures, of intermediates in their formation, and of how these properties are affected by the binding of ligands or protein cofactors, facilitates better understanding of their functions and their structures. Many single-molecule experiments have been done in this line of research; here, we discuss only two systems as examples.

3.1 Hairpin Ribozyme

The hairpin ribozyme derived from tobacco ring spot virus satellite RNA is a self-cleaving four-helix-junction RNA structure containing two internal loops (Wilson et al. 2005). The tertiary structure of this ribozyme involves docking loop A with loop B, which facilitates the site-specific reversible cleavage and ligation reactions in loop A. This ribozyme has been extensively studied by smFRET (Ha 2004; Bokinsky and Zhuang 2005). A minimal hairpin ribozyme, with loop A and loop B domains connected by a six-nucleotide (nt) bulge, was made for both bulk and single-molecule studies. The kinetics were obtained from the lifetime distributions of undocked and docked conformations (Ditzler et al. 2008). By fitting the lifetime distributions to multiple exponential functions, smFRET revealed that there were multiple populations of molecules with distinct undocking kinetics (Ditzler et al. 2008). Surprisingly, the molecules with relatively slower undocking kinetics retained their undocking rates even after separation of the two strands of the minimal hairpin ribozyme by denaturing gel electrophoresis. The authors proposed that the intra-strand structure of the S-turn or loop E motif within loop B might remain through extensive denaturation and annealing processes (Ditzler et al. 2008). It is also possible that subpopulations of hairpin ribozymes are kinetically trapped in misfolded secondary structures in other parts of the ribozyme. Secondary structure dynamics can be followed by characteristic fluorescent properties that depend directly on self-cleavage/ligation reactions of the hairpin ribozyme (Nahas et al. 2004; Liu et al. 2007).

Directly monitoring RNA global conformational changes can be facilitated by constructing multiple FRET pairs at different positions. Three-color smFRET methods have been developed to probe global structural dynamics of a Holliday junction (Hohng et al. 2004). Sequential and simultaneous multiple-color smFRET experiments will be useful in directly revealing the hidden folding dynamics of hairpin ribozymes and other larger RNA molecules (Qu et al. 2008; Steiner et al. 2008).

Mechanical force has also been applied to perturb and map the docking/undocking pathways and kinetics, and thus provide new insight into molecular heterogeneity and catalytic function. Using a combined instrument of optical tweezers and smFRET, the kinetics of the conformational switch of a Holliday junction was found to be affected by mechanical force of less than 0.5 pN (Hohng et al. 2007). Applying forces in different directions revealed structures of the transient species during the conformation changes of the Holliday junction. The presence of native and near-native "quasi-docked" tertiary structures of a hairpin ribozyme may be revealed by their distinct force-dependent undocking kinetics. Force may also be used to directly test whether the heterogeneity observed in the hairpin ribozyme persists throughout mechanical denaturation.

3.2 Pseudoknots and Frameshifting

All the secondary and tertiary structures in the coding region of an mRNA have to be unfolded to single strand to be translated by the ribosome. Pseudoknot structures (see Fig. 2A for a typical pseudoknot containing two stems and two loops) stimulate programmed −1 ribosomal frameshifting at an upstream slippery sequence of the form X XXY YYZ (0 frame) to XXX YYY Z (−1 frame), in which X is any three identical nucleotides, Y is either AAA or UUU, and in eukaryotes Z is usually not G. With the mRNA slippery sequence located at the aminoacyl (A) and

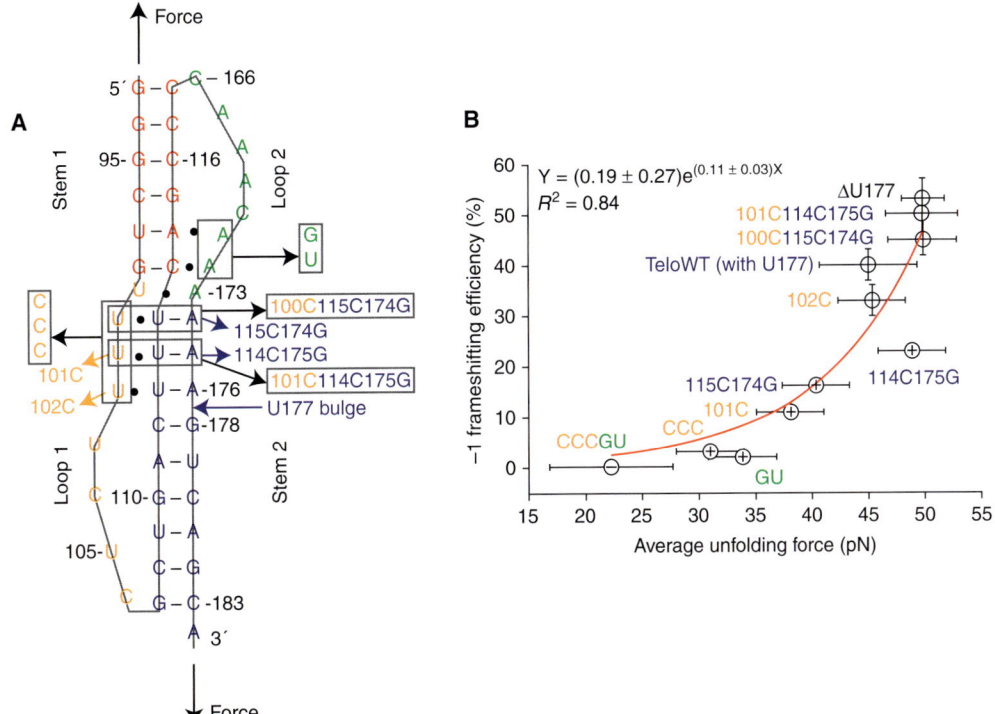

Figure 2. Pseudoknots and −1 ribosomal frameshifting. (*A*) Pseudoknots used for bulk frameshift assays and single-molecule studies. The pseudoknots contain stem 1 (red), loop 1 (yellow), stem 2 (blue), and loop 2 (green). All the mutants were made based on ΔU177. In mutant CCCGU, all base triples are disrupted (two in stem 1-loop 2 and three in stem 2-loop 1). In TeloWT, the single nucleotide bulge U177 in stem 2 is added. The directions of applied mechanical force by optical tweezers are shown with black arrows. (*B*) Correlation observed between bulk −1 frameshifting efficiency and average unfolding force. Error bars are standard deviations from bulk and single-molecule experiments. Adapted from Chen et al. 2009.

peptidyl (P) ribosome sites, a downstream pseudoknot structure provides resistance to ribosomal helicase activity and stimulates −1 ribosomal frameshifting.

Mechanical unfolding of mRNA structures using optical tweezers (Fig. 1A) may mimic the helicase activity of ribosomes and provide insight into the molecular determinants of pseudoknot mechanical stability and −1 ribosomal frameshifting. Mechanical unfolding of frameshifting stimulatory pseudoknots from infectious bronchitis virus (IBV) revealed that mechanical stability and frameshifting efficiency are affected by the sequence and length of their stems (Hansen et al. 2007; Green et al. 2008).

Bulk frameshifting assays suggested that in addition to Watson-Crick base pairing interactions, minor-groove stem 1-loop 2 interactions (Fig. 2A) are also important in stimulating −1 frameshifting in some IBV pseudoknots (Liphardt et al. 1999). A crystal structure of a pseudoknot in beet western yellows virus (BWYV) revealed the presence of extensive minor-groove base triples important for stimulating −1 frameshifting (Kim et al. 1999; Su et al. 1999).

More recently, both major-groove and minor-groove base triples were found to enhance mechanical stability and increase −1 frameshifting efficiency in 11 pseudoknots studied (Fig. 2) (Chen et al. 2009). Excellent correlation was found between mechanical stability and frameshifting efficiency. The results indicate that −1 frameshifting is stimulated by: (1) stabilizing stem 1 by forming minor-groove stem 1-loop 2 base triples; and (2) increasing torsional resistance to unfolding by forming major-groove stem 2-loop 1 base triples. The combined mutational and single-molecule studies suggested that the folding intermediate pseudoknot structures (with partial formation of stem 2 and no base triples formed) unfold in one step or two steps at low force, whereas native pseudoknots unfold in one step at high force. The folding intermediate pseudoknot structures probably do not induce high-efficiency −1 frameshifting, because the unfolding force is typically below 35 pN (see Fig. 2B) (Chen et al. 2009).

Mechanical unfolding allows direct measurement of how base triple formation stabilizes a Watson-Crick duplex structure. Remarkably, codon-anticodon recognition is enhanced by two base triples formed at the minor-groove of the first two Watson-Crick pairs of the decoding A-site codon-anticodon duplex (Ogle et al. 2001). It will be interesting to see how −1 frameshifting is affected by the A-site base triples and other interactions within the ribosome complex.

4 MOLECULAR MOTORS

Molecular motors include a wide range of biological devices usually powered by ATP or GTP. Some motors deal with proteins, such as myosins, which walk along actin, and kinesins, which carry cargo along microtubules. Other motors are vital in gene replication, transcription, and translation (Seidel and Dekker 2007). Much detailed information on the dynamics of molecular motors has been obtained by following the motors' action in real time. We will concentrate here on four motors with RNA as their substrates.

4.1 HIV Reverse Transcriptase

The infection of human immunodeficiency virus (HIV) relies on the conversion of its single-stranded RNA genome into double-stranded DNA, which is later incorporated into the host genome. This process involves three major steps: RNA-directed DNA synthesis (the synthesis of minus-strand DNA using the HIV RNA genome as template), DNA-directed RNA hydrolysis (the cleavage of the RNA template at multiple places), and DNA-directed DNA synthesis (the synthesis of the plus-strand DNA using the minus-strand DNA as the template). Amazingly, these different reactions are done by a single enzyme, the HIV reverse transcriptase (RT). RT consists of two subunits, one of which contains DNA polymerase and RNase H domains. The different nucleic acid substrates regulate the RT activities, but the mechanism of the regulation was poorly understood. The DNA directed DNA synthesis activity of RT has been studied by single-molecule experiments with or without force (Lu et al. 2004; Ortiz et al. 2005; Kim et al. 2007b).

Using smFRET techniques (see Fig. 3 caption for experimental details) (Abbondanzieri et al. 2008), Zhuang and her colleagues found that RT binds to the DNA template with a DNA or RNA primer in two opposite orientations, with either the DNA polymerase domain or the RNase H domain close to the 3′ end of the primer. This correlates nicely with RT's function of DNA-directed DNA synthesis or DNA-directed RNA hydrolysis on such substrates. RT has both DNA polymerase and RNase H function on other substrates, such as the chimeric DNA/RNA primer, a primer with two special 15-nt RNA purine sequences called polypurine tracts (PPT), and the primer with PPT and a few nucleotides of DNA extension. On such substrates, RT was observed to occupy both the DNA-polymesase-competent and RNase-H-competent orientations and to dynamically flip between the two orientations. The measured rate of primer extension correlated with the fraction of time for which the RT enzyme bound in the polymerase-competent orientation. Thus, these experiments suggest that RT can distinguish between different substrates and binds differently, and that the binding orientation determines the subsequent function. Small-molecule ligands, specifically dNTP and two clinically approved anti-HIV

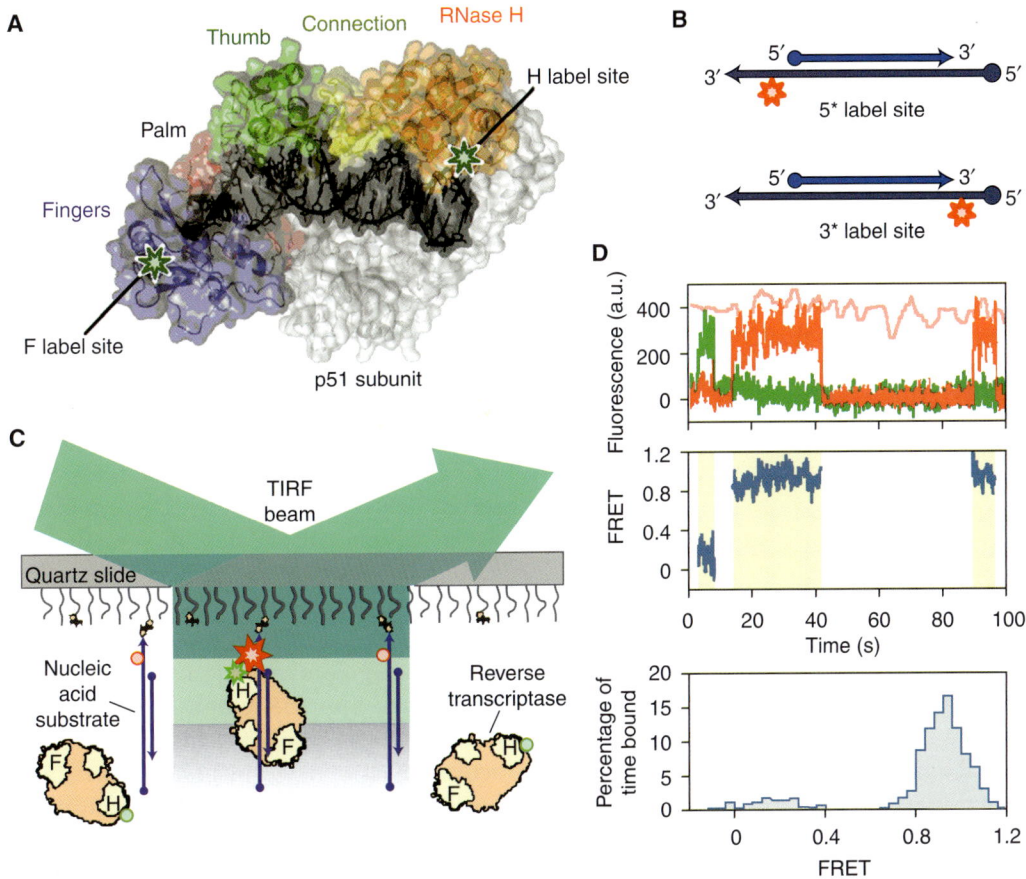

Figure 3. Single-molecule FRET assay for probing the binding mechanism of HIV-1 reverse transcriptase (RT) on nucleic acid substrates. (A) The structure of HIV-1 RT bound to a substrate. RT is labeled with Cy3 at one of the two labeling sites (green stars). (B) The substrate consists of a 19–21-nt primer strand (DNA or RNA) annealed to a 50-nt DNA template strand to mimic the different binding substrates that RT encounters in vivo. The template is labeled with a Cy5 dye (red star) at either the 5′ or 3′ end. (C) One end of the substrate is immobilized on the coverslip surface for single-molecule detection. The RT is free to diffuse in solution. Under 532 nm excitation wavelength in a total internal reflection fluorescence (TIRF) microscospe, no fluorescence is detected without RT binding; when an RT binds to the substrate, the donor dye is excited by the evanescent laser illumination and the FRET signal of the dye pair can be observed until the RT dissociates from the substrate or the dyes photobleach. The stars and spheres indicate dyes that do and do not emit fluorescence, respectively. (D) An example of smFRET analysis. *Top*: fluorescence time traces from Cy3 (green) and Cy5 (red) under excitation at 532 nm, and that from Cy5 (pink) under excitation at 635 nm (to confirm that Cy5 has not photobleached). *Middle*: FRET value calculated over the duration of the binding events (yellow shaded regions). *Bottom*: FRET distribution histogram created for the binding events. The observed FRET value and its change over time report the RT binding orientation, and the dynamics of RT orientation change, respectively. The same experiment was repeated using different FRET labeling sites on the RT and/or substrate to strengthen the conclusions. (Reprinted, with permission, from Abbondanzieri et al. 2008 [© MacMillan].)

drugs (nevirapine and efavirenz), were also studied and found to greatly affect the equilibrium and flipping dynamics of RT binding orientation, and therefore, to regulate the RT activity.

RT has very low processivity for DNA synthesis (a few to a few hundred nucleotides), but the whole HIV genome is ~10 kb long. Thus, having a highly efficient searching mechanism to locate the target site is crucial for RT function. Using smFRET, the Zhuang group observed that RT slides between the two ends of the substrate when the substrate is considerably longer than 19 nt (the length that RT covers on binding) (Liu et al. 2008). And this sliding motion was shown to be thermally driven. Very interestingly, it was also observed that when RT locates the target site on the substrate, it can flip its binding orientation to place the correct functional domain (DNA polymerase domain for DNA synthesis or the RNase H domain for RNA hydrolysis) close to the target site. The combination of sliding and flipping provides a very efficient searching mechanism. Furthermore, the nontemplate strand displacement capability of RT during DNA synthesis was investigated. It was found that in RNA strand displacement synthesis, RT was able to extend the primer with a few nucleotides before termination; but in DNA strand displacement synthesis, RT was able to complete a long stretch of primer extension before termination. The observed difference is consistent with the fact that DNA•RNA hybrids are usually more thermodynamically stable than DNA•DNA duplexes. This also shows that RT is not a very powerful motor in regard to strand displacement.

4.2 RNA Helicases

RNA helicases use nucleotide triphosphates (NTP) to unwind RNA duplexes and are involved in many viral and cellular RNA metabolism processes (Jankowsky and Fairman 2007; Pyle 2008). Some RNA helicases actually function as translocases without unwinding RNA structures. Understanding how RNA unwinding/translocating correlates with NTP binding, hydrolysis, and helicase conformational change is central to the understanding of the mechanism of helicase function. Even though single-molecule techniques have been successfully applied to many DNA helicases (Hopfner and Michaelis 2007) and other nucleic acid motors (Seidel and Dekker 2007), there have been only a few single-molecule studies of RNA helicases (Dumont et al. 2006; Marsden et al. 2006; Cheng et al. 2007; Yang et al. 2007; Myong et al. 2009).

Marsden et al. (Marsden et al. 2006) used AFM to stretch a single RNA hairpin in the presence of RNA helicases: eIF4A or Ded1p, which are involved in eukaryotic translation. Both helicases lower the unfolding force of the hairpin during AFM pulling, suggesting that both helicases weaken the RNA hairpin stability. Yang et al. (Yang et al. 2007) applied smFRET techniques to study Ded1p. However, instead of studying the molecular motor properties of Ded1p, the authors focused on how Ded1p facilitates RNA conformational changes, which is another aspect of Ded1p function.

The nonstructural protein 3 (NS3) in hepatitis C virus (HCV) is a key component of the viral replication machinery for RNA directed RNA synthesis. It has both protease and helicase domains and can unwind both DNA and RNA duplexes. Dumont et. al (Dumont et al. 2006) used optical tweezers (Fig. 1) to study the translocation and unwinding mechanism of NS3 with an RNA hairpin substrate containing a short single-strand region for loading the NS3 monomer, and a hairpin region for the detection of the unwinding activity. It was found that NS3 unwinding of, and translocation on, RNA are coordinated by ATP in discrete steps of 11 ± 3 base pairs, and that each step is composed of three rapid substeps of 3.6 ± 1.3 base pairs, also triggered by ATP binding. Force does not affect the NS3 monomer unwinding step size, but force does increase NS3 processivity. The coupled duplex unwinding and translocation rate is much faster on A•U stretches than on G•C stretches, and the dependence on the base pair free energy suggests that NS3 actively destabilizes the RNA duplex to facilitate unwinding (Cheng et al. 2007). NS3 has a higher tendency to dissociate when encountering the barrier created by a G•C stretch following an A•U stretch, and NS3 processivity is affected up to six bases before the barrier. Using smFRET without applying force, Myong et al. observed that NS3 unwinds the DNA duplex with one base pair as the fundamental step, but several one base-pair steps accumulate tension on the NS3-DNA complex, which is relieved in a burst of three base pairs (Myong et al. 2009).

RIG-I is a cytosolic protein that detects pathogen-associated molecular patterns on viral RNA, and elicits an antiviral immune response. Wild-type RIG-I is composed of two amino-terminal tandem CARDs (caspase activation and recruitment domains), a central DExH box RNA helicase domain, and a carboxy-terminal regulatory domain (RD). Myong et al. (Myong et al. 2009) used protein-induced fluorescence enhancement (PIFE) (Fischer et al. 2004) to study RIG-I. In PIFE, the substrate is labeled with a dye, whose fluorescence emission is affected by the proximity of a protein (not fluorescently labeled). Two mutants were also studied: RIGh and svRIG, which have complete or partial deletions of the CARDs, respectively. Without ATP, all three forms of RIG-I bind steadily to the RNA substrate until dissociation. On a double-stranded RNA substrate, all three forms of RIG-I proteins translocate repetitively on the RNA substrate dependent on ATP concentration, temperature, and length of the RNA duplex region. The authors proposed that the repetitive shuttling of RIG-I on dsRNA regions of the viral genome might induce conformational changes in RIG-I important for antiviral immune response signaling.

4.3 Translation

Following movement of the ribosome along the mRNA using optical tweezers. Optical tweezers in the mode shown in Figure 1A were used to monitor the motion of a single *E. coli* ribosome on a single messenger RNA during translation (Wen et al. 2008). A hairpin mRNA is used so that as the ribosome translates the 5′-side of the hairpin, double-stranded base pairs are converted to single strands leading to an increase in end-to-end distance of the RNA. A trajectory is seen in Figure 1C; it shows that a translating ribosome moves by a series of translocation-pause-translocation steps. Each translocation corresponds to translation of three nucleotides—one codon (with extension increase of ~3 nm), with a mean time of approximately 25 ms (X. Qu, unpubl.). Each pause is on the order of seconds; during this time, all the biochemistry for the translation of one codon occurs.

The ability to see individual steps of translation and to separate the process into pauses and translocations, opens the door to learn how each component (either intrinsic or extrinsic) of the translation machinery affects each substep of the process of translation. Detailed mechanisms of the fidelity of translation can be obtained: amino acid misincorporation, frameshifting, read-through of stop codons, or premature termination. The effects of codon sequence, of mRNA structures such as hairpins and pseudoknots, of elongation factors and release factors, of antibiotics, etc., can all be assessed. Single-molecule methods will allow us to find answers that cannot be obtained by ensemble methods.

Transfer RNA and ribosome dynamics by smFRET.
The Puglisi-Chu group pioneered the application of smFRET to observe the motion of fluorescently labeled tRNAs on the ribosome during translation (Blanchard et al. 2004a; Blanchard et al. 2004b; Kim et al. 2007a; Lee et al. 2007; Uemura et al. 2007). They found that tRNAs fluctuate between a classical state AA/PP (the anticodons are in the A-site and P-site of the 30S subunit, and the amino acid acceptor sites are in the A-site and P-site of the 50S subunit) and a hybrid state AP/PE (the anticodons remain in the A- and P-sites, but the amino acid acceptor sites move to the P- and E-sites of the 50S subunit). The hybrid state is favored by formation of the peptide bond. More recent results showed that there were two hybrid states and suggested that global conformational changes in the ribosome induced the different tRNA states (Munro et al. 2007).

smFRET measurements on fluorescently labeled ribosomes directly showed the conformational dynamics during the elongation phase of translation. The Noller and Ha groups (Cornish et al. 2008) labeled protein L9 of the 50S subunit with Cy3 (donor) and proteins S6 or S11 of the 30S subunit with Cy5 (acceptor). Spontaneous intersubunit fluctuations were seen consistent with the intersubunit rotation (racheting) seen by cryo-electron microscopy (Frank and Agrawal 2000), and the kinetics correspond to the tRNAs kinetics. The authors conclude that the intersubunit rotations correspond to the classical and hybrid states of the tRNAs. The fluctuations are thermally driven; they do not require EF-G, but the hybrid state is stabilized by EF-G binding. After translocation, the ribosome is left in the classical state.

Further characterization of ribosome and tRNA dynamics was obtained by measuring FRET between the L1 stalk of the 50S subunit and the tRNAs (Fei et al. 2008). Labeling two proteins of the large subunit, L1 and L33 (Cornish et al. 2009) or L1 and L9 (Fei et al. 2009), allows direct observation of the opening and closing of the L1 stalk, and its correlation with the classical-hybrid states of the tRNAs. The effects of the different tRNAs (initiator and elongator), the state of acylation of the tRNAs, and of EF-G on the dynamics have been measured. A detailed picture of the motions of the tRNAs, of the subunits, of individual proteins in the subunits, and of translocation dynamics along the mRNA is emerging.

It is clear that smFRET can provide useful information about relative motions, in the range of 1 to 10 nm, of the components of molecular machines. The kinetics of the motions can be accurately measured, with lifetimes in the range from milliseconds to minutes of each conformation. The long time limit is determined by the lifetime of the fluorophore; the short time limit depends on instrumental parameters.

5 CONCLUSIONS

Even though the ability to study individual molecules in solution or in biological cells has only recently become possible, its applications to biological questions are growing rapidly. Perhaps the detailed, step-by-step analysis of the mechanism of complex molecular motors, like the ribosome, will be an early success. In this article, we selected a very limited number of RNA-related studies as examples to show the capability of single-molecule techniques. Although we described only two specific single-molecule methods, optical tweezers and FRET, single-molecule techniques are actually very versatile and are evolving rapidly. We encourage the readers to explore further in single-molecule research; new methods are being discovered constantly, and some will surely help solve their problems.

ACKNOWLEDGMENTS

We thank Dr. Jin-Der Wen and Dr. Fei Liu for careful reading of the paper and for making helpful suggestions. Our work was supported by National Institute of Health Grant GM10840 and by the Human Frontiers Science Program.

REFERENCES

Abbondanzieri EA, Bokinsky G, Rausch JW, Zhang JX, Le Grice SF, Zhuang X. 2008. Dynamic binding orientations direct activity of HIV reverse transcriptase. *Nature* **453:** 184–189.

Blanchard SC, Gonzalez RL Jr, Kim HD, Chu S, Puglisi JD. 2004a. tRNA selection and kinetic proofreading in translation. *Nat Struct Mol Biol* **11**: 1008–1014.

Blanchard SC, Kim HD, Gonzalez RL Jr, Puglisi JD, Chu S. 2004b. tRNA dynamics on the ribosome during translation. *Proc Natl Acad Sci* **101**: 12893–12898.

Bokinsky G, Zhuang X. 2005. Single-molecule RNA folding. *Acc Chem Res* **38**: 566–573.

Carrion-Vazquez M, Oberhauser AF, Fowler SB, Marszalek PE, Broedel SE, Clarke J, Fernandez JM. 1999. Mechanical and chemical unfolding of a single protein: A comparison. *Proc Natl Acad Sci* **96**: 3694–3699.

Cecconi C, Shank EA, Bustamante C, Marqusee S. 2005. Direct observation of the three-state folding of a single protein molecule. *Science* **309**: 2057–2060.

Chen G, Chang KY, Chou MY, Bustamante C, Tinoco I Jr, 2009. Triplex structures in an RNA pseudoknot enhance mechanical stability and increase efficiency of -1 ribosomal frameshifting. *Proc Natl Acad Sci* **106**: 12706–12711.

Chen G, Wen JD, Tinoco I Jr, 2007. Single-molecule mechanical unfolding and folding of a pseudoknot in human telomerase RNA. *RNA* **13**: 2175–2188.

Cheng W, Dumont S, Tinoco I Jr, Bustamante C. 2007. NS3 helicase actively separates RNA strands and senses sequence barriers ahead of the opening fork. *Proc Natl Acad Sci* **104**:, 13954–13959.

Collin D, Ritort F, Jarzynski C, Smith SB, Tinoco I Jr, Bustamante C. 2005. Verification of the Crooks fluctuation theorem and recovery of RNA folding free energies. *Nature* **437**: 231–234.

Cornish PV, Ermolenko DN, Noller HF, Ha T. 2008. Spontaneous intersubunit rotation in single ribosomes. *Mol Cell* **30**: 578–588.

Cornish PV, Ermolenko DN, Staple DW, Hoang L, Hickerson RP, Noller HF, Ha T. 2009. Following movement of the L1 stalk between three functional states in single ribosomes. *Proc Natl Acad Sci* **106**: 2571–2576.

Crooks GE. 1999. Entropy production fluctuation theorem and the nonequilibrium work relation for free energy differences. *Phys Rev E* **60**: 2721–2726.

Ditzler MA, Rueda D, Mo J, Hakansson K, Walter NG. 2008. A rugged free energy landscape separates multiple functional RNA folds throughout denaturation. *Nucleic Acids Res* **36**: 7088–7099.

Dumont S, Cheng W, Serebrov V, Beran RK, Tinoco I Jr, Pyle AM, Bustamante C. 2006. RNA translocation and unwinding mechanism of HCV NS3 helicase and its coordination by ATP. *Nature* **439**: 105–108.

Fei J, Bronson J, Hofman JM, Srinivas RL, Wiggins CH, Gonzalez RL Jr, 2009. Allosteric collaboration between elongation factor G and the ribosomal L1 stalk direct tRNA movements during translation. *Proc Natl Acad Sci* **106**: 15702–15707.

Fei J, Kosuri P, MacDougall DD, Gonzalez RL Jr, 2008. Coupling of ribosomal L1 stalk and tRNA dynamics during translation elongation. *Mol Cell* **30**: 348–359.

Fischer CJ, Maluf NK, Lohman TM. 2004. Mechanism of ATP-dependent translocation of *E. coli* UvrD monomers along single-stranded DNA. *J Mol Biol* **344**: 1287–1309.

Frank J, Agrawal RK. 2000. A ratchet-like inter-subunit reorganization of the ribosome during translocation. *Nature* **406**: 318–322.

Green L, Kim CH, Bustamante C, Tinoco I Jr, 2008. Chacterization of the mechanical unfolding of RNA pseudoknots. *J Mol Biol* **375**: 511–528.

Greenleaf WJ, Frieda KL, Foster DA, Woodside MT, Block SM. 2008. Direct observation of hierarchical folding in single riboswitch aptamers. *Science* **319**: 630–633.

Ha T. 2004. Structural dynamics and processing of nucleic acids revealed by single-molecule spectroscopy. *Biochemistry* **43**: 4055–4063.

Hansen TM, Reihani SN, Oddershede LB, Sorensen MA. 2007. Correlation between mechanical strength of messenger RNA pseudoknots and ribosomal frameshifting. *Proc Natl Acad Sci* **104**: 5830–5835.

Hohng S, Joo C, Ha T. 2004. Single-molecule three-color FRET. *Biophys J* **87**: 1328–1337.

Hohng S, Zhou R, Nahas MK, Yu J, Schulten K, Lilley DM, Ha T. 2007. Fluorescence-force spectroscopy maps two-dimensional reaction landscape of the holliday junction. *Science* **318**: 279–283.

Hopfner KP, Michaelis J. 2007. Mechanisms of nucleic acid translocases: Lessons from structural biology and single-molecule biophysics. *Curr Opin Struct Biol* **17**: 87–95.

Iqbal A, Arslan S, Okumus B, Wilson TJ, Giraud G, Norman DG, Ha T, Lilley DM. 2008. Orientation dependence in fluorescent energy transfer between Cy3 and Cy5 terminally attached to double-stranded nucleic acids. *Proc Natl Acad Sci* **105**: 11176–11181.

Jankowsky E, Fairman ME. 2007. RNA helicases - one fold for many functions. *Curr Opin Struct Biol* **17**: 316–324.

Jarzynski C. 1997. Nonequilibrium equality for free energy differences. *Phys Rev Lett* **78**: 2690–2693.

Kim HD, Puglisi JD, Chu S. 2007a. Fluctuations of transfer RNAs between classical and hybrid states. *Biophys J* **93**: 3575–3582.

Kim S, Blainey PC, Schroeder CM, Xie XS. 2007b. Multiplexed single-molecule assay for enzymatic activity on flow-stretched DNA. *Nat Methods* **4**: 397–399.

Kim YG, Su L, Maas S, O'Neill A, Rich A. 1999. Specific mutations in a viral RNA pseudoknot drastically change ribosomal frameshifting efficiency. *Proc Natl Acad Sci* **96**: 14234–14239.

Lee TH, Blanchard SC, Kim HD, Puglisi JD, Chu S. 2007. The role of fluctuations in tRNA selection by the ribosome. *Proc Natl Acad Sci* **104**: 13661–13665.

Li PTX, Tinoco I Jr, 2009. Mechanical unfolding of two DIS RNA kissing complexes from HIV-1. *J Mol Biol* **386**: 1343–1356.

Li PTX, Bustamante C, Tinoco I Jr, 2007. Real-time control of the energy landscape by force directs the folding of RNA molecules. *Proc Natl Acad Sci* **104**: 7039–7044.

Li PTX, Vieregg J, Tinoco I Jr, 2008. How RNA Unfolds and Refolds. *Ann Rev Biochem* **77**: 77–100.

Liphardt J, Dumont S, Smith SB, Tinoco I Jr, Bustamante C. 2002. Equilibrium information from nonequilibrium measurements in an experimental test of Jarzynski's equality. *Science* **296**: 1832–1835.

Liphardt J, Napthine S, Kontos H, Brierley I. 1999. Evidence for an RNA pseudoknot loop-helix interaction essential for efficient -1 ribosomal frameshifting. *J Mol Biol* **288**: 321–335.

Liphardt J, Onoa B, Smith SB, Tinoco I Jr, Bustamante C. 2001. Reversible unfolding of single RNA molecules by mechanical force. *Science* **292**: 733–737.

Liu S, Abbondanzieri EA, Rausch JW, Le Grice SF, Zhuang X. 2008. Slide into action: Dynamic shuttling of HIV reverse transcriptase on nucleic acid substrates. *Science* **322**: 1092–1097.

Liu S, Bokinsky G, Walter NG, Zhuang X. 2007. Dissecting the multistep reaction pathway of an RNA enzyme by single-molecule kinetic "fingerprinting". *Proc Natl Acad Sci* **104**: 12634–12639.

Lu H, Macosko J, Habel-Rodriguez D, Keller RW, Brozik JA, Keller DJ. 2004. Closing of the fingers domain generates motor forces in the HIV reverse transcriptase. *J Biol Chem* **279**: 54529–54532.

Marsden S, Nardelli M, Linder P, McCarthy JEG. 2006. Unwinding single RNA molecules using helicases involved in eukaryotic translation initiation. *J Mol Biol* **361**: 327–335.

Munro JB, Altman RB, O'Connor N, Blanchard SC. 2007. Identification of two distinct hybrid state intermediates on the ribosome. *Mol Cell* **25**: 505–517.

Myong S, Cui S, Cornish PV, Kirchhofer A, Gack MU, Jung JU, Hopfner KP, Ha T. 2009. Cytosolic viral sensor RIG-I Is a 5′-triphosphate-dependent translocase on double-stranded RNA. *Science* **323**: 1070–1074.

Nahas MK, Wilson TJ, Hohng S, Jarvie K, Lilley DMJ, Ha T. 2004. Observation of internal cleavage and ligation reactions of a ribozyme. *Nat Struct Mol Biol* **11**: 1107–1113.

Ogle JM, Brodersen DE, Clemons WM Jr, Tarry MJ, Carter AP, Ramakrishnan V. 2001. Recognition of cognate transfer RNA by the 30S ribosomal subunit. *Science* **292:** 897–902.

Onoa B, Dumont S, Liphardt J, Smith SB, Tinoco I Jr, Bustamante C. 2003. Identifying kinetic barriers to mechanical unfolding of the *T. thermophila* ribozyme. *Science* **299:** 1892–1895.

Ortiz TP, Marshall JA, Meyer LA, Davis RW, Macosko JC, Hatch J, Keller DJ, Brozik JA. 2005. Stepping statistics of single HIV-1 reverse transcriptase molecules during DNA polymerization. *J Phys Chem B* **109:** 16127–16131.

Pereira MJ, Nikolova EN, Hiley SL, Jaikaran D, Collins RA, Walter NG. 2008. Single VS ribozyme molecules reveal dynamic and hierarchical folding toward catalysis. *J Mol Biol* **382:** 496–509.

Pyle AM. 2008. Translocation and unwinding mechanisms of RNA and DNA helicases. *Ann Rev Biophys* **37:** 317–336.

Qu X, Smith GJ, Lee KT, Sosnick TR, Pan T, Scherer NF. 2008. Single-molecule nonequilibrium periodic Mg^{2+}-concentration jump experiments reveal details of the early folding pathways of a large RNA. *Proc Natl Acad Sci* **105:** 6602–6607.

Rasnik I, Myong S, Ha T. 2006. Unraveling helicase mechanisms one molecule at a time. *Nucleic Acids Res* **34:** 4225–4231.

Roy R, Hohng S, Ha T. 2008. A practical guide to single-molecule FRET. *Nat Methods* **5:** 507–516.

Seidel R, Dekker C. 2007. Single-molecule studies of nucleic acid motors. *Curr Opin Struct Biol* **17:** 80–86.

Selvin PR, Ha T. 2008. *Single-molecule techniques, a laboratory manual.* Cold Spring Harbor Laboratory.

Steiner M, Karunatilaka KS, Sigel RK, Rueda D. 2008. Single-molecule studies of group II intron ribozymes. *Proc Natl Acad Sci* **105:** 13853–13858.

Stone MD, Mihalusova M, O'Connor C M, Prathapam R, Collins K, Zhuang X. 2007. Stepwise protein-mediated RNA folding directs assembly of telomerase ribonucleoprotein. *Nature* **446:** 458–461.

Su L, Chen L, Egli M, Berger JM, Rich A. 1999. Minor groove RNA triplex in the crystal structure of a ribosomal frameshifting viral pseudoknot. *Nat Struct Biol* **6:** 285–292.

Tinoco I Jr, Li PTX, Bustamante C. 2006. Determination of thermodynamics and kinetics of RNA reactions by force. *Q Rev Biophys* **39:** 325–360.

Uemura S, Dorywalska M, Lee TH, Kim HD, Puglisi JD, Chu S. 2007. Peptide bond formation destabilizes Shine-Dalgarno interaction on the ribosome. *Nature* **446:** 454–457.

Walther KA, Grater F, Dougan L, Badilla CL, Berne BJ, Fernandez JM. 2007. Signatures of hydrophobic collapse in extended proteins captured with force spectroscopy. *Proc Natl Acad Sci* **104:** 7916–7921.

Wen JD, Lancaster L, Hodges C, Zeri AC, Yoshimura SH, Noller HF, Bustamante C, Tinoco I Jr. 2008. Following translation by single ribosomes one codon at a time. *Nature* **452:** 598–603.

Wilson TJ, Nahas M, Ha T, Lilley DM. 2005. Folding and catalysis of the hairpin ribozyme. *Biochem Soc Trans* **33:** 461–465.

Yang QS, Fairman ME, Jankowsky E. 2007. DEAD-box-protein-assisted RNA structure conversion towards and against thermodynamic equilibrium values. *J Mol Biol* **368:** 1087–1100.

Aptamers and the RNA World, Past and Present

Larry Gold[1,2], Nebojsa Janjic[2], Thale Jarvis[2], Dan Schneider[2], Jeffrey J. Walker[2], Sheri K. Wilcox[2], and Dom Zichi[2]

[1]Department of Molecular, Cellular, and Developmental Biology, University of Colorado, Boulder, Colorado 80309
[2]SomaLogic, Boulder, Colorado 80301

Correspondence: lgold@somalogic.com

SUMMARY

Aptamers and the SELEX process were discovered over two decades ago. These discoveries have spawned a productive academic and commercial industry. The collective results provide insights into biology, past and present, through an in vitro evolutionary exploration of the nature of nucleic acids and their potential roles in ancient life. Aptamers have helped usher in an RNA renaissance. Here we explore some of the evolution of the aptamer field and the insights it has provided for conceptualizing an RNA world, from its nascence to our current endeavor employing aptamers in human proteomics to discover biomarkers of health and disease.

Outline

1. Introduction
2. The history of SELEX and aptamers
3. Proteomics: Driving SELEX to the best possible "winners"
4. SOMAmer specificity
5. SOMAmer structure
6. Quantitative probing of the plasma proteome at high content and low limits of detection
7. Back to the RNA world
8. Conclusions: What other functions might oligonucleotides have?

References

1 INTRODUCTION

Aptamers, the output of the SELEX process (Systematic Evolution of Ligands by EXponential enrichment), have now had 20 years to "show their stuff." They have done so admirably, and in so doing have allowed us to wonder about what else oligonucleotides might be doing that we have yet to discover as we poke around the biosphere. Deeply embedded within the concepts of the RNA World are questions regarding present biology as well as how we got here. A famous Dobzhansky quote echoed often by Carl Woese suffices—"Nothing in biology makes sense except in the light of evolution" (Dobzhansky 1973). Aptamers open our eyes to some of the possibilities.

2 THE HISTORY OF SELEX AND APTAMERS

Let us first provide a little history. Craig Tuerk was finishing his PhD thesis at the University of Colorado, and had taken on the task of more deeply understanding the nature of the "translational operator" within the bacteriophage T4 gene 43 mRNA. A hairpin and the Shine and Dalgarno domain just 5′ to the initiating AUG of the gene 43 mRNA is the RNA motif that is bound by the gene 43 protein (the replicative enzyme encoded by T4) to repress further synthesis of the polymerase when the level of replication is appropriate. Craig decided to mutate completely the hairpin loop within that motif. Those eight nucleotides were the focus of the first SELEX experiment. That first SELEX experiment (Tuerk and Gold 1990) yielded two winning hairpins among the ∼65,000 sequences of length eight—the wild type T4 sequence and another containing four changes (a quadruple "mutation" over eight nucleotides). Those four changes appeared to reduce the loop size from eight nucleotides to four nucleotides, even though the two hairpins bound with the same affinities to the gene 43 protein.

These experiments defined the SELEX process. The resulting ligands were coined "aptamers" (derived from the Greek word *aptus*; "to fit") by Andy Ellington and Jack Szostak in independent work that devised the same general strategy (Ellington and Szostak 1990; see also Green et al. 1990). The surprising data on the gene 43 mRNA motif drove us to generalize from the SELEX method to useful (in a commercial sense) single-stranded oligonucleotide shapes that could be identified through SELEX.

At some point shortly after Craig's paper, we expanded the randomized domain from eight nucleotides to 30 or 40 [and, later, at NeXstar, 50 to prove a point (Jellinek et al. 1993)], reasoning that one must access 30 randomized nucleotides or more to provide sufficient length to generate hairpins, G-quartets, bulges, and pseudo-knots. We thought then, and largely think today (but see later), that the helical regions of aptamers provide stable secondary structures that allow loops and other single-stranded regions to "collapse" into whatever three-dimensional shape is most likely. This thinking was influenced by CUUCGG hairpins (Tuerk et al. 1988), data on other common "tetra-loops" (Woese et al. 1983), and the extraordinary structures within the loops of tRNAs (Robertus et al. 1974; Suddath et al. 1974). We studied many proteins quickly, as targets, and reached the conclusion that SELEX would yield aptamers to many (if not all) proteins. Before the word "aptamer" became widely used, Craig had chosen the words "nucleic acid antibodies" (meaning antibodies made *out of* nucleic acids and NOT antibodies to nucleic acids—we are all happy that the word "aptamer" survived).

The history continued with the creation of NeXagen in 1992; NeXagen became, after some biotech stuff, a company called NeXstar. NeXagen and NeXstar were dedicated to the development of aptamers as therapeutic agents, exactly analogous to antibodies or antibody mimics. Many good aptamers were identified, some of which are in clinical development today. The first aptamer taken into the clinic was NX1838 (now called Macugen), a modified RNA aptamer with a low K_d for Vascular Endothelial Growth Factor and an activity that prevented $VEGF_{165}$ from binding to its high affinity receptors. NX1838 is a VEGF antagonist, and thus an angiogenesis inhibitor (Ruckman et al. 1998). NX1838 was tested against the wet form of age-related macular degeneration (ARMD), and, in the midst of that trial, NeXstar was acquired by Gilead (Gragoudas et al. 2004; Gonzales 2005). Shortly thereafter the therapeutic rights to NX1838 were licensed to Eyetech who finished the clinical development, renamed the compound Macugen, and after FDA approval started selling the drug in about January, 2005 (Doggrell 2005). Macugen would have been a commercial as well as financial success except that Macugen was selected specifically to target the VEGF isoform $VEGF_{165}$ and does not antagonize isoform $VEGF_{121}$ because the binding site of Macugen is missing in the shorter protein. Lucentis (and Avastin), two slightly different antibodies aimed at $VEGF_{121}$ and also $VEGF_{165}$ beat Macugen in the market based on more rapid and complete clinical response. Although $VEGF_{165}$ was clearly the predominant and most active isoform, the role of $VEGF_{121}$ in ARMD was not fully elucidated when Macugen was taken into the clinic (Kaiser 2006). Nevertheless, as concerns emerge that pan-VEGF inhibition is associated with serious cardiovascular and CNS events, it is worth noting that selective inhibition of only $VEGF_{165}$ may still be a useful option for long-term therapy of ARMD, since some VEGF activity is now known to be required for maintenance of normal blood vessels and retinal neurons (Nishijima et al. 2007; Saint-Geniez et al. 2009).

Several other aptamers are now in clinical development: AS1411 from Antisoma that targets nucleolin for acute myeloid leukemia and renal cell carcinoma; REG1 from Regado that targets Factor IX for coronary artery bypass graft and percutaneous coronary intervention; ARC1779 from Archemix that targets VWF for thrombotic microangiopathies and thrombocytopenic purpura; NU172 from ARCA that targets thrombin for coronary artery bypass graft and percutaneous coronary intervention; E10030 from Ophthotech that targets PDGF-B for ARMD and diabetic retinopathy; ARC1905 from Archemix that targets C5 for ARMD; and NOX-E36 from Noxxon that targets MCP-1 for kidney disease. Several companies including SomaLogic are engaged in developing different versions of aptamers. For example, Archemix and Ophthotech are working on RNA aptamers and NOXXON Pharma is working on spiegelmers, which are mirror-image L-RNA aptamers (Klussmann et al. 1996). Most, but not all of the aptamers in clinical development are modified RNA molecules (modified so as to have slow degradation rates from endogenous human RNases). The present therapeutic market for monoclonal antibodies (with which aptamers would compete, both being aimed largely at extracellular target molecules) is about $35B, and within that growing market there are opportunities for highly specific antagonists. The dominant patent position, staked out in a patent (Gold and Tuerk 1993) approved by the US Patent Office in 1993, will end over the next few years.

3 PROTEOMICS: DRIVING SELEX TO THE BEST POSSIBLE "WINNERS"

In the last few years at NeXstar, one of us became convinced that a major aptamer value was proteomics. We thought extensively about the use of ELISAs for measuring single analytes, and understood the value of using a sandwich of two good monoclonal antibodies to measure a rare protein in plasma, for example. That value, simply stated, is that one can multiply the specificity of each monoclonal for the intended analyte and thus ignore far more abundant proteins toward which the two monoclonal antibodies have higher K_d's. A great ELISA will quantify a nonabundant analyte below 1 pM (and sometimes at 10 fM) in plasma. Although not often stated explicitly, the issue in any assay in complex matrices is noise, not signal, and sandwich assays are intended to reduce noise (Zichi et al. 2008). Nucleic acid biochemists understand this concept deeply—"nested" PCR using contiguous primer pairs is a form of a "sandwich assay" in which specificities can be multiplied.

We were not concerned about the speed of doing SELEX. The SELEX literature contains methods to do SELEX in fewer rounds than we have found to be optimal (Tok and Fischer 2008; Lou et al. 2009), with some methods aiming to find good aptamers in a single round. We doubt that such protocols can work—if the best molecules in a library are present at a frequency of 10^{-9} to 10^{-13} (Gold 1995) or even lower, it is very difficult to discard all losers in a single round. It is, however, entirely possible to lose interesting sequences that may initially exist as a single copy, especially in the first round, because starting random libraries of 10^{14} to 10^{15} molecules are typically used for practical reasons. Furthermore, the work one does after selection to characterize aptamers is so vast compared to the selections themselves (which we do anyway in 96-well plates with many targets processed in parallel), we see no serious value in speeding up the nonrate limiting piece of the work. The limitation is aptamer quality—the equilibrium and kinetic properties needed to achieve high specificity.

We defined a great aptamer as one that would provide the specificity (in plasma, for example) of a pair of great antibodies. A low K_d was not going to be sufficient *if one wants to use a single binding reagent*. One can see this through a concocted example. Imagine that albumin in blood is present at 1 mM and that one wants to measure some analyte present at 1 pM. But then imagine that the capture agent (the monoclonal antibody or the aptamer) has a 1 pM K_d (which would be an exceptionally good K_d) for the intended analyte and binds albumin with the horrible affinity of 1 mM (merely a kiss in time). In that situation all measurements of the intended protein would actually measure half intended protein and half albumin. Let us call this the "albumin" problem (and later we will mention the "growth factor" problem, which is even more interesting). One sees immediately why ELISA formats were developed.

Thus we needed a second element of specificity intrinsic to that single capture aptamer, along with equilibrium-based discrimination. That is, we wanted the qualities of a sandwich in a format that used but a single reagent. We examined the original SELEX process exhaustively. The details are published within a patent application (Zichi et al. 2009) as well as papers in press and winding their way through the review process (Keeney et al. 2009; Ostroff et al. 2009). We have learned how to drive aptamer selection to the lowest possible K_d for a given oligonucleotide library, using five-position modified pyrimidines in the libraries. Modified pyrimidines have played an enormous role in the successful enhancement of aptamers. Bruce Eaton's original five-position modifications (Dewey et al. 1995) have been expanded to include both amino acid side-chain adducts (Vaught et al. 2004; Vaught et al. 2010) and more recently adducts that resemble "fragment-based pharmacophores" from the drug industry (Congreve et al. 2008;

J. Rohloff, personal communication). We have found many examples of so-called recalcitrant proteins that have yielded lovely aptamers using new oligonucleotide libraries (Zichi et al. 2009).

We also explored second elements of specificity for those modified aptamers because the "albumin" problem is not solved by lower K_ds alone, even though we routinely obtain K_ds between 10 pM and 100 pM. We have added as a second element of specificity a kinetic component on top of the equilibrium component. Many years ago Hopfield and Ninio independently elaborated the idea of kinetic proofreading (Hopfield 1974; Ninio 1975; Hopfield et al. 1976).

The central idea of kinetic proofreading was primarily concerned with specificity that was enhanced by using ATP or GTP hydrolysis as a method to separate two equilibria events from each other so that one could (almost) use the same binding differentiation twice. We used a version of this thinking: A component of the binding reaction (slow dissociation) can be used after using equilibrium discrimination. We have been able to select aptamers with remarkably slow dissociation rate constants, allowing us to do "kinetic challenges" during SELEX (both by simple dilution and also by incubations with alternative polyanions such as dextran sulfate at high concentration) (Schneider et al. 2009; Zichi et al. 2009).

These new aptamers are so important to the applications we study (biomarker discovery in blood, pathology, and in vivo imaging) that we have renamed them SOMAmers (SOMA; Slow Off-rate Modified Aptamers). The new name helps us distinguish and compare our data with a huge prior literature (Famulok et al. 2007; Mayer 2009).

4 SOMAMER SPECIFICITY

The requirement for our applications, including for the development of "magic bullet" therapeutics, is little to no off-target binding when both equilibrium and kinetic challenges are employed. We have shown that SOMAmers pull down largely the intended analytes from very complex biological matrices. From only these data it is clear that we have solved the "albumin" problem outlined above—in fact, abundant weak-binding proteins in plasma are the IgM's (whose concentrations in plasma are about micromolar, and which are found as multimers, and which have patches of lysines and arginines that bind nucleic acids nonspecifically). Gratifyingly IgMs are largely lost in the pull downs that are a part of the proteomics protocols.

But there are other proteins to fear in biological matrices. Growth factors almost always have heparan sulfate binding sites (even more concentrated patches of lysines and arginines than are present within the IgMs)—those sites are used by growth factors to bind loosely to the external surfaces of cells (which are negatively charged) from which point two-dimensional diffusion on the surface of the cell allows growth factors to find their high affinity receptors (Lieleg et al. 2009). This two-step method for binding was first understood by Peter von Hippel to be the "reduction of the dimensionality of the search" (a phrase he used to describe the diffusion along double-stranded DNA by the *lac* repressor as it sought its operator) (Von Hippel et al. 1982). Based on the work of Five Prime (Lin et al. 2008) we have an estimate of the number of such secreted proteins in humans, and the number is large: 3400! These proteins represent a SOMAmer-friendly collection of low abundant proteins, but in sum they represent a sink to which SOMAmers or other polyanions might bind. We call this the "growth factor" problem.

The heparan sulfate binding sites of the secreted human proteome have quite different structures from each other because the low affinity binding to cells requires nothing more than weak ionic interactions. That is, the quite different heparan sulfate binding sites on a few thousand human proteins are likely to have randomly disposed positive charges because the target (the outside surface of a cell) has an enormous local concentration of flexible, negative charges. Some of these proteins have significant binding to nontarget aptamers and SOMAmers, but most of those proteins dissociate from noncognate SOMAmers during kinetic challenge. That is, the kinetic challenge step removes the unintended "growth factor" molecules from binding to SOMAmers nonspecifically.

We have published some thoughts about the difficulties with antibodies for biomarker discovery using arrays (Zichi et al. 2008). Our exhaustive attack on the specificity problem has made possible reagents that can be used alone in a "conceptual" sandwich, solving both the albumin and the "growth factor" problems simultaneously.

5 SOMAMER STRUCTURE

Our studies include an X-ray crystal structure for one SOMAmer bound to its protein target (Janjic and Jarvis, personal communication and manuscript in preparation). The SOMAmer was identified from a library of single-stranded DNAs in which every "T" was substituted with 5-benzyl-dUMP (Fig. 1) (Vaught et al. 2010). The SOMAmer has elements in its structure that have not been observed in single-stranded nucleic acids (which of course do not have access to the hydrophobic benzenes for whatever intramolecular folds might be needed, or for interactions with the target protein). In the structure are elements of benzene—amino acid interactions, benzene stacking on other nucleotides, and even a compact

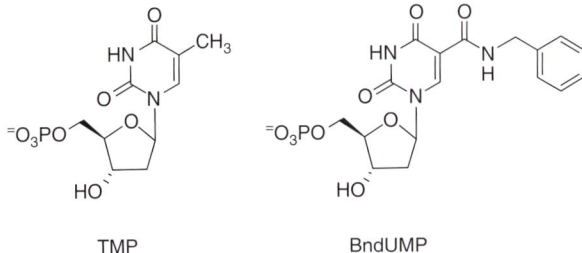

Figure 1. Thymadine-monophosphate (TMP) and the modified nucleotide 5-benzylaminocarbonyl-deoxyuridine-monophosphate (BndUMP).

hydrophobic "turn" that is remarkable. Apparently small modifications to standard nucleotide chemistry open up a new world of possible structures, an idea alluded to years ago by Harold B. White III (White 1976). Modification of the pyrimidines of oligonucleotide libraries is a continuing piece of the efforts to make better and better SOMAmers.

6 QUANTITATIVE PROBING OF THE PLASMA PROTEOME AT HIGH CONTENT AND LOW LIMITS OF DETECTION

The heart of our work is to find novel biomarkers to use for drug development and for diagnostics. We have devised an assay (Fig. 2) that allows simultaneous measurements of human proteins in plasma or serum (or other matrices) in a manner analogous to mRNA or microRNA profiling from cells or even blood (Derisi et al. 1996; Calin et al. 2004). We have adopted the tools for RNA profiling (and SNP and CNV profiling) for our assay—we measure the SOMAmers themselves after a set of simple biochemical steps that discard the unbound input SOMAmers and allow us to quantify only those SOMAmers that stay bound (through their slow dissociation rate constants) to their cognate proteins. We converted the proteomic exercise required for Biomarker Discovery into a hybridization or QPCR measurement.

This body of work (manuscripts submitted) yields array data that look exactly like the data obtained on a "DNA chip" constructed by, for example, Agilent, Affymetrix, Illumina, or NimbleGen/Roche because we use custom chips that have been printed with the complements of our SOMAmers.

We have found novel biomarkers for a variety of cancers, cardiovascular conditions, degenerative diseases, and more. The content today is "only" >800 SOMAmers (hence we have the capacity to measure >800 human proteins, simultaneously, using only about 15 μl of sample). We easily measure proteins at levels below 1 pM in plasma. Thus we achieved the primary objective we set for ourselves,

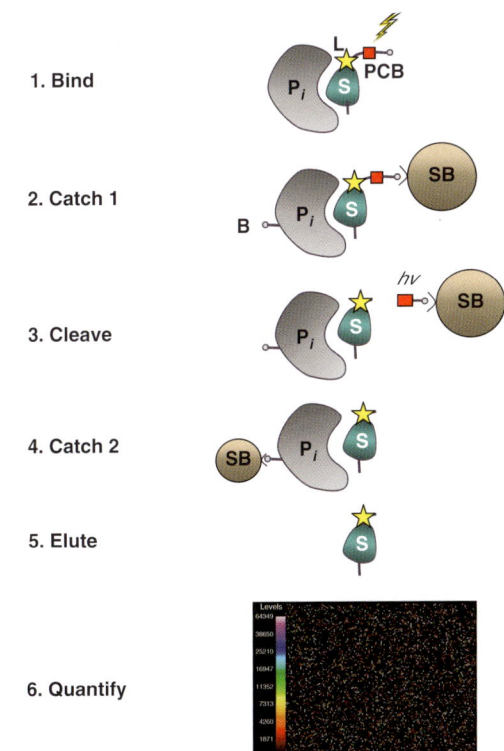

Figure 2. Overview of the SomaLogic proteomics assay. In step 1, the specific protein to be measured (P_i) binds tightly to its cognate SOMAmer binding molecule (S), which includes a photo-cleavable biotin (PCB) and fluorescent label (L) at the 5' end. In step 2, bound protein-SOMAmer complexes are captured onto streptavidin coated beads (SB) by the photo-cleavable biotin on the SOMAmer. Unbound proteins are washed away. Bound proteins are tagged with NHS-biotin (B). In step 3, the photo-cleavable biotin is cleaved by UV light ($h\nu$) and the protein-SOMAmer complexes are released into solution. In step 4, the protein-SOMAmer complexes are captured onto streptavidin coated magnetic beads and the SOMAmer are eluted into solution and recovered for quantification in step 6, hybridization to a custom DNA microarray. Each probe spot contains DNA with sequence complementary to a specific SOMAmer, and the fluorescent intensity of each probe spot is proportional to the amount of SOMAmer recovered, and thus directly proportional to the amount of protein present in the original sample.

which was to allow biomarker discovery in a way that could enhance drug development and medicine. We shall see over the next years if the effort we made will be as valuable for patients as we hoped when we started.

7 BACK TO THE RNA WORLD

We appreciate having a chance to tell our molecular biology friends what we have been up to for all these years. What started as an accident (Craig Tuerk's PhD thesis, and the isolation of what we called the major variant but which was in fact the first aptamer) and then became general (the small cottage industry, in universities and companies, doing SELEX) has the chance to be truly useful for healthcare, the fundamental reason we all enjoyed NIH funding over the years.

But of course there are lessons in our work for thinking about evolution. The most important lessons, now supported by a huge published literature (we include the patent literature, a literature that seems a bit obscure to most scientists), are that aptamers (or at least SOMAmers) can be identified for virtually any target molecule, large or small, protein or other.

Data on aptamers support only weakly notions of ubiquitous RNA-networks (e.g., Mattick 2003), which will require more data. Detailed studies of many creatures show that no region of any genome is entirely silent (Nagalakshmi et al. 2008). Although it is tempting to imagine functions for everything we find in a creature, and although it is obvious that evolution grabs noise *over time* and makes something useful, when we think about fancy networks it is a good idea to remember that noise is real in real things (Thattai and Van Oudenaarden 2001; Raser and O'Shea 2005).

But the aptamer literature is huge and compelling—one can find short single-stranded oligonucleotides that will bind to almost anything; and from binding one can imagine function. An oligonucleotide world seems sensible as a step during evolution, and we always think that early short oligonucleotides as drivers of evolution can be RNA-based, DNA-based, or even modified-oligo-based. Years ago the papers of Harold B. White (White 1976) sensibly posited a lovely idea that it was wrong to assume that the present nucleotides in RNA and DNA were the ones that were (in Woese's and Dobzhansky's way of thinking) the players during early evolution. Perhaps our use of SOMAmers, with many different pyrimidine adducts in our libraries, is analogous to White's idea that a variety of alternative nucleotides existed before simple genetics won so that life could cross Woese's Darwinian Threshold (Woese 2002).

The aptamer literature says "give me 30-40 random nucleotides within an oligo (of whatever chemistry) and binders are there if the number of molecules is large." The deepest question is why is this true, given the relatively uninteresting chemical qualities of modern nucleotides and the tiny (but with profound effect) pyrimidine adducts we have used. One way of thinking about aptamers and SOMAmers is as analogues of the conotoxins, those wonderful (and frightening) small peptides with remarkable affinities and specificities for various protein targets (Mondal et al. 2005; Halai and Craik 2009). Entropic cost upon binding is a problem solved by using non-mobile participants in a binding pair, and the conotoxins have solved the entropy problem through the use of several disulfide bonds to limit the flexibility of the peptides. Aptamers use base-pairing as the conotoxin disulfide bond equivalents, interactions within the aptamer that reduce flexibility and also provide a chance for the hairpin loops and bulges to settle into energetically favored structures. SOMAmers, with their modified nucleotides, use components more commonly thought to be from the domain of proteins.

8 CONCLUSIONS: WHAT OTHER FUNCTIONS MIGHT OLIGONUCLEOTIDES HAVE?

What have we missed in our studies of present day life? The problem is what to seek.

A huge new area, anticipated by aptamer and ribozyme research (Gold et al. 1997a; Gold et al. 1997b; Winkler et al. 2002), is that of the riboswitches (Winkler et al. 2002; Roth and Breaker 2009). From the aptamer-based capacities for binding to several small molecules, Breaker and colleagues imagined and found that bacteria use aptamers and ribozymes to do feedback regulation. They did have a huge advantage, which we want to mention here.

Bacteria have had more than three billion years of life on earth, which is much more than enough time for *every single base pair of every bacterial genome to be tested exhaustively for selective advantages*. Imagine that bacteria divided through the ages with a one day generation time, and that they had the present mutation rate of one base pair change per 10^6 base pairs per replication. Three billion years contain about 1×10^{12} d, and thus enough time for about 1×10^{12} bacterial replications. Thus bacteria may have (each) experienced 1×10^6 *mutations per base pair* since their beginnings. The word one might use is "hammered"—bacterial genomes have been hammered. Each base pair has been like a slow acting metronome—first a G:C to A:T transition, then (perhaps) a reversion to the original G:C, then a transversion, then...and so on, such that, *every single base pair in a bacterium has been changed to something else many many times*.

One sees immediately the power of conserved RNA sequences and structures in bacteria. If all bacterial genomes are hammered, what remains in common must be important. This simple thought provoked Carl Woese to make his outstanding contributions (Woese and Fox 1977; Woese 1987, 1998, 2000, 2002). Breaker and his colleagues expanded this idea with their discovery of small conserved RNA sequences that also do something profound—in fact one would argue that the number 1×10^6 mutations per base pair is the underlying power behind all searches for useful RNA sequences using genomic comparisons in bacteria (Eddy 2005).

Are there additional undiscovered functions for RNAs (or even single-stranded DNAs) that we might imagine? The successful creation of aptamers has made every graduate student audience since 1990 say, at aptamer seminars, "why doesn't the immune system use aptamers instead of antibodies?" Of course this is an example of a question that can only be answered with an evolutionary perspective. If mammals carried the SELEX libraries we use as their information content for an oligonucleotide-based immune system (the genomic-size of a typical SELEX experiment is 10^{15} times 40 base pairs—that is a lot of genome to use to fight off pathogens), that immune system would require a larger genome than that of any creature we know today (by about a factor of a million). We ought to give credit to the present protein-based immune system for the cleverness with which diversity and binding selectivity is built!

However, somewhere on this planet there might well be a creature who responds with a highly mutable expression cassette of little "aptamers" to bind to and inactivate some invading creature. Experiments have been done to test artificial variations on this theme—so-called decoys with aptamers, expressed inside cells to "immunize" against viruses (Tuerk et al. 1992; Kumar et al. 1997), although no natural examples have been found in the biosphere.

One might also imagine that eukaryotic mRNAs will contain aptamers on their 5′ or 3′ ends (or even internally, using codon choices to co-evolve aptamers) that help localize the proteins expressed by the ribosomes and those ribosomes to specific areas on the inside of cells. It is also just another way to "reduce the dimensionality of the search"—a common problem in biology (Von Hippel et al. 1982). The IRES sequences allow internal translational initiations on polycistronic mRNAs (Kieft et al. 2002; Pfingsten et al. 2006).

One might also imagine that all those proteins we think of as nonspecific nucleic acid binding proteins will nucleate on specific aptamer-like sequences. Aptamers have been selected against many such proteins [ribosomal protein S1 from *Escherichia coli* was an early example (Tuerk and Gold 1990)]. In principle any protein that is a nonspecific binder to RNA or DNA, single or double stranded, has to prefer some sequence over others. Even a perfect "backbone binder" will be influenced by the precise interphosphate distances, which will in turn be influenced by the bases themselves. We proposed "genomic SELEX" as a way to get to unexpected biology (Gold et al. 1997a)—we never embarked seriously on that work because of our focus on the medical potential. And so it goes.

Finally, and most strikingly, consider the extraordinary work from Eaton and Feldheim (Gugliotti et al. 2004, 2005; Liu et al. 2006; Feldheim and Eaton 2007). In a paper published in Science, using "catalytic aptamers" made from modified RNA libraries, these authors extended the reach of oligonucleotides to an entirely new realm (Gugliotti et al. 2004). The selected catalytic aptamers were able to recruit metals from solution, to nucleate crystal growth of specific crystal forms, and to do so in a sequence dependent manner. Should this work generalize, one might imagine a future research area of *intracellular nanotechnology microfabrication* catalyzed by oligonucleotides. Perhaps the Eaton-Feldheim work will lead us down the pathway toward an organic-inorganic fusion within modern fabrication and even biology.

Our collective thesis has two components. First, evolutionary time is long, and only in the hammered genomes of bacteria do we see the general proposition that what we have learned from in vitro evolution experiments such as SELEX is common in biology. But some of what we see today in our test tubes will have had that key stochastic accidental moment and will have been frozen and survived. Our tasks include wondering how to find those phenomena without the insights that flow from comparative bacterial genome sequences. But we argue that just because the easier discovery methods are unlikely to work, there are examples of present day oligonucleotide uses that will continue to be uncovered.

Second, modified pyrimidines change the entire reach of SELEX. As long as one can solve the replication problem (which has been done because many DNA polymerases are forgiving with respect to pyrimidine nucleotides modified at their five-positions with rather small adducts), one can change the characteristics and qualities of aptamers toward SOMAmers and so provide high quality reagents for many medical applications.

ACKNOWLEDGMENTS

We thank all of our colleagues who participated in the genesis of aptamer science and building the aptamer biotechnology industry over the past decades. They have contributed immeasurably to this endeavor and made countless insights about aptamers and the nature of biology. We especially thank past and present colleagues at the University of Colorado, NeXstar Pharmaceuticals, and SomaLogic. Finally, we thank the editors for their patient and skilled guidance in the preparation of this manuscript.

REFERENCES

Calin GA, Liu CG, Sevignani C, Ferracin M, Felli N, Dumitru CD, Shimizu M, Cimmino A, Zupo S, Dono M, et al. 2004. MicroRNA profiling reveals distinct signatures in B cell chronic lymphocytic leukemias. *Proc Natl Acad Sci U S A* **101:** 11755–11760.

Congreve M, Chessari G, Tisi D, Woodhead AJ. 2008. Recent developments in fragment-based drug discovery. *J Med Chem* **51:** 3661–3680.

Derisi J, Penland L, Brown PO, Bittner ML, Meltzer PS, Ray M, Chen Y, Su YA, Trent JM. 1996. Use of a cdna microarray to analyse gene expression patterns in human cancer. *Nat Genet* **14:** 457–460.

Dewey T, Mundt A, Crouch G, Zyniewski M, Eaton B. 1995. New uridine derivatives for systematic evolution of rna ligands by exponential enrichment. *J Am Chem Soc* **117:** 8474–8475.

Dobzhansky T. 1973. Nothing in biology makes sense except in the light of evolution. *Am Bio Teach* **35:** 125–129.

Doggrell SA. 2005. Pegaptanib: The first antiangiogenic agent approved for neovascular macular degeneration. *Expert Opin Pharmaco* **6:** 1421–1423.

Eddy SR. 2005. A model of the statistical power of comparative genome sequence analysis. *PLoS Biol* **3:** 95–102.

Ellington AD, Szostak JW. 1990. In vitro selection of RNA molecules that bind specific ligands. *Nature* **346:** 818–822.

Famulok M, Hartig JS, Mayer G. 2007. Functional aptamers and aptazymes in biotechnology, diagnostics, and therapy. *Chem Rev* **107:** 3715–3743.

Feldheim DL, Eaton BE. 2007. Selection of biomolecules capable of mediating the formation of nanocrystals. *ACS Nano* **1:** 154–159.

Gold L. 1995. Oligonucleotides as research, diagnostic, and therapeutic agents. *J Biol Chem* **270:** 13581–13584.

Gold L, Tuerk C. 1993. Methods for identifying nucleic acid ligands. US5270163.

Gold L, Brown D, He Y, Shtatland T, Singer BS, Wu Y. 1997a. From oligonucleotide shapes to genomic selex: Novel biological regulatory loops. *Proc Natl Acad Sci* **94:** 59–64.

Gold L, Singer B, He YY, Brody E. 1997b. Selex and the evolution of genomes. *Curr Opin Genet Dev* **7:** 848–851.

Gonzales CR. 2005. Enhanced efficacy associated with early treatment of neovascular age-related macular degeneration with pegaptanib sodium: An exploratory analysis. *Retina* **25:** 815–827.

Gragoudas ES, Adamis AP, Cunningham ET Jr, Feinsod M, Guyer DR. 2004. Pegaptanib for neovascular age-related macular degeneration. *N Engl J Med* **351:** 2805–2816.

Green R, Ellington AD, Szostak JW. 1990. In vitro genetic analysis of the tetrahymena self-splicing intron. *Nature* **347:** 406–408.

Gugliotti LA, Feldheim DL, Eaton BE. 2004. Rna-mediated metal-metal bond formation in the synthesis of hexagonal palladium nanoparticles. *Science* **304:** 850–852.

Gugliotti LA, Feldheim DL, Eaton BE. 2005. Rna-mediated control of metal nanoparticle shape. *Journal of the Americane Chemical Society* **127:** 17814–17818.

Halai R, Craik DJ. 2009. Conotoxins: Natural product drug leads. *Nat Prod Rep* **26:** 526–536.

Hopfield JJ. 1974. Kinetic proofreading: A new mechanism for reducing errors in biosynthetic processes requiring high specificity. *Proc Natl Acad Sci* **71:** 4135–4139.

Hopfield JJ, Yamane T, Yue V, Coutts SM. 1976. Direct experimental evidence for kinetic proofreading in amino acylation of trnaile. *Proc Natl Acad Sci* **73:** 1164–1168.

Jellinek D, Lynott CK, Rifkin DB, Janjic N. 1993. High-affinity RNA ligands to basic fibroblast growth factor inhibit receptor binding. *Proc Natl Acad Sci* **90:** 11227–11231.

Kaiser PK. 2006. Antivascular endothelial growth factor agents and their development: Therapeutic implications in ocular diseases. *Am J Ophthalmol* **142:** 660–668.

Keeney T, Kraemer S, Walker JJ, Bock C, Vaught J, Nikrad M, Stewart A, Lollo B, Stanton M, Gold L. 2009. Automation of the somalogic proteomics assay: A platform for biomarker discovery. *J Assoc Lab Automat* **14:** 360–366.

Kieft JS, Zhou K, Grech A, Jubin R, Doudna JA. 2002. Crystal structure of an RNA tertiary domain essential to hcv ires-mediated translation initiation. *Nat Struct Biol* **9:** 370–374.

Klussmann S, Nolte A, Bald R, Erdmann VA, Furste JP. 1996. Mirror-image RNA that binds d-adenosine. *Nat Biotechnol* **14:** 1112–1115.

Kumar PK, Machida K, Urvil PT, Kakiuchi N, Vishnuvardhan D, Shimotohno K, Taira K, Nishikawa S. 1997. Isolation of RNA aptamers specific to the ns3 protein of hepatitis c virus from a pool of completely random RNA. *Virology* **237:** 270–282.

Lieleg O, Baumgartel RM, Bausch AR. 2009. Selective filtering of particles by the extracellular matrix: An electrostatic bandpass. *Biophys J* **97:** 1569–1577.

Lin H, Lee E, Hestir K, Leo C, Huang M, Bosch E, Halenbeck R, Wu G, Zhou A, Behrens D, et al. 2008. Discovery of a cytokine and its receptor by functional screening of the extracellular proteome. *Science* **320:** 807–811.

Liu D, Gugliotti LA, Wu T, Dolska M, Tkachenko AG, Shipton MK, Eaton BE, Feldheim DL. 2006. Rna-mediated synthesis of palladium nanoparticles on au surfaces. *Langmuir* **22:** 5862–5866.

Lou X, Qian J, Xiao Y, Viel L, Gerdon AE, Lagally ET, Atzberger P, Tarasow TM, Heeger AJ, Soh HT. 2009. Micromagnetic selection of aptamers in microfluidic channels. *Proc Natl Acad Sci* **106:** 2989–2994.

Mattick JS. 2003. Challenging the dogma: The hidden layer of non-protein-coding RNAs in complex organisms. *Bioessays* **25:** 930–939.

Mayer G. 2009. The chemical biology of aptamers. *Angew Chem Int Ed Engl* **48:** 2672–2689.

Mondal S, Vijayan R, Shichina K, Babu RM, Ramakumar S. 2005. I-superfamily conotoxins: Sequence and structure analysis. *In Silico Biol* **5:** 557–571.

Nagalakshmi U, Wang Z, Waern K, Shou C, Raha D, Gerstein M, Snyder M. 2008. The transcriptional landscape of the yeast genome defined by RNA sequencing. *Science* **320:** 1344–1349.

Ninio J. 1975. Kinetic amplification of enzyme discrimination. *Biochimie* **57:** 587–595.

Nishijima K, Ng YS, Zhong L, Bradley J, Schubert W, Jo N, Akita J, Samuelsson SJ, Robinson GS, Adamis AP, et al. 2007. Vascular endothelial growth factor-a is a survival factor for retinal neurons and a critical neuroprotectant during the adaptive response to ischemic injury. *Am J Pathol* **171:** 53–67.

Ostroff R, Foreman T, Keeney TR, Stratford S, Walker JJ, Zichi D. 2009. The stability of the circulating human proteome to variations in sample collection and handling procedures measured with an aptamer-based proteomics array. *Journal of Proteomics*.

Pfingsten JS, Costantino DA, Kieft JS. 2006. Structural basis for ribosome recruitment and manipulation by a viral ires rna. *Science* **314:** 1450–1454.

Raser JM, O'Shea EK. 2005. Noise in gene expression: Origins, consequences, and control. *Science* **309:** 2010–2013.

Robertus JD, Ladner JE, Finch JT, Rhodes D, Brown RS, Clark BF, Klug A. 1974. Structure of yeast phenylalanine tRNA at 3 a resolution. *Nature* **250:** 546–551.

Roth A, Breaker RR. 2009. The structural and functional diversity of metabolite-binding riboswitches. *Annu Rev Biochem* **78:** 305–334.

Ruckman J, Green LS, Beeson J, Waugh S, Gillette WL, Henninger DD, Claesson-Welsh L, Janjic N. 1998. 2′-fluoropyrimidine rna-based aptamers to the 165-amino acid form of vascular endothelial growth factor (vegf165). Inhibition of receptor binding and vegf-induced vascular permeability through interactions requiring the exon 7-encoded domain. *J Biol Chem* **273:** 20556–20567.

Saint-Geniez M, Kurihara T, Sekiyama E, Maldonado AE, D'amore PA. 2009. An essential role for rpe-derived soluble vegf in the maintenance of the choriocapillaris. *Proc Natl Acad Sci* **106:** 18751–18756.

Schneider D, Nieuwlandt D, Eaton B, Stanton M, Gupta S, Kraemer S, Zichi D, Gold L. 2009. Multiplexed analyses of test samples. US2009/0042206.

Suddath FL, Quigley GJ, Mcpherson A, Sneden D, Kim JJ, Kim SH, Rich A. 1974. Three-dimensional structure of yeast phenylalanine transfer RNA at 3.0angstroms resolution. *Nature* **248:** 20–24.

Thattai M, Van Oudenaarden A. 2001. Intrinsic noise in gene regulatory networks. *Proc Natl Acad Sci* **98:** 8614–8619.

Tok JB, Fischer NO. 2008. Single microbead selex for efficient ssdna aptamer generation against botulinum neurotoxin. *Chem Commun (Camb)* 1883–1885.

Tuerk C, Gold L. 1990. Systematic evolution of ligands by exponential enrichment: RNA ligands to bacteriophage t4 DNA polymerase. *Science* **249:** 505–510.

Tuerk C, Macdougal S, Gold L. 1992. RNA pseudoknots that inhibit human immunodeficiency virus type 1 reverse transcriptase. *Proc Natl Acad Sci* **89:** 6988–6992.

Tuerk C, Gauss P, Thermes C, Groebe DR, Gayle M, Guild N, Stormo G, D'aubenton-Carafa Y, Uhlenbeck OC, Tinoco I Jr, et al. 1988. Cuucgg hairpins: Extraordinarily stable RNA secondary structures associated with various biochemical processes. *Proc Natl Acad Sci* **85:** 1364–1368.

Vaught JD, Dewey T, Eaton BE. 2004. T7 RNA polymerase transcription with 5-position modified utp derivatives. *J Am Chem Soc* **126:** 11231–11237.

Vaught J, Bock C, Carter J, Fitzwater T, Otis M, Schneider D, Rolando J, Waugh S, Wilcox SK, Eaton B. 2010. Expanding the chemistry of DNA for in vitro selection. *J Am Chem Soc* **132:** 4141–4151.

Von Hippel PH, Kowalczykowski SC, Lonberg N, Newport JW, Paul LS, Stormo GD, Gold L. 1982. Autoregulation of gene expression. Quantitative evaluation of the expression and function of the bacteriophage t4 gene 32 (single-stranded DNA binding) protein system. *J Mol Biol* **162:** 795–818.

White HB 3rd. 1976. Coenzymes as fossils of an earlier metabolic state. *J Mol Evol* **7:** 101–104.

Winkler W, Nahvi A, Breaker RR. 2002. Thiamine derivatives bind messenger RNAs directly to regulate bacterial gene expression. *Nature* **419:** 952–956.

Woese CR. 1987. Bacterial evolution. *Microbiol Rev* **51:** 221–271.

Woese CR. 1998. The universal ancestor. *Proc Natl Acad Sci* **95:** 6854–6859.

Woese CR. 2000. Interpreting the universal phylogenetic tree. *Proc Natl Acad Sci* **97:** 8392–8396.

Woese CR. 2002. On the evolution of cells. *Proc Natl Acad Sci* **99:** 8742–8747.

Woese CR, Fox GE. 1977. Phylogenetic structure of the prokaryotic domain: The primary kingdoms. *Proc Natl Acad Sci* **74:** 5088–5090.

Woese CR, Gutell R, Gupta R, Noller HF. 1983. Detailed analysis of the higher-order structure of 16s-like ribosomal ribonucleic acids. *Microbiol Rev* **47:** 621–669.

Zichi D, Eaton B, Singer B, Gold L. 2008. Proteomics and diagnostics: Let's get specific, again. *Curr Opin Chem Biol* **12:** 78–85.

Zichi D, Wilcox SK, Bock C, Schneider DJ, Eaton B, Gold L. 2009. Method for generating aptamers with improved off-rates. US2009/0004667.

In Vivo RNAi: Today and Tomorrow

Norbert Perrimon[1,2], Jian-Quan Ni[1,2], and Lizabeth Perkins[3]

[1]Department of Genetics, Harvard Medical School, Boston, Massachusetts 02115
[2]Howard Hughes Medical Institute, Harvard Medical School, Boston, Massachusetts 02175
[3]Pediatric Surgical Research Labs, Massachusetts General Hospital, Harvard Medical School, Boston, Massachusetts 02114

Correspondence: perrimon@receptor.med.harvard.edu

SUMMARY

RNA interference (RNAi) provides a powerful reverse genetics approach to analyze gene functions both in tissue culture and in vivo. Because of its widespread applicability and effectiveness it has become an essential part of the tool box kits of model organisms such as *Caenorhabditis elegans*, *Drosophila*, and the mouse. In addition, the use of RNAi in animals in which genetic tools are either poorly developed or nonexistent enables a myriad of fundamental questions to be asked. Here, we review the methods and applications of in vivo RNAi to characterize gene functions in model organisms and discuss their impact to the study of developmental as well as evolutionary questions. Further, we discuss the applications of RNAi technologies to crop improvement, pest control and RNAi therapeutics, thus providing an appreciation of the potential for phenomenal applications of RNAi to agriculture and medicine.

Outline

1. Introduction
2. RNAi reagents for in vivo screening
3. In vivo RNAi screening in *C. elegans*
4. In vivo RNAi screening in *Drosophila*
5. Applications of in vivo RNAi to the developmental biology of emerging systems and evo-devo
6. In vivo RNAi in vertebrate models
7. In vivo RNAi in plants: applications toward crop improvement
8. In vivo RNAi applications toward virus and pest control
9. In vivo RNAi applications in medicine
10. Concluding remarks

References

1 INTRODUCTION

Genetic screening is one of the most powerful methods available for gaining insights into complex biological processes. Indeed, much of what we have learned from model organisms such as *Saccharomyces cerevisiae*, *Caenorhabditis elegans*, and *Drosophila melanogaster* can be traced back to genetic screens designed to identify sets of mutations that perturb specific processes. For example, screens in yeast have led to the identification of key regulators of the cell cycle (Hartwell et al. 1974); screens in *C. elegans* have identified the genetic regulation of programmed cell death (Horvitz et al. 1999); and screens for mutations that cause embryonic lethality in *Drosophila* have elucidated the logic of body patterning in a multicellular organism (Nusslein-Volhard and Wieschaus 1980; St Johnston and Nusslein-Volhard, 1992).

Over the years many improvements and tools for genetic manipulation have become available, and as a result there now exist powerful "tool-boxes" for each model organism (Nagy et al. 2003; Venken and Bellen 2005; Kaletta and Hengartner 2006). Sophistication in approaches and tools facilitates the ease of genetic screening as well as the identification of genetic alteration(s) and requisite follow-up analyses of mutant phenotypes. For example, innovations such as mosaic analysis and tissue-specific expression of transgenes have allowed researchers to study gene function in a wider variety of tissues, stages, and contexts.

Soon after the initial discovery by Fire and Mello (Fire et al. 1998) that double-stranded RNAs (dsRNAs) can be used to knockdown the activity of individual genes, many RNA interference (RNAi)-based methods were (and continue to be) added to the tool-boxes of various organisms. These methods have truly revolutionized the field of functional genomics because of their relative ease, and most significantly, because RNAi, unlike more traditional genetic screening methods, provides a powerful reverse genetic approach, especially for organisms in which genetics is difficult, as is the case with mammalian systems. Importantly, the power of RNAi-based methods for genetic analyses became fully realized when the genome sequences of various organisms were completed (*C. elegans* Sequencing Consortium, 1998; Adams et al. 2000; Venter et al. 2001; Waterston et al. 2002; The Rat Genome Sequencing Project Consortium, 2004). Thus, the identification of all genes in the *C. elegans*, *Drosophila*, mouse, rat, and human genomes has led to the construction of numerous genome-wide RNAi resources, allowing reverse genetic screens either in tissue culture or in vivo. Today, genome-wide RNAi screening is possible in vivo in *C. elegans*, in tissue culture cells and in vivo in *Drosophila*, and in cell lines from mice, rats, and humans.

RNAi is a well-established tool for studies in tissue culture and, following the first genome-wide RNAi screen performed in *Drosophila* cells (Boutros et al. 2004), RNAi high-throughput screening (HTS) has become routine both in *Drosophila* and mammalian cells. Cell-based screening has been extensively reviewed in the past (Echeverri and Perrimon 2006; Perrimon and Mathey-Prevot, 2007; Boutros and Ahringer 2008; Mohr et al. 2011). In this review, we focus on in vivo methods and applications of RNAi. In most organisms, methods for in vivo RNAi are still in development and we discuss the state of the field, what has been learned so far, and future development. In particular, we describe the application of in vivo RNAi to characterize the function of pleiotropic genes and discuss its impact for the study of organisms for which genetic tools are either nonexistent or poorly developed.

2 RNAi REAGENTS FOR IN VIVO SCREENING

Four different types of RNAi reagents are used for in vivo studies: synthetic siRNAs, small hairpin RNAs (shRNAs), small hairpin microRNAs (shmiRNAs), and long dsRNAs (reviews by Echeverri and Perrimon 2006; Lee and Kumar 2009) (Fig. 1).

Synthetic siRNAs are small RNA duplexes composed of 19 complementary base pairs (bps) and 2-nucleotide 3′ overhangs. They are transfected into cells or injected into animals. On entering cells one strand of the siRNA duplex is incorporated into the multi-subunit ribonucleoprotein complex (RISC) and directs RISC to the target mRNA by complementary base-pairing, resulting in mRNA degradation. The effects of the siRNAs are transient, especially in actively dividing cells.

In contrast, shRNA and shmiRNA-synthesizing vectors allow for controlled or continuous expression of small transcripts in the cell that contain both the sense and antisense strand complementary to the selected mRNA target. They are either transfected into cells as plasmid DNAs or delivered using viral particles, and are maintained as extra-chromosomal copies or stably integrated in the genome as transgenes. The 50–70 bps single-stranded RNA transcripts fold back to form a stem-loop structure. ShRNAs are processed in the cytoplasm by the ribonuclease Dicer to generate siRNAs. ShmiRNAs are a variation of shRNAs in that sequences for the silencing trigger are embedded in an endogenous miRNA expression cassette. ShmiRNAs therefore exploit the endogenous microRNA pathway for the biogenesis and subsequent loading of siRNAs into RISC, and are usually more effective in knocking down target mRNAs than shRNAs.

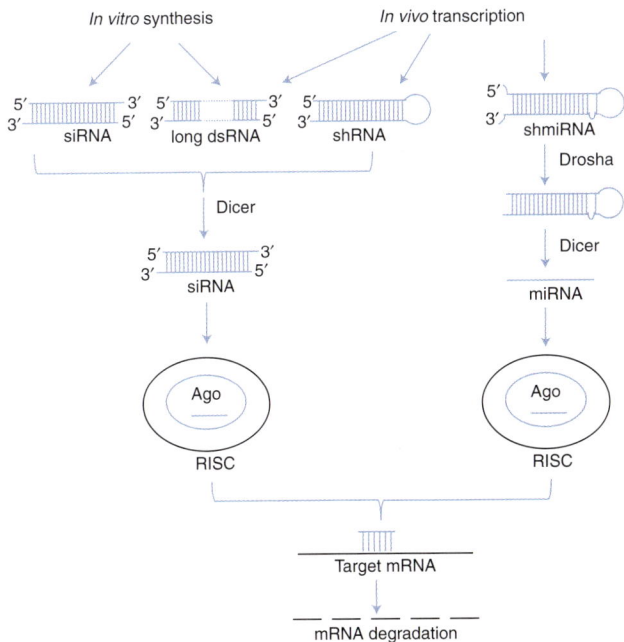

Figure 1. SiRNAs, shRNAs, shmiRNAs, and long dsRNAs pathways.

For use as RNAi reagents, long dsRNAs are usually 200–500 nucleotides (nts) in length. They can be injected into animals and in many cases, into eggs; delivered via bacteria (see sections on *C. elegans* and Planaria later); expressed as transgenes (see sections on *C. elegans*, *Drosophila* and mouse); or delivered into cultured cells by transfection or bathing (*Drosophila*). With the exception of the esiRNA (endoribonuclease-prepared siRNAs) method (Yang et al. 2002), whereby long dsRNAs are used to produce a pool of small, diced siRNAs that is then transfected into cells, long dsRNAs are not used in mammalian systems as they trigger an unwanted interferon response that can mask gene-specific effects.

3 IN VIVO RNAi SCREENING IN *C. ELEGANS*

The discovery that a dsRNA introduced into the nematode *C. elegans* is able to degrade a specific mRNA (Fire et al. 1998) marked the beginning of the revolution in in vivo RNAi. Importantly, RNAi in *C. elegans* is both systemic and transitive. First, injection, or expression, of a dsRNA into one tissue can lead to gene silencing in other tissues (Fire et al. 1998; Winston et al. 2002). Genetic analysis of this systemic effect has identified a number of genes involved in the phenomenon, including the multispan transmembrane protein SID-1, which is sufficient to confer cellular uptake of dsRNA to cells (Feinberg and Hunter 2003). Second, RNAi in *C. elegans* is transitive, whereby an RNA-dependent RNA polymerase (RdRP) is involved in an amplification step of RNAi and as a result, siRNAs that are derived from regions upstream of the original dsRNA sequences are produced (Alder et al. 2003).

The relative ease of methods required to perform RNAi experiments in *C. elegans* makes this genetically amenable model organism a logical choice for the development of technologies to study gene function on a genome-wide scale (review by Sugimoto et al. 2004). DsRNAs can be introduced into the nematode by simply soaking the animal in a solution of dsRNA, by feeding the worms bacteria that express long dsRNAs, by injection of dsRNA, or by generating transgenic hairpin-expressing animals.

Many genome-wide RNAi screens have been performed in the past 10 yr in *C. elegans* to interrogate a large variety of biological questions in developmental biology, cell signaling, aging, metabolic regulation, and neurodegenerative diseases, to name a few. These screens have been performed either in a wild-type strain or specific mutant backgrounds and either by injection or by feeding (see reviews by Kaletta and Hengartner 2006; Boutros and Ahringer 2008). In a landmark study based on injection into eggs, Sönnichsen et al. (2005) performed a genome-wide screen to identify all genes required for the first two rounds of cell division by examining embryonic phenotypes using time-lapse microscopy. However, the method of choice is large-scale RNAi screening by feeding worms bacteria that produce dsRNAs (Timmons and Fire 1998) because first, the method is less tedious by far, and second, RNAi libraries in bacteria that cover most of the 20,000 *C. elegans* genes are available (Fraser et al. 2000; Kim et al. 2005; Boutros and Ahringer 2008). For example, Ashrafi et al. (2003) screened 16,757 genes for their roles in fat storage in living worms using Nile Red staining of tissue lipids. They isolated 305 genes that when knocked down, lead to reduced body fat and 112 genes that lead to increased fat storage, representing a core set of fat regulatory genes as well as pathway-specific fat regulators.

Importantly, RNAi screening in *C. elegans* can easily be performed in various combinations, either in mutant backgrounds or by using multiple RNAs, to identify synthetic phenotypes. Such screens are a powerful means to gain an understanding of the structure of signaling networks, disease susceptibility, and identification of new drug targets. (Lehner et al. 2006), for example, systematically tested approximately 65,000 pairs of genes for their abilities to interact genetically and identified 350 genetic interactions between components of the EGF/Ras, Notch, and Wnt pathways.

Finally, an important issue with large-scale RNAi screening (also discussed later) is the rate of false positive and negative results associated with the method. False positives that occur when novel unexpected phenotypes are associated with RNAi lines appear to be a minor contributor to false discovery in *C. elegans* (Sönnichsen et al. 2005). False negatives on the other hand, because of the variability of knockdown associated with the feeding techniques, can be more of an issue (depending on the screen) and may account for the 10%–30% variability observed between screens even if they are performed in the same laboratory (Simmer et al. 2003).

4 IN VIVO RNAi SCREENING IN *DROSOPHILA*

In *Drosophila*, feeding methods for RNAi delivery, as in *C. elegans*, do not appear to work; however, RNAi reagents can be delivered either by injection into precellular blastoderm embryos or as transgenes. Importantly, although there have been reports that systemic and transitive RNAi may occur in *Drosophila* (Saleh et al. 2009; Lipardi and Patterson 2009), this does not appear to occur when the dsRNA is produced from a transgene (Roignant et al. 2003).

Injection of dsRNAs as short as 200 bps and as long as 2000 bps, as well as short 21–22 nts siRNAs injected into embryos, have been shown to have potent interfering activities (Kennerdell and Carthew 1998; Misquitta and Paterson 1999; Williams and Rubin 2002; Misquitta et al. 2008). This approach has been used, for example, to clarify the role of the MyoD-related gene *nautilus* in embryonic somatic muscle formation (Misquitta and Paterson, 1999), and the roles of both the Frizzled1 and Frizzled2 receptors in Wingless signaling (Kennerdell and Carthew 1998). RNAi injection has been used systematically to screen more than 5000 genes for cardiogenic and embryonic nervous system phenotypes. For the heart screen, dsRNA-injected embryos that carry the *D-mef2-lacZ* transgene to detect cardiac cells were examined. For the nervous system screen, embryos were stained using the 22C10 antibody that detects the entire peripheral nervous system and a subset of central nervous system neurons. This approach led to the identification of many new genes involved in either heart or neural development (Kim et al. 2004) (http://flyembryo.nhlbi.nih.gov/).

RNAi by injection has somewhat limited applications as this approach is restricted to studies of gene function during embryonic development and maternally loaded proteins may mask embryonic phenotypes. Transgenic RNAi, on the other hand, has been widely used to study gene function in somatic tissues. Importantly, and unlike in *C. elegans*, in *Drosophila* RNAi is cell-autonomous, and because of this, targeted expression of RNAi constructs using the Gal4/UAS system (Brand and Perrimon 1993) can be used for cell- or tissue-specific interrogation of gene function. Indeed, this approach has been used extensively (Fig. 2). To date, transgenic RNAi lines have been shown to be potent in all somatic tissues, including neurons and muscles. However, for unknown reasons they do not appear to be effective in the female germ line.

Several groups, working independently, have developed vectors that have been used to generate transgenic RNAi fly strains. The initial vectors were based on transgenes having an inverted–repeat configuration, driven from either a single promoter or symmetrically transcribed from opposing promoters (Lam and Thummel 2000; Fortier and Belote 2000; Martinek and Young 2000; Kennerdell and Carthew 2000; Giordano et al. 2002). Because these vectors generated variable RNAi silencing effects, a number of modifications were introduced based on the observation in plants that intron-spliced hairpin RNAs are more efficient at gene silencing than the hairpin loop RNA (Smith et al. 2000). Thus, a number of groups designed vectors that include intron sequences from genes such as *mub* (Reichhart et al. 2002), *white* (Lee and Carthew 2003), *Ret* (Pili-Floury et al. 2004), or *fushi-tarazu* (*ftz*) (Kondo et al. 2006), as well as genomic/cDNA hybrids (Kalidas and Smith 2002). Additionally, the position of the *ftz* intron within the construct, e.g., located to the end of the hairpin structure, was tested (Fig. 3) (Dietzl et al. 2007). Altogether, these intron-containing vectors gave more robust RNAi phenotypes than the inverted–repeat configuration, most likely because of the enhanced formation of duplex dsRNAs following the splicing event and/or enhanced export of the processed mRNAs from the nucleus. Finally, in addition to the RNAi vectors that generate long dsRNAs, small hairpin microRNA-based (shmiRNA) RNAi constructs that generate a single siRNA have been shown to work as *Drosophila* transgenes (Chen et al. 2007; Haley et al. 2008). Although these vectors appear to work effectively in the soma, their overall effectiveness, especially compared with long dsRNAs, has not been tested systematically.

A major source of variability between these first generation vectors is caused by the method of transgenesis used, in which the constructs are integrated into the genome at random positions using P-element based transformation. Indeed, Dietzl et al. (2007) estimated that only 60% of their RNAi lines are effective with this most likely because of position effects associated with a large number of the random insertions. To solve this source of variability, a series of vectors, the "VALIUM" series, were generated. These rely on the phiC31-mediated site-specific integration approach (Groth et al. 2004) and the RNAi constructs were strategically integrated into sites in the genome that had been preselected for optimal expression (Ni et al. 2008; Ni et al. 2009). Specifically, a number of genomic sites were identified for which high levels of induced, Gal4-driven gene expression is observed, and importantly, low basal

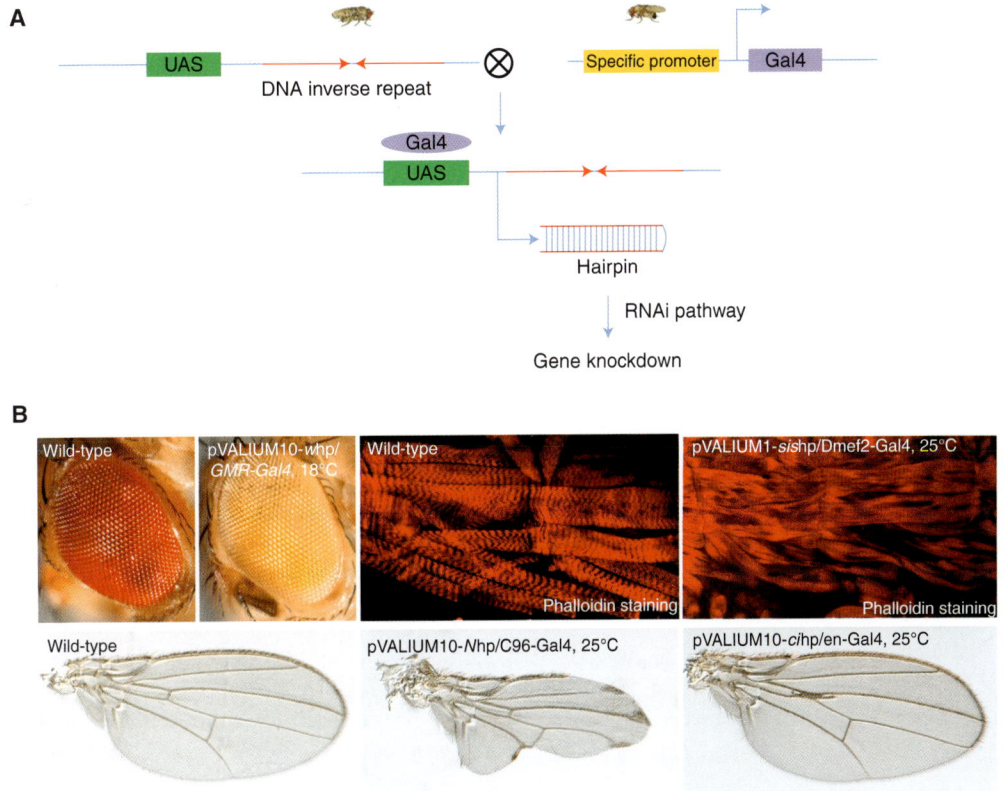

Figure 2. Transgenic RNAi in *Drosophila*. (*A*) Tissue expression of the transgenic RNAi construct is achieved following a cross between a UAS-hairpin and a Gal4 driver line. The main advantage of this method, in addition to its relatively simple design and fast execution time, is that it allows spatial and temporal control of the knockdown construct, which is essential for characterizing genes with pleiotropic functions. As thousands of Gal4 lines are available, appropriate Gal4 drivers are basically available for most questions to be addressed in the intact animal. (*B*) Examples of tissue specific RNAi phenotypes generated in the eye (knockdown of the *white* gene in the eye using the GMR-Gal4 driver), muscle (knockdown of the *sallimus* (*sis*) gene in the eye using the Dmef2-Gal4 driver), and wings (knockdown of the *Notch* (*N*) gene and *cubitus interruptus* (ci) genes in the wing using the C96-Gal4 and en-Gal4 drivers, respectively).

levels are seen in the absence of the Gal4 driver (Markstein et al. 2008). Furthermore, a series of related VALIUM vectors were built and tested for their ability to produce optimal RNAi effects. From these analyses, one optimal vector, VALIUM10, proved excellent for somatic RNAi (Figs. 2, 3) (Ni et al. 2009).

In flies, transgenic RNAi is particularly applicable to studies relevant to human biology such as cancer and metastasis, inflammation and wound healing, metabolic disorders, immunity, aging, and central nervous system disorders. This is exemplified by the hundreds of fly lines generated by individual laboratories, which can be identified either from the published literature or in the *Drosophila* database Flybase (http://flybase.bio.indiana.edu/). Building on the proven strength of transgenic RNAi, three independent efforts have already generated large-scale resources, such that RNAi lines that cover most of the *Drosophila* 13,929 protein-encoding genes (Tweedie et al. 2009) are now available (Fig. 3).

Two recently published large-scale screens, both using the Dietzl et al. (2007) library, illustrate that in flies complex developmental processes can be dissected on a genome-wide level using transgenic RNAi. First, Mummery-Widmer et al. (2009) screened for novel components of the Notch pathway by examining the effect of RNAi lines on external sensory organ development. In particular, they identified six new genes involved in asymmetric cell division and 23 novel genes regulating Notch signaling. Among the many interesting genes identified as Notch regulators were genes involved in nuclear import and the COP9 signallosome. In the second whole-genome study, Cronin et al. (2009) screened the RNAi lines for their ability to be resistant or susceptible to the ingestion of pathogenic Gram-negative bacteria *Serratia marcescens*. The initial

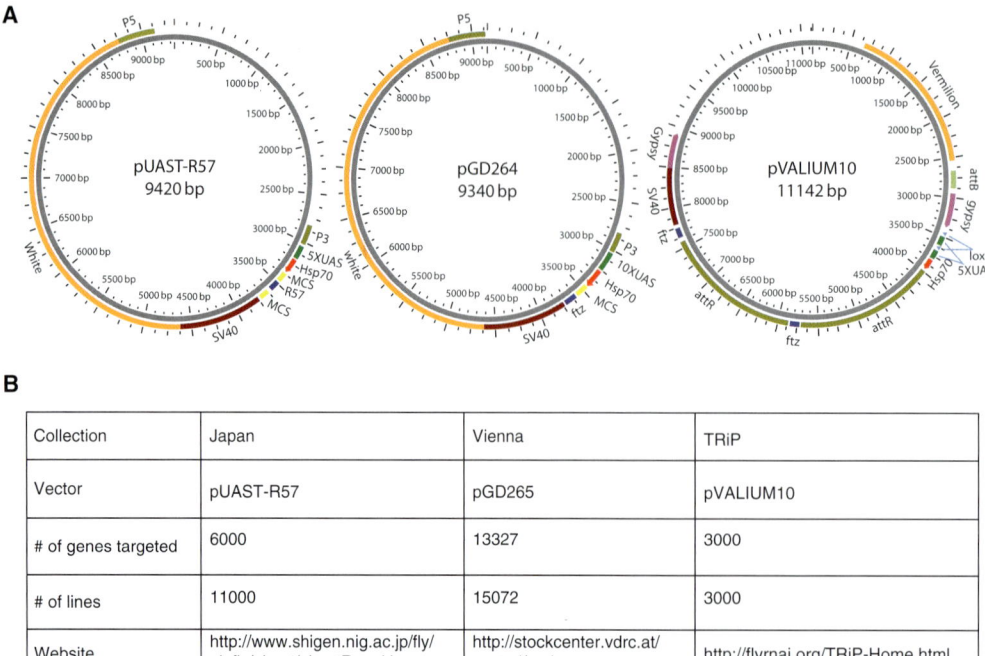

Figure 3. Structure of the vectors (A) and transgenic RNAi resources (B) available in *Drosophila*.

Collection	Japan	Vienna	TRiP
Vector	pUAST-R57	pGD265	pVALIUM10
# of genes targeted	6000	13327	3000
# of lines	11000	15072	3000
Website	http://www.shigen.nig.ac.jp/fly/nigfly/about/aboutRunai.jsp	http://stockcenter.vdrc.at/control/main	http://flyrnai.org/TRiP-Home.html

screen identified 95 resistant and 790 susceptible candidates that were subsequently analyzed using different Gal4 drivers to determine the site of action of the identified genes. A number of genes involved in intracellular processes, the immune system, the stress response, as well as genes associated with stem cell proliferation, growth, and cell death were shown to be required in the gut. Others, involved in phagocytosis and the stress response, were required in macrophages. Building on these observations, the authors characterized a requirement for the JAK/STAT pathway in response to intestinal *Serratia* infection.

An important issue with regard to *Drosophila* RNAi screens in tissue culture concerns false positives that occur from sequence specific off-target effects (OTEs) (Kulkarni et al. 2006; Ma et al. 2006; review by Perrimon and Mathey-Prevot 2007). OTEs can be avoided by selecting sequences that do not contain 19 nts or longer cross-hybridizing stretches to other genes or tri-nucleotide CAN (CA[AGCT]) repeats. In this regard, a number of software tools are available for identifying the most common off-target sequences so that they can be excluded from RNAi constructs. These include E-RNAi from the German Cancer Research Center (http://www.dkfz.de/signaling2/ernai/; Arziman et al. 2005) and SnapDragon from the *Drosophila* RNAi Screening Center (http://www.flyrnai.org/cgi-bin/RNAi_find_primers.pl; Flockhart et al. 2006). In vivo, although it is difficult to fully evaluate the rates of false positives and negatives in general, as it depends on the reagents used, the specific Gal4 driver used, and the temperature at which the flies are screened, the consensus is that OTEs appear negligible if sequences that avoid potentially problematic sequences are used (Dietzl et al. 2007; Ni et al. 2008; Ni et al. 2009). The rate of false negatives, however, in the fly screens, as discussed in the case of *C. elegans*, may be more of an issue. For example, Dietzl et al. (2007) estimate that 40% of their RNAi lines may not generate phenotypes because of low transgene expression. In some cases the effectiveness of individual RNAi lines can be improved by co-expressing Dicer2 (Dietzl et al. 2007) but in general, the newer generation of RNAi lines based on targeted insertion (Ni et al. 2009) are more likely to significantly decrease the overall rate of false negatives in RNAi screens.

Although false positives associated with transgenic RNAi appear to be less of an issue than RNAi in tissue culture, results derived from a single fly line still have to be taken with caution. To validate a transgenic RNAi phenotype, two simple follow-up experiments can be performed. First, the result can be confirmed with a second independent line, which becomes easier as more transgenic RNAi lines are generated. Second, and most conclusive, the RNAi induced phenotype can be rescued via expression of a transcript that can confer gene activity but evades the RNAi treatment, such as by having a divergent nucleotide sequence or exogenous 3′UTR (Stielow et al. 2008). Using genomic DNA of closely related species, Kondo et al. (2009)

have established a cross-species RNAi rescue method useful to rescue RNAi phenotypes. This straightforward and reliable method, based on genomic DNA fragments contained in fosmids, can be used to quickly build the construct needed to generate the transgenic flies harboring genomic DNA of a sibling *Drosophila* species that can confer activity but avoid knockdown.

5 APPLICATIONS OF IN VIVO RNAi TO THE DEVELOPMENTAL BIOLOGY OF EMERGING SYSTEMS AND EVO-DEVO

The application of RNAi to non-model and/or emerging model organisms in which few, if any, genetic tools are available is providing biologists a powerful means to characterize the roles of specific genes throughout development and evolution. As transformation methods are available in only a handful of organisms, RNAi is being delivered to emerging models either by feeding or following injection of RNAi reagents.

A large-scale RNAi screen for gene function has been performed in the planarian, *Schmidtea mediterranea*, an organism not previously accessible to extensive genetic manipulation. As in *C. elegans*, RNAi is delivered to a Planarian by feeding. In a landmark study, Reddien et al. (2005) screened 1065 genes, 5%–7% of the total, and described phenotypes associated with 240 of them. Many of these showed specific defects in regeneration, and in particular, defects were observed during stem cell/neoblast proliferation in amputated animals. Many biological insights are emerging from this work; for example, a recent study implicates the Wnt/beta-catenin pathway in anteroposterior polarity of the blastema during regeneration (Petersen and Reddien 2008).

RNAi is being used to address evo-devo questions in jellyfish, wasps, beetles, crickets, spiders, etc. For example, in the long germ band parasitic wasp *Nasonia vitripennis*, injection of pupae with dsRNAs (Lynch and Desplan 2006) has been used to examine the logic of anteroposterior patterning, and in particular mechanisms that differ from the short germ band patterning of *Drosophila*. Further, these in vivo RNAi studies have clarified the ancestral roles of the *bicoid* and *caudal* genes as patterning organizers, and helped to elucidate how these functions have evolved in higher dipterans such as *Drosophila* (Olesnicky et al. 2006; Brent et al. 2007).

6 IN VIVO RNAi IN VERTEBRATE MODELS

RNAi-based methods are now a common tool for gene perturbation in mammalian tissue culture cells (mouse, rat, monkey, and human). Beyond cell culture screens (see recent review by Mohr et al. 2011), RNAi is being used in a number of in vivo studies in which the RNAi reagents are delivered topically and directed to specific tissues or organs, such as the retina, brain, or muscles; using ex vivo, for example in hematopoietic cells; or delivered as transgenes (Sandy et al. 2005).

RNAi can be achieved locally by delivering synthetic siRNAs or using shRNAs delivered via viral particles or following transfection of plasmid DNAs. Many vectors have been built, based on either shRNA or shmiRNA designs, to optimize the level of expression of the RNAi reagent. Furthermore, much effort has been devoted to the development of methods for conditional RNAi that include irreversible and reversible approaches (Sandy et al. 2005; Lee and Kumar 2009). For example, a number of vectors based on the Cre/loxP and FLP-FRT systems have been used to induce RNAi in an irreversible way. In addition, Tet-, Ecdysone-, LacR, HIV-1 tat-, and HIV-1 LTR-, based systems have been explored for reversible conditional RNAi systems. In addition, vectors have been built for generating transgenic animals that can be either inserted into the genome at random or at targeted sites to ensure expression. To date, most of the published studies are still at the proof of principle stage (Sandy et al. 2005; Lee and Kumar 2009). Importantly, more studies are needed to evaluate technical aspects of RNAi effects, such as the level of knockdown in various cell types, variability because of the insertion site, potential epigenetic silencing of the construct, etc. Regardless of these limitations, from intense ongoing efforts will emerge many exciting applications for RNAi-based methods in the coming years (see review by Lee and Kumar 2009).

To date, and unlike the mouse, RNAi in *Xenopus* and zebra fish has not had a great impact, due in part to mixed results on the efficacy of some of the RNAi reagents, the prevalent use of the well-established method of antisense oligonucleotide morpholinos, and the lack of effective methods for controlled gene expression. In *Xenopus*, injection of siRNAs or long dsRNAs into oocytes and early blastomeres appears to work well (Zhou et al. 2002; Nakano et al. 2002), and gene silencing via transgenesis has been shown, although some difficulties have been observed in the silencing of genes at later stages of development (Li and Rohrer 2006). Similarly, in zebra fish, although a few studies have shown that dsRNAs, shRNAs and siRNAs can be effective for gene knockdown, a number of studies report that unexplained morphological abnormalities can be associated with RNAi-injected embryos (Wargelius et al. 1999; Skromne and Prince 2008). Altogether, it is not clear to what extent, at least in the absence of major technical advances, RNAi-based methods will become mainstream in fish or *Xenopus*.

Finally, in chick embryos, electroporation of siRNAs, as well as delivery using Replication Competent Avian Splice (RCAS) retroviruses to introduce hairpins intotissues, have been used successfully (Harpavat and Cepko, 2006). The RCAS approach is of particular interest as it is long lasting and transmissible because infected cells release more virus that spread to neighboring cells.

7 IN VIVO RNAi IN PLANTS: APPLICATIONS TOWARD CROP IMPROVEMENT

In plants, as in *C. elegans*, RNAi is both systemic and heritable. The siRNAs move between cells through channels in cell walls, thus enabling communication and transport throughout the plant. In addition, methylation of promoters targeted by RNAi confers heritability, as the new methylation pattern is copied in each new generation of the cell (Jones et al. 2001). Interestingly, in plants, endogenously encoded miRNAs rather than inhibiting translation are nearly or perfectly complementary to their target genes and induce mRNA cleavage by interaction with RISC.

The focus of in vivo RNAi applications in plants is directed toward the improvement of plant productivity and/or nutritional value (see reviews by Kusaba, 2004; Tang et al. 2007; Hebert et al. 2008). Among the exciting applications in which RNAi could have a major impact in agriculture is the improvement of essential food crops such as corn and rice. In addition, RNAi could be used to engineer food plants rendering them rich in dietary protein; for example, lowering the levels of natural plant toxins in cotton seeds could make this abundant plant appropriate for human consumption. Although we are still far from seeing RNAi-modified plant products in agriculture, especially considering the controversies and concerns surrounding growing genetically modified plants for human consumption, a number of successful applications have already emerged, particularly the ability of RNAi to confer resistance to common plant viruses (Zadeh and Foster 2004) and fortification of plants such as tomatoes with dietary antioxidants (Niggeweg et al. 2004).

8 IN VIVO RNAi APPLICATIONS TOWARD VIRUS AND PEST CONTROL

RNAi may have important agricultural applications as illustrated by ongoing attempts to use RNAi approaches to remedy the colony collapse disorder (CCD) in European honeybees. In recent years, millions of beehives have disappeared, most likely because of the spread through bee colonies of a lethal virus, the Israeli acute paralysis virus (IAPV). The current working hypothesis is that IAPV infection, together with poor nutrition and exposure to pesticides, weakens bee colonies to the extent that they simply disappear. One RNAi-based strategy being explored to fight IAPV infection is to feed bees siRNAs targeting specific IAPV sequences such that, following viral entry into bee cells, translation of viral proteins is blocked (Cox-Foster and vanEngelsdorp 2009).

RNAi is also becoming an important tool to combat insect pests, in particular *Anopheles gambia*, the vector for Plasmodium, the protozoan responsible for malaria. RNAi reagents are being used to dissect host-pathogen interactions and have already provided fundamental insight into the insect defense mechanisms to control the protozoan, such as the identification of the pattern-recognition receptor TEP1 in host defense (Blandin et al. 2004).

Methods to disseminate RNAi expressing transgenes that may confer resistance to a pathogen within a population are also being explored. One of the strategies being considered for the control of pathogen-laden pests is to rapidly convert a pathogen-bearing insect population to a genetically modified population that is resistant to the pathogen. For example, if wild mosquito populations could be replaced with malaria or dengue-resistant ones, this alone may provide an effective means to control these devastating diseases. The challenge with such an approach is to develop a method for rapid replacement of the wild population. In one clever demonstration, Chen et al. (2007) reported an RNAi-based method in *Drosophila* that achieves the selfish drive of a genetic element into a population. Although the approach is still at an early stage in development, the results of this study show the feasibility of RNAi-based population replacement.

9 IN VIVO RNAi APPLICATIONS IN MEDICINE

RNAi has the potential to offer more specificity and flexibility than traditional drugs to silence gene expression. In addition, because any protein that causes or contributes to a disease is susceptible to RNAi, previous disease targets considered "undruggable" are now accessible. Not surprisingly, RNAi has become a major focus for biotechnology and pharmaceutical companies, which are now in the early stages of developing RNAi therapeutics, mostly based on siRNAs, to target viral infection, cancer, hypercholesterolemia, cardiovascular disease, macular degeneration, and neurodegenerative diseases (Sah et al. 2006).

Critical issues with RNAi as a therapeutic are delivery, specificity and stability of the RNAi reagents. Delivery is currently considered the biggest hurdle as the introduction of siRNAs systemically into body fluids can result in their degradation, off-target effects, and immune detection and subsequent reactions (see for example Zimmermann et al. 2006). Thus, many efforts are focused on developing ways to modify an RNAi or attach them to delivery agents that will protect them until they reach their therapeutic destinations. These include delivery as particles or complexes using lipid nanoparticles that encapsulate the siRNA or combining siRNA molecules with peptide-based polymers. Additional advances in therapeutic applications are likely to come from chemical modifications or other approaches to improving the specificity and potency of RNAi reagents.

10 CONCLUDING REMARKS

RNAi-based methods are providing unprecedented tools useful to address fundamental questions in the biology of living organisms. As exemplified by in vivo screens in *C. elegans* and *Drosophila* as reviewed here, these tools are enhancing and/or replacing more classical genetic approaches and manipulations. Further, as most organisms possess the cellular machinery for RNAi, this near-universal approach makes loss-of-function studies approachable in organisms in which genetic tools do not exist. Finally, with the growing appreciation for the fundamental potential of RNAi and a burgeoning collection of RNAi technologies and reagents, the diversity in in vivo applications to biology, medicine, and agriculture is seemingly limitless.

ACKNOWLEDGMENTS

We thank Rui Zhou and Stephanie Mohr for comments on the manuscript and Luping Liu, Donghui Yang-Zhou and Martha Reed assistance on the figures. This work is supported by GM084947 to N.P. and the HHMI.

REFERENCES

Adams MD, Celniker SE, Holt RA, Evans CA, Gocayne JD, Amanatides PG et al. 2000. The Genome sequence of *Drosophila melanogaster*. *Science* **287**: 2185–2195.

Alder MN, Dames S, Gaudet J, Mango SE. 2003. Gene silencing in *Caenorhabditis elegans* by transitive RNA interference. *RNA* **9**: 25–32.

Arziman Z, Horn T, Boutros M. 2005. E-RNAi: A web application to design optimized RNAi constructs. *Nucleic Acids Res* **33**: W582–588.

Ashrafi K, Chang FY, Watts JL, Fraser AG, Kamath RS, Ahringer J, Rukun G. 2003. Genome-wide RNAi analysis in Caenohabitis elegans fat regulatory genes. *Nature* **421**: 268–272.

Blandin S, Shiao SH, Moita LF, Janse CJ, Waters AP, Kafatos FC, Levashina EA. 2004. Complement-like protein TEP1 is a determinant of vectorial capacity in the malaria vector *Anopheles gambiae*. *Cell* **116**: 661–670.

Boutros M, Ahringer J. 2008. The art and design of genetic screens: RNA interference. *Nat Rev Genet* **9**: 554–566.

Boutros M, Kiger AA, Armknecht S, Kerr K, Hild M, Koch B, Haas SA, Paro R, Perrimon N. 2004. Genome-wide RNAi analysis of growth and viability in *Drosophila* cells. *Science* **303**: 832–835.

Brand AH, Perrimon N. 1993. Targeted gene expression as a means of altering cell fates and generating dominant phenotypes. *Development* **118**: 401–415.

Brent AE, Yucel G, Small S, Desplan C. 2007. Permissive and instructive anterior patterning rely on mRNA localization in the wasp embryo. *Science* **315**: 1841–1843.

C. elegans Sequencing Consortium. 1998. Genome sequence of the nematode *C. elegans*: A platform for investigating biology. *Science* **282**: 2012–2018.

Chen CH, Huang H, Ward CM, Su JT, Schaeffer LV, Guo M, Hay BH. 2007. A synthetic maternal-effect selfish genetic element drives population replacement in *Drosophila*. *Science* **316**: 597–600.

Cox-Foster D, vanEngelsdorp D. 2009. Saving the honeybee. *Sci Am* **300**: 40–47.

Cronin SJ, Nehme NT, Limmer S, Liegeois S, Pospisilik JA, Schramek D, Leibbrandt A, Simoes Rde M, Gruber S, Puc U, et al. 2009. Genome-wide RNAi screen identifies genes involved in intestinal pathogenic bacterial infection. *Science* **325**: 340–343.

Dietzl G, Chen D, Schnorrer F, Su KC, Barinova Y, Fellner M, Gasser B, Kinsey K, Oppel S, Scheiblauer S et al. 2007. A genome-wide transgenic RNAi library for conditional gene inactivation in *Drosophila*. *Nature* **448**: 151–156.

Echeverri CJ, Perrimon N. 2006. High-throughput RNAi screening in cultured cells: A user's guide. *Nat Rev Genet* **7**: 373–384.

Fraser AG, Kamath RS, Zipperlen P, Martinez-Campos M, Sohrmann M, Ahringer J. 2000. Functional genomic analysis of *C. elegans* chromosome I by systematic RNA interference. *Nature* **408**: 325–330.

Feinberg EH, Hunter CP. 2003. Transport of dsRNA into cells by the transmembrane protein SID-1. *Science* **301**: 1545–1547.

Fire A, Xu S, Montgomery MK, Kostas SA, Driver SE, Mello CC. 1998. Potent and specific genetic interference by double-stranded RNA in *Caenorhabditis elegans*. *Nature* **391**: 806–811.

Flockhart I, Booker M, Kiger A, Boutros M, Armknecht S, Ramadan N, Richardson K, Xu A, Perrimon N, Mathey-Prevot B. 2006. FlyRNAi: The *Drosophila* RNAi screening center database. *Nucleic Acids Res* **34**: D489–494.

Fortier E, Belote JM. 2000. Temperature-dependent gene silencing by an expressed inverted repeat in *Drosophila*. *Genesis* **26**: 240–244.

Giordano E, Rendina R, Peluso I, Furia M. 2002. RNAi triggered by symmetrically transcribed transgenes in *Drosophila melanogaster*. *Genetics* **160**: 637–648.

Groth AC, Fish M, Nusse R, Calos MP. 2004. Construction of transgenic *Drosophila* by using the site-specific integrase from phage phiC31. *Genetics* **166**: 1775–1782.

Haley B, Hendrix D, Trang V, Levine M. 2008. A simplified miRNA-based gene silencing method for *Drosophila melanogaster*. *Dev Biol* **321**: 482–490.

Harpavat S, Cepko CL. 2006. RCAS-RNAi: A loss-of-function method for the developing chick retina. *BMC Developmental Biol* **6**: doi:10.1186/1471-213X-6-2.

Hartwell LH, Culotti J, Pringle JR, Reid BJ. 1974. Genetic control of the cell division cycle in yeast. *Science* **183**: 46–51.

Hebert CG, Valdes JJ, Bentley WE. 2008. Beyond silencing–engineering applications of RNA interference and antisense technology for altering cellular phenotype. *Curr Opin Biotechnol* **19**: 500–505.

Horvitz HR. 1999. Genetic control of programmed cell death in the nematode *Caenorhabditis elegans*. *Cancer Res* **59**: 1701s–1706s.

Jones L, Ratcliff F, Baulcombe DC. 2001. RNA-directed transcriptional gene silencing in plants can be inherited independently of the RNA trigger and requires Met1 for maintenance. *Curr Biol* **11**: 747–757.

Kaletta T, Hengartner MO. 2006. Finding function in novel targets: *C. elegans* as a model organism. *Nat Rev Drug Discov* **5**: 387–398.

Kalidas S, Smith DP. 2002. Novel genomic cDNA hybrids produce effective RNA interference in adult *Drosophila*. *Neuron* **33**: 177–184.

Kennerdell JR, Carthew RW. 1998. Use of dsRNA-mediated genetic interference to demonstrate that frizzled and frizzled 2 act in the wingless pathway. *Cell* **95**: 10171026.

Kennerdell JR, Carthew RW. 2000. Heritable gene silencing in *Drosophila* using double-stranded RNA. *Nat Biotechnol* **18**: 896–898.

Kim JK, Gabel HW, Kamath RS, Tewari M, Pasquinelli A, Rual JF, Kennedy S, Dybbs M, Bertin N, Kaplan JM et al. 2005. Functional genomic analysis of RNA interference in *C. elegans*. *Science* **308**: 1164–1167.

Kim YO, Park SJ, Balaban RS, Nirenberg M, Kim Y. 2004. A functional genomic screen for cardiogenic genes using RNA interference in developing *Drosophila* embryos. *Proc Natl Acad Sci* **101**: 159–164.

Kondo S, Booker M, Perrimon N. 2009. Cross-species RNAi rescue platform in *Drosophila melanogaster*. *Genetics* Aug 31. [Epub ahead of print]

Kondo T, Inagaki S, Yasuda K, Kageyama Y. 2006. Rapid construction of *Drosophila* RNAi transgenes using pRISE, a P-element-mediated transformation vector exploiting an in vitro recombination system. *Genes Genet Syst* **81**: 129–134.

Kulkarni MM, Booker M, Silver SJ, Friedman A, Hong P, Perrimon N, Mathey-Prevot B. 2006. Evidence of off-target effects associated with long dsRNAs in *Drosophila melanogaster* cell-based assays. *Nat Methods* **3**: 833–838.

Kusaba M. 2004. RNA interference in crop plants. *Curr Opin Biotechnol* **15**: 139–143.

Lam G, Thummel CS. 2000. Inducible expression of double-stranded RNA directs specific genetic interference in *Drosophila*. *Curr Biol* **10**: 957–963.

Lee YS, Carthew RW. 2003. Making a better RNAi vector for *Drosophila*: Use of intron spacers. *Methods* **30**: 322–329.

Lee SK, Kumar P. 2009. Conditional RNAi: Towards a silent gene therapy. *Adv DrugDeliv Rev* **61**: 650–664.

Lehner B, Crombie C, Tischler J, Fortunato A, Fraser AG. 2006. Systematic mapping of genetic interactions in *Caenorhabditis elegans* identifies common modifiers of diverse signaling pathways. *Nat Genet* **38**: 896–903.

Lipardi C, Paterson BM. 2009. Identification of an RNA-dependent RNA polymerase in Drosophila involved in RNAi and transposon suppression. *Proc Natl Acad Sci* **106**: 15645–15650.

Lipardi C, Baek HJ, Wei Q, Paterson BM. 2005. Analysis of short interfering RNA function in RNA interference by using *Drosophila* embryo extracts and schneider cells. *Methods Enzymol* **392**: 351–371.

Li M, Rohrer B. 2006. Gene silencing in *Xenopus laevis* by DNA vector-based RNA interference and transgenesis. *Cell Res* **16**: 99–105.

Lynch JA, Desplan C. 2006. A method for parental RNA interference in the wasp *Nasonia vitripennis*. *Nat Protoc* **1**: 486–494.

Ma Y, Creanga A, Lum L, Beachy PA. 2006. Prevalence of off-target effects in *Drosophila* RNA interference screens. *Nature* **443**: 359–363.

Markstein M, Pitsouli C, Villalta C, Celniker SE, Perrimon N. 2008. Exploiting position effects and the gypsy retrovirus insulator to engineer precisely expressed transgenes. *Nat Genet* **40**: 476–483.

Martinek S, Young MW. 2000. Specific genetic interference with behavioral rhythms in *Drosophila* by expression of inverted repeats. *Genetics* **156**: 1717–1725.

Misquitta L, Paterson BM. 1999. Targeted disruption of gene function in *Drosophila* by RNA interference (RNA-i): A role for nautilus in embryonic somatic muscle formation. *Proc Natl Acad Sci* **96**: 1451–1456.

Misquitta L, Wei Q, Paterson BM. 2008. Collection of *Drosophila* Embryos for RNA Interference (RNAi). *Cold Spring Harb Protoc* 10.1101/pdb.prot4917.

Mohr S, Bakal C, Perrimon N. 2011. RNAi: Results and Challenges. *Ann Rev Biochem* (in press).

Mummery-Widmer JL, Yamazaki M, Stoeger T, Novatchkova M, Bhalerao S, Chen D, Dietzl G, Dickson BJ, Knoblich JA. 2009. Genome-wide analysis of Notch signalling in *Drosophila* by transgenic RNAi. *Nature* **458**: 987–92.

Nagy A, Perrimon N, Sandmeyer S, Plasterk R. 2003. Tailoring the genome: The power of genetic approaches. *Nat Genet* **33**: 276–284.

Nakano H, Amemiya S, Shiokawa K, Taira M. 2002. RNA interference for theorganizer-specific gene Xlim-1 in *Xenopus* embryos. *Biochem Biophys Res Commun* **274**: 434–439.

Ni JQ, Liu LP, Binari R, Hardy R, Shim HS, Cavallaro A, Booker M, Pfeiffer BD, Markstein M, Wang H, et al. 2009. A *Drosophila* resource of transgenic RNAi lines for neurogenetics. *Genetics* **182**: 10891100.

Ni JQ, Markstein M, Binari R, Pfeiffer B, Liu LP, Villalta C, Booker M, Perkins L, Perrimon N. 2008. Vector and parameters for targeted transgenic RNA interference in *Drosophila melanogaster*. *Nat Methods* **5**: 49–51.

Niggeweg R, Michael AJ, Martin C. 2004. Engineering plants with increased levels of the antioxidant chlorogenic acid. *Nat Biotechnol* **22**: 746–754.

Nusslein-Volhard C, Wieschaus E. 1980. Mutations affecting segment number and polarity in *Drosophila*. *Nature* **287**: 795–801.

Olesnicky EC, Brent AE, Tonnes L, Walker M, Pultz MA, Leaf D, Desplan C. 2006. A caudal mRNA gradient controls posterior development in the wasp *Nasonia*. *Development* **133**: 3973–3982.

Perrimon N, Mathey-Prevot B. 2007. Applications of high-throughput RNA interference screens to problems in cell and developmental biology. *Genetics* **175**: 7–16.

Petersen CP, Reddien PW. 2008. Smed-betacatenin-1 is required for anteroposterior blastema polarity in planarian regeneration. *Science* **319**: 327–330.

Pili-Floury S, Leulier F, Takahashi K, Saigo K, Samain E, Ueda R, Lemaitre B. 2004. In vivo RNA interference analysis reveals an unexpected role for GNBP1 in the defense against Gram-positive bacterial infection in *Drosophila* adults. *J Biol Chem* **279**: 12848–12853.

Rat Genome Sequencing Project Consortium. 2004. Genome sequence of the Brown Norway rat yields insights into mammalian evolution. *Nature* **428**: 493–521.

Reddien PW, Bermange AL, Murfitt KJ, Jennings JR, Sanchez Alvarado A. 2005. Identification of genes needed for regeneration, stem cell function, and tissue homeostasis by systematic gene perturbation in planaria. *Dev Cell* **8**: 635–649.

Reichhart JM, Ligoxygakis P, Naitza S, Woerfel G, Imler JL, Gubb D. 2002. Splice-activated UAS hairpin vector gives complete RNAi knockout of single or double target transcripts in *Drosophila melanogaster*. *Genesis* **34**: 160–164.

Roignant JY, Carre C, Mugat B, Szymczak D, Lepesant JA, Antoniewski C. 2003. Absence of transitive and systemic pathways allows cell-specific and isoform-specific RNAi in *Drosophila*. *RNA* **9**: 299–308.

Sah DW. 2006. Therapeutic potential of RNA interference for neurological disorders. *Life Sci* **79**: 1773–1780.

Saleh MC, Tassetto M, van Rij RP, Goic B, Gausson V, Berry B, Jacquier C, Antoniewski C, Andino R. 2009. Antiviral immunity in *Drosophila* requires systemic RNA interference spread. *Nature* **458**: 346–50.

Sandy P, Ventura A, Jacks T. 2005. Mammalian RNAi: A practical guide. *Biotechniques*. **39**: 215–224.

Simmer F, Moorman C, van der Linden AM, Kuijk E, van den Berghe PV, Kamath RS, Fraser AG, Ahringer J, Plasterk RH. 2003. Genome-wide RNAi of *C. elegans* using the hypersensitive rrf-3 strain reveals novel gene functions. *PLoS Biol* **1**: E12.

Skromne I, Prince VE. 2008. Current perspectives in zebrafish reverse genetics: Moving forward. *Dev Dyn* **237**: 861–882.

Smith NA, Singh SP, Wang MB, Stoutjesdijk PA, Green AG, Waterhouse PM. 2000. Total silencing by intron-spliced hairpin RNAs. *Nature* **407**: 319–320.

Sönnichsen B, Koski LB, Walsh A, Marschall P, Neumann B, Brehm M, Alleaume AM, Artelt J, Bettencourt P, Cassin E et al. 2005. Full-genome RNAi profiling of early embryogenesis in *Caenorhabditis elegans*. *Nature* **434**: 462–469.

St Johnston D, Nüsslein-Volhard C. 1992. The origin of pattern and polarity in the *Drosophila* embryo. *Cell* **68**: 201–219.

Stielow B, Sapetsching A, Kruger I, Kunert N, Brehm M, Suske G. 2008. Indentification of SUMO-dependant chromatin-associated transcriptional repression components by genome-wide RNAi screen. *Mol Cell* **29:** 742–754.

Sugimoto A. 2004. High-throughput RNAi in *Caenorhabditis elegans*: Genome-wide screens and functional genomics. *Differentiation* **72:** 81–91.

Tang G, Galili G, Zhuang X. 2007. RNAi and microRNA: Breakthrough technologies for the improvement of plant nutritional value and metabolic engineering. *Metabolomics* **3:** 357–369.

Timmons L, Fire A. 1998. Specific interference by ingested dsRNA. *Nature* **395:** 854.

Tweedie S, Ashburner M, Falls K, Leyland P, McQuilton P, Marygold S, Millburn G, Osumi-Sutherland D, Schroeder A, Seal R et al. 2009. FlyBase: Enhancing *Drosophila* Gene ontology annotations. *Nucleic Acids Res* **37:** D555559.

Venken KJ, Bellen HJ. 2005. Emerging technologies for gene manipulation in *Drosophila melanogaster*. *Nat Rev Genet* **6:** 167–178.

Venter JC, Adams MD, Myers EW, Li PW, Mural RJ, Sutton GG, Smith HO, Yandell M, Evans CA, Holt RA, et al. 2001. The sequence of the human genome. *Science* **291:** 1304–1351.

Wargelius A, Ellingsen S, Fjose A. 1999. Double-stranded RNA induces specific developmental defects in zebrafish embryos. *Biochem Biophys Res Commun* **263:** 156–161.

Waterston RH, Lindblad-Toh K, Birney E, Rogers J, Abril JF, Agarwal P, Agarwala R, Ainscough R, Alexandersson M, An P, et al. 2002. Initial sequencing and comparative analysis of the mouse genome. *Nature* **420:** 520–562.

Williams RW, Rubin GM. 2002. ARGONAUTE1 is required for efficient RNA interference in *Drosophila* embryos. *Proc Natl Acad Sci* **99:** 6889–6894.

Winston WM, Molodowitch C, Hunter CP. 2002. Systemic RNAi in *C. elegans* requires the putative transmembrane protein SID-1. *Science* **295:** 2456–2459.

Yang D, Buchholz F, Huang Z, Goga A, Chen CY, Brodsky FM, Bishop JM. 2002. Short RNA duplexes produced by hydrolysis with *Escherichia coli* RNase III mediate effective RNA interference in mammalian cells. *Proc Natl Acad Sci* **99:** 9942–9947.

Zadeh AH, Foster GD. 2004. Transgenic resistance to tobacco ringspot virus. *Acta Virol* **48:** 145–152.

Zhou Y, Ching YP, Kok KH, Kung HF, Jin DY. 2002. Post-transcriptional suppression of gene expression in *Xenopus* embryos by small interfering RNA. *Nucleic Acids Res* **30:** 1664–1669.

Zimmermann TS, Lee AC, Akinc A, Bramlage B, Bumcrot D, Fedoruk MN, Harborth J, Heyes JA, Jeffs LB, John M et al. 2006. RNAi-mediated gene silencing in nonhuman primates. *Nature* **441:** 111–114.

Index

A

Adenine, abiotic synthesis, 33
S-Adenosylhomocysteine (SAH), riboswitch ligand, 69–71, 85
S-Adenosylmethionine (SAM)
  riboswitch ligand, 69–71
  tandem riboswitches, 72–74
Adenovirus, noncoding RNA function, 166, 169
AFM. *See* Atomic force microscopy
Age-related macular degeneration (ARMD), aptamer therapy, 334–335
Ago. *See* Argonaute
Alanyl nucleic acid (ANA)
  abiotic synthesis, 37
  polymer structure, 36
Alu RNA, RNA polymerase II loading mediation, 285–287
6-Amino-5-nitropyridin-2-one, six letter polymerase chain reaction, 15–16
ANA. *See* Alanyl nucleic acid
Ancient RNA
  conservation and persistence, 45
  evidence, 44–45
  persistence, 45
Anisomycin, peptidyl transferase reaction inhibition, 131–132
Aptamer. *See* SELEX; SOMAmers
Arabinoside, abiotic synthesis, 33–34
ArcZ, small RNA regulator function, 219–220
Argonaute (Ago)
  ancestral function, 250–251
  phylogenetic analysis in three clades, 244–245
  Piwi, 250, 258
  RNA binding modes, 247–248
  RNA interference role, 244, 257–258
  small RNA-guided cleavage, 248–250
  structure and function, 245–247
ARMD. *See* Age-related macular degeneration
Assemble program, interactive molecular modeling, 314–315, 317
Atomic composition, CHNOPS consistency, 45
Atomic force microscopy (AFM), RNA folding studies, 323–324, 328
Azithromycin, peptidyl transferase reaction inhibition, 131–132

B

Bacterial small RNA regulators (sRNAs). *See also* CRISPRs
  abundance of types, 216
  antisense small RNA regulators, 217
  base pairing with limited complementarity
    functions
      outer membrane protein synthesis repression, 219
      overview, 219–220
      transcription factor synthesis modulation, 219
    Hfq function, 218
    Hfq-independent base pairing, 218–219
    mechanisms, 217–219
    outcomes, 217–218
  evolution
    capture of random transcription for purpose of regulation, 226
    recent evolution
      conserved gene neighborhood, 225
      conserved regulation, 225
      gene duplication, 225
      horizontal transfer, 225–226
      target-imposed constraints, 225
    sRNA-mRNA connections, 226
    sRNA-tRNA connections, 226
  functional overview, 216
  mimics in regulation, 221–222
  mRNA dual function, 221
  prospects for study, 227
  protein activity modulation
    CsrB, 223
    discovery, 223
    6S RNA, 222–223
  riboswitches, 221
  tmRNA intrinsic functions, 223–224
Bayes's theorem, RNA world applicability, 45
Borate, premetabolic cycle, 14–15
Brr2
  small nuclear ribonucleoprotein structure, 193–194
  spliceosome function, 191

C

Cas. *See* CRISPRs
CCND1
  histone modification mediation, 284–285
  signal sensing, 288
Cellular life
  membranes as compartment boundaries, 52–53
  phosphoramidate nucleic acids, 57–59
  protocell
    assembly, 59
    encapsulated templated replication, 59–60
    prospects for complete model, 60–61
  RNA-catalyzed RNA replication, 56–57
  vesicle division pathways, 55–56
  vesicle growth pathways, 53–55

chbBC, small RNA regulator function, 221–222
Chirality, early life, 35
Chloramphenicol, peptidyl transferase reaction inhibition, 131
Chromatin. *See* Heterochromatin
Cid12, heterochromatic RNA interference at fission yeast centromeres, 259
Class I ligase ribozyme. *See* RNA ligase ribozyme
Clusters of regularly interspersed short palindromic repeats. *See* CRISPRs
CMT3, heterochromatic RNA interference, 262
CMV. *See* Cytomegalovirus
Contemporary RNA world, overview, 3–4
COSMIC LOPER, overview, 48
CRISPRs
    Cas protein functions, 235
    examples, 233
    functional overview, 232
    loci and Cas genes, 232, 234
    mechanism of action
        integration of new spacers, 235–236
        target interference, 237–238
        transcription and processing, 236–237
    prospects for study, 240
    RNA interference homology, 238–240
CsrB, small RNA regulator function, 223
Ctcf, X-chromosome inactivation role, 271
CyaR, small RNA regulator function, 226
Cytomegalovirus (CMV), noncoding RNA, 174

D

DDM1, heterochromatic RNA interference, 262
DEAD-box proteins
    group II intron association, 113
Ded1p, helicase mechanism studies, 328
DHFR. *See* Dihydrofolate reductase
Dicer, RNA interference role, 257–258
Dihydrofolate reductase (DHFR), long noncoding RNA regulation, 287–288
DsrA, small RNA regulator function, 219–220
Duplicator RNA, origins of ribosomal decoding site, 148–150

E

EBER. *See* Epstein-Barr virus-encoded RNA
eIF4A, helicase mechanism studies, 328
Electron density map, fitting of atomic models, 314
Elongation factors
    homology between species, 10–11
    temperature optimum in eubacteria, 11
Elongation factors, structure and function, 136–137
Epstein-Barr virus-encoded RNA (EBER)
    discovery, 174–175
    functions, 168, 170
    microRNAs, 173–174
    ribonucleoprotein complexes, 168, 170
    transcripts, 166, 168

Eri1, silencing on chromosome arms, 261
Error threshold, RNA replicase, 26–27
Erythromycin, peptidyl transferase reaction inhibition, 131
Evf2, histone modification mediation, 283–284

F

FAD
    continuity with RNA world, 47–48
    initial darwinian ancestor modern descendant, 46
fbp1+, RNA polymerase II loading mediation, 287
Fidelity
    early RNA replicase, 26–27
    translation, 133
Flavin mononucleotide (FMN), riboswitch ligand, 67–68, 85
Fluorescence resonance energy transfer (FRET)
    folding pathways of purine riboswitch aptamer, 83–84
    single-molecule studies of RNA reactions, 324–329
    vesicle growth studies, 53–54
FMN. *See* Flavin mononucleotide
Folding. *See* Modules, RNA; Secondary structure, RNA; Tertiary structure, RNA
Frameshifting. *See* Ribosome

G

GlcN6P. *See* Glucosamine-6-phosphate
glmS ribozyme
    catalytic mechanism, 98–99
    glucosamine-6-phosphate cofactor, 98–99
    structural overview, 95–96
GlmY, small RNA regulator function, 221–222
Glucosamine-6-phosphate (GlcN6P)
    glmS ribozyme cofactor, 98–99
    riboswitch ligand, 66–67
Glyceraldehyde-2-phosphate, abiotic synthesis, 33
Glycoaldehyde phosphate, abiotic synthesis, 33
Glycol nucleic acid (GNA)
    abiotic synthesis, 37
    polymer structure, 36
GNA. *See* Glycol nucleic acid
Group II intron
    associated proteins
        DEAD-box proteins, 113
        intron-encoded protein component, 104, 106–108, 111–112
        LtrA, 112
        recruitment, 113
    catalytic mechanism, 108–109
    degenerate intron splicing, 106
    evolution, 116–117
    lineages, 108
    mobility
        DNA target site recognition, 115
        retrohoming
            mechanisms, 115
            reverse splicing into DNA, 113–115
        retrotransposition into new sites, 115–116
        targetrons, 116

phylogenetic distribution, 104
RNA structure, 104–106
three-dimensional structure, 109–111
trans-splicing, 106
twintron formation, 106

H

Hairpin ribozyme
    catalytic mechanism, 96–97
    fluorescence resonance energy transfer studies, 324–325
    RNase H homology, 97
    structural overview, 95–96
Hammerhead ribozyme
    catalytic mechanism, 98
    structural overview, 95–96, 98
HDV ribozyme. *See* Hepatitis delta virus ribozyme
Hepatitis delta virus (HDV) ribozyme, structural overview, 95–96, 99–100
Herpes simplex virus-1 (HSV-1), noncoding RNA, 174
Herpesvirus saimiri U RNA (HSUR)
    function, 171
    structure, 170–171
    transcripts, 170–171
Heterochromatin
    definition, 256
    histone modification mediation by long noncoding RNA
        CCND1, 284–285
        Evf2, 283–284
        HOTAIR, 285
        SRA, 283
    histone modifications, 256–257
    position effect variegation, 256
    RNA interference
        filamentous fungi, 262–263
        fission yeast centromeres, 258–261
        silencing on chromosome arms, 261–262
Hfq, small RNA regulator base pairing with limited complementarity role, 218
Histones. *See* Heterochromatin
HIV. *See* Human immunodeficiency virus
HOTAIR, histone modification mediation, 285
HOX genes, *HOTAIR* regulation, 285
Hrr1, heterochromatic RNA interference at fission yeast centromeres, 259
HSUR. *See* Herpesvirus saimiri U RNA
HSV-1. *See* Herpes simplex virus-1
Human immunodeficiency virus (HIV), reverse transcriptase RNA-binding studies, 326–328
Hydrolysis, susceptibility of RNA bonds, 12

I

IAPV. *See* Israeli acute paralysis virus
IDA. *See* Initial darwinian ancestor
Imidazo(1,2-c)pyrimidin-5(1H)-one, six letter polymerase chain reaction, 15–16

Initial darwinian ancestor (IDA)
    AMP-containing enzymatic cofactors as modern descendants, 45–46
    origin of life context, 47
    time vantage, 44
Initiation factors, deletion studies, 142
Internal transesterification reaction, small self-cleaving ribozymes, 94–95
Internally transcribed spacer (ITS), phylogenetic reconstruction, 303
Intron-encoded protein. *See* Group II intron
Israeli acute paralysis virus (IAPV), RNA interference targeting, 350
ITS. *See* Internally transcribed spacer

K

Kaposi's sarcoma-associated herpesvirus (KSHV)
    microRNA, 173
    PAN RNA function, 171–172
Kasugamycin, resistance mechanisms, 142
Khps1, DNA methylation mediation, 285
KSHV. *See* Kaposi's sarcoma-associated herpesvirus

L

L1 ligase. *See* RNA ligase ribozyme
L22, Epstein-Barr virus-encoded RNA binding, 170
La, Epstein-Barr virus-encoded RNA binding, 168
Last universal common ancestor (LUCA), RNA fragments, 11
Leakage, premetabolic cycle, 15
Life. *See also* Cellular life
    definition, 8
    origin study approaches, 8–9
    RNA-first view on origin
        nonenzymatic replication of RNA, 24–25
        nucleotide biosynthesis, 31–32
        polynucleotide abiotic synthesis, 22–24
        RNA replicase, 25–31
    RNA-later view on origin
        alternate genetic systems, 35–38
        nucleotide synthesis, 32–35
Linezolid, peptidyl transferase reaction inhibition, 131
Long noncoding RNA. *See also* Polycomb proteins; X-chromosome inactivation
    advantages in function
        sequence specificity, 275–276
        transcription site localization, 275
    cell structure integrity role, 288
    classification, 280
    expression regulation, 281
    functional overview, 280
    processing, 281–282
    promoter-associated RNA functions, 287–288
    prospects for study, 288
    protein complex recruitment, 282–283
    signal sensing, 288
    strandedness, 280

Long noncoding RNA (*Continued*)
  subcellular localization, 280–281
  transcriptional boundary marking, 288
  transcriptional regulation mechanisms
    DNA methylation, 285
    histone modification, 283–285
    RNA polymerase II loading, 285–287
    transcription factors, 287
    transcription interference, 287
LtrA, group II intron association, 112
LUCA. *See* Last universal common ancestor

M

M1CB, herpesvirus targeting, 173
Macugen, historical perspective, 334
Malaria, RNA interference targeting, 350
Manip program, interactive molecular modeling, 314
Meiotic silencing of unpaired DNA (MSUD), RNA interference pathway, 251
Membrane
  compartment boundaries, 52–53
  vesicles
    division pathways, 55–56
    growth pathways, 53–55
1-Methyl-adenine derivatives, polymerization, 24
2-Methylimidazolide derivatives, polymerization, 24
MHV-68. *See* Murine herpesvirus-68
MicA, small RNA regulator function, 219–220
MicroRNA, viral function, 172–174
Modules, RNA
  crystal structure comparisons, 315–316
  examples, 313
  large assembly modeling, 313
  searching, 312–313
Montmorillonite, nucleoside 5′-phosphorimidazole polymerization, 24
mRNA
  splicing. *See* Spliceosome
  translation. *See* Ribosome
MSUD. *See* Meiotic silencing of unpaired DNA
Murine herpesvirus-68 (MHV-68), noncoding RNA, 175

N

NAD/P
  abiotic synthesis of NAD, 46
  continuity with RNA world, 47–48
  initial darwinian ancestor modern descendant, 45–47
  replication chemistry, 48–49
NEAT RNAs, cell structure integrity role, 288
NMR. *See* Nuclear magnetic resonance
NRON, transcription factor mediation, 287
NS3, helicase mechanism studies, 328
Nuclear magnetic resonance (NMR), ligand-induced conformational change studies in riboswitches, 86
Nucleolin, Epstein-Barr virus-encoded RNA binding, 170
Nucleoside 5′-phosphorimidazoles, polymerization, 23–24

Nucleotides
  abiotic synthesis, 32–35
  activation in polymerization, 20–21, 25, 32
  synthetic ribozymes, 31–32

O

OmrB, small RNA regulator function, 219–220
OmtA, small RNA regulator function, 219–220
Optical tweezers
  force measurements, 322–323
  ribosome-mRNA movement studies, 328–329
OxyS, small RNA regulator function, 219–220

P

P15AS, DNA methylation mediation, 285
PABPC1. *See* Poly(A)-binding protein C1
PAN RNA. *See* Polyadenylated nuclear RNA
PCR. *See* Polymerase chain reaction
Peptide nucleic acid (PNA)
  abiotic synthesis, 37
  polymer structure, 36
Peptidyl transferase center. *See* Ribosome
Perchlorate, solar system distribution, 15
PEV. *See* Position effect variegation
Phosphoramidate nucleic acids, cellular life origins, 57–59
5-Phosphoribosyl-1-pyrophosphate (PRPP), ribozyme synthesis limitations, 32
piRNA, functions, 250–251, 258
Piwi. *See* Argonaute
PKR. *See* Protein kinase R
PNA. *See* Peptide nucleic acid
Poly(A)-binding protein C1 (PABPC1), PAN RNA association, 172
Polyadenylated nuclear (PAN) RNA, Kaposi's sarcoma-associated herpesvirus function, 172
Polycomb proteins
  long noncoding RNA advantages in regulation, 274–276
  RNA recruitment, 272–274, 282
Polymerase chain reaction (PCR), six letter polymerase chain reaction, 15–16
Position effect variegation (PEV), heterochromatin silencing, 256
Pre-mRNA splicing. *See* Spliceosome
Primordial RNA world, overview, 2–3
p-RNA, structure, 35–36
Protein kinase R (PKR), Epstein-Barr virus-encoded RNA binding, 168
Protein synthesis. *See also* Ribosome
  duplicator RNA and origins of ribosomal decoding site, 148–150
  evolutionary driving force for translation from RNA world, 150–151, 162–163
  RNA role, 8, 10, 22, 126–127, 142–144, 157–158
  stop tRNAs and type I release factor evolution, 148
  translation steps, 156–157
Protocell. *See* Cellular life
Prp2, spliceosome function, 191

Prp5
    heterochromatic RNA interference at fission yeast centromeres, 260
    spliceosome function, 190–191
Prp8
    heterochromatic RNA interference at fission yeast centromeres, 260
    spliceosome function, 191, 195
Prp10, heterochromatic RNA interference at fission yeast centromeres, 260
Prp16, spliceosome function, 191
Prp19, spliceosome function, 188, 191
Prp22, spliceosome function, 191–192
Prp28, spliceosome function, 191
Prp43, spliceosome function, 192
PRPP. See 5-Phosphoribosyl-1-pyrophosphate
Pseudoknot
    frameshifting studies, 325–326
    telomerase RNA, 208–209

R

R2 retrotransposon, RNA structure elucidation, 297–298
RCNMV. See Red clover necrotic mosaic virus
Rdp1, heterochromatic RNA interference at fission yeast centromeres, 259
Red clover necrotic mosaic virus (RCNMV), noncoding RNA, 175
Release factors, stop tRNAs and type I release factor evolution, 148
Release factors, tRNA mimics, 129–130
RepA
    Polycomb protein recruitment, 274–275
    signal sensing, 288
    X-chromosome inactivation role, 269–270
Ribose, instability, 12
Ribosome. See also Protein synthesis
    aminoacyl-tRNA selection, 144–146
    antibiotic inhibition of peptidyl transferase reaction, 130–133
    complexes with both A- and P-site substrates bound, 127–128
    crystal structures, 125–126, 156
    duplicator RNA and origins of ribosomal decoding site, 148–150
    elongation factor structure and function, 136–137
    evolutionary driving force for translation from RNA world, 150–152, 162–163
    fidelity of translation, 133
    induced-fit activation of peptide synthesis, 128–129
    mRNA movement studies with optical tweezers, 328–329
    peptidyl transferase center
        metal ions, 128
        peptide bond formation mechanism, 126–127, 160
        peptide release, 160–161
        ribozyme activity, 143–144, 157–158
        RNA components, 126
    peptidyl-tRNA protection from hydrolysis, 128–129
    protein structures and functions, 148
    pseudoknot frameshifting studies, 325–326
    release factors as tRNA mimics, 129–130
    rRNA
        role in mRNA decoding, 134–136
        tRNA interactions in P site, 146–147
    subunits and functions, 124–125
    transition state intermediate and stabilization, 128
    translation steps, 156–157
    translocation
        E site conservation and role, 162
        RNA molecular mechanics, 147
        subunit interactions, 161–162
        tRNA energy storage, 162
    tRNA
        binding sites in small ribosomal subunit, 133–134
        CCA end binding to E site of large subunit, 130
        decoding
            evolutionary role of minor groove recognition, 158–159
            minor groove recognition by RNA, 158
            structural changes, 159
        dynamics analysis with fluorescence resonance energy transfer, 329
    stop tRNAs and type I release factor evolution, 148
Riboswitch
    S-adenosylmethionine riboswitch, 69–71
    classes, 64–65, 68–69
    domains, 80
    effector molecule recognition, 84–85
    functional overview, 64, 80
    gene regulation
        mechanisms, 65–66, 80
        temporal sensitivity, 86–87
    ligand binding affinity and kinetics of function, 67–68
    origins, 75
    ribozyme control mechanisms, 67
    RNA world functions, 89
    small RNA regulators, 221
    structure
        examples, 80–82
        folding pathways of purine riboswitch aptamer, 82–84
        ligand-induced conformational change, 85–86
        models for structural switching, 87–89
    tandem switches, 71–75
RIG-I, helicase mechanism studies, 328
RITS. See RNA-induced initiation of transcriptional silencing complex
RNA III, small RNA regulator function, 221
RNAi. See RNA interference
RNA-induced initiation of transcriptional silencing complex (RITS), heterochromatic RNA interference at fission yeast centromeres, 259–260
RNA interference (RNAi). See also Argonaute
    CRISPR homology, 238–240
    eukaryote distribution, 257
    heterochromatic RNA interference
        filamentous fungi, 262–263
        fission yeast centromeres, 258–261
        silencing on chromosome arms, 261–262

RNA interference (RNAi) (*Continued*)
   immune function, 250
   mechanisms, 257–258
   pest control applications, 350
   screening in vivo
      Caenorhabditis elegans, 345–346
      developmental biology applications, 349
      Drosophila melanogaster, 346–349
      plants and crop improvement, 350
      reagents, 344–345
      vertebrate models, 349–350
   therapeutic prospects, 350–351
RNA ligase ribozyme
   magnesium dependence, 30
   polymerization reaction and optimization, 30
   structures, 29–30
RNA polymerase II
   heterochromatic RNA interference at fission yeast centromeres, 260
   long noncoding RNA mediation of loading, 285–287
RNA–protein complexes, modeling, 313
RNA replicase ribozyme
   accuracy and survival, 26–27
   cellular life origins, 56–57
   chicken-and-egg paradox, 27–29
   ligase evolution, 29–31
   RNA-first view of life, 25
RNA technology, as third RNA world, 4
RprA, small RNA regulator function, 219–220
rRNA. *See* Ribosome
RybB, small RNA regulator function, 219–220
RyhB, small RNA regulator function, 219–220

S

SAH. *See* S-Adenosylhomocysteine
SAM. *See* S-Adenosylmethionine
Secondary structure, RNA
   comparative sequence analysis
      automation
         align then fold approach, 300
         fold and align approach, 299
         fold then align approach, 300
         programs, 300
         overview, 311–312
   crystal structure comparisons, 315–316
   evolution studies, 302–304
   folding space restraints
      combined methods for determination, 297–298
      comparative structures, 297
      experimental findings, 296–297
      overview, 295–296
   free energy minimization, 294
   functional RNA discovery, 300–301
   overview of analysis, 310–311
   partition functions and probabilities, 294–295
   phylogenetic reconstruction, 303
   prediction programs, 295
   prospects for study, 301–302

SELEX
   clinical applications of aptamers, 334–335
   historical perspective, 334–335
   proteomics for aptamer optimization, 335–336
   RNA function discovery, 338–339
   RNA world applications of aptamers, 337–338
   SOMAmers
      plasma proteome probing at high throughput and specificity, 337
      specificity, 336
      structure, 336
SF1, spliceosome function, 190
SF3 proteins, spliceosome function, 188, 190
SgrS, small RNA regulator function, 221
Six letter polymerase chain reaction, 15–17
Small nuclear ribonucleoproteins (snRNPs), spliceosome structures
   electron microscopy, 193–194, 197–199
   high-resolution structures, 194–195
   U1 small nuclear ribonucleoprotein, 195–196
Small RNA regulators. *See* Bacterial small RNA regulators; CRISPRS
Small self-cleaving ribozymes
   active site mechanisms, 96
   glmS ribozyme, 98–99
   hairpin ribozyme, 96–98
   hammerhead ribozyme, 98
   hepatitis delta virus ribozyme, 99–100
   internal transesterification reaction, 94–95
   structural overview, 95–96
snRNPs. *See* Small nuclear ribonucleoproteins
SOMAmers
   plasma proteome probing at high throughput and specificity, 337
   specificity, 336
   structure, 336
Spliceosome
   cis-acting elements and catalytic steps of splicing, 182–184
   conformational two-state model for catalytic center, 187
   intron- and exon-defined assembly pathways, 183, 185
   overview of pre-mRNA splicing, 182
   protein
      composition dynamics, 187–189
      flexibility of metazoan spliceosomes, 189–190
      posttranslational modifications, 192–193
      RNA footprinting studies, 197
      splice site recognition role, 190
      structural rearrangement facilitation, 190–192
   RNA–RNA interactions, 185–187
   small nuclear ribonucleoprotein structures
      electron microscopy, 193–194, 197–199
      high-resolution structures, 194–195
      U1 small nuclear ribonucleoprotein, 195–196
   U2-dependent spliceosome mechanism, 183–184
SRA, histone modification mediation, 283
SreA, small RNA regulator function, 221
SreB, small RNA regulator function, 221
sRNAs. *See* Bacterial small RNA regulators
Sub2, spliceosome function, 190–191
SUNH4/KYP, heterochromatic RNA interference, 262
Sutherland's synthesis, prebiotic origins of RNA, 13

Systematic evolution of ligands by exponential enrichment. *See* SELEX; SOMAmers

T

Targetron, group II intron, 116
Telomerase
    evolutionary origins of components, 211–212
    functional overview, 206
    protein and RNA component interplay, 206, 210
    RNA
        evolutionary divergence, 206–207
        fine-tuning of repeat synthesis activity, 209–210
        holoenzyme biogenesis and regulation motifs, 210–211
        motif gain-of-function in ribonucleoprotein context, 212
        nuclear assembly and addressing in vertebrates, 211
        pseudoknot, 208–209
        stem terminus element, 209
        template 5′-boundary enforcement, 210
        template, 207–208
        TERT-binding motifs, 210
        yeast function, 211
Template-directed replication
    multi-stem-loops structures, 28
    nonenzymatic replication of RNA, 24–25
TER. *See* Telomerase
Tertiary structure, RNA
    constraint extraction, 312
    crystal structure comparisons, 315–316
    large assembly modeling, 313
Thiamin pyrophosphate (TPP), riboswitch ligand, 66–67, 85, 86
Thioester peptide nucleic acid (tPNA)
    abiotic synthesis, 37
    polymer structure, 36–37
4-Thiouridylate, ribozyme synthesis, 31–32
Threose
    instability, 12
    ribose substitution, 15
    threose nucleotide polymers, 35–36
tmRNA, functions, 223–224
tPNA. *See* Thioester peptide nucleic acid
TPP. *See* Thiamin pyrophosphate
Translation. *See* Protein synthesis; Ribosome
tRNA. *See* Ribosome
Tsix
    Polycomb protein binding, 274
    signal sensing, 288
    X-chromosome inactivation role, 270–272

U

Unnatural nucleic acids, potential in RNA world, 15–16

V

Vascular endothelial growth factor (VEGF), Macugen inhibition, 334
VEGF. *See* Vascular endothelial growth factor
Vesicle. *See* Membrane
Viral noncoding RNA
    adenovirus, 166, 169
    cytomegalovirus, 174
    Epstein-Barr virus, 166, 168, 170, 174–175
    herpes simplex virus-1, 174
    herpesvirus saimiri, 170–171
    Kaposi's sarcoma-associated herpesvirus, 171–172
    microRNAs, 172–174
    murine herpesvirus-68, 175
    overview, 166–168
    prospects for study, 175
    red clover necrotic mosaic virus, 175
    West Nile virus, 175

W

West Nile virus (WNV), noncoding RNA, 175
WNV. *See* West Nile virus
Work, Gibbs free energy change, 323

X

X-chromosome inactivation (XCI)
    long noncoding RNA advantages, 274–276
    noncoding RNA regulators, 269–270
    overview, 268
    RNA in pairing, counting, and choice, 270–272
    signal sensing by RNA, 288
    symmetry break by Tsix RNA, 272
    unique features, 268–269
XCI. *See* X-chromosome inactivation
Xist
    Polycomb protein regulation, 272–274
    signal sensing, 288
    X-chromosome inactivation role, 269–270
Xite, X-chromosome inactivation role, 271